T0334528

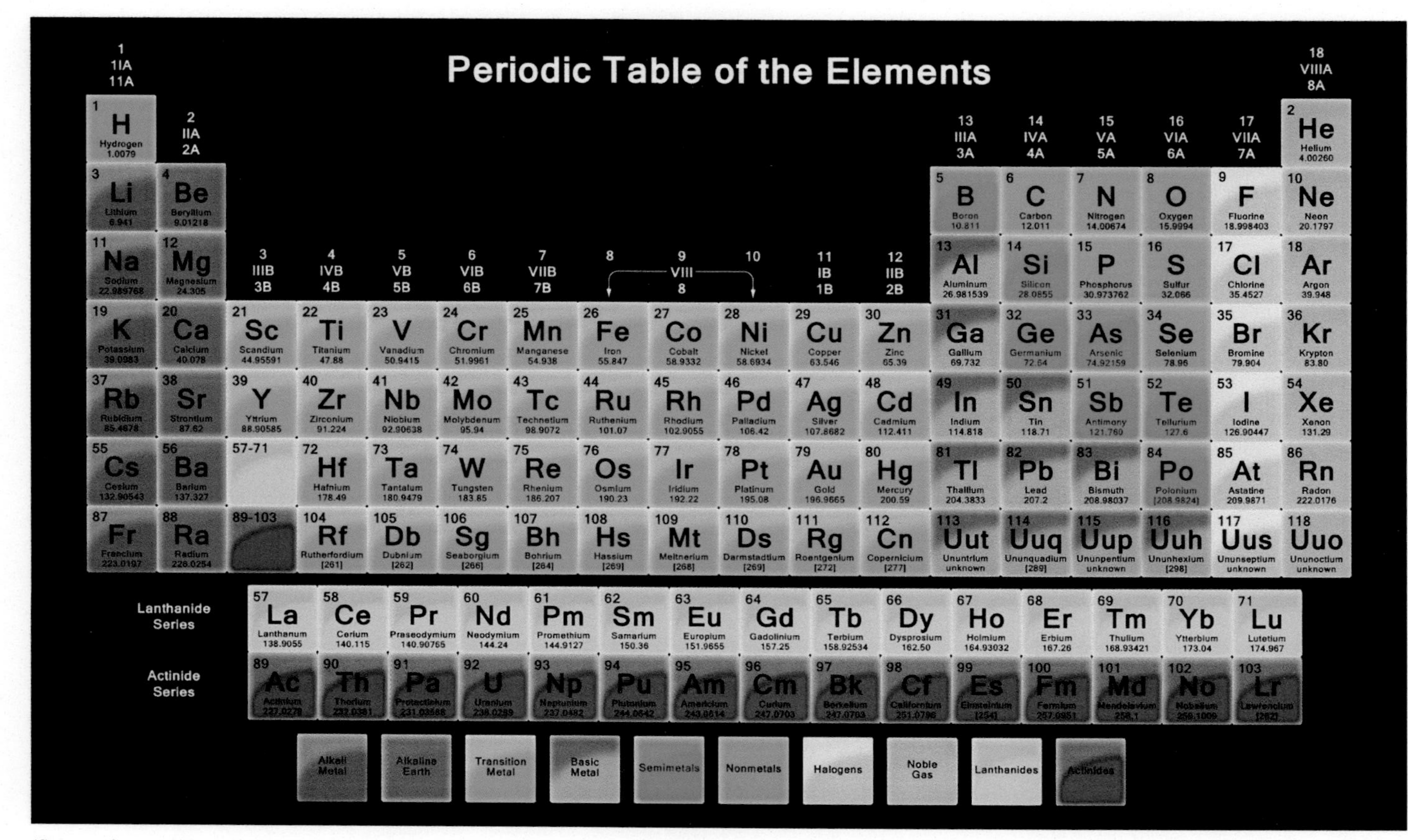

(Source: Science Notes http://sciencenotes.org)

Air Pollution Calculations

Air Pollution Calculations

Quantifying Pollutant Formation, Transport, Transformation, Fate and Risks

Daniel A. Vallero

Elsevier
Radarweg 29, PO Box 211, 1000 AE Amsterdam, Netherlands
The Boulevard, Langford Lane, Kidlington, Oxford OX5 1GB, United Kingdom
50 Hampshire Street, 5th Floor, Cambridge, MA 02139, United States

Notices
Knowledge and best practice in this field are constantly changing. As new research and experience broaden our understanding, changes in research methods, professional practices, or medical treatment may become necessary.

Practitioners and researchers must always rely on their own experience and knowledge in evaluating and using any information, methods, compounds, or experiments described herein. In using such information or methods they should be mindful of their own safety and the safety of others, including parties for whom they have a professional responsibility.

To the fullest extent of the law, neither the Publisher nor the authors, contributors, or editors, assume any liability for any injury and/or damage to persons or property as a matter of products liability, negligence or otherwise, or from any use or operation of any methods, products, instructions, or ideas contained in the material herein.

Library of Congress Cataloging-in-Publication Data
A catalog record for this book is available from the Library of Congress

British Library Cataloguing-in-Publication Data
A catalogue record for this book is available from the British Library

ISBN: 978-0-12-814934-8

Publisher: Candice Janco
Acquisition Editor: Laura S. Kelleher
Editorial Project Manager: Emily Thomson
Production Project Manager: Omer Mukthar
Designer: Matthew Limbert

Typeset by SPi Global, India

Dedication

"To my wife, Janis. To paraphrase the Hollies, you have been the air that I breathe."

Daniel A. Vallero

Contents

Preface

There are three basic types of nonfiction books that deal with environmental subject matter: descriptive, quantitative, and quantitatively descriptive. Descriptive books usually have the broadest readership, often for a general readership. They are devoid of equations and symbols. Instead, they rely on narratives about particular events and policies. Barry Commoner's, Rachel Carson's, and Al Gore's books fall into this category. They often have a few major points they want to make, such as the need to take actions to prevent cancer, protect endangered species, or restore and prevent the loss of habitat.

In descriptive books, the authors have information and knowledge that they want to share in order to build or enhance the reader's awareness of a particular environmental or public health problem, such as global warming or asbestos in consumer products. They have the luxury of addressing a single or just a few topics and do not worry about all the other aspects involved. In this sense, they are often better reads at the beach or over the morning coffee. In my opinion, the principal driving force behind this type of environmental book is awareness.

Quantitative books are most popular with scientists and, especially, engineers. They are also descriptive but describe phenomena and processes using numbers, symbols, equations, and technical jargon. In addition to the attention to detail, they are usually much more expansive in their treatment of environmental subject matter. I believe this is because books are not the principal means by which specific scientific information is shared. This is mainly done by journal articles. As such, quantitative environmental books are the beneficiaries of the massive amount of scientific findings in peer-reviewed journals. Air pollution books, for example, do not usually present the results of the author's own experiments for the first time. If they do present such results, the author has already published them in a peer-reviewed journal article. And, quite likely, most of the results presented are those of other scientists. The book author has chosen them to illustrate a point and to use the equations and coefficients derived by these legions of researchers and professionals who have made the effort to submit their findings to the scrutiny of the scientific community.

In this sense, the author of a quantitative environmental book is a messenger. The principal driving force for a quantitative book is the need to explain how things work. The author, hopefully, has much experience in applying the methods and is sharing this with the readership. To do so, it requires that the reader begins with a threshold of knowledge, usually much higher than what is required of the descriptive environmental book reader. I liken this to a prerequisite college course. Taking an air pollution modeling course based on a particular software would be very difficult if the student had not already completed the software course. Similarly, reading a quantitative atmospheric book would be very difficult without a modicum of knowledge of algebra and physics. Usually, however, the book will build knowledge, not only of the subject matter but also of new applications, both mathematical and physical. That is, the atmospheric science book not only improves one's knowledge of the atmosphere but also enhances the general mathematical and scientific acumen of the reader. I believe the best way to learn science is experiential and the quantitative book is a virtual way to experience environmental phenomena and processes.

The third type of the book, the hybrid, combines narrative descriptions with quantitation. A phenomenon is described but is explained further using symbolic and technical tools. For example, completely describing a disaster requires descriptive information, for example, time lines, historical events, key players, and adverse outcomes. Not only describing the toxic plume in Bhopal, India, requires this type of information, but also it needs chemical information, for example, physical and chemical factors of the released chemical, methyl isocyanate, its biological toxicological information, geographic doses, and other quantitative information.

My book, *Fundamentals of Air Pollution*, falls into this category. Actually, the book had three previous editions before mine. When I reviewed the third edition, I found it to be an excellent documentation of the progress in addressing air pollution in the United States. It certainly had a wealth of quantitative information, but much of the book was highly descriptive. I attempted to add quantitation in the fourth edition but still thought it needed substantial quantitative information to explain the many air pollution processes, including the diseases and other outcomes, air pollution control technologies, and spatial and temporal complexities. I completely revamped the book in the fifth edition, including a substantial amount of quantitative information.

After this effort, it became clear to me that explaining air pollution problems requires more than a cursory combination of descriptive and quantitative answers to questions at the end of each chapter of an air pollution book. In fact, instructors who use my book asked specifically for quantitative answers, which I take to mean a dedicated resource. In my opinion, I suspect those of others who teach air pollution courses and who practice in field that addresses air pollution would appreciate such a resource. This suspicion led to the book you are reading. Not only did I hope it would be a companion to *Fundamentals of Air Pollution*, but also I have striven to make it a companion to any air pollution book. In this sense, its primary aim is to make the concepts and topics more "real" to the reader.

I am told that most people, including engineers and scientists, are predominantly experiential learners. We learn very little by simply reading about or "seeing" something and even less hearing about it. We are empathic readers who must insert ourselves into a situation. We learn more effectively by experiencing, at least virtually, than by merely watching. And since mathematics and quantification are the tools of engineering and science, the examples in this book provide a step toward experience in the matters of air pollution. The experience is vicarious, but the problems presented are exactly the types being confronted daily by decision-makers, scientists, and engineers. Indeed, it is possible that I have borrowed, and hopefully acknowledged one or a few of your own experiences. If so, I am grateful to your contributions to clean air. Even if reading this book is your first step, I look forward to your future contributions to clean air, with a hope that this book will have added to your success.

Daniel A. Vallero

Chapter 1

Introduction

1.1 WHAT IS AIR POLLUTION?

Air pollution can be studied from numerous perspectives. Researchers often choose a very specific aspect of air pollution and delve deeply into its meaning. They need to understand mathematics and the basic sciences to learn what has already known and to extend this knowledge to their needs. Air pollution is the presence of contaminants or substances in the air that interfere with human health or welfare or produce other harmful environmental effects [1].

This definition requires context, that is, something is an air pollutant if it causes conditions to deviate from a desired state. Thus, an air pollutant interferes a benchmark of a desired condition that, at a minimum, provides for air quality that supports human and other life. However, the definition goes beyond health and extends to welfare, including ecological condition and societal well-being. From a system perspective, the sum of these desired states is, in fact, a single desired state. Pope Francis put this succinctly in his encyclical letter, *Laudato sí* [2]:

> *Some forms of pollution are part of people's daily experience. Exposure to atmospheric pollutants produces a broad spectrum of health hazards, especially for the poor, and causes millions of premature deaths. People take sick, for example, from breathing high levels of smoke from fuels used in cooking or heating. There is also pollution that affects everyone, caused by transport, industrial fumes, and substances that contribute to the acidification of soil and water, fertilizers, insecticides, fungicides, herbicides, and agrotoxins in general.*

Thermodynamically speaking, an air pollutant or any pollutant is the result of inefficiency. The pollutant is mass or energy that is exiting a control volume. In trying to produce something of value, a waste product is released to the atmosphere. Not all scientists are researchers. Indeed, some of the best scientists are those who apply what researchers have discovered. Such scientists work in national, state, provincial, and local government agencies with missions to protect the environment. They write regulations, rules, and guidance and issue permits and licenses on emissions and ambient air quality protections.

Often in combating air pollution, the researchers and regulatory scientists and engineers work closely together. For example, in the United States, the Environmental Protection Agency (EPA) has three basic functions: (1) to conduct research, (2) to enforce environmental rules and regulations, and (3) to provide funding for other entities to help with the first two functions. EPA's air pollution enforcement and compliance work is conducted by the Office of Air Quality Planning and Standards (OAQPS) and its 10 regional offices, and its research is conducted by the Office of Research and Development (ORD). To demonstrate the collaboration between research, development, technology transfer, and enforcement of air pollution laws, OAQPS and ORD's air programs are conducted on the same campus in Research Triangle Park, North Carolina. ORD funds and collaborates on air quality research with numerous universities and institutions.

There are other perspectives in addition to research and enforcement, including teaching, journalism, and policy making. Those who educate others and write about and promulgate polices and laws must understand air pollution in different ways but with a modicum of science.

Understanding the principles of air pollution and the calculations involved is something we all share, although the extent and specificity of these calculations vary. For example, a research scientist may need to completely explain all the reactions in a combustion facility that lead to the generation of a toxic compound, for example, benzo(a)pyrene (BaP). The air emissions permit writer does not ordinarily need to know all the reactions to the extent of the researcher, but enough to know which ones may be most difficult to change. The high school chemistry teacher may need to know only the most basic chemical reactions, but the air pollution engineering college professor may need to know just as much as the researcher to tailor the lectures and homework properly to the target student. The journalist who can "do the math" and grasp the scientific principles has an advantage over her colleagues that only know the essence of the science. The local air pollution authority manager may need to know as much as the permit writer, since he supervises the permit writers. The politician who serves on a technical committee responsible for writing air pollution legislation will need to know more than the other

Air Pollution Calculations. https://doi.org/10.1016/B978-0-12-814934-8.00001-6

members of the legislature, perhaps even grasp the math and science, as well as the permit writer and the researcher, at least to the extent of knowing whether to increase or decrease funding.

Therefore, every air pollution perspective requires an aptitude and appreciation for calculations. This book attempts to provide examples of most of the types of calculations one may encounter, whether to succeed in one's career or simply to trust what one reads in a report or article.

1.2 A BIT OF HISTORY

Not long ago, making environmental decisions based on and underpinned by sound physical and biological science was not the norm compared with those made for other reasons [3]. At first, the weight of environmental factors was based on theory and limited, isolated studies. Indeed, science-based air pollution decision-making was tracked directly with the experiences of contagious diseases. Even after germ theory took hold in the biological research community, there was substantial time before public health decisions became predominantly based on contagion data. A rather large body of theory testing occurred before John Snow was able to make the famous decision to intervene and insist on infection controls by banning the use of contaminated Thames River water to abate the outbreak of the "Broad Street" cholera epidemic in the mid-19th century [4, 5].

Concerns about air pollution predated Snow's work. For centuries, people have known intuitively that something was amiss when their air was filled with smoke or when they smelled an unpleasant odor. But, for most pollutants, those that were not readily sensed without the aid of sensitive equipment, a baseline had to be set to begin to act. One way to look at the interferences mentioned in the definition is to place them within the context of "harm." Harm can be acceptable, so long as it is sufficiently unavoidable and arises only due to obtaining a more favored good or service. The objects of the harm have received varying levels of interests. In the 1960s, the perception of harm to ecosystems grew as larger number of people perceived that the very survival of certain biological species was threatened and that these threats extended to humans. Indeed, the direct harm to humans was perceived earlier than those to ecosystems, especially in terms of diseases directly associated with obvious episodes, such as respiratory diseases and even death associated with combinations of weather and pollutant releases [6].

Myriad pollutant emissions have accompanied the industrial and technological eras. Nuclear power plants are associated with the possibilities of meltdown and the release of airborne radioactive materials. Burning fossil fuels releases products of incomplete combustion, for example, polycyclic aromatic hydrocarbons. Even complete combustion releases the greenhouse gas, carbon dioxide (CO_2). Leaks from chemical, pesticide, and other manufacturing facilities may cause exposures to toxic substances, such as the release of methyl isocyanate in Bhopal, India. In the last quarter of 20th century, these apprehensions led to the public's growing wariness about "toxic" chemicals added to the more familiar "conventional" pollutants like soot, carbon monoxide, and oxides of nitrogen and sulfur. Toxic chemical exposures were increasingly linked to cancer and threats to hormonal systems in humans and wildlife, neurotoxicity (notably from lead and mercury exposure in children), and immune system disorders.

Some substances thought to be almost exclusively water or food contaminants were found in the air. For example, the source blood lead (blood-Pb) levels in children were initially thought to be pica, that is, eating chips of lead paint. However, the Pb concentrations declined precipitously in children living near roadways following the decreased use of tetraethyl lead and other Pb organometallic compounds in gasoline, which may have accounted for more than half of blood-Pb in the 1970s [7].

Growing numbers of studies have provided evidence linking disease and adverse effects to extremely low levels of certain particularly toxic substances. For example, exposure to dioxin at almost any level above what science could detect could be associated with numerous adverse effects in humans.

At the threshold of the new millennium, the need to consider air pollution within the context of larger-scale environmental systems became obvious, including the loss of aquatic diversity in lakes due to deposition of acidic precipitation, so-called acid rain. Acid deposition was also being associated with the corrosion of materials, including some of the most important human-made structures, such as the pyramids in Egypt and monuments throughout the world. Presently, global pollutants have become the source of public concern, such as those that seemed to be destroying the stratospheric ozone layer or those that appeared to be affecting the global climate. This escalation of awareness of the multitude of pollutants complicated matters.

To set the stage for air pollution calculations, let us consider to some air pollution events to begin to understand why we need sound approaches that have evolved from lessons learned over time [8]. In addition to the need to hunt and gather, early people were nomadic because they wanted to exit the stench of the animal, vegetable, and human wastes that they generated. Within enclosed dwellings, air pollutants accumulated. For centuries, the open-fire emissions created smoked that lingered. Thus, among the first air pollution control technology was the chimney, which helped to remove the combustion products and cooking odors from the living quarters. This was not only a boon to indoor air quality but also an early lesson in the cumulative sources' impact on ambient air. This was aptly stated in AD 61 by the Roman philosopher Seneca:

As soon as I had gotten out of the heavy air of Rome and from the stink of the smoky chimneys thereof, which, being stirred, poured forth whatever pestilential vapors and soot they had enclosed in them, I felt an alteration of my disposition [9].

Eleanor of Aquitaine, the wife of King Henry II of England, moved from her castle in the year 1157 because wood smoke was generating so much air pollution [6]. In 1306, Edward I issued a royal proclamation enjoining the use of *sea coal* in furnaces in London. Elizabeth I barred the burning of coal in London when Parliament was in session. Perhaps as an early lesson in the difference between promulgating air pollution law and the ability to enforce it, coal continued to be burned despite these royal edicts. As evidence, by 1661, London's air had become sufficiently polluted that John Evelyn submit to King Charles II and Parliament a brochure entitled, "Fumifugium: or the Inconvenience of the Aer, and Smoak of London Dissipated (together with some remedies humbly proposed)" [10]. This lays out several proposed actions that are viable and used today [11].

The principal industries associated with the production of air pollution in the centuries preceding the industrial revolution were metallurgy, ceramics, and preservation of animal products. In the bronze and iron ages, villages were exposed to dust and fumes from many sources. Native copper and gold were forged, and clay was baked and glazed to form pottery and bricks before 4000 BCE. Iron was in common use, and leather was tanned before 1000 BCE. Most of the methods of modern metallurgy were known before AD 1. They relied on charcoal rather than coal or coke. However, coal was mined and used for fuel before AD 1000, although it was not made into coke until about 1600; and coke did not enter metallurgical practice significantly until about 1700. These industries and their effluents as they existed before 1556 are best described in the book "De Re Metallica" published in that year by Georg Bauer, known as Georgius Agricola. It was translated into English by President (and engineer) Herbert Clark Hoover and his wife [12]. Combustion in kilns generated air pollution even before the industrial revolution. For example, ceramics [13] and animal product preservation industries [14] burned various materials to heat kilns.

The industrial revolution soon began after it was found that harnessing of steam could provide power to pump water and to move heavy objects. Early in the eighteenth century, Savery, Papin, and Newcomen designed their pumping engines and culminated in 1784 in Watt's reciprocating engine. The reciprocating steam engine was ubiquitous in industry until it was displaced by the steam turbine in the twentieth century. During most of the nineteenth century, coal was the principal fuel used to make steam, although some oil was used for this purpose late in the century. Thus, the predominant air pollution problem of the nineteenth century was smoke and ash from the burning of coal or oil in the boiler furnaces of stationary power plants, locomotives, marine vessels, and in-home heating fireplaces and furnaces. Great Britain took the lead in addressing this problem, and in the words of Sir Hugh Beaver [15],

By 1819, there was sufficient pressure for Parliament to appoint the first of a whole dynasty of committees "to consider how far persons using steam engines and furnaces could work them in a manner less prejudicial to public health and comfort." This committee confirmed the practicability of smoke prevention, as so many succeeding committees were to do, but as was often again to be experienced, nothing was done.

In 1845, during the height of the great railway boom, an act of Parliament required that locomotives consume their own smoke and 2 years later applied the same requirement to factory furnaces. These and later laws were passed with limited success. Air pollution from the emerging chemical industry was considered a separate matter and was made the responsibility of the Alkali Inspectorate created by the Alkali Act of 1863. As opposed to the kingdom-wide proclamations in Britain, the United States considered smoke abatement (as air pollution control was then known) be a municipal responsibility. No federal or state smoke abatement laws or regulations were promulgated until the next century. The first municipal ordinances and regulations limiting the emission of black smoke and ash appeared in the 1880s and were directed toward industrial, locomotive, and marine rather than domestic sources. By the end of the 19th century, smelting of sulfide ores was identified as the source of crop damage.

Air pollution control technology in the 19th century began to look like those in use today, including improved fuel feed technologies, especially stokers for mechanical firing of coal, scrubbers to remove acid gases from effluent gas streams, and cyclones and fabric filters (baghouses) to collect particulate matter (PM). Also, with the growth of scientific knowledge, physical and chemical principles began to replace or at least augment "trial and error" and intuitive processes in pollution control design.

The first quarter of the 20th century experienced dramatic changes in the technology of both the production of air pollution and its engineering control. However, this was not accompanied by substantial changes in legislation, regulations, understanding of the problem, or public attitudes toward the air quality. As cities and factories grew, the severity of the pollution problem increased.

A big technological change was the replacement of the steam engine by the electric motor as the means of operating machinery and pumping water. This substantially transferred the smoke and ash emission from the boiler house of the factory to the boiler house of the electric generating station. At the start of this period, coal was hand-fired in the boiler house; by the middle of the period, it was mechanically fired by stokers, but by the end of the period, pulverized coal, oil, and gas firing had begun to dominate. Each form of firing produced its own characteristic emissions to the atmosphere.

Steam locomotives started to enter central business districts in the heart of the larger cities. Later, the urban terminals of many railroads had been electrified, thereby transferring much air pollution from the railroad right-of-way to the electric generating station. The replacement of coal by oil in many applications decreased ash emissions from those sources. There was rapid technological change in industry. However, the most significant change was the rapid increase in the number of automobiles from almost none at the turn of the century to millions by 1925. The principal technological changes in the engineering control of air pollution were the perfection of the motor-driven fan, which allowed large-scale gas-treating systems to be built; the invention of the electrostatic precipitator, which made particulate control in many processes feasible; and the development of a chemical engineering capability for the design of process equipment, which made the control of gas and vapor effluents feasible.

The first large-scale surveys of air pollution were undertaken—Salt Lake City, Utah (1926) [16]; New York City (1937) [17]; and Leicester, England (1939) [18]. By mid-20th century, major air pollution became almost ubiquitous in urban areas. Public health disasters, including air pollution-related deaths, occurred in the Europe and North America. The Meuse Valley, Belgium, episode occurred in 1930 [19]; the Donora, Pennsylvania, episode occurred in 1948 [20]; and the Poza Rica, Mexico, episode in 1950 [21]. Smog appeared in Los Angeles in the 1940s. In response, the first National Air Pollution Symposium in the United States was held in Pasadena, California in 1949 [22], and the first US Technical Conference on Air Pollution was held in Washington, DC, in 1950 [23].

Building of natural gas pipelines resulted in rapid displacement of coal and oil as home heating fuels resulting in dramatic improvement in air quality; witness the much-publicized decrease in black smoke in Pittsburgh (Fig. 1.1) and St. Louis [6]. The diesel locomotive began to displace the steam locomotive, thereby slowing the pace of railroad electrification. The internal combustion engine bus started its displacement of the electrified streetcar. The love for the automobile continued to grow, and the numbers on the road proliferated.

During this period, substantial national air pollution legislation or regulations were not adopted anywhere in the world until the first state air pollution law in the United States was adopted by California in 1947.

FIG. 1.1 (A) Pittsburgh after the decrease in black smoke. (B) Pittsburgh before the decrease in black smoke. *(Source: (A) Allegheny County, Pennsylvania.)*

1.2.1 Black smoke

Great Britain experienced a major air pollution disaster in London in 1952, which led to the passage of the Clean Air Act in 1956 and an expansion of the authority of the Alkali Inspectorate [24, 25]. The new law target home heating, which had been mainly done by burning soft coal on grates in separate fireplaces in each room. This led to a successful substitute of much lower-smoke fuels and central or electric heating instead of fireplace heating.

The outcome was a decrease in "smoke" concentration. This was determined by measuring the interference of visible light, referred to as "blackness" of paper filters that collect air. The opacity of the filter is proportional to the amount of soot and dust on the filter. These opacity readings improved from $175\,\mu g\,m^{-3}$ in 1958 to $75\,\mu g\,m^{-3}$ in 1968 [26]. Today, the filters are weighed before and after passage of air, but the percentage of black carbon is still of interest. Thus, variations on this British method, now known as the black smoke (BS) method, are still in use. The updated BS method, known as the filter smoke number (FSN), draws air through a pipe at constant temperature (see Fig. 1.2) [27]. It typically requires 26 impacting samples from a 4.5 μm inlet, through which the air to be measured flows onto white filter paper, which is stained. The stained filter is analyzed for blackness of the stain is 27 measured by light absorption. The blackness or darkness of the stain is measured using a reflectometer. Smoke particles composed of elemental carbon (EC) typically contribute the most stain darkness.

The mix of ambient particles varies widely spatially and seasonally. Thus, the correlation between BS measurements and PM mass is highly specific to the place and time of measurement, making comparisons difficult [29]. Thus, the conversions of the opacity readings mentioned above are conversions from blackness on paper filters to mass per volume of soot and dust. However, before the end of the 20th century, opacity readings were mainly used to quantify PM. As mentioned, PM is now mainly quantified using monitors that differentiate particle diameters, especially 10 and 2.5 μm, by mass. So, this begs the question, how were these early BS method results converted to specific values of $175\,\mu g\,m^{-3}$ in 1958 to $75\,\mu g\,m^{-3}$ in 1968?

The answer is that empirically derived conversions were applied. For reflectometer readings of 40–99, the following formula is used [30]:

$$C=\frac{F}{V}\times\left(91,679.22-3332.0460\,\mathrm{R}+49.618884\,R^2-0.35329778\,R^3+0.0009863435\,R^4\right)$$

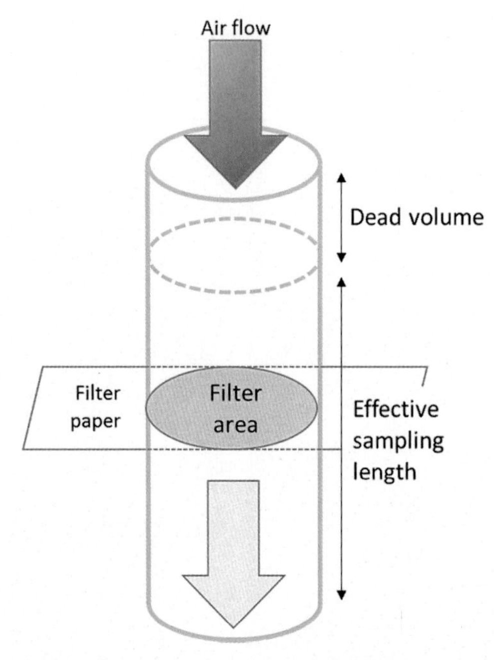

FIG. 1.2 Apparatus for black smoke (BS) method. The effective sampling length is 405 mm. The dead volume is the volume of gas away from the sampling point in the pipe up to the filter paper, which includes the probe and lines. The opacity of the filter area is measured using a reflectometer. *(Modified from A.L. GmbH, Smoke Value Measurement with the Filter Paper Method, Application Notes, Graz, Austria, AVL1007E (2005), vol. Revision 2.)*

where C = concentration in μg m^{-3}; V = volume of air sampled, in cubic feet (m^3 is converted to cubic feet by multiplying by 35.315); R = reflectometer reading; and F = a factor relating to the sampler clamp size:

0.288 for 0.5 in. clamp
1.00 for 1 in. clamp
3.68 for 2 in. clamp
12.80 for 4 in. clam

This formula represents the calibration curve to within ±1.3% over the range of reflectometer readings between 40 and 90. When used to calculate concentrations from reflectometer readings between 91 and 98, the results may be underestimated by as much as 6%. For darker stains with reflectometer readings between 40 and 20, the formula used is $C = \frac{F}{V} \times (214{,}245.1 - 15{,}130.512\,R + 508.181\,R^2 - 8.831144\,R^3 + 0.0628057\,R^4)$. Thus, the 175 and 75 values could have been reported as 175 ± 2 and 75 ± 1, respectively. Such ratios are important because we need to compare readings over decades, or even centuries, during which time measurement technologies continue to improve and methods are replaced. Thus, there have been several studies attempting to compare BS method results with later, direct PM measurements [27, 31–33].

1.2.2 Recent history

In the middle of the 20th century, almost every country in Europe and Japan, Australia, and New Zealand experienced serious air pollution in its larger cities, with each enacting national air pollution control legislation. By 1980, air pollution institutions emerged, including Warren Springs Laboratory, Stevenage, England; the Institut National de la Santé et de las Recherche Medicale at Le Visinet, France; the Rijksinstituut Voor de Volksgezondheid, Bilthoven and the Instituut voor Gezondheidstechniek-TNO, Delft, the Netherlands; the Statens Naturvardsverk, Solna, Sweden; the Institut für Wasser-Bodenund Luft-hygiene, Berlin; and the Landensanstalt für Immissions und Bodennutzungsshutz, Essen, Germany, along with many in Japan.

Smog in the United States continued to worsen, notably in Los Angeles, but extending to large cities throughout the nation. In 1955, the first federal air pollution legislation was enacted, which funded research, training, and technical assistance. The Public Health Service (PHS) of the US Department of Health, Education, and Welfare oversaw these efforts until 1970, when it was transferred to the newly formed EPA. The initial federal legislation was amended and extended several times between 1955 and 1980, greatly increasing federal authority, particularly in emissions control [34]. The automobile continued to proliferate, peaking in 2005 and declining substantially in 2009 and 2010, likely due to the economic recession. However, the rest of the world has continued to increase vehicle ownership. Global production of motor vehicles increased from approximately 54 million in 1997 to 95 million in 2016. In fact, China produced twice as many vehicles as either the United States or Japan in 2016 (over 28 million compared with 12 million in the United States and 9 million in Japan) [35].

Air pollution technologies grew rapidly from the 1950s to the 1970s, focusing predominantly on automotive air pollution and its control, as well as oxides of sulfur (SO_x) pollution and its control by sulfur oxide removal from flue gases and fuel desulfurization, and on control oxides of nitrogen, especially nitrogen dioxide (NO_2) produced in combustion processes [6].

Air pollution meteorology became a separate discipline. By 1980, mathematical models of the pollution of the atmosphere were being energetically developed. A start had been made in elucidating the photochemistry of air pollution. Air quality monitoring systems became operational throughout the world. A wide variety of measuring instruments became available, many of which are very similar to those being used today.

Environmental protection became almost universally accepted in the 1970s. Organizationally, this has taken the form of departments or ministries of the environment in governments at all levels throughout the world. In addition to national agencies, states, provinces, counties and cities took on greater responsibility for air and water quality, sold waste sanitation, noise abatement, and control of the hazards associated with radiation and the use of pesticides. This is paralleled in industry, where formerly diffuse responsibility for these areas is increasingly the responsibility of environmental protection departments. Similar changes became evident in research and education institutions.

More recently, attention was directed to the problems caused by the buildup of greenhouse gases in the atmosphere, depletion of the stratospheric ozone layer by chlorofluorocarbons (CFCs), long-range transport of pollution (especially persistent organic pollutants, i.e., POPs), Prevention of Significant Deterioration (PSD), and acidic deposition. Disasters have recently highlighted the importance of air pollution. The meltdown of the Fukushima nuclear power facility in Japan and the potential for attendant migration of radioisotopes created concern on the West Coast of North America. Like the

PBT pollution of Inuit peoples, Fukushima served as another reminder that pollutants can travel thousands of miles in winds aloft. Wildfires in North and biomass burning in South America have released large amounts of PM and other air pollutants. Even what would be considered water disasters, the huge oil spills in the Gulf of Mexico and around the world had to include measurements of volatile organic compounds and other potential airborne contaminants.

As we shall see when discussing combustion efficiencies, one of the indicators of success is the production of carbon dioxide (CO_2). The presence of water and CO_2 means we have complete combustion and have avoided producing many very toxic and persistent air pollutants, such as carbon monoxide (CO) and polycyclic aromatic compounds (PACs), known as products of incomplete combustion (PICs). More recently, however, the product of complete combustion, CO_2, is also a greenhouse gas.

There are many successes, but industrial pollution continues to be a worldwide and growing problem. Hopefully, developing nations will not repeat the 20th century problems experienced by Western nations and will apply lessons without having to suffer the consequences of side effects from economic development.

Air pollution problems are not solely the result of industry. In fact, a problem that originally plagued residences in London and other developed nations centuries ago remains among the worst air pollution problems worldwide. Open-fire and other solid-fuel cookstoves may be responsible for four million premature deaths annually, especially in developing nations [36].

1.3 THE SCIENCE OF AIR POLLUTION

Understanding air pollution begins with the physical sciences. The atmosphere is one of the spheres of importance to environment. It surrounds the globe and interacts with and connects the earth's surface to space. As such, all physical processes are at work in the atmosphere, including gravity, pressure, friction, and exchanges of matter and energy. The atmosphere is the major system by which substances are transported and transformed into either essential or detrimental compounds that we breath.

Like most air pollution texts, chemical air pollutants dominate this book. However, pollutants can also be physical and biological. The energy from ultraviolet (UV) light is an example of a physical stressor. Although exposure is to the physical contamination, that is, energy at this wavelength, the exposure has been indirectly increased by chemical contamination. For example, the release of chemicals into the atmosphere, in turn react with ozone in the stratosphere, decreasing the ozone concentration and increasing the amount of UV radiation at the earth's surface. This has meant that the mean UV dose in the temperate zones of the world has increased. This has been associated with an increase in the incidence of skin cancer, especially the most virulent form, melanoma.

Air pollutants may also be biological, as when bacteria and viruses are released to the atmosphere from medical facilities. Other biological air pollutants include irritants and allergens, such as pollen and molds (i.e., bioaerosols).

Thus, calculations related to the structure of the atmosphere are crucial to science that underpins air pollution decisions. The troposphere, that is, the lowest atmospheric layer, is the major location of most air pollution. However, the stratosphere is also important, since it contains the ozone layer, which absorbs much of the incoming ultraviolet (UV) light. When the stratospheric ozone (O_3) concentrations fall, exposures to UV increase, which damages skin cells and can lead to chronic problems, notably increased incidence of melanoma, the most virulent form of skin cancer.

This book approaches air pollution from a question-and-answer perspective. Questioning is the very essence of air quality. Some of the questions go back to the earliest times of human existence, like "why does the air smell so bad" or "when I burn different things, why does the smoke look different?" Modern questions are often different only in details, such as "what chemicals make the smoke smell this way" or "what chemicals are in the fuel that lead to smog?"

There are myriad questions about air pollution. Some can be readily and completely answered. Most can be answered only under stipulated scientific conditions. Some have more than one answer. Some have answers that will differ depending on whom one asks. Each of these types of questions is asked in this book.

1.4 THE ATMOSPHERE

The concentration of gases is often the first expression of atmosphere. At the global scale, the gas concentrations are usually described by volume rather than by weight. The presence of water is always a consideration in environmental sciences and engineering. This is no different for atmospheric science. Water in its various physical states, that is, solid, liquid, or gas, is found at some concentration in most of the lower atmosphere. The vapor phase of water in the atmosphere is commonly referred to as humidity. Given the varying amounts of humidity, meteorologists and atmospheric scientists often express gas concentrations as dry volume, that is, no humidity. By dry volume, 99.997% of the atmosphere consists of four gases,

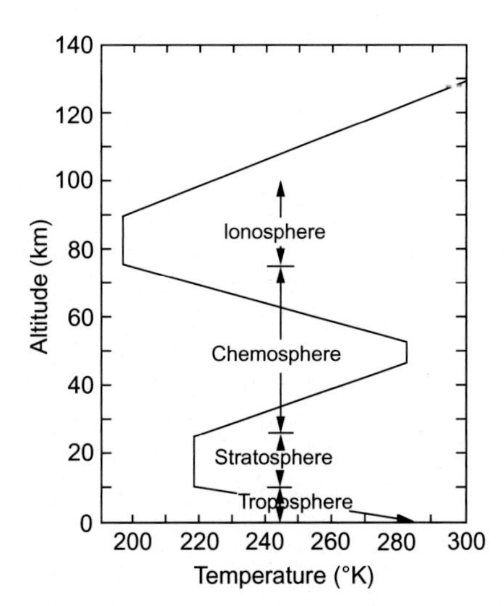

FIG. 1.3 Temperature profile of the Earth's atmosphere.

molecular nitrogen and oxygen (N_2 and O_2, respectively), argon (Ar), and carbon dioxide (CO_2). Chemically, Ar is inert (nonreactive) since it is a noble gas. The other three compounds are also very stable and nonreactive under atmospheric conditions of temperature and pressure, so they remain very stable components of the atmosphere. Approximately 99% of the mass of the atmosphere lies within 50 km (km) of the Earth's surface, that is, in the troposphere (lowest level) and stratosphere. This is where air pollution occurs.

The earth's atmospheric temperature varies with altitude (Fig. 1.3), as does the density of the substances comprising the atmosphere. The earth is warmed when incoming solar radiation is absorbed at or near the earth's surface and reradiated at longer electromagnetic wavelengths (infrared). In general, the air grows progressively less dense with increasing altitude moving upward from the troposphere through the stratosphere and the chemosphere to the ionosphere. In the upper reaches of the ionosphere, the gaseous molecules are few and far between as compared with the troposphere.

The ionosphere and chemosphere are of interest to space scientists because they must be traversed by space vehicles *en route* to or from the moon or the planets, and they are also regions in which satellites travel in the earth's orbit. These regions are also of interest to communication scientists because of their influence on radio communications. However, these layers are of interest to air pollution scientists primarily because of their absorption and scattering of solar energy, which influence the amount and spectral distribution of solar energy and cosmic rays reaching the stratosphere and troposphere.

The stratosphere is of interest to a wide range of scientists and engineers. Aeronautical scientists and engineers see it as a layer to traverse when exiting the earth's orbit but also because it has been traversed by airplanes like the supersonic transport (SST) planes in the late 20th century. The stratosphere is also the venue of communication scientists who must consider ionization and other factors to provide reliable radio, television, and satellite communications.

Air pollution scientists and engineers must also be familiar with the stratosphere for numerous reasons. Chemical compounds are formed and transformed there. Persistent chemicals can move thousands of miles via global transport. For example, debris from aboveground nuclear tests and volcanic eruptions can be suspended for protracted periods of time within the stratosphere given the intense rates of absorption and scattering of solar energy. Arguably, the best-known reason for air pollution experts to care about the stratosphere is that the lower layers contain the ozone layer, which absorbs harmful ultraviolet (UV) solar radiation. The layer has been threatened by chemicals, especially the chlorofluorocarbons (CFCs), several of which have been banned as a result. Global change scientists are also concerned about CFCs and other gases released at the earth's surface or by high-altitude aircraft.

1.4.1 Tropospheric physics

Most air pollution occurs in the troposphere, so except for stratospheric ozone depletion and climate change, air pollution texts, including this one, almost exclusively address pollution near the earth's surface. Thus, we need to understand the mechanisms responsible for gas and particle concentrations and mixing in the troposphere. These mechanisms not only are keys to estimating where pollutants form and move but also help to explain the likelihood that humans and ecosystems will be exposed to air pollutants. We can begin by posing and answering a few questions about atmospheric gases.

TABLE 1.1 Mixing ratio of principal gases in the earth's troposphere

Gas	Mixing ratio (Dry air) ($mol\,mol^{-1}$)
Nitrogen (N_2)	0.78
Oxygen (O_2)	0.21
Argon (Ar)	9.3×10^{-3}
Carbon dioxide (CO_2)	3.7×10^{-4}
Neon (Ne)	1.8×10^{-5}
Ozone (O_3)	$0.01–1.0 \times 10^{-5}$
Helium (He)	5.2×10^{-6}
Methane (CH_4)	1.7×10^{-6}
Krypton (Kr)	1.1×10^{-6}

Data from R.M. Goody, Y.L. Yung, Atmospheric Radiation: Theoretical Basis. Oxford University Press, 1995; and D.J. Jacob, Introduction to Atmospheric Chemistry, Princeton University Press, 1999.

What is the mixing ratio? What advantage is there to using this as opposed to concentrations of gases in the atmosphere?
The general definition is the ratio of the amount of a single substance to the total amount of all substances in a mixture. In air pollution, this can be either a mole ratio (r_i) or mass ratio (ζi):

$$r_i = \frac{n_i}{n_{total} - n_i} \tag{1.1}$$

where n_i = the amount (e.g., moles) of the single substance and n_{total} = the total amount of the mixture.

$$\zeta = \frac{m_i}{m_{total} - m_i} \tag{1.2}$$

where m_i = the mass of the single substance and m_{total} = the total mass of the mixture.

So, then, why use something seemingly more complicated than simple chemical concentration to express the amount of gases in the atmosphere? The principal reason is that the mixing ratio has the advantage of remaining constant with changing air density, as opposed to listing concentrations (e.g., ppm or mg m^{-3}). The mixing ratios of some important tropospheric gases are provided in Table 1.1.

Why is the mixing ratio below 100 km altitude in the atmosphere so highly variable for most substances other than molecular nitrogen and the noble gases?
Noble gases have their outer shell filled with electrons so are chemically nonreactive. Thus, their concentrations depend on sources entirely, and do not diminish due to chemical reactions in the upper atmosphere like most other substances. Similarly, N_2 is nonreactive except under extreme conditions, for example, very high temperatures and pressures like those in internal combustion engines, where it is oxidized to nitric oxide:

$$N_2 + O_2 \rightarrow 2NO$$

Why do the mixing ratios for methane and nitrous oxide decrease beginning at about 20 km of altitude?
This is the flip side of the previous question. These are much more chemically reactive. Thus, with increasing temperature and photochemistry with altitude, they increasingly degrade.

Why do concentrations of H_2O vapor and CO_2 vary so much in the atmosphere compared with most other substances?
Water vapor concentrations vary because of continual phase changes caused by changes of temperature. Carbon dioxide variability is related to its uptake and release by plants and, to a lesser degree, its emissions from animals and other organisms from respiration and anthropogenic activities.

What are common atmospheric gases that absorb incoming (shortwave) solar radiation?
Ozone, O_2, CO_2, and H_2O absorb portions of the solar spectrum.

Which gases absorb outgoing thermal (infrared) radiation?
Carbon dioxide, H_2O, O_3, N_2O, CH_4, and chlorofluorocarbons absorb portions of the thermal emission spectrum. (Note that ozone and water vapor absorb in both spectral ranges, so they can both shield insolation and hold in heat (see Fig. 1.4).

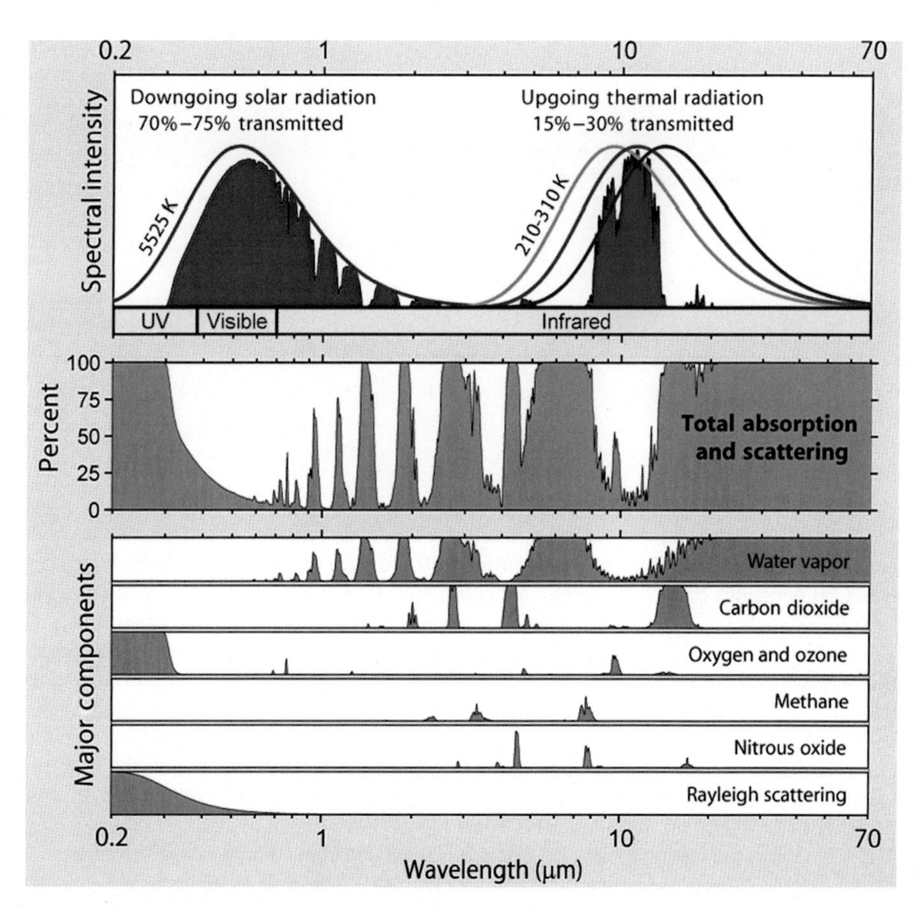

FIG. 1.4 Radiation transmitted to the atmosphere. *(Source: R. Rohde, Radiation Transmitted by the Atmosphere (2007). Based on Data from L. Gordley, B. Marshall, C. Allen, Linepak: Algorithms for Modeling Spectral Transmittance and Radiance 3 (22) (1994) 2016, https://commons.wikimedia.org/wiki/File:Atmospheric_Transmission.png.)*

Why does such a small amount of the atmosphere's gases not drift off into space?
The force of gravity.

Which principal atmospheric gas changes the most in concentration with altitude?
Ozone concentrations increase with height, reaching peak levels in the middle stratosphere, and then decrease up to ~ 50 km. High O_3 concentrations are produced in the stratosphere by photodecomposition of O_2 and recombination to form O_3.

How does the temperature change with altitude in the troposphere and stratosphere?
In the troposphere, temperature
a) decreases with increasing altitude up to ~12–15 km,
b) stays the same (i.e., isothermal conditions) within most of the stratosphere,
c) increases with height up to the highest part of the stratosphere,
d) becomes isothermal above the stratosphere before it,
e) decreases with height up to ~ 80 km, and
f) becomes isothermal before it again increases with height.

These changes result from electromagnetic spectral processes that cause warming and cooling, in different regions of the atmosphere, that is, in the troposphere, warming due to reradiation of incoming solar radiation as infrared from the earth's surface and solar heating and, modulating heating by the stratospheric ozone layer by absorption of UV light, and seasonal changes in the intensity of incoming sunlight and length of daylight due to increasing angles in summer (heating) and decreasing angles in winter (cooling).

What is the spectral range of thermal radiation?
4–30 μm.

At what spectral range is the atmosphere transparent to IR energy (i.e., heating)?
The so-called atmospheric window is the spectral region of ~7–14 μm. The atmosphere in this spectral region is, generally, transparent to the emission of infrared energy.

What are the major sources of CH_4 and CO_2 in the environment? How do these two compounds relate to each other?
>60% of the world's CH_4 emissions come from human activities. It is emitted from industries, agriculture, and waste treatment. Natural gas and petroleum systems are the largest source of CH_4 emissions from industry in the United States. It is the primary component of natural gas, so CH_4 is emitted to the atmosphere during the production, processing, storage, transmission, and distribution of natural gas. Because gas and crude oil commonly exist in the same or adjacent geologic strata, the production, refinement, transportation, and storage of petroleum is a source of CH_4 emissions. Domestic livestock (e.g., cattle, buffalo, sheep, goats, and camels) produce large amounts of CH_4 as part of their normal digestive process (e.g., rumination). In addition, when animal manure is stored, CH_4 is produced by anaerobic digestion. Agriculture is the primary source of CH_4 emissions. Landfills are the third largest US source of CH_4 emissions. Methane is also emitted from natural sources. Wetlands are the largest source, emitting CH_4 from bacteria that degrade organic materials in the absence of oxygen. Termites, oceans, sediments, volcanoes, and wildfires are other natural sources of CH_4.

Carbon dioxide is the redox mirror image to methane; that is, it is produced from complete oxidation of organic compounds. CO_2 is naturally present in the atmosphere as part of the carbon biogeochemical cycle. Human activities are altering the carbon cycle—both by adding more CO_2 to the atmosphere and by influencing the ability of natural sinks, like forests, to remove CO_2 from the atmosphere. While CO_2 emissions come from a variety of natural sources, human-related emissions are responsible for the increase that has occurred in the atmosphere since the industrial revolution.

Why has the preindustrial concentration of N_2O changed so little as of 1998 compared with that of methane and carbon dioxide?
Anthropogenic sources have increased at a much higher rate for CH_4 and CO_2 than for N_2O.

What were the sources of perfluoromethane before the industrial revolution? Why has it doubled since then?
This is sort of a trick question, since there is much debate about the sources of CF_4 before modern times. The most likely sources are crustal emissions due to weathering, volcanic activity, etc. The very long half-life means that whatever amount is emitted will remain in the atmosphere for tens of thousands of years. This and the additional anthropogenic sources, especially aluminum refining, are the reason for the increase. In an aluminum smelter, alumina is reduced in a reduction cell to form elemental aluminum (Al^0) and carbon dioxide:

$$Al_2O_3 + 3C \rightarrow 4Al + 3CO_2$$

At very high temperatures needed to melt Al (i.e., about 1150°C), these reduction cells can produce perfluorocarbons (CF_4 and C_2F_6):

$$4Na_3AlF_6 + 3C \rightarrow 4Al + 12NaF + 3CF_4$$

$$4Na_3AlF_6 + 4C \rightarrow 4Al + 12NaF + 2\,C_2F_6$$

What are some of the possible air pollutants that can be released by confined animal feeding operations? What can be done to address them?
The major airborne pollutants from confined feeding operations are various forms of nitrogen. Organic compounds will breakdown ultimately to CO_2 in oxidized parts of the system, for example, lagoons, whereas CH_4 will be released from reduced environments. The nitrogen can be released in the gas phase as nitric oxide (NO), where oxidation is occurring (known as nitrification), or as ammonia (NH_3), in reduced (denitrification) environments, even including waste treatment process; for example, nitrates in water can be aerosolized by aeration of farm lagoons. If lagoons become anaerobic, the sulfur-reducing bacteria (e.g., *Desulfovibrio spp.*) will release H_2S. Other pollutants can be released, for example, pesticides and other farm chemicals. In addition, biological pollutants can be released as or on aerosols into the air from the animal waste (e.g., spores, cysts, bacteria, fungi, and viruses).

1.4.2 Atmospheric pressure

Atmospheric pressure is the force exerted on an object by the weight of the atmosphere above it. It can be visualized as a column atop the object on the earth's surface, extending all the way to the top of the atmosphere. Atmospheric pressure results from the transfer of momentum by the random motion of air molecules.

Unequal (differential) heating of the earth's surface makes the air move because of temperature and pressure differences. The laws of potentialities state that movement is from higher potentials to lower potentials (e.g., height, energy, and pressure).

1.4.3 Standard conditions

Standardization is crucial to good science, since sound and credible science depends on repeatability. Diversity is a wonderful thing in most human enterprises. "Variety is the spice of life." However, if each scientist uses a unique method, generalizations are impossible. Science depends on logical progressions. Most commonly, science is deductive, that is, knowledge is inferred from the general to specifics. Scientific laws (the general) dictate the applications (the specific).

Of course, even these laws are subject to change, given enough countervailing evidence, for example, inductive reasoning by a sufficient body of specifics to override outdated scientific principles. Such paradigm shifts come only after much evidence and run into resistance from the *status quo* [39, 40]. But to make science repeatable, we need standards.

Thus, to compare the corpus of air pollution literature and knowledge and to apply meaning to atmospheric research and investigations, we must begin with what air pollution scientists mean by "the atmosphere."

The standard atmosphere is unit of pressure. One standard atmosphere (atm) = 101,325 Pa (Pa) = 101.325 kilopascals (kPa) = 1013.25 hectopascals (hPa) = 1013.25 millibars (mb). Air pollution publications use these and other pressure units (e.g., mm Hg). American engineers use a mix of pressure units, depending on the task, for example, designing an air mover, installing a pump, or testing indoor air. Meteorologists not only use atm but also use other units like hPa, inches Hg, and mm Hg. Air pollution texts and manuals may prefer using kPa most often because this unit is commonly used in environmental and engineering literature to express vapor pressure and partial pressures. Some environmental partitioning coefficients include Pa units, for example, Henry's law coefficients, but the Pa is usually too large to be used in most environmental literature, given the frequently that air pollutants thankfully have low partial pressures and concentrations in the atmosphere. Also, where air pollution precursors are at very high concentrations, for example, a stack or reactor, chemical concentrations may be expressed as in Pa units for a stack test, for example. The key is that secondary data sources useful to air pollution calculations vary in the units used for pressure and other factors, so conversions are often needed.

Standard temperature and pressure (STP) is 273.15 K and 1 atm. It is used to allow for consistent comparisons of results from scientific studies. For many engineering studies and experiments, normal temperature and pressure (NTP) is preferred, because NTP is 293.15 K (20°C), which is closer to "room temperature." Studies conducted at different temperatures and pressures need to be adjusted to these standards. For example, if a study was done at 290 K and 1.5 atm, the results must be corrected to 273.15 K and 1 atm for STP. Other factors are particularly important for emissions, especially moisture content and oxygen levels. In stack tests, corrections to STP are a means of ensuring that the amount of any pollutant that can leave a stack is consistently applied to all similar stacks in a jurisdiction. In testing air movers and fans, corrections to NTP allow for consistent comparisons for indoor air studies.

For air pollution studies specifically, NTP is often used. For example, EPA's standard conditions for air pollution measurements are 20°C and 1 atm.

1.5 RECENT TRENDS

Air pollution engineers, scientists, and managers have done a remarkable job in the past half-century. Public health has been the principal driver for assessing and controlling air contaminants, but air pollution abatement laws and programs throughout the world have also recognized that effects beyond health are also important, especially *welfare* protection. One of the main welfare considerations is that ecosystems are important *receptors* of contamination. Another welfare concern is that contaminants impact structures and other engineered systems by corrosion. Thus, from an air pollution perspective, there is a cascade of hazards from human health to ecosystems to abiotic (i.e., nonliving) systems.

The US Congress enacted the 1970 amendments to the Clean Air Act to provide a comprehensive set of regulations to control air emissions from area, stationary, and mobile sources. This law authorized the US EPA to establish National Ambient Air Quality Standards (NAAQS) to protect public health and the environment from the "conventional" (in contrast to "toxic") pollutants: carbon monoxide, particulate matter (PM), oxides of nitrogen, oxides of sulfur, and photochemical oxidant smog or ozone. The metal lead (Pb) was later added as the sixth NAAQS pollutant. The original goal was to set and to achieve NAAQS in every state by 1975. These new standards were combined with charging the 50 states to develop state implementation plans (SIPs) to address industrial sources in the state.

The ambient atmospheric concentrations are measured at over 4000 monitoring sites across the United States. The ambient levels have continuously decreased, as shown in Table 1.2. The Clean Air Act Amendments of 1977 mandated new dates to achieve attainment of NAAQS (many areas of the country had not met the prescribed dates set in 1970). Other amendments were targeted at types of air pollution that had not previously been addressed sufficiently, including acidic deposition, tropospheric ozone pollution, depletion of the stratospheric ozone layer, and a new program for air toxics, known as the National Emission Standards for Hazardous Air Pollutants (NESHAPS).

1.6 SPATIAL SCALE

Characterizing and controlling air pollution varies by the size of the problem. Air is polluted at all geographic scales, from the individual receptor, for example, a person, to local, to global. A town engineer or city manager, for example, may rely on air pollution data and reports at the urban or regional scale. These can be subdivided into five ascending scales: local, urban, regional, continental, and global. For situations at finer scales, for example, truck exhaust exposure during road repairs, this information may need to be augmented with data on the concentration of certain air pollutants at microenvironmental scale (e.g., indoor air in a home, office, or garage) or personal scale (i.e., at the individual's breathing zone). The spheres of

TABLE 1.2 Percentage decrease in ambient concentrations of National Ambient Air Quality Standard Pollutants from 1990 to 2014

Pollutant and averaging time	Decrease in ambient concentration
Carbon monoxide (CO) 8 hour	77%
Lead (Pb) 3-month average	99%
Nitrogen dioxide (NO_2) annual	56%
Nitrogen dioxide (NO_2) 1 hour	50%
Ozone (O_3) 8 hour	22%
Particulate matter 10 μm (PM_{10}) 24 hour	39%
Particulate matter 2.5 μm ($PM_{2.5}$) annual	42%
Particulate matter 2.5 μm ($PM_{2.5}$) 24 hour	44%
Sulfur dioxide (SO_2) 1 hour	85%

Source: U.S. Environmental Protection Agency. Our Nation's Air—Status and Trends Through 2016. (2017). Available: https://gispub.epa.gov/air/trendsreport/2017/#highlights.

influence of the air pollutants themselves range from molecular (e.g., gases and nanoparticles) to planetary (e.g., diffusion of greenhouse gases throughout the troposphere). The local scale is up to about 5 km of the earth's surface. The urban scale extends to the order of 50 km. The regional scale is from 50 to 500 km. Continental scales range from several 1000 km. Of course, the global scale is planetwide [11].

1.6.1 Local scale

Local air pollution problems are usually characterized by one or several large emitters or many relatively small emitters. The local scale is the focus of most waste managers. Indeed, the waste management facility is often a source of air pollution, for example, fugitive dust from a landfill or stack emissions from an incinerator. Thus, the waste manager, like any reputable environmental professional, must find ways to ameliorate air emissions.

Lowering the release height of a source, for example, an incinerator, spreads the potential local impact for a given release. Examples include carbon monoxide and hydrocarbon released from tailpipes or oxides of sulfur released from power plants. In fact, this phenomenon is the reason that certain controls have led to larger-scale problems, that is, solving a local air pollution problem by increasing the height of the point of release (i.e., tall stacks) allowed pollutants to travel long distances from the source. Any ground-level source, such as evaporation of volatile organic compounds from a gasoline station, will produce the highest concentrations near the source, with concentrations generally diminishing with distance. This phenomenon is known as a concentration gradient.

Large sources that emit high above the ground through stacks can also cause local problems, especially under unstable meteorological conditions that cause portions of the plume to reach the ground in high concentrations.

There are many releases of pollutants from relatively short stacks or vents on the top of one- or two-story buildings. Under most conditions, such releases are caught within the turbulent downwash downwind of the building. This allows high concentrations to be brought to the ground surface. Many different pollutants can be released in this manner, including compounds and mixtures that can cause odors. Usually, the effects of accidental releases are confined to the local scale.

1.6.2 Urban scale

Waste managers are part of the decision-making structure of urban areas. They may represent individual towns, cities, or counties in a metropolitan area or be part of a multijurisdictional planning association [42]. Air pollution problems in urban areas generally result from either primary or secondary pollutants. Primary pollutants are released directly from sources, whereas secondary pollutants form through chemical reactions of the primary pollutants in the atmosphere after release. Air pollution problems can be caused by individual sources on the urban scale and the local scale. For gaseous pollutants that are relatively nonreactive, such as carbon monoxide (CO), and relatively slowly reactive, such as sulfur dioxide (SO_2), the contributions from individual sources combine to yield high concentrations. Since motor vehicles are the major source of CO in the ambient air, "hot spots" of high concentration can occur especially near intersections with high traffic. The emissions are especially high from idling vehicles. The hot spots are exacerbated if high buildings surround the intersection, since the volume of air in which the pollution is contained is severely restricted. The combination of these factors results in high concentrations.

Particulate matter (PM) is highly variable in its composition. By mass, the chemical reactivity of the individual particle may range. The liquids and solids that make of the PM often consist of chemically reactive substances, especially those adsorbed onto particle surfaces.

Tropospheric ozone (O_3) is a dominant air pollution problem in many urban areas. It forms from reactions among secondary pollutants in the presence of sunlight's ultraviolet range, that is, photochemical oxidant smog. Many metropolitan areas in the United States are in nonattainment for ozone, that is, they are not meeting the air quality standards. In the United States, the Clean Air Act Amendments (CAAA) of 1990 classifies the various metropolitan areas to be in nonattainment according to the severity of the problem for that attainment with the National Ambient Air Quality Standards (NAAQS).

Oxides of not only nitrogen, principally nitric oxide (NO), but also nitrogen dioxide (NO_2) are emitted from automobiles and from combustion processes. Hydrocarbons are emitted from many different sources. The various species have widely varying reactivities. Determining the emissions of these chemical species from myriad sources as a basis for pollution control programs can be difficult, but methods continue to improve.

1.6.3 Regional scale

Three problem types contribute to air pollution problems on the regional scale. The first is the blend of urban oxidant problems at the regional scale. Many major metropolitan areas are merging spatially and continue to grow. Urban geographers refer to some of the larger urban aggregations as "megalopolises." As a result, the air from one metropolitan area, containing both secondary pollutants formed through reactions and primary pollutants, flows on to the adjacent metropolitan area. The pollutants from the second area are then added on top of the "background" from the first.

The second problem involves the release of relatively slow-reacting primary air pollutants that undergo reactions and transformations during lengthy transport times. Protracted transport times result over regional scales, which allows for sufficient time for transformation not only transforming the parent compounds but also the potential to generate numerous transformation by-products that form along the way. The distance between an emission and the harm it causes can be large. To illustrate, acid rain can result when the gas, sulfur dioxide (SO_2), is released primarily through combustion of fossil fuels (especially from coal and oil) and is then oxidized during long-distance transport to sulfur trioxide (SO_3):

$$2SO_2 + O2 \rightarrow 2SO_3 \tag{1.3}$$

Although SO_2 is a gas, both gas-phase and liquid-phase oxidation of SO_2 occurs in the troposphere. The SO_3 in turn reacts with water vapor to form sulfuric acid:

$$SO3 + H_2O \rightarrow H_2SO_4 \tag{1.4}$$

Sulfuric acid reacts with numerous compounds to form sulfates. These are fine (submicrometer) particulates. Nitric oxide (NO) results from high-temperature combustion, both in stationary sources such as power plants or industrial plants in the production of process heat and in internal combustion engines in vehicles. The NO is oxidized in the atmosphere, usually rather slowly, or more rapidly if there is ozone present, to nitrogen dioxide (NO_2), which also reacts further with other constituents, forming nitrates, which is also in fine particulate form.

The sulfates and nitrates existing in the atmosphere as fine particulates, generally in the size range $<1\ \mu m$ (aerodynamic diameter), can be removed from the atmosphere by several processes. "Rainout" occurs when the particles serve as condensation nuclei that lead to the formation of clouds. If the droplets grow to a sufficient size, they fall as raindrops with the particles in suspension or in solution. Another mechanism, known as "washout," also involves rain, but the particles in the air are captured by raindrops falling through the air. Both mechanisms contribute to "acid rain," which results in the sulfate and nitrate particles reaching lakes and streams and increasing their acidity. As such, acid rain is both a regional and continental problem [11].

A third type of regional problem is visibility, which may be reduced by specific plumes or by the regional levels of particulate matter that produce various intensities of haze. The fine sulfate and nitrate particulates just discussed are largely responsible for reduction of visibility. This is especially problematic in locations of natural beauty, where it is desirable to keep scenic vistas as free of obstructions to the view as possible. Regional haze is a type of visibility impairment that is caused by the emissions of air pollutants from numerous sources across a broad region. The CAAA provides special protections for such areas; the most restrictive denoted as mandatory Federal Class I areas that cover over 150 national parks and wilderness areas in the United States [43]. Decreased visibility can also impair safety, especially concerning aviation.

1.6.4 Continental scale

In Europe, there is little difference between what would be considered regional scale and continental scale. However, on larger continents, there would be a substantial difference. Perhaps, of greatest concern on the continental scale is that the air pollution policies of a nation are likely to create impacts on neighboring nations. Acid rain in Scandinavia has been considered to have had impacts from Great Britain and Western Europe. Japan has considered that part of their air pollution problem, especially in the western part of the country, has sources in China and the Korean peninsula. For decades, Canada and the United States have cooperated in studying and addressing the North American acid rain problem. Likewise, tall stacks in Great Britain and the lowland countries of continental Europe have contributed t acid rain in Scandinavia. British industries for some years simply built increasingly tall stacks as a method of air pollution control, reducing the immediate ground-level concentration but emitting the same pollutants into the higher atmosphere. The local air quality improved but at the expense of acid rain at other parts of Europe.

1.6.5 Global scale

The release of radioactivity from the accident at Chernobyl would be considered primarily a regional or continental problem. However, elevated levels of radioactivity were detected in the Pacific Northwest part of the United States soon after the accident. Likewise, persistent organic pollutants, such as polychlorinated biphenyls (PCBs), have been observed in Arctic mammals, thousands of miles from their sources. These observations demonstrate the effects of long-range transport.

In this and other instances, the gases are released from numerous locations on the earth's surface but when added together can change the temperature and other climatological features of the troposphere. In the case of chlorofluorocarbons (CFCs), their release results in free chlorine (Cl) atoms that attack ozone (O_3) in the stratosphere, which is ordinarily quite stable and resists vertical air exchange between layers. However, halogen compounds can be transferred from the troposphere into the stratosphere by injection through the tops of thunderstorms that occasionally penetrate the tropopause, the boundary between the troposphere and the stratosphere. Some transfer of stratospheric air downward also occurs through occasional gaps in the tropopause. Since the ozone layer is considerably above the troposphere, the transfer of chlorofluorocarbons upward to the ozone layer is expected to occur gradually. Thus, there is a lag from the first release of these gases until an effect is seen, that is, the so-called thinning of the O_3 layer at the poles. Similarly, with the cessation of the use of these materials worldwide, there has been a commensurate lag between in the restoration of the O_3 layer.

Other important planetary scale pollutants are the attendant global problem of climate change that is generated by excessive amounts of radiant gases (commonly known as greenhouse gases (GHGs)), especially methane (CH_4) and CO_2, which is not normally considered an air pollutant. A portion of radiation from the earth's surface is intercepted by the carbon dioxide in the air and is reradiated both upward and downward. That that is radiated downward keeps the ground from cooling rapidly. As the carbon dioxide concentration continues to increase, the earth's temperature is expected to increase [44, 45].

1.7 STRESSOR-RECEPTOR PARADIGM

The main aim in controlling air pollution is to prevent adverse responses to receptor categories exposed to stressors in the atmosphere: human, animal, plant, and material. These adverse responses, shown in Table 1.3, have characteristic response times: short term (i.e., seconds or minutes), intermediate term (i.e., hours or days), and long term (i.e., months or years). To elicit no adverse responses, the pollutant concentration in the air must be lower than the concentration level at which these responses occur. Fig. 1.5 illustrates the relationship between these concentration levels. This figure displays response curves, which remain on the concentration duration axes because they are characteristic of the receptors, not of the actual air quality to which the receptors are exposed. The odor response curve, for example, to hydrogen sulfide, shows that an inhalation event of one second can establish the presence of the odor but that, due to odor fatigue, the ability to continue to recognize that odor can be lost in a matter of minutes. Nasopharyngeal and eye irritations, for example, by ozone, are similarly subject to acclimatization due to tear and mucus production. The three visibility lines correlate with the concentration of suspended particulate matter in the air. Attack of metal, painted surfaces, or nylon hose is shown by a line starting at one second and terminating in a matter of minutes, when the acidity of the droplet is depleted by the material attacked [46].

TABLE 1.3 Examples of receptor category characteristic response times for exposures to air pollutants

Examples of receptor category characteristic response times			
	Characteristic response times		
Receptor category	**Short term (from seconds to minutes)**	**Intermediate term (from hours to days)**	**Long term (months to years)**
Human	Odor, visibility, nasopharyngeal, and eye irritation	Acute respiratory disease	Chronic respiratory disease and lung cancer
Animal, vegetation	Field crop loss and ornamental plant damage	Field crop loss and ornamental plant damage	Fluorosis of livestock, decreased fruit, and forest yield
Material	Acid droplet pitting and nylon hose destruction	Rubber cracking, silver tarnishing, and paint blackening	Corrosion, soiling, and material deterioration

Source: A.C. Stern, R. Boubel, Turner, and DB, Fox, Donald L, Fundamentals of Air Pollution, 1984.

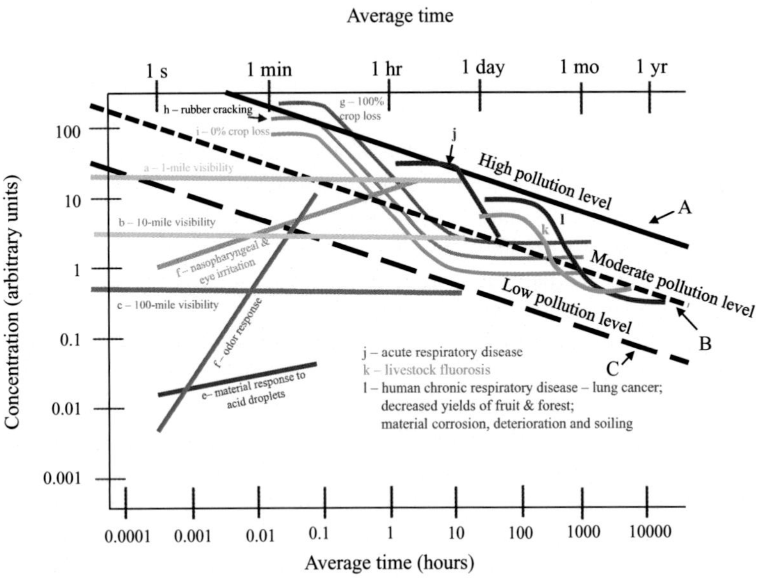

FIG. 1.5 Adverse responses to various pollution levels. *(Data from A.C. Stern, R. Boubel, Turner, and DB, Fox, Donald L, Fundamentals of Air Pollution, 1984.)*

Plant damage can be measured biologically or socioeconomically. Biological damage can be measured as stress on growth and survival. Socioeconomic damage can be measured as 0% loss when there is no loss of the sale value of the crops or ornamental plants but a 100% loss if the crop is damaged to the extent that it cannot be sold. Such responses are related to dose, that is, concentration times duration of exposure, as shown by the percent loss curves on the chart. Many manifestations of material damage, for example, rubber cracking by ozone, require an exposure duration long enough for the adverse effects to be economically significant. That is, attack for just a few seconds or minutes will not affect the utility of the material for its intended use, but attack for several days will.

The biological response line for acute respiratory disease is a dose-response curve that for a constant concentration becomes a duration-response curve.

The shape of such a curve reflects the ability of the human body to cope with short-term, ambient concentration respiratory exposures and the overwhelming of the body's defenses by continued exposure. For example, fluorosis of livestock is not induced until there has been a long enough period of deposition of a high enough ambient concentration of fluoride to increase the level of fluoride in the forage. Since the forage is either consumed by livestock or cut for hay at least once during the growing season, the duration of deposition ends after the growing season. The greater the duration of the season, the greater the time for deposition; hence, the shape of the line is labeled "fluorosis." Long-term vegetation responses—decreased yield of fruit and forest—and long-term material responses—corrosion, soiling, and material deterioration—are shown on the chart as having essentially the same response characteristics as human chronic respiratory disease and lung cancer.

The relationship of these response curves to ambient air quality is shown by lines A, B, and C, which represent the maximum or any other chosen percentile line from a display such as Fig. 1.5, which shows actual air quality. Where the air quality is poor (line A), essentially, all the adverse effects displayed will occur. Where the air quality is good (line C), most of the intermediate and long-term adverse effects displayed will not occur. Where the air quality is between good and poor, some of the intermediate and long-term adverse effects will occur but in an attenuated form compared with those of poor air quality.

Air pollutants, like most pollutants, under other circumstances would be "resources," such as compounds of nitrogen. This book spends time addressing the biogeochemical cycle of nitrogen and other nutrients and the cycles of water, carbon, and metals. In the soil, nitrogen compounds are essential nutrients. As air pollutants, however, certain chemical species of nitrogen can cause respiratory problems directly or, in combination with hydrocarbons and sunlight indirectly, can form ozone and smog. Thus, simply "removing" pollutants is often less feasible, if even possible, than managing systems to ensure that optimal conditions for health and environmental quality exist. The oxidation state of nitrogen and other elements often determines the difference between an essential and toxic compound. Even an essential compound in excess can become a pollutant. The quantity and location are determinants of whether a substance is a pollutant.

The system view also helps to inform societal decision-making. For example, the term "zero emission" has been applied to vehicles, as the logical next step following low-emission vehicles (LEVs) and ultralow-emission vehicles (ULEVs). The emissions of certain pollutants from the vehicle are indeed approaching zero. However, the function of moving people using vehicles is not "zero emission." Presently, for example, electric cars are not emission-free, but are a type of emission trading from a mobile source to a stationary source. That is, the electricity is generated at a power plant that is emitting pollutants as it burns fossil fuels or has the problem of radioactive wastes if it is a nuclear power plant. Even hydrogen, solar, and wind systems are not completely pollution-free since manufacture, transport, and installation of the parts and assemblages require energy and materials, some of which find their way to the atmosphere.

1.8 CALCULATION BASICS

We end this chapter with some examples of basic calculations. Mathematics is the language of science. Air pollution engineering and science is explained through a series of equations, reactions, statistics, indices, and models. In fact, arithmetic is the source of many errors in engineering in general. The author is guilty of these mistakes. As evidence, in demonstrating how to calculate the amount of carbon monoxide (CO) released using a simple model at the beginning of the book, *Environmental Contaminants* [47], I used the wrong sign for an exponent. The exponent was supposed to be +3 but was shown as −3. Thus, in this hypothetical example, an error of six orders of magnitude is carried through the example, underestimating the amount of carbon monoxide in the air by a factor of a million! In other words, if you make the same mistake I did, you would have what is known as a "false negative," that is, appearing not to have a problem with CO in your basement, but there is a million times more CO in the basement air than the model shows.[1]

Certainly, the calculator helps with the more common mistakes but can introduce other types of errors, especially regarding significant figures. Like my CO model mistake, the calculator or computer will simply do what you command. You must decide what is the proper level of uncertainty and validity. The tool you use will obediently string out the answer as far as its programming will allow. If that means 21 figures to the right of the decimal point, that is what you will get.

Thus, understanding and explaining air pollution requires how to not only apply the wealth of mathematical tools but also be vigilant about their limitations. As such, a modicum of mathematical knowledge is needed before we venture into the physical, chemical, and biological properties and mechanisms involved in air pollution. Thus, we begin our air pollution quest with some sample calculations of the chemical principles and the laws of motion and thermodynamics.

---

1. I included an errata sheet with the book, since I discovered this error shortly after printing. If you happen to have the book, please make sure that it includes this sheet.

1.8.1 Mathematics and statistics

Here are a few questions and answers to "prime the pump" on the proper use of math and statistics in air quality decision-making.

Engineers often use scientific notation because we often work with very small and very large numbers. A number is product of a coefficient (with a value between 1 and <10) and an exponent of 10 (e.g., 10^4 or 10^{-11}). To demonstrate, at a rate of 1 atom per second, how many years would it take to count the number of atoms in 12 g of carbon? What if your computer could count 1 million atoms per second?

Time is equal to the total number of atoms (N), divided by the counting rate (R):

$$t = \frac{N}{R}$$

$$N = \frac{N_A}{12g} \times 1g = 5.02 \times 10^{22} \text{ atoms}$$

$$\therefore t = \frac{N}{R} = \frac{5.02 \times 1022 \text{ atoms}}{1\ atom^{-}\text{sec}} = 5.02 \times 10^{22} \text{ sec} = 1.59 \times 10^{15} \text{ years}$$

So, it would take you 1,590,000,000,000,000 years, or about 100,000 times the estimated age of the universe. However, your computer would take.

$$\frac{1.59 \times 10^{15}}{10^6} = 1.59 \times 10^9 \text{ years, or } only\ 1,590,000,000 \text{ years.}$$

The confidence and validity a calculation is crucial in any scientific and engineering endeavor. MarMar is trying to determine how much water has been added to the soil on her land, so she asks each of her three grandchildren to measure and record the rainfall for a week. The results are shown below. What is the total rainfall?

Child	Days	Rainfall amount (cm)
Samuel	1–7	2
Alexander	8–14	3.009
Chloe	15–21	4.10002

Using a calculator, MarMar's result shows the total to be 9.10902 cm. However, she recognizes that this seems to be overly precise and that this information's usefulness depends on its validity, the degree to which these values in fact show the amount of water reaching the soil. She knows that Samuel, the youngest, probably was not nearly as careful as his older relatives and that Chloe and Alexander may have been a bit too precise in their interpretation of the rain in the gauge. So, she decides to apply the rules of significant figures:

Rule 1. When adding or subtracting, the answer's significance is only as high as the lowest value in the set. In this case, given Samuel's rainfall result, the answer is simply *9 cm*.

MarMar was happy with the results but, like a good scientist, decided that next time, the team would implement a standard operating procedure (SOP). She trained her "staff" on how to read the gauge and improved the way the results were recorded, for example, waterproof tablets. The results for the next period are shown below. What is the total rainfall?

Child	Days	Rainfall amount (cm)
Samuel	22–28	8.2
Alexander	29–35	5.05
Chloe	36–42	3.24

The calculator gives a total of 16.49 cm. Applying Rule 1, she knows that once again Samuel's results, albeit more precise this time, is the least significant figure. However, she knows that she cannot simply truncate the "9." Thus, she applies Rule 2:

To round off, a significant figure stays the same if the next figure is <5 but is increased by 1, if the next figure is >5. Thus, she should report that, for this 3-week period, the total rainfall was 16.5 cm.

MarMar would like to know the total rainfall on her property for the 6-week period. If she has 1.579 ha (ha) of land, of which 12% is impervious surfaces (e.g., buildings and concrete drives), and assuming the rainfall results are the same over the entire area, how much water reached the soil?

Each hectare is 10^4 m^2, so we can multiply the rainfall height by the area receiving the rainfall to ascertain the volume of water. We also need to convert our rainfall results from cm to m, that is, divide by 100, to get an answer in cubic meters. Assuming that the water is diverted and will not reach the soil under the impervious surfaces, we set up the problem as.

$$\frac{(1.579 \times 0.88)\text{m}^2 \times 10^4 \times (9 \times 16.5)\text{m}}{100} = 20634.372\,\text{m}^3$$

However, we know that this is an overly precise number. The third rule of significant figures states that when multiplying or dividing, the significant figures of the answer are the same as the quantity in the calculation with the fewest significant figures. This means that since Samuel's first answer has but one significant figure, so should our answer. Using the three rules, we should report the answer as $2 \times 10^4\ m^3$.

MarMar now wants to estimate the amount of nitrate (NO_3) that is reaching the soil from rainfall. Using the map in Fig. 1.6, she decides that there must be about 0.5 mg L^{-1} NO_3 in the average rainfall in her county. How much NO_3 reached her soil in during the 6 weeks?

Concentration of an air pollutant is expressed as mass per volume. This is also common for water pollutants, but soil and sediment pollutants are often expressed as mass per mass, for example, mass of nitrate per mass of soil. So, since the source of the concentration data is the National Atmospheric Deposition Program, NO_3 is considered an air pollutant here. To convert this concentration to a total mass, we must multiply the concentration by the total volume of water that we found in the previous question:

$$\text{Concentration} \times \text{Volume} = \text{Total Mass of Pollutant} \tag{1.5}$$

Since the concentration is expressed in liters, we must also convert L to m^3.

$$\left(0.5\,\text{mg}\,\text{L}^{-1}\,\text{NO}_3\right) \times \left(2 \times 10^4\,\text{m}^3\right) \times \left(10^3\,\text{L}\,\text{m}^{-3}\right) = 10^7\,\text{mg} = 10^4\,\text{g} = 10\,kg\,NO_3$$

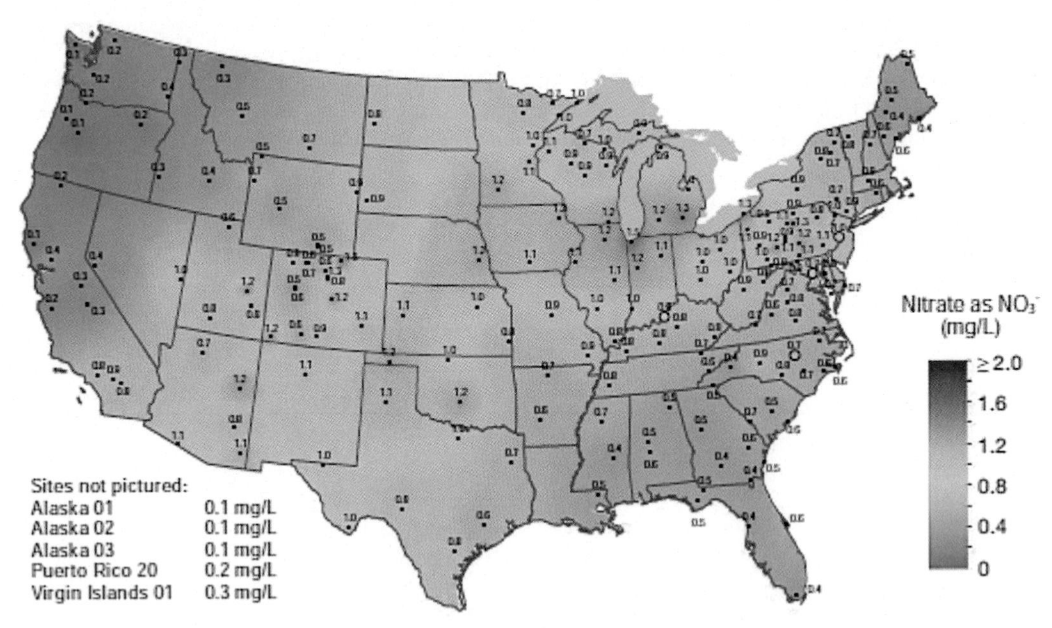

FIG. 1.6 Nitrates in precipitation. *(Source: Illinois State Water Survey. University of Illinois at Urbana-Champaign. (2013). NADP Data Report 2013-01, National Atmospheric Deposition Program: 2012 Annual Summary.)*

Air pollution is best described as a concentration and often as a suite of pollutants, such as the so-called criteria pollutants or certain hazardous air pollutants, for example, those associated with a common source or those that lead to similar health effects. However, it can be also described using an index. The air quality index (AQI), shown in Table 1.4, was developed by the US Environmental Protection Agency (EPA) for five air pollutants: particulate matter (PM), ground-level ozone (O_3), carbon monoxide (CO), sulfur dioxide (SO_2), and nitrogen dioxide (NO_2). A wildfire has been burning upwind of your town. The city engineer measured the fine particles ($PM_{2.5}$)[2] in the air and found the average for the last 24hs to be 22μg per cubic meter ($\mu g\,m^{-3}$). What category would this be in the AQI?

The AQI calculates the quality of the air for each pollutant on a scale from 0 to 500. Decreasing values indicate improving air quality. However, each pollutant is reported separately, so one pollutant can be good (i.e., green), whereas another can be

TABLE 1.4 Air quality index

Air quality index	Who needs to be concerned?	What should I do?
Good 0–50	It's a great day to be active outside for just about everyone	
Moderate 51–100	Some people who may be unusually sensitive to particle pollution	**Unusually sensitive people:** *Consider reducing* prolonged or heavy exertion. Watch for symptoms such as coughing or shortness of breath. These are signs to take it easier
		Everyone else: It's a good day to be active outside
Unhealthy for sensitive groups 101–150	Sensitive groups include **people with heart or lung disease, older adults, children and teenagers**	**Sensitive groups:** *Reduce* prolonged or heavy exertion. It's OK to be active outside but take more breaks and do less intense activities. Watch for symptoms such as coughing or shortness of breath
		People with asthma should follow their asthma action plans and keep quick relief medicine handy
		If you have heart disease: Symptoms such as palpitations, shortness of breath, or unusual fatigue may indicate a serious problem. If you have any of these, contact your health-care provider

Continued

2. $PM_{2.5}$ is particulate matter that has a diameter less than or equal to 2.4 μm.

TABLE 1.4 Air quality index—cont'd

Air quality index	Who needs to be concerned?	What should I do?
Unhealthy 151–200	Everyone	**Sensitive groups:** *Avoid* prolonged or heavy exertion. Move activities indoors or reschedule to a time when the air quality is better
		Everyone else: *Reduce* prolonged or heavy exertion. Take more breaks during all outdoor activities
Very unhealthy 201–300	Everyone	**Sensitive groups:** *Avoid all* physical activity outdoors. Move activities indoors or reschedule to a time when air quality is better
		Everyone else: *Avoid* prolonged or heavy exertion. Consider moving activities indoors or rescheduling to a time when air quality is better
Hazardous 301–500	Everyone	**Everyone:** *Avoid* all physical activity outdoors
		Sensitive groups: Remain indoors and keep activity levels low. Follow tips for keeping particle levels low indoors

hazardous (i.e., maroon) at same time and place. The AQI is a piecewise linear function of the concentration of each pollutant, that is, a function of real numbers with a graph of straight line segments. Since we know that the concentration is 22 μg m^{-3}, we can use the linear function to find the index value (I):

$$I = \frac{I_{high} - I_{low}}{C_{high} - DI_{low}} \times (C - C_{low}) + I_{low} \quad (1.6)$$

where C = pollutant concentration, C_{low} = concentration breakpoint $\leq C$, C_{high} = concentration breakpoint $\geq C$, I_{low} = index breakpoint corresponding to C_{low}, and I_{high} = index breakpoint corresponding to C_{high}.

Table 1.5 shows the breakpoints for the AQI pollutants.

We can see from Table 1.5 that is 22 μg m^{-3} falls within the moderate category. To find the actual AQI value, we use Eq. (1.6):

$$I = \frac{100 - 51}{35.4 - 12.1} \times (22 - 12.1) + 51 \cong 72$$

Thus, the fires seem to be lowering the air quality, but there is still about 28 AQI points before $PM_{2.5}$ concentrations are unhealthy for sensitive groups, like those with asthma or babies.

TABLE 1.5 Breakouts for air quality index (AQI)

These breakpoints							Equal this AQI	And this category
O_3 (ppm) 8 h[b]	O_3 (ppm) 1 h*	$PM_{2.5}$ (μg/m^3) 24 h	PM_{10} (μg/m^3) 24 hour	CO (ppm) 8 h	SO_2 (ppb) 1 h	NO_2 (ppb) 1 hour	AQI	
0.000–0.054	–	0.0–12.0	0–54	0.0–4.4	0–35	0–53	0–50	Good
0.055–0.070	–	12.1–35.4	55–154	4.5–9.4	36–75	54–100	51–100	Moderate
0.071–0.085	0.125–0.164	35.5–55.4	155–254	9.5–12.4	76–185	101–360	101–150	Unhealthy for sensitive groups

TABLE 1.5 Breakouts for air quality index (AQI)—cont'd

These breakpoints							Equal this AQI	And this category
O_3 (ppm) 8h	O_3 (ppm) 1h*	$PM_{2.5}$ (μg/m^3) 24h	PM_{10} (μg/m^3) 24hour	CO (ppm) 8h	SO_2 (ppb) 1h	NO_2 (ppb) 1hour	AQI	
0.086–0.105	0.165–0.204	(55.5–150.4)[c]	255–354	12.5–15.4	(186–304)[d]	361–649	151–200	Unhealthy
0.106–0.200	0.205–0.404	(150.5–250.4)[c]	355–424	15.5–30.4	(305–604)[d]	650–1249	201–300	Very unhealthy
[c]See notes	0.405–0.504	(250.5–350.4)[c]	425–504	30.5–40.4	(605–804)[d]	1250–1649	301–400	Hazardous
[ab]See notes	0.505–0.604	(350.5–500.4)[c]	505–604	40.5–50.4	(805–1004)[d]	1650–2049	401–500	Hazardous

[a]*Areas are generally required to report the AQI based on 8 hour ozone values. However, there are a small number of areas where an AQI based on 1 hour ozone values would be more precautionary. In these cases, in addition to calculating the 8 hour ozone index value, the 1 hour ozone value may be calculated, and the maximum of the two values may be reported.*
[b]*Eight hour O_3 values do not define higher AQI values (≥ 301). AQI values of 301 or higher are calculated with 1 hour O_3 concentrations.*
[c]*If a different SHL for $PM_{2.5}$ is promulgated, these numbers will change accordingly.*
[d]*One hour SO_2 values do not define higher AQI values (≥ 200). AQI values of 200 or greater are calculated with 24 h SO_2 concentrations.*

Source: Technical Assistance Document for the Reporting of Daily Air Quality-the Air Quality Index (AQI): US Environmental Protection Agency, 2016. EPA-454/B-16-002.

Air quality measurements are often taken at discrete times. We are interested in the concentrations of volatile organic compounds (VOCs) in the air of a school parking lot every six hours. What flow would we need to fill 6 L (L) canisters to represent two days? Give answer both per L and per m^3?

At this rate, we would need eight canisters. Thus, the flow is 6 L per 6 h, or 1 L hr. 1 or about 1.7×10^{-2} L min^{-1}. Since 1 L = 0.001 = 10^{-3}, the flow rate is 1×10^{-3} L hr.$^{-1}$ or 1.7×10^{-5} L min^{-1}.

Calculate the daily VOC averages for the parking lot if the gas chromatograph readings are.

Day	Start time	VOC concentrations (μg m^{-3})
1	0:00	13
	6:00	54
	12:00	38
	18:00	35
2	0:00	15
	6:00	29
	12:00	33
	18:00	19

The arithmetic mean is calculated as

$$\overline{x} = \frac{\sum_{i=1}^{n} x_i}{n} \tag{1.7}$$

Eq. (1.6) simply states that we need to sum (Σ) all the values (x_1 through x_n) and divide by the number of entries (n). For day 1, this is.

$$\overline{x} = \frac{x_1 + x_2 + x_3 + x_4}{n} = \frac{13 + 54 + 38 + 35}{4} = \frac{120}{4} = 30\,\mu g\,m^{-3}.$$

For day 2, this is.

$$\overline{x} = \frac{x_1 + x_2 + x_3 + x_4}{n} = \frac{15 + 29 + 33 + 19}{4} = \frac{96}{4} = 24\,\mu g\,m^{-3}.$$

We will revisit and extend our discussion of statistics and decision science with calculations throughout this book. These approaches are particularly useful in data analysis and interpretation related to public health and ecosystem risk assessment, as well as in modeling.

1.9 SYSTEMS THINKING AND SUSTAINABILITY

Before venturing into the details of air pollution physics, chemistry, and biology, we need to consider how the approach has changed in addressing air pollution and environmental problems in general.

As mentioned, the 20th century saw an explosion of control technologies developed and deployed to address the air pollution that accompanied worldwide industrial expansion. The predominant technologies were specific to the extent of pollution and type of pollutant, ranging from no controls to increasingly sophisticated technologies designed to address specific pollutants. As evidence, electrostatic precipitation and fabric filter technologies improved and continue to improve in the efficiency and effectiveness of removing particulate matter (PM) from stack and vent emissions. At first, these technologies collected all size ranges of PM, but as research demonstrated that smaller particles were more respirable and could penetrate the lungs more deeply [50, 51], engineers designed equipment to remove these smaller particles [11].

Traditionally, air pollution has been addressed mainly as an engineering problem, focusing on the source, for example, a stack or vent from a factory or from the exhaust of an automobile. Engineering solutions were designed to fit the emission. The factory is an example of a stationary source and the automobile an example of a mobile source. The releases from these sources had to be connected to the ambient air quality of a neighborhood, town, or larger area. We knew that the individual contributions of each source contributed to the mass balance of pollutants in the ambient air, although we did not and still do not completely understand the mechanisms. There are many "black boxes" in air pollution engineering. Thus, thermodynamics indicated that if the concentration of a pollutant emitted from a stack or tailpipe could be decreased, the air quality would be expected to improve correspondingly. However, early on, air pollution experts knew that the atmosphere was not a neutral sink into which pollutants entered, but a dynamic system that sometimes degraded these emissions but other times exacerbated them.

The central goal of air pollution work has been to decrease harm. The metric for harm is risk. The health risk from an air pollutant is proportional to the amount of the pollutant reaching a person, that is, exposure. Location is a key factor in air pollutant exposure. Time spent and activities undertaken in different locations are important determinants of air pollution exposure. Air pollution monitoring may provide reliable estimates of ambient concentrations of air pollutants at various sites in a region. However, most people spend the majority of their time in indoor locations [52], with distinctions among ambient (often defined further as outdoor), indoor, and personal-scale exposures. The locations where people spend their time, such as in a kitchen, bedroom, vehicle, or garage, are known as microenvironments. The microenvironmental connections between outdoor and indoor and between these and personal-scale concentrations can vary substantially by the type of pollutant and mechanisms. For example, indoor concentrations of fine particulate matter do not correlate closely with outdoor concentrations [53]. Elevated concentrations of ozone (O_3) in the outdoor air do not typically penetrate indoor environments, but can enter dwellings more effectively by mechanical and open windows [54].

Air pollution managers and engineers are increasingly applying systems thinking and looking for greener, more sustainable approaches. We still rely heavily on proved air pollution control, measurement, and models but are increasing adding system tools, such as design for the environment (DfE), design for disassembly, human exposure modeling, and multicriterion decision analysis (MCDA). One of the means of achieving this is by expert elicitation, where professional judgment is used as a first step in screening for potential exposure and risk of large numbers of chemicals that are either air pollutants or which may be transformed into pollutants [55].

System thinking begins with an appreciation of how science and engineering is changing to address air quality needs. To date, air pollutant, especially the so-called air toxics, has been addressed using best or maximally achievable control technologies. However, personal exposure to most chemicals predominantly occurs within near-field exposure scenarios, for example, exposure while using a product that contains the chemical or contact with an article treated with the chemical [56]. This is leading to new screening models and tools, including multicriterion decision analysis (MCDA) and expert elicitation exposure-based chemical prioritization [57].

System tools, like MCDA, allow for the consideration of numerous variables, from various information sources. For near-field air pollution, the two main categories of variables are (1) those associated with product use and (2) the physical and chemical properties of the pollutant.

Thus, one of the important gaps in data and tools needed for life cycle and risk-based decision-making to address and to prevent air pollution is improved exposure predictions. For example, human health characterization factors (CFs) in life cycle assessments (LCAs) have benefited from improved hazard, especially toxicity, information [58]. The CFs are weighting factors used in an important component of the LCA process, that is, the life cycle impact assessment (LCIA). An air pollution LCIA converts emissions into impact scores for various impact categories [59].

An impact score is a weighted sum of the damage due to all air pollutant emissions to one of these or other impact categories [60]. LCAs generally and LCIAs specifically are important means of taking a systematic view of the costs and benefits of an entire process rather than a single stage of the life cycle. For air pollution, LCAs allow the engineer or process designer to compare among various alternatives. For example, a chemical compound may appear to be the best choice for manufacturing of product, but the LCIA may indicate that it will produce a toxic air pollutant in later stages, which will have to be treated. The treatment will add costs and risks that can be prevented by changes in process design or selection of safer chemicals that, if substituted, may obviate or greatly reduce the treatment costs. Likewise, the LCA may indicate that a choice of a substance may increase air and other pollutants in earlier stages, such as the extraction of ores, which may not be necessary if another substance with less extraction-related pollution is chosen.

Until recently, LCAs have underweighted exposure and relied more heavily on inherent toxicity estimates to indicate risks posed by substances in the human health CFs [61]. Knowing how toxic an air pollutant is necessary, but not sufficient, for air pollution decision-making. Reliable exposure predictions are also needed. Unlike the inherent physical and chemical properties of a pollutant, exposure predictions rely on numerous factors that differ among air pollutants. These factors go beyond physical and biological sciences and must also incorporate the social sciences, for example, the use of products or daily activities that increase exposures [55, 62].

1.10 CONCLUSIONS

With this introduction to scientific and mathematics commonly used in air pollution science, engineering, technology, and management, we are now ready to discuss the calculations involved in how an air pollutant is formed. In subsequent chapters, we will address how these change after being emitted, how they are exchanged between the atmosphere and other environmental media, how they are physically transported, and their environmental fate.

We are now ready to consider the various ways that pollutants form.

REFERENCES

[1] U.S. Environmental Protection Agency, Terms of environment, 9 May 2018. Available: https://ofmpub.epa.gov/sor_internet/registry/termreg/searchandretrieve/glossariesandkeywordlists/search.do?details=&vocabName=Terms%20of%20Env%20(2009)#formTop, 2018.

[2] P. Francis, Laudato si: On care for our common home, Our Sunday Visitor, 2015.

[3] R.M. Alexander, Social Aspects of Environmental Pollution, Air Resources Center, Water Resources Research Institute, Oregon State University, Corvallis, OR, 1971.

[4] S.J. Snow, Commentary: Sutherland, Snow and water: the transmission of cholera in the nineteenth century, Int. J. Epidemiol. 31 (5) (2002) 908–911.

[5] S. Newsom, Pioneers in infection control: John Snow, Henry Whitehead, the Broad Street pump, and the beginnings of geographical epidemiology, J. Hosp. Infect. 64 (3) (2006) 210–216.

[6] A.C. Stern, H.C. Wohlers, R.W. Boubel, W.P. Lowery, Fundamentals of Air Pollution, Academic Press, New York and London, 1973.

[7] J. Schwartz, H. Pitcher, The relationship between gasoline lead and blood lead in the United States, J. Off. Stat. 5 (4) (1989) 421.

[8] D.A. Vallero, Excerpted and Revised From: Fundamentals of Air Pollution, fifth ed., Elsevier Academic Press, Waltham, MA, 2014. p. 999 pages cm.

[9] W. Boubel, D. Fox, D. Turner, A. Stern, Sources of air pollution, in: Fundamentals of Air Pollution, 1994, pp. 72–96.

[10] J.P. Lodge, J. Evelyn, R. Barr, The Smoake of London: Two Prophecies: Fumifugium: or, The Inconvenience of the Aer and Smoake of London Dissipated, Maxwell Reprint Co, 1969.

[11] D.A. Vallero, Fundamentals of Air Pollution, fifth ed., Elsevier Academic Press, Waltham, MA, 2014. p. 999 pages cm.

[12] H. Hoover, L.H. Hoover, De re metallica, Courier Corporation, 1950.

[13] C. Piccolpasso, Li tre libri dell'arte del vasaio, 1557.

[14] M. Duhamel du Monceau, Traité général des pesches, et histoire des poissons qu'elles fournissent, tant pour la subsistance des hommes, que pour plusieurs autres usages qui on rapport aux arts et au commerce. Seconde partie, (2016).

[15] S.H. Beaver, The growth of public opinion, in: F.S. Mallette (Ed.), Problems and Control of Air Pollution, Reinhold Publishing Corp, New York, 1955, pp. 1–11.

[16] Smoke-abatement investigations at Salt Lake City, Utah, in 1912 to 1920 are reported. Recommendations for a program of air pollution control are included, 1926.

[17] S. Pincus, A.C. Stern, A study of air pollution in New York City, Am. J. Public Health Nations Health 27 (4) (1937) 321–333.

[18] G. Britain, Department of Scientific and Industrial Research: Atmospheric Pollution in Leicester: A Scientific Survey, Atmospheric Pollution Research, Tech. Paper, No. 1, 1945.

[19] J. Firket, Sur les causes des accidents survenus dans la vallée de la Meuse, lors des brouillards de décembre 1930, Bull. Acad. R. Med. Belg. 11 (5) (1931) 683–741.

[20] H.H. Schrenk, H. Heimann, G.D. Clayton, W. Gafafer, H. Wexler, Air Pollution in Donora, Pa. Epidemiology of the Unusual Smog Episode of October 1948, Preliminary Report, No. 306, 1949.

[21] L.C. McCabe, G. Clayton, Air Pollution by hydrogen sulfide in Poza Rica, Mexico. An evaluation of the incident of Nov. 24, 1950, Arch. Indust. Hyg. Occupational Med. 6 (3) (1952) 199–213.

[22] J. Schueneman, Proceedings of the First National Air Pollution Symposium, American Public Health Association, 1951.

[23] L. McCabe, Air Pollution, US Technical Conference on Air Pollution Proceedings 1950, McGraw-Hill Book Co., New York, 1952.

[24] E. Wilkins, Air pollution and the London fog of December, 1952, J. Roy. Sanitary Inst. 74 (1) (1954) 1–21.

[25] Ministry of Health, Mortality and morbidity during the London fog of December 1952: Reports on Public Health and Medical Subjects, Ministry of Health London, 1954.

[26] Royal Commission of Environmental Pollution, First report, HMSO, London, 1971.

[27] P. Lakshminarayanan, S. Aswin, Estimation of Particulate Matter from Smoke, Oil consumption and Fuel Sulphur, SAE Technical Paper 0148-7191, 2016.

[29] R. Damberg, A. Vasu, M. Schmdt, J. Langstaff, V. Sandiford, K. Martin, Review of the national ambient air quality standards for particulate matter: Policy assessment of scientific and technical information. OAQPS staff paper. Final report, Environmental Protection Agency, Office of Air Quality Planning and Standards, Research Triangle Park, North Carolina, 2005. EPA-452/R-05-005a.

[30] A. Loader, D. Mooney, R. Lucas, UK smoke and sulphur dioxide network, Networks (1998).

[31] T. Götschi, et al., Comparison of black smoke and PM2. 5 levels in indoor and outdoor environments of four European cities, Environ. Sci. Technol. 36 (6) (2002) 1191–1197.

[32] P. Quincey, A relationship between black smoke index and black carbon concentration, Atmos. Environ. 41 (36) (2007) 7964–7968.

[33] M.R. Heal, P. Quincey, The relationship between black carbon concentration and black smoke: a more general approach, Atmos. Environ. 54 (2012) 538–544.

[34] A.C. Stern, E. Professor, History of air pollution legislation in the United States, J. Air Pollut. Control Assoc. 32 (1) (1982) 44–61.

[35] P. Statistics, The International Organization of Motor Vehicle Manufacturers, 2018.

[36] A.L. Northcross, N. Hwang, K. Balakrishnan, S. Mehta, Exposure to smoke from the use of solid fuels and inefficient stoves for cooking and heating is responsible for approximately 4 million premature deaths yearly. As increasing investments are made to tackle this important public health issue, there is a need for identifying and providing guidance on best practices for exposure and stove performance monitoring, particularly for public health research, EcoHealth 12 (1) (2015) 196–199.

[39] T. Kuhn, What Are Scientific Revolutions, MIT, Boston, 1981.

[40] T.S. Kuhn, D. Hawkins, The structure of scientific revolutions, Am. J. Phys. 31 (7) (1963) 554–555.

[42] C. Hornsby, M. Ripa, C. Vassillo, S. Ulgiati, A roadmap towards integrated assessment and participatory strategies in support of decision-making processes. The case of urban waste management, J. Clean. Prod. 142 (2017) 157–172.

[43] U.S. Environmental Protection Agency, List of Areas Protected by the Regional Haze Program (40 CFR PART 81 ed.), 9 May 2018. Available: https://ofmpub.epa.gov/sor_internet/registry/termreg/searchandretrieve/glossariesandkeywordlists/search.do?details=&vocabName=Terms%20of%20Env%20(2009)#formTop, 2018.

[44] Intergovernmental Panel on Climate Change, IPCC, in: Climate Change, 2014.

[45] D.A. Vallero, Engineering aspects of climate change, in: Climate Change, second ed., Elsevier, 2015, pp. 547–568.

[46] A.C. Stern, R. Boubel, Turner, and DB, Fox, Donald L, in: Fundamentals of Air Pollution, 1984.

[47] D.A. Vallero, Environmental Contaminants: Assessment and Control, Elsevier Academic Press, Amsterdam; Boston, 2004. pp. xxxix, 801 p.

[50] International Commision on Radiological Protection Task Force on Lung Dynamics and Task Group on Lung Dynamics, Deposition and retention models for internal dosimetry of the human respiratory tract, Health Phys. 12 (2) (1966) 173.

[51] American Lung Association, Health Effects of Air Pollution, American Lung Association, New York, NY, 1978.

[52] N.E. Klepeis, et al., The National Human Activity Pattern Survey (NHAPS): a resource for assessing exposure to environmental pollutants, J. Expo. Anal. Environ. Epidemiol. 11 (3) (2001) 231–252.

[53] R. Meier, et al., Differences in indoor versus outdoor concentrations of ultrafine particles, PM2. 5, PM absorbance and NO_2 in Swiss homes, J. Exposure Sci. Environ.l Epidemiol. (2015).

[54] D. Lai, P. Karava, Q. Chen, Study of outdoor ozone penetration into buildings through ventilation and infiltration, Build. Environ. 93 (2015) 112–118.

[55] D.A. Vallero, Air pollution monitoring changes to accompany the transition from a control to a systems focus, Sustainability 8 (12) (2016) 1216.

[56] J.F. Wambaugh, et al., High-throughput models for exposure-based chemical prioritization in the ExpoCast project, Environ. Sci. Technol. 47 (15) (2013) 8479–8488.

[57] M.D. Wood, K. Plourde, S. Larkin, P.P. Egeghy, A.J. Williams, V. Zemba, I. Linkov, D.A. Vallero, Advances on a decision analytic approach to exposure-based chemical prioritization. Risk Anal. (2018), https://doi.org/10.1111/risa.13001.

[58] P.A. Schulte, et al., Occupational safety and health, green chemistry, and sustainability: a review of areas of convergence, Environ. Health 12 (1) (2013) 1.

[59] J. Bare, TRACI 2.0: the tool for the reduction and assessment of chemical and other environmental impacts 2.0, Clean Techn. Environ. Policy 13 (5) (2011) 687–696.

[60] R. van Zelm, et al., European characterization factors for human health damage of PM 10 and ozone in life cycle impact assessment, Atmos. Environ. 42 (3) (2008) 441–453.

[61] J.C. Bare, D.A. Vallero, Incorporating exposure science into life-cycle assessment, in: AccessScience. McGraw-Hill Education; Yearbook of Science & Technology, 2014.

[62] D.A. Vallero, C. Brasier, Sustainable Design: The Science of Sustainability and Green Engineering, John Wiley & Sons, 2008.

FURTHER READING

[63] A. L. GmbH, Smoke Value Measurement with the Filter Paper Method, Application Notes, Graz, Austria, AVL1007E, vol. Revision 2, 2005.

[64] R.M. Goody, Y.L. Yung, Atmospheric Radiation: Theoretical Basis, Oxford University Press, 1995.

[65] R. Rohde, Radiation Transmitted by the Atmosphere, in: Based on Data from Gordley, L., Marshall, B., Allen, C. (1994) Linepak: Algorithms for Modeling Spectral Transmittance and Radiance, 2007. vol. 3, no. 22, p. 2016, https://commons.wikimedia.org/wiki/File:Atmospheric_Transmission.png.

[66] U.S. Environmental Protection Agency, Our Nation's Air—Status and Trends Through 2016, Available: https://gispub.epa.gov/air/trendsreport/2017/#highlights, 2017.

[67] Illinois State Water Survey. University of Illinois at Urbana-Champaign, NADP Data Report 2013-01, National Atmospheric Deposition Program: 2012 Annual Summary, 2013.

[68] Technical Assistance Document for the Reporting of Daily Air Quality-the Air Quality Index (AQI), US Environmental Protection Agency, 2016 EPA-454/B-16-002.

Chapter 2

Characterizing air pollutants

Biological agents, like viruses, bacteria, and pollen, are air pollutants, as well as physical agents, like electromagnetic fields. However, by and large, most air pollutants are chemical compounds. The basic math examples in Chapter 1 introduced the concept of concentration. This is a chemical and a physical concept, that is, it is another way of saying that a pollutant is dissolved or suspended in a fluid, that is, air. Therefore, air pollution calculations almost always involve chemistry. There are certainly many air pollution chemists, but those in other fields must also have a strong chemistry underpinning. An engineer designing monitoring equipment or installing air pollution controls is indeed applying chemistry along with physics. A meteorologist concerned about acid rain or climate change is engaging in chemical reactions. The veterinarian studying the effects of air pollution on the lungs of rats must apply chemistry. This is a theme throughout this book, so let us begin with a few straightforward air pollution chemistry example calculations.

2.1 THE PERIODIC TABLE

Air pollutants are chemical compounds that form from reactions of elements. Thus, we will start our consideration of pollutant formation from the periodic table of elements (inside cover).

What is an "amu"?

The nucleus of an atom, consisting of protons and neutrons, accounts for the entirety of the atomic mass, or the atomic mass unit (amu). The element is the simplest form of matter and what appears in the periodic table of elements (see inside cover). They are not completely immutable, since under conditions of extreme energy in nuclear reactions, subatomic particles are emitted. These particles are worrisome sources of pollution, for example, releases from leaking and breached nuclear power plants. An atom is the smallest part of an element that can enter into a chemical reaction. The molecule, which may also be an atom, is the smallest subdivision of an element that can exist as a natural state of matter. As mentioned, the nucleus of an atom accounts for the entirety of the atomic mass, that is, the amu. These two particles comprising the atom's nucleus are sometimes referred to as nucleons. An amu is defined as one-twelfth of the mass of carbon (C^{12}), which is 1.66×10^{-27} kg. The atomic weight of an element listed in most chemistry handbooks is the relative atomic weight, that is, the total number of nucleons in the atom. So, for example, oxygen (O) has an atomic mass equal to 16. The atomic number (*Z*) is the number of protons in the nucleus. The chemical nomenclature for atomic weight *A* and the number of element *E* appears as

$$^{A}_{Z}E \tag{2.1}$$

Often, *Z* is understood and not shown; for example, the most stable form of C is usually shown as ^{12}C, rather than $^{12}_{6}C$.

2.1.1 Isotopes

The same element may have two or more atomic weights, owing to different numbers of neutrons. It is the same element because the number of electrons and protons of stable atoms are the same. Elements having different atomic weights are known as isotopes. Each different weight is a different isotope. Stable isotopes do not undergo natural radioactive decay, but radioactive isotopes undergo spontaneous radioactive decay, that is, their nuclei disintegrate. As a result, new isotopes or new elements are formed. A radiogenic isotope is the stable product of an element's radioactive decay. For example, lead (Pb; $Z=82$) has four naturally occurring isotopes: ^{204}Pb, ^{206}Pb, ^{207}Pb, and ^{208}Pb. Only the isotope ^{204}Pb is stable. The isotopes ^{206}Pb and ^{207}Pb are progeny from the radioactive decay of uranium (U). However, ^{208}Pb is a product from thorium (Th) decay. The radioactive decay results in the heavier isotopes of lead increasing in abundance compared with ^{204}Pb.

Air Pollution Calculations. https://doi.org/10.1016/B978-0-12-814934-8.00002-8

2.2 EMPIRICAL CALCULATIONS

Let us consider how a chemical formula is a first indication of how an air pollutant is formed.

A citizen used a 6 L stainless steel canister to collect air for 1 min at 6 L min^{-1} and brought the canister to the Townsville Fire Department. The fire department delivered the canister to the State Environmental Laboratory. The citizen said that the sample was taken near a site where a rusty 55 gal drum was found by some children in the creek in the neighborhood. The children and neighbors reported an unpleasant smell near the site where the drum was found. After accounting for what is normally expected in the air (mainly N_2, O_2, and water vapor), the gravimetric analysis of the gas in the canister indicated the following elemental compositions for an unknown compound:

Carbon: 40.0%
Hydrogen: 6.7%
Oxygen: 53.3%

Based on this evidence, what do you think was in the air and in the canister?

This calls for developing a chemical formula empirically. The first step is to divide the elemental percentages according to the respective atomic weights:

$$C: \frac{40.0}{12} = 3.3$$

$$H: \frac{6.7}{1} = 6.7$$

$$O: \frac{53.3}{16} = 3.3$$

Next, each ratio is divided by the smallest quotient; in this case, it is 3.3 and shows result as an integer:

$$C: \frac{3.3}{3.3} = 1$$

$$H: \frac{6.7}{3.3} = 2$$

$$O: \frac{3.3}{3.3} = 1$$

This yields the formula: CH_2O, which is *formaldehyde*.

It is now up to the air pollution experts to decide if they correctly ruled out other compounds, for example, their assumptions about normal air and water vapor. If it is indeed formaldehyde, they must engage the process of "source apportionment," that is, how much of the formaldehyde came from the drum. Formaldehyde can exist in various states of matter.

Does that mean that the drum contains formaldehyde?

Given it was stored in a drum, we must assume it was originally mainly in liquid state. If dissolved in water, it will become a hydrate, $H_2C(OH)_2$, known as methanediol. The drum could also have contained a mixture, such as the aqueous solution of formaldehyde and methanol, CH_3OH, the simplest alcohol. If this mixture is 40% formaldehyde by volume, it is known as formalin. In addition, formaldehyde readily converts to other compounds, which means that first responders may also need to test for these degradation products.

The challenge of source apportionment is ruling out other potential sources. This is particularly difficult for formaldehyde, given it is produced by many atmospheric processes, including combustion during fires and exhaust from cars and trucks. For example, if the person taking the sample was smoking during the time the canister was filling, this could create a positive bias, that is, higher formaldehyde concentrations than was actually present, since environmental tobacco smoke (ETS) contains formaldehyde. The smoker could be emitting greater than 200 μg m^{-3} formaldehyde [1]. Indeed, human breath contains between about 1 and 70 ppb formaldehyde [1], so even if not smoking, the person could be biasing the sample. If the sample were taken in a wooden area, the decomposition of the trees and other organic matter is also a source.

Thus, to answer this question, the contents of the barrel should also be analyzed. In addition, good scientific method requires quality assurance, which includes controls, for example, additional air samples at distances from the location of the drums.

It is important to note that this example merely introduces empirical formula development. It is by no means how air pollution analysis is normally conducted. The actual approach applies chromatography and detection, for example, using gas chromatography-mass spectroscopy (GC-MS).

Many earth and atmospheric science texts and references express the larger constituents of the atmosphere in percent volume of air. Percentage is a unitless express, that is, volume of constituent per volume of air. So long as the same units are being used, percentage can be used for any calculation. For example, it can be used for mass, volume, number of fibers, or people, as in 20% carbon in an organic solution (mass), 20% benzene in a gasoline mixture (volume), 20% asbestos fibers in a rock formation (fibers), and 20% people who work in a mine that contract lung diseases. Percentage should not be used when units differ, such as the earlier examples of $mg L^{-1}$ and $mg m^{-3}$, which are mass per volume.

How are the chemical concentrations of air pollutants reported?

Percentage involves relatively large numbers. For example, 1 cm^3 (cc or cm^3) of a substance in 100 cm^3 of air means that substance has a concentration of 1% by volume. Since the common air pollution unit is cubic meters (m^3), we would convert the 100 cm^3 to m^3. Given that 1 cm^3 is 10^{-6} m^3, our 1% concentration is 10^4 of the substance per 10^6 air, by volume. A review of the scientific notation indicates that this is indeed a percentage. Given that the two numbers have the same base, in this case 10, we divide by subtracting the exponents. Thus, the quotient of 10^6 and 10^4 is 10^2, so this must be a percentage because, by definition, a percent is a fraction per 100.

As mentioned, it is common to see the major gases of the atmosphere be expressed as percentages, for example, 78.08% nitrogen (N_2), 20.95% oxygen (O_2), and 0.93% argon (Ar) in dry air. Water vapor is also often reported as a percentage, for example, relative humidity of X%. Such percentages work well for larger numbers like these, but they become awkward as concentrations decrease. In the same atmospheric tables that describe N_2, O_2, and Ar, other lower concentration gases may also be reported in percentage, for example, 0.0002% methane (CH_4) and 0.000001% O_3. Social scientists, especially economists, use percentages like 0.005% when describing changes in the marketplace. This is the same as a change of 0.000005. In the physical sciences and engineering, we might use scientific notation to express this same number, that is, 5×10^{-6}. However, there are much better ways to express concentration of air pollutants that, thankfully, most often exist as very small fractions of the air. Notably, gases of very low concentrations are more commonly reported in parts per million or billion (ppm and ppb, respectively) or as milligram or microgram per cubic meter of air ($mg\ m^{-3}$ or $\mu g\ m^{-3}$, respectively).

As will be explained in Chapter 10, density is important to reporting the concentrations of air pollutants. Often, it is necessary to convert mass to volume, for example, $mg m^{-3}$ to ppm. At a particular temperature and pressure, density can be found for a chemical species. When a concentration is expressed simply as ppm, it is unclear whether a volume or weight basis is intended. To avoid confusion caused by different units, air pollutant concentrations in this book are generally expressed as micrograms per cubic meter of air ($mg m^{-3}$) at an environmental temperature (e.g., 20° or 25°C) and atmospheric pressure of 1 atm (atm), which is 760 mm mercury (mmHg). To convert from units of ppm (vol.) to $mg\ m^{-3}$, we assume that the ideal gas law is accurate under ambient conditions. A generalized formula for the conversion of a gas-phase substance at 25°C and 760 mmHg is

$$1\,\text{ppm (vol) pollutant} = \frac{1\,\text{L pollutant}}{10^6\text{L air}} = \frac{\left(\frac{1\text{L}}{22.4}\right) \times \text{MW} \times 10^6 \times \mu g \times g \times m^{-1}}{10^6\text{L} \times 298\text{K}^{\circ} \times 10^{-3}\text{m}^3 \times \text{L}^{-1}} = 40.9 \times \text{MW}\,\mu g\,m - 3 \quad (2.2)$$

where MW is the molecular weight in grams per mole ($g mol^{-1}$).

Thus, at 25°C and 1 atm, 180 ppb of ozone (O_3), which is three atoms of 16 MW oxygen, that is, MW = 48, would be.

Gas-phase contaminant concentration ($\mu g m^{-3}$) = contaminant concentration (ppm) × molecular weight of the contaminant × 40.9

$$\text{Thus, } ?\mu g\,m^{-3}\,O_3 = 0.180 \times 48 \times 40.9 = 353\,\mu g\,m^{-3}.$$

2.3 HEALTH CHARACTERIZATION

Human health is the primary reason for controlling air pollution, with ecological and material damage also being very important. Chapter 11 provides examples of risk calculations, but let us consider a couple examples here in the context of pollutant formation.

Explain why the bars for acute toxicity (LC_{50}) are so much taller than those for chronic toxicity (minimum risk level (MRL) and cancer risk).

See Fig. 2.1.

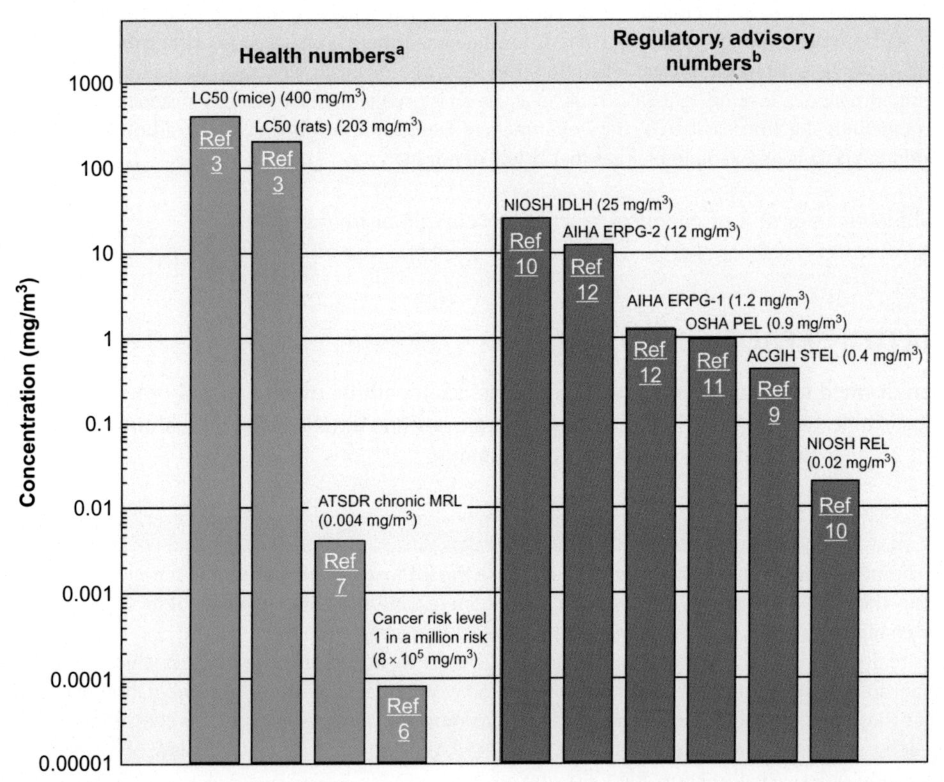

FIG. 2.1 Health data for inhalation exposure to formaldehyde as of 1999. To convert concentrations in air (at 25°C) from ppm to mg m^{-3}: mg m^{-3} ¼ (ppm) × (molecular weight of the compound)/(24.45). For formaldehyde: 1 ppm = 1.23 mg m^{-3}. American Industrial Hygiene Association's emergency response planning guidelines (AIHA ERPG). ERPG 1 is the maximum airborne concentration below which it is believed nearly all individuals could be exposed up to 1 h without experiencing other than mild transient adverse health effects or perceiving a clearly defined objectionable odor; ERPG 2 is the maximum airborne concentration below which it is believed nearly all individuals could be exposed up to 1 h without experiencing or developing irreversible or other serious health effects that could impair their abilities to take protective action. American Conference of Governmental and Industrial Hygienists' short-term exposure limit (ACGIH STEL) expressed as a time-weighted average exposure, the concentration of a substance that should not be exceeded at any time during a workday; lethal concentration 50 (LC_{50}), a calculated concentration of a chemical in air to which exposure for a specific length of time is expected to cause death in 50% of a defined experimental animal population. This is a common expression of acute hazard. National Institute of Occupational Safety and Health's immediately dangerous to life or health (NIOSH IDLH) limit, NIOSH recommended exposure limit to ensure that a worker can escape from an exposure condition that is likely to cause death or immediate or delayed permanent adverse health effects or prevent escape from the environment; NIOSH's recommended exposure limit (NIOSH REL), NIOSH recommended exposure limit for an 8 or 10 h time-weighted average exposure and/or ceiling; Occupational Safety and Health Administration's permissible exposure limit (OSHA PEL) expressed as a time-weighted average, the concentration of a substance to which most workers can be exposed without adverse effect averaged over a normal 8 h workday or a 40 h workweek. [a]Health numbers are toxicological values derived from animal testing or risk assessments developed by the US government. [b]Regulatory values have been incorporated in US government regulations, whereas advisory values are nonregulatory but based on advice from the US government or other groups as advised. OSHA numbers are regulatory, whereas NIOSH, ACGIH, and AIHA numbers are advisory. *(Source: U.S. Environmental Protection Agency. Formaldehyde - CAS 50-00-0. 25 April 2018. Available: https://www.epa.gov/sites/production/files/2016-09/documents/formaldehyde.pdf.)*

Acute effects occur shortly after exposure to a pollutant, so the dose must be high enough to elicit the effect. Chronic effects occur after a much longer time. Usually, a chronic effect is associated with a chronic exposure, that is, a continuous exposure for years. However, a chronic effect could occur even after long time periods without an exposure, for example, exposure to radiation or asbestos for months or years, followed by years without these exposures. With no apparent effect for several years, persons can contract cancer and other chronic diseases from these earlier exposures. This time between first exposures and onset of the disease is known as the latency period.

Explain why a first responder needs to know the density of substances that may be in a building? Why do the different densities between air and those of carbon disulfide and ammonia matter for a first responder?

Density is an indicator of where a substance will be. If it is heavier than air, it would move downward (e.g., in a lower level of a building, near the floor in a room, or in a lower compartment of a vehicle (e.g., a boat's hull)). If it is lighter than air, it will tend to move upwardly. In either situation, if it reaches a threshold, for example, a lower explosive limit (LEL) or a flammable gas finds an ignition source; it places the responder and others at risk. In either case, there is also a good chance that the air will be displaced by the moving gases, leading to asphyxiation and inhalation risks (e.g., people in a lower compartment below the leaking heavy gas or in an upper compartment of a leaking light gas). Thus, evacuation is usually the first step (the distance of evacuation is specified in first responder guidelines, e.g., those issued by the Department of Transportation and NIOSH). Carbon disulfide is an example of the heavier gas, and ammonia is an example of a gas that is less dense than air.

Thus, air pollutants are a mix of chemical reactions occurring at myriad rates.

2.4 FORMATION REACTIONS

Air pollutants are formed under combinations of reactions, each with its own rates and orders of reactions. Air pollution seldom occurs in a single system. We will discuss how the rate law and kinetics affect transformation in the next chapter, but for now, we will address the initial formation of air pollutants.

Consider the chemical plant manager who wants to increase the intensity of heat added to a boiler. She wants to know which option increases the heat most quickly. His options include increasing the concentration of molecular oxygen, adding more diesel fuel, or changing the fuel mix. What ingredient, that is, reactant, is the best option?

The manager is really asking which reactant can she add to give the highest rate of the reaction, that is, combustion. What is the reaction mechanism? So, she must understand how combustion works, that is, what is the combination of reactants, in concentrations, to produce the flame? Thus, this is a rate law question. We know that the rate of reaction is affected by the concentration of the reactants. Rate is the change over a time period, that is, $A+B \rightarrow \frac{\Delta A}{t}$. We know that the rate equals $k[A]^m$, where k is the rate constant and m is the reaction order. We also know the following:

If $m=0$, the reaction is zero order, meaning that the rate is independent of the concentration of A, that is, [A].

If $m=1$, the reaction is first order, so the rate is directly proportional to [A].

If $m=2$, the reaction is second order, so the rate is proportional to the square of [A], that is, $[A]^2$.

Unfortunately for the plant manager, the order of the reaction can only be determined by experimentation. The plant manager doubles the amount of diesel fuel per minute and finds that the oxidation does not change. However, if the diesel is mixed with an additive, the reaction is first order. If O_2 is increased, the rate is also not affected, that is, the flame is receiving sufficient oxygen and is not O_2-limited. Thus, of these options, changing the mix with the additive is the one that gives the highest rate of combustion.

Indeed, a better way to consider these mixed systems is pseudo-orders in the rate laws. Let us consider carbon dioxide, which is a product of complete combustion.

Assume the average weight of a passenger vehicle is 1800 kg and the average gas mileage is 10 km l^{-1} and that gasoline weighs an average of 0.7 kg l^{-1}, with 85% of this weight as carbon. How much CO_2 does a passenger vehicle generate from combusting gasoline, which is a mixture of several organic compounds?

The question does not give the number of kilometers traveled per year. In 2011, the average US driver logged 21,688 km (US Department of Transportation), so the carbon in the total amount of gasoline combusted annually for the vehicle would be

$$\frac{21688\,\text{km}}{\text{year}} \times \frac{1\,\text{l}}{10\,\text{km}} \times \frac{0.7\,\text{kg}}{\text{l}} \times 0.85 \cong \frac{1290\,\text{kg C}}{\text{year}}$$

The carbon dioxide emitted is

$$\frac{1290\,\text{kg C}}{\text{year}} \times \frac{44\,\text{kg CO}_2}{12\,\text{kg C}} = \frac{4730\,\text{kg CO}_2}{\text{year}}$$

Thus, the average US passenger vehicle emits much more (3.7 times as much) CO_2 than its weight every year.

Using the data from the previous question, if there are 1 billion vehicles in the world, estimate the annual global carbon emissions from vehicles. Give your answer in tons, tonnes, and teragrams.

$$4.7 \times 10^3\ \text{kg}\ CO_2 \times 10^9 = 4.7 \times 10^{12}\ \text{kg} = 4.7 \times 10^{15}\ \text{g}$$

1 t (metric ton, t) = 1000 kg; thus, vehicles would emit about *4.7× 10^9 t CO_2 per year.*

1 teragram (Tg) = 10^9 g; thus, vehicles would emit about *4700 Tg per year.*

Incidentally, this is probably an overestimate, since the US driver is likely to log many more miles than the world average, especially compared with Europe.

How is the formation of CO_2 different in respiration versus combustion?

Both processes produce carbon dioxide and water, but combustion requires a heat source to oxidize organic compounds (exothermic), whereas respiration is a biochemical process that converts O_2 to CO_2 and H_2O endogenously (i.e., cellular metabolism, often catalyzed by enzymes). Both processes follow the general oxidation reaction; for example, for methane, this is.

$$CH_4(g) + 2O_2(g) \rightarrow CO_2(g) + 2H_2O(g) \tag{2.3}$$

For all hydrocarbons, the general stoichiometry is

$$C_xH_y + zO_2 \rightarrow xCO_2 + \frac{y}{2}H_2O. \tag{2.4}$$

If the reactants in the chemical reaction are heated, the Greek symbol delta (Δ) is often shown above the arrow. Thus, the combustion reaction would be

$$C_xH_y + zO_2 \xrightarrow{\Delta} xCO_2 + \frac{y}{2}H_2O \tag{2.5}$$

As mentioned, the respiration process is often catalyzed, so the reaction could denote the enzyme, for example, cytochrome P-450:

$$C_xH_y + zO_2 \overset{cyp450}{\rightarrow} xCO_2 + \frac{y}{2}H_2O \tag{2.6}$$

We hear a lot about how animal operations emit the greenhouse gas, CH_4. What are some other air pollutants that can be released by confined animal feeding operations? What can be done to address them?

The major airborne pollutants from confined feeding operations are various forms of nitrogen. Organic compounds will breakdown ultimately to CO_2 in oxidized parts of the system, for example, lagoons, whereas CH_4 will be released from reduced environments. The nitrogen can be released in the gas phase as nitric oxide (NO), where oxidation is occurring (known as nitrification), or as ammonia (NH_3), in reduced (denitrification) environments, even including waste treatment process; for example, nitrates in water can be aerosolized by aeration of farm lagoons. If lagoons become anaerobic, the sulfur-reducing bacteria (e.g., *Desulfovibrio* spp.) will release H_2S. Other pollutants can be released, for example, pesticides and other farm chemicals. In addition, biological pollutants can be released as or on aerosols into the air from the animal waste (e.g., spores, cysts, bacteria, fungi, and viruses).

2.4.1 Formation kinetics

With a basic understanding of air pollutant concentrations and measurement, we can now address the theories underlying the processes that form the pollutants. Chemical kinetics describes the rate of a chemical reaction, that is, the rate at which the reactants are transformed into products. Since a rate is a change in quantity that occurs with time, this describes air pollutant formation:

$$\text{Reaction rate} = \frac{\text{change in product concentration}}{\text{corresponding change in time}} \tag{2.7}$$

and

$$\text{Reaction rate} = \frac{\text{change in reactant concentration}}{\text{corresponding change in time}} \tag{2.8}$$

When a compound breaks down, the change in product concentration decreases proportionately with the reactant concentration. Thus, for substance A, the kinetics is

$$\text{Rate} = -\frac{\Delta(A)}{\Delta t} \tag{2.9}$$

The negative sign denotes that the reactant concentration (the parent contaminant) is decreasing. It stands to reason then that the degradation product C resulting from the concentration will be increasing in proportion to the decreasing concentration of the contaminant A, and the reaction rate for Y is

$$\text{Rate} = \frac{\Delta(C)}{\Delta t} \tag{2.10}$$

Thus, the rate of reaction at any time is the negative of the slope of the tangent to the concentration curve at that specific time (see Fig. 2.2).

For a reaction to occur, the molecules of the reactants must collide. High concentrations of a substance are more likely to collide than low concentrations. Thus, the reaction rate must be a function of the concentrations of the reacting substances. The mathematical expression of this function is known as the "rate law." The rate law can be determined experimentally for any contaminant. Varying the concentration of each reactant independently and then measuring the result will give a concentration curve. Each reactant has a unique rate law (this is one of a contaminant's physicochemical properties).

In a reaction of reactants, A and B to yield product C (i.e., $A+B \rightarrow C$), the reaction rate increases in accordance with the increasing concentration of either A or B. If the amount of A is tripled, then the rate of this whole reaction triples. Thus, the rate law for such a reaction is

$$\text{Rate} = k[\text{A}][\text{B}] \tag{2.11}$$

The rate law for the different reaction $X+Y \rightarrow Z$, in which the rate is only increased if the concentration of X is increased (changing the Y concentration has no effect on the rate law), must be

$$\text{Rate} = k[\text{X}] \tag{2.12}$$

Eqs. (2.11) and (2.12) indicate that the concentrations in the rate law are the concentrations of reacting chemical species at any specific point in time during the reaction. The rate is the velocity of the reaction at that time. The constant k is the rate constant, which is unique for every chemical reaction and is a fundamental physical constant for a reaction, as defined by environmental conditions, including pH, temperature, pressure, and the type of solvent.

The rate constant is the rate of the reaction when all reactants are present in a 1 M (M) concentration. Accordingly, the rate constant k is the rate of reaction under conditions standardized by a unit concentration. By drawing a concentration

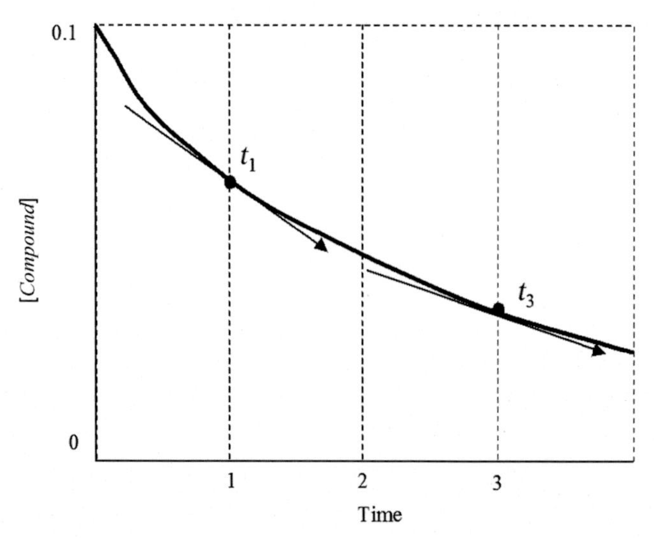

FIG. 2.2 The kinetics of the formation and/or transformation of a compound. The rate of reaction at any time is the negative of the slope of the line tangent to the concentration curve at that time. The rate is higher at t_1 than at t_3. This rate is concentration-dependent (first order). *(Sources: D.A. Vallero, Fundamentals of Air Pollution. 5th ed. Elsevier Academic Press Waltham, MA, 2014, p. 999 pages cm; J.N. Spencer, G.M. Bodner, L.H. Rickard, Chemistry: Structure and Dynamics. John Wiley & Sons, 2010.)*

curve for a contaminant that consists of an infinite number of points at each instant of time, an instantaneous rate can be calculated along the concentration curve. At each point on the curve, the rate of reaction is directly proportional to the concentration of the compound at that moment in time. This is a physical demonstration of kinetic order. The overall kinetic order is the sum of the exponents (powers) of all the concentrations in the rate law. For the rate $k[A][B]$, the overall kinetic order is 2. Such a rate describes a second-order reaction because the rate depends on the concentration of the reactant raised to the second power. Other decomposition rates are like $k[X]$ and are first-order reactions because the rate depends on the concentration of the reactant raised to the first power.

The kinetic order of each reactant is the power that its concentration is raised in the rate law. So, $k[A][B]$ is first order for each reactant, and $k[X]$ is first order for X and zero order for Y. In a zero-order reaction, compounds degrade at a constant rate and are independent of reactant concentration.

A "parent compound" is transformed into "chemical daughters" or "progeny." For example, pesticide kinetics often concerns itself with the change of the active ingredient in the pesticide to its "degradation products."

Air pollution physics must be considered mutually with air pollution chemistry. Any complete discussion of the physical process of solubility, for example, must include a discussion of chemical phenomenon *polarity*. Further, any discussion of polarity must include a discussion of electronegativity. Likewise, discussions of sorption and air-water exchanges must consider both chemical and physical processes. Such is the nature of environmental science; all concepts are interrelated.

Air pollutant formation processes are a function of both the chemical characteristics of the compartment (e.g., air) and those of the contaminant. The inherent properties of the air pollutant are influenced and changed by the extrinsic properties of the air and other media in which the pollutant resides in the environment.

2.5 REACTION TYPES

Air pollution involves five categories of chemical reactions. Various types of reaction occur within boilers and other industrial operations, in stacks, in the plume, in the microenvironments and ecosystems, and within organisms. These categories include the following:

Synthesis or combination:

$$A + B \rightarrow AB \tag{2.13}$$

In combination reactions, two or more substances react to form a single substance. Two types of combination reactions are important in environmental systems, that is, formation and hydration.

Formation reactions are those where elements combine to form a compound. Examples include the formation of ferric oxide and the formation of octane:

$$4Fe\ (s) + 3O_2\ (g) \rightarrow 2Fe_2O_3\ (s) \tag{2.14}$$

$$8C\ (s) + 9H_2\ (g) \rightarrow C_8H_{18}\ (l) \tag{2.15}$$

The g, l, and s in parentheses indicate gas, liquid, and solid, respectively.

Hydration reactions involve the addition of water to synthesize a new compound, for example, when calcium oxide is hydrated to form calcium hydroxide and when phosphate is hydrated to form phosphoric acid:

$$CaO_2\ (s) + H_2O\ (l) \rightarrow Ca(OH)_2\ (s) \tag{2.16}$$

$$P_2O_5\ (s) + 3H_2O\ (l) \rightarrow 2H_3PO_4\ (aq) \tag{2.17}$$

The aq in parentheses indicates that product is aqueous.

Decomposition:

$$AB \rightarrow A + B \tag{2.18}$$

Decomposition is often referred to as "degradation" when applied to the breakdown of organic compounds. In decomposition, one substance breaks down into two or more new substances, such as in the decomposition of carbonates. For example, calcium carbonate breaks down into calcium oxide and carbon dioxide:

$$CaCO_3\ (s) \rightarrow CaO\ (s) + CO_2\ (g) \tag{2.19}$$

Single replacement:

$$A + BC \rightarrow AC + B \tag{2.20}$$

Single replacement (or single displacement) commonly occurs when one metal ion in a compound is replaced with another metal ion, such as when trivalent chromium replaces monovalent silver:

$$3AgNO_3\,(aq) + Cr\,(s) \rightarrow Cr(NO_3)_3\,(aq) + 3\,Ag\,(s) \tag{2.21}$$

Double replacement (also metathesis or double displacement):

$$AB + CD \rightarrow AD + CB \tag{2.22}$$

In double replacement reactions, metals are exchanged between the salts. These newly formed salts have different chemical and physical characteristics from those of the reagents and are commonly encountered in metal precipitation reactions, such as when lead is precipitated (indicated by the "(s)" following $PbCl_2$), as in the reaction of a lead salt with an acid like potassium chloride:

$$Pb(ClO_3)_2\,(aq) + 2KCl\,(aq) \rightarrow PbCl_2\,(s) + 2KCLO_3\,(aq) \tag{2.23}$$

Complete Combustion:

Complete or efficient combustion (thermal oxidation) occurs when an organic compound, in this case methane, is oxidized in the presence of heat (indicated by Δ):

$$a(CH)_x + bO_2 \xrightarrow{\Delta} cCO_2 + dH_2O \tag{2.24}$$

Combustion is the combination of O_2 in the presence of heat (as in burning fuel), producing CO_2 and H_2O during complete combustion of organic compounds, such as the combustion of octane:

$$C_8H_{18}\,(l) + 17O_2\,(g) \xrightarrow{\Delta} 8CO_2\,(g) + 9H_2O\,(g) \tag{2.25}$$

Complete combustion may also result in the production of molecular nitrogen (N_2) when nitrogen-containing organics are burned, such as in the combustion of methylamine:

$$4CH_3NH_2\,(l) + 9O_2\,(g) \xrightarrow{\Delta} 4CO_2\,(g) + 10H_2O\,(g) + 2N_2\,(g) \tag{2.26}$$

Respiration is similar to these reactions but at much lower temperatures.

Incomplete combustion is particularly important in air pollutant formation. Among the most are the polycyclic aromatic hydrocarbons (PAHs), dioxins, furans, and CO. At least two observations can be made about these categories. First, all are kinetic, as denoted by the one-directional arrow (→). Second, in the environment, many processes are incomplete, such as the common problem of incomplete combustion and the generation of new compounds in addition to carbon dioxide and water.

Indeed, many equilibrium reactions take place. However, getting to equilibrium requires a kinetic phase. Upon reaching equilibrium, the kinetic reactions (one-way arrows) would be replaced by two-way arrows (↔). Changes in the environment or in the quantities of reactants and products can invoke a change back to kinetics. Kinetics will be explored more deeply in Chapter 3.

Arguably, the most important aspect of equilibrium reactions for environmental science and air pollution, specifically, is partitioning. How and why a pollutant is transformed and transported in the environment is dictated by equilibria. It determines the affinity of a pollutant, or any substance, for certain components of a system. The system can be an ecosystem; for example, the inherent properties of the air pollutant, where it is found, and the conditions of environment, determine its form and its location. For example, even an air pollutant with a high vapor pressure may not necessarily move into the atmosphere if other conditions give it more affinity for the soil, for example, adsorption to the soil and/or bioaccumulation in the soil microbes.

Partitioning and equilibria will be discussed in much detail in Chapter 4.

2.6 POLLUTANT FORMATION PROCESSES: INTRODUCTION TO CHEMODYNAMICS

Before a pollutant is formed, numerous processes take place. The inherent properties of a substance are the first determinant of where it is likely to be found, how rapidly it will move and change, and what kinds of damage it may cause. Numerous conditions are needed to complement the inherency, especially the physical, chemical, and biological processes within each compartment (e.g., substrate, air, or water) as they affect the substance residing in that compartment. The inherent properties of the substance are influenced and changed by the extrinsic properties of the media (including the cell, tissue, or organism) in which the pollutant resides in the environment. Table 2.1 briefly highlights some of these important processes. Note that every process listed requires interactions between the inherent properties of the substance and the conditions of the environmental medium where the substance resides.

TABLE 2.1 Physical, chemical, and biological processes involved in air pollutant formation

Process	Description	Physical phases involved	Major mechanisms at work	Outcome of process	Factors included in process
Advection	Transport by turbulent flow, mass transfer	Aqueous, gas	Mechanical	Transport due to mass transfer	Concentration gradients, porosity, permeability, hydraulic conductivity, circuitousness, or tortuosity of flow paths
Dispersion	Transport from source	Aqueous, gas	Mechanical	Concentration gradient-driven transport	Concentration gradients, porosity, permeability, hydraulic conductivity, circuitousness, or tortuosity of flow paths
Molecular diffusion	Fick's law (concentration gradient)	Aqueous, gas, solid	Mechanical	Concentration gradient-driven transport	Concentration gradients
Liquid separation	Various fluids of different densities and viscosities are separated within a system	Aqueous	Mechanical	Recalcitrance to due formation of separate gas and liquid phases (e.g., gasoline in water separates among benzene, toluene, and xylene)	Polarity, solubility, K_d, K_{ow}, K_{oc}, coefficient of viscosity, density
Density stratification	Distinct layers of differing densities and viscosities	Aqueous	Physical/ chemical	Recalcitrance or increased mobility in the transport of lighter fluids (e.g., light nonaqueous phase liquids (LNAPLs)) that float at water table in groundwater or at atmospheric pressure in surface water	Density (specific gravity)
Migration along flow paths	Faster through large holes and conduits, for example, path between sand particles in an aquifer	Aqueous, gas	Mechanical	Increased mobility through fractures	Porosity, flow path diameters
Sedimentation	Heavier compounds settle first	Solid	Chemical, physical, mechanical, varying amount of biological mechanisms	Recalcitrance due to deposition of denser compounds	Mass, density, viscosity, fluid velocity, turbulence (R_N)

Filtration	Retention in mesh	Solid	Chemical, physical, mechanical, varying amount of biological mechanisms	Recalcitrance due to sequestration, destruction, and mechanical trapping of compounds in soil micropores	Surface charge, soil, particle size, sorption, polarity
Volatilization	Phase partitioning to vapor	Aqueous, gas	Physical	Increased mobility as vapor phase of contaminant migrates to soil-gas phase and atmosphere	P^0, concentration of contaminant, solubility, temperature
Dissolution	Cosolvation, attraction of water molecule shell	Aqueous	Chemical	Various outcomes due to the formation of hydrated compounds (with varying solubilities, depending on the species)	Solubility, pH, temperature, ionic strength, activity
Fugacity	Escape from one type of environmental compartment to another	All phases	Physical but influenced by chemical and biological mechanisms	Fleeing potential	All partitioning conditions affect fugacity
Absorption	Retention on solid surface	Solid	Chemical, physical, mechanical, varying amount of biological mechanisms	Partitioning of lipophilic compounds into soil organic matter	Polarity, surface charge, van der Waals attraction, electrostatics, ion exchange, solubility, K_d, K_{ow}, K_{oc}, coefficient of viscosity, density
Adsorption	Retention on solid surface	Solid	Chemical, physical, varying amount of biological mechanisms	Recalcitrance due to ion exchanges and charge separations	Polarity, surface charge, van der Waals attraction, electrostatics, ion exchange, solubility, K_d, K_{ow}, K_{oc}, coefficient of viscosity, density
Chemisorption	Retention on surface wherein strength of interaction is stronger than solely physical adsorption, resembling chemical bonding	Solid	Chemical and biochemical in addition to the physical mechanisms	Recalcitrance due to ion exchanges and charge separations	Polarity, surface charge, van der Waals attraction, electrostatics, ion exchange, solubility, K_d, K_{ow}, K_{oc}, coefficient of viscosity, density, the presence of biofilm

Continued

TABLE 2.1 Physical, chemical, and biological processes involved in air pollutant formation—cont'd

Process	Description	Physical phases involved	Major mechanisms at work	Outcome of process	Factors included in process
Ion exchange	Cations attracted to negatively charged particle surfaces or anions are attracted to positively charged particle surfaces, causing ions on the particle surfaces to be displaced	Solid	Chemical and biochemical in addition to the physical mechanisms	Recalcitrance due to ion exchanges and charge separations	Polarity, surface charge, van der Waals attraction, electrostatics, ion exchange, solubility, K_d, K_{ow}, K_{oc}, coefficient of viscosity, density, the presence of biofilm
Complexation	Reactions with matrix (e.g., soil compounds like humic acid) that form covalent bonds	Solid	Chemical, varying amount of biological mechanisms	Recalcitrance and transformation due to reactions with soil organic compounds to form residues (bound complexes)	Available oxidants/ reductants, soil organic matter content, pH, chemical interfaces, available O_2, electric interfaces, temperature
Oxidation/ reduction	Electron loss and gain	All	Chemical, physical, varying amount of biological mechanisms	Destruction or transformation due to mineralization of simple carbohydrates to CO_2 and water from respiration of organisms	Available oxidants/ reductants, soil organic matter content, pH, chemical interfaces, available O_2, electric interfaces, temperature
Ionization	Complete cosolvation leading to separation of compound into cations and anions	Aqueous	Chemical	Dissolution of salts into ions	Solubility, pH, temperature, ionic strength, activity
Hydrolysis	Reaction of water molecules with contaminants	Aqueous	Chemical	Various outcomes due to the formation of hydroxides (e.g., aluminum hydroxide) with varying solubilities, depending on the species	Solubility, pH, temperature, ionic strength, activity
Photolysis	Reaction catalyzed by electromagnetic energy (sunlight)	Gas (major phase)	Chemical, physical	Photooxidation of compounds with hydroxyl radical upon release to the atmosphere	Free radical concentration, wavelength and intensity of EM radiation
Bioavailability	Fraction of the total mass of a compound present in a compartment that has the potential of being absorbed by the organism	All phases	Biological and chemical	Uptake, absorption, distribution, metabolism, and elimination	*Bioaccumulation* is the process of uptake into an organism from the abiotic compartments. *Bioconcentration* is the concentration of the pollutant within an organism above levels found in the compartment in which the organism lives

Biodegradation	Microbially mediated, enzymatically catalyzed reactions	Aqueous, solid	Chemical, biological	Various outcomes, including destruction and formation of daughter compounds (degradation products) intracellularly and extracellularly	Microbial population (count and diversity), pH, temperature, soil moisture, acclimation potential of available microbes, nutrients, appropriate enzymes in microbes, available and correct electron acceptors (i.e., oxygen for aerobes and others for anaerobes)
Cometabolism	Other organic compounds metabolized concurrently by microbes that are degrading principal energy source	Aqueous but wherever biofilm comes into contact with organic compounds	Biochemical	Coincidental degradation of organic compounds	Enhanced microbial activity, the presence of a good energy source (i.e., successful acclimation) and production of enzymes in metabolic pathways
Activation	Metabolic, detoxification process that renders a compound more toxic	Aqueous, gas, solid, tissue	Biochemical	Phase 1 or 2 metabolism, for example, oxidation may for epoxides on aromatics	
Cellular respiration	Conversion of nutrients' biochemical energy into adenosine triphosphate (ATP), with release of waste products	Catabolic reactions, oxidation of one molecule and reduction of another	Biochemical	Microbial respiration, along with metabolism, degrades organic compounds	Aerobic respiration involves oxygen as final electron acceptor, whereas anaerobic respiration has another final electron acceptor
Fermentation	Cellular energy derived from oxidation of organic compounds	Endogenous electron acceptor, that is, usually an organic compound. Differs from cellular respiration where electrons are donated to an exogenous electron acceptor, (e.g., O_2) via an electron transport chain	Biochemical	Organic compounds degraded to alcohols and organic acids, ultimately to methane and water	Often an anaerobic process
Enzymatic catalysis	Cell produces biomolecules (i.e., complex proteins) are that speed up biochemical reactions	Enzymes reactive site binds substrates by noncovalent interactions	Biological (intracellular in single-celled and multicelled organisms)	Catalyzed reaction follows three steps: substrate fixation, reaction, desorption of the product	Noncovalent bonding includes hydrogen bonds, dipole-dipole interactions, van der Waals or dispersion forces, stacking interactions, hydrophobic effect

Continued

TABLE 2.1 Physical, chemical, and biological processes involved in air pollutant formation—cont'd

Process	Description	Physical phases involved	Major mechanisms at work	Outcome of process	Factors included in process
Metal catalysis	Reactions sped up in the presence of certain metallic compounds (e.g., noble metal oxides in the degradation of nitric acid)	Aqueous, gas, solid, and biotic	Chemical (especially reduction and oxidation)	Same chemical reaction but faster	Chemical form of metal, pH, temperature
Pharmacokinetics/ toxicokinetics	Rates at which uptaken substances are absorbed, distributed, metabolized, and eliminated	Absorption, distribution, metabolism, and elimination of a substance by the body, as affected by uptake, distribution, binding, elimination, and biotransformation	Biochemical	Mass balance of substance after uptake	Available detoxification and enzymatic processes in cells
Pharmacodynamics/ toxicodynamics	Effects and modes of action of chemicals in an organism	Uptake, movement, binding, and interactions of molecules at their site of action	Biochemical	Fate of compound or its degradates	Affinities of compounds to various tissues

Source: D. Vallero, Environmental Biotechnology: A Biosystems Approach. Elsevier Science, 2015.

Air pollution chemodynamics addresses the processes by which chemicals move and change not only in the atmosphere but also in other environmental media; for example, they may form in the soil or water. Their fate can indeed by any medium, water, land, and organisms. Thus, it is important to take a system view of pollutant transport, transformation, and fate [2]. One recurring theme in environmental science and engineering is the need to understand threats to public health and ecological resources. For example, all engineering professions, including environmental engineers, are required to "hold paramount the public's safety, health, and welfare" [3]. Scientists are concerned with adding knowledge, and engineers apply this best available information and knowledge to address societal needs. Environmental engineers are particularly interested in protecting public health and ecosystem conditions.

2.7 NONCHEMICAL CONCENTRATIONS

Air pollutants are not only chemical substances, but also be microbiological or physical. These, however, can also be expressed as concentrations.

A microbiological concentration is often the number of colonies of a microbe in a sample of medium, usually agar. The units of airborne microbial concentrations are reported as colony-forming units (CFU) per air volume, that is, $CFU\,m^{-3}$ [4, 5].

Examples of physical concentrations include radiation and particulates. Radiation concentrations can be waves, for example, X-rays and electromagnetic fields (EMF). Radiation concentrations can be expressed as fluence of particles, for example, photons, or energy. Fluence is actually a flux, rather than a true concentration. That is, it is the number of particles or amount of energy over a defined surface area, for example, photons cm^{-2} skin surface or millijoules energy cm^{-2} skin surface ($mJ\,cm^{-2}$).

The other main types of physical concentration are solid and liquid suspensions of matter, that is, aerosols (see Chapter 10). We will give much attention to aerosols, known in air pollution literature as particulate matter (PM). Aerosol concentrations can be expressed in the same way as chemical concentrations, that is, mass per air volume, $\mu g\,m^{-3}$ PM.

Depending on their shape, aerosols may be counted and reported as counts per volume, for example, fibers m^{-3}. A fiber is simply a stretched-out particulate, that is, it has an aspect ratio greater than 3:1. The aspect ratio is the ratio of length to width. A fiber of particular importance to air pollution is asbestos.

Why was asbestos used so commonly in the 20th century?

In addition to fireproofing, asbestos had numerous uses, including heat resistance and insulation, providing strength to building materials (flooring, roofing tiles, etc.), noise insulation, and corrosion resistance and keeping vermin from gnawing through piping and other materials.

Governmental bodies establish action levels for asbestos. These are usually aimed at longer-term exposures; for example, in the United Kingdom, occupational action levels are calculated over a continuous 12-week period. They are also designed to be worst-case scenarios; for example, any single employee's exposure can be assumed to apply to all employees. Chrysotile's 4 h control limit is 0.2 fibers per mL (f mL^{-1}) and 0.9 f mL^{-1} for 10 min [6]. The action level is 72 fiber hours per mL. All other forms of asbestos are more protective, that is, 0.2 f mL^{-1} for 4 h, 0.6 f mL^{-1} for 10 min, and an action level of 48 f- hrs mL^{-1}. If a person who works 5 days per week is exposed to 0.3 f mL^{-1} chrysotile for 3 h every work day, what is his cumulative exposure? Does it exceed the action level?

Exposure (E) can be calculated using [6]

$$E = t_d \times t_w \times t_c \tag{2.27}$$

where t_d is the daily exposure (hours), t_w is the length of work, and t_c is the cumulative exposure duration (weeks). Thus,

$$E = 0.3 \times 3\text{ hours} \times 5\text{ days} \times 12\text{ weeks} = 54\, f - hrs\, mL^{-1}$$

Therefore, since this is less than 72 f- hrs mL^{-1}, it does not exceed the action level. However, if it were any other asbestos fiber, it would have exceeded the action level. This means additional protections and controls would have to be put in place immediately.

A worker is exposed to 2 f mL^{-1} chrysotile for a continuous 2 h period each week for 10 weeks. In each of the following 2 weeks, the worker is exposed to an additional exposure of 1 f mL^{-1} for 10 h. What is the cumulative exposure? Does it exceed the action level?

$$E = [(2 \times 2\,\text{hrs}) \times 10\,\text{weeks}] + [(1 \times 10\,\text{hrs}) \times 2\,\text{weeks} = 60\,f - hrs\,mL^{-1}$$

Again, the exposure is below the action limit. However, workers are also protected by control limits. For chrysotile, the given 1 f mL^{-1} exceeds the 0.9 f mL^{-1}; control limit for exposures averaged over 10 min. This is an example of meeting an action level but still requiring controls [6].

Now that we have placed air pollution in a historical context and introduced the most basic atmospheric scientific concepts, we can begin to understand the thermodynamic and fluidic systems involved in air pollution. We can begin with how pollutants are formed. The following chapters will address what happens after these substances are transformed and transported.

The physical and chemical principles that apply to chemical formation also apply to transformation kinetics after a substance is emitted to atmosphere and released into the other components of the environment. These pollutant transformations are the subject of the next chapter.

REFERENCES

[1] T. Salthammer, S. Mentese, R. Marutzky, Formaldehyde in the indoor environment, Chem. Rev. 110 (4) (2010) 2536–2572.
[2] D. Vallero, Environmental Biotechnology: A Biosystems Approach, Elsevier Science, 2015.
[3] National Society of Professional Engineers, NSPE Code of Ethics for Engineers, 26 January 2016. Available:http://www.nspe.org/resources/ethics/code-ethics, 2016.
[4] J.E. Schillinger, T. Vu, P. Bellin, Airborne fungi and bacteria, J. Environ. Health 62 (2) (1999) 9.
[5] D. Hospodsky, N. Yamamoto, J. Peccia, Accuracy, precision, and method detection limits of quantitative PCR for airborne bacteria and fungi, Appl. Environ. Microbiol. 76 (21) (2010) 7004–7012.
[6] HSE, Asbestos: The Analysts' Guide for Sampling, Analysis and Clearance Procedures, (2005).

FURTHER READING

[7] D.R. Bullard, et al., APTI Course SI:409: Basic Air Pollution Meteorology: Student Guidebook, U.S. Environmental Protection Agency, Air Pollution Training Institute, Research Triangle Park, NC, 1982.
[8] D.A. Kaden, C. Mandin, G.D. Nielsen, P. Wolkoff, Formaldehyde, (2010).
[9] W. H. Organization, WHO Guidelines for Indoor Air Quality: Selected Pollutants, (2010).
[10] U.S. Environmental Protection Agency, Formaldehyde, 25 April 2018. Available: https://www.epa.gov/formaldehyde, 2018.
[11] Toxicological Profile for Formaldehyde (Draft), (1997).
[12] Agency for Toxic Substances and Disease Registry, Formaldehyde and Your Health, 15 April 2018. Available: https://www.atsdr.cdc.gov/formaldehyde/index.html, 2018.
[13] The National Institute for Occupational Safety and Health, Formaldehyde, 25 April 2018. Available: https://www.cdc.gov/niosh/topics/formaldehyde/, 2018.
[14] An Update on Formaldehyde: 2013 Revision, Available: https://www.cpsc.gov/s3fs-public/pdfs/AN%2520UPDATE%2520ON%2520FORMALDEHYDE%2520final%25200113.pdf, 2013.
[15] D.A. Vallero, Fundamentals of Air Pollution, 5th ed., Elsevier Academic Press, Waltham, MA, 2014 p. 999 pages cm.
[16] C. Li, et al., Long-term persistence of polychlorinated dibenzo-p-dioxins and dibenzofurans in air, soil and sediment around an abandoned pentachlorophenol factory in China, Environ. Pollut. 162 (2012) 138–143.
[17] Emissions Monitoring Guidance (Note AG2), Available: http://www.epa.ie/pubs/advice/air/emissions/AG2%20Air%20Emissions%20Monitoring%20Guidance%20Note_rev3.pdf, 2015.
[18] Continuous Emission Manual, Available: https://digital.osl.state.or.us/islandora/object/osl%3A20999/datastream/OBJ/view, 1992.
[19] EPA-454/R-07-001, Urban Air Toxics Monitoring Program (UATMP), (2005).
[20] U.S. Environmental Protection Agency, Formaldehyde - CAS 50-00-0, 25 April 2018. Available: https://www.epa.gov/sites/production/files/2016-09/documents/formaldehyde.pdf, 2018.
[21] J.N. Spencer, G.M. Bodner, L.H. Rickard, Chemistry: Structure and Dynamics, John Wiley & Sons, 2010.

Chapter 3

Pollutant transformation

3.1 INTRODUCTION

The substances that form from human activities and natural phenomena are released into myriad parts of the environment. After their release, they move, change, and end up in various places on earth. These processes are collectively known as transformation, transport, and fate. Some, maybe most, air pollution texts discuss the movement, that is, transport, immediately after discussing air pollutant formation (see Chapter 2):

Formation → Transport → Transformation → Fate

Others prefer the following:

Formation → Transformation → Transport → Fate

Either is fine, but no simple flow is 100% correct. The challenge is deciding what is being transported. For example, consider an air pollutant that forms during combustion but changes into other compounds even before exiting the stack and then forms more compounds in the atmosphere as the plume travels downwind. In this case, it may be better to discuss transformation before transport, as we do in this book.

After a substance is released, it may change into other substances. As shown in Fig. 3.1, the parent compound (A) can be not only an air pollutant but also a precursor for other compounds that may also be air pollutants (A′ and A″). In fact, A could be the precursor for both A′ and A″, or it could be the precursor for A′, which is the precursor for A″.

Transport occurs everywhere in Fig. 3.1. The precursors and parent compound are moving within the reactor, for example, mixing by turbulence from convection. Transformation occurs almost everywhere once the chemical is formed. So, many air pollution experts link the terms as transport and fate, with fate being understood to include kinetics and equilibriums.

As discussed in Chapter 2, the formation of an air pollutant depends on the inherent properties of a substance. Some substances readily change under certain conditions, whereas others resist change under the same conditions. Furthermore, the rates of change differ for these reactions. A compound not only may form under high heat and pressure but also may form under ambient atmospheric conditions, albeit the former's rate of reaction is much faster than the latter.

The processes that take place during the formation of the air pollutant also occur after the pollutant is released into the atmosphere, water bodies, soil, sediment, and biota, for example, forests. A compound's inherent properties are the first determinant of how rapidly they will change. As shown in Table 2.1 in Chapter 2, the physical, chemical, and biological processes within the air, water, soil, or sediment, as well as those within the organism after uptake, determine whether a substance will be chemically transformed.

It is tempting, given that our subject is air pollution, to limit our discussions of chemical transformation to those that take place in the atmosphere. Indeed, substances change in myriad ways due to atmospheric conditions. However, this is only part of the transformation story. Chemicals change anywhere in the environment. Indeed, some of the same processes that transform a chemical in the air also do so in the soil, sediment, and water.

Furthermore, the same processes also take place inside an organism, including the cells of the human body, albeit at different rates, with different catalysts, and with different impacts. For example, an organic compound like benzene is released from a stack. Some of the benzene is degraded in the atmosphere, some falls to the earth, and some deposits in the soil. In the soil, bacteria use the benzene via metabolism, breaking it down to carbon dioxide and water, which is used by plants in photosynthesis. Cattle eat the plants, and the rumen in the anaerobic chambers of the cow's digestive system reduces the plant biomass to methane, which is released to the atmosphere. The benzene is transported to many other places, where abiotic and biotic processes transform it to other compounds. Indeed, every process mentioned in Table 2.1 is important to air pollution.

Air Pollution Calculations. https://doi.org/10.1016/B978-0-12-814934-8.00003-X

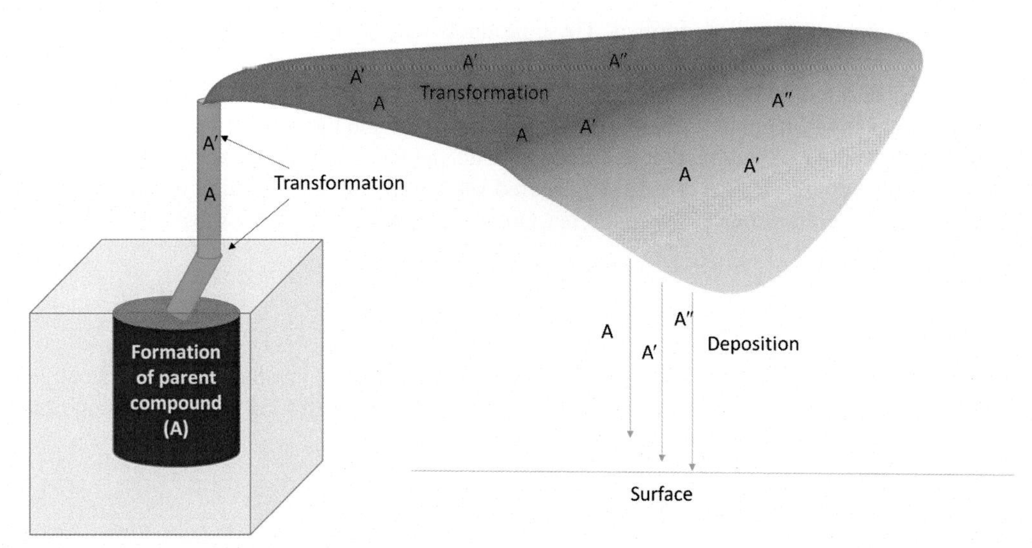

FIG. 3.1 Typical flow from air pollutant formation to contact to deposition to the earth's surface. Note that after parent compound (A) is formed, it is transformed to A′ in the stack and continues to be transformed in the plume. Downwind, a third transformation product is formed (A″) in the plume from the transformation of A and/or A′.

3.2 THE COMPOUND

With a few exceptions, air pollutants are chemical compounds. The exceptions are as follows:

- Elements, especially metals like lead (Pb), cadmium (Cd), and mercury (Hg)
- Particulate matter (PM)
- Microbes and other biological substances, known collectively as "bioaerosols"

Even these three exceptions consist of compounds. For example, metals exist in various oxidation states, such as metallic compounds, for example, the toxic salt, the lead oxide (PbO), or the organometallic tetraethyl lead ($(CH_3CH_2)_4Pb$). Particulate matter is an agglomeration of various compounds, depending on where it originated, for example, mineral oxides from the earth's crust, sodium chloride (NaCl) and other salts from sea spray, and organic compounds from incomplete combustion sources. Microbes, pollen, and other bioaerosols consist of organic chemicals, including biomolecules, for example, proteins, carbohydrates, lipids, nucleic acids, and metabolites [1].

Some compounds are considered air pollutants, and others are not. The distinction is arbitrary. If one were to ask a group of 100 randomly selected adults if carbon monoxide (CO) is an air pollutant, it is likely that all 100 would answer "yes." If the same group were asked if water is an air pollutant, most would likely answer "no." However, if one were to ask if carbon dioxide (CO_2) is a pollutant, there would be some disagreement. For example, some may say, correctly, that CO_2 is required for photosynthesis and, therefore, essential to life on earth, so it must not be an air pollutant. Others would argue that this essentiality does not preclude CO_2 from being a pollutant. After all, it is a greenhouse gas. Incidentally, water is also a greenhouse gas in vapor form. So, perhaps, the best definition of an air pollutant is situational. It is essential in some amount and in certain places, but it is harmful in others. Therefore, for these compounds, there exists an optimal range between too little and too much.

Notwithstanding the situational aspect of air pollution, certain compounds are inherently pollutants upon entering the atmosphere, particularly those that cause direct harm to the environment and human health. The unanimity of our 100 respondents that CO is always an air pollutant is correct.

3.3 AIR POLLUTION KINETICS

In Chapter 2, we considered kinetics that are important to the formation of an air pollutant. It is worthwhile to extend this discuss to the transformation after the air pollutant or an air pollutant precursor is emitted.

The individual chemical compounds that participate in reactions are called reactants, and the compounds that are generated are called products. Any chemical reaction requires that molecules collide. As the number of collisions increases, the larger the number of products that will be formed. These products may be new air pollutants, less toxic products from previous air pollutants, or even more toxic products from previous air pollutants.

A kinetic reaction is denoted by the one-directional arrow (→). In the environment, many processes are incomplete, such as the common problem of incomplete combustion in an incinerator and the generation of new compounds other than the two that result from complete oxidation, that is, carbon dioxide and water. Recall from Reactions (2.24) and (2.25) that when a compound breaks down in nature or in an engineered system like an air pollution control device, the change in product concentration drops proportionately with the reactant concentration. Thus, for substance A, the kinetics is.

$$\text{Rate} = -\frac{d[A]}{dt} \tag{3.1}$$

The brackets indicate "concentration." The minus and plus superscripts indicate these are ions, in which the minus sign indicates an anion and the plus sign indicates a cation. The negative sign denotes that the reactant concentration, that is, the parent compound in the atmosphere, is decreasing. The degradation product C resulting from the concentration increases in proportion to the decreasing concentration of the contaminant A:

$$\text{Rate} = +\frac{d[C]}{dt} \text{ or simply, } \frac{d[C]}{dt} \tag{3.2}$$

Recall that the rate of reaction at any time is the negative of the slope of the tangent to the concentration curve at that time [2, 3]. The intensity of molecular collisions is expressed as the "rate law." For example, when reacting A and B to yield product C (i.e., $A + B \rightarrow C$), the reaction rate increases in accordance with the increasing concentration of either A or B. If the amount of A increases fivefold, then the rate of this whole reaction increases fivefold. Restating Eq. (3.1) here, the rate law for such a reaction is.

$$\text{Rate} = k[\text{A}][\text{B}] \tag{3.3}$$

For the reaction $X + Y \rightarrow Z$, where the rate is only increased if the concentration of X is increased (changing the Y concentration has no effect on the rate law), the rate law must be

$$\text{Rate} = k[\text{X}] \tag{3.4}$$

The concentrations in the rate law are the concentrations of reacting chemical species at any specific point in time during the reaction. The rate expresses how fast reaction is happening at that time. The constant k in the equations is the rate constant, which is unique for every chemical reaction and is a fundamental physical constant for a reaction, depending on the environmental conditions (e.g., pH, temperature, pressure, and type of solvent).

3.3.1 Rate order and half-life

A key characteristic of an air pollutant is the time it takes for it to become transformed into other compounds, that is, reaction rate. This information can also be used to estimate how long the air pollutant takes to break down, that is, its persistence as indicated by its half-life. The rate constant is the rate of the reaction when all reactants are present in a 1 M (M) concentration. Accordingly, the rate constant k is the rate of reaction under conditions standardized by a unit concentration. By drawing a concentration curve for a contaminant that consists of an infinite number of points at each instant of time, an instantaneous rate can be calculated along the concentration curve. At each point on the curve, the rate of reaction is directly proportional to the concentration of the compound at that moment in time. This is a physical demonstration of kinetic order.

A reaction rate can be plotted against reactant concentrations. The rate is first order if the relationship is linear, that is, a reaction that proceeds at a rate (in units of moles per time, e.g., M sec^{-1}) that depends linearly on a single reactant concentration.

The overall kinetic order is the sum of the exponents (powers) of all the concentrations in the rate law. Consequently, in the rate k[A][B], the overall kinetic order is 2. Such a rate describes a second-order reaction because the rate depends on the concentration of the reactant raised to the second power. Other decomposition rates are like k[X] and are first-order reactions because the rate depends on the concentration of the reactant raised to the first power.

The kinetic order of each reactant is the power that its concentration is raised in the rate law. So, k[A][B] is first order for each reactant, and k[X] is first order X and zero order for Y. In a zero-order reaction, compounds degrade at a constant rate and are independent of reactant concentration.

What are the rates when hydrogen peroxide and ionic iodide react?
The reaction is

$$H_2O_2 + 2H^+ + 3I^- \rightarrow 2H_2O + I_3^- \tag{3.5}$$

Thus, the decreasing rates of reactants are.

$$-\frac{d[H_2O_2]}{dt}$$
$$-\frac{d[H^+]}{dt}$$
$$-\frac{d[I^-]}{dt}$$

The increasing rates of products are

$$\frac{d[H_2O]}{dt}$$

$$\frac{d[I_3^-]}{dt}$$

What is the stoichiometry of this reaction?
The number of moles (the number in front of the compound in a reaction) gives the relative quantities of substances taking part in a reaction, i.e., the stoichiometry. So, Reaction (3.5) tells us [4]

$$-\frac{d[H_2O_2]}{dt} = -\frac{1}{2}\frac{d[H^+]}{dt} = -\frac{1}{3}\frac{d[I^-]}{dt} = \frac{1}{2}\frac{d[H_2O]}{dt} = \frac{d[I_3^-]}{dt}$$

The rate for this reaction is governed by the stoichiometry:

$$\text{Rate} = -\frac{d[H_2O_2]}{dt} = -\frac{1}{2}\frac{d[H^+]}{dt} = -\frac{1}{3}\frac{d[I^-]}{dt} = \frac{1}{2}\frac{d[H_2O]}{dt} = \frac{d[I_3^-]}{dt}$$

Can this be generalized?
Yes. Let us express this type of reaction as [4]

$$aA + bB \rightarrow cC + dD \tag{3.6}$$

Using this notation, the generalized reaction says that the rate may be expressed by any one of the following:

$$\text{Rate} = -\frac{1}{a}\frac{d[A]}{dt} = -\frac{1}{b}\frac{d[B]}{dt} = \frac{1}{c}\frac{d[C]}{dt} = \frac{1}{d}\frac{d[D]}{dt} \tag{3.7}$$

Acetyl chloride (CH_3COCl) is hydrolyzed at 23°C in water to produce acetic acid (CH_3COOH) and hydrochloric acid, (HCl):

$$H_3C\text{–}C(=O)\text{–}Cl + H_2O \longrightarrow HO\text{–}C(=O)\text{–}CH_3 + HCl \tag{3.8}$$

If the beginning concentration of CH_3COCl is 0.1 M, what is the concentration of water at the beginning and completion of this reaction? What happens to the rate order in this reaction?
Reaction (3.8) is in the structural format. The reaction can also be shown in a line notation form, for example, the simplified molecular-input line-entry system (SMILES) format:

$$CH_3COCl + H_2O \rightarrow CH_3COOH + HCl \tag{3.9}$$

Experiments show that this reaction's rate law to be

$$-\frac{d[CH_3COCl]}{dt} = k[CH_3COCl]\,[H_2O] \tag{3.10}$$

This is a demonstration of a variation on the general rate law that we have been discussing, that is, the differential rate law (DRL), which expresses the concentration of one or more reactants, that is, Δ[R], over a specific time interval Δ*t*. DRLs help to describe processes during the reaction at the molecular scale. The DRL format is

$$\text{DRL: Rate} = -\frac{d[A]}{dt} = k[A]^n \tag{3.11}$$

This differs from the integrated rate law (IRL), which expresses the rate in terms of the initial concentration, that is, $[R_0]$, and the concentration of the reactant(s), that is, [R], after a given time period, *t*. Thus, the IRL is calculated by integrating the DRL. The IRL provides a way to determine rate constants and the reaction order from experimentally derived data. The IRL format is

$$\text{IRL: } \ln[A] = -kt + \ln[A]_0$$

where $[A]_0$ is the initial concentration of the reactant(s) after a specific time period (*t*) and [*A*] is the concentration after *t*.
Table 3.1 shows the equations for the DRL and IRL rate order of a reaction, which can be displayed as a linear plot and indicates the amount of time before half of the compound is transformed, that is, its half-life ($t_{1/2}$).
To make the math easier, let us also assume that the reaction occurs in 1 L solution of water. The room temperature assumption allows us to assume that the solution (water) density is approximately 1 g mL^{-1} and the mass is 1000 g. Therefore, H_2O is 18 g mol^{-1}:

$$\text{Number of moles } H_2O \text{ at } t=0 = \frac{1000g}{18g\ mol^{-1}} = 55.6\,M \text{ at the beginning of this reaction.}$$

Reaction (3.9) gives the following stoichiometry:

$$1CH_3COCl + 1H_2O \rightarrow 1CH_3COOH + 1HCl$$

Since we know that there is 0.1 mol of CH_3COCl, then there must be
55.6–0.1 = *55.5 M* CH_3COCl at the completion of this reaction.
Because water is in excess, we can use the mean of the beginning and ending concentrations to represent the constant, that is, 55.55 M. Therefore, $-\frac{d[CH_3COCl]}{dt} = k[CH_3COCl]\ [H_2O]$ can be simplified to

$$-\frac{d[CH_3COCl]}{dt} = k'[CH_3COCl] = k(55.55M)$$

where $k' = k[H_2O] = k(55.55)$
This means the overall order of the reaction has been lowered from second to first, or what is known as a pseudo first-order reaction.
If **k** ***=*** $1.2\times10^{-3}\ M^{-1}\ s^{-1}$***, when would*** $[CH_3COCl]$ ***equal 0.05 M?***
$-\frac{d[CH_3COCl]}{dt} = k'[CH_3COCl]$ can be rearranged to
$-\frac{d[CH_3COCl]}{[CH_3COCl]} = k'dt$, which can be integrated to give

$$-\ln\frac{[CH_3COCl]_t}{[CH_3COCl]_0} = k't$$

Since we started with 0.1 M and then 0.05 M, then $[CH_3COCl]_t$ is the $t_{1/2}$. So, using the information in Table 3.1, we find

$$t_{1/2} = \frac{1}{k'} \times -\ln\left(\frac{0.05M}{0.1M}\right) = \frac{\ln 2}{k'} = \frac{0.693}{k'}$$

From the earlier calculation, $k' = k\ (55.55\,M) = (1.2\times10^{-3}\ M^{-1}\ s^{-1})\ (55.55\,M) = 0.067\,s^{-1}$
Therefore, $t_{1/2} = \frac{0.693}{0.067\ sec^{-1}} = \underline{10.3\,s}$
Thus, assuming this *k* value, half of the acetyl chloride would have been transformed to acetic acid and hydrochloric acid in 10.3 s. Incidentally, the *k* value is experimentally derived. In this case, the reported value [5] for this equation was rounded to 1.2×10^{-3} M^{-1} s^{-}1 from the reported value of $1.16\times10^{-3}\ M^{-1}\ s^{-1}$. Had we used the reported value, the $t_{1/2}$ would have been 10.7 s.

TABLE 3.1 Relationships between first- and second-order reactions

Reaction order	Differential rate law	Integrated rate law	Linear plot	Half-life ($t_{1/2}$)
First	$\frac{d[A]}{dt} = k[A]$	$[A] = [A]_0e^{-kt}$	ln[*A*] vs *t* Slope = −*k*	$\frac{\ln(2)}{k}$
Second	$\frac{d[A]}{dt} = k[A]^2$	$\frac{1}{[A]} - \frac{1}{[A]_0} = kt$	$\frac{1}{[A]}$ vs *t* Slope = *k*	$\frac{1}{k[A]_0}$

What if we were not told the temperature. Does that matter?

Yes, it matters very much. Sometimes, assumptions must be made. For example, if not told otherwise, we often assume that a solution is in water. There may often be times when we are fairly certain that standard temperature and pressure are assumed. However, it is not wise to make these types of assumptions when addressing air pollution, since temperature can range widely; for example, combustion and stack temperatures are very high, habitations are about room temperature, and refrigerated units are quite low. Therefore, one can make these assumptions only if they are reported as such. For example, the solution to a calculation should have statement something akin to the following: "Since not specified, this calculation assumed the temperature to be X° and pressure to be Y atm."

3.3.2 Atmospheric kinetics

The acetyl chloride hydrolysis example in the previous section could happen in many places. Obviously, this could occur in an industrial reactor to synthesize organic compounds, including dyes. This could allow for releases to the air in manufacturing and occupational settings. There are hundreds of thousands of such chemicals in the marketplace.

Some chemical compounds are much more likely be released into the atmosphere and undergo transformation there. For example, oxides of nitrogen (NO_X) and hydrocarbons that are emitted from vehicles and other sources will be transformed. Some of these transformations eventually lead to the production of ozone (O_3) in the troposphere. These are predominantly photochemical transformations.

Why should we care about anything other than photochemistry, since it is so important to air pollution transformation?

Certainly, many air pollution experts devote much of their research and interest to photochemistry, especially those concerned with oxidative processes that form smog and other important pollutants. However, other processes, including reduction-oxidation and hydrolysis, also play roles in atmospheric transformations. There are two prominent reasons to consider all of the processes. First, air pollutants often form in media, before moving to the atmosphere. We shall discuss Henry's law and vapor pressure in the next chapter.

The second reason is that the air contains water and particulate matter (PM), in which and on which chemical reactions occur. Indeed, air pollution chemistry must address these and other processes because they occur in water droplets in clouds and in and on PM. The release of relatively slow-reacting primary air pollutants undergoes reactions and transformations during lengthy transport times. These protracted transport times allow pollutants to move over long distances, during which the parent compounds but of numerous transformation by-products form (see Fig. 3.2). For example, sulfur dioxide (SO_2) released as a gas primarily through combustion of fossil fuels is oxidized during long-distance transport to sulfur trioxide (SO_3):

$$2SO_2 + O_2 \rightarrow 2SO_3 \tag{3.12}$$

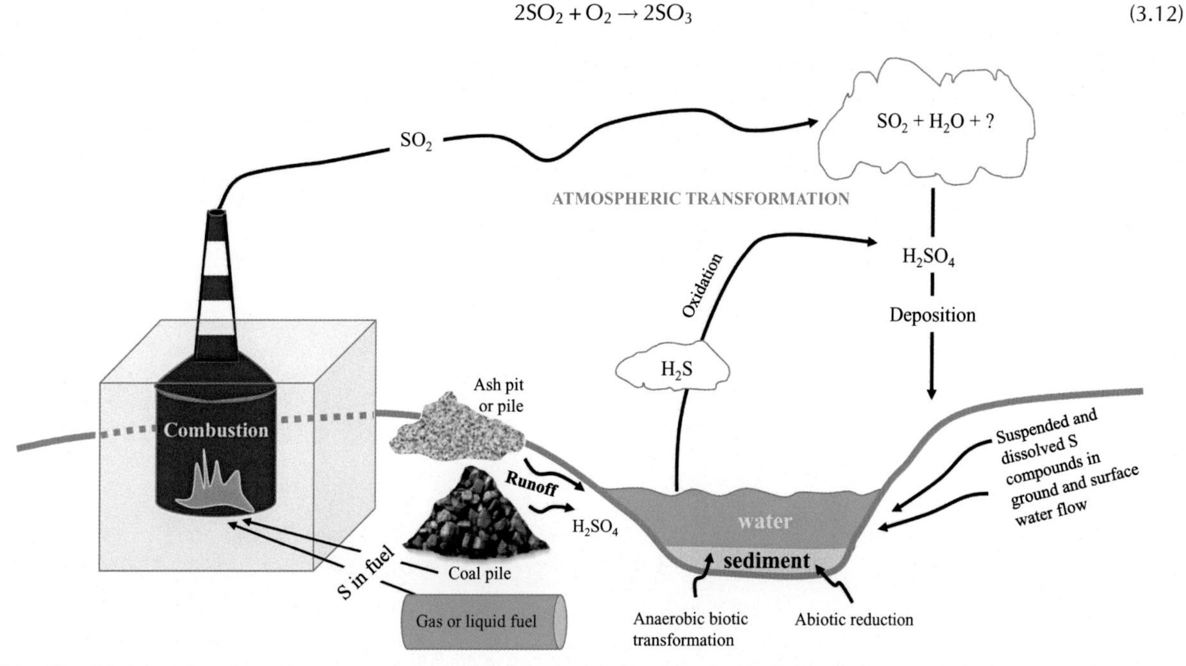

FIG. 3.2 Simplified depiction of transformation pathways leading to atmospheric sulfur dioxide. Sulfur dioxide not only is formed during combustion and emitted to the atmosphere but also is transformed from the oxidation of hydrogen sulfide and other less oxidized sulfur species biotically, followed by oxidation in water droplets in the atmosphere.

Although SO_2 is a gas, both gas-phase and liquid-phase oxidation of SO_2 occur in the troposphere. The SO_3 in turn reacts with water vapor to form sulfuric acid:

$$SO_3 + H_2O \rightarrow H_2SO_4 \quad (3.13)$$

The sulfuric acid reacts with numerous compounds to form sulfates. These are fine (submicrometer) aerosols. Nitric oxide (NO) results from high-temperature combustion, both in stationary sources such as power plants or industrial plants in the production of process heat and in internal combustion engines in vehicles. The NO is oxidized in the atmosphere, usually rather slowly or more rapidly if there is ozone present to nitrogen dioxide (NO_2), which also reacts further with other constituents, forming nitrates, which is also in fine particulate form.

The sulfates and nitrates existing in the atmosphere as aerosols, generally in the size range $<1\,\mu m$ aerodynamic diameter, can be removed from the atmosphere by several processes. "Rainout" occurs when the particles serve as condensation nuclei that lead to the formation of clouds. If the droplets grow to a sufficient size, they fall as raindrops with the particles in suspension or in solution. Another mechanism, known as "washout," also involves rain, but the particles in the air are captured by raindrops falling through the air. Both mechanisms contribute to "acid rain," which results in the sulfate and nitrate particles reaching lakes and streams and increasing their acidity [2].

During the atmospheric residence time, chemical compounds generally become oxidized. Gases in reduced states undergo step reactions to form ionic substances, which are in turn washed out by rain and other precipitation, known as deposition. For example, hydrogen sulfide (H_2S) is a reduced species of sulfur. After emission to the atmosphere, it is dissolved in water vapor and is oxidized to form sulfate compounds or anions:

$$H_2S\,(g) + H_2O\,(l) \rightarrow SO_4^{2-} \quad (3.14)$$

While we have been discussing nonmetals, metals undergo similar transformations. For example, mercury (Hg) can undergo chemical transformations in the atmosphere that change its properties, including its likelihood to accumulate in organisms. Mercury will react in gas, liquid, and solid phases. Reactions with ozone [6] and OH radicals [7] are the most important gas-phase oxidation pathways. Zerovalent or elemental mercury, that is, Hg^0, is oxidized to divalent mercury (Hg^{2+}), which is less volatile and more likely to condense on surfaces, including particulate matter and biota. When Hg is dissolved in water, it can be oxidized, for example, by O_3 and OH radicals [6, 8–10].

The Hg aqueous-phase reaction rates are much faster than those in the gas phase. However, the low aqueous solubility of Hg^0 and the low water concentrations in the atmosphere make for about equal amounts of transformations for gas and liquid phases. This is another example of how a compound's transformation can be dictated by chemistry and physics, that is, rate laws and solubility.

These and other reactions can also reduce visibility and form haze. The fine sulfate- and nitrate-laden PM is largely responsible for reduction of visibility. Decreased visibility can impair safety, especially concerning aviation, but is especially problematic in locations of natural beauty, where it is desirable to keep scenic vistas as free of obstructions to the view as possible. The United States and other countries require special protections for such areas. For example, the United States denotes the most restrictive areas as mandatory Federal Class I areas that cover over 150 national parks and wilderness areas [11].

What is the role of ionization in pollutant transformation?

Processes like ionization within a droplet or on an aerosol will set the stage for the kinetics in the atmosphere (see Fig. 3.3).

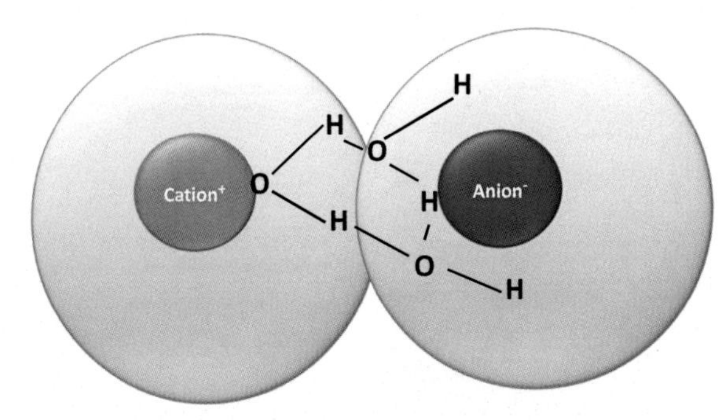

FIG. 3.3 Ion pairs in a water droplet. *Modified from D.A. Vallero, Fundamentals of Air Pollution, 5th ed., Elsevier Academic Press, Waltham, MA, 2014, p. 999 pages cm;and V. Evangelou, Environmental Soil and Water Chemistry: Principles and Applications. John Wiley Sons, Inc., Canada, 1998.*

The key atmospheric condition driving many reactions is the available light. Photochemistry distinguishes atmospheric chemical reactions, that is, dark versus photochemical. Dark reactions are often also thermal reactions, that is, they need the heat energy to form products from the reactants. Of course, both dark and light reactions occur in the atmosphere, given that sunlight is only available during the day (see Fig. 3.4).

The dominant type of photochemical reactions in the atmosphere is photodissociation. Photodissociation is a chemical reaction that degrades compounds by bombarding the molecule with photons. Photolysis is the type of photodissociation wherein a compound is decomposed by light or other forms of radiant energy. During photooxidation in the atmosphere, photodissociation combines with oxidation to break down chemicals in the presence of sunlight. The oxidation reactions involve the exchange of electrons between the pollutant molecule and other reactive compounds.

If the only concern for air pollutants were the pollutant's residence time in the atmosphere, the degradation processes would mainly be photolysis and other types of photooxidation. However, formation and transformation occur in most environmental compartments, so other processes occur before, during, and after atmospheric residence time, especially redox, biodegradation, and hydrolysis [2].

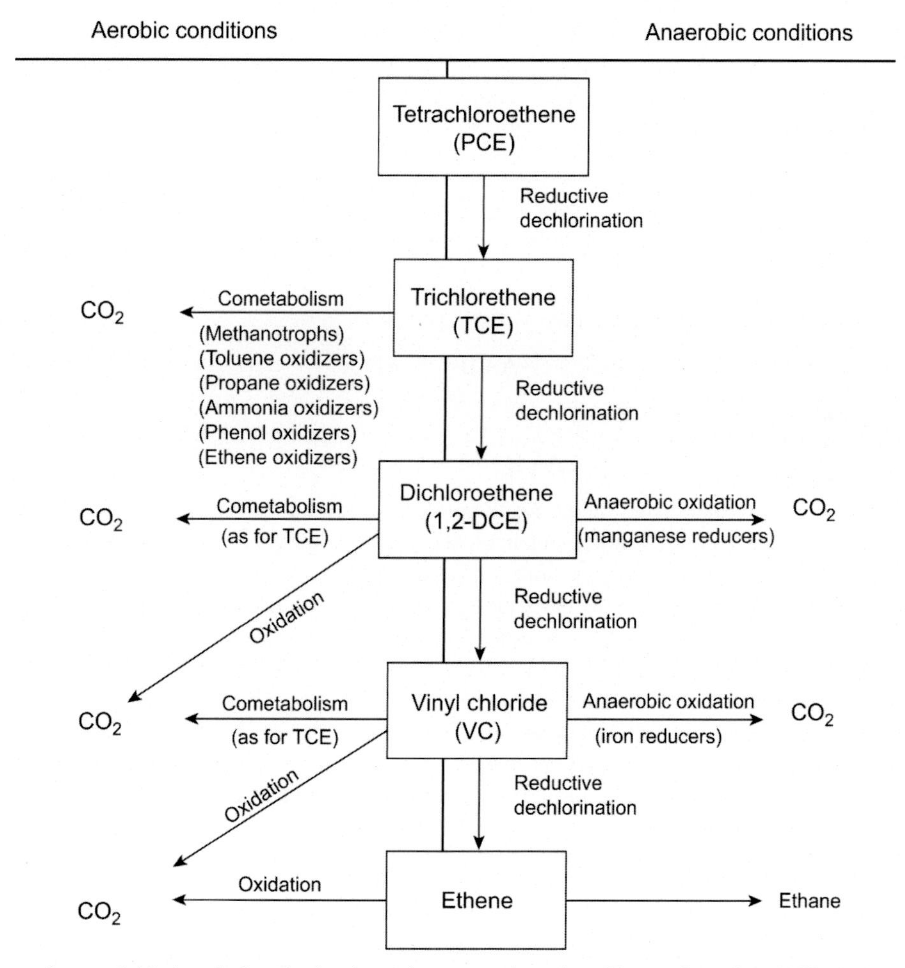

FIG. 3.4 Degradation pathways of chlorinated ethers that involve various types of reactions. The transformation of different chemical species varies in reaction kinetics. Note that before any of these compounds can be oxidized, including photodissociation, they undergo processes in water (e.g., reductive dichlorination) and in microbes, especially cometabolism. The toxic air pollutant, vinyl chloride, accumulates mainly from anaerobic digestion, that is, reductive dichlorination of a degradation product, that is, 1,2-DCE. Vinyl chloride is degraded either by direct oxidization to carbon dioxide or first by reductive dichlorination to ethene, which is then oxidized to carbon dioxide. These oxidative steps can occur completely or in part by photodissociation in the atmosphere. *Courtesy D.A. Vallero, Fundamentals of Air Pollution, 5th ed., Elsevier Academic Press, Waltham, MA, 2014, p. 999 pages cm; adapted from C. GeoSyntec, Bioaugmentation for remediation of chlorinated solvents: technology development, status, and research needs. Environ. Secur. Technol. Certif Program (ESTCP), 122 (2005).*

So, atmospheric transformation is exclusively oxidative, given that about 20% is molecular oxygen, right?
Wrong, reduction also occurs in the atmosphere. Remember, the 20% O_2 is for dry air. Real air contains water in all phases, that is, solid, liquid, and gas. Just as in surface water, an airborne droplet can contain ions and radicals that will reduce Hg^{2+} to Hg^0. For example, Hg^{2+} can be reduced to Hg^0 if sulfur dioxide (SO_2) is dissolved in water vapor or HO_2 radicals are present.

Atmospheric chemical reactions can be either photochemical or thermal (or dark). Often, photochemical reactions occur first in sunlight, by interactions of photons with chemical species to form products, but these newly formed compounds may undergo further light or dark chemical reactions. The subsequent thermal reactions can be reductive or oxidative.

From this, we can surmise that atmospheric transformations can also be differentiated as either homogeneous or heterogeneous. If the reaction only occurs in the gas phase, the reaction is homogeneous. If the reaction involves an interface with a surface of an aerosol or within a liquid droplet, it is heterogeneous. In the series of transformations, the numerous step reactions, are either homogeneous (e.g., all reactants are in the gas phase) or heterogeneous (e.g., gas phase and aqueous phase reactions at the interface between air and droplet).

3.4 ENERGY IN ATMOSPHERIC TRANSFORMATIONS [12]

Almost every reaction in the atmosphere is driven by photochemistry. There are numerous types of atmospheric reactions. Up to now, we have mainly been discussing mass, for example, moles of compounds, when discussing kinetics. But as the term kinetics implies, all chemical transformations require energy. A unimolecular reaction involves a single reactant. Many atmospheric unimolecular reactions require light energy via the photon ($h\nu$). These include the following:

- $A + h\nu \rightarrow C$
- Luminescence: $A + h\nu \rightarrow AB^* \rightarrow AB + h\nu$ (where AB^* is an intermediate, short-lived excited product)
- Quenching: $A + h\nu \rightarrow AB^* \rightarrow AB + M$ (where M is a third-body molecule)
- Photodissociation: $A + h\nu \rightarrow AB^* \rightarrow A + B$

Bimolecular reactions have two reactants, such as

$$A + B \rightarrow C + D$$

Termolecular reactions have three reactants:

$$A + B + M \rightarrow C + D + M$$

The sun provides the energy that drives the atmosphere. After reaching the earth's atmosphere and surface, the solar radiation is converted to various forms, including mechanical energy that moves the molecules in the atmosphere, chemical energy via photosynthesis, and potential energy stored in fossil fuels.

The physics of radiative transfer provides an early clue about how compounds are formed and transformed in the atmosphere. Sunlight covers almost the entire electromagnetic radiation (EMR) spectrum shown in Fig. 3.5. The EMR that makes it through the atmosphere is converted to new wavelengths by the earth's surface, for example, from shorter-wave light to longer-wave infrared (heat), which are reradiated back to the atmosphere. The atmosphere's gases and aerosols absorb and filter this energy by emission, absorption, and scattering.

Physicists apply the term blackbody to an object that is a perfect emitter and absorber of radiation at all wavelengths. Although no such object exists in nature, the properties describable by theory are useful for comparison with materials found in the real world. The amount of radiation, or radiant flux over all wavelengths (F), from a unit area of a blackbody is dependent on the temperature of that body and is given by the Stefan-Boltzmann law:

$$F = \sigma T^4 \tag{3.15}$$

where σ is the Stefan-Boltzmann constant and equals $8.17 \times 10^{-11}\,\text{cal}\,\text{cm}^{-2}\,\text{min}^{-1}\,\text{deg.}^{-4}$ and T is the temperature in degrees K. Radiation from a blackbody ceases at a temperature of absolute zero, 0°K.

Absorptivity is the amount of radiant energy absorbed as a fraction of the total amount that falls on the object. Absorptivity depends on both frequency and temperature; for a blackbody, it is 1. Emissivity is the ratio of the energy emitted by an object to that of a blackbody at the same temperature. It depends on both the properties of the substance and the frequency. Kirchhoff's law states that for any substance, its emissivity at a given wavelength and temperature equals its absorptivity. Note that the absorptivity and emissivity of a given substance can vary substantially for different frequencies [2].

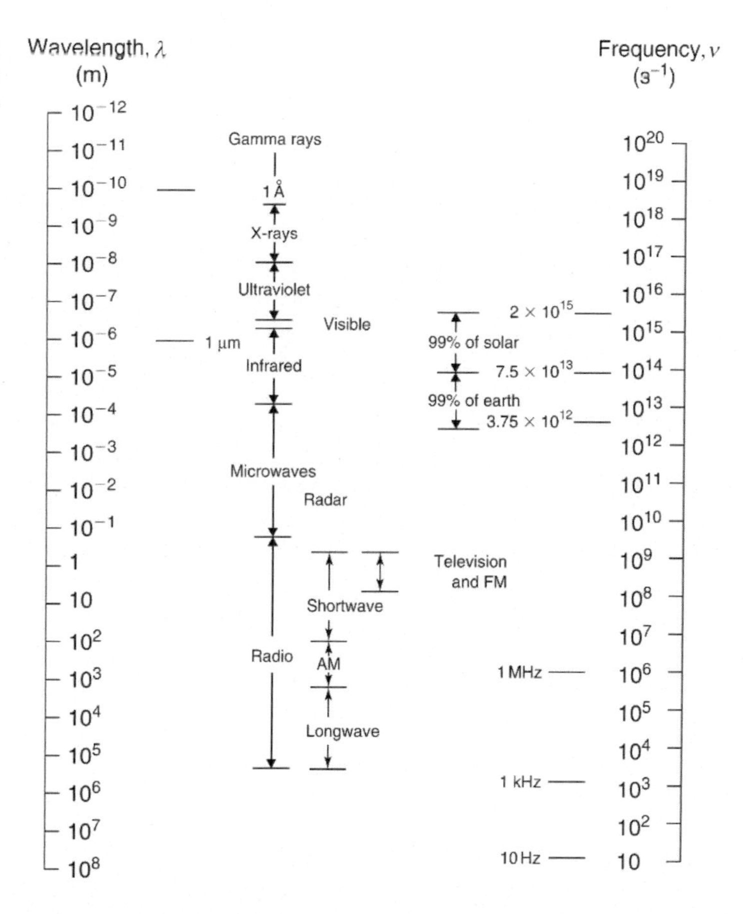

FIG. 3.5 Electromagnetic spectrum, denoting the regions of solar and earth radiation. *Source: D.A. Vallero, Fundamentals of Air Pollution, 5th ed., (Elsevier Academic Press, Waltham, MA, 2014, p. 999 pages cm.)*

Eq. (3.15) shows that the total radiation from a blackbody depends on the fourth power of its absolute temperature. The frequency of the maximum intensity of this radiation is also related to temperature through Wien's displacement law, which is derived from Planck's law:

$$\nu_{max} = 1.04 \times 10^{11} T \tag{3.16}$$

where frequency ν is in s^{-1} and the constant is in s^{-1} K^{-1}.

The radiant flux can be determined as a function of frequency from Planck's distribution law for emission:

$$E_\nu d\nu = c_1 \nu^3 \left[\exp\left(c_2 . \nu \cdot T^{-1}\right) - 1\right]^{-1} d\nu \tag{3.17}$$

where

$c_1 = 2\pi \cdot h \cdot c^{-2}$
$h = 6.62 \times 10^{-27}$ erg s (Planck's constant)
$c = 3 \times 10^8 \, \mathrm{m\,s^{-1}}$ (speed of light)
$c_2 = h \cdot k^{-1}$

and

$$k = 1.37 \times 10^{-16} \, \mathrm{erg\,K^{-1}} \text{ (Boltzmann's constant)}$$

The radiation from a blackbody is continuous over the electromagnetic spectrum. The term blackbody is misleading, since numerous nonblack materials can behave like blackbodies. The sun behaves almost like a blackbody; snow radiates in the infrared nearly as a blackbody. At some wavelengths, water vapor radiates very efficiently. Unlike solids and liquids, many gases absorb and reradiate energy selectively in discrete wavelength bands, rather than smoothly over a continuous spectrum [2].

The sun radiates approximately as a blackbody, with an effective temperature of about 6000°K. The total solar flux is 3.9×10^{26} W. Using Wien's law, it has been found that the frequency of maximum solar radiation intensity is 6.3×10^{14} s^{-1}

(wavelength (λ) = 0.48 μm), which is in the visible part of the spectrum; 99% of solar radiation occurs between the frequencies of $7.5 \times 10^{13}\ s^{-1}$ ($\lambda = 4\ \mu m$) and $2 \times 10^{15}\ s^{-1}$ ($\lambda = 0.15\ \mu m$) and about 50% in the visible region between $4.3 \times 10^{14}\ s^{-1}$ ($\lambda = 0.7\ \mu m$) and $7.5 \times 10^{14}\ s^{-1}$ $\lambda = 4\ \mu m$). The intensity of this energy flux at the distance of the earth is about $1400\ W\ m^{-2}$ on an area normal to a beam of solar radiation. This value is known as the solar constant. Due to the eccentricity of the earth's orbit as it revolves around the sun once a year, the earth is closer to the sun in January (perihelion) than in July (aphelion). This results in about a 7% difference in radiant flux at the outer limits of the atmosphere between these two times.

Since the area of the solar beam intercepted by the earth is πE^2, where E is the radius of the earth, and the energy falling within this circle is spread over the area of the earth's sphere, $4\pi E^2$, in 24 h, the average energy reaching the top of the atmosphere is $338\ W\ m^{-2}$. This average radiant energy reaching the outer limits of the atmosphere is depleted as it attempts to reach the earth's surface. Ultraviolet radiation with a wavelength $<0.18\ \mu m$ is strongly absorbed by molecular oxygen in the ionosphere 100 km above the earth; shorter X-rays are absorbed at even higher altitudes above the earth's surface. At 60–80 km above the earth, the absorption of 0.2–0.24 μm wavelength radiation leads to the formation of ozone; below 60 km, there is so much ozone that much of the 0.2–0.3 μm wavelength radiation is absorbed. This ozone layer in the lower mesosphere and the top of the stratosphere shields life from much of the harmful ultraviolet radiation. The various layers warmed by the absorbed radiation reradiate in wavelengths dependent on their temperature and spectral emissivity. Approximately 5% of the total incoming solar radiation is absorbed above 40 km. Under clear sky conditions, another 10%–15% is absorbed by the lower atmosphere or scattered back to space by the atmospheric aerosols and molecules; as a result, only 80%–85% of the incoming radiation reaches the earth's surface. With average cloudiness, only about 50% of the incoming radiation reaches the earth's surface, because of the additional interference of the clouds.

The inorganic chemical species most important to atmospheric kinetics and transformation reactions can be grouped into five classifications [13]:

1. Odd oxygen species
2. Odd hydrogen species
3. Reactive nitrogen species
4. Reactive sulfur species
5. Reactive halogen species

Odd chemical species have odd subscripts. Odd oxygen species include O_3, atomic O, and O(1D), which is an O atom in an excited singlet state. Odd hydrogen species include OH, HO_2, and atomic H. Reactive nitrogen species include NO, NO_2, NO_3, HNO_2, HNO_3, and NH_3. Reactive sulfur species include SO_2, SO_3, H_2SO_3, H_2SO_4, and H_2S. Reactive halogen species include F, FO, Cl, ClO, Br, BrO, I, IO, HCl, and Cl_2.

3.4.1 Atmospheric photochemistry

Let us consider what happens when a photon from sunlight attacks a molecule.

What is a photon?

The foregoing discussion may leave the reader wondering whether EMR is a wave or a particle. The answer is "yes." This conundrum is known as the "wave-particle duality." As a Roman Catholic Christian, I am familiar with the term "duality," as in my belief that Christ is wholly human and wholly divine.

However useful this term is for the metaphysical, it is rather confusing for the physical. Some aspects of waves and particles are mutually exclusive. Particles travel well-defined paths, whereas waves bend around corners (i.e., diffraction) and propagate differently than particles. They also differ in how they exchange energy, that is, particles by collision at specific points in time in space, but waves disperse in space and deposit continuously as the wave front encounters matter. Albert Einstein and Leopold Infeld [14] summed up the duality this way:

But what is light really? Is it a wave or a shower of photons? There seems no likelihood for forming a consistent description of the phenomena of light by a choice of only one of the two languages. It appears we must use sometimes the one theory and sometimes the other, while at times we may use either. We are faced with a new kind of difficulty. We have two contradictory pictures of reality; separately neither of them fully explains the phenomena of light, but together they do.

Indeed, Eq. (3.18) and Fig. 3.5 show that the EMR spectrum varies by wavelength (λ), so it must behave as a wave. However, the change in energy between the two states corresponds to a quantum or "photon" of solar radiation. So, a photon is an amount, that is, a bundle, of light energy (*E*). The frequencies (*ν*) of absorption are expressed by Planck's law:

$$E = h\nu = \frac{hc}{\lambda} \tag{3.18}$$

where h is Planck's constant; c is the speed of light; and ν and λ are the frequency and wavelength of the light of the photon, respectively. The photon is represented as $h\nu$.

Molecules and atoms interact with photons of solar radiation under certain conditions to absorb photons of light of various wavelengths. Solar radiation initiates the formation of free radicals. Molecules interact with solar radiation by absorbing light energy, causing the molecule to undergo a transition from the ground electronic state to an excited state.

What is the photon energy at the extremes of visible light?

The range of λ for visible light is 400–700 nm.

Since $E = \frac{hc}{\lambda}$

and Planck's constant $(h) = 6.626 \times 10^{-34}\,\text{J·s} = 4.136 \times 10^{-15}\,\text{eV·s}$

then, $hc = (4.136 \times 10^{-15}\,\text{eV·s})\,(2.997 \times 10^{8}\,\text{m s}^{-1}) = 1.240 \times 10^{-6}\,\text{eV·m} = 1240\,\text{eV·nm}$

$$\text{At}\ \lambda = 400\,\text{nm},\ E = \frac{1240\,\text{eV}\cdot\text{nm}}{400\,\text{nm}} = 3.10\,\text{eV}$$

$$\text{At}\ \lambda = 700\,\text{nm},\ E = \frac{1240\,\text{eV}\cdot\text{nm}}{700\,\text{nm}} = 1.77\,\text{eV}$$

The energy of the photons in visible light ranges from 1.77 to 3.10 eV.

3.4.2 Energy and bonds

Energy is often defined as the capacity to do work. Changes in ecosystems are manifestations of the law of conservation of mass and of the laws of thermodynamics. The former states that the total mass remains constant during a chemical change or reaction or in any isolated process. We shall shortly apply it to material flow.

Energy must be supplied to break a chemical bond. Indeed, atoms bond together to form compounds so that they can gain lower energies than they have as individual atoms. Energy, especially heat, is released when this happens. The amount of energy is equal to the difference between the energies of the bonded atoms and the energies of the released atoms. The bonded atoms have a lower energy than the individual atoms do. When atoms combine to form a compound, energy is always released, so the compound has a lower overall energy [15]. Each molecule absorbs solar radiation at its own range of wavelengths (see Table 3.2).

TABLE 3.2 Bond energy and wavelengths for important bonds in air pollutant molecules. The likelihood of a photolytic reaction depends on the probability that a compound will absorb a specific wavelength of light or on the probability that the excited molecular species will undergo a reaction

Bond	Bond energy (kJ mol^{-1})	Wavelength (nm)
O—H	465	257
H—H	436	274
C—H	415	288
N—H	390	307
C—O	360	332
C—C	348	344
C—Cl	339	353
Cl—Cl	243	492
Br—Br	193	630
O—O	146	820

Source: A. Felsot, Abiotic/biotic degradation & transformation, in: Richland (Ed.), Environmental Attenuation of Contaminants, Washington State University, Washington, 2005.

TABLE 3.3 Mean bond energies ($kJ\,mol^{-1}$)

Single bonds						Multiple bonds	
H—H	432	N—H	391	I—I	149	C=C	614
H—F	565	N—N	160	I—Cl	208	C≡C	839
H—Cl	427	N—F	272	I—Br	175	O=O	495
H—Br	363	N—Cl	200			C=O[a]	745
H—I	295	N—Br	243	S—H	347	C≡O	1072
		N—O	201	S—F	327	N=O	607
C—H	413	O—H	467	S—Cl	253	N=N	418
C—C	347	O—O	146	S—Br	218	N≡N	941
C—N	305	O—F	190	S—S	266	C≡N	891
C—O	358	O—Cl	203			C=N	615
C—F	485	O—I	234	Si—Si	340		
C—Cl	339			Si—H	393		
C—Br	276	F—F	154	Si—C	360		
C—I	240	F—Cl	253	Si—O	452		
C—S	259	F—Br	237				
		Cl—Cl	239				
		Cl—Br	218				
		Br—Br	193				

Note: — is a single bond, = is a double bond, and ≡ is a triple bond.
[a]*C == O* (CO_2)*=799.*
Data from K. Song, Donald Le. Bond Energies. (2017). 24 May 2018. Available: https://chem.libretexts.org/Core/Physical_and_Theoretical_Chemistry/Chemical_Bonding/Fundamentals_of_Chemical_Bonding/Bond_Energies.

In the simple example of water reacting to become hydrogen and oxygen gases, the bonds of two water molecules break:

$$2H_2O \rightarrow 2H_2 + O_2 \tag{3.19}$$

An excited molecule can follow several pathways, including fluorescence, collisional deactivation, direct reaction, and photodissociation.

Note that the data in Table 3.2 are quite specific, that is, a particular bond energy is associated with a particular wavelength of light. Certainly, each molecule has a characteristic bond energy, and the exact value of a bond energy depends on the particular molecule. However, bond energies need to be generalized so that we can make air pollution reaction calculations. For example, all O—H bonds have a bond energy of roughly the same value because they are all O—H bonds. The same applies to all the bonds in Table 3.2. Thus, about 100 kcal of energy is needed to break 1 mol of C—H bonds, which allows us to say that the bond energy of a C—H bond is about $100\,kcal\,mol^{-1}$. A C—C bond has an approximate bond energy of $80\,kcal\,mol^{-1}$, whereas the bond energy of a double-bond carbon to carbon (C=C) is about $145\,kcal\,mol^{-1}$. Therefore, with all the reactions taking place in the atmosphere, the mean bond energies of a specific bond in different molecules can be used to generalize bond energy. These are provided in Table 3.3.

3.4.3 Photoabsorption

Energy levels of a molecule are described by quantum mechanics as a discrete sequence, depending on the spectroscopic properties of the molecule.[1] Based on information from Ilčev [16] and Sportisse [17], let us consider an example of energy transition.

1. This is a key concept to spectroscopy, which is discussed in Chapter 9.

Consider the discrete sequence, (En)n, specific to a molecule trapped in an energy potential, metaphorically, "in a well." If the well corresponds to a one-dimensional interval, (0.1), and x is the spatial variable, what are the possible energy levels?

We can only estimate the particle location using a probability density function, $p(x)$, derived from the wave function $f(x)$ as $p(x)=|f(x)|^2$. The Schrödinger equation governs this function:

$$-\frac{h^2}{2m}\frac{d^2f}{dx^2}+V(x)f=Ef \quad (3.20)$$

where m is the particle mass; $V(x)$ is the energy potential that defines the well; E corresponds to the particle energy; and h is Planck's constant, that is, 6.63×10^{-34} J s. The particle can move freely within the well ($V=0$) but is "trapped," which means that its probability density function is null at the well's boundaries.

The governing equation for f is

$$\frac{d^2f}{dx^2}=\frac{2mE}{h^2}f,\ f(0)=f(1)=0 \quad (3.21)$$

Thus, the solutions are in the form $f(x)\sim\sin\left(\sqrt{\frac{2mE}{h^2}}x\right)$, where $\sqrt{\frac{2mE}{h^2}}=n\pi$ with n being a positive integer.

Solution: This results in a discrete spectrum of energy levels in which $E_n=n^2\cdot\frac{h^2}{2m}\pi^2$.

The emission of a photon by a compound with an energy level of E_1 will correspond to transition from E_1 to a lower energy state, for example, E_2. Therefore, $E_2 < E_1$. Conversely, when a photon is gained, that is, absorbed by the compound, the transition is to a higher energy level, that is, $E_1 < E_2$. Planck's law (Eq. 3.18) requires that the photon's wavelength be fixed by the energy transition. That is, a photon may only be emitted or absorbed if its wavelength corresponds to a transition that is possible (see Fig. 3.6).

The energy gap is inversely proportional to wavelength, that is, a smaller λ EMR contains more energy. Thus, UV light contains much more energy than infrared EMR (see Fig. 3.5).

Bond breakage calculations must follow rules. The amount of energy needed to break a bond increases with bond strength. Bond order and bond length indicate the type and strength of covalent bonds between atoms, when they share pairs of electrons. The number of chemical bonds between a pair of atoms is known as bond order, which is the indicator of bond stability, that is, a triple bond is stronger than a double bond, which is stronger than a single bond. The most plentiful gas in the atmosphere is molecular nitrogen, N_2, which has a triple bond between the nitrogen atoms. Thus, N≡N has a bond order that is 3. Thus, it requires much energy to break that bond, so N_2 behaves as if it were inert under standard environmental conditions. However, the energy in an internal combustion engine or an electric generating power plant is sufficient to break the bond, allowing the N to react, for example, to be oxidized to NO_2. The bond order also indicates where a molecule may be attacked, for example, by microbial degradation or heat. For example, in acetylene, H—C≡C—H, the C—H bonds (order 1) are much weaker than the C≡C bond (order 3), so the molecule will first break at the C—H bonds. Bond order and length are inversely proportional, that is, the longer the bond, the weaker the bond.

3.5 ATMOSPHERIC KINETICS CALCULATIONS

A resistant compound's half-life is longer than a reactive compound's. However, such statements are not robust, so we need to quantify them. We can use energy and bond information to estimate kinetics. Let us first consider bond length, which is inversely related to molecular resistance to degradation.

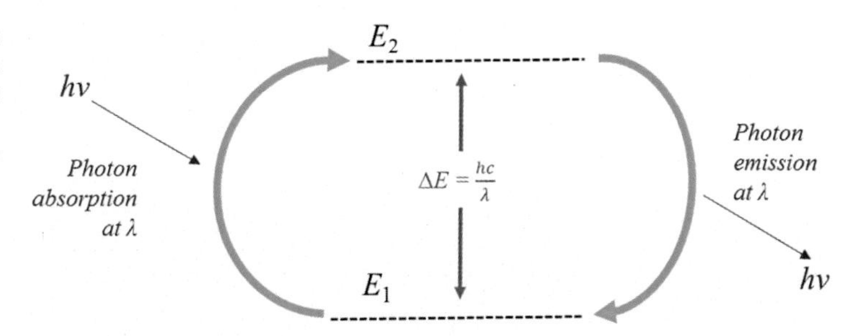

FIG. 3.6 Planck's law and energy transition. When a photon (hv) is absorbed or emitted, it causes the transition between energy states E_1 and E_2 at a specific wavelength (λ). Note that h is Planck's constant and c is the speed of light. *(Based on B. Sportisse, Atmospheric Pollution. From Processes to Modelling, 2008.)*

What is the carbon-oxygen bond length for carbon dioxide?
We need a bit more information to answer this question, that is, the covalent radius of this bond (see Table 3.4).

TABLE 3.4 Covalent radii

		Covalent radius (1×10^{-12} m)		
Atomic number	**Element**	**Single bond**	**Double bond**	**Triple bond**
1	H	32		
2	He	46		
3	Li	133	124	
4	Be	102	90	85
5	B	85	78	73
6	C	75	67	60
7	N	71	60	54
8	O	63	57	53
9	F	64	59	53
10	Ne	67	96	
11	Na	155	160	
12	Mg	139	132	127
13	Al	126	113	111
14	Si	116	107	102
15	P	111	102	94
16	S	103	94	95
17	Cl	99	95	93
18	Ar	96	107	96
19	K	196	193	
20	Ca	171	147	133
21	Sc	148	116	114
22	Ti	136	117	108
23	V	134	112	106
24	Cr	122	111	103
25	Mn	119	105	103
26	Fe	116	109	102
27	Co	111	103	96
28	Ni	110	101	101
29	Cu	112	115	120
30	Zn	118	120	
31	Ga	124	117	121
32	Ge	121	111	114
33	As	121	114	106
34	Se	116	107	107

Continued

TABLE 3.4 Covalent radii—cont'd

		Covalent radius (1×10^{-12} m)		
Atomic number	Element	Single bond	Double bond	Triple bond
35	Br	114	109	110
36	Kr	117	121	108
37	Rb	210	202	
38	Sr	185	157	139
39	Y	163	130	124
40	Zr	154	127	121
41	Nb	147	125	116
42	Mo	138	121	113
43	Tc	128	120	110
44	Ru	125	114	103
45	Rh	125	110	106
46	Pd	120	117	112
47	Ag	128	139	137
48	Cd	136	144	
49	In	142	136	146
50	Sn	140	130	132
51	Sb	140	133	127
52	Te	136	128	121
53	I	133	129	125
54	Xe	131	135	122
55	Cs	232	209	
56	Ba	196	161	149
57	La	180	139	139
58	Ce	163	137	131
59	Pr	176	138	128
60	Nd	174	137	
61	Pm	173	135	
62	Sm	172	134	
63	Eu	168	134	
64	Gd	169	135	132
65	Tb	168	135	
66	Dy	167	133	
67	Ho	166	133	
68	Er	165	133	
69	Tm	164	131	
70	Yb	170	129	

TABLE 3.4 Covalent radii—cont'd

		Covalent radius (1×10^{-12} m)		
Atomic number	**Element**	**Single bond**	**Double bond**	**Triple bond**
71	Lu	162	131	131
72	Hf	152	128	122
73	Ta	146	126	119
74	W	137	120	115
75	Re	131	119	110
76	Os	129	116	109
77	Ir	122	115	107
78	Pt	123	112	110
79	Au	124	121	123
80	Hg	133	142	
81	Tl	144	142	150
82	Pb	144	135	137
83	Bi	151	141	135
84	Po	145	135	129
85	At	147	138	138
86	Rn	142	145	133
87	Fr	223	218	
88	Ra	201	173	159
89	Ac	186	153	140
90	Th	175	143	136
91	Pa	169	138	129
92	U	170	134	118
93	Np	171	136	116
94	Pu	172	135	
95	Am	166	135	
96	Cm	166	136	
97	Bk	168	139	
98	Cf	168	140	
99	Es	165	140	
100	Fm	167		
101	Md	173	139	
102	No	176	159	
103	Lr	161	141	
104	Rf	157	140	131
105	Db	149	136	126
106	Sg	143	128	121

Continued

TABLE 3.4 Covalent radii—cont'd

		Covalent radius (1×10^{-12} m)		
Atomic number	**Element**	**Single bond**	**Double bond**	**Triple bond**
107	Bh	141	128	119
108	Hs	134	125	118
109	Mt	129	125	113
110	Ds	128	116	112
111	Rg	121	116	118
112	Cn	122	137	130
113	Uut	136		
114	Uuq	143		
115	Uup	162		
116	Uuh	175		
117	Uus	165		
118	Uuo	157		

Source of data: P. Pekka, A. Michiko, Molecular double-bond covalent radii for elements Li–E112. Chem. Eur. J. 15 (46) (2009) 12770–12779.

This molecule has double bonds between the carbon atom and the two oxygen atoms. From the table, we see the following:

		Covalent radius (1×10^{-12} m)		
Atomic number	Element	Single bond	Double bond	Triple bond
and				
6	C	75	67	60
8	O	63	57	53

Therefore, the C double bond 67 is picometer (1×10^{-12} m) long and that an O double bond is 57 picometer long. Thus, the sum gives us the bond length of a C=O bond, which is approximately *124 picometers.*

How strong are the bonds of a carbon tetrachloride (tetrachloromethane) molecule compared with carbon dioxide?
This is CCl4. Thus, it has four single C—Cl bonds. From Table 3.4, we find the following:

		Covalent radius (1×10^{-12} m)		
Atomic number	Element	Single bond	Double bond	Triple bond
and				
6	C	75	67	60
17	Cl	99	95	93

Thus, the bond length is 75 + 99 = 174 picometers. So, all four bonds have this length. The bond is weaker than that of CO2, which was found to be 124 picometers. This should not be a surprise, given that CCl_4 is single-bonded and CO_2 is double-bonded. Thus, in the atmosphere, under the same conditions, CCl_4 will be degraded more easily.

Another rule regards resonance, wherein electrons do not occupy a fixed position in a molecule. They move, for example, back and forth, in a molecule. Hence, they resonate. Theoretical chemistry uses probability to indicate where the bonds will exist in the orbitals. A normal, nonresonant bond can be described as an integer. For example, a normal bond configuration can be likened to wearing a glove on one hand or two gloves on both hands. However, a resonant bond is likened to the 75% probability of between 1.5 and 1.8 hands wearing a glove. Therefore, the bond order does not have to be an integer for resonant bonds.

What is the bond order for the nitrate ion?

The nitrate anion, NO_3^-, is important in environmental science and air pollution engineering. It is a key component of the nitrogen cycle. Particulate matter may contain it, along with other ions. Before deciding on the bond order, we should discuss Lewis structures. The Lewis electron dot structures (LEDS) are diagrams that depict the bonding in any covalently bonded molecule or ion [18]. The total number of electrons equals the sum of the numbers of valence electrons on each atom. Lewis structures do not indicate nonvalenced electrons, since we are concerned about bond strength and breakage. After determining the total number of available electrons, the electrons must be positioned into the structure. They are first placed as lone pairs, that is, one pair of dots for each pair of electrons available. Lone pairs are first positioned on outer atoms, other than hydrogen, until each outer atom has eight electrons in bonding pairs and lone pairs. Next, the extra lone pairs are positioned on the central atom. Lone pairs should first be placed on more highly electronegative atoms.

This first positioning indicates bonding. For example, after the lone pairs of electrons are placed, some atoms may not have an octet of electrons. These are usually the central atoms. This means that the atoms must form a double bond. That is, a lone pair of electrons is moved to form a second bond between the two atoms. Sharing the bonding pair between the two atoms means that the atom that originally had the lone pair keeps its octet. Thus, the other atom must have two more electrons in its valence shell.

Therefore to diagram nitrate, we take the following steps [19]:

1. Nitrogen is less electronegative than oxygen, so it is the central atom by multiple criteria. Determine the number of valence electrons. Nitrogen has five valence electrons; each oxygen has six, for a total of $(6 \times 3)+5=23$. The anion has a charge of -1; thus, we have an extra electron, making the total number of electrons 24.
2. Draw the ion's skeleton:

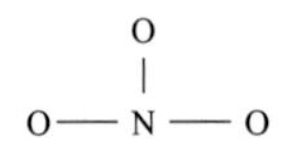

O
|
O — N — O

3. Of the 24 valence electrons in NO_3^-, six were needed to make the skeleton. Place the remaining 18 electrons to fill the octets of as many atoms as possible, beginning with the most electronegative atoms first then proceeding to the atoms with higher electronegativity:

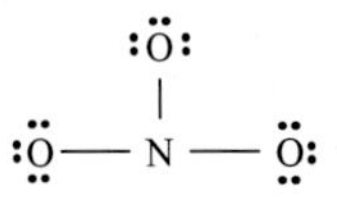

:Ö:
|
:Ö — N — Ö:

4. If the octets are not yet filled for all the atoms, fill them by making multiple bonds and making a lone pair of electrons on the higher electronegative atom into a bonding pair of electrons shared with the lesser electronegative atom:

:Ö: | :Ö — N — Ö: ⟶ :Ö: | :Ö — N ═ Ö:

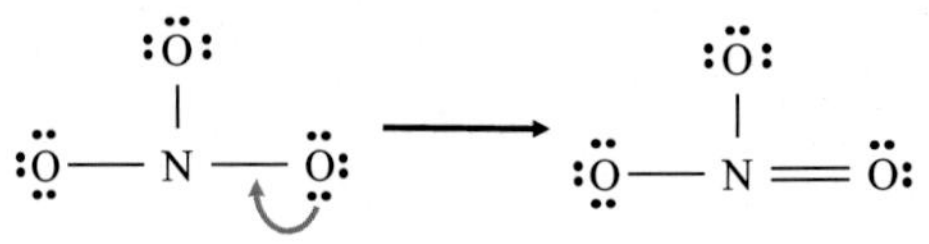

So, the Lewis structure is

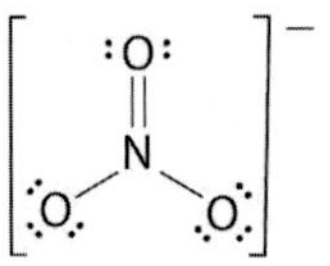

$[O{=}N(O)(O)]^-$

We see that there are a total of four bonds and three bond groups between the individual atoms. Thus, we divide the number of bonds between the individual atoms by the total number of bonds, that is, $4/3=1.33$.

Therefore, nitrate's bond order is *1.33*.

Which molecule is more resistant to breaking down in the atmosphere, carbon dioxide or carbon monoxide?

The LEDS for the two molecules are

$:\ddot{O}=C=\ddot{O}: \quad :C\equiv O:$

Thus, we know from the bond order that the triple bond in CO will be more difficult to break than the double bond in CO_2. From Table 3.4, we can calculate the bond lengths:

CO_2, 67 + 57 = 124 picometers; CO, 60 + 53 = 113 picometers.

Thus, under the same conditions, *carbon dioxide's half-life is shorter than that of carbon monoxide.*

3.5.1 Free radicals

Atmospheric reactions cannot be understood without an appreciation for free radicals, that is, an atom, molecule, or ion that has an unpaired valence electron. As shown in Fig. 3.6, when the photon, $h\nu$, is absorbed, it adds energy to compounds at point "a" in the atmosphere, in this case NO_2:

$$NO_2 + h\nu \rightarrow NO + O \tag{3.22}$$

Similarly, molecular oxygen is cleaved into two oxygen atoms by the photons delivered from ultraviolet light, that is, the O_2 is photolyzed:

$$O_2 + h\nu \rightarrow 2O\cdot \tag{3.23}$$

This breaks off atomic oxygen (O), which reacts very rapidly with the ubiquitous O_2 to form ozone (O_3):

$$O\cdot + O_2 + M \rightarrow 2O_3 + M \tag{3.24}$$

The third-body molecule (M) can be any molecule that takes the excess energy from the reaction and dissipates it as heat. However, the two most ubiquitous third-body molecules are N_2 and O_2. Neither N_2 nor O_2 is limiting to tropospheric reactions, since they are the most abundant and are available everywhere in the troposphere.

Photons also generate free radicals when they react with hydrocarbons, such as the hydroperoxyl radical, $HO_2^{\cdot}$, depicted in Fig. 3.7. These radicals oxidize the NO, so that NO_2 is again produced at point "b," starting the cycle again.

Like any reaction, atmospheric reactions are proportional to concentration, such as the reaction of ozone and nitric acid to produce nitrogen dioxide and molecular oxygen:

$$O_3 + NO \rightarrow NO_2 + O_2 \tag{3.25}$$

Note that this is an example of one air pollutant (ozone) reacting with another air pollutant (nitric acid) to form another air pollutant (nitrogen dioxide) and an essential molecule (molecular oxygen).

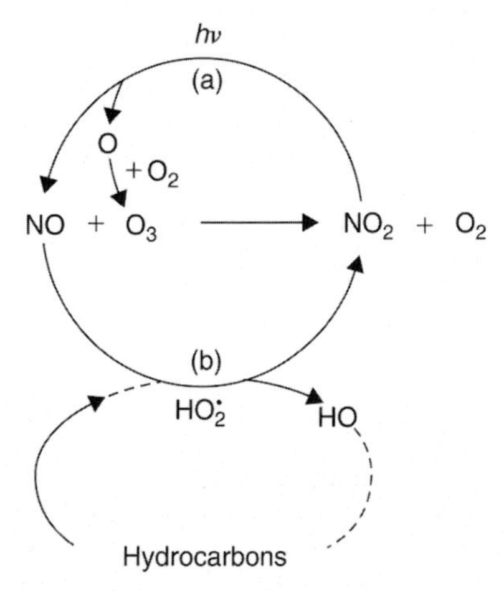

FIG. 3.7 Simplified photochemical cycle in the atmosphere. *Source: D.A. Vallero, Fundamentals of Air Pollution, 5th ed., Elsevier Academic Press, Waltham, MA, 2014, p. 999 pages cm.*

What is the rate order and rate constant for reaction (3.25)?

Reactions in the environment and the atmosphere range from very fast to extremely slow. Recall that the rate (*r*) of a reaction depends on the concentration of the reactants:

$$r = k[A]^m [B]^n \tag{3.26}$$

where *A* is reactant 1, *B* is reactant 2, and *m* and *n* are the reaction orders. The sum of reaction orders equals the overall reaction order. The proportionality rate constant *k* is specific to a reaction at a given temperature and any catalyst being used. The exponents are derived experimentally and not from stoichiometric coefficients.

The values of the exponents give the order of the reaction for each chemical species and the overall order of the whole reaction. Thus, from our discussion of rate order and rate laws, we know that we can derive experimentally the reaction rates to determine the kinetic velocity and half-lives in Reaction (3.12). First, based on the stoichiometry and Table 3.1, we see that this is a second-order reaction:

$$\frac{d[O_3]}{dt} = \frac{d[NO]}{dt} = -[O_3] \cdot [NO] \cdot k \tag{3.27}$$

Therefore, fitting reaction Eqs. (3.25) and (3.26), this must be a *second-order reaction*. According to *Fundamentals of Air Pollution* [2], the second-order rate constant *k* for this reaction is $24.2\,ppm^{-1}\,min^{-1}$.

3.6 TRANSFORMATION THERMODYNAMICS

This chapter's focus on kinetics could be an excuse for waiting to address thermodynamics until the next chapter's discussion of equilibriums. However, since thermodynamics is all about systems, it is important to both kinetics and equilibriums together. We need to discuss energy systematically, that is, both the light and the heat, if you will. So, consider this section to be a brief segue from air pollutant transformation to pollutant transport and fate.

3.6.1 Entropy and enthalpy

The first law of thermodynamics is the law of conservation of mass and energy. The thermodynamic expression is

$$\Delta U = q + w \tag{3.28}$$

where ΔU is the change of internal energy of a system, q is the heat that passes through the boundary into the system, and w is the work done on the system. If the heat and work were lost by the system, the signs of q and w would be negative. This highlights the equivalence of heat and work. Internal energy of an isolated system is constant (conservation of energy).

If an electric motor produces $200\,kJ\,s^{-1}$ of work by lifting a load and loses $50\,kJ\,s^{-1}$ of heat to the surrounding air, then the change of internal energy of the motor per second is

$$\Delta U = -2 - 15 = -17\,kJ$$

The quantities q and w have negative signs as the system (motor) is losing work and heat to the surroundings.

The conservation of mass and energy requires that every input and output be included. Energy or mass can be neither created nor destroyed, only altered in form. For any system, the transfer of energy or mass is associated with mass and energy crossing the control boundary within the control volume (Fig. 3.8). If mass does not cross the boundary, but work and/or heat do, the system is a "closed" system. If mass, work, and heat do not cross the boundary, the system is an isolated system. Too often, open systems are treated as closed, or closed systems include too small of a control volume.

Although we are tempted to infer from the first law that heat can be completely converted into work, we should not. We need another thermodynamics law to explain these conversions. The second law is less direct and less obvious than the first. In all energy exchanges, if no energy enters or leaves the system, the potential energy of the state will always be less than that of the initial state. The tendency toward disorder, that is, entropy, requires that external energy is needed to maintain any energy balance in a control volume, such as a heat engine, a waterfall, or an ethanol processing facility. Entropy is ever present. Losses must always occur in conversions from one type of energy (e.g., mechanical energy of an oil refinery's equipment ultimately to chemical energy of the produced fuel).

The mass and energy coming in and going out across two-dimensional surfaces in these systems are known as fluxes. These fluxes are measured and yield energy balances within a region in space through which a fluid travels. This region, that is, the control volume, is where balances occur and can take many forms. With any control volume, the calculated mass balance is

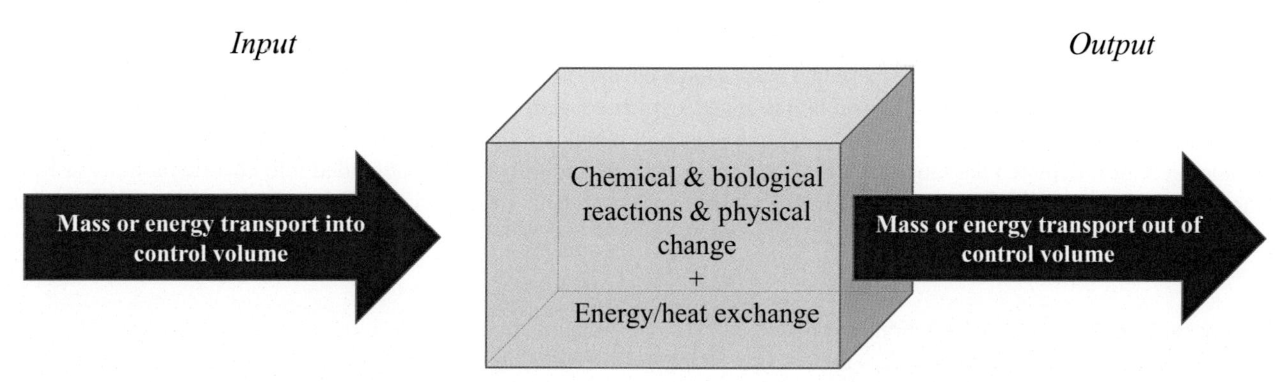

FIG. 3.8 Control volume showing input, change, and output. The process applies to both mass and energy balances.

$$\begin{bmatrix} \text{Quantity of} \\ \text{mass per unit volume} \\ \text{in a medium} \end{bmatrix} = [\text{Total flux of mass}] + \begin{bmatrix} \text{Rate of production or loss} \\ \text{of mass per unit volume} \\ \text{in a medium} \end{bmatrix} \tag{3.29}$$

This can be restated as

$$\frac{dM}{dt} = M_{in} - M_{out} \tag{3.30}$$

where M = mass and t = specified time interval.

Note that such conservation equations commonly express net inputs, net outputs, and net system losses. The quantity of energy is expressed as balance equations for the entire process, with losses at every step. If we are concerned about a specific chemical, for example, losing good ones, like oxygen, or forming bad ones, like the toxic air pollutants, the equation needs a reaction term (R):

$$\frac{dM}{dt} = M_{in} - M_{out} \pm R \tag{3.31}$$

Environmental systems are composed of interrelationships among *abiotic* (nonliving) and *biotic* (living) components of the environment. Organisms live according to the transfer of mass and energy via the concept of "trophic state." Humans live within this interconnected network or web of life, so they are affected, positively or negatively, by the condition of these components. Disasters are major affronts to these finely balanced relationships [20].

There are many ways to express the second law of thermodynamics, but one that adds directly to the first law is

…that no process is possible in which heat absorbed from a reservoir can be completely converted into work.

The second law introduces the concept of spontaneous change and in so doing introduces two vital properties in thermodynamics, namely, entropy (S) and the Gibbs free energy (G). The Gibbs free energy allows us to determine whether a process is spontaneous. Another way of expressing the second law is that the entropy, S, of a system and its surroundings increases during a spontaneous change:

$$\mathrm{To}_{\mathrm{tal}} > 0 \tag{3.32}$$

A system's entropy is an indication of disorder. It is possible to calculate entropy values from theory and to determine it indirectly, but there is no way of measuring it as we do heat:

$$\mathrm{q} = \mathrm{m.SH.\Delta T} \tag{3.33}$$

where SH is the specific heat and ΔT is the change of temperature (using a thermometer). Entropy values reflect its quantitative description as being a measure of disorder, and the standard entropy (S°m) of a solid such as a diamond (C) is very low, $2.4\,\mathrm{J\,K^{-1}\,mol^{-1}}$, as it is a very ordered solid, while the value for CO_2 is large, $213.7\,\mathrm{J\,K^{-1}\,mol^{-1}}$. This reflects the fact that gases are a very disordered state of matter.

An example of the second law is that of an expanding gas. Consider a high-pressure gas (say, from a cylinder) expanding into a low-pressure region (atmosphere). The gas changes from reasonably well-ordered gas to a chaotic disordered gas, and the entropy changes from a small value to a large value:

$$\Delta S = \Delta_{Sfinal} - \Delta_{Sinitial} > 0 \tag{3.34}$$

The process as we know from experience is spontaneous, and yes, the entropy is greater than zero.

Another way of expressing the second law is in any spontaneous process; the change is always accompanied by a dispersal of energy into a more disordered state. For instance, if a ball is dropped from a height, it will bounce and continue bouncing until it depleted the kinetic energy. The energy has been dissipated and degraded into chaotic motions of floor molecules that have been heated by each bounce. Thus, to predict whether a chemical reaction will be spontaneous, we can apply the following:

$$\Delta_r G = \Delta_r H - T\Delta_r S \tag{3.35}$$

where subscript "r" refers to "reaction" or process and H is the enthalpy, that is, the heat (q) at constant pressure. $\Delta_f G$ is the Gibbs free energy of formation, ΔrS is the entropy change of reaction, and T is in Kelvin. The process is spontaneous if Gibbs free energy of formation is negative, that is, $\Delta_r G < 0$. Alternatively, if $\Delta_r G$ is positive, the reaction does not go forward and is termed nonspontaneous.

How does enthalpy change when glucose is oxidized at 37°C (mean human body temperature)?

$$C_6H_{12}O_6 + 6O_2 = 6H_2O + (6CO_2)$$

The enthalpy change for this reaction is $-2807.8\,\text{kJ mol}^{-1}$, and the entropy change is $182.4\,\text{J K-L mol}^{-1}$.
So, the $\Delta_r G$ for this reaction is

$$\Delta_r G = [-2807.8 - (310.15 \times 182.4)/1000]\,\text{kJmol}^{-1} = -2864.4\,\text{kJmol}^{-1}$$

This oxidation is thermodynamically the same as the biological process that takes place when a person digests a mole (180.2 g) of glucose. This is the energy we use when we exercise. So, if 18 g of glucose is consumed, the muscles would have at most 286 kJ of energy. Equating this to the energy involved in lifting a weight through a height *h* (i.e., $E = mgh$), a 70 kg person could use this energy to climb 417 m.

Can we block CO_2 from entering the atmosphere by using it in the manufacture of useful chemicals?

Let us consider one possible useful reaction that of using CO_2 to produce ethanol, a possible liquid fuel for the future:

$$2CO_2(g) + 3H_2(g) = C_2H_5OH(l)$$

To see if the $\Delta_r G$ is spontaneous, we must look up data in a table of thermodynamic properties—see Table 3.5. By using $\Delta_f G$ for each compound or element in the reaction (the formation refers to the formation from its elements), then,

$$\Delta_r G = \Delta_f G(\text{products}) - \Delta_f G(\text{reactants}) \tag{3.36}$$

$$\Delta_r G = [-174.8 - 3(0) - 2(-394.4)] = +614\,\text{kJ}\left(\text{mol}^{-1}\right)$$

The Gibbs free energy is positive, so the reaction is not spontaneous. Too bad!

An important lesson from this query is that because the Gibbs free energy of formation of CO_2 is so very large and negative, it will be virtually impossible to use it to make new chemicals. All is not lost, however, because we could emulate nature, by coupling, that is, linking, reactions to involve the CO_2 in making useful chemicals. Research in this area is ongoing in many laboratories, and CO_2 is being used in living processes involving algae to make hydrocarbon chemicals including diesel.

From data like those in Table 3.5, it is possible to determine whether any reaction you can possibly imagine will or will not happen spontaneously.

By using thermodynamic properties, compounds with apparently unrelated properties can be compared. We have seen this in determining whether a reaction will go spontaneously or not from enthalpy data and from entropy data—apparently unrelated properties. There are many other such examples in thermodynamics. Notably, Maxwell's relations are a particularly good set of examples, such as

$$\left(\frac{dS}{dV}\right)_T = \left(\frac{d\rho}{dT}\right)_V \tag{3.37}$$

With this, the entropy change of a gas with respect to volume at constant temperature can be found by knowing the change of pressure (ρ) with temperature at constant volume for the gas. The right-hand side (RHS) of the equation is

TABLE 3.5 Thermodynamic data—enthalpies and Gibbs free energies of formation and standard entropy (at 298 K and 1 bar)

Compound	$\Delta_f H$ (kJ mol^{-1})	$\Delta_f G$ (kJ mol^{-1})	$S^o m$ (kJ K^{-1} mol^{-1})
CO_2	−393.5	−394.4	213.7
CH_4 (methane)	−74.8	−50.7	186.3
C_2H_6 (ethane)	−84.7	−32.8	229.6
C_4H_{10} (butane)	−126.2	−17.0	310.2
C_6H_6 (benzene)	+49.0	+124.3	173.3
CH_3OH (methanol)	−238.7	−166.3	126.8
C_2H_5OH (ethanol)	−277.7	−174.8	160.7
$C_6H_{12}O_6$ (glucose)	−1268	−910	212
H_2O	−285.8	−237.1	69.9
H_2	0	0	130.7

Source of data: D.A. Vallero, T. M. Letcher, Unraveling Environmental Disasters. Newnes, 2012.

relatively easy to measure, but left-hand side (LHS) is impossible. The Maxwell relationship makes it possible to determine the LHS of the equation.

The change in free energy associated with the movement of the solute from one compartment to another is directly proportional to the difference in chemical potential between the compartments. In all energy exchanges in an isolated system, if no energy enters or leaves the system, the potential energy of the state will always be less than that of the initial state. This, as we have seen, is related to the entropy so that external energy is needed to keep things going—in a control volume, refrigerator, heat engine, waterfall, etc.

Physicists look to these laws to determine how mass and energy are distributed. Further, ecologists consider these distributions of mass and energy when investigating the complex interrelationships between and within the compartments of the food webs and chains and consider humans to be among the consumers [21]. Food chains illustrate the complexity and vulnerability of environmental systems (see Fig. 3.9). Species at a higher trophic level are predators of lower-level species, so materials and energy flow downward. The transfer of mass and energy upwardly and downwardly between *levels of biological organization* can be measured and predicted, given certain initial and boundary conditions. However, the types and abundance of species and interaction rates vary in time and space. Stress from an air pollutant deposited on an ecosystem can change these trophic interrelationships.

The substance of all species consists mainly of molecular arrangements of the elements carbon, oxygen, hydrogen, and most contain nitrogen. These four biophile elements have an affinity for each other to form complex organic compounds. The smallest organisms, for example, the viruses, bacteria, and other microbes, are quite efficient in finding and using organic material as sources of energy and carbon, but for much of human history, the systems within microbes have been the agents of epidemics. However, the cause of the epidemics involves numerous interrelationships among the trophic states. For example, when habitats are changed to allow advantages to certain species (e.g., rats carrying plague microbes), the disaster cannot simply be attributed to the particular disease agent, but to the overall changes to the environment.

We touched on thermodynamics in Chapter 2 in the discussion of ideal gases. Thermodynamics are at work in the chemical transformation of pollutants and their precursors. For example, raising temperature increases the kinetics of a system. Other factors affect the rates of transformation, e.g., presence of water. If an air pollutant is highly reactive (i.e., fast reaction rate) in water vapor that is dispersed and remains in a plume for days, much or all of this compound will have been transformed within the water vapor. However, if a compound has a very slow reaction rate in water vapor, the volume of the parent compound would have changed very little. Similarly, a reactive compound sorbed to particulate matter will more likely be transformed during transport than a less reactive one. Of course, since water vapor is rather ubiquitous in the atmosphere, aqueous reactions will occur in and on these aerosols as a result of condensation.Since this atmospheric residence time is often expressed as its atmospheric half-life, reaction rates also give clues as to how a pollutant will behave in the environment. This is the concern of equilibrium and partitioning in the environment (see Chapter 4).

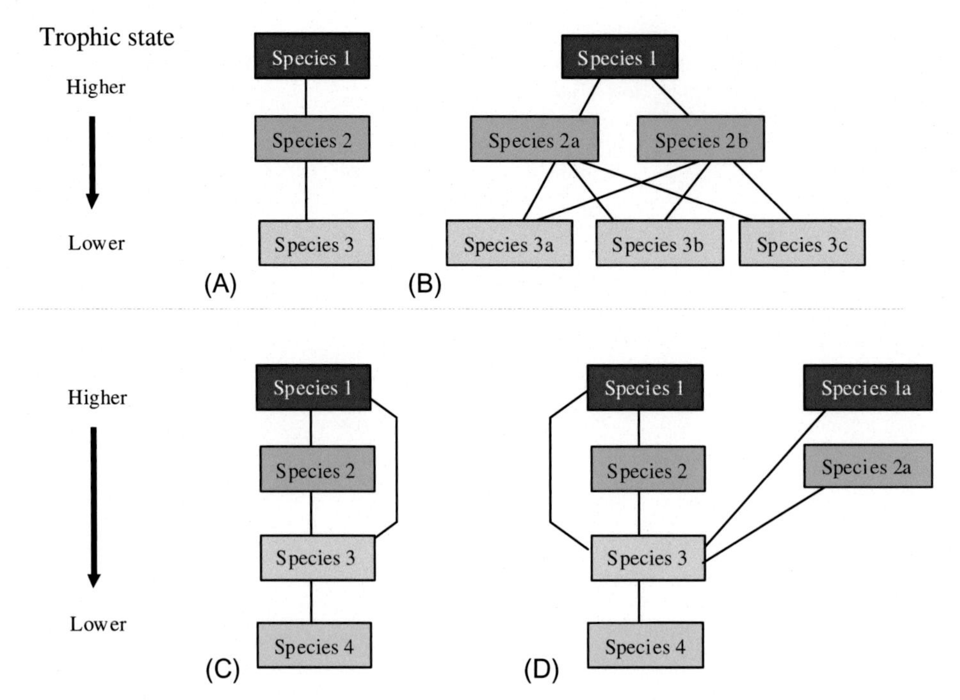

FIG. 3.9 Energy and matter flow in environmental systems from higher trophic levels to lower trophic levels. Lines represent interrelationships among species. (A) Linear biosystem, (B) multilevel trophic biosystem, (C) omnivorous biosystem, and (D) multilevel biosystem with predation and omnivorous behaviors. An interference at any of these levels or changes in mass and energy flow can lead to disasters. *Based on T.E. Graedel, On the concept of industrial ecology. Annu. Rev. Energy Environ. 21 (1) (1996) 69–98.*

Elemental mercury vapor (Hg^0), for example, does not readily react with other atmospheric constituents. In addition, Hg^0 has a very low aqueous solubility. These properties make for long atmospheric residence time (approximately 1 year) [22, 23]. However, Hg^0 can be transformed to more reactive species, with much higher aqueous solubility.

Chemical transformation is explained by air pollution thermodynamics. Ecosystems and individual organisms and their components are systems. Unfortunately, the English language has numerous connotations of "systems." Even scientists have various definitions. For example, a more general understanding of scientists and technicians is that a "system" is a method of organization, for example, from smaller to larger aggregations. The "ecosystem" and the "organism" are examples of both types of systems. They consist of physical phases and order (e.g., producer-consumer-decomposer, predator-prey, individual-association-community, or cell-tissue-organ-system). They are also a means for understanding how matter and energy move and change within a parcel of matter. Within the context of thermodynamics, a system is a sector or region in space or some parcel of a sector that has at least one substance that is ordered into phases. Reactors, stack gases, plumes, the open atmosphere, microenvironments, organisms, and cells have qualities of both closed and open systems. A closed system does not allow material to enter or leave the system (engineers refer to a closed system as a "control mass"). The open system allows material to enter and leave the systems (such a system is known as a control volume).

Air pollution results from complex interactions of substances in various states of matter, in not only the atmosphere but also the many environmental media where substances are transformed. These transformations may render a compound that was previously not very toxic into a very toxic substance. Conversely, what may have been a precursor to an air pollutant may have been changed into a less toxic compound or compounds. During airborne transport, an air pollutant may also undergo chemical changes. These changes may form toxic compounds or other types of problems; for example, they may become stronger greenhouse gases. After deposition, chemical reactions occur in the soil, water, and biota.

As mentioned previously, $NO_x = NO + NO_2$. If the average vehicle decreases its NO_x emissions from 17 to 0.64 kg year^{-1} and its emissions of volatile organic compounds from 35 to 1 kg year^{-1}, how would the global emissions of these compounds change? What effect might this have on smog?

The higher numbers are the estimated emissions for the year 2000, and the lower numbers are the changes required under the Tier 2 emission limits for mobile sources that went into effect in 2007 in the United States [24].

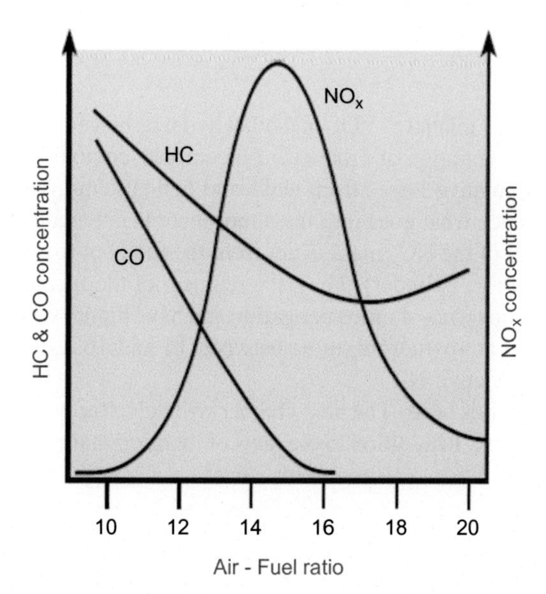

FIG. 3.10 Stoichiometry of carbon monoxide (CO), hydrocarbons (HC), and oxides of nitrogen (NO_x) at different air-to-fuel ratios. *Source: U.S. Environmental Protection Agency, Principles and Practices of Air Pollution – Student Manual: APTI Course 452, 3rd ed., U.S. Environmental Protection Agency, Research Triangle Park, North Carolina, 2003. [Online]. Available: http://www.4cleanair.org/APTI/452combined.pdf.*

Photochemical oxidant smog is formed from NO_x and VOCs in sunlight. The rate-limiting step for ozone formation is either the amount of NO_x or the concentrations of hydrocarbons in the troposphere. If NO_x is the rate-limiting substance, the region is said to be NO_x-limited. If the VOC concentration is the limiting substance to ozone formation, the region is said to be hydrocarbon or VOC-limited.

A heavily treed area that releases high concentrations of VOCs would not seem as a dramatic effect from mobile source reductions of VOCs compared with those that lack trees and are usually congested with high vehicular activity. Smog is prevalent in areas especially during the summer time during peak temperatures and sunlight. Metropolitan areas with large numbers of vehicles and with large wooded areas can vary between being NO_x-limited and VOC-limited. Emissions from plant life can overcompensate for the number of vehicles that are the leading source of NO_x.

Whether a region is NO_x- or VOC-limited affects the types of actions needed to attain ozone standards. For example, if there are naturally high concentrations of VOCs in an area, pollution controls and preventive activities to lower NO_x concentrations are more likely to be implemented.

What would the answers to the three previous questions be if the average gas mileage were 15 and 25 km l^{-1}? Based on these calculations, make a statement about the relationship between air pollution and efficiency.

At 15 km l^{-1}, the carbon in the total amount of gasoline combusted annually for the vehicle would be

$$\frac{21688\,\text{km}}{\text{year}} \times \frac{1\,\text{L}}{15\,\text{km}} \times \frac{0.7\,\text{kg}}{\text{L}} \times 0.85 \cong \frac{1012\,\text{kg C}}{\text{year}}$$

The carbon dioxide emitted is

$$\frac{1012\,\text{kg C}}{\text{year}} \times \frac{44\,\text{kg CO}_2}{12\,\text{kg C}} = \frac{3711\,\text{kg CO}_2}{\text{year}}$$

At 25 km l^{-1}, the carbon in the total amount of gasoline combusted annually for the vehicle would be

$$\frac{21688\,\text{km}}{\text{year}} \times \frac{1\,\text{L}}{25\,\text{km}} \times \frac{0.7\,\text{kg}}{\text{L}} \times 0.85 \cong \frac{607\,\text{kg C}}{\text{year}}$$

The carbon dioxide emitted is

$$\frac{607\,\text{kg C}}{\text{year}} \times \frac{44\,\text{kg CO}_2}{12\,\text{kg C}} \cong \frac{2227\,\text{kg CO}_2}{\text{year}}$$

Thus, even with these improved fuel efficiencies (59 mpg), the average US passenger vehicle still emits more CO_2 than its weight every year. Indeed, an 1800 kg car would have to get 31 kg l^{-1} (73 mpg) to emit less than its weight in CO_2 each year.

Generally, fuel efficiency is indirectly related to VOC and other hydrocarbon emission rates. The more thermodynamically efficient the engine, the lower the concentration of products of incomplete combustion, including VOCs. Other efficiencies must also improve, for example, evaporative losses from tanks and fuel systems, pass-through during start-ups, etc. These losses also decrease fuel efficiency since what goes into the atmosphere is not available for combustion to turn the wheels.

NO_x emissions are different. Most of the NO_x mass is not from the fuel, but from the molecular nitrogen (N_2) in the air brought into the combustion reactor (e.g., cylinder). Thus, the richness of the fuel-air mix will affect NO_x emissions. It is possible to have better fuel efficiency in terms of miles per gallon and have higher rates of NO_x. Thus, as shown in Fig. 3.10, the highest NO_x emissions are found at air-fuel mixtures between 14 and 16. Combustion at very high temperatures is needed to oxidize the relatively nonreactive N_2.

We will revisit kinetics throughout this book. The next chapter will delve further into thermodynamics as they pertain to air pollutant equilibrium and partitioning. In addition to the laws of thermodynamics, the laws of motions will be applied to pollutant transport phenomena.

REFERENCES

[1] J.W. Schaeffer, et al., Size, composition, and source profiles of inhalable bioaerosols from Colorado dairies, Environ. Sci. Technol. 51 (11) (2017) 6430–6440.

[2] D.A. Vallero, Fundamentals of Air Pollution, 5th ed., Elsevier Academic Press, Waltham, MA, 2014. p. 999 pages cm.

[3] J.N. Spencer, G.M. Bodner, L.H. Rickard, Chemistry: Structure and Dynamics, John Wiley & Sons, 2010.

[4] IUPAC, Compendium of Chemical Terminology—The Gold Book, (2014).

[5] J.J. BelBruno, Chemistry 6: General Chemistry, 22 May 2018. Available: https://www.dartmouth.edu/~genchem/0405/spring/6belbruno/, 2005.

[6] B. Hall, The gas phase oxidation of elemental mercury by ozone, in: Mercury as a Global Pollutant, Springer, 1995, pp. 301–315.

[7] J. Sommar, K. Gårdfeldt, D. Strömberg, X. Feng, A kinetic study of the gas-phase reaction between the hydroxyl radical and atomic mercury, Atmos. Environ. 35 (17) (2001) 3049–3054.

[8] J. Munthe, K. Kindbom, O. Kruger, G. Petersen, J. Pacyna, Å. Iverfeldt, Examining source-receptor relationships for mercury in Scandinavia modelled and empirical evidence, Water Air Soil Pollut. Focus 1 (3–4) (2001) 299–310.

[9] J. Munthe, K. Kindbom, O. Kruger, G. Petersen, J. Pacyna, Å. Iverfeldt, Emission, deposition, and atmospheric pathways of mercury in Sweden, Water Air Soil Pollut. (2001).

[10] P. Nelson, C. Peterson, A. Morrison, Atmospheric emissions of mercury-sources and chemistry, Clean Air Environ. Qual. 38 (4) (2004) 48.

[11] U.S. Environmental Protection Agency, List of Areas Protected by the Regional Haze Program (40 CFR PART 81 ed.), 9 May 2018. Available: https://ofmpub.epa.gov/sor_internet/registry/termreg/searchandretrieve/glossariesandkeywordlists/search.do?details=&vocabName=Terms%20of%20Env%20(2009)#formTop, 2018.

[12] D.A. Vallero, Excerpted and Revised From: Fundamentals of Air Pollution, 5th ed., Elsevier Academic Press, Waltham, MA, 2014. p. 999 pages cm.

[13] J. Liang, Chemical Modeling for Air Resources: Fundamentals, Applications, and Corroborative Analysis, Academic Press, 2013.

[14] L. Infeld, The Evolution of Physics, CUP Archive, 1971.

[15] K. Song, D. Le, Bond Energies, 24 May 2018. Available: https://chem.libretexts.org/Core/Physical_and_Theoretical_Chemistry/Chemical_Bonding/Fundamentals_of_Chemical_Bonding/Bond_Energies, 2017.

[16] S.D. Ilčev, Global Satellite Meteorological Observation (GSMO) Theory, Springer, 2018.

[17] B. Sportisse, Atmospheric Pollution. From Processes to Modelling, (2008).

[18] G.N. Lewis, The atom and the molecule, J. Am. Chem. Soc. 38 (4) (1916) 762–785.

[19] I. Hunt, How to draw Lewis diagrams, 24 May 2018. Available: http://www.chem.ucalgary.ca/courses/351/Carey5th/Ch01/ch1-3depth.html, 2018.

[20] D. Vallero, Excerpt: Environmental Biotechnology: A Biosystems Approach, Elsevier Science, 2015.

[21] T.E. Graedel, On the concept of industrial ecology, Annu. Rev. Energy Environ. 21 (1) (1996) 69–98.

[22] F. Slemr, W. Seiler, C. Eberling, P. Roggendorf, The determination of total gaseous mercury in air at background levels, Anal. Chim. Acta 110 (1) (1979) 35–47.

[23] O. Lindqvist, et al., Mercury in the Swedish environment—recent research on causes, consequences and corrective methods, Water Air Soil Pollut. 55 (1–2) (1991) xi–261.

[24] Wikipedia, Exhaust gas, 17 July 2018. Available: https://en.wikipedia.org/wiki/Exhaust_gas, 2018.

FURTHER READING

[25] V. Evangelou, Environmental Soil and Water Chemistry: Principles and Applications, John Wiley Sons, Inc, Canada, 1998.

[26] C. GeoSyntec, Bioaugmentation for remediation of chlorinated solvents: technology development, status, and research needs, Environ. Secur. Technol. Certif Program (ESTCP) (2005) 122.

[27] A. Felsot, Abiotic/biotic degradation & transformation, in: Richland (Ed.), Environmental Attenuation of Contaminants, Washington State University, Washington, 2005.
[28] P. Pekka, A. Michiko, Molecular double-bond covalent radii for elements Li–E112, Chem. Eur. J. 15 (46) (2009) 12770–12779.
[29] D. Vallero, Environmental Biotechnology: A Biosystems Approach, Elsevier Science, 2015.
[30] D.A. Vallero, T.M. Letcher, Unraveling Environmental Disasters, Newnes, 2012.
[31] U.S. Environmental Protection Agency (Ed.), Principles and Practices of Air Pollution – Student Manual: APTI Course 452, 3rd ed., U.S. Environmental Protection Agency, Research Triangle Park, North Carolina, 2003 [Online]. Available:http://www.4cleanair.org/APTI/452combined.pdf.

Chapter 4

Environmental partitioning

4.1 INTRODUCTION

The kinetics discussions in Chapters 2 and 3 set the stage for understanding pollutant transport. We are now ready to discuss equilibrium and the partitioning of chemical species within the different compartments of the environment.

A good place to start a discussion of air pollutant partitioning is with vapor pressure and solubility. Vapor pressure is an expression of the maximum amount of a chemical species that can be held in the gas phase, and solubility is an expression of the maximum amount of a chemical species that can be held in the liquid phase. Thus, both are "saturation properties" [1]. These two properties combine with others to create the partition coefficients.

4.1.1 Vapor pressure

As vapor pressure is the pressure exerted by a vapor in a confined space, vaporization is the change of a liquid or solid to the vapor phase. The vapor pressure of a chemical species is the portion of the total atmospheric pressure due to that chemical species in its vapor phase. Thus, vapor pressure is on one of the molecular characteristics of a compound that determines its physical, chemical, and biological behavior. A compound's structure includes the geometry of the molecule, its electronic characteristics, and its arrangement in space (e.g., in a crystalline liquid or solid) [2]. This is obvious for some features, such as lower molecular weight will mean higher vapor pressures and lower boiling points.

The vapor pressure of a chemical species is a unique, inherent property. For example, volatile organic compounds (VOCs) reach the ambient air predominantly in the gas phase since their vapor pressures are usually greater than 10^{-2} kilopascals (kP) or 9.9×10^{-4} atm [3].

For VOCs, their vapor pressure fairly well predicts that a substantial amount of these compounds will be in the atmosphere. Although there is much debate on what distinguishes a VOC, it is a compound with a relatively low boiling point and relatively high vapor pressure (see Table 4.1). For example, the European Union defines a VOC as any organic compound with an initial boiling point less than or equal to 250°C at 1 atm [4]. Some compounds are so volatile that they are designated very volatile organic compounds (VVOCs). The key here is that compounds vary under the same environmental conditions, in the number of molecules that move from the liquid phase to the vapor phase.

A 2007 study [5] measured the composition of conventional gasoline (containing <10% of ethanol), E85 (85% ethanol and 15% gasoline), ultra-low-sulfur diesel (ULSD), and B20 (20% soy biodiesel and 80% ULSD) from a fuel station in Michigan, the United States (see Table 4.2). You have two fuel containers holding 4 L of gasoline and 4 L of E85 fuel in your garage but forgot to put the caps on both. You smell odors in the morning and, after checking, replace the caps. Which one will have lost the most toluene and ethanol to the air in the garage?

The vapor pressure of toluene is about 4 kP and that of ethanol is about 8 kP at 25°C. The gasoline contains 1.5×10^4 mg L^{-1} toluene and <10% (<10^5 mg L^{-1}). The E85 contains 4.1×10^4 mg L^{-1} toluene and 15% (1.5×10^5 mg L^{-1}).

Therefore, the gasoline will have moved more toluene molecules into the air in the garage. The exact amount will be determined by the size of the spout, the temperature range overnight, and the effect of mixtures (e.g., Raoult's law, which is discussed later in this chapter). The spout size is important because it determines the flux. That is, if the containers each have spout the same size, we may disregard this variable. However, if one spout has an opening twice as large as the other, the flux of toluene and ethanol would double.

At the other end of the vapor pressure spectrum are the relatively nonvolatile organic compounds (NVOCs). These have vapor pressures <10^{-5} kilopascals. These vapor pressures are predominantly found in and on particles, unless significant energy is added to increase their volatility. Thus, NVOCs can be estimated to be nearly equal to the total amount of the air pollutant in particulate matter collected, since so few molecules escape to the atmosphere. Between the extremes, the

Air Pollution Calculations. https://doi.org/10.1016/B978-0-12-814934-8.00004-1

TABLE 4.1 Categories of organic compounds according to vapor pressure

Category	Abbreviation	Boiling point rang (°C)	Example compounds
Very volatile organic compounds	VVOC	From <0 to 50–100	Propane, butane, methyl chloride
Volatile organic compounds	VOC	From 50–100 to 240–260	Formaldehyde, d-limonene, toluene, acetone, ethanol (ethyl alcohol) 2-propanol (isopropyl alcohol), hexanal
Semivolatile organic compounds	SVOC	From 240–260 to 380–400	Pesticides dichlorodiphenyltrichloroethane (DDT), chlordane, phthalates, fire retardants (including polybrominated biphenyls, PBBs), polychlorinated biphenyls, several polycyclic aromatic hydrocarbons (PAHs), chlorinated dioxins and furans.

Sources of data: U.S. Environmental Protection Agency, Technical Overview of Volatile Organic Compounds, General Definition and Classification. 5 June 2018. Available: https://www.epa.gov/indoor-air-quality-iaq/technical-overview-volatile-organic-compounds; WHO Guidelines for Indoor Air Quality: Selected Pollutants. World Health Organization, 2010; 0143-2060, Indoor air quality: organic pollutants (1989).

semivolatile organic compounds (SVOCs) with vapor pressures between 10^{-2} and 10^{-5} kilopascals can exist in substantial concentrations in both the gas and particle phases in the ambient air. The SVOCs include some of the most important pollutants, including dioxins, polychlorinated biphenyls (PCBs), polycyclic aromatic hydrocarbons (PAHs), and numerous other persistent organic pollutants (POPs).

This is an oversimplification. An inherent property like vapor pressure certainly is part of the prediction of where a compound will reside in the environment and its propensity to move, in this case, to the atmosphere. However, vapor pressure alone will not explain the partitioning and transport. All inherent properties of a chemical species and those of the environmental compartment are constantly affecting one another under real-world conditions. Vapor pressure interacts with other factors, like the chemical species' solubility in water and its likely sorption to matter. For example, even a substance with a very high vapor pressure may remain in the water if it is highly water soluble or in the soil if it has a strong affinity to a soil particle. These features will be discussed respectively in our considerations of Henry's law and sorption, respectively.

4.1.2 Solubility

As mentioned, the second saturation property of a chemical species is its solubility. Often, this means solubility in water, that is, aqueous solubility. However, other solvents are also important in air pollution, for example, solubility of an organic compound in octanol, as we shall consider in our discussion of octanol-water partitioning.

If a compound has high aqueous solubility, that is, it is easily dissolved in water under normal environmental conditions of temperature and pressure, it is hydrophilic. That is, it is very likely to remain dissolved in water unless and until something changes. If, conversely, a substance is not easily dissolved in water under these conditions, it is hydrophobic, and many of its molecules will exit the water and enter compartments with substances of which the chemical has more affinity, for example, hydrophobic compounds in sediment or tissues in organisms. In these instances, the inherent property, aqueous solubility, is a factor of partitioning. However, like vapor pressure, this one property does not fully predict the amount of a chemical species that will be in each compartment. Such prediction is afforded by the partitioning coefficients discussed in this chapter.

Vapor pressure and aqueous solubility do provide the first clues to where an air pollutant may reside and its likelihood to move to other environmental compartments. For example, the very hydrophobic compounds known as polychlorinated biphenyls (PCBs) and many polycyclic aromatic hydrocarbons (PAHs) would be expected to be found in soil and sediment, with less in water, given their low aqueous solubility and low vapor pressures. However, despite these properties, they are often found in water. Again, this is predicted with partitioning; for example, some of the PCB and PAH molecules are sorbed to suspended particles, or they are absorbed within the food chain.

As evidenced, substantial amounts of PCBs are found in the ocean, not only due to other partitioning discussed below but also because of simple arithmetic. The enormous abundance of water and biota in the ocean means that many molecules of PCBs will be dissolved [6]. For example, at saturation, 0.45 mg of Aroclor 1232 (molecular weight, 232.2) will be dissolved in water at 20°C [7]. If this is dissolved in 1 million liters of water, at saturation, this would be 450 grams of this PCB. Studies show, however, that that the concentrations are much lower than saturation, for example, less than one picogram per liter for several PCB congeners in the North Atlantic and Arctic Oceans [8].

TABLE 4.2 Chemical composition of volatile organic compounds in neat fuels from a Michigan filling station in 2007

Fuel	Gasoline	E85	E85/gasoline	Diesel	B20	B20/diesel
Unit	($mg\,L^{-1}$)	($mg\,L^{-1}$)	(%)	($mg\,L^{-1}$)	($mg\,L^{-1}$)	(%)
Aromatics						
Benzene	6140	862	14	67	37	55
Toluene	15,400	4110	27	238	214	90
Ethylbenzene	3080	1990	65	124	186	150
p-Xylene, m-xylene	9120	6980	76	420	496	118
o-Xylene	4610	2790	60	185	212	115
Isopropylbenzene	351	156	45	44	70	160
n-Propylbenzene	2110	665	32	115	167	146
p-Isopropyltoluene	88	29	33	112	83	75
4-Ethyl toluene	8380	2650	32	400	464	116
2-Ethyltoluene	3460	928	27	194	264	136
1,3,5-Trimethylbenzene	4060	1030	25	202	150	74
1,2,4-Trimethylbenzene	10,600	3270	31	720	575	80
1,2,3-Trimethylbenzene	3950	975	25	2120	961	45
sec-Butylbenzene	159	65	41	90	117	130
n-Butylbenzene	822	198	24	375	111	30
Styrene	14	4	32	<0.02	<0.02	–
Naphthalene	2240	378	17	3000	1220	41
Alkanes						
n-Heptane	12,800	3330	26	174	138	79
n-Octane	2870	1550	54	481	612	127
n-Nonane	1790	1050	59	7020	4120	59
n-Decane	1390	262	19	7690	4310	56
n-Undecane	1120	121	11	7730	4560	59
n-Dodecane	822	85	10	8370	5000	60
n-Tridecane	644	80	12	13,400	7620	57
n-Tetradecane	213	19	9	10,100	7130	71
n-Pentadecane	62	<0.02	–	9030	7580	84
n-Hexadecane	18	<0.02	–	9300	5550	60
Cyclohexane	9830	880	9	191	69	36
Methyl cyclohexane	8280	778	9	426	270	63
Total measured VOCs	114,000	34,900	31	80,700	51,600	64

Source of data: J.-Y. Chin, S. A. Batterman, VOC composition of current motor vehicle fuels and vapors, and collinearity analyses for receptor modeling, Chemosphere, 86 (9) (2012) 951–958.

Three compounds are found to be leaking into the groundwater from an underground tank. Compound A has a vapor pressure of 12 kPa and density of 1.6 g mL^{-1}. Compound B has a vapor pressure of 59 kPa and density of 0.7 g mL^{-1}. Compound C has a vapor pressure of 0.003 kPa and density of 1.1 g mL^{-1}. These values are all at 20°C. Based on these data alone, give the order of each compound likely to be emitted to the atmosphere through the soil.

These approximate the values for neat solvents:

A = tetrachloromethane or carbon tetrachloride (CCl_4).

B = diethyl ether [$(C_2H_5)_2O$].

C= diethylene glycol [$(HOCH_2CH_2)_2O$]. Based only on the density and vapor pressure, the order would be that C is the least likely to be emitted given the relatively low vapor pressure and density less than that of water. The most likely to partition to the atmosphere is B, with a very high vapor pressure and low density. Thus, the order would be B, A, and C.

However, A and C are each heavier than water, so they could settle toward the bottom of the aquifer if they have low aqueous solubilities, that is, they do not dissolve in the water.

4.2 EQUILIBRIUM

After the kinetic stage, the chemistry of the air pollutant considers chemical species at equilibrium. This explains how a pollutant will behave in the environment, including the extent to which it will reach the atmosphere and how long it will remain there. Consider the reversible reaction:

$$aA + bB \leftrightarrow cC + dD \tag{4.1}$$

Note that the left side of Fig. 4.1 includes reactions from Chapter 3, that is, kinetics. To the right side, however, the reaction changes to one represented by Eq. (4.2), that is, an equilibrium reaction. Initially (t_0), the rate of the forward reaction is high, and the rate of the reverse reaction is low. As the reaction proceeds, the rate of the forward reaction decreases, while the rate of the reverse reaction increases. This will change when the reaction rates approach equality, that is, at t_{equil}, when the rate of the forward reaction equals the rate of the reverse reaction. At t_{equil}, the system has reached equilibrium.

4.2.1 Solubility constant

The discussion of equilibrium and partitioning can begin by delving a bit further into solubility. When a solid phase of a chemical species dissolves in water, after the kinetic stage, that chemical species has solubility product constant (K_{sp}), that is, an equilibrium constant. The K_{sp} represents the point at which that solute dissolves in solution. The greater the aqueous solubility of a chemical species, the larger the K_{sp} will be.

Consider the aqueous reaction:

$$aA_{(s)} \leftrightarrow cC_{(aq)} + dD_{(aq)} \tag{4.2}$$

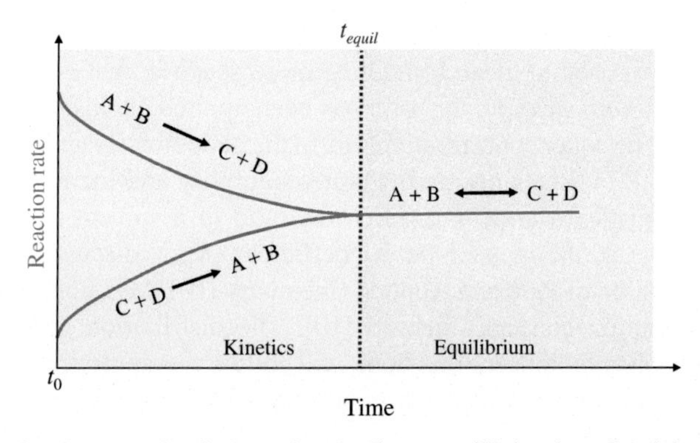

FIG. 4.1 The change in rate of forward and reverse chemical reactions leading to equilibrium in a closed thermodynamic system.

By multiplying the molarities or concentrations of the products (cC and dD), the K_{sp} of a reaction can be found. The result is raised by the power of the coefficients:

$$K_{sp} = [C]^c [D]^d \tag{4.3}$$

Since the concentrations of solid-phase species, for example, $aA_{(s)}$, do not change the expression, they are not included in equilibrium constant calculations; any change in their concentrations is negligible. That is, K_{sp} is the maximum extent that a solid-phase species can be dissolved in solution.

What is the K_{sp} for cadmium fluoride?

This is a salt that ionizes in water. The solubility of a salt is that amount of salt added to water to reach the solution's equilibrium. So, the equilibrium reaction for cadmium fluoride is

$$CdF_{2(s)} \leftrightarrow Cd^{2+}_{(aq)} + F^{-}_{(aq)}$$

Thus,

$$K_{sp} = \left[Cd^{2+}_{(aq)}\right]\left[F^{-}_{(aq)}\right]^2$$

Thus, K_{sp} is an example of an equilibrium or partitioning coefficient, that is, from a solid-phase chemical species to aqueous product, we can now extend and generalize Eq. (4.2) to other equilibria.

Fig. 4.1's abruptness does not accurately and completely represent reality. In the environment or a small vial for that matter, the kinetics and equilibria are net phenomena. Indeed, the forward and reverse reactions continue at a microscopic scale, so it more properly stated that the system has reached dynamic equilibrium.

Recall from the kinetic discussions in Chapter 3 that the reaction has a forward rate:

$$k_f [A]^a [B]^b \tag{4.4}$$

The reaction also has a backward rate:

$$k_b [C]^c [D]^d \tag{4.5}$$

Thus, at equilibrium, the forward equals the backward rate:

$$k_f [A]^a [B]^b = k_b [C]^c [D]^d \tag{4.6}$$

Thus, an equilibrium constant, K_{eq}, can be found for this reaction:

$$K_{eq} = \frac{k_f}{k_b} = \frac{[C]^c [D]^d}{[A]^a [B]^b} \tag{4.7}$$

4.2.2 The partition coefficient

The ratio of the concentrations of solute in two media at equilibrium is known as a partition coefficient, P. However, the term "partition coefficient" has been used to mean something more specific, that is, the ratio of the concentrations of a solute between two solvents. Even more specific, the term has been applied exclusively to the concentrations of a solute in a nonpolar solvent and in water. The water concentrations are in the denominator and the nonpolar solvent concentrations in the numerator. Thus, increasing P indicates decreasing water solubility and increasing lipophilicity.

In biomedical and pharmacological parlance, P is often the ratio of a solvent in unionized water and in unionized *n*-octanol [9], rendering it the same as the octanol-water coefficient (K_{ow}), discussed later in this chapter. In fact, the chemical authority, International Union of Pure and Applied Chemistry (IUPAC), considers the term, partition coefficient, to be obsolete, preferring "distribution constant" instead [10]. The distribution coefficient differs from the partition coefficient by including ionized solvents. Indeed, this book will follow the approach of air pollution and environmental venues, which considers the partition coefficient to an inclusive term for any type of partitioning at equilibrium, and K_{ow} to be specific to octanol-water partitioning.

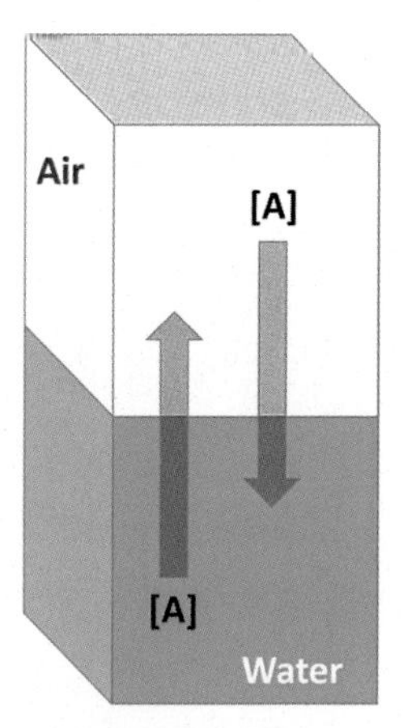

FIG. 4.2 Partitioning of compound A between two phases at equilibrium. In this case, the fugacity (*f*) at equilibrium is $f_{A,water} = f_{A,air}$.

4.2.3 Fugacity [11]

Recall for Chapter 3 that the chemical potential of a compound is a form of energy that can be absorbed or released during a reaction because of a change in the particle number of that compound. When temperature and pressure are held constant, chemical potential is the partial molar Gibbs free energy (see discussion in Chapter 3). At chemical equilibrium, the sum of the product of chemical potentials and stoichiometric coefficients is zero, since free energy is minimized. This is also true for equilibrium between physical phases, for example, environmental compartments.

The fugacity of a substance is its potential to escape from one compartment to another. Fugacity is similar to chemical potential. However, fugacity is proportional to concentration, whereas chemical potential is not [12]. In the vapor or gas phase, substances vary in their tendency to escape from one compartment to another, for example, from the water to the air (see Fig. 4.2). At this fleeing indicates fugacity has units of pressure. When equilibrium is reached, the chemical compound has a common fugacity in all phases. For example, when the fugacity of toluene in water is equal to its fugacity in air, then a toluene equilibrium is achieved for air and water. This is not the same as having equal concentrations in both compartments. Indeed, equilibrium between air and water, for example, would have very different concentrations in these two compartments, given their densities, temperature, and other conditions.

Fugacity is dynamic, for example, if the fugacity in water is higher than the fugacity in air, toluene will evaporate, that is, volatilize, until establishing a new equilibrium. The common units for fugacity calculations are mol m^{-3}, which are air concentration units. This necessitates unit conversions for the fleeing between other "nonair" compartments. For example, water concentrations expressed in mg L^{-1} and soil or sediment concentrations expressed in mg kg^{-1} would have converted to mol m^{-3}. The dry bulk density of soil is the dry weight of the soil divided by soil volume. So, the same soil that is loose will have a much smaller bulk density than a highly compacted soil. That is why chemical concentration in solid matrices, like soil, is most often expressed as mass of chemical species per mass of matrix, as opposed to concentrations in air and water, which are usually expressed as mass per volume.

Therefore, the densities of water and the solid media would have to be specified; for example, bulk densities of different soils vary.

Chemical concentration and fugacity directly relate to one another via the fugacity capacity constant, known as the *Z* value [13]:

$$C_i = Z_i \cdot f \tag{4.8}$$

where

Ci = concentration of substance in compartment i (mass per volume)
Z_i = fugacity capacity (time^2 per length^2)
f = fugacity (mass per length per time^2)

The *Z* value depends on pressure, temperature, physicochemical properties of the compound, and the characteristics of the environmental compartment where the compound resides. These factors are usually more sensitive than concentration, since concentration dependence is very weak in typical environmental conditions, that is, highly diluted [12]. When *Z* is highest, it reaches the highest concentrations in an environmental compartment.

The Z value expresses the capacity of a phase or environmental medium for a given chemical species. When the compound is readily soluble in a phase, that is, the phase can hold a large quantity of the compound, Z is large. Small Z values indicate that the phase can accept only a small quantity of the compound. Deriving Z values for each chemical in each phase usually begins in the air phase. In the air, the ideal gas law is applied (see Chapter 2):

$$PV = nRT \tag{4.9}$$

where P is pressure, which can be substituted with fugacity here; V is the volume of the air; n is the number of moles of the compound; R is the gas constant $\left(8.314 \times \frac{(\mathrm{Pa{\cdot}m^3})}{\mathrm{mol{\cdot}K}}\right)$; and T is absolute temperature (K). Since $\mathrm{C} = \frac{n}{V}$ and $\mathrm{C} = \mathrm{Z} \times f$, Eq. (4.5) can be rewritten in fugacity terms:

$$Z_A = \frac{1}{RT} = \frac{C_A}{f_A} \tag{4.10}$$

where the subscript A refers to the air phase. Z_A under standard conditions, then, is about $4 \times 10^{-4}\,\mathrm{mol\,m^{-3}} \cdot \mathrm{Pa}$ for all compounds in air. Given that a partition coefficient is the ratio of the concentrations in two environmental media at equilibrium, the partition is the ratio of the Z values of the two media. For example, the air-water partition coefficient (K_{AW}), which will be discussed in detail later, is

$$K_{AW} = \frac{C_A}{C_W} = \frac{Z_A f_A}{Z_W f_W} \tag{4.11}$$

Because K_{AW} is measured when f_A is equal to f_W, then,

$$K_{AW} = \frac{Z_A}{Z_W} \tag{4.12}$$

And we can generalize the partition coefficient K beyond air to be the fleeing between Phase 1 and Phase 2:

$$K_{1,2} = \frac{Z_1}{Z_2} \tag{4.13}$$

The logic of Eq. (4.8) and Fig. 4.2 can be extended from the strict definitions of physical phases to environmental compartments. Fugacity can relate linearly to concentrations in the gaining and losing compartments:

$$C_{i,j} = Z_{i,j} \cdot f_{i,j} \tag{4.14}$$

where

Cij = concentration of ith substance in jth compartment (mol $\mathrm{m^{-3}}$)
$Z_{i,j}$ = fugacity capacity of ith substance in jth compartment $\left(\frac{\mathrm{mol}}{\mathrm{m^3\ pascal}}\right)$
$f_{i,j}$ = fugacity (pascal, Pa)

The Z constants are known as half-partition coefficients since, at equilibrium, half of the capacity is in each compartment. Thus, at equilibrium, the fugacity of the two environmental compartments (e.g., air and water) will be equal:

$$f_{\mathrm{A,water}} = f_{\mathrm{A,air}} \tag{4.15}$$

In addition, the ratio of concentration to fugacity capacity in the two compartments (i.e., 1 and 2) will be equal:

$$\frac{C_{A,1}}{Z_{A,1}} = \frac{C_{A,2}}{Z_{A,2}} \tag{4.16}$$

The rate constants can also be related to the concentration and fugacity capacities:

$$k_{1,2} = \frac{C_{A,2}}{C_{A,1}} = \frac{Z_{A,2}}{Z_{A,1}} \tag{4.17}$$

To derive the Z constants, we can use Raoult's law, which states that the partial vapor pressure of each component of an ideal mixture of liquids is equal to the vapor pressure of the pure component multiplied by its mole fraction in the mixture. An ideal mixture is one in which all intermolecular interactions are the same. Ideal conditions can often be assumed for ambient environmental conditions, since the concentrations of air pollutants are very dilute, that is, the pollutant concentration accounts for only small number of molecules compared to the number of air molecules. This means that the relative lowering of vapor pressure of a dilute solution containing a nonvolatile solute is equal to the mole fraction of solute in the solution. Thus, Raoult's law for a single component is

$$p_i = p_i^* x_i \tag{4.18}$$

where p_i is the partial pressure of compound i in the gaseous mixture in the compartment above the liquid (e.g., headspace), p_i^* is the inherent vapor pressure of pure compound, and x_i is the mole fraction of compound I in the mixture in the liquid.

Thus, Eq. (4.5) can be modified according to Raoult's law to derive Z constants:

$$f_{i,j} = C_{i,j} \cdot v_j \cdot \gamma_{i,j} \cdot f_R \tag{4.19}$$

where v_j is the molar volume of jth compartment (mol m^{-3}), $\gamma_{i,j}$ is activity coefficient in the jth compartment, and f_R is the reference fugacity (Pa). The $\gamma_{i,j}$ activity coefficient is the ratio of the compound's fugacity to the ideal solution's f_R and can be derived from the phase density. Typical phase densities used on air models are shown in Table 4.3.

Knowledge about the affinities of a compound for each phase enables predictions of the amount and rate of transformation, transport, and fate. This means that the likely place for a pollutant in the environment can be predicted by the partition coefficients. These coefficients can be viewed as a potential. That is, at the time when equilibrium is achieved among all phases and compartments, the chemical potential in each compartment has been reached.

4.3 PARTITIONING AND TRANSPORT

We can apply the fugacity concepts to estimate the affinity of a chemical species for various environmental compartments. This can be a first determinant of the amount of the compound that will be transported. If a chemical species has a particularly strong affinity for a compartment, it is less likely to escape, or put another way, less mass of the compound will partition to another compartment. Indeed, at equilibrium, the fugacity of the system of all environmental compartments is

$$f = \frac{M_{total}}{\sum_i (Z_i \cdot V_i)} \tag{4.20}$$

where M_{total} = total number of moles of a substance in all of the environmental system's compartments and V_i = volume of compartment i where the substance resides. M commonly represents molarity, that is, the number of moles of solute in a volume, for example, number of liters of solution.

4.3.1 Volatilization

If a chemical substance obeys the ideal gas law, then fugacity capacity is the reciprocal of the gas constant (R) and absolute temperature (T). Recall that the ideal gas law described in Eq. (4.5) can be arranged as

$$\frac{n}{V} = \frac{P}{RT} \tag{4.21}$$

TABLE 4.3 Phase density default values used in air pollution models

Phase	Density (kg m^{-3})
Air	1.19
Water	1000
Sediment	1500
Colloids	1000
Biota	1000

Source of data: D. Mackay, S. Paterson, Evaluating the multimedia fate of organic chemicals: a level III fugacity model, Environ. Sci. Technol. 25 (3) (1991) 427–436.

where

n = number of moles of a substance
P° = substance's vapor pressure

Then,

$$P^\circ = \frac{n}{V} \cdot RT = f \quad (4.22)$$

And

$$C_i = \frac{n}{V} \quad (4.23)$$

Therefore,

$$Z_{air} = \frac{1}{RT} \quad (4.24)$$

This relationship allows for predicting the behavior of the substance in the gas phase. The substance's affinity for other environmental media can be predicted by relating the respective partition coefficients to the Henry's law constants (K_H). For water, the fugacity capacity (Z_{water}) can be found as the reciprocal of K_H:

$$Z_{water} = \frac{1}{K_H} \quad (4.25)$$

This is the dimensioned version of the Henry's law constant (length^2 per time^2). Note that Henry's law is a special case of Raoult's law.

4.3.2 Compartmental affinity and fugacity

From the foregoing discussion, some chemical compounds have more affinity for the air than for water or other compartments, for which Henry's law can be conveniently applied. To estimate the amount of a substances that escapes and remains, then, fugacity must be calculated for every environmental compartment. This is important for air pollutants that have sources throughout the environment. For example, many persistent organic compounds spend part of their life cycles in sediment reservoirs beneath surface waters. As evidence, substances like polychlorinated aromatic compounds (e.g., PCBs) that have been banned and not used for decades continue to be emitted to the atmosphere. In sediment, the fugacity capacity is directly proportional to the contaminant's sorption potential, expressed as the solid-water partition coefficient (K_d) and the average sediment density ($\rho_{sediment}$). Sediment fugacity capacity is indirectly proportional to the chemical substance's Henry's law constant:

$$Z_{sediment} = \frac{\rho_{sediment} \cdot K_d}{K_H} \quad (4.26)$$

Fugacity explains the partitioning that allows that to happen, but it also explains how the pollutant enters and exits biota in ecosystems. For biota, particularly fauna and especially fish and other aquatic vertebrates, the fugacity capacity is directly proportional to the density of the fauna tissue (ρ_{fauna}) and the chemical substance's bioconcentration factor (BCF) and inversely proportional to the contaminant's Henry's law constant:

$$Z_{fauna} = \frac{\rho_{fauna} \cdot BCF}{K_H} \quad (4.27)$$

As in the case of the sediment fugacity capacity, a higher bioconcentration factor means that the fauna's fugacity capacity increases and the actual fugacity decreases. Again, this is logical, since the organism is sequestering the contaminant and keeping it from leaving if the organism has a large BCF. This is a function of both the species of organism and the characteristics of the contaminant and the environment where the organism resides. So, factors like temperature, pH, and ionic strength of the water and metabolic conditions of the organism will affect BCF and Z_{fauna}. This also helps to explain why published BCF values may have large ranges.

The total biochemodynamic partitioning of the environmental system is merely the aggregation of all of the individual compartmental partitioning. So, the moles of the contaminant in each environmental compartment (M_i) are found to be the product of the fugacity, volume, and fugacity capacity for each compartment:

$$M_i = Z_i \cdot V_i \cdot f \tag{4.28}$$

To consolidate the preceding discussions of partitioning and to demonstrate multicompartmental fugacity, consider a hypothetical, three-compartment lake ecosystem of air, water, and fish. Assume that 100 kg of dichloromethane (CH_2Cl_2) is accidentally spilled into this lake. The airshed contains 10 billion cubic meters of air, over 7 billion liters of water that contains 3.5 cubic meters of fish (see Fig. 4.3). What is the partitioning of dichloromethane (CH_2Cl_2) among these three compartments?

First, we must establish compatible units, so we need to convert the water volume from liters to m^3, that is, $7 \times 10^9\,L = 7 \times 10^6\,m^3$.

Next, we must decide on a bioconcentration factor. There are several reported BCF values for CH_2Cl_2. For example, the US Environmental Protection Agency [14] reports one experimentally derived unitless BCF as 22.9 and predicted BCF values ranging from 2.63 to 15.1. The highest reported BCF by European Chemicals Agency [15] is $40\,L\,kg^{-1}$. We will use the European value, but it is important to investigate which reported values are best and relevant. This requires exploring the metadata, that is, the information about the data.

Partitioning is temperature-dependent, so let us assume it to be 25°C.

Also, the densities of biota vary considerably. For example, the size of fish affects the specific gravity of their tissue; for example, a study of mesopelagic fish in the Pacific Ocean off the northeastern US coastline [16] found mean specific gravities between 1.039 and $1.062\,g\,cm^{-3}$. Fish are generally a bit denser than water, so we will assume that the fish density to be $1.05\,g\,cm^{-3}$. However, it is preferable to establish the mean or median body density of the species of interest and use these in the calculations.

The vapor pressure at 25°C is 0.46 atm. Henry's law constant is $3.25 \times 10^{-3}\,\frac{atm \cdot m^3}{mol}$. By converting the CH_2Cl_2 mass to moles,

$$100\,kg \times \frac{1000\,g}{kg} \times \frac{1\,mol}{84.93\,g} = 1177\,mol$$

For air partitioning, we plug our data into Eq. (4.20):

$$Z_{air} = \frac{1}{RT} = \frac{1}{0.0821\,L \cdot \frac{atm}{mol \cdot K} \cdot 298K} \times \frac{1000\,L}{1\,m^3} = 40.9\,\frac{mol}{atm \cdot m^3}$$

For partitioning to water, we use Eq. (4.21):

$$Z_{water} = \frac{1}{K_H} = \frac{1}{3.25 \times 10^{-3}\,\frac{atm \cdot m^3}{mol}} = 308\,\frac{mol}{atm \cdot m^3}$$

And for partitioning to the fish tissue, we use Eq. (4.23):

$$Z_{fauna} = \frac{\rho_{fauna} \cdot BCF}{K_H} = \frac{1000\frac{g}{m^3} \cdot 40\frac{L}{kg}}{3.25 \times 10^{-3}\,\frac{atm \cdot m^3}{mol}} \times \frac{1\,kg}{1000\,g} \times \frac{1\,m^3}{1000\,L} = 12.3\,\frac{mol}{atm \cdot m^3}$$

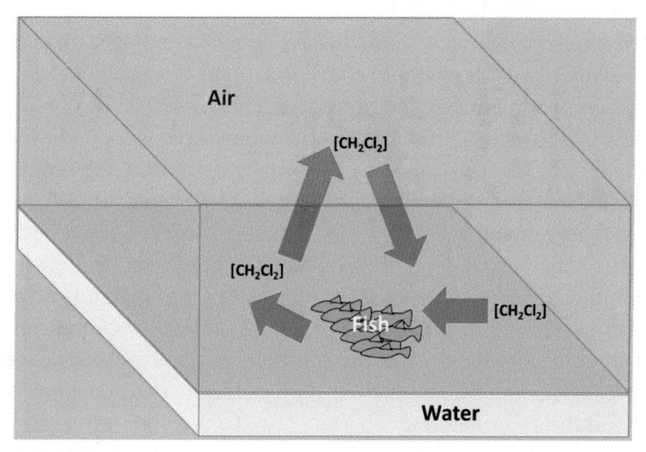

FIG. 4.3 Hypothetical three-compartment ecosystem.

To calculate the system fugacity, we divide the mass of CH_2Cl_2 entering the system by the sum of the three compartment's partitions and volumes:

$$f = \frac{1177\,\text{mol}}{(40.9 \times 10^{10}) + [308 \times (7 \times 10^{6})] + (12.3 \times 3.5)} = 2.9 \times 10^{-9}\ \text{atm}.$$

So, to find the moles of CH_2Cl_2 in each compartment, we use Eq. (4.24):

$$M_{air=}\ 2.9 \times 10^{-9} \times 10^{10} \times 40.9 = 1171\ \text{mol}$$

$$M_{water=}\ 2.9 \times 10^{-9} \times (7 \times 10^{6}) \times 308 = 6.2\ \text{mol}$$

$$M_{fish} =\ 2.9 \times 10^{-9} \times 3.5 \times 12.3 = 1.2 \times 10^{-7}\ \text{mol}$$

From the above calculations, what can be said about the affinity of CH_2Cl_2 for environmental media?

In terms of partitioning at equilibrium, dichloromethane would appear to have the greatest affinity for air. Indeed, the mass of CH_2Cl_2 fleeing to the air is over two orders of magnitude higher than water and 10 orders of magnitude higher than fish.

However, the *Z* values and chemical concentrations tell a different story. Given the volumes, the highest CH_2Cl_2 concentrations are in the water, which has the highest fugacity capacity (i.e., $Z = 308$). The concentration in water is 6.2 moles in 7 million m^3 of water, that is, 8.9×10^{-7} mol m^{-3}.

The air concentration is 1171 moles in 10^{10} m^3 = 1.2×10^{-7} mol m^{-3}.

The fish tissue concentration is 1.2×10^{-7} moles in 3.5 m^3 = 3.4×10^{-8} mol m^{-3}.

All things considered, however, the CH_2Cl_2 is evenly distributed among the three phases.

Assume that a later study showed that the air volume in which the CH_2Cl_2 is mixed was overstated, that is, the mixing layer is half the altitude, so that the volume of air should have been 5×10^9 m^3. How would it affect the partitioning?

Note that none of the *Z* values are affected by volume. The air value is based on the gas laws, so it is not constrained by the volume of air nor by the inherent properties of the chemical species. The water *Z* value is not affected by water volume, but is the reciprocal of the Henry's law coefficient, which is unique to each chemical species. Similarly, the fauna *Z* value is not dependent on air or water volume, but is inversely proportional to the Henry's law coefficient and directly proportional to the BCF and the tissue density for each chemical species.

However, the concentration calculations depend on volume. Thus, the air concentration would double if the air volume is halved. The 1171 moles fleeing to the atmosphere would be diluted in half the volume of air. In this case, the rounded air concentration is $2.3 \times 10^{-7}\ mol\,m^{-3}$.

Comparing the respective fugacity capacities for each phase or compartment in an environmental system is useful for air pollution and environmental calculations. First, if one compartment has a very high fugacity (thus, a low fugacity capacity) for a contaminant and the source of the contaminant no longer exists, then one would expect the concentrations in that compartment to decrease rather precipitously with time under certain environmental conditions. Conversely, if a compartment has a very low fugacity, the contaminant would have a strong affinity for that compartment and only after much energy is added would it move away (the PCB example above illustrates this problem) and decrease the chemical concentration of the contaminant in that compartment. Second, if a continuous source of the contaminant exists and a compartment has a high fugacity capacity (and low fugacity), this compartment may serve as a conduit for delivering the contaminant to other compartments with relatively low fugacity capacities. Third, by definition, the higher relative fugacities of one set of compartments compared with another set in the same ecosystem allow for comparative analyses and estimates of sources and sinks (or "hot spots") of the contaminant, which is an important part of fate, transport, exposure, and risk assessments.

Fugacity-based, multicompartmental environmental models take these relationships into account. The movement of a contaminant through the environment can be expressed regarding how equilibrium is achieved in each compartment. The processes driving this movement can be summarized into transfer coefficients or compartmental rate constants, known as *D* values [17]. So, by first calculating the Z values, as we did for toluene in the previous examples, and then equating inputs and outputs of the contaminant to each compartment, we can derive D value rate constants. The actual transport process rate (N) is the product of fugacity and the D value:

$$N = D \cdot f \tag{4.29}$$

And since the contaminant concentration is $Z{\cdot}f$, we can substitute and add a first-order rate constant k to give us a first-order rate D value (D_R):

$$N = V[c]k = (V \cdot Z \cdot k) \cdot f = D_{R} \cdot f \quad (4.30)$$

Although the concentrations are shown as molar concentrations (i.e., in brackets), they may also be represented as mass per volume concentrations.[1]

Diffusive and nondiffusive transport processes follow Fick's laws, that is, diffusive processes. They can also be expressed with their own D values (D_D), which is related to the mass transfer coefficient (K) applied to area A:

$$N = KA[c] = (K \cdot A \cdot Z.) \cdot f = D_{D} \cdot f \quad (4.31)$$

Nondiffusive transport (bulk flow or advection) within a compartment with a flow rate (G) has a D value (D_A) and is expressed as

$$N = G[c] = (G \cdot Z \cdot)f = D_{A} \cdot f \quad (4.32)$$

This means that a substance that is moving through the environment, during its residence time in each phase, is affected by numerous physical transport and chemical degradation and transformation processes. The processes are addressed by models with the respective D values, so that the total rate of transport and transformation is expressed as

$$f \cdot (D_1 + D_2 + \ldots D_n) \quad (4.33)$$

Very fast processes have large D values, and these are usually the most important when considering the contaminant's behavior and change in the environment.

Models, though imperfect, are important tools for estimating the movement of contaminants in the environment. They do not obviate the need for sound measurements. In fact, measurements and models are highly complementary. Compartmental model assumptions must be verified in the field. Likewise, measurements at a limited number of points depend on models to extend their meaningfulness. Understanding of the basic concepts of a contaminant transport model allows for improved exploration of principle mechanisms for the movement of contaminants throughout the environment.

4.4 INTEGRATING INHERENT PROPERTIES AND SUBSTRATE CHARACTERISTICS

The measure of the amount of chemical that can dissolve in a liquid is called solubility. It is usually expressed in units of mass of solute (that that is dissolved) in the volume of solvent (that that dissolves). Solubility may also be expressed in mass per mass or volume per volume, represented as parts per million (ppm), parts per billion (ppb), or parts per trillion (ppt). Occasionally, solubility is expressed as a percent or in parts per thousand; however, this is uncommon for contaminants and is usually reserved for nutrients and essential gases (e.g., percent carbon dioxide in water or ppt water vapor in the air).

The solubility of a compound is very important to environmental transport, including atmospheric transport. The diversity of solubilities in various solvents is a fairly reliable indication of where one is likely to find the compound in the environment. For example, the various solubilities of the most toxic form of dioxin, tetrachlorodibenzo-*para*-dioxin (TCDD), are provided in Table 4.4. Based on these solubility differences, if a bioreactor has been operating and releasing dioxins, one would expect TCDD to have a much greater affinity for sediment, organic particles, and the organic fraction of soils. The low water solubilities indicate that dissolved TCDD in the water column should be at only extremely low concentrations. But, as will be seen in the discussion regarding cosolvation, for example, other processes may override any single process, for example, dissolution, in an environmental system.

Polarity is an important physicochemical characteristic of a substance that determines its solubility. The polarity of a molecule is its unevenness in charge. Since the water molecule's oxygen and two hydrogen atoms are aligned so that there is a slightly negative charge at the oxygen end and a slightly positive charge at the hydrogen ends and since "like dissolves like," polar substances have an affinity to become dissolved in water, and nonpolar substances resist being dissolved in water.

Increasing temperature, that is, increased kinetic energy, in a system increases the velocity of the molecules, so that intermolecular forces are weakened. With increasing temperature, the molecular velocity becomes sufficiently large to overcome all intermolecular forces, so that the liquid boils (vaporizes). Intermolecular forces may be relatively weak or strong. The weak forces in liquids and gases are often called van der Waals forces.

1. Throughout this text, bracketed values indicate molar concentrations, but these may always be converted to mass per volume concentration values.

TABLE 4.4 Solubility of tetrachlorodibenzo-*para*-dioxin in water and organic solvents

Solvent	Solubility ($mg\,L^{-1}$)
Water	1.93×10^{-5}
Water	6.90×10^{-4} (25°C)
Methanol	10
Lard oil	40
n-Octanol	50
Acetone	110
Chloroform	370
Benzene	570
Chlorobenzene	720
Orthochlorobenzene	1400

If a compound has high aqueous solubility, that is, it is easily dissolved in water under normal environmental conditions of temperature and pressure, it is hydrophilic. If, conversely, a substance is not easily dissolved in water under these conditions, it is said to be hydrophobic. Since many contaminants are organic (i.e., consist of molecules containing carbon-to-carbon bonds and/or carbon-to-hydrogen bonds), the solubility can be further differentiated as to whether under normal environmental conditions of temperature and pressure the substance is easily dissolved in organic solvents. If so, the substance is said to be lipophilic (i.e., readily dissolved in lipids). If, conversely, a substance is not easily dissolved in organic solvents under these conditions, it is said to be lipophobic.

Although many possible outcomes can occur after a substance is released into the environment, the possibilities can fall into three basic categories:

1. The chemical may remain where it is released and retain its physicochemical characteristics (at least within a specified time).
2. The substance may be transported to another location.
3. The substance may be changed chemically, known as the transformation of the chemical.

This is a restatement of the conservation law. Every molecule of mass moving into and out of the control volume must be accounted for and any chemical changes to the contaminant that take place within the control volume. A control volume may be a simple cube (Fig. 4.4A) through which contaminant fluxes are calculated. However, a control volume can also be a cell within an organism, the organism itself (e.g., taking in substances, which it absorbs, metabolizes, distributes, and eliminates), or an entire ecosystem (Fig. 4.4B).

The first law of thermodynamics requires that any change in storage of a substance's mass in a control volume must equal the difference between the mass of the chemical transported into the system less the mass of the chemical transported out of the system. Given the transformations discussed in Chapter 3, the actual chemical species transported in may be different from what was initially released and what has entered the control volume. Thus, the mass balance equation may be written as

$$\text{Accumulation or loss of contaminant A} = \text{Mass of A transported in} - \text{Mass of A transported out} \pm \text{Reactions} \quad (4.34)$$

The reactions may be either those that generate chemical A (i.e., sources) or those that destroy chemical A (i.e., sinks).

The entering mass transported equals the inflow to the system that includes pollutant discharges, transfer from other control volumes and other media (e.g., if the control volume is soil, the water and air may contribute mass of chemical A), and formation of chemical A by abiotic chemistry and biological transformation. Conversely, the outflow is the mass transported out of the control volume, which includes uptake, by biota; transfer to other compartments (e.g., volatilization to the atmosphere); and abiotic and biological degradation of chemical A.

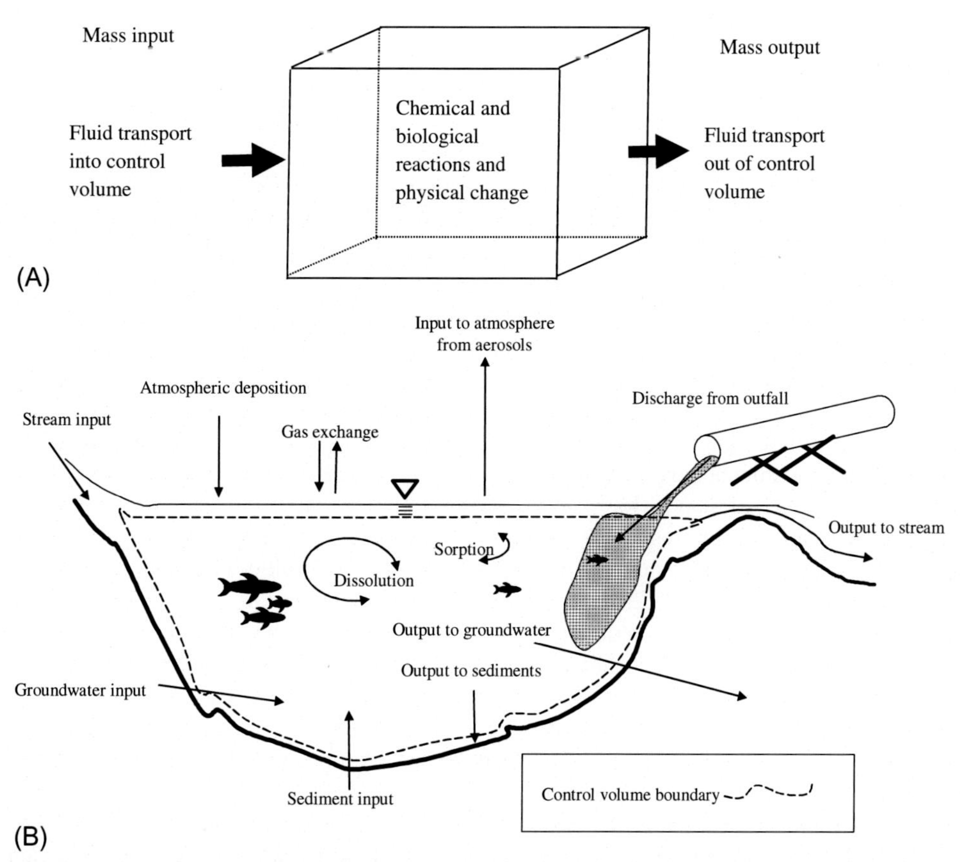

FIG. 4.4 (A) Control volume of an environmental matrix (e.g., soil, sediment, or other unconsolidated materials) or fluid (e.g., water, air, or blood). (B) A pond. Both volumes have equal masses entering and exiting, with transformations and physical changes taking place within the control volume.

The rate of change of mass in a control volume is equal to the rate of chemical A transported in less the rate of chemical A transported out, plus the rate of production from sources, and minus the rate of elimination by sinks. Stated as a differential equation, the rate of change of contaminant A is

$$\frac{d[A]}{dt} = -v \cdot \frac{d[A]}{dx} + \frac{d}{dx}\left(D \cdot \frac{d[A]}{dx}\right) + r \tag{4.35}$$

where

v = fluid velocity
$\frac{d[A]}{dx}$ = concentration gradient of chemical A
r = internal sinks and sources within the control volume

These rates operate at various scales. For example, Eq. (4.35) can be applied from the cell to the planet. As evidence, it is the basis for pharmacokinetic and pharmacodynamic modeling. A basic concept of air pollution is that substances tend to have affinity for certain compartments in abiotic and biotic systems. Thus, the specific partitioning relationships that control the "leaving" and "gaining" compartments must be known. Chemical compounds exit and enter the atmosphere from clouds, water bodies, soil, and biota and move within soil and sediment matrices, in and on particles, within the atmosphere and within organic tissues. Furthermore, these partitioning relationships can be applied to estimate and to model where a substance will go within a designed system (e.g., a scrubber) and after it is released (e.g., from the air to the water). Basic relationships between sorption, solubility, volatilization, and organic carbon-water partitioning are respectively expressed by coefficients of sorption (distribution coefficient, K_D, or solid-water partition coefficient, K_p), dissolution or solubility coefficients, air-water partitioning (often expressed as K_{AW} or Henry's law (K_H) constant), octanol-water partitioning (K_{ow}), and organic carbon-water (K_{oc}).

In biochemodynamics, the environment can be subdivided into finite compartments. Recall that the mass of the contaminant entering and the mass leaving a control volume must be balanced by what remains within the control volume. Environmental systems are a cascade of control volumes. Within each control volume, an individual compartment may be a gainer or loser of a chemical species' mass, but the overall mass must balance.

The common approach is to address each compartment where a contaminant is found in discrete phases of air, water, soil, sediment, and biota. However, a complicating factor in environmental chemodynamics is that even within a single compartment, a contaminant may exist in various phases (e.g., dissolved in water and sorbed to a particle in the solid phase). Interphase reactions or the physical interactions of the contaminant at the interface between each compartment determine the amount of any substance in the environment. Within a compartment, a contaminant may remain unchanged for a designated time period, or it may move physically, or it may be transformed chemically into another substance.

Indeed, interactions of mechanisms occur within a control volume or environmental compartment. A mass fraction will remain unmoved and unchanged. Another fraction remains unchanged but is transported to a different compartment. Another fraction becomes chemically transformed with all remaining products staying in the compartment where they were generated. And a fraction of the original contaminant is transformed and then moved to another compartment. Thus, upon release from a source, the contaminant moves because of thermodynamics. If a substance is dissolved in water and the water comes into contact with another substance, for example, octanol, the substance will tend to move from the water to the octanol. Its octanol-water partitioning coefficient reflects just how much of the substance will move until the aqueous and organic solvents (phases) will reach equilibrium. So, for example, in a spill of equal amounts of the polychlorinated biphenyl decachlorobiphenyl (log K_{ow} of 8.23) and the pesticide chlordane (log K_{ow} of 2.78), the PCB has much greater affinity for the organic phases than does the chlordane (more than five orders of magnitude). This does not mean that a great amount of either of the compounds is likely to stay in the water column, since they are both hydrophobic, but it does mean that they will vary in the time and mass of each contaminant moving between phases. The rate (kinetics) is different, so the time it takes for the PCB and chlordane to reach equilibrium will be different. This can be visualized by plotting the concentration of each compound with time (see Fig. 4.5). When the concentrations plateau, the compounds are at equilibrium with their phase.

Air pollution and other health studies increasingly rely on biomarkers as indicators of concentrations that occur in control volumes where humans and organisms live. Physicians have used markers for decades, for example, cholesterol, fat, and blood sugar. However, environmental toxicologists have adopted numerous chemicals to indicate pollution and its effect. Some of these chemicals are the parent substances to which the organism, including a person, has been exposed. A good example of this is the metal lead (Pb). The amount of lead in a child's blood or hair indicates the extent to which the child has been exposed to Pb. In this case, the total mass of Pb is a reliable indicator of the time and activities of the child within various control volumes, for example, microenvironments in the home (see Fig. 4.6).

Given the r term in Eq. (4.35), however, the biomarkers are compounds other than the parent. Thus, biomarkers are often metabolites of the compounds. For example, smokers inhale nicotine, much of which is metabolized to cotinine (see Fig. 4.7). Thus, cotinine biomarkers can be measures of exposure to tobacco smoke.

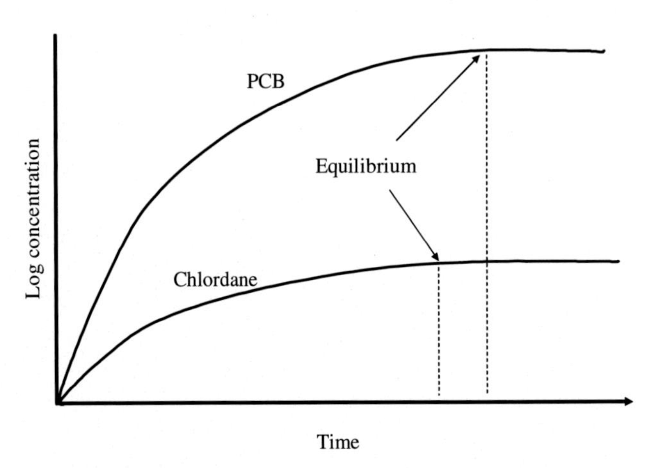

FIG. 4.5 Relative concentrations of a polychlorinated biphenyl (PCB) and chlordane in octanol with time.

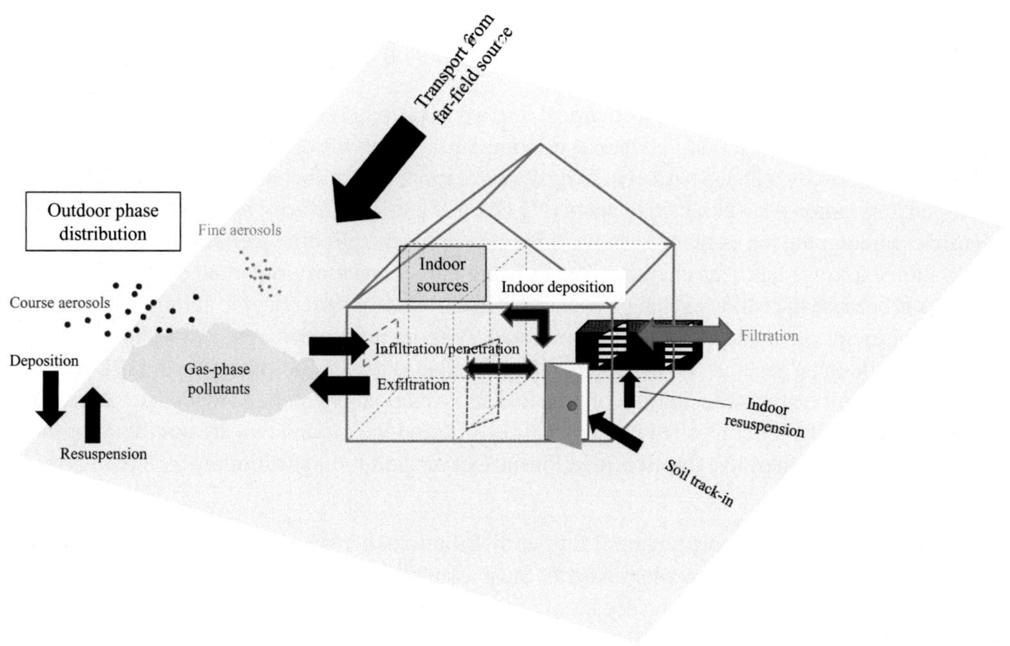

FIG. 4.6 The home as a microenvironmental control volume. *(Data from US Department of Energy, Lawrence Berkeley Laboratory (2003). http://eetd.lbl.gov/ied/ERA/CalEx/partmatter.html (Accessed 2004); and D.A. Vallero, Environmental Contaminants: Assessment and Control, Elsevier Academic Press, Burlington, MA, 2004.)*

FIG. 4.7 Chemical structures of (**1**) S(−)-nicotine (when X = H_2) or S(−)-cotinine (when X = O) and (**2**) S(−)-nicotine N1-glucuronide (when X = H_2) or S(−)-cotinine N1-glucuronide (when X = O). *(Source: O. Ghosheh, E.M. Hawes, N-glucuronidation of nicotine and cotinine in human: formation of cotinine glucuronide in liver microsomes and lack of catalysis by 10 examined UDP-glucuronosyltransferases, Drug Metab. Dispos. 30 (9) (2002) 991–996.)*

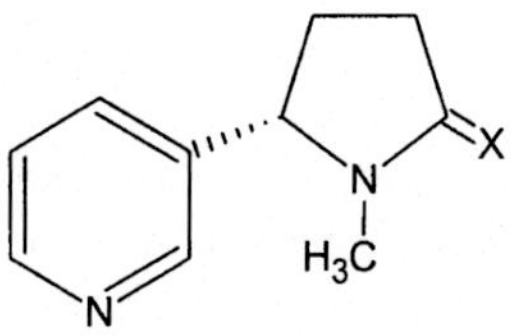

4.5 MOVEMENT FROM LIQUID TO SOLID PHASE

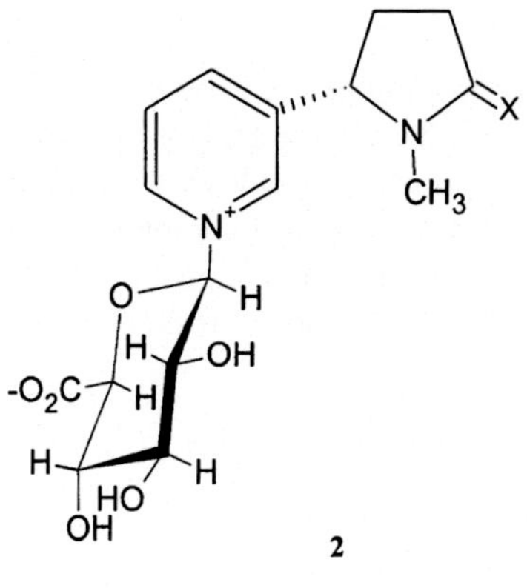

Sorption is a very important transfer process that determines how bioavailable or toxic a compound will be in surface waters, but it is also very important for atmospheric transport. Surface water-atmospheric exchanges are a means by which large quantities of air pollutants reach the atmosphere. Also, sorption occurs at the microscopic scale in water droplets. In addition, chemical transport can occur to and from a solid particle via any fluid, including air and water. The physicochemical transfer [18] of chemical is an interaction of the solute (i.e., the chemical being sorbed) with the surface of a solid surface that can be complex and dependent upon the properties of the chemical and the water. Other fluids are often of such small concentrations that they do not determine the ultimate solid-liquid partitioning. While it is often acceptable to consider "net" sorption, let us consider briefly the four basic types or mechanisms of sorption.

Adsorption is the process wherein the chemical in solution attaches to a solid surface, which is a common sorption process in clay and organic constituents in soils. This simple adsorption mechanism can occur on clay particles where little carbon is available, such as in groundwater.

Absorption is the process that often occurs in porous materials so that the solute can diffuse into the particle and be sorbed onto the inside surfaces of the particle. This commonly results from short-range electrostatic interactions between the surface and the contaminant.

Chemisorption is the process of integrating a chemical into a porous material surface via a chemical reaction. In soil, this is usually the result of a covalent reaction between a mineral surface and the contaminant.

Ion exchange is the process by which positively charged ions (cations) are attracted to negatively charged particle surfaces or negatively charged ions (anions) are attracted to positively charged particle surfaces, causing ions on the particle surfaces to be displaced. Particles undergoing ion exchange can include soils, sediment, airborne particulate matter, or even biota, such as pollen particles. Cation exchange has been characterized as being the second most important chemical process on earth, after photosynthesis. This is because the cation exchange capacity (CEC), and to a lesser degree anion exchange capacity (AEC) in tropical soils, is how nutrients are made available to plant roots. Without this process, the atmospheric nutrients and the minerals in the soil would not come together to provide for the abundant plant life on planet earth [19].

These four types of sorption are a combination of physical and chemical phenomena that occur at surfaces. The first two types of sorption are predominantly controlled by physical factors, and the second two are combinations of chemical reactions and physical processes. Generally, sorption reactions affect air pollution partitioning in three ways, by affecting the following [20]:

1. The chemical contaminant's transport potential due to distributions between the aqueous phase and particles
2. The aggregation and transport of the contaminant because of electrostatic properties of suspended solids
3. Surface reactions such as dissociation, surface catalysis, and precipitation of the chemical contaminant

When a chemical species reaches a soil matrix, some of the chemical remains in soil solution, and the remainder is adsorbed onto the surfaces of the soil particles. Sometimes, this sorption is strong due to cations adsorbing to the negatively charged soil particles. Other times, the attraction is weak. Sorption of chemicals on solid surfaces, like those in sediment, soil, and unconsolidated materials, is a means by which the chemical species is retained in that matrix. Thus, sorption determines the residence time in the matrix. Strongly sorbed compounds will not be allowed to move freely between a particle and the surrounding liquid or gas. For example, sorption slows that rate at which substances move downwardly through the soil profile. It is also one of the challenges for remediation, for example, how to desorb toxic substances from soil or other contaminated materials.

Chemical compounds eventually establish a balance between the mass on the solid surfaces and the mass that is in solution. Molecules will migrate from one phase to another to maintain this balance. The properties of both the chemical and the soil (or other matrix) will determine how and at what rates the molecules partition into the solid and liquid phases. These physicochemical relationships, known as sorption isotherms, are found experimentally. Fig. 4.8 illustrates three isotherms for pyrene from experiments using different soils and sediments.

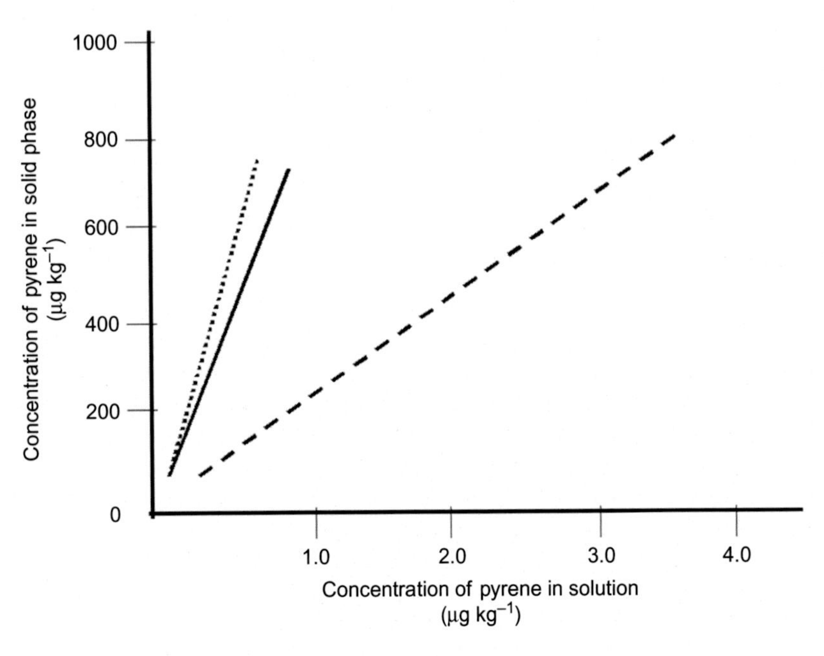

FIG. 4.8 Sorption isotherms for the polycyclic aromatic hydrocarbon, pyrene. *(Source: J.J. Hassett, W.L. Banwart, The sorption of nonpolar organics by soils and sediments, Reactions and Movement of Organic Chemicals in Soils, no. reactionsandmov, (1989), pp. 31–44.)*

The x-axis in Fig. 4.8 shows the concentration of pyrene dissolved in water, and the y-axis shows the concentration in the solid phase. Each line represents the relationship between these concentrations for a single soil or sediment. A straight-line segment through the origin represents the data well for the range of concentrations shown. Not all portions of an isotherm are linear, particularly at high concentrations of the contaminant. Linear chemical partitioning can be expressed as

$$S = K_D \cdot C_W \tag{4.36}$$

where S = concentration of contaminant in the solid phase (mass of solute per mass of soil or sediment), C_W = concentration of contaminant in the liquid phase (mass of solute per volume of pore water), and K_D = partition coefficient (volume of pore water per mass of soil or sediment) for this contaminant in this soil or sediment.

For many chemicals and the substrates (e.g., soils), the partition coefficient can be estimated using

$$K_D = K_{OC} \cdot C_{org} \tag{4.37}$$

where K_{OC} = organic carbon partition coefficient (volume of pore water per mass of organic carbon) and C_{org} = substrate organic matter (mass of organic carbon per mass of substrate, e.g., soil).

This relationship is a very useful tool to estimate a chemical's unknown K_D from the known K_{OC} of the contaminant and the organic carbon content of the soil horizon of interest. K_{oc} values are readily available and published in handbooks and manuals. Users of these report values must keep in mind that these, like several other partitioning coefficients, are substrate-dependent. Thus, be certain that the substrate being used in these calculations is from experiments using the same substrate (e.g., carbon source = anthracite) or otherwise normalized. This is but another example of the importance of metadata. If the substrate is not reported, this may preclude the use of these values or, at a minimum, report that the substrate is unknown.

The actual derivation of K_D is

$$K_D = C_S (C_W)^{-1} \tag{4.38}$$

where C_S is the equilibrium concentration of the solute in the solid phase and C_W is the equilibrium concentration of the solute in the water.

Therefore, K_D is a direct expression of the partitioning between the aqueous and solid (soil or sediment) phases. A strongly sorbed chemical like a dioxin or the banned pesticide DDT can have a K_D value exceeding 10^6. Conversely, a highly hydrophilic, miscible substance like ethanol, acetone, or vinyl chloride will have K_D values less than 1. This relationship between the two phases demonstrated by Eq. (4.38) and Fig. 4.9 is the Freundlich adsorption isotherm:

$$C_{sorb} = K_F \cdot C^n \tag{4.39}$$

where C_{sorb} is the concentration of the sorbed contaminant, that is, the mass sorbed at equilibrium per mass of sorbent; K_F is the Freundlich isotherm constant; and n is the exponent of the chemical concentration curve of the function of C_{sorb} and C_W in Fig. 4.9. The exponent determines the linearity or order of the reaction. Thus, if $n = 1$, then the isotherm is linear, meaning

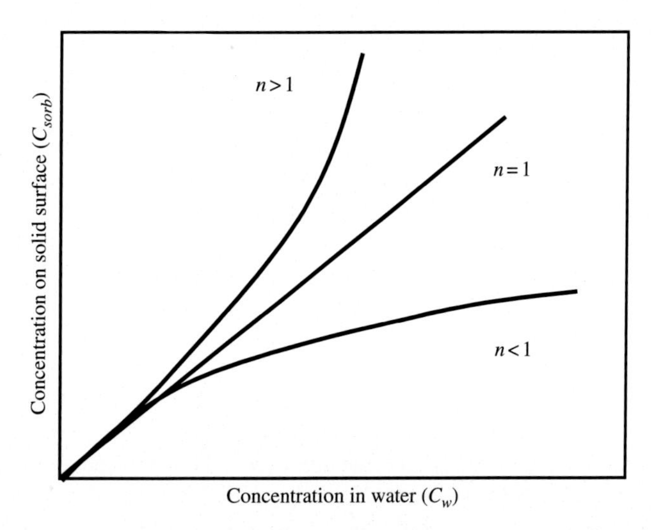

FIG. 4.9 Hypothetical Freundlich isotherms with exponents (n) less than, equal to, and greater than 1, as applied to the equation $C_{sorb} = K_F \cdot C^n$. *(Sources: H.F. Hemond, E.J. Fechner, Chemical Fate and Transport in the Environment, Elsevier, 2014; P.M. Gschwend, Environmental Organic Chemistry, John Wiley & Sons, 2016.)*

the more of the contaminant in solution, the more would be expected to be sorbed to surfaces. For values of $n<1$, the amount of sorption is in smaller proportion to the amount of solution and, conversely, for values of $n>1$, a greater proportion of sorption occurs with less contaminant in solution. Also note that if $n=1$, then Eq. (4.25) and the Freundlich adsorption isotherm are identical.

Increasing organic matter content is elevated in substrates, that is, soil and sediment, and is accompanied by an increase in the amount of a contaminant that is sorbed, that is, sorption is directly proportional to the soil/sediment organic matter content. This allows for a conversion of K_D values from those that depend on specific soil or sediment conditions to those that are soil-/sediment-independent sorption constants, K_{OC}:

$$K_{OC}=K_D\cdot(f_{OC})^{-1} \tag{4.40}$$

where f_{OC} is the dimensionless weight fraction of organic carbon in the soil or sediment. The K_{OC} and K_D have units of mass per volume. Table 4.5 provides the log K_{OC} values that are calculated from chemical structure and those measured empirically for several organic compounds and compares them with the respective K_{ow} values.

Of principal interest in air pollution, after a chemical has moved through water, soil, sediment, or biota, it must move into the atmosphere. As discussed in Chapter 5, volatilization is a function of the concentration of a contaminant in solution and the contaminant's partial pressure.

4.6 THE OCTANOL-WATER COEFFICIENT

The octanol-water partition coefficient (K_{ow}) is arguably the most commonly applied partitioning coefficient for water systems, but it also has major applications in air and other media. Since many air toxics are organic (i.e., consisting of molecules containing carbon-to-carbon bonds and/or carbon-to-hydrogen bonds), the solubility can be further differentiated as to whether under normal environmental conditions of temperature and pressure the substance is easily dissolved in organic solvents. If so, the substance is said to be lipophilic (i.e., readily dissolved in lipids). If, conversely, a substance is not easily dissolved in organic solvents under these conditions, it is said to be lipophobic.

This affinity for either water or lipids underpins an important indicator of environmental partitioning, that is, the octanol-water partition coefficient (K_{ow}). The K_{ow} is the ratio of a substance's concentration in octanol ($C_7H_{13}CH_2OH$) to the substance's concentration in water at equilibrium (i.e., the reactions have all reached their final expected chemical composition in a control volume of the fluid). Octanol is a surrogate for lipophilic solvents in general because it has degrees of affinity for both water and organic compounds, that is, octanol is amphiphilic. Since the ratio forming the K_{ow} is $[C_7H_{13}CH_2OH]/[H_2O]$, then the larger the K_{ow} value, the more lipophilic the substance. Values for solubility in water and K_{ow} values of some important environmental compounds, along with their densities, are shown in Table 4.6.

Table 4.6 elucidates some additional aspects of solubility and organic/aqueous phase distribution. Water solubility is somewhat inversely related to *Kow*, but the relationship is uneven. This results from the fact that various organic compounds are likely to have affinities for neither, either, or both the organic and the aqueous phases. Most compounds are not completely associated with either phase, that is, they have some amount of amphiphilicity.

The relationship between density and organic/aqueous phase partitioning is very important to pollutant transport. For example, a nonaqueous-phase liquid (NAPL) will partition at different rates in groundwater through the vadose (unsaturated) zone, depending on its density. A lighter compound may stay near the surface and be more likely to volatilize into the atmosphere than a denser compound, even if they have equal high K_{ow} values and water solubility.

This affinity for either water or lipids underpins an important indicator of environmental partitioning. The K_{ow} is the ratio of a substance's concentration in octanol ($C_7H_{13}CH_2OH$) to the substance's concentration in water at equilibrium (i.e., the reactions have all reached their final expected chemical composition in a control volume of the fluid). Octanol is a surrogate for lipophilic solvents in general because it has degrees of affinity for both water and organic compounds, that is, octanol is amphiphilic. Since the ratio forming the K_{ow} is $[C_7H_{13}CH_2OH]/[H_2O]$, then the larger the K_{ow} value, the more lipophilic the substance.

Also, what seem to be minor structural changes to a molecule can make quite a difference in phase partitioning and in density. Even the isomers (i.e., same chemical composition with a different arrangement) vary in their K_{ow} values and densities (note that the "1,1" vs "1,2" arrangements of chlorine atoms on 1,1-dichloroethane and 1,2-dichloroethane cause the former to have a slightly decreased density but twice the K_{ow} value than the latter). The location of the chlorine atoms alone accounts for a significant difference in water solubility in the two compounds.

As mentioned, chemical transformation and movement within other environmental media often occurs prior to a pollutant's release to the atmosphere. Most air pollution is emitted from stacks, vents, or ruptured containers above ground, but

TABLE 4.5 Calculated and experimental organic carbon coefficients (K_{OC}) compared with octanol-water coefficients (K_{ow}) for selected contaminants found at hazardous waste sites

		Calculated		Measured	
Chemical	**log K_{ow}**	**log K_{oc}**	**K_{oc}**	**log K_{oc}**	**K_{oc} (geomean)**
Benzene	2.13	1.77	59	1.79	61.7
Bromoform	2.35	1.94	87	2.10	126
Carbon tetrachloride	2.73	2.24	174	2.18	152
Chlorobenzene	2.86	2.34	219	2.35	224
Chloroform	1.92	1.60	40	1.72	52.5
Dichlorobenzene, 1,2- *(o)*	3.43	2.79	617	2.58	379
Dichlorobenzene, 1,4- *(p)*	3.42	2.79	617	2.79	616
Dichloroethane, 1,1-	1.79	1.50	32	1.73	53.4
Dichloroethane, 1,2-	1.47	1.24	17	1.58	38.0
Dichloroethylene, 1,1-	2.13	1.77	59	1.81	65
Dichloroethylene, *trans* -1,2-	2.07	1.72	52	1.58	38
Dichloropropane, 1,2-	1.97	1.64	44	1.67	47.0
Dieldrin	5.37	4.33	21,380	4.41	25,546
Endosulfan	4.10	3.33	2,138	3.31	2,040
Endrin	5.06	4.09	12,303	4.03	10,811
Ethylbenzene	3.14	2.56	363	2.31	204
Hexachlorobenzene	5.89	4.74	54,954	4.90	80,000
Methyl bromide	1.19	1.02	10	0.95	9.0
Methyl chloride	0.91	0.80	6	0.78	6.0
Methylene chloride	1.25	1.07	12	1.00	10
Pentachlorobenzene	5.26	4.24	17,378	4.51	32,148
Tetrachloroethane, 1,1,2,2-	2.39	1.97	93	1.90	79.0
Tetrachloroethylene	2.67	2.19	155	2.42	265
Toluene	2.75	2.26	182	2.15	140
Trichlorobenzene, 1,2,4-	4.01	3.25	1,778	3.22	1,659
Trichloroethane, 1,1,1-	2.48	2.04	110	2.13	135
Trichloroethane, 1,1,2-	2.05	1.70	50	1.88	75.0
Trichloroethylene	2.71	2.22	166	1.97	94.3
Xylene, *o-*	3.13	2.56	363	2.38	241
Xylene, *m-*	3.20	2.61	407	2.29	196
Xylene, *p-*	3.17	2.59	389	2.49	311

Sources: EPA/540/R-96/018, Soil Screening Guidance: User's Guide (1996); EPA/540/R-95/128 PB96-963502, Soil Screening Guidance: Technical Background Document (1996).

TABLE 4.6 Solubility, octanol-water partitioning coefficient, and density values for some environmental pollutants

Chemical	Water solubility (mg L^{-1})	*Kow*	Density (kg m^{-3})
Atrazine	33	724	
Benzene	1780	135	879
Chlorobenzene	472	832	1110
Cyclohexane	60	2754	780
1,1-Dichloroethane	4960	62	1180
1,2-Dichloroethane	8426	30	1240
Ethanol	Completely miscible	0.49	790
Toluene	515	490	870
Vinyl chloride	2790	4	910
Tetrachlorodibenzo-*para*-dioxin (TCDD)	1.9×10^{-4}	6.3×10^{6}	

Source of data: H.F. Hemond, E.J. Fechner, Chemical Fate and Transport in the Environment, Elsevier, 2014; TCDD data sources: National Toxicology Program, NTP 11th Report on Carcinogens, in: Report on carcinogens: carcinogen profiles. vol. 11 (2004) 1; NTP Report on Carcinogens: Background Document for 2,3,7,8-Tetrachlorodibenzo-p-Dioxin (TCDD). Available: https://ntp.niehs.nih.gov/ntp/newhomeroc/other_background/tetrachl_tcdd_4apps_508.pdf (1999); EPA 811-F-95-0031-T, National Primary Drinking Water Regulations: Dioxin (2,3,7,8-TCDD) (1995).

there are also sources that have their beginnings below ground. Leaking pipelines, underground tanks and contaminated soil, porous media, and water can be surreptitious sources of air pollutants. As mentioned, the relationship between density and organic-/aqueous-phase partitioning is very important to this type of pollutant transport, as shown in Fig. 4.10. The transport of the lipophilic compound through the vadose zone assumes that the compound has extremely high K_{ow} values and extremely low water solubility. That is, they have a greater affinity for lipids than for water. As the aqueous solubility of a substance increases, its flow will increasingly follow the water flow lines.

When a dense, miscible fluid seeps into the zone of saturation, the dense contaminants move downward. When these contaminants reach the bottom of the aquifer, the shape dictates their continued movement and slope of the underlying bedrock or other relatively impervious layer, which will likely be in a direction other than the flow of the groundwater

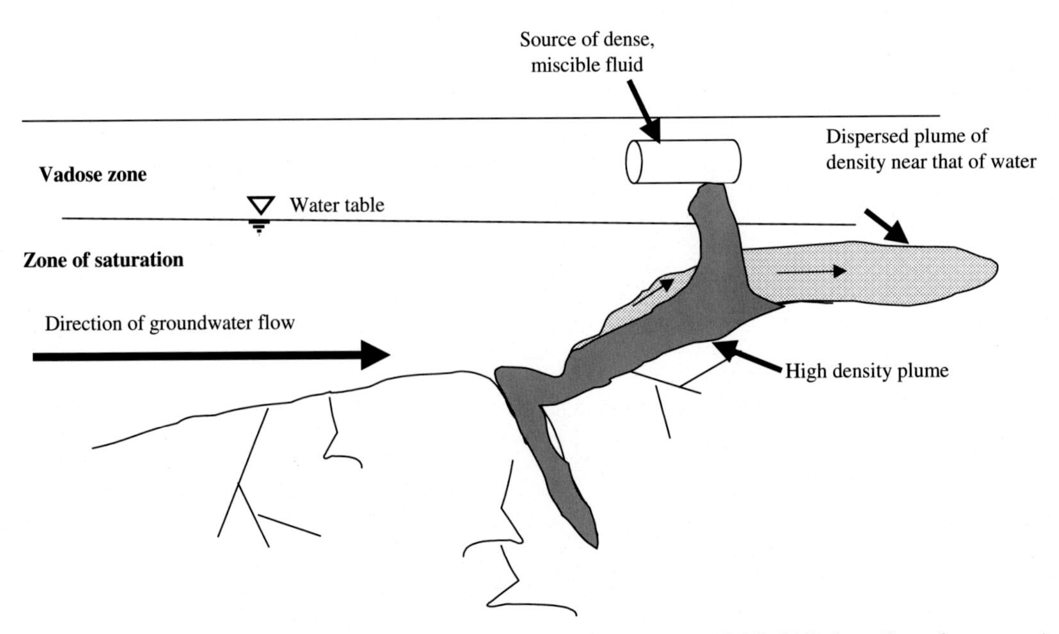

FIG. 4.10 Hypothetical plume of dense, highly hydrophilic fluid. *(Based on information provided by M.N. Sara, Ground-water monitoring system design, in: Practical Handbook of Ground-Water Monitoring, DM Nielsen Lewis Publishers, USA, 1991, pp. 17–68.)*

in the aquifer. Solution and dispersion near the boundaries of the plume will have a secondary plume that will generally follow the general direction of groundwater flow. The physics of this system points how that deciding where the plume is heading will entail more than the fluid densities, including solubility and phase partitioning. So, monitoring wells will need to be installed upstream and downstream from the source.

Another possibility is that the water may contain an amphiphilic compound (e.g., an alcohol) or a surfactant, which dissolves both hydrophilic and lipophilic compounds. As such, even a highly lipophilic compound may be dissolved by this solvent or surfactant that is carried with the water and ultimately is volatilized into the atmosphere.

If a source consists entirely of a light, hydrophilic fluid, the plume may be characterized as shown in Fig. 4.11. Low-density organic fluids, however, often are highly volatile, that is, their vapor pressures are sufficiently high to change phases from liquid to gas. So, let us now consider another physicochemical property of environmental fluids, vapor pressure, which must be considered along with density and solubility.

4.6.1 Co-solvation

An important process in plume migration is that of cosolvation, the process where a substance is first dissolved in one solvent and then the new solution is mixed with another solvent. As mentioned, with increasing aqueous solubility, a pollutant will travel along the flow lines of the ground or surface water. However, even a substance with low aqueous solubility can follow the flow under certain conditions. Even a hydrophobic compound like a chlorinated benzene (called a dense nonaqueous-phase liquid, DNAPL), which has very low solubility in pure water, can migrate into and within water bodies if it is first dissolved in an alcohol or an organic solvent (e.g., toluene). So, a DNAPL will migrate downward because its density is less than that of water and is transported in the DNAPL that has undergone cosolvation with the water. Likewise, the ordinarily lipophilic compound can be transported in the vadose zone or upper part of the zone of saturation where it undergoes cosolvation with water and a light nonaqueous-phase liquid (LNAPL), for example, toluene.

Contaminants are often mixtures. The constituents in these mixtures will partition within an environmental compartment at different rates. For example, the dense substance that moves toward the bottom of the aquifer in Fig. 4.11 with time may partition between lighter and denser constituents. The lighter constituents will be more likely to migrate upward and could become air pollutants via volatilization than the denser constituents of the mixture, which will likely remain below the surface. However, this should be considered along with other inherent properties and environmental conditions. For example, even the denser constituents may migrate if they become sorbed to particles that migrate. In addition, some of the constituents may differentially migrate because of cosolvation. Others may resist dissolution and continue to migrate downwardly. The bottom line is that K_{ow} must be applied to each constituent and that partitioning is highly specific to environmental conditions.

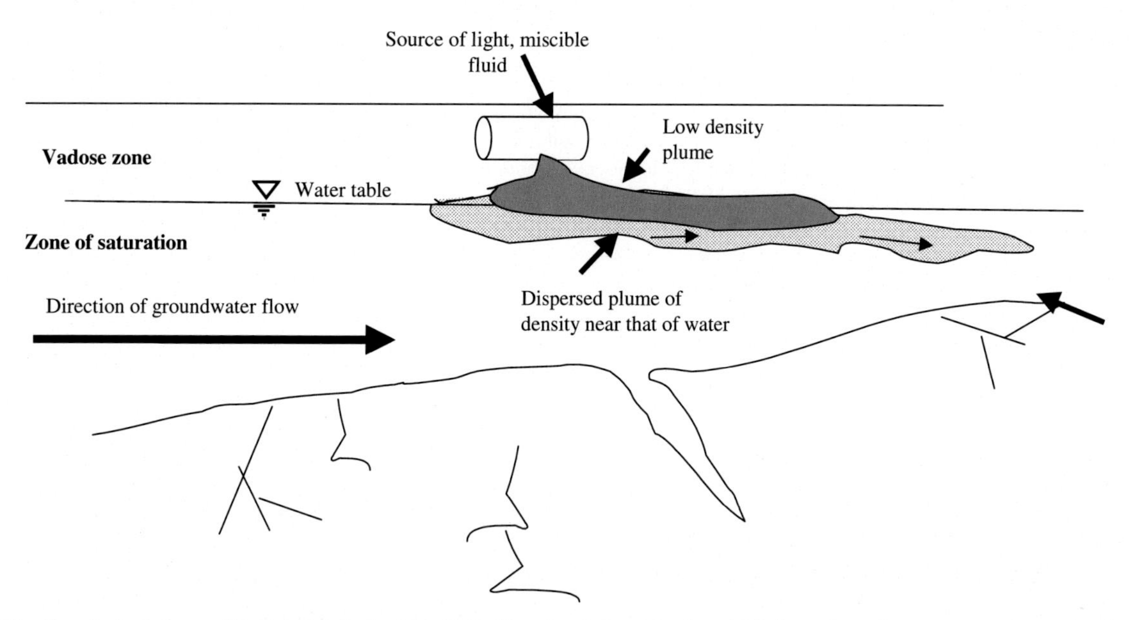

FIG. 4.11 Hypothetical plume of light, highly hydrophilic fluid. *(Based on information from M.N. Sara, Ground-water monitoring system design, in: Practical Handbook of Ground-Water Monitoring, DM Nielsen Lewis Publishers, USA, 1991, pp. 17–68.)*

4.7 PARTITIONING BETWEEN AIR AND TISSUE

The principal interest in partitioning between an organic substrate and water, that is, the *Kow*, was mainly the need of water pollution. Air pollution also needs a means of characterizing this partitioning for organic air pollutants in terms of sources likely to release the organic air pollutants. However, organic solvent and water partitioning is not completely useful for understanding the movement of substances among organic substrates and the air, especially within the lungs. Although both systems derive O_2 that is dissolved in a substrate (water or air), K_{ow} is very useful for gills but much less useful for the lungs. Thus, the octanol-air partitioning coefficient (K_{oa}) was established.

Models have been developed to combine K_{ow} and Henry's law constant (K_H) to estimate the equilibrium between air and organic tissues. Henry's law constant and the K_{OA} are physical-chemical properties that are often not measured but are needed to understand the environmental fate and transport of chemicals. When measured values of the movement from the water to air are not available or to estimate potential long-range transport or terrestrial bioaccumulation is of air pollutants, both K_H and K_{oa} can be estimated.

As mentioned, Henry's law constant is the air-water partition coefficient, which is the ratio of vapor pressure to solubility in water. It is commonly reported with units of atm-m^3/mole, or if it is a unitless value, it is reported as the air-water partitioning coefficient (K_{AW}). The conventional Henry's law constant is calculated as

$$K_H = \frac{P^{\circ}\,\mathrm{x\,MW}}{760 \times C_{sol}} \tag{4.41}$$

where P° = vapor pressure (torr), MW = molecular weight (g mol^{-1}), 760 = conversion factor (1 atm = 760 torr), and C_{sol} = water solubility (mg L^{-1}). The resulting units for K_H are atm-m^3/mole. The units of mg and liter (L) are canceled since there are 1000 mg in 1 g and 0.001 m^3 in 1 L.

The unitless Henry's law constant *(K_{AW})* is calculated as

$$K_{AW} = \frac{C_{air}}{C_{water}} = \frac{K_H}{R \cdot T} \tag{4.42}$$

where R is the ideal gas constant of 8.205746 × 10^{-5} atm-m^3/K-mole, T is absolute temperature expressed in Kelvin (K), and K_H is expressed in atm-m^3/mole. Use parameters measured or calculated for the same temperature (e.g., 25°C).

Earlier, we considered three compounds that are leaking into the groundwater from an underground tank. Recall that compound A has a vapor pressure of 12 kPa and density of 1.6 g mL^{-1}. Compound B has a vapor pressure of 59 kPa and density of 0.7 g mL^{-1}. Compound C has a vapor pressure of 0.003 kPa and density of 1.1 g mL^{-1}. These values are all at 20°C. Based on these data alone, we found B to be the highest and C to be the lowest in the amount emitted from soil to the air, but since A and C are each heavier than water, they could settle toward the bottom of the aquifer. What other physicochemical properties could change this order?

Aqueous solubility and sorption can certainly change the order of affinity for environmental compartments. If the compounds completely dissolve in water, then even denser compounds may not settle and may be transported with the water in a solution (e.g., a droplet containing the solvent). At 20°C, the aqueous solubility of CCl_4 is 80 g kg^{-1}. The aqueous solubility of $(C_2H_5)_2O$ is 75000 g kg^{-1}. $(HOCH_2CH_2)_2O$ is completely miscible. So, even though the diethylene glycol is denser than water, it will be dissolved in the water. To a lesser extent, a portion of the ether will also likely be dissolved, given its relatively high water solubility. The tetrachloromethane will not likely be dissolved, so a fraction could settle and not be released, even given its high vapor pressure.

Sorption is also a complicating factor, since the clay, organic matter, and other substances between the troposphere and the aquifer could hold various compounds. Sorption varies considerably depending on substrate and other conditions. The K_{oc} of CCl_4 has been found to be about 71 mL g^{-1}. $(C_2H_5)_2O$ is about 14. Diethylene glycol is about 2 (see http://www.epa.gov/opptintr/rsei/pubs/tech_app_b.pdf). Thus, they all are within the same order of magnitude for this particular type of sorption, that is, to organic matter, so this may not be a comparatively limiting step for these three compounds. However, for many semivolatile compounds, sorption can be very important in terms of half-life and transport (e.g., increasing K_{oc} may mean more affinity for PM, and transport with these particles becomes a major route). Higher K_{oc} may also retard the movement through the vadose zone, thus decreasing the amount moving from soil to air.

4.7.1 Calculating the octanol-air partition coefficient

For air pollutants, partitioning between organic compounds and water, that is, *Kow*, is often insufficient. Air is another fluid from which compounds partition. Thus, the potential air pollutant transport and bioaccumulation call for another partitioning between air and the organic phase, analogous to the partitioning between the water and organic phase. This is the octanol-air coefficient (K_{OA}) for a chemical compound. The K_{OA} is the ratio of the concentration of a compound in an organic phase (n-octanol) and air at equilibrium $\frac{C_{air}}{C_{water}} = \frac{C_{octanol}}{C_{air}}$, where $C_{octanol}$ and C_{air} represent equilibrium concentrations in air and n-octanol with the same units. The K_{OA} has been used to describe partitioning between air and aerosol particles, air and foliage, and air and soil [21–23]. Handbook values are available, but measured K_{OA} values may be substantially different than calculated values, so measured values are preferred. In the absence of data, K_{OA} values calculated using a quantitative structural activity relationship (QSAR) calculator, such as EPI Suite and the following equations [24]:

$$K_{OA} = \frac{K_{ow} \cdot R \cdot T}{K_H} \tag{4.43}$$

where K_{ow} is the n-octanol-water partition coefficient (unitless), R is the ideal gas constant $= 8.205746 \times 10^{-5}$ atm-m^3/K-mole, T is absolute temperature expressed in Kelvin (K), and K_H is Henry's law constant expressed in atm-m^3/mole.

The ideal gas law assumptions can simplify the equation to a relationship between water and air partitioning, that is, K_{ow} and K_{OA}:

$$K_{OA} = \frac{K_{ow}}{K_{AW}} \tag{4.44}$$

K_{AW} is the air-water partition coefficient (unitless).

When a measured K_{OA} is not available and long-range transport or terrestrial bioaccumulation may be a concern, the K_{OA} can be calculated by using Eq. (4.21) when measured K_{ow} and vapor pressure or Henry's law constant is available. If these are not available, a QSAR approach is recommended (e.g., EPI Suite). The published values for these key partitioning coefficients are given in Table 4.7.

Note that partitioning coefficients in Table 4.7 are reported as log values. This is because the ranges are so large that log-log relationships are commonly used, including for comparing chemicals' affinities. Thus, Eq. (4.44) can be restated as

$$\log K_{\mathrm{OA}} = \log K_{ow} - \log K_{\mathrm{AW}} \tag{4.45}$$

The values in Table 4.7 demonstrate the variability of estimates depending on the method used. For example, the published log K_{OA} for methyl isocyanate (MIC) is 2.112, so taking its antilog gives a $K_{OA} = 129$. However, calculating K_{OA} from Eq. (4.22) gives an estimated value:

$$\log K_{OA} = \log K_{ow} - \log K_{AW} = 0.79 - (-1.4222) \approx 1.5 \tag{4.46}$$

The calculated value is 0.7 less than the measured or published value. The difference means that the calculated K_{OA} is underreported by about one-third. Thus, if the calculated value is used, the affinity for tissue is understated. That is, the MIC is less likely to exit from tissue into air than the calculated value indicates.

The published log K_{OA} for formaldehyde is 2.77. Calculating an estimate for K_{OA} from Eq. (4.22) gives

$$\log K_{OA} = \log K_{ow} - \log K_{AW} = 0.35 - (-4.861) \approx 4.81$$

This difference is a large discrepancy, that is, two orders of magnitude between the two methods. As for MIC, the calculated method underestimates formaldehyde's affinity for tissue and overestimates its likelihood to move into the air.

These differences may be explained somewhat by the fact that the calculation is a rule of thumb. Each partitioning coefficient is measured differently. For example, K_{ow} is actually not usually calculated as partitioning between pure octanol and pure water; rather, the measurement reflects octanol saturated with water and water saturated with octanol [1, 21]. Whenever the inherent properties like solubility and vapor pressure are incorrect, derived properties like Kow, K_{AW}, and K_{OA} will propagate and amplify such error.

These differences also appear in the organic carbon partitioning coefficients (K_{oc}), that is, the two right columns in Table 4.7, with large discrepancies between QSAR and K_{ow}-based K_{oc} values. This is a reminder that octanol is merely a surrogate or indicator of the organic phase. This problem emphasizes that an actual, published K_{OA} based on credible methods is preferred whenever possible.

4.7.2 Dynamics within an organism

To this point, we have discussed the kinetics and equilibria of the processes leading to the release, behavior in environmental compartments, and movement among compartments. These same physical and chemical processes take place within

TABLE 4.7 Key air, water, and organic tissue partitioning coefficients for hazardous air pollutants calculated from quantitative structural activity relationships (QSAR) and from published estimates

Chemical name	Log K_{ow} from QSAR	Log K_{oa} from QSAR	Log K_{aw}	Log K_{oc} from QSAR	Log K_{oc} estimated from K_{ow}
Acetaldehyde	−0.17	2.387	−2.564	−0.2676	0.5078
Acetamide	−1.16	5.179	−6.339	0.475	0.2244
Acetonitrile	−0.15	2.753	−2.851	0.6693	1.1292
Acetophenone	1.67	5.067	−3.371	1.7147	1.9946
2-Acetylaminofluorene	3.12	11.225	−8.105	3.3436	2.6255
Acrolein	0.19	3.025	−2.302	−0.007	0.6903
Acrylamide	−0.81	5.808	−7.158	0.755	0.5507
Acrylic acid	0.44	5.368	−4.82	0.1583	0.3493
Acrylonitrile	0.21	2.459	−2.249	0.93	1.4556
Allyl chloride	1.93	1.786	−0.347	1.5977	1.6746
4-Aminobiphenyl	2.84	8.064	−5.224	3.3928	2.4854
Aniline	1.08	5.19	−4.083	1.8465	1.4013
o-Anisidine	1.16	6.495	−4.422	1.662	1.6121
Asbestos					
Benzene (including benzene from gasoline)	1.99	2.647	−0.644	2.1637	1.8482
Benzidine	1.92	10.595	−8.675	3.0757	1.6231
Benzotrichloride	3.9	5.873	−1.973	3.0005	3.3844
Benzyl chloride	2.79	3.858	−1.774	2.6495	1.9958
Biphenyl	3.76	5.531	−1.9	3.71	3.4799
Bis(2-ethylhexyl)phthalate (DEHP)	8.39	11.707	−4.957	5.0776	4.9977
Bis(chloromethyl)ether	0.57	2.929	−0.749	0.9867	1.1498
Bromoform	1.79	4.125	−1.66	1.5027	2.0826
1,3-Butadiene	2.03	1.527	0.478	1.5977	1.7267
Calcium cyanamide	−0.81	2.057	−7.975	0.6693	0.8637
Caprolactam (see modification)	0.66	6.645	−5.985	1.3892	1.2646
Captan	2.74	9.467	−6.543	2.4018	2.4663
Carbaryl	2.35	9.242	−6.874	2.55	2.1344
Carbon disulfide	1.94	1.848	−0.23	1.337	1.6833
Carbon tetrachloride	2.44	2.424	0.052	1.6424	2.4558
Carbonyl sulfide	−1.33	−1.634	1.397	−0.2676	−0.0398
Catechol	1.03	9.653	−7.309	2.3899	1.7455
Chloramben	1.9	10.966	−8.801	1.3299	1.1851
Chlordane	6.26	8.802	−2.702	4.8296	5.3979
Chlorine	0.85	None	None	1.1211	0.7373
Chloroacetic acid	0.34	5.443	−6.422	0.1583	0.2774

Continued

TABLE 4.7 Key air, water, and organic tissue partitioning coefficients for hazardous air pollutants calculated from quantitative structural activity relationships (QSAR) and from published estimates—cont'd

Chemical name	Log K_{ow} from QSAR	Log K_{oa} from QSAR	Log K_{aw}	Log K_{oc} from QSAR	Log K_{oc} estimated from K_{ow}
2-Chloroacetophenone	1.93	5.779	−3.849	1.9952	2.1882
Chlorobenzene	2.64	3.427	−0.896	2.369	2.4644
Chlorobenzilate	3.99	9.265	−5.529	3.1872	3.0699
Chloroform	1.52	2.401	−0.824	1.5027	1.7094
Chloromethyl methyl ether	0.32	2.226	−1.906	0.7261	1.0115
Chloroprene	2.53	2.169	0.361	1.7832	2.1954
Cresols/cresylic acid (isomers and mixture)	2.06	6.657	−4.388	2.4777	2.165
o-Cresol	2.06	6.657	−4.309	2.4865	2.1706
m-Cresol	2.06	6.657	−4.456	2.4777	2.1761
p-Cresol	2.06	6.657	−4.388	2.4777	2.165
Cumene	3.45	3.817	−0.328	2.8438	3.1761
2,4-D, salts, and esters	2.62	9.044	−5.839	1.4717	1.7659
DDE	6	8.842	−2.769	5.0701	5.6496
Diazomethane	2	1.861	0.139	1.1211	1.7354
Dibenzofurans	4.05	6.825	−2.775	3.9619	3.5146
1,2-Dibromo-3-chloropropane	2.68	4.285	−2.221	2.0636	2.5686
Dibutyl phthalate	4.61	8.912	−4.131	3.0635	3.283
1,4-Dichlorobenzene(p)	3.28	4.197	−1.006	2.5743	2.9852
3,3-Dichlorobenzidine	3.21	12.145	−8.935	3.5038	2.8234
Dichloroethyl ether (bis(2-chloroethyl)ether)	1.56	3.672	−3.158	1.508	1.548
1,3-Dichloropropene	2.29	2.289	−0.838	1.8583	1.7614
Dichlorvos	0.6	5.055	−4.63	1.7321	1.8194
Diethanolamine	−1.71	7.085	−8.801	−0.6817	−0.7323
N,N-Diethyl aniline (N,N-dimethylaniline)	2.17	4.625	−2.634	1.8958	2.1377
Diethyl sulfate	1.14	4.867	−3.6	1.4502	1.7171
3,3-Dimethoxybenzidine	2.08	13.211	−11.131	2.7066	1.9949
Dimethyl aminoazobenzene	4.29	9.309	−5.019	3.3071	3.8243
3,3′-Dimethyl benzidine	3.02	11.61	−8.59	3.5038	2.1762
Dimethyl carbamoyl chloride	−0.72	3.809	−4.529	−0.0527	0.254
Dimethylformamide	−0.93	4.59	−5.52	−0.247	0.0936
1,1-Dimethyl hydrazine	−1.19	4.356	−5.546	1.0773	0.2233
Dimethyl phthalate	1.66	6.698	−5.094	1.4996	1.6789
Dimethyl sulfate	0.16	4.133	−3.786	0.9289	1.175

TABLE 4.7 Key air, water, and organic tissue partitioning coefficients for hazardous air pollutants calculated from quantitative structural activity relationships (QSAR) and from published estimates—cont'd

Chemical name	Log K_{ow} from QSAR	Log K_{oa} from QSAR	Log K_{aw}	Log K_{oc} from QSAR	Log K_{oc} estimated from K_{ow}
4,6-Dinitro-o-cresol and salts	2.27	8.174	−4.242	2.8776	2.7083
2,4-Dinitrophenol	1.73	7.678	−5.454	2.6635	2.4538
2,4-Dinitrotoluene	2.18	7.602	−5.656	2.7601	2.4585
1,4-Dioxane (1,4-diethyleneoxide)	−0.32	3.297	−3.707	0.4205	0.5946
1,2-Diphenylhydrazine	3.06	9.806	−4.709	3.1775	2.5081
Epichlorohydrin (l-chloro-2,3-epoxypropane)	0.63	3.269	−2.906	0.9959	1.0834
1,2-Epoxybutane	0.86	2.922	−2.133	0.9959	1.3102
Ethyl acrylate	1.22	3.519	−1.858	1.0273	1.5896
Ethylbenzene	3.03	3.521	−0.492	2.6495	2.7335
Ethyl carbamate (urethane)	−0.02	5.648	−5.58	1.0837	0.7596
Ethyl chloride (chloroethane)	1.58	1.931	−0.343	1.337	1.2407
Ethylene dibromide (dibromoethane)	2.01	3.285	−1.576	1.5977	1.7007
Ethylene dichloride (1,2-dichloroethane)	1.83	2.136	−1.317	1.5977	1.2841
Ethylene glycol	−1.2	4.071	−5.61	−1.0381	−0.65
Ethylene imine (aziridine)	−0.28	3.243	−3.306	0.9563	0.7266
Ethylene oxide	−0.05	2.259	−2.218	0.5102	0.6686
Ethylene thiourea	−0.49	4.372	−4.862	1.1128	0.8169
Ethylidene dichloride (1,1-dichloroethane)	1.76	2.066	−0.639	1.5027	1.5531
Formaldehyde	0.35	2.77	−4.861	−0.4836	0.8894
Heptachlor	5.86	8.003	−1.92	4.6155	4.747
Hexachlorobenzene	5.86	7.298	−1.158	3.792	4.2389
Hexachlorobutadiene	4.72	5.075	−0.376	2.9269	4.1482
Hexachlorocyclopentadiene	4.63	5.694	0.043	3.1473	4.3738
Hexachloroethane	4.03	4.797	−0.799	2.294	3.5927
Hexamethylene-1,6-diisocyanate	3.2	5.907	−2.707	3.6829	2.7769
Hexamethylphosphoramide	−0.22	9.315	−6.087	0.3474	1.0526
Hexane	3.29	1.445	1.867	2.119	3.3844
Hydrazine	−1.47	4.76	−6.23	1.1211	−1.797
Hydrochloric acid	0.54	1.04	−0.5	1.1211	0.4683
Hydrogen fluoride (hydrofluoric acid)	0.23	1.181	−0.951	1.1211	0.1992
Hydrogen sulfide (see modification)	0.23	0.679	−0.449	1.1211	−1.1981
Hydroquinone	1.03	9.653	−8.714	2.3811	1.5851

Continued

TABLE 4.7 Key air, water, and organic tissue partitioning coefficients for hazardous air pollutants calculated from quantitative structural activity relationships (QSAR) and from published estimates—cont'd

Chemical name	Log K_{ow} from QSAR	Log K_{oa} from QSAR	Log K_{aw}	Log K_{oc} from QSAR	Log K_{oc} estimated from K_{ow}
Isophorone	2.62	5.188	−3.566	1.8139	2.061
Lindane (all isomers)	4.26	None	None	3.4482	3.5927
Maleic anhydride	1.62	5.414	−3.794	−0.8957	1.3626
Methanol	−0.63	3.128	−3.73	−0.1968	0.0877
Methoxychlor	5.67	11.069	−5.081	4.4296	3.8468
Methyl bromide (bromomethane)	1.18	1.64	−0.523	1.1211	1.0324
Methyl chloride (chloromethane)	1.09	1.565	−0.443	1.1211	0.7894
Methyl chloroform (1,1,1-trichloroethane)	2.68	3.438	−0.153	1.6424	2.1607
Methyl ethyl ketone (2-butanone) (see modification)	0.26	2.83	−2.633	0.6542	1.2811
Methyl hydrazine	−1	4.889	−5.889	1.1243	0.3225
Methyl iodide (iodomethane)	1.59	2.238	−0.667	1.1211	1.3101
Methyl isobutyl ketone (hexone)	1.16	3.484	−2.249	1.1003	1.8453
Methyl isocyanate	0.79	2.212	−1.422	1.5977	0.6852
Methyl methacrylate	1.28	3.507	−1.885	0.961	1.6228
Methyl tert-butyl ether	1.43	2.513	−1.62	1.0631	1.3544
4,4-Methylene bis(2-chloroaniline)	3.47	12.341	−8.871	3.7557	3.0446
Methylene chloride (dichloromethane)	1.34	1.768	−0.877	1.337	1.0845
Methylene diphenyl diisocyanate (MDI)	5.22	9.657	−4.437	5.4548	4.53
4,4′-Methylenedianiline	2.18	10.791	−8.611	3.3275	1.7614
Naphthalene	3.17	4.837	−1.745	3.1887	2.8637
Nitrobenzene	1.81	4.87	−3.008	2.3549	2.1675
4-Nitrobiphenyl	3.57	7.746	−4.176	3.9012	3.2571
4-Nitrophenol	1.91	8.954	−7.77	2.4636	2.3675
2-Nitropropane	0.87	3.388	−2.313	1.4886	1.6586
N-Nitroso-N-methylurea	−0.52	7.872	−8.392	1.0414	0.8765
N-Nitrosodimethylamine	−0.64	3.435	−4.128	1.3577	0.5662
N-Nitrosomorpholine	−0.43	5.809	−5.999	1.3524	0.5475
Parathion	3.73	8.647	−4.914	3.3842	3.1709
Pentachloronitrobenzene (Quintobenzene)	5.03	8.742	−2.743	3.7779	3.8551
Pentachlorophenol	4.74	10.032	−5.999	3.6954	4.0684
Phenol	1.51	6.149	−4.866	2.2724	1.8995
p-Phenylenediamine	−0.39	7.17	−7.56	1.5293	0.7159
Phosgene	−0.71	−0.272	−0.166	−0.102	0.3031

TABLE 4.7 Key air, water, and organic tissue partitioning coefficients for hazardous air pollutants calculated from quantitative structural activity relationships (QSAR) and from published estimates—cont'd

Chemical name	Log K_{ow} from QSAR	Log K_{oa} from QSAR	Log K_{aw}	Log K_{oc} from QSAR	Log K_{oc} estimated from K_{ow}
Phosphine	−0.27	−0.269	−0.001	1.1211	−0.2347
Phosphorus	−0.27	−0.269	−0.001	1.1211	−0.2347
Phthalic anhydride	2.07	5.656	−6.176	0.1469	1.3515
Polychlorinated biphenyls (Aroclors)	7.62	10.173	−2.553	5.3208	5.1682
1,3-Propane sultone	−0.28	3.736	−4.016	0.9737	−0.9317
beta-Propiolactone	−0.8	1.7	−2.5	0.5508	0.417
Propionaldehyde	0.33	2.764	−2.523	−0.007	1.0222
Propoxur (Baygon)	1.9	8.763	−7.233	1.7778	1.7256
Propylene dichloride (1,2-dichloropropane)	2.25	2.432	−0.938	1.7832	1.718
Propylene oxide	0.37	2.554	−2.546	0.7155	0.8511
1,2-Propylenimine (2-methyl aziridine)	0.13	3.53	−3.388	1.1616	0.9534
Quinoline	2.14	6.691	−4.166	3.1887	1.7614
Quinone	0.25	7.552	−1.708	1.6148	1.9355
Styrene	2.89	3.838	−0.949	2.6495	2.5599
Styrene oxide	1.59	4.992	−3.19	2.0565	1.725
2,3,7,8-Tetrachlorodibenzo-p-dioxin	6.92	10.761	−2.689	5.3964	4.8308
1,1,2,2–Tetrachloroethane	2.19	3.399	−1.824	1.9775	2.0739
Tetrachloroethylene (perchloroethylene)	2.97	3.141	−0.14	1.9775	2.9505
Titanium tetrachloride	1.47	None	None	1.6424	1.2754
Toluene	2.54	3.154	−0.566	2.369	2.369
2,4-Toluene diamine	0.16	7.677	−7.517	1.7434	0.9593
2,4-Toluene diisocyanate	3.74	7.083	−3.343	3.8706	3.2455
o-Toluidine	1.62	5.686	−4.092	2.0606	1.6336
Toxaphene (chlorinated camphene)	6.75	9.556	−2.806	4.8178	5.8579
1,2,4-Trichlorobenzene	3.93	4.978	−1.236	3.1322	3.2931
1,1,2-Trichloroethane	2.01	2.768	−1.473	1.7832	1.6399
Trichloroethylene	2.47	2.497	−0.395	1.7832	2.0999
2,4,5-Trichlorophenol	3.45	8.481	−4.179	3.2497	3.294
2,4,6-Trichlorophenol	3.45	8.481	−3.973	3.2497	3.2774
Triethylamine	1.51	3.967	−2.215	1.7059	1.6618
Trifluralin	5.31	7.372	−2.376	4.2147	4.2518
2,2,4-Trimethylpentane	4.09	2	2.094	2.3808	3.5493

Continued

TABLE 4.7 Key air, water, and organic tissue partitioning coefficients for hazardous air pollutants calculated from quantitative structural activity relationships (QSAR) and from published estimates—cont'd

Chemical name	Log K_{ow} from QSAR	Log K_{oa} from QSAR	Log K_{aw}	Log K_{oc} from QSAR	Log K_{oc} estimated from K_{ow}
Vinyl acetate	0.73	2.05	−1.68	0.7469	1.2633
Vinyl bromide	1.52	1.819	−0.299	1.337	1.3622
Vinyl chloride	1.62	1.361	0.056	1.337	1.4056
Vinylidene chloride (1,1-dichloroethylene)	2.12	2.005	0.028	1.5027	1.8482
Xylenes (isomers and mixture)	3.09	3.662	−0.532	2.5743	2.7769
o-Xylenes	3.09	3.662	−0.674	2.5831	3.7422
m-Xylenes	3.09	3.662	−0.532	2.5743	2.7769
p-Xylenes	3.09	3.662	−0.55	2.5743	2.7335

an organism after uptake. The importance of the factors can be very different within the organism. For example, the degradation or activation of a pollutant may be much faster in the presence of organic catalysts, that is, enzymes.

The reaction rates and equilibria differ, as do residence times and half-lives of chemical processes. However, the laws of thermodynamic and motion apply at every level of interest to air pollution, from cells to the planet.

4.7.3 Microbial partitioning

Microbes in soil, water, sediment, and air, including fungi, bacteria, and viruses, use an array of organic compounds as carbon and energy sources. Indeed, engineers take advantage of this to treat wastes that contain organic pollutants. Before the metabolism can happen, however, the contaminants and O_2 must first move from the vapor phase to the aqueous phase where they can be metabolized by the microorganisms. Thus, even though these are volatile compounds, they are commonly treated exclusively in the aqueous phase. The volumetric mass transfer rate (mol m^{-3} s^{-1}) of gaseous substrates (e.g., compounds to be treated, oxygen, and nutrients) to the aqueous phase is

$$K_{1a_{G/A}}\left(\frac{S_G}{K_{G/A}} - S_A\right) \tag{4.47}$$

where $K_{1a_{G/A}}$ is the global volumetric mass transfer coefficient (hr^{-1}); S_G and S_A are the substrate (e.g., benzene) concentrations ($mol\,m^{-3}$) in the bulk gas and aqueous phases, respectively; and $K_{G/A}$ is the substrate partition coefficient (dimensionless) between the gaseous and aqueous phases. $K_{G/A}$ is calculated as follows:

$$K_{G/A} = \frac{S_G}{S_A^*} \tag{4.48}$$

where S_A^* is the substrate concentration at the gas/aqueous interface ($mol\,m^{-3}$).

In addition to the vapor phase and aqueous phase, there is also a nonaqueous (e.g., lipids) phase in the substrate (see Fig. 4.12). Reactions can only occur when reagents come into contact, which in air pollution often requires that an agent move at least two phases (Fig. 4.12B). This translates into a film profile, with sectors based on the octanol-water coefficient (K_{ow}) and $K_{G/A}$. This partitioning can also occur between organisms and substrate, for example, microbes, as depicted in Fig. 4.13.

The inherent properties of air pollutants do not fully explain a substance's affinity for an environmental compartment. As mentioned, the vapor pressure and molecular weight, for example, will give clues to the behavior of an air pollutant, but are not definitive. The physicochemical properties must be combined with environmental conditions to characterize and estimate the fate of the pollutant. After entering the air, water, soil, and biota, a chemical compound will either remain in a compartment or move to another. Indeed, this is seldom an exclusive residence, as some of the chemical remain and the rest will move to one or more other compartments. The propensity for this distribution is known as partitioning.

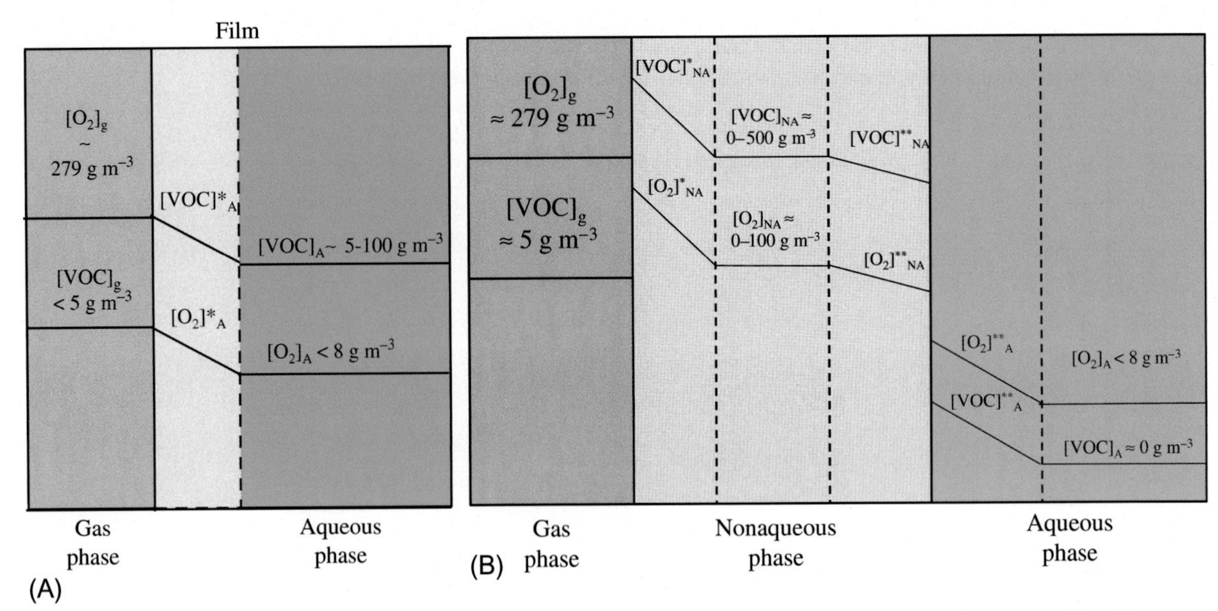

FIG. 4.12 Concentration profiles of lipophilic substrates (volatile organic compounds—VOCs—and O_2) in a single-phase system (A) and in a two-phase partitioning bioreactor (B). [VOC] and $[O_2]$ = concentrations of volatile organic compounds and O2, respectively, in the treated gas phase ([ ]g), aqueous phase ([ ]A), nonaqueous phase ([ ]NA); [ ]* and [ ]** represent the equilibrium concentrations at the gaseous/nonaqueous and nonaqueous/aqueous interfaces, respectively. Note: All concentrations are based on air contaminated with 5 g VOC m^{-3}. *(Data from R. Muñoz, S. Villaverde, B. Guieysse, S. Revah, Two-phase partitioning bioreactors for treatment of volatile organic compounds, Biotechnol. Adv. 25 (4) (2007) 410–422.)*

Phase partitioning is also sometimes called "phase distribution." Partitioning certainly is influenced by inherency. If a compound has high aqueous solubility, that is, it is easily dissolved in water under normal environmental conditions of temperature and pressure, it is hydrophilic. That is, it is very likely to remain dissolved in water. If, conversely, a substance is not easily dissolved in water under these conditions, it is said to be hydrophobic, and many of its molecules will exit the water and enter compartments with substances of which the chemical has more affinity, for example, hydrophobic compounds in sediment or tissues in organisms. Thus, aqueous solubility is a factor of partitioning yet does not fully predict the amount of a chemical that will be in each compartment. Such prediction is afforded by partitioning coefficients.

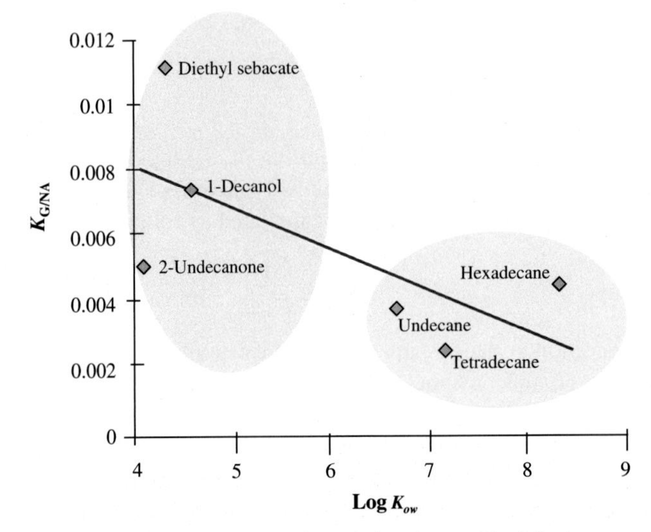

FIG. 4.13 Gaseous/nonaqueous hexane partition coefficient ($K_{G/NA}$) in organic solvents. The left cluster represents solvents toxic to the fungus, *Fusarium solani*. The right cluster shows biocompatible solvents that were biodegraded by *F. solani*. Note: Silicone oil ($K_{G/NA}$ = 0.0034; unknown log K_{ow}) was the only nonaqueous-phase substance tested showing both biocompatible and nonbiodegradable characteristics. *(Sources: R. Muñoz, S. Villaverde, B. Guieysse, S. Revah, Two-phase partitioning bioreactors for treatment of volatile organic compounds, Biotechnol. Adv. 25 (4) (2007) 410–422; S. Arriaga, R. Muñoz, S. Hernández, B. Guieysse, S. Revah, Gaseous hexane biodegradation by* Fusarium solani *in two liquid phase packed-bed and stirred-tank bioreactors. Environ. Sci. Technol. 40 (7) (2006) 2390–2395.)*

4.8 CONCLUSIONS

Partitioning among the environmental compartments begins the pollutant transport process. This chapter has applied the laws of thermodynamics, including fugacity, to the transport of air pollutants. The next chapter adds the physical transport via the laws of motion. Together, these explain how a pollutant moves in time and space.

REFERENCES

[1] D. Mackay, W.-Y. Shiu, K.-C. Ma, S.C. Lee, Handbook of Physical-Chemical Properties and Environmental Fate for Organic Chemicals, CRC Press, 2006.

[2] S. Hilal, S. Karickhoff, L. Carreira, Prediction of chemical reactivity parameters and physical properties of organic compounds from molecular structure using SPARC, National Exposure Research Laboratory, Office of Research and Development, US Environmental Protection Agency, Research Triangle Park, NC, 2003.

[3] R. Lewis, S. Gordon, Sampling for organic chemicals in air, Environmental Protection Agency, Research Triangle Park, NC, United States, 1994 Atmospheric Research and Exposure Assessment Lab.

[4] U.S. Environmental Protection Agency, Technical Overview of Volatile Organic Compounds, General Definition and Classification, 5 June. Available: https://www.epa.gov/indoor-air-quality-iaq/technical-overview-volatile-organic-compounds, 2018.

[5] J.-Y. Chin, S.A. Batterman, VOC composition of current motor vehicle fuels and vapors, and collinearity analyses for receptor modeling, Chemosphere 86 (9) (2012) 951–958.

[6] E. Jurado, R. Lohmann, S. Meijer, K.C. Jones, J. Dachs, Latitudinal and seasonal capacity of the surface oceans as a reservoir of polychlorinated biphenyls, Environ. Pollut. 128 (1) (2004) 149–162.

[7] O. Faroon, J.N. Olson, Toxicological Profile for Polychlorinated Biphenyls (PCBs), 2000.

[8] R. Gioia, et al., Polychlorinated biphenyls in air and water of the North Atlantic and Arctic Ocean, J. Geophys. Res. Atmos. 113 (D19) (2008).

[9] Y. Kwon, Handbook of Essential Pharmacokinetics, Pharmacodynamics and Drug Metabolism for Industrial Scientists, Springer Science & Business Media, 2001.

[10] A.D. McNaught, A.D. McNaught, Compendium of Chemical Terminology, Blackwell Science, Oxford, 1997.

[11] D.A. Vallero, Excerpted and Revised From: Fundamentals of Air Pollution, fifth ed., Elsevier Academic Press, Waltham, MA, 2014. p. 999 pages cm.

[12] D. Mackay, S. Paterson, Evaluating the multimedia fate of organic chemicals: a level III fugacity model, Environ. Sci. Technol. 25 (3) (1991) 427–436.

[13] H.F. Hemond, E.J. Fechner, Chemical Fate and Transport in the Environment, Elsevier, 2014.

[14] U.S. Environmental Protection Agency, Chemistry Dashboard, 21 April 2017. Available: https://www.epa.gov/chemical-research/chemistry-dashboard, 2017.

[15] European Chemicals Agency, Dichloromethane, 31 May 2018. Available: https://echa.europa.eu/registration-dossier/-/registered-dossier/15182/2/3, 2018.

[16] P. Davison, The specific gravity of mesopelagic fish from the northeastern Pacific Ocean and its implications for acoustic backscatter, ICES J. Mar. Sci. 68 (10) (2011) 2064–2074.

[17] D. Mackay, L. Burns, G. Rand, Fate modeling, in: Fundamentals of Aquatic Toxicology: Effects, Environmental Fate, and Risk Assessment, Taylor & Francis, Philadelphia, PA, 1995, pp. 563–585.

[18] W. Lyman, Transport and transformation processes, in: Rand GM Fundamentals of Aquatic Toxicology. Effects, Environmental Fate, and Risk Assessment, second ed., Taylor and Francis, Washington DC, 1995, pp. 449–492.

[19] D. Richter, D.A. Vallero (Eds.), Ion Exchange, Duke University, Durham, North Carolina, 1996.

[20] J. Westfall, Adsorption mechanisms in aquatic surface chemistry, in: W. Stumm (Ed.), Aquatic Surface Chemistry: Chemical Processes at the Particle-Water Interface, vol. 87, John Wiley & Sons, 1987.

[21] W.-Y. Shiu, D. Mackay, K.-C. Ma, S.C. Lee, Handbook of Physical-Chemical Properties and Environmental Fate for Organic Chemicals, CRC Press, 2006.

[22] D. Mackay, W. Shiu, K. Ma, Physical-Chemical Properties and Environmental Fate and Degradation Handbook, Chapman & Hall CRCnetBASE, CRC Press LLC, Boca Raton, FL, 1999.

[23] T. Harner, M. Shoeib, Measurements of octanol-air partition coefficients (K OA) for polybrominated diphenyl ethers (PBDEs): predicting partitioning in the environment, J. Chem. Eng. Data 47 (2) (2002) 228–232.

[24] U.S. Environmental Protection Agency, Guidance for reporting on the environmental fate and transport of the stressors of concern in problem formulations, Available: https://www.epa.gov/pesticide-science-and-assessing-pesticide-risks/guidance-reporting-environmental-fate-and-transport, 2017.

FURTHER READING

[25] WHO Guidelines for Indoor Air Quality: Selected Pollutants, World Health Organization, 2010.

[26] 0143-2060, Indoor air quality: organic pollutants, 1989.

[27] D.A. Vallero, Fundamentals of Air Pollution, fifth ed., Elsevier Academic Press, Waltham, MA, 2014. p. 999 pages cm.

[28] O. Ghosheh, E.M. Hawes, N-glucuronidation of nicotine and cotinine in human: formation of cotinine glucuronide in liver microsomes and lack of catalysis by 10 examined UDP-glucuronosyltransferases, Drug Metab. Dispos. 30 (9) (2002) 991–996.

[29] J.J. Hassett, W.L. Banwart, The sorption of nonpolar organics by soils and sediments, in: Reactions and Movement of Organic Chemicals in Soils, 1989, pp. 31–44. no. reactionsandmov.

[30] P.M. Gschwend, Environmental Organic Chemistry, John Wiley & Sons, 2016.

[31] EPA/540/R-96/018, Soil Screening Guidance: User's Guide, 1996.

[32] EPA/540/R-95/128 PB96-963502, Soil Screening Guidance: Technical Background Document, 1996.

[33] National Toxicology Program, NTP 11th Report on Carcinogens, in: Report on carcinogens: carcinogen profiles, vol. 11, 2004, p. 1.

[34] NTP Report on Carcinogens: Background Document for 2,3,7,8-Tetrachlorodibenzo-p-Dioxin (TCDD), Available: https://ntp.niehs.nih.gov/ntp/newhomeroc/other_background/tetrachl_tcdd_4apps_508.pdf, 1999.

[35] EPA 811-F-95-0031-T, National Primary Drinking Water Regulations: Dioxin (2,3,7,8-TCDD), 1995.

[36] M.N. Sara, Ground-water monitoring system design, in: D.M. Nielsen (Ed.), Practical Handbook of Ground-Water Monitoring, Lewis Publishers, USA, 1991, pp. 17–68.

[37] R. Muñoz, S. Villaverde, B. Guieysse, S. Revah, Two-phase partitioning bioreactors for treatment of volatile organic compounds, Biotechnol. Adv. 25 (4) (2007) 410–422.

[38] S. Arriaga, R. Muñoz, S. Hernández, B. Guieysse, S. Revah, Gaseous hexane biodegradation by *Fusarium solani* in two liquid phase packed-bed and stirred-tank bioreactors, Environ. Sci. Technol. 40 (7) (2006) 2390–2395.

Chapter 5

Air partitioning

5.1 INTRODUCTION

Transport begins with the likelihood of a chemical escaping from its first location, which is just after it is formed and to which it has been released. For air pollutants, this is usually an emission to the air. This first location may be the ambient air or a smaller compartment, such as a room in a home. However, air pollutants also are transformed from substances that are first released to other environmental media, for example, a compound that is discharged to water and is transformed into a more volatile compound, which is the air pollutant. Fig. 5.1 depicts both types of air pollutants.

There can be several steps before a parent compound, which may or may not be harmful, becomes an air pollutant. That is, a water, soil, or sediment pollutant can become an air pollutant, but an inherently harmless substance can also become an air pollutant if it reacts with other substances after reaching the environment. Spilled onto soil is transformed by microbes on the soil particle surfaces and in the soil pore water, becoming a compound with a higher vapor pressure than the spilled pesticide, so that the transformation product is subsequently released to the air.

The previous chapter addresses the affinities of chemical compounds for the various environmental compartments and, conversely, their propensity to flee a compartment, that is, environmental partitioning, which begins the transport process. This chapter applies the environmental partitioning principles specifically to the partitioning to and from the air.

5.2 HENRY'S LAW REVISITED

We introduced vapor pressure and Henry's law constants in our discussion of multicompartmental partitioning in Chapter 4. This set the stage for the movement of molecules into and out of the atmosphere. This occurs at all scales, from molecular and global. Henry's law constants are based on aqueous solubility and partial pressure, so they can be calculated for a chemical species from its saturation vapor pressure ($p°$) and its solubility (S°):

$$K_H = \frac{p^{\circ}_{\text{liquid}}}{S^{\circ}_{\text{liquid}}} \text{ and } \frac{p^{\circ}_{\text{solid}}}{S^{\circ}_{\text{solid}}} \tag{5.1}$$

Note that K_H is calculated for the same physical state of the chemical species, either liquid or solid. In theory, at constant partial pressure, gas solubility in liquids, including water, increases with decreasing temperature. Gas molecules are increasingly driven from the liquid with rising heat, due to kinetic molecular energy. As mentioned, Henry's law states that the partial pressure (p) of a solute (A) in equilibrium in a solution is directly proportional to the solute's mole fraction (x_A) in the solution, with K_H as the proportionality constant:

$$p[\text{A}] = K_H \cdot x_A \tag{5.2}$$

When a gas contacts a liquid, a small quantity of the gas will be dissolved in the liquid by diffusion. After time, the liquid-gas arrangement reaches equilibrium, and eventually, every gas contacting the liquid will reach a saturation concentration. This exchange follows thin-film theory, which is discussed later in this chapter.

Henry's law tells us that, at equilibrium, p for any gas in a mixture will be proportional to its x. In addition to mole fraction, Henry's law constants can also be expressed according to the concentration of a chemical species, A:

$$p = K_H \cdot [\text{A}] \tag{5.3}$$

However, concentration is usually expressed as mass per volume, which means that the concentration-based K_H is no longer unitless, but is expressed in units of $\frac{\text{volume-pressure}}{\text{mass}}$, often $\frac{\text{L-atm}}{\text{mol}}$ or $\frac{\text{Pa-m}^3}{\text{mol}}$.

Each chemical species has its own K_H, that is, it is an inherent property of the chemical. Also, given that it is derived from partial pressures at equilibrium, it is also temperature-dependent. Tables in handbooks often give K_H values at 298 K.

Air Pollution Calculations. https://doi.org/10.1016/B978-0-12-814934-8.00005-3

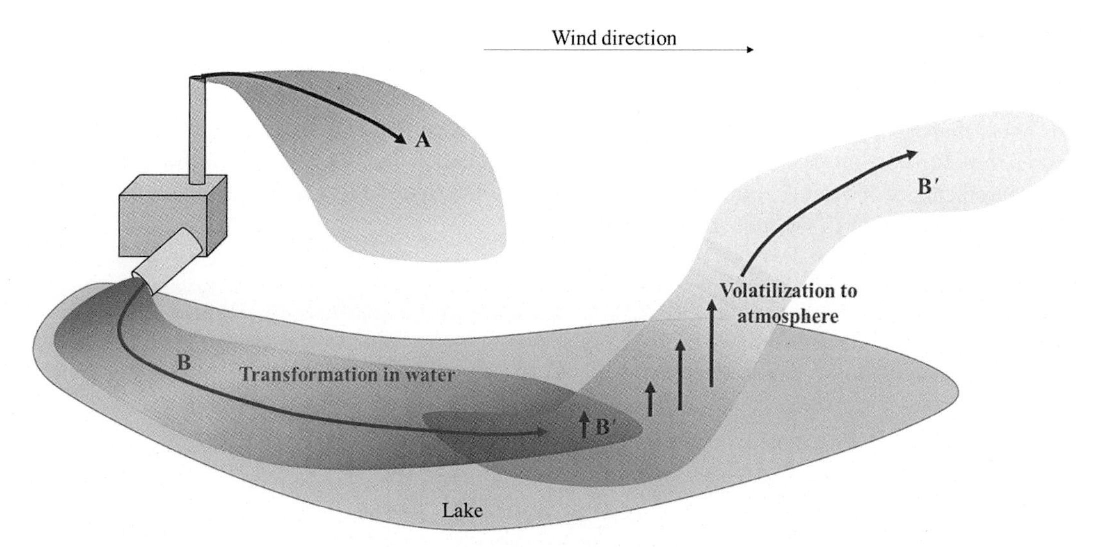

FIG. 5.1 Simplified depiction of direct (A) and indirect (B and B′) air pollutants. In the indirect air pollution scenario, the parent compound (B) is first released to water and is subsequently transformed abiotically or biotically in the water to a more volatile compound (B′), which is then released to the atmosphere. In this case, B is the air pollutant precursor, and B′ is the air pollutant.

For example, at this temperature, the K_H for Henry's law constants is derived as an infinite-dilution limit, that is, for solute A:

$$K_H = \lim_{x_A \to 0} \left(\frac{f_A}{x_A} \right) \tag{5.4}$$

Therefore, we are reminded that K_H is not, in fact, a "constant," since it has nonlinear temperature dependence. Interestingly, empirical studies show that at low temperatures, K_H values tend to increase until reaching a maximum, which varies with each solute-solvent pairing. After reaching the maximum K_H value, the measured K_H falls [1]. As mentioned, many K_H experiments have been conducted at 298 K, meaning that as the actual temperature varies from this, some amount of error will be introduced (see Fig. 5.2). This can be a crucial limitation for air pollution control equipment and other devices where temperatures can be quite high.

Henry's law constant applications work well for dilute solutions that, thankfully, are usually encountered in air pollution. For examples, the approach has been applied successfully in the design of air-stripping columns [2]. Certain chemical mixtures, for example, those encountered at hazardous waste facilities, may deviate from Henry's law due to

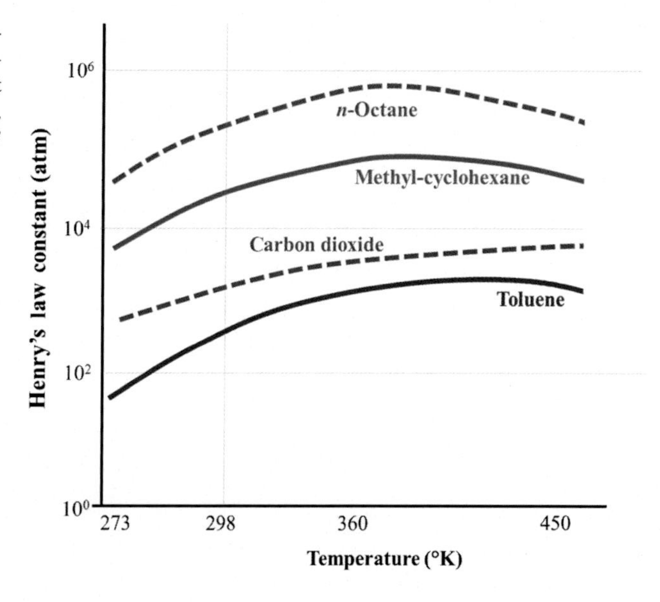

FIG. 5.2 Henry's law constant-temperature relationship for four compounds. Note that the constants' values continue to increase with temperature after the 298 K, the temperature at which the constants are most commonly reported. *(Source of information: F.L. Smith, A.H. Harvey, Avoid common pitfalls when using Henry's law, Chem. Eng. Progr. 103 (9) (2007) 33–39.)*

component interactions or concentrations that are too high or too low to reflect reported values. Error should be on the side of being environmentally conservative, that is, the actual gas-phase concentration is not likely to be underestimated. For example, concentrated mixtures of organic compounds, such as in a separate oil layer, would likely call for Raoult's law assumptions rather than Henry's law. Unfortunately, for equilibrium partitioning that can be applied generally to hazardous waste mixtures, reliable data are sparse [2]. Preferably, where certainty tolerances are tight, for example, public health concerns, measurements of the equilibrium partitioning between the liquid and gas phase would decrease uncertainties and avoid propagation of error in subsequent calculations.

An industrial plant manager wants to remove hydrogen sulfide (H_2S) from a water solution with 0.195 M. He has asked an air pollution expert to calculate K_H for H_2S and its aqueous solubility at a partial pressure (p) of 25.5 mm Hg at STP, that is, 760 mm Hg = 1 atm.

The expert begins with Eq. (4.25) to find a rough calculation of $K_{H\text{-}H2S}$:

$p = K_{H\text{-}H2S}: \cdot [H_2S]$

$$K_{H-H2S} = \frac{[H2S]}{p} = \frac{0.195}{1\text{ atm}} = 0.195\frac{\text{M}}{\text{atm}}$$

To be more exact, we may want to employ mole fractions. Since most of the solution is water, we may assume that 1 L weighs 1 kg. So, the number of moles of water is total water mass divided by molar mass of water:

$$\frac{1000\text{g}}{18} = 55.56\,\text{mol}$$

Therefore, the molar fraction (x) of H_2S is the number of moles divided by the total number of moles in the solution:

$$\text{Mole fraction of } H_2S = \frac{x_{H_2S}}{\text{total number of moles}} = \frac{0.195}{0.195 + 55.56} = 0.0035$$

At STP, pressure is 1 atm, so

$$K_{H-H2S} = \frac{p_{H_2S}}{x_{H_2S}} = \frac{1}{0.0035} = 286\,\text{atm} = 2.17 \times 10^5\,\text{mm Hg}.$$

To find solubility, we solve for $[H_2S]$:

$$[H_2S] = K_{H-H_2S} : \cdot p = \left(0.195\frac{\text{M}}{\text{atm}}\right)\left(\frac{25.5\text{mm Hg}}{\frac{760\text{mm Hg}}{\text{atm}}}\right) = 6.54 \times 10^{-3}\,\text{M}$$

The air pollution expert may want to ask why the calculation data were for STP, given that this is 0°C. Perhaps, the removal equipment will be refrigerated. If not, the gas solubility in water will decrease with increasing temperature (see Fig. 5.3). This can be important if the polluted water is stored at higher temperatures before treatment, that is, H_2S gas will be released faster, that is,

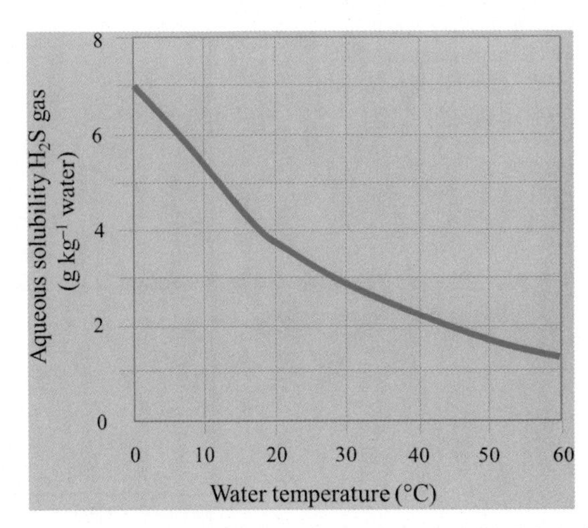

FIG. 5.3 Solubility of hydrogen sulfide in water. *(Source of data: Engineering Toobox. Solubility of Gases in Water. (2001). 2 June 2018. Available: https://www.engineeringtoolbox.com/gases-solubility-water-d_1148.html.)*

becomes less soluble by nearly a factor of 5, at 20 and 0°C. Conversely, during treatment, higher temperatures would lead to higher transfer rates from the tank to the gas collection equipment. Thus, K_H can be a friend or foe, depending on whether it is better or worse for the contaminant gas to remain in solution.

According to the ideal gas law, a chemical species vapor's partial pressure is directly proportional to the number of moles of that vapor per volume of gas. This means that Henry's law constants can be dimensionless for a given temperature, that is, no units, as the ratio of the chemical species' vapor concentration (C_g) and the chemical species' aqueous-phase concentration (C_{aq}):

$$K_H = \frac{C_g}{C_{aq}} \tag{5.5}$$

Thus, the dimensionless K_H, also known as the air-water coefficient (K_{AW}), can be calculated by rearranging K_H with units and the gas law factors for a chemical species A:

$$K_H\,(\text{dimensionless}) = \frac{K_H\,(\text{units})}{R \times T} = \frac{\dfrac{\dfrac{\text{atm}_\text{A}}{\text{mol}_\text{A}}}{\text{liter}_\text{water}}}{R\left(\dfrac{\text{liter}_\text{air} \cdot \text{atm}_\text{A}}{\text{liter}_\text{air}\text{K}}\right) \cdot T} = \frac{\dfrac{\text{mol}_\text{A}}{\text{liter}_\text{air}}}{\dfrac{\text{mol}_\text{A}}{\text{liter}_\text{water}}} \tag{5.6}$$

where R is the universal gas constant and T is temperature in K.

Soil pores contain both air and water. The carbon dioxide concentration in soil air in a landfill was recently analyzed at 20°C and found in equilibrium with the soil water to be 50 micromoles (μM). What is the corresponding CO_2 concentration in air?

Using the dimensionless K_H, we find

$$50\left(\frac{\mu\text{mol CO}_2}{\text{liter}_\text{water}}\right)\mu\text{M/liter}_\text{water} \times 44\left(\frac{\dfrac{\mu\text{mol CO}_2}{\text{liter}_\text{air}}}{\dfrac{\mu\text{mol CO}_2}{\text{liter}_\text{water}}}\right) = 2200\,\frac{\mu\text{mol CO}_2}{\text{liter}_\text{air}}$$

Convert the above dimensionless K_H above to units of atm · m³/mol.

At 20°C and converting liters to m³,

$$K_H = \frac{\dfrac{44\text{ mol}}{\text{liter}_\text{air}}}{\dfrac{1\text{ mol}}{\text{liter}_\text{water}}} \times RT \times \left(\frac{1\text{m}^3{}_\text{water}}{1000\text{ liter}_\text{water}}\right) = \frac{0.082\text{ liter atm }(20+273)\text{K}}{\text{mol K}} \times \frac{1\text{m}^3{}_\text{water}}{1000\text{ liter}_\text{water}}$$

$K_H = 1.06$ atm · m³/mol

During this study, MegaTox (MT), a chlorinated pesticide's concentration, was also analyzed at 20°C, with the soil water concentration of 5 x 10^{-5} μM found in equilibrium with the soil air. What is the corresponding pesticide concentration in air? Is this a proper conversion?

At 20°C and converting liters to m³, the corresponding MT concentration in air is

$$5 \times 10^{-5}\left(\frac{\mu\text{mol MT}}{\text{liter}_\text{water}}\right)\mu\text{M/liter}_\text{water} \times 232\left(\frac{\dfrac{\mu\text{mol MT}}{\text{liter}_\text{air}}}{\dfrac{\mu\text{mol MT}}{\text{liter}_\text{water}}}\right) = 0.01\,\frac{\mu\text{mol MT}}{\text{liter}_\text{air}}$$

There are 10^3 μM per mol, so the measurement was 5×10^{-8} mol of MT in the air:

$$K_H = \frac{\frac{5 \times 10^{-8}\,\text{mol}}{\text{liter}_{\text{air}}}}{\frac{1\,\text{mol}}{\text{liter}_{\text{water}}}} \times RT \times \left(\frac{1\,\text{m}^3{}_{\text{water}}}{1000\,\text{liter}_{\text{water}}}\right)$$

$$K_H = \frac{\frac{5 \times 10^{-8}\,\text{mol}}{\text{liter}_{\text{air}}}}{\frac{1\,\text{mol}}{\text{liter}_{\text{water}}}} \times \frac{0.082\,\text{liter atm}\,(20+273)\text{K}}{\text{mol K}} \times \frac{1\,\text{m}^3{}_{\text{water}}}{1000\,\text{liter}_{\text{water}}}$$

$K_H = 1.2 \times 10^{-6}$ atm · m^3/mol

This is likely not a proper conversion, since we are assuming that the pesticide vapor is behaving as an ideal gas. However, given the low concentration, this assumption may be tolerable.

Recently, MegaTox (MT) was reformulated and renamed SoluPest (SP), which has a molecular weight of 201. The analysis at 20° C found the soil water concentration of 50 M at equilibrium with the soil air. The partial pressure of SP was found to be 3×10^{-5} atm. What is its Henry's law constant?

$$K_H = \frac{p}{[SP]} = \frac{3 \times 10^{-5}}{50} = 6 \times 10^{-7}\,\text{atm} \cdot \text{m}^3/\text{mol}$$

5.3 MOVEMENT INTO THE ATMOSPHERE

A principal means of a substance being released as a plume is volatilization, which is a function of the concentration of a contaminant in solution and the contaminant's partial pressure. That is, the proportionality between solubility and vapor pressure can be established for any chemical.

Whether a chemical finds its way to the atmosphere begins with the chemical's vapor pressure. Indeed, one of the key characteristics of a chemical compound is its volatility, which is often used to categorize air pollutants (see Discussion Box "Classifying Chemical Species by Volatility"). However, vapor pressure is not a complete predictor of how much of the chemical species will migrate to the atmosphere. Given that Henry's law is a function of a substance's solubility in water and its vapor pressure, this is an expression of the proportionality between the concentration of a dissolved contaminant and its partial pressure in the open atmosphere at equilibrium. That is, Henry's law constant is an equilibrium constant, that is, the ratio of concentrations when chemical equilibrium is reached in a reversible reaction, the time when the rate of the forward reaction is the same as the rate of the reverse reaction.

Classifying Chemical Species by Volatility

As discussed in previous chapters, vapor pressure is commonly used to categorize pollutants. In particular, air pollution scientists and engineers categorize organic compounds into three major classes: volatile, semivolatile, and nonvolatile. Organic compounds are ubiquitous in the atmosphere, widely ranging in toxicity, hazard, and the potential for exposure and risks to humans and other species. Indeed, most of the attention to organic compounds ranges from very large areas, for example, the role of hydrocarbons in the formation of ozone in large urban areas, to small microenvironments, for example, indoor air in homes and offices.

Volatile organic compounds (VOCs) readily move to the headspace of a container or to the atmosphere. The vapor pressure of a chemical species in the liquid or solid phase is the pressure that is exerted by its vapor when the liquid and vapor are in dynamic equilibrium (see Fig. 5.4).

Vapor pressure increases in direct proportion to temperature (Fig. 5.5). Vapor pressure is the inherent property that provides a clue as to whether a compound is an air pollutant and the rate of volatilization to the atmosphere. Substances of low molecular weight and certain molecular structures have sufficiently high vapor pressures that they can exist in either the liquid or gas phases under environmental conditions.

Vapor pressure is indeed an inherent property, but data sets report vapor pressure as an expression of the partial pressure of a chemical substance in a gas phase that is in equilibrium with the nongaseous phases. Thus, it is a partitioning property. For many

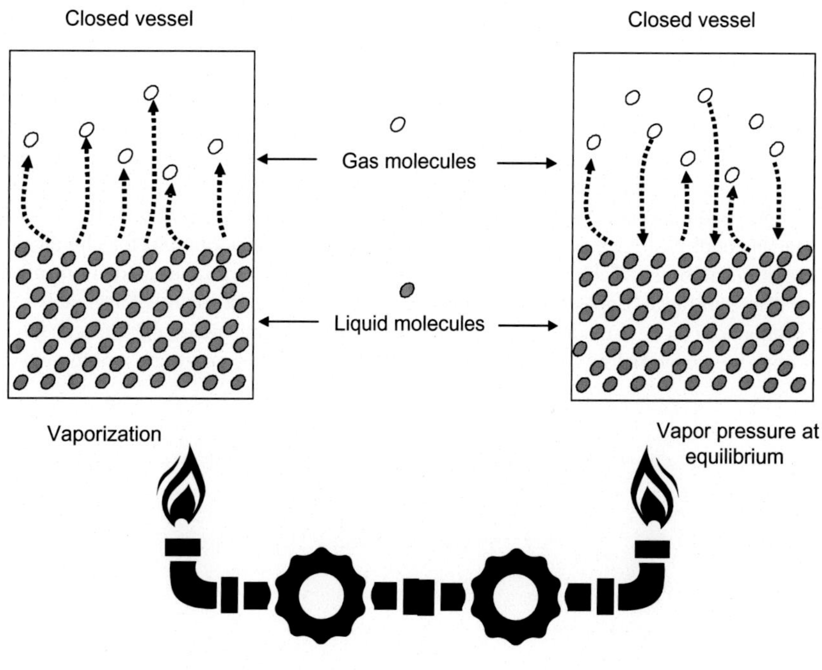

FIG. 5.4 Vapor pressure of equal amounts of hypothetical fluid during vaporization and at equilibrium in two identical, evacuated, and closed vessels with limited headspace at equal temperatures. The pressure in the space above the liquid increases from zero and eventually stabilizes at a constant value. *(Source: D. Vallero, Environmental Contaminants: Assessment and Control. Academic Press, 2010.)*

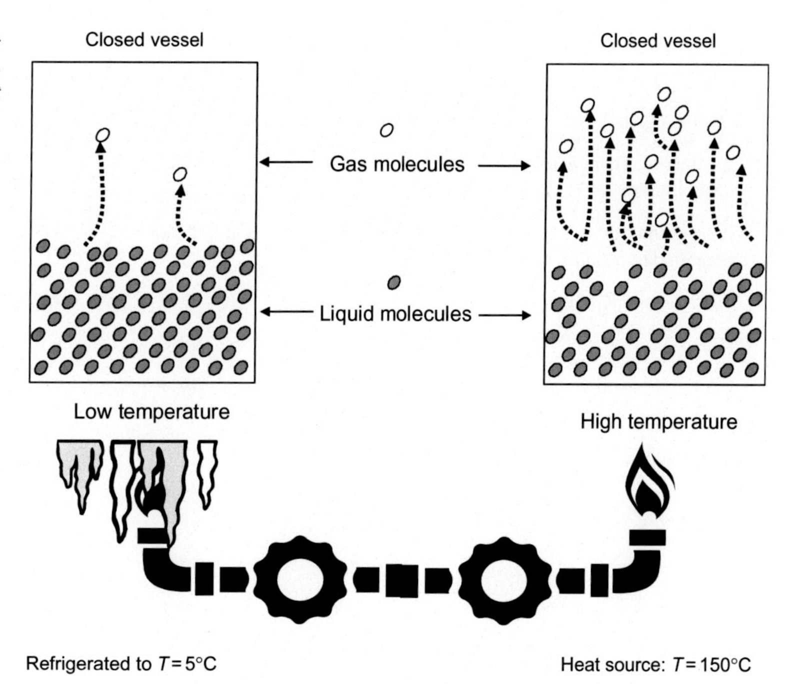

FIG. 5.5 Vapor pressure increases in direct proportion with increasing temperature. *(Source: D. Vallero, Environmental Contaminants: Assessment and Control. Academic Press, 2010.)*

vapors under environmental conditions, a fraction of the molecules will be in liquid phase and a fraction in the gaseous phase. Thus, under identical conditions, a molecule with a vapor pressure higher than another molecule, the first molecule's liquid-to-gas ratio will be less than the second molecule's liquid-to-gas ratio.

It is also important to keep in mind that above 0 K, all compounds, even solids, have vapor pressures above zero. Thus, under environmental conditions, no substance is completely "nonvolatile." However, there are so few molecules at equilibrium that reach the air for the so-called NVOCs that they are often ignored.

Generally, VOCs have vapor pressures greater than 10^{-2} kilopascals [3]. The US EPA and the World Health Organization use boiling point as the factor for determining volatility (see Table 4.1 in Chapter 4), adding another category, very volatile organic compound (VVOC), which exists almost entirely in the gas phase under most environmental conditions [4]. This is logical and an expression of the same principles, since boiling point and vapor pressure are directly related, that is, the temperature at which a compound becomes a vapor. At its boiling point, a liquid's vapor pressure is equal to the external pressure. Other ways to classify VOCs have been proposed, such as evaporation rates over time, for example, what remains after six months [5].

In practice, such chemical classifications can be subjective and poorly defined in the literature. These inconsistencies have recently been addressed by evaporation studies under realistic conditions. To this end, the South Coast (California) Air Quality Management District measured the rate of volatilization (evaporation) for chemicals having a wide range of published vapor pressure values (see Table 5.1) [5]. Fig. 5.6 shows the highly variable predictions of volatility.

Any substance, depending upon the temperature, can exist in any phase. However, in many environmental contexts, many compounds can be excluded from certain phases. For example, molecular nitrogen (N_2) certainly exists in a liquid phase; however, it must be cooled below its boiling point (-196°C). Nowhere on earth do ambient conditions meet this temperature requirement,

TABLE 5.1 Chemical compounds included in evaporation studies conducted by the South Coast (California) Air Q

Compound	Manufacturer	Purity	Carbon atoms	Boiling point (°C)	Vapor pressure (mm Hg @ 25°C)
Isopropyl alcohol (IPA)	EMD Millipore	≥99.99%	3	83	33
C9-C16 Hydrotreated light distillate (light distillate)	Calumet Specialty Products	UNK	9–16	216–278	<0.1–0.3
Ethyl lactate	JT Baker	99%	5	154	1.2
N-methylpyrrolidinone (NMP)	Sigma-Aldrich	99%	5	202	0.5
Benzyl alcohol	Sigma-Aldrich	≥99%	6	205	0.14
Dodecane (C12)	Sigma-Aldrich	≥99%	12	216	0.3
Propylene glycol	SAFC Global	≥99.5%	3	187	0.13
Ethylene glycol	Aldrich	99.8%	2	198	0.09
2,2,4-Trimethylpentanediol diisobutyrate	Aldrich	98.5%		280	0.004
2,2,4-Trimethyl-1,3-pentanediol monoisobutyrate	Aldrich	99%	12	255	0.01
Pentadecane (C15)	Sigma-Aldrich	99+%	15	270	0.01
Dipropylene glycol	Aldrich	99%	6	232	<0.01
2-Methyl hexadecane	MP biomedicals	90%–100%	17	291	<0.01
Hexadecane (C16)	Sigma-Aldrich	99%	16	287	0.005
Naphthenic oil (Hynap N60HT)	San Joaquin Refinery	UNK	16–20	279	<0.001
Heptadecane (C17)	Aldrich	99%	17	302	<0.001
Naphthenic-based metal working fluid (MWF)	W.S. Dodge Oil	UNK	UNK	UNK	UNK
Alkyl alkanolamine	Taminco Higher Amines, Inc.	>99%	8	283	<0.01
Methyl palmitate	Fluka	≥99.0%	17	332	<0.001
Soy oil	W. R. Meadows	UNK	14–18	>250	<0.001
Glycerol	EM Science	99.5+%	3	290	<0.001

UNK, information not known.

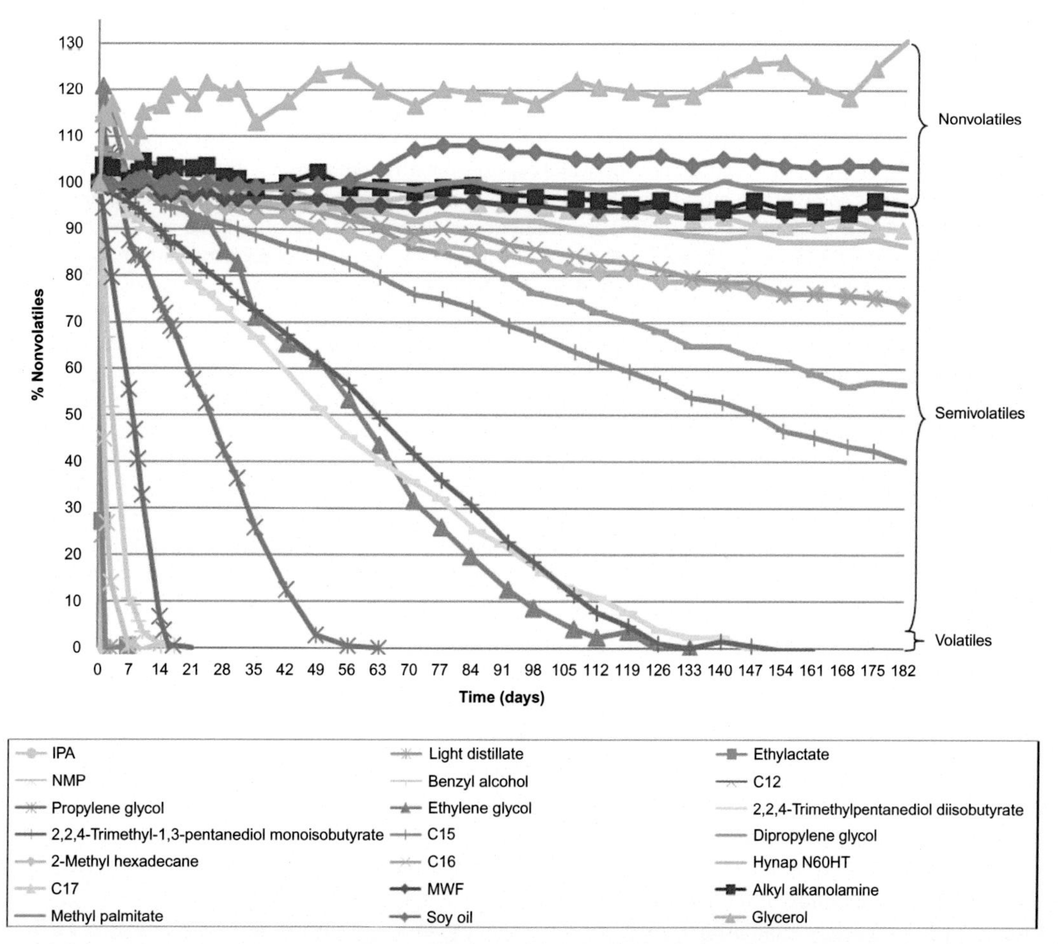

FIG. 5.6 Evaporation profiles for compounds described in Table 5.1, which were selected by California's South Coast Air Quality Management District to demonstrate three broad classifications of volatility. *(Source: Nonvolatile, semivolatile, or volatile: Redefining volatile for volatile organic compounds. (2012). Available: http://www.aqmd.gov/docs/default-source/planning/architectural-coatings/redefining-volatile-for-vocs.pdf.)*

so N2 will be considered an atmospheric gas. Conversely, a vapor is a substance that is in its gas phase but under typical environmental conditions exists as a liquid or solid under a given set of conditions. Although the pressure in the closed container in Fig. 5.4 is constant and below the boiling point, molecules of the vapor will continue to condense into the liquid phase, and molecules of the liquid will continue to evaporate into the vapor phase. However, the rate of these two processes is equal, meaning no net change in the amount of vapor or liquid. This is an example of dynamic equilibrium or equilibrium vapor pressure.

Vapor pressure is also an important consideration in air pollution chemical analysis. The vapor pressure of a substance increases as boiling points at a given pressure decrease. Thus, volatility is a measure of the propensity of a substance to form concentrations of vapor above the liquid. This is an important aspect of gas chromatography. For example, some gas chromatographs (GC) have a capillary column, that is, a silica tube coated on the inside with a film that holds the sample, in an oven. When a sample is injected into the GC, it is held in the film while the column is heated. The lower-boiling-point compounds, that is, the VOCs, usually exit the column first, move with a carrier gas (e.g., helium), and are the first to reach the detector, for example, mass spectrometer (MS). Thus, the first peaks on the MS chart are the compounds with lowest boiling points. As the GC continues to increase the temperature of the column, the compounds with increasingly higher boiling points are detected. Vapor pressure and boiling points do not predict the order of detection invariably, as other chemical factors like halogenation and polarity affect residence time on the column. However, a compound's boiling point is usually a good first indicator of residence time on a column and, as such, will determine which capillary column to use, for example, one for VOCs and one for SVOCs.

Polarity and vapor pressure are not independent qualities of a substance. The boiling point of a compound frequently is related to polarity, especially if the compound is in an aqueous or other highly polar solvent. Recall from the discussion of Henry's law that the greater the solubility, this less fugacity is expected because the compound has an affinity for the water.

Vapor pressure is also a crucial factor for emergency responders and in environmental cleanup. For example, chemical compounds stored in tanks and other vessels can present problems to first responders if a volatile compound is also flammable or explosive and otherwise hazardous. These hazardous substances can exit rapidly if a tank is disturbed, for example, a buried drum containing an explosive VOC. Also, even an imperceptible leak in tank containing a VOC can fill an enclosed space rapidly. This is one of the reasons that lower explosive limits (LELs) are so important for first responders, since volatile compounds that may have been released can concentrate in the air (e.g., inside a confined space or room) to a point where they are dangerously close to exploding or at concentrations that lead to the loss of consciousness of workers.

The reason for specificity is that the controls and prevention methods differ depending on a substance's volatility. Engineers and managers in industrial, commercial, and governmental facilities generating or handling volatile compounds must take steps to prevent these compounds from moving to the atmosphere and exposing nearby communities. The semivolatile organic compound (SVOC) category comprises some of the most toxic, bioaccumulating, and persistent environmental contaminants, which have vapor pressures between 10^{-5} and 10^{-2} kilopascals. These values correspond to classifications that are based upon observations of the compounds' behaviors during air sampling. [3] A particularly important aspect of SVOCs is that they may be transported from soil in the gas phase or as aerosols [6]. Thus, if SVOCs are present in the waste stream or storage, they must be monitored as both vapor and particulate matter.

Often, VOCs and SVOCs are released to the atmosphere during thermal processes, such as manufacturing and waste incineration. In addition, ash and residues contain products of incomplete combustion, such as polycyclic aromatic hydrocarbons, dioxins, furans, and hexachlorobenzene, as well as numerous pesticides, phenolics, solvents, and pharmaceuticals, which fall under the SVOC chemical classification. Proper management of these chemicals requires an understanding of the factors that lead to the release, movement, and degradation of these compounds.

Incidentally, volatility classification is also applied to inorganic substances, albeit less commonly than organic. These include volatile inorganic compounds (VICs), semivolatile inorganic compounds (SVICs), and nonvolatile inorganic compounds (NVICs). The metrics can be the same as those for organic compounds, for example, vapor pressure and boiling points.

In the environment, volatility acts with other inherent properties of a substance, which affects the likelihood that a substance will move to the atmosphere. For example, ground water can be contaminated with nonaqueous-phase liquids (NAPLs) of varying densities. Light NAPLs (LNAPLs) are more likely to remain in the upper reaches of aquifers than dense NAPLs (DNAPLs). The contaminated soil and ground water can become an air pollution source if an LNAPL includes a relatively insoluble substance that distributes between liquid and gas phases (see Fig. 5.7). Because of its hydrophobicity, the LNAPL can infiltrate and move along the water table near the top of the zone of saturation, just above the capillary fringe. However, some of the contaminant fluid lags the plume and slowly solubilizes in the pore spaces of the soil and unconsolidated material. The more soluble forms of the fluid find their way to the zone of saturation and move with the general groundwater flow, that is, advectively. The higher vapor pressures of portions of the plume will lead to upward movement of volatile compounds in the gas phase.

This illustrates how a water contamination scenario becomes an air pollution scenario because of the interplay of inherent properties, that is, vapor pressure, solubility, and density of the fluid components. Of course, the extent and severity of the pollution is also modulated or worsened by other environmental conditions, including temperature,

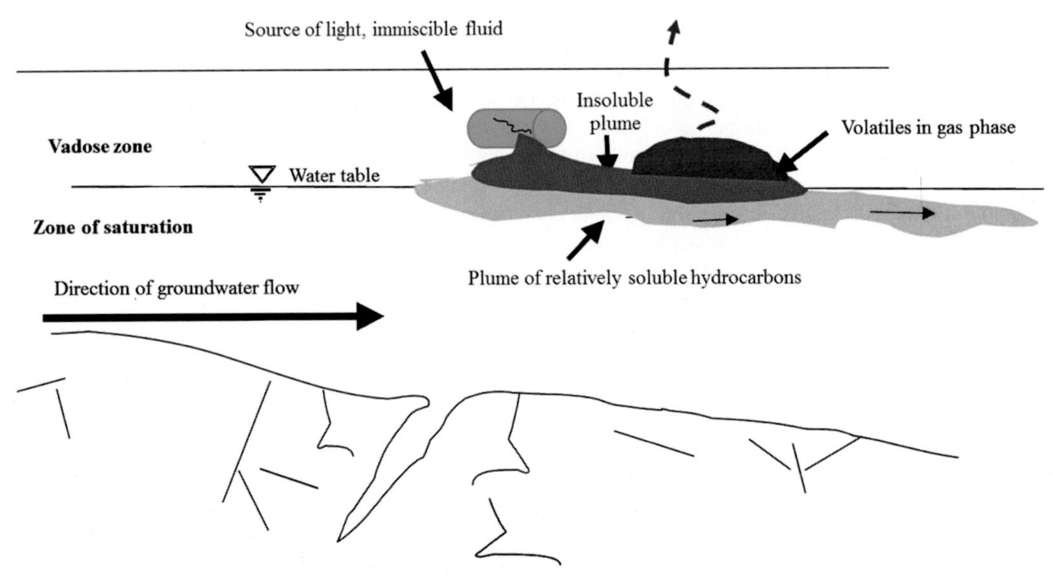

FIG. 5.7 Hypothetical plume of low-density, hydrophobic fluid. *(Reproduced with permission from D. Vallero, Translating Diverse Environmental Data Into Reliable Information: How to Coordinate Evidence from Different Sources, Academic Press, 2017; adapted from D.A. Vallero, Fundamentals of Air Pollution, 5th ed., Elsevier Academic Press, Waltham, MA, 2014, p. 999 pages cm; M.N. Sara, Ground-water monitoring system design, in: D.M. Nielsen (Ed.), Practical Handbook of Ground-Water Monitoring, Lewis Publishers, USA, 1991, pp. 17–68.)*

atmospheric pressure, relative humidity, soil type (e.g., more permeable soil allows increased contaminant transfer), concentration of pollutant, and the presence of other substances. If surfactants are present, cosolvation can occur, that is, the surfactant's polar head likes the water, and its nonpolar tail likes the NAPL. This results in the NAPL behaving more like a soluble substance and is less likely to escape to the atmosphere.

As previously discussed, Henry's law states that the concentration of a dissolved gas is directly proportional to the partial pressure of that gas above the solution:

$$p_a = K_H[C] \tag{5.7}$$

where

K_H = Henry's law constant,
p_a = partial pressure of the gas,
$[C]$ = molar concentration of the gas.

Recall that concentration is the quantity of solute in a gas or liquid solution or the quantity of a substance in a solid matrix or substrate, for example, in soil. Other letters besides C are also commonly used to denote the concentration of a chemical species, that is, the solute, including i, x, and A, which we used when introducing fugacity and Henry's law. We will use C in this discussion because many of the derivations of partition coefficients in the scientific literature use it for concentrations in the various environmental compartments.

If the compound is in water, the proportionality is

$$p_A = K_H C_W \tag{5.8}$$

where C_W is the concentration of gas in water. Note that from this point forward, we will drop the square brackets although they often signal concentration; for example, the concentration of molecular oxygen is $[O_2]$, the concentration of hydrogen ions is $[H^+]$, and the concentration of pentachlorophenol is $[C_6HCl_5O]$.

5.3.1 Net exchange

Combining the concentration of a dissolved contaminant and its partial pressure in the headspace (including the open atmosphere) at equilibrium is a reliable means of estimating the likelihood that a chemical will move into the atmosphere. As mentioned, the dimensionless version of K_H partitioning is like that of sorption, except that instead of the partitioning between solid and water phases, it is between the air and water phases (K_{AW}):

$$K_{AW} = \frac{C_A}{C_W} \tag{5.9}$$

where C_A is the concentration of gas A in the air.

The relationship between the air/water partition coefficient and Henry's law constant for a substance is

$$K_{AW} = \frac{K_H}{RT} \tag{5.10}$$

where R is the gas constant (8.21×10^{-2} L atm mol^{-1} K^{-1}) and T is the temperature (°K).

The lake manager has returned from a meeting of stakeholders who asked about a legacy pesticide called toxaphene. They had heard that pesticide was still being deposited on Lake Superior. The manager asks the town engineer if that is true. Calculate the net exchange of toxaphene based on reported values of 918 nanograms (ng) m^{-3} in water and 1.2 ng m^{-3} in air [7,8].

To answer, we use the reported measurements of C_w and C_A to calculate the concentration ratio:

$$\frac{C_W}{C_A} = \frac{918}{0.02} = 4.59 \times 10^4$$

Next, we look up K_H and find that Bidleman and Jantunen [7–9] report it to be

$$\text{Log } K_H = 10.42 - \frac{3309}{295} = -0.458 \frac{\text{Pa} \cdot \text{m}^3}{\text{mol}}$$

Thus, $K_H = 0.348 \frac{Pa \cdot m^3}{mol}$

Then, compare the ratio $\frac{C_W}{C_A}$ at equilibrium, that is, when $\frac{1}{K_{AW}} = \frac{RT}{K_{AW}}$ to see whether toxaphene is being deposited or is volatilizing.

That is,

If $\frac{C_W}{C_A} = \frac{1}{K_{AW}}$, then the net exchange of toxaphene is 0, that is, no net exchange.

If $\frac{C_W}{C_A} > \frac{1}{K_{AW}}$, then the net exchange of toxaphene is from the water to the air, that is, net volatilization of toxaphene.

If $\frac{C_W}{C_A} < \frac{1}{K_{AW}}$, then the net exchange of toxaphene is from the air to the water, that is, net deposition of toxaphene.

We choose R with compatible units, that is, $R = 8.314 \frac{m^3 \cdot Pa}{mol \cdot T}$, with T in K.

$$\frac{1}{K_{AW}} = \frac{RT}{K_{AW}} = \frac{8.314 \times 295}{0.348} = 7.05 \times 10^3.$$

$$4.59 \times 10^4 > 7.05 \times 10^3.$$

Since $\frac{C_W}{C_A} > \frac{1}{K_{AW}}$, the water is over saturated, that is, toxaphene is volatilizing from Lake Superior to the atmosphere.

However, this is not really what the stakeholders asked. They wanted to know if toxaphene was still being deposited. The answer is "yes," but less is being deposited than volatilizing. Another caveat that should be shared with the lake manager is that these calculations are assumed to reflect the entire lake. Of course, this is not the case. They are the mean of a few sites. So, every other location of this large lake system will have C_w values that are higher and lower, possibly substantially so, than these "grab samples." Also, the measurements were taken more than a decade ago, so they must be different now.

Recently, the C_w for toxaphene was modeled [10] for the year 2011 and found to be 0.493 ng L^{-1}. If C_A has not changed, how does this change the results reported above?

The calculations used volume units of m^3. 1 m^3 = 1000 L; therefore,

0.493 ng L^{-1} = 493 ng m^{-3}; and therefore,

$\frac{C_W}{C_A} = \frac{493}{0.02} = 2.47 \times 10^4$, which is still larger than $\frac{1}{K_{AW}}$, that is, 7.05×10^3.

Therefore, the lake continues to be a source of toxaphene. That is, net exchange from the water to the atmosphere is higher than the net exchange from the atmosphere to the water. However, the exchanges are approaching equilibrium. It is likely that since the water concentration decreased, so would the air, for example, if the toxaphene is being degraded by microbes or hydrolysis. Therefore, our assumption that C_A stayed the same is likely incorrect. If we assume that C_A has fallen to 0.15, our calculation would be

$\frac{C_W}{C_A} = \frac{493}{0.015} = 3.29 \times 10^4$, which again means net volatilization.

However, there are conditions that could decrease C_w while increasing C_A, for example, cloud effects and aerosols, where the toxaphene is less vulnerable to degradation and/or physical and meteorologic processes allow for increasing air concentrations. So, if C_A were to rise to 0.08 ng m^{-3}, our net exchange would be

$\frac{C_W}{C_A} = \frac{493}{0.08} = 6.16 \times 10^3$, which is less than $\frac{1}{K_{AW}}$, that is, 7.05×10^3.

Thus, the situation where the mass of toxaphene decreases and the concentration of air increases would change the system to net deposition, that is, the air is adding toxaphene to the lake water.

The lake manager presented the engineer's results at the next meeting and has returned with a new question. The stakeholders noticed that all the calculations were based on $T = 295$ K, or about 22°C. They have heard about global warming and wanted to know what would happen to the toxaphene exchanges if the temperature increased by 5°.

Henry's law constant is not reported for 300 K, but Fig. 5.8 shows the values from Jantunen and Bidleman [9]. Thus, at 300 K, K_H for toxaphene is about 0.6 Pa · m^3 · mol^{-1}.

$$\frac{1}{K_{AW}} = \frac{RT}{K_{AW}} = \frac{8.314 \times 300}{0.6} = 4.16 \times 10^3$$

If we use the first calculations,

$$\frac{C_W}{C_A} = \frac{918}{0.02} = 4.59 \times 10^4$$

$$4.59 \times 10^4 > 4.16 \times 10^3$$

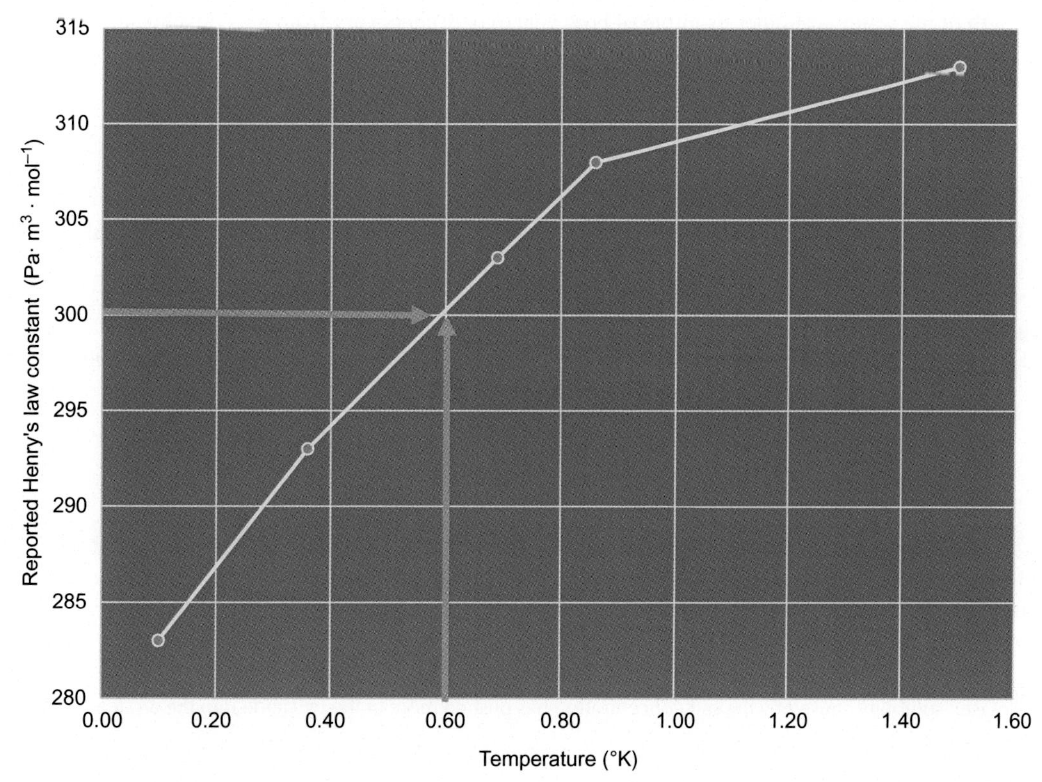

FIG. 5.8 Temperature dependence of Henry's law constants for toxaphene. *(Source of data: L.M. Jantunen, T.F. Bidleman, Temperature dependent Henry's law constant for technical toxaphene, Chemosphere-Glob, Change Sci. 2 (2) (2000) 225–231.)*

Since $\frac{C_W}{C_A} > \frac{1}{K_{AW}}$, this means that the net exchange of toxaphene remains from the water to the air, that is, net volatilization. Therefore, the water continues to be the source and the air the sink for toxaphene.

5.3.2 Henry's law complications

Henry's law relationships work well for most environmental conditions, representing a limiting factor for systems where a substance's partial pressure is approaching zero. At very high partial pressures (e.g., 30 pascals) or at very high contaminant concentrations (e.g., >1000 ppm), Henry's law assumptions cannot be met. Such vapor pressures and concentrations are seldom seen in ambient environmental situations but may be seen in industrial and other source situations. Thus, in modeling and estimating the tendency for a substance's release in vapor form, Henry's law is a good metric and is often used in compartmental transport models to indicate the fugacity from the water to the atmosphere.

Henry's law constants are highly dependent upon temperature, since both vapor pressure and solubility are also temperature-dependent. So, when using published K_H values, one must compare them isothermally. Figs. 5.2 and 5.4 indicate, however, that K_H values and temperature are not linear relationships.

The single partition coefficient uncertainties are further complicated by numerous other factors when combining different partitioning coefficients in a model or study. Therefore, it is important either to use only values derived at the same temperature (e.g., sorption, solubility, and volatilization all at 20°C) or to adjust them accordingly. For example, general adjustment is an increase of a factor of 2 in K_H for each 8°C temperature increase.

Reported K_H values, like other partition coefficients, are based on neat chemical species. That is, the experimental measurements are usually conducted with a single pure compound in deionized water. However, in the environment, the chemical species exists in the presence of many other chemicals. In fact, for real aqueous solutions, the K_H constants are reported as an air-water distribution ratio. Water in the real world is not deionized. In addtion, water often contains organic solvents and alcohols that can increase the concentrations of otherwise insoluble organic compounds by cosolvation. For example compared to other organic compounds, benzene is relatively soluble in water. So, the benzene may dissolve an organic chemical species that otherwise has very low aqueous solubility, resulting in higher water concentrations than would be expected for pure water. In this cosolvation example, the water dissolves the benzene, and the benzene dissolves the organic chemical species. This would substantially change the $S^{\circ}_{\text{liquid}}$ value in Eq. (5.1).

Ionic strength of the water is another example of how neat experiments vary from K_H constants reported as an air-water distribution ratio. Gas solubility generally decreases with increasing salinity, known as "salting out." This effect is estimated with the Sechenov equation [11]:

$$\log\left(\frac{K_{H0}^{bp}}{K_H{}^{bp}}\right) = k_s \times b(\text{salt}) \tag{5.11}$$

where K_{H0}^{bp} is Henry's law constant in pure, deionized water; $K_H{}^{bp}$ is Henry's law constant in the salt solution; k_s is the Sechenov constant, which is based on molality; and b(salt) is the molality of the salt.

Increased salt concentrations decrease the aqueous solubility of organic compounds. Thus, fresh waters more readily dissolve organic compounds compared with marine waters. If k_s and b(salt) are known for a chemical species vapor, the real-world Henry's law constant can be approximated by the product of pure Henry's law constant and 10 to the power of the product of the Sechenov constant and the salt molality:

$$K_{H0}^{bp} = K_H{}^{bp} \times 10^{[k_s s\, b(salt)]} \tag{5.12}$$

A pesticide has a reported Henry's law constant of 0.74 Pa m³ mol⁻¹ at 25°C. The reported k_s for this pesticide is 0.40 L mol⁻¹. If salt concentration is 0.5 mol L⁻¹, what is the salt-adjusted Henry's law constant?

$$K_{H0}^{bp} = 1.17 \times 10^{(0.40\times 0,\,5)} = 1.49\,\text{Pa}\cdot\text{m}^3\cdot\text{mol}^{-1}$$

The decreased solubility caused by the salt concentration has pushed more of the pesticide into the air, leading to an increased Henry's law constant.

Unfortunately, Sechenov constants and others are not available for many species [12].

Any sorbed or otherwise bound fraction of the chemical species will not exert a partial pressure, so this fraction should not be included in calculations of partitioning from water to air. For example, it is important to distinguish the mass of the contaminant in solution (available for the K_{AW} calculation) from that in the suspended solids (unavailable for K_{AW} calculation). This is crucial for many hydrophobic organic contaminants, most of which will not to be dissolved in the water column (except as cosolutes mentioned earlier), with the largest mass fraction in the water column being sorbed to particles.

The relationship between K_H and K_{ow} is also important. It is often used to estimate the environmental persistence, as reflected the chemical half-life ($t_{1/2}$) of a contaminant. However, many other variables determine the actual persistence of a compound after its release. Note, in the table, for example, that benzene and chloroform have nearly identical values of K_H and K_{ow} yet benzene is far less persistent in the environment. We will consider these other factors in the next chapters, when we discuss abiotic chemical destruction and biodegradation.

With these caveats in mind, however, relative affinity for a substance to reside in air and water can be used to estimate the potential for the substance to partition not only between water and air but also more generally between the atmosphere and biosphere, especially when considering the long-range transport of contaminants (e.g., across continents and oceans). Such long-range transport estimates make use of both atmospheric $t_{1/2}$ and K_H. Also, the relationship between octanol-water and air-water coefficients can be an important part of predicting a contaminant transport. For example, Fig. 5.9 provides some general classifications according to various substances' K_{AW} and K_{ow} relationships. In general, chemicals in the upper left-hand group have a great affinity for the atmosphere, so unless there are contravening factors, this is where to look for them. Conversely, substances with relatively low K_{AW} and K_{ow} values are less likely to be transported long distance in the air. Since K_{AW} is proportional to K_H, these groupings also apply to Henry's law constants.

5.4 INTERFACES

Although our discussions of partitioning and equilibria have shown direct molecular exchanges between compartments, this is seldom the case. Instead, there are one or more intermediate zones between the compartments. Indeed, for exchanges involving biota, there is almost never a direct exchange, but passage through a film. In water-air exchanges, the molecular transport of many chemical species traverses through thin films at the interface between the water and air.

The film can be envisioned as a stagnant layer at interface of each compartment. Chapter 6 delves into the theoretical structure of a liquid-gas exchange (see Figs. 6.8 and 6.9 in Chapter 6). That is, air pollution and other chemical transport

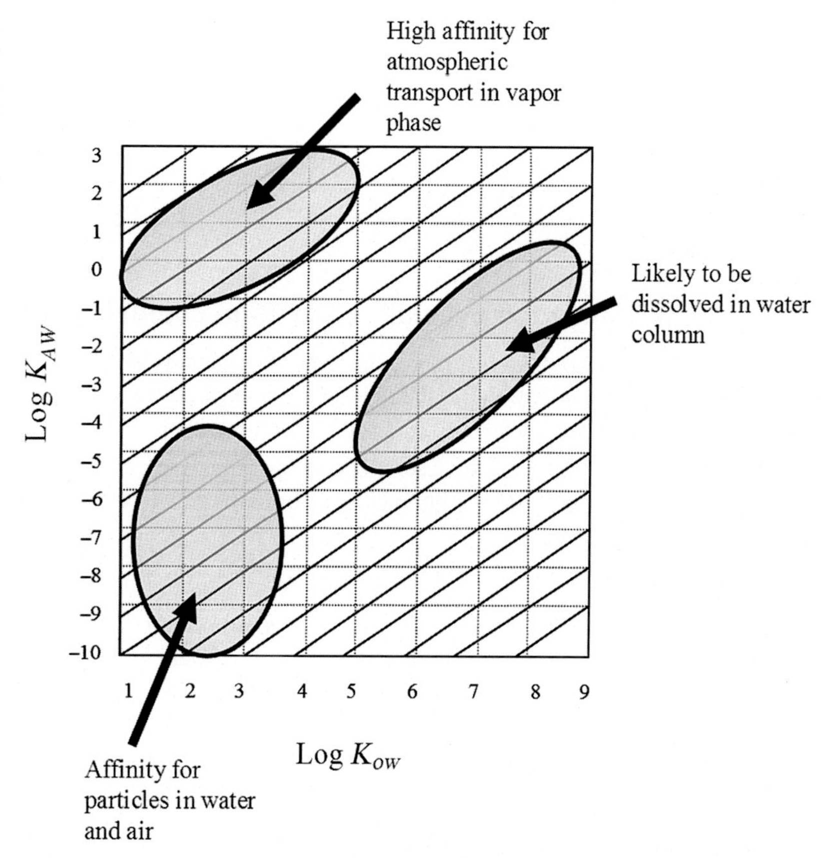

FIG. 5.9 Relationship between air-water partitioning and octanol-water partitioning and affinity of classes of contaminants for certain environmental compartments under standard environmental conditions. *(Source of information: D. van de Meent, et al., Persistence and transport potential of chemicals in a multimedia environment, in: SETAC Pellston Workshop, Fairmount Hot Springs, British Columbia, Canada, 1998, pp. 14–18.)*

models often assume that the volume of gas above the gas-side film is turbulent, whereas the gas in the gas film is assumed to be quiescent or stagnant.

Dispersion is generally a combination of diffusion and advection, that is, the transport of a chemical species along with the movement of air. By distance transported, advection is the primary mechanism by which air pollutants are transported, but the pollutants would not find their way to be advected without first diffusing into the atmosphere.

5.5 PARTITIONING BETWEEN AIR AND TISSUE

The principal interest in partitioning between an organic substrate and water, that is, the *Kow*, was mainly the need of water pollution. Air pollution also needs a means of characterizing this partitioning for organic air pollutants in terms of sources likely to release the organic air pollutants. However, organic solvent and water partitioning is not completely useful for understanding the movement of substances among organic substrates and the air, especially within the lungs. Although both systems derive O_2 that is dissolved in a substrate (water or air), K_{ow} is very useful for gills but much less useful for lungs. Thus, the octanol-air partitioning coefficient (K_{oa}) was established.

Models have been developed to combine K_{ow} and Henry's law constant (K_H) to estimate the equilibrium between air and organic tissues. Henry's law constant and the K_{OA} are physical-chemical properties that are often not measured but are needed to understand the environmental fate and transport of chemicals. When measured values are not available for an air pollutant, both K_H and K_{oa} can be estimated.

As mentioned, Henry's law constant is the air-water partition coefficient, which is the ratio of vapor pressure to solubility in water. It is commonly reported with units of atm-m^3/mole, or if it is a unitless value, it is reported as the air-water partitioning coefficient (K_{AW}). The conventional Henry's law constant is calculated as

$$K_H = \frac{P^{\circ} \text{ x MW}}{760 \times C_{sol}} \tag{5.13}$$

where P° = vapor pressure (torr), MW = molecular weight (g mol^{-1}), 760 = conversion factor (1 atm = 760 torr), and C_{sol} =and water solubility (mg L^{-1}). The resulting units for K_H are atm-m^3/mole. The units of mg and liter (L) are canceled since there are 1000 mg in one gram and 0.001 m^3 in one liter.

The unitless Henry's law constant (K_{AW}) is calculated as

$$K_{AW} = \frac{C_{air}}{C_{water}} = \frac{K_H}{R \cdot T} \tag{5.14}$$

where R is the ideal gas constant of 8.205746×10^{-5} atm-m^3/K-mole, T is absolute temperature expressed in Kelvin (K), and K_H is expressed in atm-m^3/mole. Use parameters measured or calculated for the same temperature (e.g., 25°C).

5.5.1 Calculating the octanol-air partition coefficient

In discussed in Chapter 4, air is a fluid from which compounds partition, as expressed by the octanol-air coefficient (K_{OA}) for a chemical compound, which is the ratio of the concentration of a compound in an organic phase (n-octanol) and air at equilibrium ($\frac{C_{air}}{C_{water}} = \frac{C_{octanol}}{C_{air}}$, where $C_{octanol}$ and C_{air} represent equilibrium concentrations in air and n-octanol with the same units). The K_{OA} has been used to describe partitioning between air and aerosol particles, air and foliage, and air and soil [13–15]. K_{OA} can be calculated from other partitioning coefficients and the gas laws:

$$K_{OA} = \frac{K_{ow} \cdot R \cdot T}{K_H} \tag{5.15}$$

where K_{ow} is the n-octanol-water partition coefficient (unitless), R is the ideal gas constant $= 8.205746 \times 10^{-5}$ atm-m^3/K-mole, T is absolute temperature expressed in Kelvin (K), and K_H is Henry's law constant expressed in atm-m^3/mole.

The ideal gas law assumptions can simplify the equation to a relationship between water and air partitioning, that is, K_{ow} and K_{OA}:

$$K_{OA} = \frac{K_{ow}}{K_{AW}} \tag{5.16}$$

K_{AW} is the air-water partition coefficient (unitless).

When a measured K_{OA} is not available and long-range transport or terrestrial bioaccumulation may be a concern, the K_{OA} can be calculated by using Eq. (5.16) when measured K_{ow} and vapor pressure or Henry's law constant is available. If these are not available, a QSAR approach is recommended (e.g., EPI Suite). The published values for these key partitioning coefficients are given in Table 4.7 in Chapter 4.

Recall that partitioning coefficients in Table 4.7 are reported as log values to account for the ranges of several orders of magnitude for partitioning coefficients. Thus, Eq. (5.16) can be restated as

$$\log K_{OA} = \log K_{ow} - \log K_{AW} \tag{5.17}$$

The values in Table 4.7 demonstrate the variability of estimates depending on the method used. Compare the published log K_{OA} for methyl isocyanate (MIC) to the calculated value from Eq. (5.17).

The published log K_{OA} is 2.112, so taking its antilog gives a $K_{OA} = 129$. However, calculating K_{OA} from Eq. (5.17) gives an estimated value:

$$\log K_{OA} = \log K_{ow} - \log K_{AW} = 0.79 - (-1.4222) \approx 1.5 \tag{5.18}$$

The calculated value is 0.7 less than the measured or published value. The difference means that the calculated K_{OA} is under-reported by about one-third. Thus, if the calculated value is used, the affinity for tissue is understated. That is, the MIC is less likely to exit from tissue into air than the calculated value indicates.

Make the same comparison for formaldehyde.

The published log K_{OA} for formaldehyde is 2.77. Calculating and estimating for K_{OA} from Eq. (5.17) give

$$\log K_{OA} = \log K_{ow} - \log K_{AW} = 0.35 - (-4.861) \approx 4.81$$

This difference is a large discrepancy, that is, two orders of magnitude between the two methods. As for MIC, the calculated method underestimates formaldehyde's affinity for tissue and overestimates its likelihood to move into the air.

These differences may be explained somewhat by the fact that the calculation is a rule of thumb. Each partitioning coefficient is measured differently. For example, K_{ow} is actually not usually calculated as partitioning between pure octanol and pure water; rather, the measurement reflects octanol saturated with water and water saturated with octanol [13]. Whenever the inherent properties like solubility and vapor pressure are incorrect, derived properties like *Kow*, K_{AW}, and K_{OA} will propagate and amplify such error.

These differences also appear in the organic-carbon partitioning coefficients (K_{oc}), that is, the two right columns in Table 4.7, with large discrepancies between QSAR- and *Kow*-based K_{oc} values. This is a reminder that octanol is merely a surrogate or indicator of the organic phase. This problem emphasizes that an actual, published K_{OA} based on credible methods is preferred whenever possible.

REFERENCES

[1] F.L. Smith, A.H. Harvey, Avoid common pitfalls when using Henry's law, Chem. Eng. Progr. 103 (9) (2007) 33–39.

[2] U. EPA, Air Emissions Models for Waste and Wastewater, US Environmental Protection Agency, Washington, DC, 1994.

[3] R. Lewis, S. Gordon, Sampling for Organic Chemicals in Air, Environmental Protection Agency, Research Triangle Park, NC, United States, 1994. Atmospheric Research and Exposure Assessment Lab.

[4] U.S. Environmental Protection Agency, Technical Overview of Volatile Organic Compounds, General Definition and Classification, 5 June. Available: https://www.epa.gov/indoor-air-quality-iaq/technical-overview-volatile-organic-compounds, 2018.

[5] U.-U.T. Võ, M.P. Morris, Nonvolatile, semivolatile, or volatile: redefining volatile for volatile organic compounds, J. Air Waste Manage. Assoc. 64 (6) (2014) 661–669.

[6] R.W. Williams, R.R. Watts, R.K. Stevens, C.L. Stone, J. Lewtas, Evaluation of a personal air sampler for twenty-four hour collection of fine particles and semivolatile organics, J. Expo. Sci. Environ. Epidemiol. 9 (2) (1999) 158.

[7] T.F. Bidleman, Air-Water Gas Exchange of Chemicals, 6 June 2018. Available: http://www.recetox.muni.cz/res/file/konference/bidleman/5-gas-exchange.pdf, 2018.

[8] L.M. Jantunen, T.F. Bidleman, Air—water gas exchange of toxaphene in Lake Superior, Environ. Toxicol. Chem. 22 (6) (2003) 1229–1237.

[9] L.M. Jantunen, T.F. Bidleman, Temperature dependent Henry's law constant for technical toxaphene, Chemosphere-Glob. Change Sci. 2 (2) (2000) 225–231.

[10] M. Knabb, The impact of climate change on air-water exchange of toxaphene in Lake Superior, University of Minnesota, 2013.

[11] J. Setschenow, Über die konstitution der salzlösungen auf grund ihres verhaltens zu kohlensäure, Zeitschrift für Physikalische Chemie 4 (1) (1889) 117–125.

[12] R. Sander, Compilation of Henry's law constants (version 4.0) for water as solvent, Atmos. Chem. Phys. 15 (8) (2015).

[13] D. Mackay, W.-Y. Shiu, K.-C. Ma, S.C. Lee, Handbook of Physical-Chemical Properties and Environmental Fate for Organic Chemicals, CRC Press, 2006.

[14] D. Mackay, W. Shiu, K. Ma, Physical-Chemical Properties and Environmental Fate and Degradation Handbook, Chapman & Hall CRCnetBASE, CRC Press LLC, Boca Raton, FL, 1999.

[15] T. Harner, M. Shoeib, Measurements of octanol-air partition coefficients (K OA) for polybrominated diphenyl ethers (PBDEs): predicting partitioning in the environment, J. Chem. Eng. Data 47 (2) (2002) 228–232.

[16] U.S. Environmental Protection Agency, Guidance for reporting on the environmental fate and transport of the stressors of concern in problem formulations, Available https://www.epa.gov/pesticide-science-and-assessing-pesticide-risks/guidance-reporting-environmental-fate-and-transport, 2017.

FURTHER READING

[17] Engineering Toobox, Solubility of Gases in Water, 2 June 2018. Available: https://www.engineeringtoolbox.com/gases-solubility-water-d_1148.html, 2001.

[18] D. Vallero, Environmental Contaminants: Assessment and Control, Academic Press, 2010.

[19] Nonvolatile, semivolatile, or volatile: Redefining volatile for volatile organic compounds, Available: http://www.aqmd.gov/docs/default-source/planning/architectural-coatings/redefining-volatile-for-vocs.pdf, 2012.

[20] D. Vallero, Translating Diverse Environmental Data Into Reliable Information: How to Coordinate Evidence from Different Sources, Academic Press, 2017.

[21] D.A. Vallero, Fundamentals of Air Pollution, fifth ed., Elsevier Academic Press, Waltham, MA, 2014. p. 999 pages cm.

[22] M.N. Sara, Ground-water monitoring system design, in: D.M. Nielsen (Ed.), Practical Handbook of Ground-Water Monitoring, Lewis Publishers, USA, 1991, pp. 17–68.

[23] D. van de Meent, et al., Persistence and transport potential of chemicals in a multimedia environment, in: SETAC Pellston Workshop, Fairmount Hot Springs, British Columbia, Canada, 1998, pp. 14–18.

Chapter 6

Physical transport of air pollutants

6.1 INTRODUCTION

The previous two chapters address the partitioning of chemical compounds among the various environmental compartments. This chapter applies the laws of diffusion and motion to pollutant transport. Together with partitioning, the movement of air pollutants can be estimated and predicted.

Air pollution is both a temporal and spatial concept. We are interested in how concentrations of pollutants change in time and space. As discussed in Chapter 3, pollutant concentrations change because of chemical reactions. They also change because the pollutants move. This can be envisioned as a thermodynamic control volume into which a chemical species moves. This physical movement increases the concentration of the chemical species. While in the volume, the chemical species may react, which decreases the concentration of the chemicals species. Finally, the chemical species may physically move out of the control volume, which further decreases the concentration.

The inflow, reaction, and outflow are continuous, so the process is dynamic. The rates of inflow, reaction, and outflow are constantly changing.

6.2 PARTITIONING MEETS THE LAWS OF MOTION

The gas laws and thermodynamics determine where the molecules of a chemical compound will be found. As discussed in previous chapters, molecules tend to flee or remain in a medium, depending on the inherent properties of the molecule and of the medium. Thus, fugacity determines the extent to which a molecule will move from one medium to another. The other concept to keep in mind is that kinetics occurs during partitioning, even though partitioning is an equilibrium property. The partitioning coefficients are calculated during equilibrium. However, in dynamic systems, like those in the environment or within an organism, temperature, pressure, and other conditions are constantly changing. During pollutant transport, transformation is also occurring. Thus, in addition to the parent chemical species, its degradation products are also being transported. These products have their own inherent properties that dictate the extent to which they are transported. For example, if the parent compound is a gas and its degradation product is a solid, they would move at different rates and via different transport mechanisms. The solid would move as an aerosol, likely by nondiffusive transport, whereas the gas would may also move by diffusive transport mechanisms. Indeed, we will consider the differences between gas and aerosol transport in detail in this chapter and in others. Notably, the main difference in air pollution control technology classifications is whether the pollutant being treated is in a vapor phase or aerosol phase.

As mentioned, fugacity-based, multicompartmental environmental models combine inherent properties of the chemical species with substrate characteristics, including the matrix and the fluids within the matrix. For air pollution, the dominant fluid is air, but water must also be addressed, for example, in droplets and air-water exchanges. Also, air exists in places other than the open atmosphere, for example, within soil interstices and building products and articles with foam and fabric. Certainly, fluids other than air are important to air pollution, including fuels, lubricants, and propellants. Even blood and bodily fluids are important when considering the effects of air pollutants on human and wildlife health.

After partitioning has moved a substance into the air, the next factor of transport is motion, that is, the dynamics of movement in the system. Recall that partitioning is driven by the laws of thermodynamics, whereas physical transport is driven by the laws of motion. For example, if the molecules are moved into the water, they will be transported hydrodynamically. If they move into the air, they will move aerodynamically. Other processes, like molecular diffusion, will be at work and are particularly important in quiescent systems like sediment. These follow the laws of diffusion, that is, Fick's laws, which define a gradient from high to low concentration of a chemical species.

Air Pollution Calculations. https://doi.org/10.1016/B978-0-12-814934-8.00006-5

6.2.1 Concentration and flux

The concept of concentration, introduced in Chapter 1, is crucial to transport. Recall that concentration is the quantity of a substance per volume or mass of substrate and can be expressed in units of mass per mass, volume per volume, or mass per volume. For air pollutants, mass per volume is usually preferred, for example, micrograms of a pollutant per cubic meter of air ($\mu g\,m^{-3}$), although volume per volume is sometimes seen, for example, parts per billion (ppb) of ozone. Concentration is more uncertain for air than for other environmental media because air is much more compressible. Boyle's law states that at constant temperature, volume is inversely proportional to pressure. Thus, changes in pressure will change the air concentration of a chemical species, even if the number of molecules remains the same (see Fig. 6.1).

The flux of a chemical species is the amount that passes through a surface perpendicular to the flow (see Fig. 6.2). Flux (q) is most easily calculated in one direction:

$$q = \frac{m}{A\Delta t} \tag{6.1}$$

where m is the mass of the chemical species, A is the cross-sectional area of the flux surface, and t is time duration of the flux. Thus, flux can be expressed as $\mu g\,m^{-2}\,s^{-1}$. Other nomenclature and symbols are used for flux; for example, J or N_x is commonly used to indicate mass flux, and j is used to indicate diffusive flux.

The quantity of a chemical species the moves through a cross-sectional surface varies by transport mechanism. For example, if the mechanism is advection, that is, the passive entrainment of a chemical species carried by the air, the flux is readily related to concentration and fluid velocity:

$$q = \frac{m}{V} \times \frac{V}{A \cdot t} = C \cdot u \tag{6.2}$$

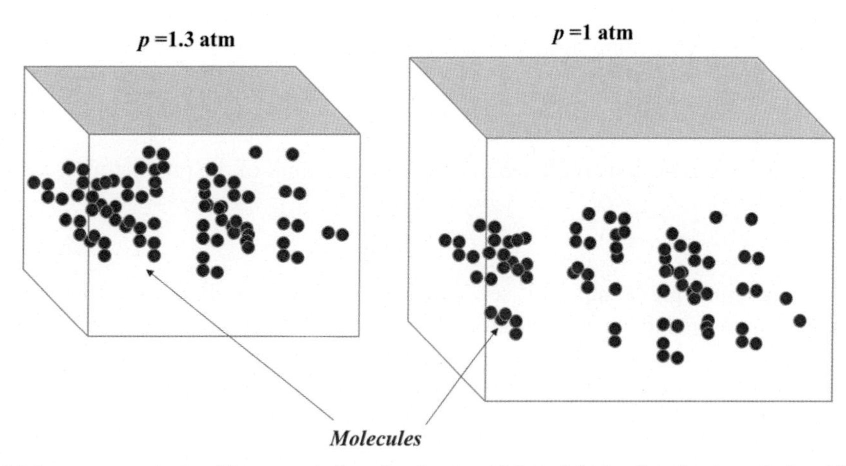

FIG. 6.1 Effect of Boyle's law on concentration. The concentration of molecules at 1.3 atm is higher than the concentration at 1 atm for the same number of molecules because the volume is larger at 1 atm. This is a theoretical, instantaneous volume increase. In reality, the molecules would spread out within the new, larger volume.

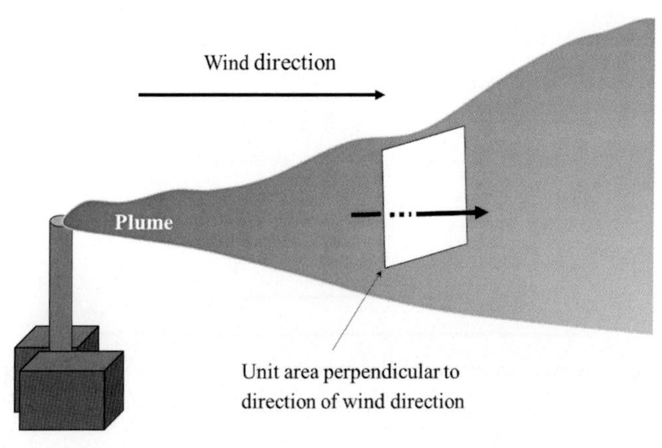

FIG. 6.2 Flux of a chemical species through of an imaginary cross-sectional area.

where C is the concentration of the chemical species and u is fluid (air) velocity.

The other principal transport mechanism is diffusion, that is, molecular motions lead to random movement of the chemical species. Commonly, advection and diffusion are the two main flux mechanisms, so the total flux would be the sum of the advective flux (Cu) and the diffusive flux (j):

$$q = Cu + j \tag{6.3}$$

Other mechanisms come into play, such as gravitational settling during air pollutant deposition from the atmosphere to the earth's surface and in air pollution control equipment. At a large scale, for example, emissions of air pollutants into the ambient air, advection is the principal transport mechanism for most chemical species. However, diffusive flux cannot be ignored for complete and accurate calculations.

6.2.2 Molecular diffusive flux

Molecular diffusion can be demonstrated with a two-compartment system (see Fig. 6.3). In compartment A, the chemical species concentration is lower than that in compartment B. C_A is the low-concentration compartment, and C_B is the high-concentration compartment. Assuming no net flow of the fluid between the two compartments, that is, no advection, the fluctuating flow (u') in one direction, x, is compensated by flow of the same magnitude in opposite direction, also x. Thus, Δx is the distance between the compartment centers. Simply, the net diffusive flux (j) from A to B equals the flux from A to B minus the flux from B to A:

$$j = C_A \cdot u' - C_B \cdot u' = u' \Delta C \tag{6.4}$$

where ΔC is the concentration difference. Eq. (6.4) can be rearranged by using Δx, the distance between the compartment centers and assuming the limit approaching infinitely small distances:

$$j = (u' \Delta x) \frac{\Delta C}{\Delta x} \tag{6.5}$$

Furthermore, we can derive a diffusion coefficient, D, which is the limit of the product $u' \Delta x$ as Δx becomes small [1]:

$$j = -D \frac{\Delta C}{\Delta x} \tag{6.6}$$

The units of j are area per time, for example, $m^2\ sec^{-1}$. Eq. (6.7) is known as Fick's law of diffusion, which shows that diffusive flux is directly proportional to the concentration gradient. The difference in concentration determines the amount of diffusion, that is, flux increases with the difference in concentration of the two compartments.

Eq. (6.7) is the discrete version of Fick's law of diffusion. The law can also be depicted in a continuous version:

$$j = -D_{O} \cdot i_c \tag{6.7}$$

where D_O is again the proportionality constant for molar concentration, $[c]$, and

$$i_c = \frac{\partial [c]}{\partial x} \tag{6.8}$$

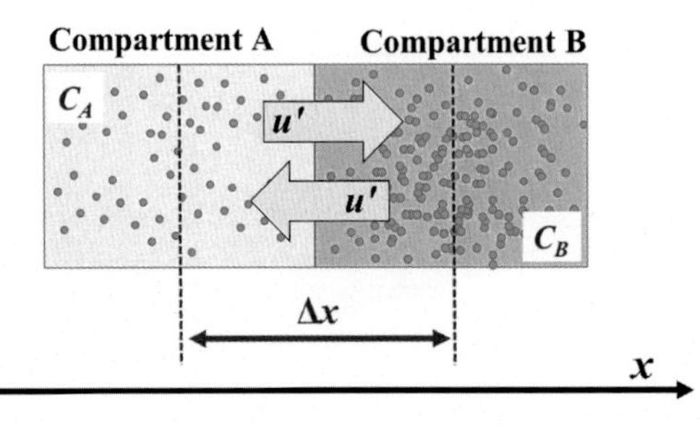

FIG. 6.3 Uneven exchange of molecules in a two-compartment system. Note: In compartment A, the chemical species concentration is lower than that in compartment B. C_A is the concentration of chemical C in compartment A, C_B is the concentration of chemical C in compartment B, and u' is the fluctuating flow in dimension x.

The negative sign denotes that the transport is in the direction of greater to lesser contaminant concentrations. Fick's second law comes into play when the concentrations are changing with time. The change of concentrations with respect to time is proportional to the second derivative of the concentration gradient:

$$\frac{\partial[c]}{\partial t} = \frac{\partial^2[c]}{\partial x^2} \tag{6.9}$$

The diffusive flux can be calculated by inserting the cross-sectional surface area:

$$j = -AD\frac{\Delta C}{\Delta x} \tag{6.10}$$

Consider the garage of an automobile repair shop with an average carbon monoxide (CO) concentration of 25 mg m^{-3}. The garage is connected to the office by a 3 m-long hallway. The hallway is 1 m wide and 2 m high. The office is ventilated so that the average CO concentration is 2.5 mg m^{-3}. If the diffusion coefficient is 8×10^{-3} m^2 sec^{-1}, calculate the CO diffusive flux between the garage and the office:

$$j = -AD\frac{\Delta C}{\Delta x} = 2\,\text{m}^2 \times \left(8 \times 10^{-3}\,\text{m}^2\,\text{sec}^{-1}\right) \times \frac{25 - 2.5\,\text{mg}\,\text{m}^{-3}}{3\,\text{m}} \approx 0.26\,\text{mg}\,\text{sec}^{-1} = 260\,\mu g\,\text{sec}^{-1}$$

If the repair shop improved the HVAC system in the garage so that the CO concentration decreased to 3 mg m^{-3}, what would you expect the office concentration (C_{office}) to be?

$$0.26\,\text{mg}\,\text{sec}^{-1} = -AD\frac{\Delta C}{\Delta x} = 2\,\text{m}^2 \times \left(8 \times 10^{-3}\,\text{m}^2\,\text{sec}^{-1}\right) \times \frac{3 - [\text{CO}_{\text{office}}]\,\text{mg}\,\text{m}^{-3}}{3\,\text{m}} = \frac{0.016 \times (3 - \text{CO}_{\text{office}})}{3}$$

$$0.26 = 0.005 - (1 - \text{CO}_{\text{office}})$$

$$1 - \text{CO}_{\text{office}} = 0.255$$

$$\text{CO}_{\text{office}} \approx 0.75\,\text{mg}\,\text{m}^{-3}$$

Thus, the office air would be expected to be improved from 2.5 to about 0.75 mg m^{-3}.

Of course, this assumes that the entire mass of CO is transported by diffusion and that the diffusion coefficient remains unchanged.

As mentioned, molecular diffusion is based on the random movement of the molecules. The diffusive flux and the random orientation of the molecules in Fig. 6.3 can be considered mathematically. Fig. 6.4 shows two adjacent stacks of molecules with their horizontal center at $x = 0$. If each molecule moves in a random manner, that is, by Brownian movement, when energy (heat) is added to the system (Fig. 6.4B), after a certain time these random movements will result in half of the molecules moving to the left and half to the right, on average. The concentration of molecules will then decrease because the molecular space has increased in size (Fig. 6.4A area is smaller than Fig. 6.4C area).

Since the concentration gradient (i_c) is the change in concentration (e.g., in units of kg m^{-3}) with length (in meters), the units of i_c are kg m^{-4}.

The concentration gradient unit, having an exponent of −4, calls for some explanation. Volume's unit is length cubed; for example, m^3 is the unit of volume. This is an expression of three-dimensional space. This is where the chemical species resides as a concentration, for example, 1 mg of chemical per cubic meter of air. Therefore, when we compare one location's concentration of a chemical species to another location's concentration of the same chemical species, we are expressing the change of number of molecules or mass per volume from the first location to the second:

$$i_c = \frac{\text{mass per llength}^3 - \text{mass per length}^3}{\text{length}} \tag{6.11}$$

So, the units represent mass per volume per distance, for example, kg m^{-4}.

These units indicate that diffusion is analogous to the physical potential field theories. The flow of matter or energy is from the direction of high potential to low potential, such as from high pressure to low pressure. This gradient is observed in all phases of matter, solid, liquid, or gas. So, molecular diffusion is a major factor of transport in porous media, such as soil or sediment, but is often ignored in other processes with bulk transport mechanisms, for example, advection. It is worth

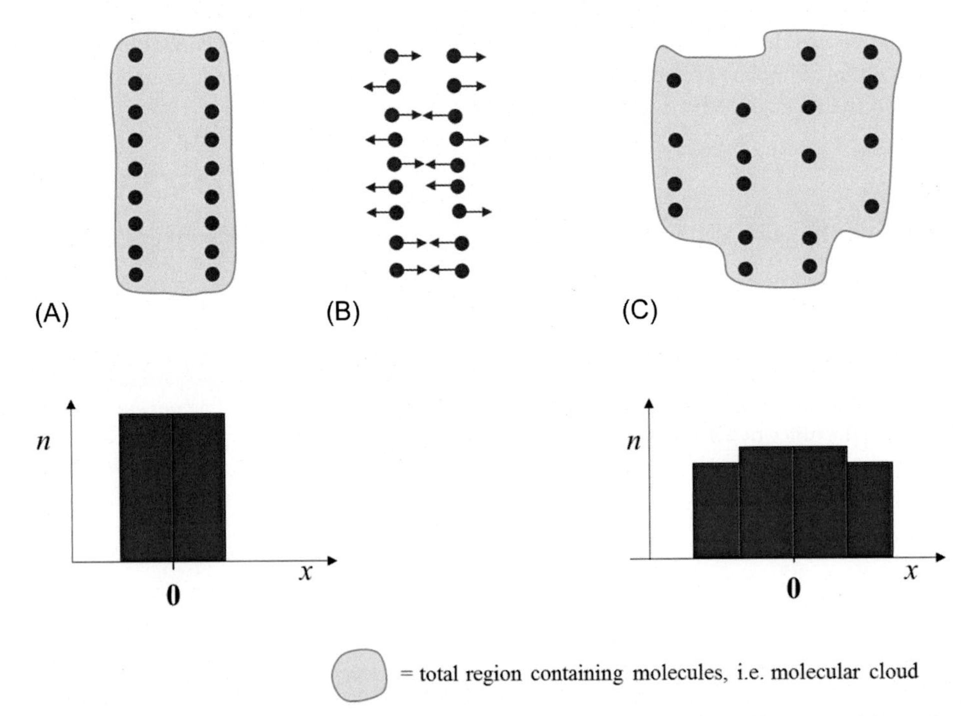

FIG. 6.4 Random motion of molecules depicting Fickian diffusion in one dimension (x). Bottom graphs are histograms of the number of molecules (n) by location on the x-axis before and after the motion. Note that the size of the molecular cloud increases. Thus, the concentration of molecules decreases because diffusion has expanded the size of the cloud. This depiction can be expanded to two dimensions by adding the y dimension and to three dimensions by adding the y and z dimensions. Air pollution transport and fate models commonly employ the z-axis. (A) Initial molecular distribution. (B) Random (Brownian) motion. (C) Final molecular distribution.

noting that even in porous media, if the flow is 2×10^{-5} m s^{-1} or higher, molecular diffusion is often ignored as a transport mechanism [2]. However, even in low-flow systems, molecular diffusion can be an important process within pockets, for example, for source characterization from locations within the porous media. It may be the principal means by which a contaminant becomes mixed in a quiescent container, such as a drum, a buried sediment, or a covered pile, or at the boundaries near clay or artificial liners in landfill systems [3].

Since molecular diffusion dominates transport only when advection and dispersion are near zero, ambient air can most often be assumed to be turbulent. If molecular diffusion does dominate, it is usually in a small pocket for only a very short time period. As mentioned, molecular diffusion should not be ignored in quiescent systems, such as within pollution control equipment or in wet soils and sediment, where concentration gradients serve as a major transport process. In these instances, diffusion is an important mechanism within air pollution sources, such as contaminants, like methane, that are generated in the reduced conditions in sediments or wetland soils and released into and held with the interstices of porous media before volatilizing or aerosolizing into the atmosphere (see Fig. 6.5). Molecular diffusion in atmospheric systems most often occurs as a transport mechanism only in a very thin boundary layer between the fluid media.

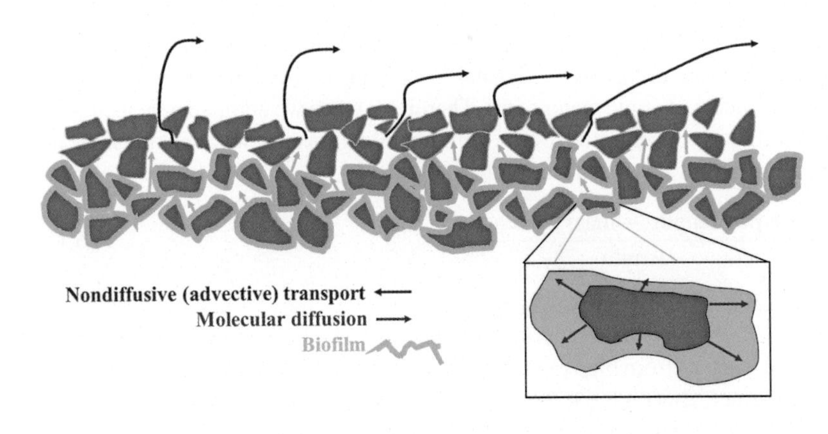

FIG. 6.5 Microscopic side view of wetland soil profile. Molecular diffusive transport occurs within the biofilm surrounding soil particles and in the void spaces, that is, interstices between particles. Nondiffusive transport becomes dominant in the nonquiescent regions, that is, air velocity increases so that the airflow becomes turbulent.

6.2.3 Eddy dispersion

Fickian processes vary by scale. A second type of diffusion is known as turbulent or eddy diffusion. Turbulent motion in fluids is characterized by the formation of eddies of various sizes. In the atmosphere, airflow is almost always turbulent. In most air pollution scenarios, laminar airflow is almost exclusively seen in engineered, controlled systems, for example, laminar flow hoods.

Even though they are more complex than the random, molecular-scale diffusion, the eddies produced by turbulence can also be modeled from the concentration gradients of Fick's first law. For this reason, turbulent dispersion is often called "eddy diffusion." However, molecular agitation at the macroscale is much less of a factor in the mixing of the air and dispersing the airborne substance. Therefore, eddy dispersion is arguably a more descriptive term. However, if the concentration gradient assumption holds, the equations applied to molecular diffusion may also be used to estimate the transport of contaminants by eddy dispersion. Like molecular diffusion, eddy diffusion can be modeled in one, two, or three dimensions. One-dimensional models assume that the diffusion coefficient (D) does not change with respect to direction. However, D must be adjusted to the model. This must be done when D is expected to vary with spatial location and time (which it always does, but if the change is not significant, it may be ignored). The coefficient may also be *anisotropic*, that is, it may vary in different directions or vertically in the air or water.

Although the equations thus far have been concerned with the concentration gradient in the horizontal direction, the gradient may occur in any of the three dimensions. For air pollutants, the vertical dimension is very important, for example, the gradient from the stack upward (z) and horizontally (x and y).

Contaminated soil is being treated in a building 200 m^2 of untreated soil. At 3 m beneath the contaminated soil, the underlying soil has a concentration of 2 μg cm^{-3} total hydrocarbons (THC). If the diffusion coefficient is 0.01 cm^2 s^{-1} in this soil, what is the flux density of the vapor and the rate of vapor release by molecular diffusion, assuming complete mixing and a one-dimensional flux?

The vertical concentration gradient (upward on z-axis):

$$\frac{dC}{dz} = (2 \times 10^{-6}\,\text{g cm}^{-3})/300\,\text{cm}) = 6.7 \times 10^{-9}\,\text{g cm}^{-4}$$

The flux density is.

$$j = -D\frac{\Delta C}{\Delta z} = (10^{-2}\,\text{cm}^2\,\text{sec}^{-1}) \times (6.7 \times 10^{-9}\,\text{g cm}^{-4}) = 6.7 \times 10^{-11}\,\text{g cm}^{-2}\,\text{sec}^{-1}$$

This tells us how much of the THC breaks through the soil surface, so applying the flux density to the 200 m^2 (2×10^6 cm^2), the penetration of the vapor into the air:

$$(6.7 \times 10^{-11}\,\text{g cm}^{-2}\,sec^{-1}) \times (2 \times 10^6\,\text{cm}^2) \times (3600\,\text{s hr.}^{-1}) \times (24\,\text{h day}^{-1}) = 11.5 g day^{-1}$$

The plant manager is told that rather than THC, she must calculate the flux density of a single hydrocarbon with a diffusion coefficient of 0.1 cm^2 s^{-1} in this soil. The underlying soil has a concentration of 0.5 μg cm^{-3} of this chemical species. Using the same assumptions, what is the flux density of the vapor and the rate of vapor release by molecular diffusion?

The vertical concentration gradient (upward on z-axis):

$$\frac{dC}{dz} = (5 \times 10^{-7}\,\text{g cm}^{-3})/300\,\text{cm}) = 1.7 \times 10^{-9}\,\text{g cm}^{-4}$$

The flux density is

$$j = -D\frac{\Delta C}{\Delta z} = (0.1\,\text{cm}^2\,\text{sec}^{-1}) \times (1.7 \times 10^{-9}\,\text{g cm}^{-4}) = 1.7 \times 10^{-10}\,\text{g cm}^{-2}\,\text{sec}^{-1}$$

Applying the flux density to the 200 m^2 (2×10^6 cm^2), the penetration of the vapor into the air:

$$(1.7 \times 10^{-10}\,\text{g cm}^{-2}\,\text{sec}^{-1}) \times (2 \times 10^6\,\text{cm}^2) \times \left(3600\,\text{sec hr}^{-1}\right) \times \left(24\,\text{hr day}^{-1}\right) = 28.8\,g\,day^{-1}$$

Obviously, something is wrong with this calculation, since the calculated amount is greater than the total hydrocarbon mass emitted. A likely culprit is the given concentration of the hydrocarbon, that is, one-fourth of THC. Another possibility is the diffusion coefficient that was 10 times higher for the single compound than for THC. It is also possible that the measurements were accurate but that the assumption of the soil being well-mixed is incorrect. Indeed, soil is highly heterogeneous and can vary in chemical concentrations even within a few centimeters.

These calculations cannot both be correct. At a minimum, additional samples must be taken and analyzed, and the literature needs to be consulted for more representative diffusion coefficients. Also, it is important to keep in mind that D is but one of the coefficients in this scenario. Others, such as sorption and K_{ow}, may be as or more important than D within the soil. Once the hydrocarbons enter the air, these other partitioning coefficients are less important than D (although Henry's law and others will be important in droplets and at water-air-aerosol interfaces).

6.2.4 Advection

From a purely physical motion perspective, the most straightforward pollutant transport process arguably is advection, that is, the transport of matter within the streamlines of a fluid, that is, with the water flow or airflow. Eddy diffusion may be seen as a transition between diffusion and bulk transport. However, eddy diffusion can be classified as a type of advection, given that eddy diffusion transport of substances is with the flow.

During advection, a chemical species is moved along with the fluid. Advection is considered a passive form of transport because the contaminant moves along with the transporting fluid. That is, the contaminant moves only because it happens to reside in the medium. Advection occurs within a single medium and among media. The rate and direction of transport are completely determined by the rate and direction of the flow of the media.

The simplest bulk transport within one environmental medium or compartment is known as homogeneous advection, where only one fluid is carrying the contaminant. Thus, in terms of time and space, most of the transport of an air pollutant is by homogeneous advection. The three-dimensional rate of homogeneous, advective transport (N) is simply the product of the fluid medium's flow rate and the concentration of the contaminant in the medium:

$$N = QC \tag{6.12}$$

where Q is the flow rate of the fluid medium (e.g., $m^3\ sec^{-1}$) and C is the concentration of the chemical contaminant being transported in the medium (e.g., $\mu g\ m^{-3}$). Therefore, the units for three-dimensional advection are mass per time (e.g., $\mu g\ sec^{-1}$).

What is the homogeneous advection of a plume in the air containing $100\ \mu g\ m^{-3}$ vinyl chloride (gas phase) flowing at $10\ m^3 sec^{-1}$ that equals to $1000\ \mu g\ sec^{-1}$ or $1\ mg\ sec^{-1}$?

$$N = 100\ \mu g\ m^{-3}\ \text{vinyl chloride} \times 10\ m^3\ sec^{-1} = 1000\ \mu g\ sec^{-1} = 1\ mg\ sec^{-1}$$

Therefore, the plume of vinyl chloride can be characterized as a $1\ mg\ m^{-3}$ plume until it is further diluted in the atmosphere.

Heterogeneous advection refers to those cases where there is a secondary phase present inside the main advective medium, for example, the presence of particulate matter (PM) carried by wind. Heterogeneous advection involves more than one transport system within the compartment.

In the vinyl chloride case above, if particulate matter (PM) is also flowing in the plume and the PM has a concentration of $100\ \mu g\ m^{-3}$ and vinyl chloride makes up 0.05 of this PM concentration, what is the homogeneous particle phase advection?
The aerosol (PM) phase of vinyl chloride in the plume $100\ \mu g\ m^{-3} \times 0.05 = 5\ \mu g\ m^{-3}$. The homogeneous particle phase advection is $5\ \mu g\ m^{-3} \times 10\ m^3\ sec^{-1} = 50\ \mu g\ sec^{-1}$. This value must be added to the gas-phase homogeneous calculation:

Thus, the heterogeneous (gas and particle phases) vinyl chloride advection would be $1000\ \mu g\ sec^{-1}$ (vapor) + $50\ \mu g\ sec^{-1}$ (PM) = $1050\ \mu g\ sec^{-1}$. Thus, not only the concentration of the dissolved fraction of the contaminant must be known but also the concentration of chemical in and on the solid particles.

Heterogeneous advection is a common transport mechanism for highly lipophilic compounds or otherwise insoluble matter that is often sorbed to particles. This may be similar to chemical transport, such as when metals form both lipophilic and hydrophilic species (e.g., ligands), depending upon their speciation. Many biomolecules and other complex organic compounds, such as the PAHs and PCBs, are relatively insoluble in water. Therefore, most of their advective transport is by attaching to particles. In fact, lipophilic organics are likely to have orders of magnitude greater than concentrations in suspended matter than is dissolved in the water (recalling the discussion of K_{ow} in Chapter 5).

Another example of advective transport is atmospheric deposition of contaminants. The sorption of contaminants to the surface of atmospheric water droplets is known as wet deposition and sorption to solid particles is known as dry deposition. The process where these contaminants are delivered by precipitation to the earth is advection.

Rather than three-dimensional transport, many advective models are represented by the one-dimensional mass *flux* equation for advection, which can be stated as.

$$J_{Advection} = \bar{v}\eta_e[c] \tag{6.13}$$

where

$\bar{v}$ = average linear velocity ($m\,s^{-1}$).
η_e = effective porosity (percent, unitless).
$[c]$ = molar concentration of the solute ($kg\,m^{-3}$).

The porosity term is crucial to unconsolidated material, like soil and underground aquifers, but is not needed for the open atmosphere airflow or water flow. Thus, for atmospheric and surface water fluxes, the equation can be stated in two dimensions:

$$J_{Advection} = \bar{v}[c] \tag{6.14}$$

Two-dimensional fluxes are an expression of the transport of a contaminant across a unit area. This rate of this transport is the flux density (see Fig. 6.2), which is the contaminant mass moving across a unit area per time. Often in air pollution, fluid velocities vary considerably in time and space (e.g., calm vs gusty wind conditions). Thus, estimating flux density for advection in a turbulent fluid usually requires a time integration to determine average concentrations of the contaminant. For example, a piece of air monitoring equipment may collect samples every minute, but the model or calculation calls for an hourly value, so the 60 values are averaged to give one integrated concentration of the air pollutant.

What is the flux density of a pesticide with the atmospheric concentration of 15 ng m^{-3} and moving at a velocity of 0.1 m sec^{-1} measured at an air monitor as the pesticide moves downwind?

$$C_{pesticide} = 15\,ng\,m^{-3}$$

$$J_{advection} = \bar{v} \cdot C_{pesticide} = (0.1\,m\,sec^{-1})(15\,ng\,m^{-3}) = 1.5\,ng\,m^{-2}$$

Four other equally spaced monitors were found to be measuring the pesticide at the same time as the above (monitor B). What is the average flux density of the pesticide at the atmospheric concentration based on the following measurements?

Site	Concentration (ng m^{-3})	Wind velocity (m sec^{-1})
A	99.5	0.2
B	15.0	0.1
C	4.1	0.08
D	0.08	0.1
E	0.03	1.3

The average concentration is $99.5+15+4.1+0.08+0.03=118.71/5 \approx 23.7\,ng\,m^{-3}$
The average wind velocity is $0.2+0.1+0.08+0.1+1.3=1.78/5 \approx 0.36\,m\,sec^{-1}$
Given these average, the mean flux is.

$$J_{advection} = \bar{v} \cdot C_{pesticide} = (0.36\,m\,sec^{-1})(23.7\,ng\,m^{-3}) \approx 8.5\,ng\,m^{-2}$$

Therefore, the additional monitors differ substantially from the single measurement in terms of concentration and flux of the pesticides. The high value at site A may indicate that the monitor is near the source, for example, a pesticide manufacturing facility or a sprayed field.

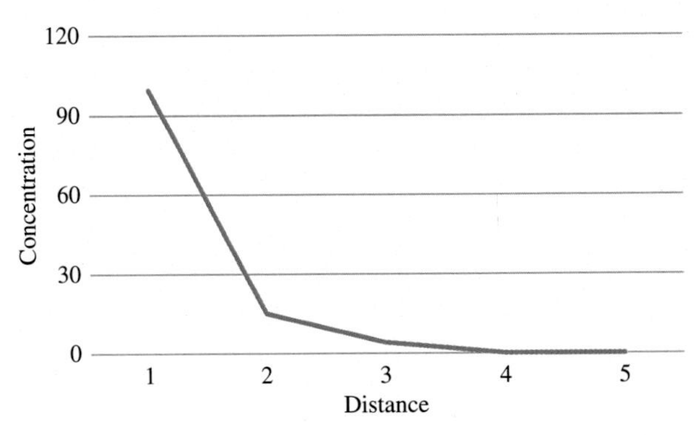

FIG. 6.6 Hypothetical comparison of pesticide concentration in air with distance.

What may be inferred if the values reported above were from equally spaced monitors along a transect instead of equally distributed in two-dimensional space as shown in Fig. 6.6?

This appears to be a concentration gradient, with the highest value possibly near the source.

6.2.5 Dispersion

Advection and diffusion provide motion for pollutant dispersion, but numerous dispersion processes exist. The dominant type of dispersion varies according to scale. Air pollutant transport literature identifies two principal types, that is, dynamic dispersion and mechanical dispersion. These are not mutually exclusive terms. In fact, mechanical dispersion is a factor in dynamic dispersion.

Computational approximations are based on first principles of motion and thermodynamics and can be applied to any physical, chemical, or biological agent. They are also useful in predicting the dispersion of agents in emergency situations.

The spread of a contaminant into multiple directions longitudinally is known as dynamic dispersion. If in air, the spreading is known as aerodynamic dispersion, and if in water, it is hydrodynamic dispersion. The scattering of molecules results from physical processes that affect the velocity of different molecules in an environmental medium. For example, in aquifers, the process is at work when the contaminant traverses the flow path of the moving groundwater. This is the result of two physical mechanisms: molecular diffusion and mechanical dispersion. Molecular diffusion can occur under both freely flowing and stagnant fluid systems, while mechanical dispersion is of most importance in flowing systems. The units of dynamic dispersion d_d are area per time (e.g., $cm^2\ sec^{-1}$).

Since dispersion is the mixing of the pollutant within the fluid body (e.g., the atmosphere), the question arises as to whether it is better to calculate the dispersion from physical principles, using a deterministic approach, or to estimate the dispersion using statistics and probabilities. The Eularian model bases the mass balance around a differential volume. A Lagrangian model applies statistical theory of turbulence, assuming that turbulent dispersion is a random process described by a distribution function. The Lagrangian model follows the individual random movements of molecules released into the plume, using statistical properties of random motions that are characterized mathematically. Thus, this mathematical approach estimates the movement of a theoretical volume of chemical from one point in the plume to another distinct point during a unit time.

Physicists may refer to this theoretical volume as a "particle," that is, a theoretical point that is followed in a fluid. The point represents the path that the pollutant is expected to take. Unfortunately, in air pollution engineering and atmospheric sciences, the term particle is often used to mean an aerosol. Particle is also commonly used to describe unconsolidated materials, such as soils and sediment. The present discussion, for example, accounts for the effects of these particles (e.g., frictional) as the fluid moves through unconsolidated material. Thus, rather than using the term particle here, we will use the term molecule. The path each molecule takes during this time is an ensemble mean field that relates to the molecule's displacement probabilities:

$$\overline{C}(x,y,z,t) = M_{total}P(Dx,t) \tag{6.15}$$

where

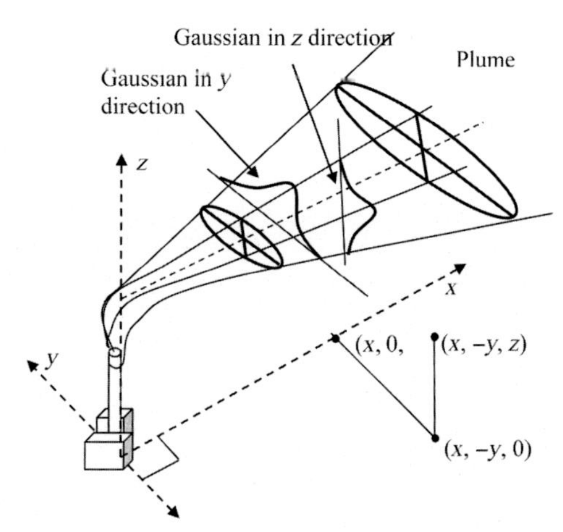

FIG. 6.7 Gaussian plume. The molecules of the airborne chemical species are assumed to be distributed vertically and horizontally in a statistically normal manner about the plume centerline away from the point of release. *(Source: D.A. Vallero, Fundamentals of Air Pollution, 5th ed., Elsevier Academic Press, Waltham, MA, 2014, p. 999 pages cm.)*

$Dx = x_2 - x_1$ = molecule displacement.
$P(Dx_2, t)$ = probability that the point x_2 will be immersed in the dispersing media at time t.
M_{Total} = total mass of molecules released at x_1.
$\overline{C}$ = mean concentration of all released molecules = mass of molecules in the plume $dx \cdot dy \cdot dz$ around x_2.

Gaussian dispersion models assume a normal distribution of the plume (see Fig. 6.7). This is a common but at best a first approximation of the actual dispersion within a biochemodynamic system.

In a deterministic approach, the dispersion includes mixing at all scales. At the larger scales, characteristics of strata and variability in the permeability of the layers must be described. So, a deterministic dispersion flux would be.

$$J_{Dispersion} = \underline{D} \cdot \text{grad}[c] \tag{6.16}$$

where

$J_{Dispersion}$ = mass flux of solute due to dispersion ($\text{kg m}^{-2}\text{ s}^{-1}$).
$\underline{D}$ = dispersion tensor (m s^{-1}).
C = concentration of chemical contaminant (kg m^{-3}).

The $\underline{D}$ includes coefficients for each direction of dispersion [3], that is, longitudinally, horizontally, and vertically (D_{xx}, D_{xy}, D_{xz}, D_{yy}, D_{yz}, and D_{zz}).

There is a lack of consistency in the usage of terms in pollutant transport and even within the air pollution science and engineering community. For example, the net transport of substances by diffusion and advection in fluids is sometimes referred to as convection. However, convection is also defined as vertical motion because of buoyancy and is contrasted with advection, which is the horizontal motion in the atmosphere driven by winds. Convection is often considered a local phenomenon, whereas advection is applied to long-range transport. Diffusion and dispersion are sometimes used synonymously. Given this lack of consistency, when addressing pollutant transport mechanisms, it is best to discuss each type of mass transport precisely.

6.2.6 Merging mass transfer and flow

Our depictions of chemical species transport, for example, Fig. 6.3, represent the macroscale between discrete compartments. However, the fluid flow and the coefficients for mass transfer and diffusion (D) must be connected.

The rate of mass transfer, that is, the flux at the interface (N_i), is proportional to the product of interfacial area and the difference in concentration at the interface and in the bulk solution. So, the mass transfer coefficient is proportionality constant (k). In a liquid, this is.

$$N_i = k(C_l - C_{l-i}) \tag{6.17}$$

The flux, N_i, is a total flux that includes diffusion and advection. Note that the mass transfer coefficient allows for changing the size of the flux surface and changes in the amount of mass being transferred, while flux per area remains constant [4]. For example, tripling the surface area between two fluid compartments will triple the mass of benzene being transferred to the adjacent compartment, but the benzene flux per area remains unchanged.

A 19.2 L, closed tank is isothermal at 25°C, so that the vapor pressure of water in the tank is 0.031 atm. The tank has 0.8 L water with 150 cm^2. In 3 min, the air is 5% saturated. Calculate the mass transfer coefficient.

Comparing the partial pressure of water to total pressure and adjusting temperature,

$$N_i = \frac{(C_{vapor} \times V_{air})}{A_{liquid} \times t} = 0.05\frac{\left(\frac{0.031\text{ atm}}{1\text{atm}} \times \frac{1\text{ mol}}{22.4\text{L}} \times \frac{273^\circ\text{K}}{298^\circ\text{K}} \times (19.2 - 0.8\text{L})\right)}{150\text{ cm}^2 \times 180\text{ sec}} = 4.3 \times 10^{-8}\text{ mol cm}^{-2}\text{ sec}^{-1}.$$

When will the tank's headspace reach 90% saturation?

The air concentration immediately at the water surface is saturated. During short time periods, the concentration in the bulk solution (i.e., liquid water) is zero. So, calculate the difference in concentration at the water surface from the concentration at the bulk solution. Applying Eq. (6.13) and solving for k,

$$4.3 \times 10^{-8}\text{ mol cm}^{-2}\text{ sec}^{-1} = k\left(\frac{0.031\text{ atm}}{1\text{ atm}} \times \frac{1\text{ mol}}{22.4 \times 10^3\text{ cm}^3} \times \frac{273^\circ\text{K}}{298^\circ\text{K}}\right) - (0)$$

$$k = \frac{4.3 \times 10^{-8}\text{ mol cm}^{-2}\text{ sec}^{-1}}{\left(\frac{0.031\text{ atm}}{1\text{ atm}} \times \frac{1\text{ mol}}{22.4 \times 10^3\text{ cm}^3} \times \frac{273^\circ\text{K}}{298^\circ\text{K}}\right)} = 0.034\text{ cm sec}^{-1}$$

At this transfer rate, the saturation percentage can be ascertained from the mass balance between accumulation in the vapor phase and the evaporation rate:

$$\frac{d}{dt}V_{C-vapor} = \text{A}N_\text{i} = \text{A}k[C_{water}(\text{sat}) - C_{water}] \tag{6.18}$$

For dry air, $t = 0$ and $C_{water} = 0$

Integrating the mass balance with these values gives.

$$\frac{C_{water}}{C_{water}(\text{sat})} = 1 - e^{-\left(\frac{kA}{V}\right)t}$$

Thus,

$$t = \frac{V}{kA}\ln\left(1 - \frac{C_{water}}{C_{water}(\text{sat})}\right) = \frac{18.4 \times 10^3\text{ cm}^3}{(0.034\text{ cm sec}^{-1}) \times 150\text{ cm}^2}\ln(1 - 0.9) = 8.3 \times 10^3\text{ sec}$$

Thus, it will take over 2 h for the headspace of the tank to be 90% saturated [5].

If the flow is between fluids that differs substantially in density, for example, gas versus liquid or aqueous-phase versus nonaqueous-phase liquid, the interface between the high- and low-concentration compartments is not a simple plane, but a boundary or interface with films on each side (see Figs. 6.8 and 6.9). As mentioned, Brownian diffusion occurs in quiescent and stagnant conditions, whereas mass transfer by advection occurs in turbulence.

The connections can be constructed from transfer theories. The two-film theory [6] represents both conditions, that is, transport of a chemical species across the films on either side of the compartment boundary is governed by molecular diffusion and the bulk of the fluid outside of the film is governed by bulk, nondiffusive transport. The concentration profile in the two-film figures assumes steady-state conditions, that is, the diffusive flux is constant. They also assume a linear concentration profile through each stagnant film, instantaneous equilibrium, and dilute solutions, that is, Henry's law applies (see Chapter 5). In addition, two-film theory assumes that transport by bulk diffusion is not limiting.

Note that the concentration gradients disappear in the bulk phases due to turbulence.

The two-film theory is the simplest for interfacial mass transfer and does not hold for many systems, since the stagnant interface layers do not exist. However, even though the physical layer may not exist, systems may behave as if they do [7].

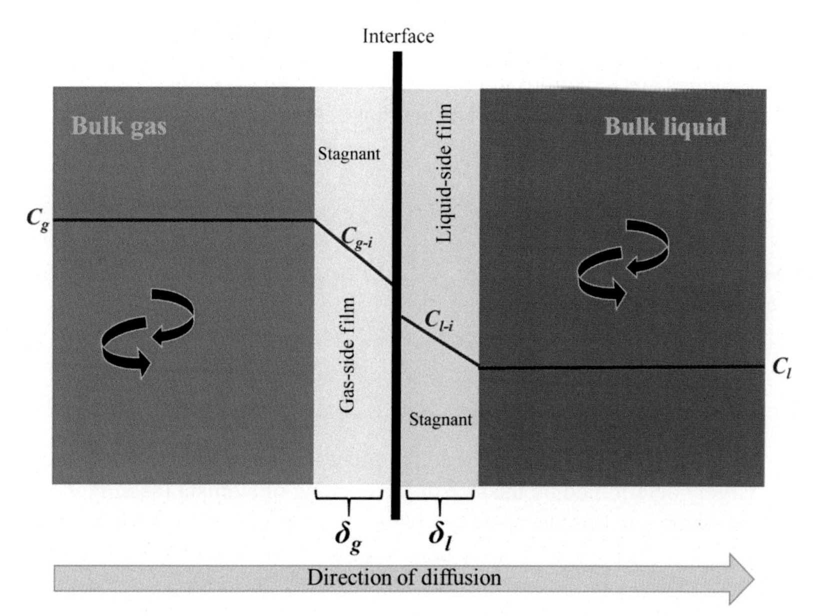

FIG. 6.8 Gas-liquid interface according to the two-film theory. *Notes*: δ_g is gas-side film thickness, δ_l is liquid-side film thickness, C_g is the bulk gas concentration, C_l is the bulk liquid concentration, $C_{g\text{-}i}$ is the gas-side film (gas interface) concentration, and $C_{l\text{-}i}$ is the liquid-side film (liquid interface) concentration. Often, in a gas-liquid interface, the gas concentrations are expressed as partial pressures (see Fig. 6.9).

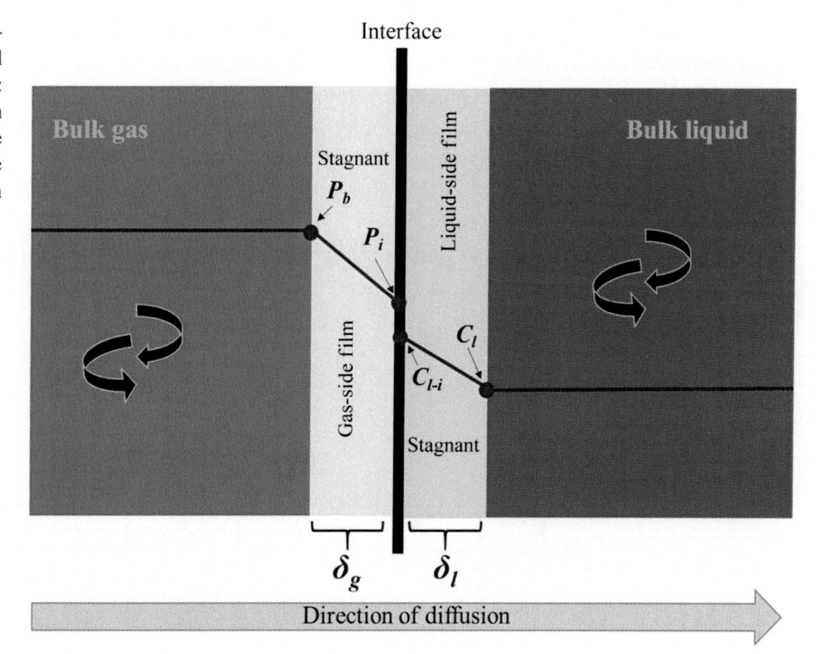

FIG. 6.9 Gas-liquid interface according to the two-film theory, based on partial pressure of chemical species in gas phase in bulk and interface layers. *Notes*: δ_g is gas-side film thickness, δ_l is liquid-side film thickness, P_b is the bulk gas partial pressure, P_i is the gas-side film (gas interface) partial pressure, C_l is the bulk liquid concentration, and $C_{l\text{-}i}$ is the liquid-side film (liquid interface) concentration.

Another theory of mass transport, penetration theory [8], more realistically assumes that diffusion is unsteady because eddies move from the bulk phase to the interface and remain there for a constant time interval of exposure, t_e. It differs from two-film theory in that it assumes the film is sufficiently thick, that is, large δ values, so that diffusion may dominate in the z-direction, but convection is the principal mechanism in the x-direction. The chemical species penetrates into an eddy during t_e by unsteady molecular diffusion. Meanwhile, the chemical diffuses into the interface fluid. Penetration theory predicts that the mass transfer coefficient is directly proportional to the square root of the molecular diffusivity (i.e., the diffusion coefficient, D_{AB}).

These and other film theories are helpful in understanding the concentration gradient, but not worthy of further application here. Empirical evidence indicates that mass transfer coefficient, k, should vary with square root of the diffusion coefficient, D, and k should also vary with the two-third power of the fluid velocity, ν. However, both of these film theories predict smaller variances [9]. This may in part be due to the nonuniform flow common in many environmental settings, especially in the troposphere. Some may also be due to the ongoing chemical kinetics that change the physical attributes of chemical compounds of interest.

6.3 DIFFUSION AND BIOLOGICAL PROCESSES

The figures and equations introduced thus far have a temporal component. Each environmental compartment acts as a reservoir for a chemical species. The amount of time the chemical remains in a compartment or reservoir is its mean residence time (MRT):

$$\tau = \frac{M}{J_{out}} \tag{6.19}$$

where τ = MRT at steady state, M is the mass of chemical species in a compartment, and J_{out} is the amount of chemical species leaving the compartment. In this sense, Fick's law can be seen as the concentration gradient against the resistance of the chemical to exit the compartment:

$$J = \frac{i_c}{R} \tag{6.20}$$

where R is the resistance of the compartment. For example, the exchange of gases between a plant's leaf and the atmosphere is largely determined by the size of the opening of stomata (see Fig. 6.10), that is, the small openings in an epidermis of a plant where gas flux occurs, especially for water vapor and carbon dioxide (CO_2). The stomata can be seen as the "gatekeepers" for all gas diffusions between the ambient air and the plant [10]. The stomata are the gates that balance gas exchange between the leaf interior and the bulk atmosphere. As such, they ideally maximize CO_2 flux into the leaf and minimize H_2O flux out of the leaf (i.e., limit transpiration). Indeed, the actual gas flux is not governed by a single R, but by a pathway consisting of a series of R values, beginning with the interface between the leaf and the air, that is, the leaf boundary layer. The dominant R between the air and the leaf interior is provided by the stomatal pore openings, with additional R provided by CO_2 as it confronts the aqueous and lipid boundaries within the leaf tissue, that is, mesophyll cell and chloroplasts, collective called "mesophyll resistance." The outward flux of H_2O moving to the atmosphere encounters all the R series, except mesophyll resistance. The stomatal pores are regulated by guard cells, which increase or decrease in size as a response to external stimuli, for example, changes in temperature and humidity. Interestingly, stomatal pores represent only about 3% of the leaf surface but account for 98% of CO_2 inward flux and H_2O outward flux [10]. See Discussion Box "Diffusion and Food."

Diffusive flux affects and is part of many biological processes, such as delivering carbon dioxide (CO_2) to a plant to begin photosynthesis.

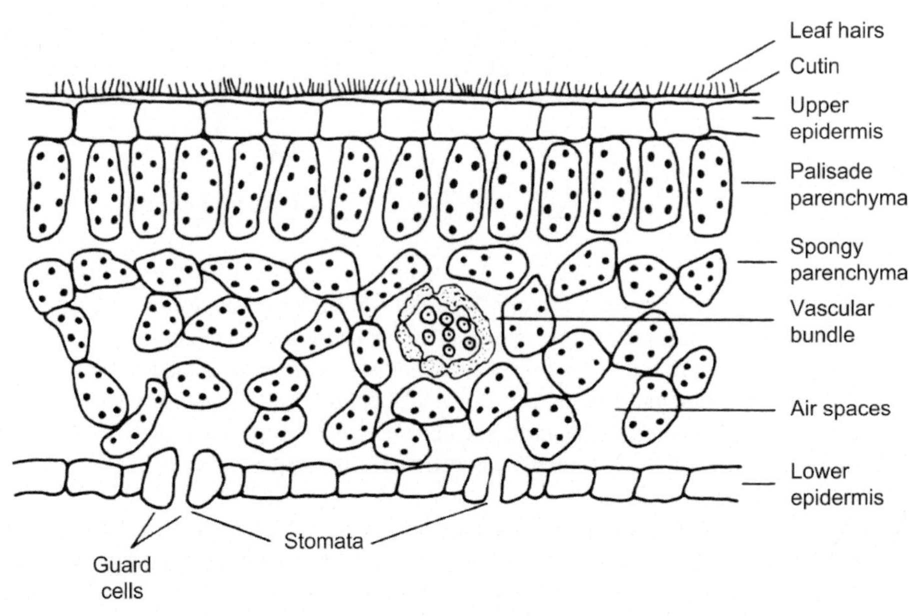

FIG. 6.10 Leaf cross section. *(Source: D.A. Vallero, Fundamentals of Air Pollution, 5th ed., Elsevier Academic Press, Waltham, MA, 2014, p. 999 pages cm.)*

If a plant leaf's CO_2 concentration ranges from about 100 μL L^{-1} inside the leaf to 400 μL L^{-1} In the atmosphere [11]. The water vapor gradient of concentration ranges from saturation inside the leaf to whatever is the vapor pressure of the atmosphere. The atmospheric vapor pressure is a function of the relative humidity and temperature of the air around the leaf. If the stomatal resistance is 5 s cm^{-1}, what is the flux of CO_2 into the leaf and water vapor exiting the leaf at 25°C and 30% relative humidity?

Calculate the amount CO_2 entering the leaf:

$$J = \frac{(400 - 100)\,\text{uL L}^{-1}}{5\,\text{sec cm}^{-1}}$$

Converting liters to cubic centimeters,

$$J = \frac{300\,\text{uL} \times 1000\,\text{cm}^{-3}}{5\,\text{sec cm}^{-1}} = 0.06\,\mu\text{L cm}^{-2}\,\text{sec}^{-1}$$

Thus, applying the gas laws, the amount of CO_2 entering the leaf would be

$$0.06\,\mu L\,\text{cm}^{-2}\,\text{sec}^{-1} \times \frac{1\,\mu\text{mol}}{22.4\,\text{umol}} = 0.0027\,\mu\text{mol CO}_2\,\text{cm}^{-2}\,\text{sec}^{-1}$$

During an ozone air pollution episode, the stomatal aperture shrinks, so that the resistance increases to 10 s cm^{-1}. What happens to the flux of CO_2 into the leaf and H_2O transpiration at 100% relative humidity and 25°C?

The increased resistance means that the flux is halved, that is, 0.0013 μmol CO_2 cm^{-2} s^{-1}.

At 100% relative humidity and 25°C, then, the air contains

$$\frac{29.8\,\text{g m}^{-3}\;1\,\text{mol water}}{18\,\text{g}} = 1.6\,\text{mol m}^{-3}.$$

Thus, if the air has only 30% relative humidity at this temperature, the water vapor content would be 33% of 1.6, that is, about 0.5 mol m^{-3}, so the leaf-to-air flux is

$$\frac{1.6 - 0.50\,\text{mol m}^{-3} \times 1\,\text{m}^3}{5\,\text{sec cm}^{-1} \times 10^6\,\text{cm}^3} = 0.23\,\mu\text{mol H}_2\text{O cm}^{-2}\,\text{sec}^{-1}.$$

Diffusion and Food

Air pollutant gas flux follows the same pathways as CO_2 and water vapor. Changes to stomatal and other biotic mechanisms can indirectly relate to many public health problems. For example, tropospheric ozone (O_3) can greatly damage crop yields by accelerating leaf senescence and detachment, that is, abscission and by decreasing stomatal aperture size. These changes lead to diminished carbon uptake and decreased photosynthetic carbon fixation (see Fig. 6.11). Elevated concentrations of O_3 depress yields of crops important to global food supply, for example, average global yield reductions as high as 5.5% for corn, 14% for wheat, and 15% for soybeans [12]. Given that climate change may lead to increasing emissions of hydrocarbon and oxides of nitrogen in many areas, the troposphere's O_3 concentrations may continue to increase in future decades, exacerbating the food supply stress. Ozone causes visible injury symptoms to foliage. Increasing ozone exposure can

- moderate biomass growth via carbon availability or more directly and can decrease translocation of fixed carbon to edible plant parts, that is, grains, fruits, pods, and roots, due to reduced availability at source, redirection to synthesis of chemical protectants, or reduced transport capabilities via phloem;
- decrease carbon transport to roots, reduce nutrient and water uptake, and affect anchorage;
- moderate or bring forward flowering and induce pollen sterility;
- induce ovule and/or grain abortion;
- diminish the ability of some genotypes to withstand other stresses such as drought, high vapor pressure deficit, and high photon flux density via effects on stomatal control.

The functioning of the stomata and other mechanisms can be adversely affected by numerous air pollutants, which results in substantial crop loss. Sulfur dioxide (SO_2), oxides of nitrogen (NOx), ozone (O_3), peroxyacetyl nitrate (PAN), and hydrocarbons

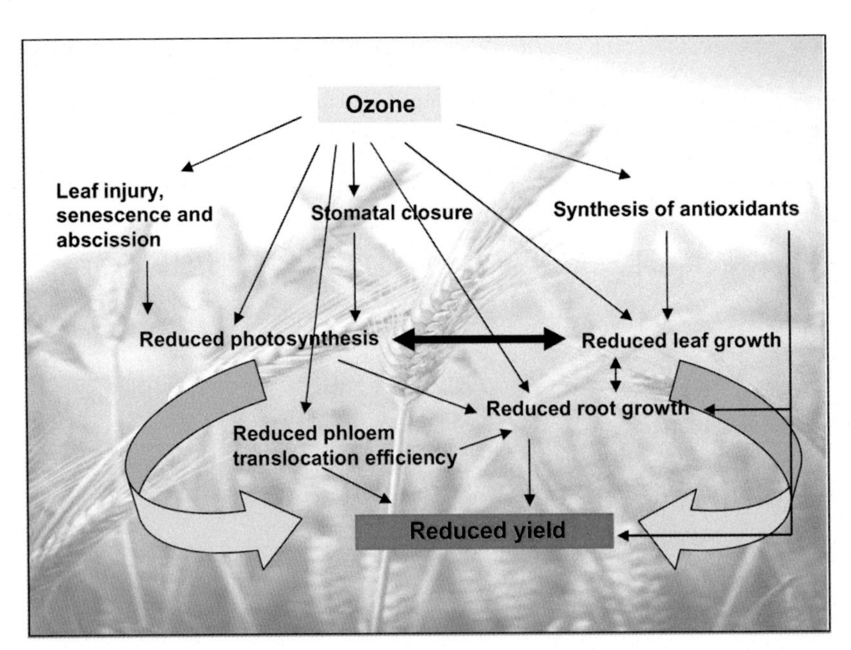

FIG. 6.11 Effects of ozone on carbon gain and carbon use that impact on crop yield. *(Source: S. Wilkinson, G. Mills, R. Illidge, W.J. Davies, How is ozone pollution reducing our food supply? J. Exp. Bot. 63 (2) (2012) 527–536.)*

are notable examples. By far, exposure to O_3 is the biggest problem, causing up to 90% of the air pollution injury to vegetation in the United States [13].

Generally, the pollutant gases follow the same diffusive pathways as CO_2 but are often followed by other mechanisms; for example, NOx dissolves in cells and generates phytotoxic nitrite ions (NO_2) and nitrate ions (NO_3^-), which affects metabolism. Exposure to SO_2 causes stomata to close, which protects against further leaf injury, but inhibits photosynthesis by diminishing CO_2 flux. Once SO_2 enters the plant, it dissolves in water and generates phytotoxic sulfite ions. However, at low concentrations, these ions will help to detoxify the plant and provide a source of the sulfur, which is a plant nutrient. This is an example of an optimal range, below which is nutrient deficiency and above which is toxicity.

Crop loss varies in time and space, and assigning the fraction due to air pollution is unclear. For example, when average daily ozone concentrations reach >50 ppb, yields of vegetables can be reduced by up to 15% [13].

Predictions are that the same atmospheric conditions that are conducive to drought formation likely will also give rise to intense oxides of nitrogen and sulfur and other precursor emission events, which may become more severe over the coming decades [3].

These calculations provide an opportunity to introduce the concept of water-use efficiency (WUE), which is the ratio of water being used for plant metabolism to water lost to transpiration:

$$\mathrm{WUE} = \frac{\left[c_{CO2-fixed}\right]}{\left[c_{H2O-lost}\right]} \tag{6.21}$$

where $[c_{CO2-fixed}]$ is the moles of CO_2 fixed in the plant and $[c_{H2O-lost}]$ is moles of H_2O lost to evapotranspiration.

If moles of water are lost, they are not available for plant processes, especially photosynthesis. Thus, a falling WUE indicates plant stress is pending or present. For most vegetation, WUE ranges from 0.86 to 1.50 and varies by environmental conditions, and WUE increases with decreasing stomatal conductance [14, 15]. Stomatal conductance is the rate of passage of CO_2 entering or water vapor exiting through the stomata of a leaf (usually measured in mmol m^{-2} s^{-1} and, therefore, is a diffusive flux).

In the previous example of an ozone air pollution episode, in which the leaf resistance increases to 10 s cm^{-1}, what happens to the flux of CO_2 into the leaf and H_2O transpiration at 100% relative humidity and 25°C?

Recall that the stomatal resistance (r) is 5 s cm^{-1}. Thus, the calculations show the water-use efficiency of this leaf when $r = 5$ s cm^{-1}; WUE is about $\frac{0.0100\ \mu mol}{\mu mo} = 10$ mmol mol^{-1}. Thus, the gradient of water vapor between the leaf and the atmosphere is so much steeper than the gradient for CO_2, meaning that the WUE is very low and plants are less efficient at holding fixed CO_2, than the atmosphere is at pulling the water from the plant by diffusion.

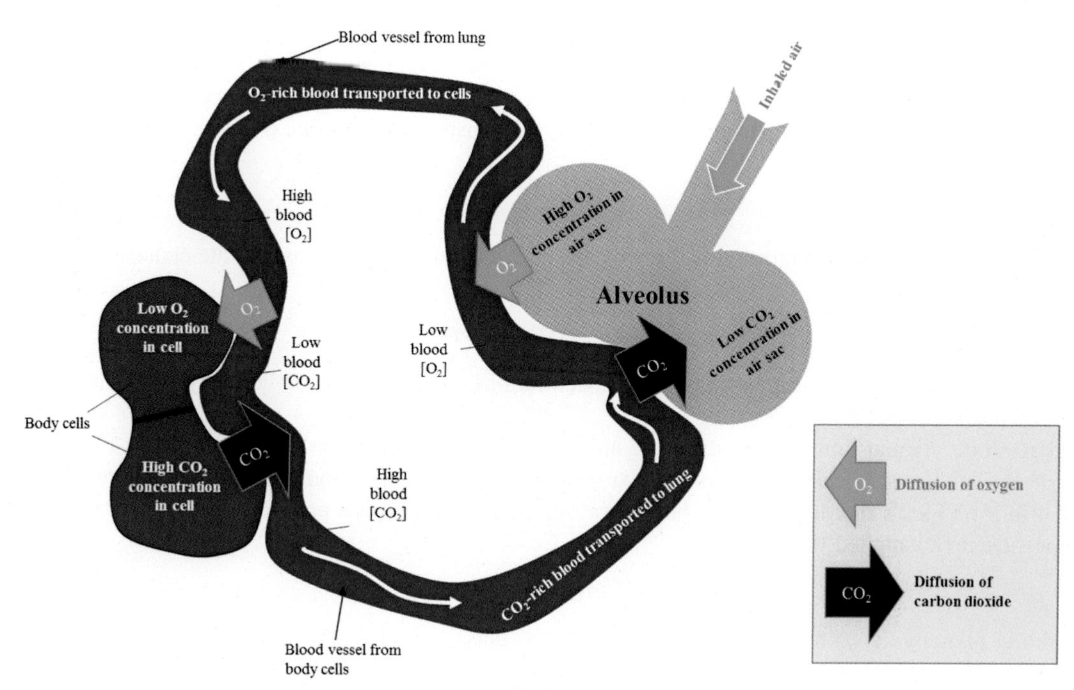

FIG. 6.12 Diffusion of oxygen and carbon dioxide between the lung, blood, and cells.

As tropospheric CO_2 rises, photosynthesis can occur at the same rate at lower stomatal conductance, which would be expected to increase.

Diffusion is also a crucial mechanism in animals. The respiratory system delivers O_2 to the bloodstream and removes CO_2. Air containing O_2 flows into the nose and/or mouth and down through the upper airway to the alveolar region, where O_2 diffuses across the lung wall to the bloodstream. The counterflow involves transfer of CO_2 from the blood to the alveolar region and then up the airways and out the nose (see Fig. 6.12). The respiratory system may be divided into three regions: the nasal, tracheobronchial, and pulmonary. The nasal region is composed of the nose and mouth cavities and the throat. The tracheobronchial region begins with the trachea and extends through the bronchial tubes to the alveolar sacs. The pulmonary region is composed of the terminal bronchi and alveolar sacs, where gas exchange with the circulatory system occurs.

The continued bifurcation of the trachea forms many branching pathways of increasingly smaller diameter by which air moves to the pulmonary region. The trachea branches into the right and left bronchi. Each bronchus divides and subdivides at least 20 times; the smallest units, bronchioles, are located deep in the lungs. The bronchioles end in about 3 million air sacs, the alveoli [3]. The diffusion of O_2 and CO_2 is example of mass transfer via the concentration gradient from high to low concentration, that is, the oxygenation of the blood is accomplished by the concentration gradient across the surface of the lung. The lung surface is the interface across which diffusion occurs. In this case, the concentration gradients of O_2 and CO_2 are in opposite directions.

After the air is channeled through the trachea to the first bronchi, the flow is divided at each subsequent bronchial bifurcation until very little apparent flow is occurring within the alveolar sacs. Mass transfer is controlled by molecular diffusion in this final region. Because of the very different flows in the various sections of the respiratory region, particulates suspended in air and gaseous air pollutants are treated differently in the lung.

For gases, solubility controls removal from the airstream. Highly soluble gases such as SO_2 are absorbed in the upper airways, whereas less soluble gases such as NO_2 and ozone (O_3) may penetrate to the pulmonary region. Irritant gases are thought to stimulate neuroreceptors in the respiratory walls and cause a variety of responses, including sneezing; coughing; bronchoconstriction; and rapid, shallow breathing. The dissolved gas may be eliminated by biochemical processes or may diffuse to the circulatory system. [3].

6.4 PARTITIONING, TRANSPORT AND KINETICS

Fugacity-based, multicompartmental environmental models account for transformation and transport relationships. The movement of a contaminant through the environment can be expressed according to the processes that lead to the equilibrium of a chemical species in each compartment. The processes driving this movement can be summarized into transfer coefficients or compartmental rate constants, that is, the previously mentioned *D* values [16]. By first calculating the *Z* values introduced in Chapters 4 and 5 and then equating the amount of a chemical species entering and exiting each compartment, we can derive *D* value rate constants. The actual transport process rate (*N*) is the product of fugacity (*f*) and the *D* value:

$$N = D \cdot f \tag{6.22}$$

Because the contaminant concentration is $Z{\cdot}f$, we can substitute and add a first-order rate constant *k* to give us a first-order rate *D* value (D_R):

$$N = V[\mathrm{c}]k = (V \cdot Z \cdot k) \cdot f = D_R \cdot f \tag{6.23}$$

Although the concentrations are shown as molar concentrations (i.e., in brackets), they may also be represented as mass per volume concentrations.[1]

As discussed throughout this chapter, diffusive and nondiffusive transport processes follow Fick's laws, that is, diffusive processes can be estimated for both molecular diffusion and eddy diffusion. Thus, the diffusive flux of a chemical species that follows Fick's first law, N_D, can be generalized and expressed as a product of its concentration and the mass transfer coefficient (*K*) applied to area A:

$$N_D = K\mathrm{A}[\mathrm{c}] \tag{6.24}$$

Substituting fugacity ($Z \cdot f$) for concentration, we find the mass transfer to be

$$N_D = (K \cdot \mathrm{A} \cdot Z) \cdot f \tag{6.25}$$

Therefore, the diffusive mass transfer is simply the product of the compartmental rate constant resulting from Fick's first law processes and fugacity:

$$N_D = D_D \cdot f$$

Nondiffusive transport (bulk flow or advection) within a compartment with a flow rate (*G*) has a *D* value (D_A) and is expressed as

$$N = G[\mathrm{c}] = (G \cdot Z \cdot)f = D_A \cdot f \tag{6.26}$$

This means that a substance is moving through the environment, during its residence time in each phase, is affected by numerous physical transport and chemical degradation and transformation processes. The processes are addressed by models with the respective *D* values, so that the total rate of transport and transformation is expressed as

$$f \cdot (D_1 + D_2 + \ldots D_n) \tag{6.27}$$

Very fast processes have large *D* values, and these are usually the most important when considering the contaminant's behavior and change in the environment.

Models, though imperfect, are important tools for estimating the movement of contaminants in the environment. They do not obviate the need for sound measurements. In fact, measurements and models are highly complementary. Compartmental model assumptions must be verified in the field. Likewise, measurements at a limited number of points depend on models to extend their meaningfulness. Understanding the basic concepts of a contaminant transport model, we are better able to explore the principal mechanisms for the movement of contaminants throughout the environment.

6.4.1 Transfer coefficients and fugacity

The movement of a contaminant through the environment can be expressed according to the equilibrium achieved in each compartment. The processes driving this movement can be summarized into transfer coefficients or compartmental rate constants, that is, the *D* values [16]. So, by first calculating the *Z* values, as we did in the previous chapter, and then equating inputs and outputs of the contaminant to each compartment, we can derive *D* value rate constants. The actual transport process rate (*N*) is the product of fugacity and the compartment's rate constant:

1. Throughout this text, bracketed values indicate molar concentrations, but these may always be converted to mass per volume concentration values.

$$N = D \cdot f \tag{6.28}$$

And, since the contaminant concentration is $Z \cdot f$, we can substitute and add a first-order rate constant k to give us a first-order rate D value (D_R):

$$N = V[c]k = (V \cdot Z \cdot k) \cdot f = D_R \cdot f \tag{6.29}$$

As mentioned, although the concentrations are shown as molar concentrations they may also be represented as mass per volume concentrations.

As mentioned, pollutant transport processes follow Fick's laws, that is, diffusive processes. They can also be expressed with their own D values (D_D), which is related to the mass transfer coefficient (K) applied to area A:

$$N = KA[c] = (K \cdot A \cdot Z \cdot) \cdot f = D_D \cdot f \tag{6.30}$$

Advection, also known as bulk flow or nondiffusive transport, within a compartment with a flow rate (G) has a D value (D_A) and is expressed as

$$N = G[c] = (G \cdot Z \cdot f) = D_A \cdot f \tag{6.31}$$

This means that a substance is moving through the environment, during its residence time in each phase, and is affected by numerous physical transport and chemical degradation and transformation processes. In the atmosphere, the pollutant is moving with the air. Thus, transport is expressed by advection to the extent to which a chemical species is part of the airflow.

The transport processes are addressed by models with the respective D values, so that the total rate of transport and transformation is expressed as

$$f \cdot (D_1 + D_2 + \ldots D_n) \tag{6.32}$$

Very fast processes have large D values, and these are usually the most important when considering the contaminant's behavior and change in the environment.

Commonly, air pollutants are transported in a dominant or preferential direction, that is, from the source, moving in the direction of the wind. Models now combine physical transport and fugacity to predict chemical concentrations as functions of time. The concentrations may be in any environmental media but may also represent mass transfer to an individual person because of potential contact with media of varying characteristics. For example, the Stochastic Human Exposure and Dose Simulation (SHEDS) model differentiates transfer rates from textured surfaces, smooth surfaces, and air. The transfer can be based on

- a simple decay/dispersion model,
- time-specific distributions (<1 day, 1–7 days, 8–30 days, and >31 days),
- user-specified time series from measurement studies or an external model.

The SHEDS fugacity module divides a microenvironment (e.g., rooms within a house), for example, rooms with vinyl flooring, carpet, air, and wall compartments (see Fig. 6.13). Based on fugacity-based model inputs (Table 6.1), the module provides output concentrations of chemicals in time series for the different compartments will be used as contacted concentrations for simulated persons in the microenvironments, that is, compartments where people undertake different activities, for example, eating, sleeping, and cleaning [17, 18]. Note that chemical exposure models like these include combinations parameters of kinetics, fluxes, mass transfer coefficients, and partitioning coefficients.

The parameters can be quite specific. For example, the parameters and default values were recently expanded in the latest version of SHEDS [19], to estimate exposures in a higher throughput manner, that is, SHEDS-HT (see Table 6.2).

Although the same principles hold for indoor and outdoor air, indoor microenvironments can be treated as a control volume consisting of a well-mixed box, with air exchange. The air exchange rate is often expressed as air exchanges per hour (ACH), that is, the number of times per hour that the complete indoor air volume (e.g., a room or an entire building) is replaced by incoming outdoor air. Therefore, for a conservative (nonreactive) chemical species with no indoor sinks, the indoor concentration (C_{indoor}) can be estimated to be

$$C_{indoor} = \frac{\sum S_1 \ldots S_n}{V_{ME} \times ACH} \tag{6.33}$$

where $s_1 \ldots s_n$ are the sources of the chemical species and V_{ME} is the volume of air in the microenvironment. Since ACH is indirectly proportional to C_{indoor}, ventilation is a means for ameliorating indoor air pollution.

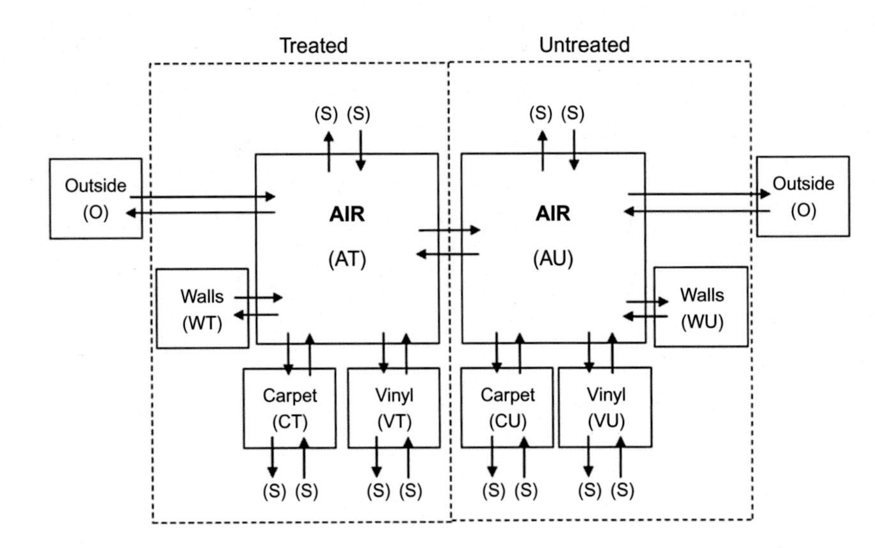

FIG. 6.13 Schematic of a microenvironment treated with a pesticide showing fugacity compartments. Each area is divided into four compartments: "air," "walls," "carpet," and "vinyl." The symbols in parentheses are the abbreviations given to each compartment in the fugacity module. The symbol "S" represents sources and/or sinks. The arrows in this diagram represent flows, either of the chemical itself (diffusive flows) or flows of particles that may carry the chemical (advective flows). *(Source: J. Xue, Planned Methodologies for Extending SHEDS-Multimedia Version 3 (Aggregate) to SHEDS-Multimedia Version 4 (Cumulative or Aggregate), 2007.)*

TABLE 6.1 Chemical-specific inputs for the SHEDS fugacity module

Variable description	Units
Chemical decay rate in air	d^{-1}
Chemical decay rate in carpet	d^{-1}
Chemical decay rate in vinyl	d^{-1}
Chemical decay rate in walls	d^{-1}
Diffusion coefficient in air	$m^2\,d^{-1}$
Octanol-water partitioning coefficient (K_{ow})	unitless
Vapor pressure	pascals
Solubility	$mol\,m^{-3}$
Molecular weight	$g\,mol^{-1}$

Note: d = day.
Source of information: J. Xue, Planned Methodologies for Extending SHEDS-Multimedia Version 3 (Aggregate) to SHEDS-Multimedia Version 4 (Cumulative or Aggregate), 2007.

On the day that a 30 m² office had new carpet installed, the formaldehyde (HCHO) was measured and found to be emitted from the carpet at a rate of 0.01 mg per hour, and the indoor air HCHO concentration was 10 ppb. Outdoor measurements did not detect HCHO in the air. The floor-to-ceiling height of the office is 2.4 m. Building management has informed the facilities department that no worker can enter the building until the HCHO concentration is 1 ppb at 23°C. At this temperature, the air density is 1.2 kg m⁻³; what ventilation rate is needed to reach this target concentration?

Since HCHO was not detect outside and assuming the carpet is the only source, we can solve for ACH:

$$\mathrm{ACH} = \frac{S_{carpet}}{V_{ME} \times C_{inside}} = \frac{\left(\dfrac{0.01\ \mathrm{mg\,HCHO}}{\mathrm{h}}\right)\cdot\left(\dfrac{\mathrm{kg}}{10^6\ \mathrm{mg}}\right)}{(30\ \mathrm{m}^2 \times 2.4\ \mathrm{m}) \times \left(\dfrac{1\ \mathrm{kg\,HCHO}}{10^9\ \mathrm{kg\,air}}\right)\cdot\left(\dfrac{1.2\ \mathrm{kg\,air}}{\mathrm{m}^3\ \mathrm{air}}\right)}$$

$$\mathrm{ACH} = 0.12\ \mathrm{h}^{-1}$$

TABLE 6.2 Parameters in the fugacity input file for SHEDS-HT

Parameter name	Description	Units	Default value in SHEDS-HT[a]
aer.out	Air exchange rate for rooms with outdoors	$L\ d^{-1}$	Lognormal (11.9, 1.7); assumed
area.sur	Total floor area of the house	m^2	Lognormal (130, 1.8); assumed
height	Height of walls	m	Uniform (2.44,3); assumed
lg.carb.f	Organic carbon fraction for large particles	–	Normal (0.15, 0.01); mean estimated from values reported for three larger particle sizes, variability assumed. Source: [17]
lg.clean.air	Cleaning removal rate for large particles in air, for example, from air filters on HVAC systems, or electrostatic filters	$L\ d^{-1}$	Uniform (0.03, 0.5); mean estimated from values reported for three larger particle sizes, variability assumed. Source: [17]
lg.clean.sur	Cleaning removal rate for large particles on surfaces, for example, from vacuuming or mopping	$L\ d^{-1}$	Uniform (0.035, 0.045); mean estimated from values reported for three larger particle sizes, variability assumed. Source: [17]
lg.depos	Air-to-floor large-particle deposition rate	$m\ d^{-1}$	Normal (387, 20); mean estimated from values reported for three larger particle sizes, variability assumed. Source: [17]
lg.load.air	Loading of large particles in air	$\mu g\ m^{-3}$	Uniform (2.2, 2.5); mean estimated from values reported for three larger particle sizes, variability assumed. Source: [17]
lg.load.sur	Loading of large particles on surfaces	$\mu g/cm^2$	Mean estimated from values reported for three larger particle sizes, variability assumed. Source: [17]
lg.resus	Surface-to-air large-particle resuspension rate	$L\ d^{-1}$	Uniform (0.0015, 0.0017); mean estimated from values reported for three larger particle sizes, variability assumed. Source: [17]
sm.carb.f	Organic carbon fraction for small particles	unitless	Normal (0.3, 0.03); mean estimated from values reported for three larger particle sizes, variability assumed. Source: [17]
sm.clean.air	Cleaning removal rate for small particles in air, for example, from air filters on HVAC systems, or electrostatic filters	$L\ d^{-1}$	Uniform (0.018, 0.22); mean estimated from values reported for three larger particle sizes, variability assumed. Source: [17]
sm.clean.sur	Cleaning removal rate for small particles on surfaces, for example, from vacuuming or mopping	$L\ d^{-1}$	Uniform (0.035, 0.045); mean estimated from values reported for three larger particle sizes, variability assumed. Source: [17]
sm.depos	Air-to-floor small-particle deposition rate	$m\ d^{-1}$	Normal (11, 1); mean estimated from values reported for three larger particle sizes, variability assumed. Source: [17]
sm.load.air	Loading of small particles in air	$\mu g\ m^{-3}$	Uniform (15, 25); mean estimated from values reported for three larger particle sizes, variability assumed. Source: [17]
sm.load.sur	Loading of small particles on surfaces	$\mu g\ cm^{-2}$	Uniform (6, 14.5); mean estimated from values reported in Bennett and Furtaw (2004) for three smaller particle sizes, variability assumed
sm.resus	Surface-to-air small-particle resuspension rate	$L\ d^{-1}$	Uniform (0.00072, 0.00082); mean estimated from values reported in Bennett and Furtaw (2004) for three smaller particle sizes, variability assumed
sm.clean.air	Cleaning removal rate for small particles in air, for example, from air filters on HVAC systems, or electrostatic filters	$L\ d^{-1}$	Uniform (0.018, 0.22); mean estimated from values reported in Bennett and Furtaw (2004) for three smaller particle sizes, variability assumed
temp	Indoor temperature	K	Normal (296, 2); assumed
thick.bou	Boundary layer thickness over surfaces	m	Uniform (0.025, 0.0275); assumed
thick.sur	Effective thickness of surfaces	m	Normal (0.0098, 0.002); assumed

[a] *Normal distributions reported as (mean, SD); lognormal as (geometric mean, geometric standard deviation); uniform as (min, max).*
Note: *d = day.*
Data from K.K. Isaacs, et al., SHEDS-HT: an integrated probabilistic exposure model for prioritizing exposures to chemicals with near-field and dietary sources, Environ. Sci. Technol. 48 (21) (2014) 12750–12759.

Thus, the air exchange rate is low, that is, only 0.12 per hour or 3 per day. Note that the existing concentration does not factor into this calculation because the equations are simply finding the amount of air exchanges needed to keep the air at 1 ppb HCHO and not the time to reach steady state [20].

Models, though imperfect, are important tools for estimating the movement of contaminants in the environment. They do not obviate the need for sound measurements. Indeed, measurements and models are highly complementary. Compartmental model assumptions must be verified in the field. Likewise, measurements at a limited number of points depend on models to extend their meaningfulness. Understanding of the basic concepts of a contaminant transport model allows for improved exploration of principle mechanisms for the movement of contaminants throughout the environment.

As mentioned in Chapter 4, fugacity concepts can be used to estimate the affinity of a chemical species for various environmental compartments, providing one of the first clues as to whether, where, and how much of the compound will be transported. A particularly strong affinity for a compartment will hinder the amount of a chemical species that will flee, that is, decrease fugacity. Therefore, the concepts and laws in this chapter should be combined with those in Chapters 4 and 5 for a more complete understanding of pollutant transport.

REFERENCES

[1] B. Cushman-Roisin, Environmental transport and fate, Thayer School of Engineering Dartmouth College, University Lecture, 2012.

[2] W.A. Tucker, L.H. Nelken, Diffusion coefficients in air and water, in: Handbook of Chemical Property Estimation Methods: Environmental Behavior of Organic Compounds, American Chemical Society, Washington, DC, 1990. p 17. 1–17. 25. 7 Tab, 28 Ref. 1990.

[3] D.A. Vallero, Fundamentals of Air Pollution, 5th ed., Elsevier Academic Press, Waltham, MA, 2014. p. 999 pages cm.

[4] E.L. Cussler (Ed.), Fundamentals of mass transfer, in: Diffusion: Mass Transfer in Fluid Systems, third ed., Cambridge University Press, Cambridge, 2009, pp. 237–273. Cambridge Series in Chemical Engineering.

[5] E.L. Cussler, Modified from "fundamentals of mass transfer", in: E.L. Cussler (Ed.), Diffusion: Mass Transfer in Fluid Systems, third ed., Cambridge University Press, Cambridge, 2009, pp. 237–273.

[6] W. Lewis, W. Whitman, Principles of gas absorption, Ind. Eng. Chem. 16 (12) (1924) 1215–1220.

[7] G. McKay, M.J. Bino, A. Altememi, External mass transfer during the adsorption of various pollutants onto activated carbon, Water Research 20 (4) (1986) 435–442.

[8] R. Higbie, Penetration theory leads to use of the contact time in the calculation of the mass transfer coefficients in the two film theory, Trans. Am. Inst. Chem. Eng. 365 (1935) 31.

[9] E.L. Cussler, Diffusion: Mass Transfer in Fluid Systems, third ed., Cambridge University Press, Cambridge, 2007, pp. 237–273.

[10] T. Lawson, M.R. Blatt, Stomatal size, speed, and responsiveness impact on photosynthesis and water use efficiency, Plant Physiol. 164 (4) (2014) 1556–1570.

[11] E.S. Bernhardt, W. Schlesinger, Biogeochemistry: An Analysis of Global Change, Elsevier Inc, Waltham, MA, USA and Oxford, UK, 2013.

[12] S. Wilkinson, G. Mills, R. Illidge, W.J. Davies, How is ozone pollution reducing our food supply? J. Exp. Bot. 63 (2) (2012) 527–536.

[13] G.E. Brurst, Air Pollution Effects on Vegetables, 25 June 2018. Available: https://extension.umd.edu/learn/air-pollution-effects-vegetables, 2007.

[14] E.S. Bernhardt, Biogeochemistry: An Analysis of Global Change, 1991.

[15] C.B. Osmond, K. Winter, H. Ziegler, Functional significance of different pathways of CO_2 fixation in photosynthesis, in: Physiological Plant Ecology II, Springer, 1982, pp. 479–547.

[16] D. Mackay, L. Burns, G. Rand, Fate modeling, in: Fundamentals of Aquatic Toxicology: Effects, Environmental Fate, and Risk Assessment, Taylor & Francis, Philadelphia, PA, 1995, , pp. 563–585.

[17] D.H. Bennett, E.J. Furtaw, Fugacity-based indoor residential pesticide fate model, Environ. Sci. Technol. 38 (7) (2004) 2142–2152.

[18] P.R. Williams, B.J. Hubbell, E. Weber, C. Fehrenbacher, D. Hrdy, V. Zartarian, An overview of exposure assessment models used by the US Environmental Protection Agency, Modell. Pollut. Complx Environ. Syst. 2 (2010) 61–131.

[19] K.K. Isaacs, et al., SHEDS-HT: an integrated probabilistic exposure model for prioritizing exposures to chemicals with near-field and dietary sources, Environ. Sci. Technol. 48 (21) (2014) 12750–12759.

[20] H.F. Hemond, E.J. Fechner, Chemical Fate and Transport in the Environment, Elsevier, 2014.

FURTHER READING

[21] J. Xue, Planned Methodologies for Extending SHEDS-Multimedia Version 3 (Aggregate) to SHEDS-Multimedia Version 4 (Cumulative or Aggregate), 2007.

Chapter 7

Water and the atmosphere

7.1 INTRODUCTION

Water not only is essential to all living systems but also is a carrier of air pollutants, reacts with them via hydrolysis and other chemical processes, and is involved in their transport and deposition back to the earth, for example, by droplet mechanisms.

This chapter specifically addresses the cycles of water and its interrelationships with the atmosphere.

7.2 THE WATER MOLECULE AS AN AGENT OF CHANGE

The water molecule (see Fig. 7.1) interacts with other molecules in the earth's spheres with profound effect. In the lithosphere, water is a key component of physical and chemical processes that shape the terrain, that is, geomorphology (see Fig. 7.2). The lithosphere affects air pollution in many ways. The geomorphology of an area can influence and even dictate where people live; for example, most cities are in coastal plains and in valleys near waterways. The terrain also determines wind patterns and air circulation; for example, inversions in urban valleys can greatly exacerbate air pollution episodes, and land-sea breezes can improve or degrade air quality depending on the direction.

An example of a physical mechanism occurs when the molecular configuration engenders its substantial expansion by volume as it freezes, that is, the change of state from liquid to solid, or when it sublimates, that is, the change of state from gas to solid. These changes in state initiate frost heaving, one of the mechanisms needed to break down rock formations. The resulting unconsolidated materials provide surfaces for chemical and biological processes that eventually create soil [1].

Chemically, this molecular configuration, especially polarity, induces the water to dissolve many chemical compounds, including acids that react with compounds in solid rock, for example, within fissures and crack, changing the geomorphology of the terrain. The alignment of a water molecule's oxygen and two hydrogen atoms gives a slightly negative charge at the oxygen end and a slightly positive charge at the hydrogen ends. Given that "like dissolves like," polar substances have an affinity to become dissolved in water, and nonpolar substances resist being dissolved in water. The hydrogen atoms form an angle of 105 degrees with the oxygen atom. The asymmetry of the water molecule leads to a dipole moment in the symmetry plane toward the more positive hydrogen atoms.

The hydrosphere is the discontinuous stratum around the earth that holds its water in many forms, including groundwater beneath the surface, surface water, water in soil and biota, and moisture in the atmosphere. The fluid properties of air and water combine in the hydrosphere, not only to make weather but also to move and transform all types of chemical compounds, including pollutants. The atmosphere holds 0.0055% of the earth's total amount of water (0.04% of fresh water) compared with the 96.5% in the oceans (see Table 7.1). However, the atmosphere accounts for much of water's activity on earth, from weather systems, to nutrient cycling, to pollutant transformation (e.g., hydrolysis), as well as pollutant transport and deposition [2].

7.2.1 Evaporation and transpiration in the hydrosphere

Note that the atmosphere is an entry as a water source in Table 7.1. Water must move to and from the hydrosphere to complete the water cycle. The principal means of exchange from the other sources to the atmosphere is by evaporation and transpiration. The underlying inherent property of water that drives these processes is vapor pressure (see Chapter 4).

Evapotranspiration is the combination of evaporation and vegetation transpiration from land and water surfaces, that is, water fluxes. A flux is the movement of energy or matter across a two-dimensional surface. This surface can be envisioned as a source. Simply stated, flux (J) is.

$$J = \frac{Q}{A} \tag{7.1}$$

Air Pollution Calculations. https://doi.org/10.1016/B978-0-12-814934-8.00007-7

FIG. 7.1 Configuration of the water molecule, showing the electronegativity (δ) at each end. The hydrogen atoms form an angle of 105 degrees with the oxygen atom. *(Source: D.A. Vallero, Fundamentals of Air Pollution, 5th ed., Elsevier Academic Press, Waltham, MA, 2014, p. 999 pages cm.)*

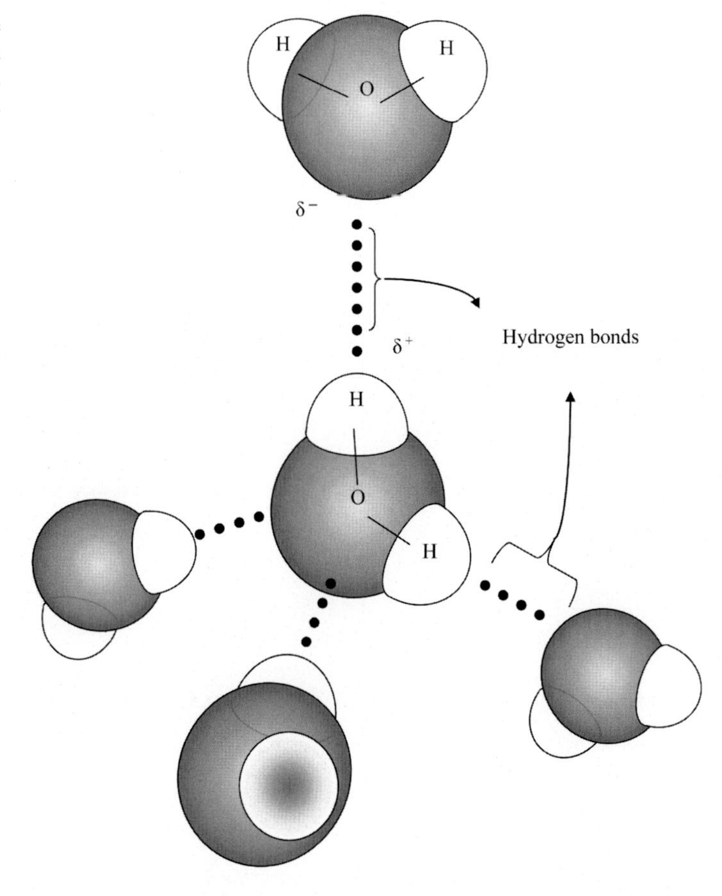

FIG. 7.2 Side view example of some of the many physical, chemical, and biological mechanisms that change terrain. (A) Initial rock formation. (B) Erosion and weather form pockets of unconsolidated material as matrices for early soils. Water flows into fissures and widens them by chemical reactions, for example, carbonic acid in limestone. (C) Early microbial and plant root systems continue to deepen the soil horizons and to smooth the terrain.

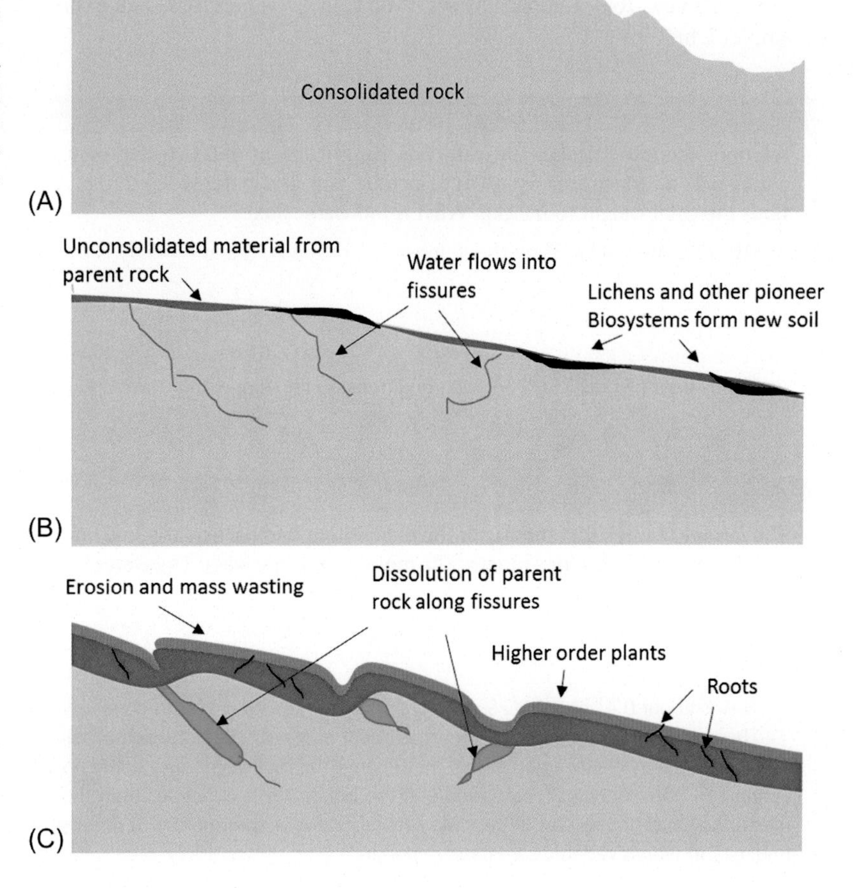

TABLE 7.1 Estimation of water volume in the hydrosphere

Water source	Water volume (× 1000 km³)	Percent of fresh water	Percent of total water
Oceans, seas, and bays	1,338,000	–	96.5
Ice caps, glaciers, and permanent snow	24,064	68.6	1.74
Groundwater	23,400	–	1.7
Fresh	10,530	30.1	0.76
Saline	12,870	–	0.93
Ground ice and permafrost	300	0.86	0.022
Lakes	176.4	–	0.013
Fresh	91	0.26	0.007
Saline	85.4	–	0.007
Soil moisture	16.5	0.05	0.001
Atmosphere	**12.9**	**0.04**	**0.001**
Swamp water	11.5	0.03	0.0008
Rivers	2.1	0.006	0.0002
Biological water	1.1	0.003	0.0001

Sources: P.H. Gleick, Water in Crisis: A Guide to the Worlds Fresh Water Resources, 1993; U.S. Geological Survey, The World's Water. (2016). 28 February 2018. Available: https://water.usgs.gov/edu/earthwherewater.html.

where Q is the flow rate and A is the surface area (units of length squared) of the source. Q can be expressed as volume or mass, for example, mg sec^{-1} or L hr.$^{-1}$. Therefore, J is expressed in mass or volume per time per area, for example, L hr^{-1} m^2.

A farmer needs to calculate the water flux from 10,000 m² soil. To test the soil, he adds 1 L of water to a 0.2 m-high soil column in a tube with a surface area of 0.01 m². The tube, soil, and water weigh 2.5 kg. After 1 day, he weighs the soil, water, and tube and finds the total weight to be 2 kg. What is the water flux?

At STP, 1 L water = 1 kg. Thus, the difference between $t=0$ and $t=1$ day is 0.5 kg, so.

$$J=\frac{0.5\,\frac{L}{\text{day}}}{0.01\,\text{m}^2}=50\,\text{L}\,\text{day}^{-1}\,\text{m}^2$$

This is about 2 L hr.$^{-1}$ m^2. So, the conditions on this day would have a water flux of about 20,000 L from his soil every hour.

The farmer realizes that the day of the experiment had particularly low humidity and high temperatures, so he repeated the experiment every day for a month. The monthly average was 0.1 kg water loss per day. What was the water flux for the month?

$$J=\frac{0.1\,\frac{L}{\text{day}}}{0.01\,\text{m}^2}=10\,\text{L}\,\text{day}^{-1}\,\text{m}^2$$

This is about 0.4 L hr.$^{-1}$ m^2. So, the conditions on this day would have a water flux of about 4000 L from his soil every hour.

There are various reasons that the experiment may not reflect the actual soil flux. For example, the sides of the tube are impervious, so this may create a barrier that may change the hydrodynamics of the system. Also, if the column of soil is narrower than the actual soil horizon, water may puddle at the bottom. If it is deeper than the horizon, then the water may migrate away from the surface and not be part of the flux. Any laboratory experiment will differ from the real world. It is up to the experimenter to understand which variables are most important.

Evaporation accounts for the movement of water to the air from various sources, whereas transpiration is exclusively biological. The biological water entry in Table 7.1 appears small, that is, 0.3% of the fresh water, but transpiration is responsible for about 10% of water that evades from the earth's surface to the troposphere [3]. Indeed, transpiration is a botanical process whereby the water absorbed by plants moves to tissue for hydration and to the leaves for photosynthesis. Animals contribute a much smaller amount of water to the atmosphere from respiration.

During evapotranspiration, a pollutant may be taken from one compartment, for example, the soil, and enter the air in a plume, that is, advectively [4]. That is, a pollutant can move from one compartment to another when it is dissolved or suspended in water, and when the water is subsequently transported to the air, the pollutant is transported with the water.

Evaporation (E) can be calculated using Shuttleworth's [5] adaptation of the empirically derived Penman formula [6]:

$$\mathrm{E} = \frac{mR_n + \gamma \cdot 6.43(1 + 0536 \times U_2) \cdot \delta e}{\lambda_v \cdot (m + \gamma)} \tag{7.2}$$

where m is the slope of the saturation vapor pressure curve ($\mathrm{kPa\,K^{-1}}$), R_n is the net irradiance ($\mathrm{MJ\,m^{-1}\,day^{-1}}$), γ is the psychrometric constant ($\mathrm{kPa\,K^{-1}}$), U_2 is wind speed ($\mathrm{m\,s^{-1}}$), δe is vapor pressure deficit (kPa), and λ_v is the latent heat of vaporization ($\mathrm{MJ\,kg^{-1}}$).

The psychrometric constant, γ, is calculated as

$$\frac{0.0016286 \cdot P}{\lambda_v} \tag{7.3}$$

where P is atmospheric pressure in kP.

Water provides an important vehicle for transporting chemical species and particulate matter in the air. For chemical species transport, vapor pressure of water is key. An aspect of vapor pressure that leads to volatilization is the vapor pressure deficit (VPD). The VPD is the difference between the amount of water vapor in the air and the amount that the air can hold:

$$\mathrm{VPD} = \overset{\circ}{p}_{saturated} - \overset{\circ}{p}_{actual} \tag{7.4}$$

and

$$\overset{\circ}{p}_{actual} = \frac{RH \times \overset{\circ}{p}_{saturated}}{100} \tag{7.5}$$

where $\overset{\circ}{p}_{saturated}$ is the saturated vapor pressure, $\overset{\circ}{p}_{actual}$ is the actual vapor pressure, and RH is the relative humidity. This has been described as the air's drying power [7], since it is a controlling factor for evapotranspiration, that is, the combination of transpiration of water vapor from plant tissue, especially leaves, and abiotic evaporation. As such, VPD is commonly calculated by farmers, landscapers, and agricultural scientists and engineers.

Saturation vapor pressure is temperature-dependent, that is, $\overset{\circ}{p}_{saturated}$ increasing with increasing air temperature (see Table. 7.2 and Fig. 7.3). By extension, VPD is also temperature-dependent.

TABLE 7.2 Saturation vapor pressure of air and temperature

Saturation vapor pressure (kP)	Temperature (C°)
0.611	0
0.657	1
0.706	2
0.758	3
0.813	4
0.872	5
0.935	6
1.002	7

TABLE 7.2 Saturation vapor pressure of air and temperature—cont'd

Saturation vapor pressure (kP)	Temperature (C°)
1.073	8
1.148	9
1.228	10
1.312	11
1.402	12
1.497	13
1.598	14
1.705	15
1.818	16
1.937	17
2.064	18
2.197	19
2.338	20
2.486	21
2.643	22
2.809	23
2.983	24
3.167	25
3.361	26
3.565	27
3.779	28
4.005	29
4.242	30
4.492	31
4.754	32
5.029	33
5.318	34
5.621	35
5.940	36
6.273	37
6.623	38
6.990	39
7.374	40
7.776	41

Data from J. Monteith, M. Unsworth, Principles of Environmental Physics: Plants, Animals, and the Atmosphere, Academic Press, 2013; Q. Cronk, Vapour Pressure Deficit Calculation. University of British Columbia. Center for Plant Research, 2018. Available online: http://cronklab.wikidot.com/calculation-of-vapour-pressure-deficit.

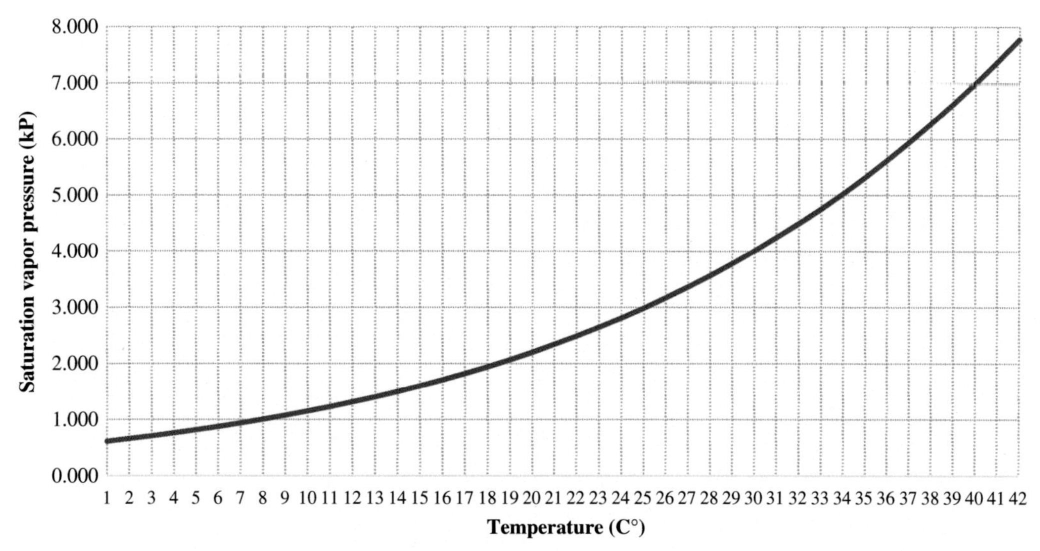

FIG. 7.3 Saturation vapor pressure of air and temperature. *(Data from J. Monteith, M. Unsworth, Principles of Environmental Physics: Plants, Animals, and the Atmosphere, Academic Press, 2013; Q. Cronk, Vapour Pressure Deficit Calculation. University of British Columbia. Center for Plant Research, 2018. Available online: http://cronklab.wikidot.com/calculation-of-vapour-pressure-deficit.)*

A professor is building an evapotranspiration algorithm in a model to predict the volatilization of pesticides and realizes that he has not accounted for the vapor pressure deficit. So, he walks outside and finds that at noon, the relative humidity is 75% and the temperature is 28°C. What is the VPD today at noon?

From Table 7.2, we see that at 28°C, the saturation vapor pressure is 3.779 kP. Therefore,

$$100 - \text{RH} = 25$$

$$25/100 = 0.25$$

$$0.25 \times 3.779\,\text{kP} = 0.94\,\text{kP}$$

Later at 7:00 p.m., the professor measures the relative humidity to be 90% and the temperature is 21°C. What is the VPD now? What might these two measurements indicate?

From Table 7.2, we see that at 21°C, the saturation vapor pressure is 2.486 kP. Therefore,

$$100 - \text{RH} = 21$$

$$21/100 = 0.21$$

$$0.21 \times 2.486\text{kP} = 0.52\,\text{kP}$$

Therefore, with the decreasing temperature and rising relative humidity, the deficit has fallen substantially, that is, the air has less drying power at 7:00 p.m. compared with noon.

Even this very small study indicates that if evapotranspiration plays a role in the volatilization of a pesticide, the drying power decline means that less volatilization is occurring with decreasing VPD, that is, the plume from the pesticide-treated area is less at 7:00 p.m. than at noon on this day. So, he may want to enhance the volatilization algorithm to account for VPD.

VPD is crucial as an indication of stress on plant life, which means that when combined with other stressors, like ozone, the dryness contributes to crop and ecosystem stresses. In addition, however, VPD is also a factor in pollutant transport. That is, when the water volatilizes, it can take along with it other substances into the atmosphere.

This is an example of the importance of translational science. The agricultural science and engineering community has studied VPD and dryness to improve crops and to protect ecosystems. Environmental and air pollution scientists and engineers can translate this knowledge into pollutant transport models and computations of how substances move into and out of the atmosphere and other compartments.

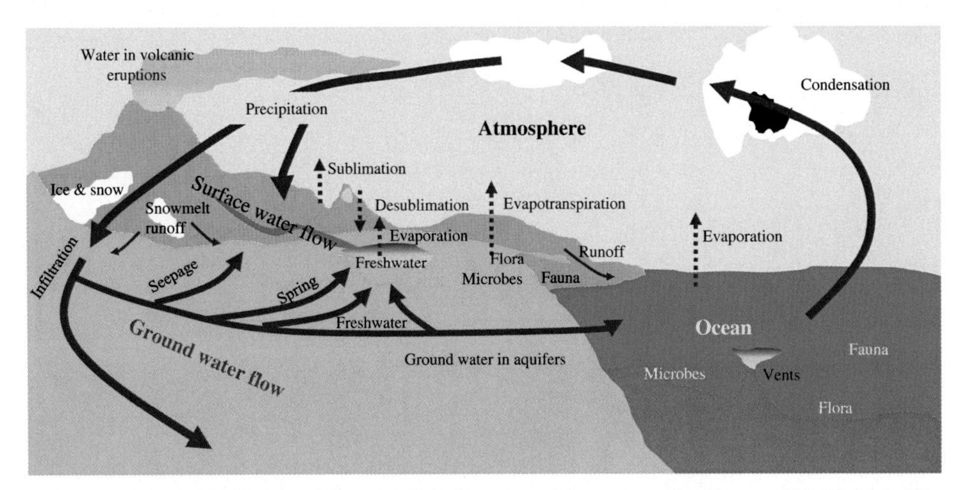

FIG. 7.4 The hydrologic (water) cycle. US Geological Survey (2013). Summary of the water cycle. *(Source: U.S. Geological Survey. Summary of the Water Cycle. (2016). 28 February 2018. Available: https://water.usgs.gov/edu/watercyclesummary.html.)*

7.3 THE HYDROSPHERE

All air pollution management takes place in the hydrosphere, whether it is the exchange of pollutants between ground and surface waters and the atmosphere, the storage of pollutants in soil water, or the release and the biota's uptake of pollutants after their release (see Fig. 7.4).

Matter and energy cycling takes place at all levels, each of which is important to air pollution management. The first law of thermodynamics dictates that these cycles be both mass and energy balanced. Energy cycling almost always includes water, and water cycling includes energy. A cycle can begin with the release of heated plumes from stacks and outfall structures, to surface and groundwater, to the atmosphere. Exchanges of matter and energy occur between the hydrosphere, atmosphere, and lithosphere, for example, soil and sediment.

Water is both a sink and source of energy. Heat reservoirs in terrestrial and aquatic systems receive added heat, which when released can alter habitats, for example, changes to freeze-thaw cycles, seasonal variations, and selectivity of certain soil bacteria genera. Some water-energy relationships are obvious and direct; for example, the thermal inversions that lead to urban air pollution are the result of differences in heat energy in water vapor at various layers in the troposphere. Some of the water-energy relationships are incremental and indirect, such as the transfer of energy between trophic states in an ecosystem, which relies in part on energy and water exchanges among biotic tissue and abiotic substrates (e.g., aqueous-phase phosphorous (K) compounds that transfer energy during photosynthesis).

Water-energy cycling occurs in both natural and anthropogenic systems. The design, selection, and operation of pollution control equipment must consider water-energy interactions. For example, at the facility scale, water and energy are often addressed together, such as the management of water and energy at a factory or power plant. Managing water is a simultaneous process with managing energy. Indeed, mismanagement of water systems at the facility level can involve trade-offs between types of pollution. For example, allowing heated water to be released in any amount, even the permitted level, would increase the overall temperature of the receiving waters. At some threshold, the added heat will adversely affect species diversity and other aspects of environmental integrity of aquatic ecosystems. Meanwhile, pollutants released to the atmosphere can be transformed by hydrolysis and other chemical reactions in water, whether in receiving water bodies, soil, roots, or droplets suspended in the atmosphere. These processes not only help to degrade the pollutants but also can make some pollutants even more toxic.

7.4 IONIZATION, ACIDITY AND ALKALINITY [8]

Biogeochemical cycles depend on changes in ionization, which determines the acidity and alkalinity of water and the production of acids and bases. The carbon, nitrogen, sulfur, and other cycles must consider the equilibriums among chemical species in terms of molecules and ions.

When a salt is dissolved in water, it dissociates, that is, becomes ionic species. With time and under the right conditions, a fraction of most substances can become dissolved in water. Ions that are dissolved in a solution can react with one another and can form solid complexes and compounds. For example, when the salts calcium sulfate and sodium chloride are added to water, it is commonly held that $CaSO_4$ and NaCl are in the water. However, what is in fact happening is that the metals and nonmetals are associated with one another in equilibrium:

$$CaSO_4(s) \leftrightarrow Ca^{2+}(aq) + SO_4^{2-}(aq) \tag{7.6}$$

and

$$NaCl(s) \leftrightarrow Na^+(aq) + Cl^-(aq) \tag{7.7}$$

Thus, all four of the dissociated ions are free and no longer associated with each other, that is, the Na^+, Ca^{2+}, Cl^-, and SO_4^- ions are "unassociated" in the water. The Na and Cl atoms are no longer linked to each other as they were before the compound was added to the water.

Even though atoms are neutral, in the process of losing or gaining electrons, they become electrically charged species, that is, ions. An atom that loses one or more electrons is positively charged and called a cation. For example, the potassium atom loses one electron and becomes the monovalent potassium cation:

$$K - e^- \rightarrow K^+ \tag{7.8}$$

The mercury atom that loses two electrons becomes the divalent mercury cation:

$$Hg - 2e^- \rightarrow Hg^{2+} \tag{7.9}$$

When the chromium atom loses three electrons, it forms the trivalent chromium cation:

$$Cr - 3e^- \rightarrow Na^{3+}$$

Conversely, an atom that gains electrons becomes a negatively charged ion, known as an anion. For example, when chlorine gains an electron, it becomes the chlorine anion:

$$Cl + e^- \rightarrow Cl^- \tag{7.10}$$

When sulfur gains two electrons, it becomes the divalent sulfide anion:

$$S + 2e^- \rightarrow S^{2-} \tag{7.11}$$

Note that the Greek prefix (mono-, di-, tri-, …) denotes the valence, which is the number of electrons that will become the ion. This is an integer that expresses the difference from neutrality. For Eq. (7.11), this difference is 2, so the anion S^{2-} is two electrons different than neutrality.

Ionic reactions frequently occur when the ions of water-soluble salts react in aqueous solution to form salts that are often nearly insoluble in water. This causes them to separate into insoluble precipitates:

$$(\text{Ions in Solution 1}) + (\text{Ions in Solution 2}) \rightarrow \text{Precipitate} + \text{Unreacted ions} \tag{7.12}$$

A salt in the solid phase is a collection of ions in a lattice, that is, the ions are surrounded by one another. However, when the salt is dissolved in water, the ions become surrounded by the water, rather than by the other ions. Each ion now has its own coordinating water envelope or "hydration sphere," that is, a collection of water molecules surrounding it. An ion-association reaction is an ion-ion interaction between ions in an ion-containing solution, that is, an electrolyte. The term comes from the Swedish chemist, Svante Arrhenius, who observed the relationship between electric and chemical properties of molecules. An electrolyte conducts electricity when dissolved in water, yet other chemicals do not.

Thus, when the salt lattice enters the water, the ions assemble into couplets of separate oppositely charged ions, that is, cations and anions, the so-called ion pairs. The pairs are held together by electrostatic attraction. Ion association is the reverse of dissociation where the ions separate from a compound into free ions. Ions or molecules surrounded by water exist in the aqueous phase.

Solvation is the interaction of solvent, like water, with its solutes, that is, the molecules and ions dissolved in the solvent. Ionic compounds that are soluble in water break apart, that is, dissociate, into their ionic components within a solvation shell, that is, the solvent interface of a chemical species that constitutes the solute. When the solvent is water, the solvation shell is referred to as a hydration shell or hydration sphere. The number of solvent molecules surrounding each unit of solute is called the hydration number of the solute. When water molecules configure around a metal ion, for example, the electronegative oxygen atom of the water molecule is electrostatically attracted to the positive charge on the metal ion. The result is a solvation shell of water molecules that surround the ion (see Fig. 7.5).

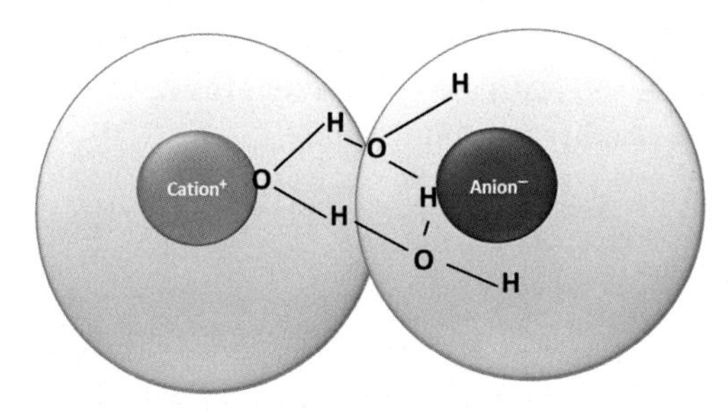

FIG. 7.5 Ion pair. *(Modified from V. Evangelou, Environmental Soil and Water Chemistry: Principles and Applications, John Wiley Sons, Inc, Canada, 1998.)*

Solutions may contain nonelectrolytes, strong electrolytes, and/or weak electrolytes. Nonelectrolytes do not ionize. They are nonionic molecular compounds that are neither acids nor bases. Sugars, alcohols, and most other organic compounds are nonelectrolytes. Some inorganic compounds are also nonelectrolytes.

Weak electrolytes only partially dissociate in water. Most weak electrolytes dissociate less than 10%, that is, greater than 90% of these substances remain undissociated. Organic acids, such as acetic acid, are generally weak electrolytes.

An example of dissociation is strontium carbonate dissolved in water:

$$SrCO_3(s) \rightarrow Sr^{2+}(aq) + CO_3^{2-}(aq) \tag{7.13}$$

Conversely, the reverse reaction forms a solid; that is, it returns from the solution to again form the lattice of ions surrounding ions. This is a precipitation reaction. In our Sr example, the carbonate species is precipitated:

$$Sr^{2+}(aq) + CO_3^{2-}(aq) \rightarrow SrCO_3(s) \tag{7.14}$$

The ionic product (Q) is a measure of the ions present in the solvent. The solubility product constant (K_{sp}) is the ionic product when the system is in equilibrium. So, solubility is often expressed as the specific K_{sp}, the equilibrium constant for dissolution of the substance in water. Since K_{sp} is another type of chemical equilibrium, it is a state of balance of opposing reversible chemical reactions that proceed at constant and equal rates, resulting in no net change in the system (hence the symbol, $\leftrightarrow$). Like sorption, Henry's law, and other equilibrium constant, solubility follows Le Chatelier's principle, which states that in a balanced equilibrium, if one or more factors change, the system will readjust to reach equilibrium. K_{sp} values and the resulting solubility calculations for some important reactions are shown in Table 7.3. K_{sp} constants for many reactions can be found in engineering, hydrogeology, and chemistry handbooks.

Since $SrCO_3$ is a highly insoluble salt (aqueous solubility $= 6 \times 10^{-3}$ mg L^{-1}), its equilibrium constant for the reaction is quite small:

$$K_{sp} = \left[Sr^{+2}\right]\left[CO_3^{2-}\right] = 1.6 \times 10^{-9} \tag{7.15}$$

The small K_{sp} value of the constant reflects the low concentration of dissolved ions. So, as the number of dissolved ions approaches zero, the compound is being increasingly insoluble in water. This does not mean that an insoluble product cannot be dissolved, but that chemical treatment is needed. In this case, strontium carbonate requires the addition of an acid to *solubilize* the Sr^{+2} ion.

TABLE 7.3 Solubility product constant versus solubility for four types of salts

Salt	Example	Solubility product, K_{sp}	Solubility, S
AB	$CaCO_3$	$[Ca_{2+}][CO_3^{2-}] = 4.7 \times 10^{-9}$	$(K_{sp})^{1/2} = 6.85 \times 10^{-5}$M
AB_2	$Zn(OH)_2$	$[Zn^{2+}][OH^-]^2 = 4.5 \times 10^{-17}$	$(K_{sp}/4)^{1/3} = 2.24 \times 10^{-6}$M
AB_3	$Cr(OH)_3$	$[Zn^{3+}][OH^-]^3 = 6.7 \times 10^{-31}$	$(K_{sp}/27)^{1/4} = 1.25 \times 10^{-8}$M
A_3B_2	$Ca_3(PO_4)_2$	$[Ca^{2+}]^3[PO_4^{3-}]^2 = 1.3 \times {}^{10-32}$	$(K_{sp}/108)^{1/5} = 1.64 \times 10^{-7}$M

Source: EM 1110-1-4012, Engineering and Design Precipitation/Coagulation/Flocculation, 2001.

To precipitate a compound, the product of the concentration of the dissolved ions in the equilibrium expression must exceed the value of the K_{sp}. The concentration of each of these ions does not need to be the same. For example, if $[Sr^{+2}]$ is 1×10^{-5} M, the carbonate ion concentration must exceed 0.0016 M for precipitation to occur because $(1 \times 10^{-5}) \times (1.6 \times 10^{-4}) = 1.6 \times 10^{-9}$ (i.e., the K_{sp} for strontium carbonate).

If 100 mL of 0.050 M NaCl is added to 200 mL of 0.020 M $Pb(NO_3)_2$, will the lead chloride that is formed precipitate?

Calculate the ion product (Q) and compare it with the K_{sp} for the following reaction:

$$PbCl_2(s) \rightarrow Pb^{2+}(aq) + 2Cl^-(aq)$$

When the two solutions are mixed, the unassociated ions are formed as follows:

$$[Pb^{2+}] = \frac{0.2\,L \times 2.0 \times 10^{-2}\,M}{0.3\,L} = 1.3 \times 10^{-2}\,M,$$

and

$$[Cl^-] = \frac{0.1\,L \times 5.0 \times 10^{-2}\,M}{0.3\,L} = 1.7 \times 10^{-2}\,M.$$

The value for the ion product is calculated as follows:

$$Q = [Pb^{2+}][Cl^-]^2 = [1.3 \times 10^{-2}][1.7 \times 10^{-2}]^2 = 3.8 \times 10^{-7}$$

The K_{sp} for this reaction is 1.6×10^{-5}.

$Q < K_{sp}$, so no precipitate will be formed. If the ion product was greater than the K_{sp}, a precipitate would have formed.

Environmental conditions affect solubility. Compared with a solution in pure water, an ion's solubility is decreased in an aqueous solution that contains a common ion, that is, one of the ions that make up the compound. This "common ion effect" allows a precipitate to form if the K_{sp} is exceeded. For example, soluble sodium carbonate (Na_2CO_3) in solution with strontium ions can cause the precipitation of strontium carbonate, since the carbonate ions from the sodium salt are contributing to their overall concentration in solution and reversing the solubility equilibrium of the "insoluble" compound, strontium carbonate:

$$Na_2CO_3\,(s) \rightarrow 2\,Na^+(aq) + CO_3^{2-}\,(aq) \tag{7.16}$$

$$SrCO_3\,(s) \rightarrow Sr^+(aq)^{2+}\,CO_3^{2-}\,(aq) \tag{7.17}$$

Also, a complexing agent, known as a ligand, may react with the cation of a precipitate, which enhances the solubility of the compound. In addition, several metal ions are weakly acidic and readily hydrolyze in solution. For example, when ferric ion (Fe^{3+}) reacts with water, that is, hydrolyzes,

$$Fe^{3+} + H_2O \rightarrow Fe(OH)^{2+} + H^+ \tag{7.18}$$

When these metal ions hydrolyze, they produce a less soluble complex. The solubility of the salt is inversely related to the pH of the solution, with solubility increasing as the pH decreases. The minimum solubility is found under acidic conditions when the concentrations of the hydrolyzed species approach zero.

The ionization of water is crucial to air pollution chemistry. The acidity and alkalinity of water are expressed as a solution's hydrogen ion $[H^+]$ concentration and hydroxide ion $[OH^-]$ concentration, respectively. Unless deionized, water is the combination of the neutral molecular water and these ions. Water has the ability to act as either an acid or a base. In fact, water chemists sometimes show molecular water as HOH, perhaps to demonstrate ionization. The dissociation in water is a reversible reaction:

$$HOH(l) \leftrightarrow H^+(aq) + OH^-(aq) \tag{7.19}$$

However, this equilibrium really involves the hydronium ion (H_3O^+), which is the chemical form of the hydrogen ion (H^+) in an aqueous solution. When hydrogen loses its only electron, all that is left is a proton, which is highly reactive, so it immediately reacts with a water molecule, generating H_3O^+. By definition, an acid donates a proton to a solution, that is, H^+, but this proton is extremely small, so it is more reasonable that it exists as hydronium. Therefore, the dissociation of water is a reversible reaction better described as

$$2HOH(l) \leftrightarrow H_3O^+(aq) + OH^-(aq) \tag{7.20}$$

This dissociation is continuous, and the number of ions dissociating at any time is known as the water dissociation constant, K_w, with the formula in the same format as any equilibrium constant discussed thus far, that is, the concentrations of products divided by the concentrations of reactants, all raised to the power of the coefficients of the balanced reaction:

$$K_w = \frac{[H_3O^+][OH^-]}{[HOH]^2} \tag{7.21}$$

However, the ions represent so small a fraction of the total mass in this reaction that the molecular water is assumed to be essentially pure water. Since pure substances are solvents without solutes, that is, concentration is meaningless, they are not included in equilibrium calculations. At 25°C, this dissociation reaction is.

$$K_w = [H_3O^+][OH^-] = 1.0 \times 10^{-14} \text{mol L}^{-1} \tag{7.22}$$

Recall that the brackets denote concentration. Because this equilibrium reaction is balanced, concentration proportions are 1:1. So, if $[H_3O^+]$ equals x, $[OH^-]$ must also be x. Thus, the equilibrium simplifies further to

$$x^2 = 1.0 \times 10^{-14}$$

Therefore, $[H_3O^+] = [OH^-] = \sqrt{1.0 \times 10^{-14}}$ mol L^{-1}, and

$$[H_3O^+] = [OH^-] = 1.0 \times 10^{-7} \text{mol L}^{-1}$$

Water chemistry refers to "p" as a negative log:

$$pX = -\log X \tag{7.23}$$

The pH of a substance is the expression of the power of hydrogen in a solution, that is, pH is a measure of the strength of an acid or base in that solution. Much of the literature describes pH as negative of the base logarithm ($\log_{10}$, or simply log) of $[H^+]$:

$$pH = -\log[H^+] \tag{7.24}$$

This is usually acceptable given the many other uncertainties in acidity calculations. However, as discussed above, pH is more accurately described as the negative of log $[H_3O^+]$:

$$pH = -\log[H_3O^+] \tag{7.26}$$

At equilibrium, $[H_3O^+] = 1.0 \times 10^{-7}$ mol L^{-1}. The negative log of 1.0×10^{-7} is 7, so water is neutral at pH=7.

Since, at equilibrium, $[OH^-]$ is also 1.0×10^{-7} mol L^{-1}, then pOH is also equal to 7. Thus, the center of the pH scale is 7. Thus, the pH and pOH scales are reciprocal to one another, with both ranging from 0 to 14. The number of hydronium ions equals the number of hydroxide at pH=7. The acidity increases logarithmically with decreasing pH (i.e., increasing number of hydronium ions, or protonation). The alkalinity increases logarithmically with increasing pH (i.e., increasing number of hydroxyl ions). This means that each pH integer represents a change in acidity by a factor of 10. A pH 5 solution is 10 times more acidic than neutral. A pH 2 solution has 100,000 times more hydronium ions than neutral water (pH 7), or $[H_3O^+] = 10^{12}$ versus $[H_3O^+] = 10^7$, respectively.

The Brønsted-Lowry model defines acids as proton donors and bases as proton acceptors. The Lewis model states that acids are electron-pair acceptors and bases are electron-pair donors. When acids react with bases, a double replacement reaction takes place, resulting in neutralization. The products are water and a salt. One mole of acid neutralizes precisely one mole of base. Being electrolytes, a strong acid dissociates and ionizes 100% into H_3O^+ and anions. These anions are the acid's specific conjugate base. A strong base also dissociates and ionizes completely. As mentioned, this ionization results in hydroxide ions and cations, known as the base's conjugate acid. Weak acids and weak bases dissociate less than 100% into the respective ions.

Numerous acids are involved in air pollution, but four strong acids are noteworthy, that is, hydrochloric acid (HCl), nitric acid (HNO_3), sulfuric acid (H_2SO_4), and perchloric acid ($HClO_4$). Many weak acids are important, including carbonic acid, acetic acid, and phosphoric acid. Strong bases include sodium hydroxide (NaOH) and potassium hydroxide (KOH). Weak bases include ammonia (NH_3), which dissolves in the surface waters to become ammonium hydroxide (NH_4OH) and organic amines, that is, compounds with the functional group, −NH.

Nonmetal oxides, such as carbonate (CO_2) and sulfate (SO_2), are generally acidic, forming carbonic acid and sulfuric acid, respectively, in water. The metal oxides like those of calcium (e.g., CaO) and magnesium (e.g., MgO) are generally basic. These two metal oxides, for example, form calcium hydroxide (CaOH) and magnesium hydroxide (MgOH) in water, respectively. These and other reactions are discussed in Chapter 8.

A laboratory has an aqueous solution of sulfuric acid, which is 0.05 M. What are the hydronium ion and chlorine ion molar concentrations and the pH of the solution?

Since this is a strong acid, it should ionize and dissociate completely. Thus,

$[H_3O^+]=0.05$ and $[Cl^-]=0.05$, so none of the associated acid remains.

What is the pH of the solution above?

Since $[H_3O^+]=0.05$, $pH=-\log[H^+]=-\log 0.05\,M=1.3$.

What is the $[OH^-]$ of this solution?

$$K_w\,[H_3O^+][OH^-]=1.0\times10^{-14}$$

So, $[OH^-]=\frac{K_w}{[H_3O^+]}=\frac{10^{-14}}{0.05}=2.0\times10^{-13}$.

Thus, even at an extremely low pH, with a very high relative concentration of hydronium ions in the acidic solution, there is still a small amount of hydroxide ion concentration.

The strength of an acid or base is irrespective of concentration, that is, the molarity of the solution. Rather, the strength is the extent to which the acid or base dissociates when it enters water. Whether at a concentration of 6.0 M or 0.00001 M, sulfuric acid will completely ionize in the water, whereas acetic acid will not completely ionize at any concentration. As evidence, Table 7.4 shows the pH for a number of acids and bases, all with the same molar concentration, 0.1 M. The strongest acids are at the top, and the strongest bases are at the bottom. The weak acids and bases are in the middle of the table. Increasing acidic and basic strength moves from distilled water (blue) upward to the table's strongest acid, hydrochloric acid (red), and downward to the table strongest base, sodium hydroxide (green).

Most environmental acid-base reactions involve weak substances. The amount of ionization in most environmental reactions, especially those in the ambient environment (as opposed to those in chemical manufacturing and laboratory reactors), is quite weak, usually well below 10% dissociation. Recall that water itself is a weak acid in that autoionizes into 10^{-14} M concentration of hydroxide and hydronium ions. At 25°C, there are 55.35 mol water per liter. Since half of the ions are hydronium ions, this means.

$$\frac{1.0\times10^{-7} M\,H_3O^+}{55.35 M\,H_2O}=1.8\times10^{-9}\text{ hydronium ions per water molecule.}$$

Even this small ratio provides enough H_3O^+ given the amount of water available in the hydrologic cycle and the highly reactive nature of each hydrogen ion.

Generally, for every 1000 molecules of a weak acid, only a few, say 50, molecules of the acid dissociate into hydronium ions in the water. The acetic acid acid-base equilibrium reaction is.

$$CH_3COOH\,(aq)+H_2O\,(l)\leftrightarrow H_3O^+\,(aq)+CH_3COO^-\,(aq) \tag{7.27}$$

The acetate ion (CH_3COO^-) is the reaction's conjugate base, and the hydronium ion is the active acid chemical species. The reaction is in equilibrium, so all species exist together. This introduces another equilibrium constant for acid reactions, that is, the acid constant (K_a):

$$K_a=\frac{[H_3O^+][CH_3COO^-]}{[CH_3COOH]} \tag{7.28}$$

At 25°C, acetic acid is 1.8×10^{-5}. If the percent dissociation in an acid reaction is known, the product of this percentage and the initial acid concentration will give the molar concentration of protons, $[H^+]$. For example, if a 0.1 M solution of cyanic acid (HOCN) is 2.8% ionized, the $[H^+]$ can be found. We know that HOCN is a weak acid because the percent ionization is less than 100. In fact, it is well below 10%. Therefore, the proton molar concentration is.

$[H^+]=2.8\%\times0.1\,M=0.0028$ or 2.8×10^{-3}. Published K_a constants show that the HOCN constant at 25°C is 3.5×10^{-3}, meaning that environmental conditions, especially temperature, are slightly affecting the pH of the solution. Remember that all equilibrium constants are temperature-dependent.

Weak bases follow exactly the same protocol as weak acids, with a base equilibrium constant, K_b. Some important acid and base equilibrium constants are provided in Tables 7.5 and 7.6. A monoprotic acid can donate a single proton per acid

TABLE 7.4 Experimentally derived pH values for 0.1 M solutions of acids and bases at 25°C

Compound	pH
HCl	1.1
H_2SO_4	1.2
H_3PO_4	1.5
CH_3COOH	2.9
H_2CO_3 (in saturated solution)	3.8
HCN	5.1
NaCl	6.4
H_2O (distilled)	7.0
$NaCH_3CO_2$	8.4
$NaSO_3$	9.8
NaCN	11.0
NH_3 (aqueous)	11.1
$NaPO_4$	12.0
NaOH	13.0

Data from J.N. Spencer, G.M. Bodner, L.H. Rickard, Chemistry: Structure and Dynamics. John Wiley & Sons, 2010.

TABLE 7.5 Equilibrium constants for selected environmentally important weak monoprotic acids and bases at 25°C

Monoprotic acid	Dissociation reaction	K_a
Hydrofluoric acid	$HF+H_2O \leftrightarrow H_3O^++F^-$	7.2×10^{-4}
Nitrous acid	$HNO_2+H_2O \leftrightarrow NO_2^-+H_3O^+$	4.0×10^{-4}
Lactic acid	$CH_3CH\ (OH)\ CO_2H+H_2O \leftrightarrow CH_3CH\ (OH)\ CO_2^-+H_3O^+$	1.38×10^{-4}
Benzoic acid	$C_6H_5CO_2H+H_2O \leftrightarrow C_6H_5CO_2^-+H_3O^+$	6.4×10^{-5}
Acetic acid	$HC_2H_3O_2+H_2O \leftrightarrow C_2H_3O_2^-+H_3O^+$	1.8×10^{-5}
Propionic acid	$CH_3CH_2CO_2H+H_2O \leftrightarrow CH_3CH_2CO_2^-+H_3O^+$	1.3×10^{-5}
Hypochlorous acid	$HOCl+H_2O \leftrightarrow OCl^-+H_3O^+$	3.5×10^{-8}
Hypobromous acid	$HOBr+H_2O \leftrightarrow OBr^-+H_3O^+$	2×10^{-9}
Hydrocyanic acid	$HCN+H_2O \leftrightarrow CN^-+H_3O^+$	6.2×10^{-10}
Phenol	$HOC_6H_5+H_2O \leftrightarrow OC_6H_5^-+H_3O^+$	1.6×10^{-10}
Base	**Dissociation reaction**	K_b
Dimethylamine	$(CH_3)_2NH+H_2O \leftrightarrow (CH_3)_2NH_2^++OH^-$	5.9×10^{-5}
Methylamine	$CH_3NH_2+H_2O \leftrightarrow CH_3NH_3^++OH^-$	7.2×10^{-4}
Ammonia	$NH_3+H_2O \leftrightarrow NH_3^++OH^-$	1.8×10^{-5}
Hydrazine	$H_2NNH_2+H_2O \leftrightarrow H_2NNH_3^++OH^-$	1.2×10^{-6}
Aniline	$C_6H_5NH_2+H_2O \leftrightarrow C_6H_5NH_3^++OH^-$	4.0×10^{-10}
Urea	$H_2NCONH_2+H_2O \leftrightarrow H_2NCONH_3^++OH^-$	1.5×10^{-14}

Data source: M. Olia, A.S. Casparian, How to Prepare for the Fundamentals of Engineering, FE/EIT Exam. Barron's Educational Series, 2000.

TABLE 7.6 Equilibrium constants for selected environmentally important polyprotic acids at 25°C

Acid	Dissociation reactions	K_{a1}	K_{a2}	K_{a3}
Sulfuric acid	$H_2SO_4 + H_2O \leftrightarrow HSO_4^- + H_3O^+$	1.0×10^3		
	$HSO_4^- + H_2O \leftrightarrow SO_4^{2-} + H_3O^+$		1.2×10^{-2}	
Hydrogen sulfide	$H_2S + H_2O \leftrightarrow HS^- + H_3O^+$	1.0×10^{-7}		
	$HS^- + H_2O \leftrightarrow S^{2-} + H_3O^+$		1.3×10^{-13}	
Phosphoric acid	$H_3PO_4 + H_2O \leftrightarrow HPO_4^- + H_3O^+$	7.1×10^{-3}		
	$HPO_4^- + H_2O \leftrightarrow HPO_4^{2-} + H_3O^+$		6.3×10^{-8}	
	$HPO_4^{2-} + H_2O \leftrightarrow PO_4^{3-} + H_3O^+$			4.2×10^{-13}
Carbonic acid	$H_2CO_3 + H_2O \leftrightarrow HCO_3^- + H_3O^+$	4.5×10^{-7}		
	$HCO_3^- + H_2O \leftrightarrow CO_3^{2-} + H_3O^+$		4.7×10^{-11}	

Data source: M. Olia, A.S. Casparian, How to Prepare for the Fundamentals of Engineering, FE/EIT Exam. Barron's Educational Series, 2000.

molecule, but polyprotic acid can donate two or more protons per molecule. For example, the monoprotic nitrous acid, HNO_2, has but one H atom available to donate, but the polyprotic phosphoric acid, H_3PO_4, has three H atoms available to donate to the reaction.

As will be discussed in Chapter 8, the atmosphere contains relatively large amounts of carbon dioxide (on average about 400 ppm), so that the CO_2 becomes dissolved in droplets suspended in the atmosphere, as well as in surface water and in soil water (CO_2 is a common soil gas). Thus, one of the most important environmental acid-base reactions[1] is the dissociation of CO_2:

$$CO_2 + H_2O \leftrightarrow H_2CO_3^* \tag{7.29}$$

The asterisk (*) denotes that this compound is actually the sum of two compounds, the dissolved CO_2 and the reaction product, carbonic acid H_2CO_3.

Since the carbonic acid that is formed is a diprotic acid, that is, it has two hydrogen atoms, an additional equilibrium step reaction occurs in water. The first reaction forms bicarbonate and hydrogen ions:

$$H_2CO_3^* \leftrightarrow HCO_3^- + H^+$$

followed by a reaction that forms carbonate and hydrogen,

$$HCO_3^- \leftrightarrow CO_3^{2-} + H^+ \tag{7.30}$$

Each of the two-step reactions has its own acid equilibrium constant, K_{a1} and K_{a2}, respectively, as shown in Table 7.5. For a triprotic acid, there would be three unique constants. Note that the constants decrease substantially with each step. In other words, most of the hydrogen ion production occurs in the first step.

Numerous air pollution reactions can be predicted from the relative strength of acids and bases, since their strength results from how well the proton via the hydronium ion is transferred from the acid and the electron is transferred via the hydroxide ion from the base. If an acid is weak, its conjugate base must be strong, and if an acid is strong, its conjugate base must be weak. Likewise, if a base is weak, its conjugate acid must be strong, and if the base is strong, its conjugate acid must be weak. The tables show actual K_a and K_b constants; however, many sources report values for pK_a and pK_b. Recall that the "p" denotes negative logarithm, so the larger the pK_a, the weaker the acid, and the larger the pK_b, the weaker the base will be.

1. For an excellent discussion of carbon dioxide equilibrium in water, see H.F. Hemond, E.J. Fechner-Levy, *Chemical Fate and Transport in the Environment*, Academic Press, San Diego, CA, 2000.

Hydrocyanic acid demonstrates how K_a is an indicator of relative strength of reactants and products:

Stronger base **Weaker base**

$$HCN + OH^- \Leftrightarrow CN^- + H_2O \quad (7...)$$

Stronger acid ($K_a = 6.2 \times 10^{-10}$) **Weaker acid** ($K_a = 1.8 \times 10^{-16}$)

The ratio of the K_a constant values of the two acids is a direct way to quantify the equilibrium. In the hydrocyanic acid instance above, the ratio is $\frac{6.2\times10^{-10}}{1.8\times10^{-16}} \cong 4 \times 10^6$.

This large quotient indicates that the equilibrium is quite far to the right. So, if HCN is dissolved in a hydroxide solution (e.g., NaOH), the resulting reaction will produce much greater amounts of the cyanide ion (CN^-) than the amount of both the hydroxide ion (OH^-) and molecular HCN. Conversely, for an aqueous solution of sodium cyanide (NaCN), the water will only react with a tiny amount of CN^-.

Water is an amphoteric compound, that is, it can act as either an acid or a base. Recall that this is a reason that water is sometimes shown as HOH. When water acts as a base, its $pK_b = -1.7$. When water acts as an acid, its $pK_a = 15.7$.

The relationship between conjugate acid-base equilibrium and pH is important to air pollution chemistry and other environmental, toxicological, and pharmacological fields. The Henderson-Hasselbach equation states this relationship:

$$pK_a = pH + \log\frac{[HA]}{[A^-]} \quad (7.31)$$

Thus, Henderson-Hasselbach states that when the pH of an aqueous solution equals the pK_a of an acidic component, the concentrations of the conjugate acids and bases must be equal (since the log of $1 = 0$). If pH is 2 or more units lower than pK_a, the acid concentration will be greater than 99%. Conversely, when pH is greater than pK_a by 2 or more units, the conjugate base concentration will account for more than 99% of the solution [9].

This has strong implications for air pollutant and other environmental treatment technologies, since it shows that mixtures of acidic and nonacidic compounds can be separated with a pH adjustment. The application of this principle is also important to the transformation of environmental contaminants in the form of weak organic acids or bases, because these compounds in their nonionized form are much more lipophilic, meaning they will be absorbed more easily through the skin than when they exist in ionized forms. As a rule, the smaller the pK_a for an acid and the larger the pK_b for a base, the more extensive will be the dissociation in aqueous environments at normal pH values and the greater compound's electrolytic nature.

7.5 POLLUTANT DEPOSITION FROM THE HYDROSPHERE

The cycle in Fig. 7.4 requires that after water reaches the atmosphere, it will return to the earth in various physical states. From an air pollution perspective, this return is either "wet" or "dry" deposition.

Air pollutants and other chemical species can be removed from the atmosphere by several processes. Both homogeneous and heterogeneous reactions occur. If the reaction only occurs in the gas phase, the reaction is homogeneous. If the reaction involves an interface with a surface of an aerosol or within a liquid droplet, it is heterogeneous. Certain chemical species of sulfur (S) and N have sufficient aqueous solubility to be dissolved by the droplet. These liquid-phase compounds are already acidic (e.g., including concentrations of H_2SO_4, H_2NO_3, and H_2CO_3*) before washout. Thus, the water droplet plays a key role in both dry and wet deposition of pollutants that lead to acidic conditions in soils and surface waters (see Fig. 7.6).

Aerosol deposition is the process by which particles collect and deposit on solid surfaces, resulting in a decrease in the air concentration of the chemical species (C_A). As mentioned, the two main subprocesses are wet and dry deposition. As discussed in Chapter 6, the rate at which this occurs is a flux density (F):

$$F = vC_A \quad (7.32)$$

where v is the deposition velocity. The velocity is the result of several mechanisms, including gravity, that is, settling velocity; concentration gradient; and eddy diffusion, that is, turbulence (see Chapter 6). Other processes include impaction, thermophoresis, and electrophoresis. These processes will be considered in detail in discussions of particulate matter control technologies, since the design of pollution control equipment is based on these physical processes. Particle size and wind speed are key factors in predicting v.

The flux density is often used for dry deposition (F_d), exclusively, which is sometimes called "dryfall":

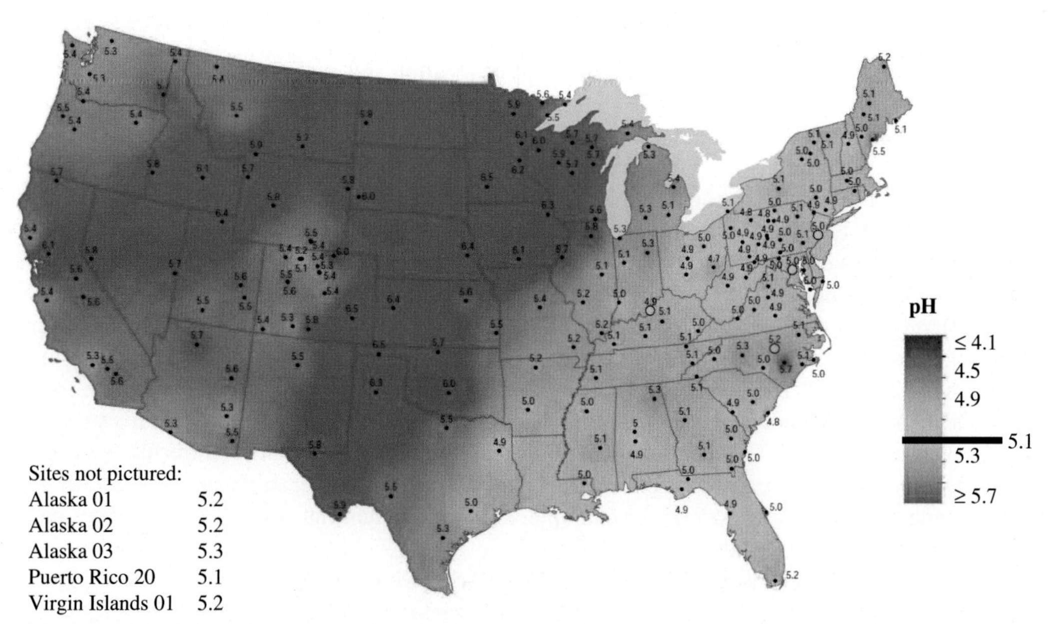

FIG. 7.6 Precipitation-weighted mean pH in wet deposition for the year 2012. *(Source: Illinois State Water Survey. University of Illinois at Urbana-Champaign. NADP Data Report 2013-01, National Atmospheric Deposition Program: 2012 Annual Summary, 2013.)*

$$F_d = v_d \cdot C_A \tag{7.33}$$

where v_v is the dry deposition flux density.

The units are length per time, that is, actual velocity units, for example, mm $\sec^{-1}$.

The deposition velocity of NH_3 in forests and other natural ecosystems has been found to be from about 1 to about 5 cm sec^{-1} [10]. If the concentration of NH_3 in the atmosphere ranges between 0.08 and 5.0 ppb [11], what would be the expected range of dry deposition of NH_3 in terrestrial ecosystems?

The volume per volume units (ppb) must be converted to mass per volume units (mg m^3). For any gas at STP, 1 ppb = 1 nanoliter L^{-1}. Using Eq. (7.7) for the lowest values ($v_d = 1$ cm sec^{-1} and $C = 0.08$ ppb),

0.08 nL NH_3/L air × 1 L air/1000 mL air × 1 mL air/ 1 cm^3 air × 1 L NH_3/10^9 nL NH_3 × 1 mol NH_3/22.4 L NH_3 × 17 g NH_3/1 mol NH_3 × 1000 mg NH_3/1 g NH_3 =

$$6.1 \times 10^{-11} \text{ mg NH}_3 \text{ cm}^{-2} \text{ sec}^{-1}.$$

Using Eq. (7.7) for the highest values ($v_d = 5$ cm sec^{-1} and $C = 5.0$ ppb),

5.0 nL NH_3/L air × 1 L air/1000 mL air × 1 mL air/1 cm^3 air × 5 L NH_3/10^9 nL NH_3 × 1 mol NH_3/22.4 L NH_3 × 17 g NH_3/1 mol NH_3 × 1000 mg NH_3/1 g NH_3 =

$$1.9 \times 10^{-8} \text{ mg NH}_3 \text{ cm}^{-2} \text{ s}^{-1}$$

Therefore, using these data, the ecosystems receive ammonia from the dry deposition at a rate ranging from 6.1×10^{-11} to 1.9×10^{-8} mg NH_3 cm^{-2} s^{-1}. So, every hectare of these ecosystems is receiving 0.0061–0.19 mg NH_3, and every km^2 is receiving 0.61–190 mg NH_3 from the atmosphere in the form of dry deposition each second.

Wet deposition depends on droplet formation. Therefore, the fraction of a chemical species that is deposited by wet deposition depends on the aqueous solubility of the chemical species. Therefore, gases and aerosols must first become dissolved or suspended in a droplet before washout from the atmosphere by precipitation and, to a lesser extent, other hydrologic processes, for example, mist and fog formation. The washout rate depends on Henry's law and air-water partition coefficient (K_{AW}) (see Chapter 5) and particle scavenging [12]. The scavenging ratio (S) is

$$S = C_p \cdot \rho \cdot C_{aerosol} - 1 \tag{7.34}$$

where C_p is the concentration of the chemical species in the precipitation (g kg^{-1}), ρ is the density of the air (about 1.2 kg m^{-3}), and $C_{aerosol}$ is the aerosol concentration of the chemical species (g m^{-3}).

The wet deposition rate depends on the type of precipitation event and the vertical distribution of the particulate matter.

7.6 INTERCOMPARTMENTAL EXCHANGE TO AND FROM THE HYDROSPHERE

The total amount of water in the hydrosphere is constant, but its location and quality vary in time and space. Air pollution managers and engineers must account for possible changes in weather, climate, and other hydrologic conditions with time. It is reasonable and prudent to expect changes over the life of a design. Mean global and local temperature increases due to increased atmospheric concentrations of global greenhouse should be factored into designs and facility citing decisions [13, 14]. For example, certain facilities should be hardened or simply not be built in areas potentially affected by rising water levels on coasts and on other large water bodies, for example, the Great Lakes in North America. In addition, the relationships between water supplies and engineered systems can be affected by changes in hydrologic and hydraulic conditions, for example, the saltwater wedge encroaching inward (see Fig. 7.7).

The density difference between fresh and salt water can be a slowly unfolding disaster for the health of people living in coastal communities and for marine and estuarine ecosystems. Salt water contains a significantly greater mass of ions than does fresh water (see Table 7.7). The denser saline water can wedge beneath fresh waters and pollute surface waters and groundwater. This phenomenon, known as saltwater intrusion, can significantly alter an ecosystem's structure and function and threaten freshwater organisms. The literature often draws abrupt distinctions between freshwater and saltwater environments, but in fact, there are large transition zones between the two. For example, an estuary is a partially enclosed coastal water body that connects a freshwater system, for example, a river, to water that is influenced by tides and within which there is a gradient between salt water and fresh water.

The effect in sensitive coastal habitats like those in South Florida would be exacerbated by the rising sea level, which could submerge low-lying areas of the Everglades (see Fig. 7.8), increasing the salinity in portions of the aquifer. The rising seawater could force salty waters upstream into coastal areas, thus threatening surface water supplies. Furthermore, siting pollution control technologies in these sensitive areas would be sources of additional contaminants, including air pollutants. Similar problems could also occur in northeastern US aquifers that are recharged by fresh portions of streams that are vulnerable to increased salinity during severe droughts. [13] Indeed, similar groundwater-surface water-seawater-atmospheric interactions could be affected across the globe. Planning, engineering, and construction in these areas require methods and tools to optimize siting alternatives, including geographic information systems (GIS) [15].

Salinity of water is a relative term. The values in Table 7.7 are at best averages and target concentrations. Note that the classifications using two dominant ions, Na^+ and Cl^-, differ by three orders of magnitude between "fresh" and "saline" waters. The same is true for bromine (Br), fluorine (F), boron (B), magnesium (Mg), and calcium (Ca). Consider the hypothetical example in Fig. 7.9 of these ionic strengths before and after saltwater intrusion. Although the water near the water supply is not "salt water" nor does it currently violate the *sec*ondary drinking water standard, the ionic concentrations are cause for concern and serve as a warning that the trend is likely to be toward even higher salinity.

Salinity is the total dissolved solids (TDS), not just ionic composition. As mentioned, the dissolved solids can be several orders of magnitude higher in salt water than fresh water, but this is not always true for the nutrients. It is true for potassium (K), which has a mean concentration of $2.3\,mg\,L^{-1}$ in fresh water but $416\,mg\,L^{-1}$ in seawater with 35% salinity. However, nitrogen (N) and phosphorus (P) do not differ substantially, with N having $0.25\,mg\,L^{-1}$ in fresh water and $0.5\,mg\,L^{-1}$ in seawater and P having $0.02\,mg\,L^{-1}$ in fresh water and $0.07\,mg\,L^{-1}$ in seawater [16].

The foregoing discussion illustrates how an indirect effect like the warming of the atmosphere from increasing concentrations of greenhouse gases can lead to hydrospheric impacts, for example, rising sea level and water contamination. The interconnectedness of the atmosphere, hydrosphere, and biosphere is complex and extensive (see Fig. 7.10). A small change in one small part of the spheres can lead to unanticipated outcomes. Thus, hydrologic cycling specifically and biogeochemical cycling generally are sensitive to initial conditions and, as such, must be treated as chaotic systems.

The basic hydrologic continuity equation states that a water system storage changes at a rate of the difference between water entering the system (I) and the amount of water exiting the system (O):

$$\frac{dS}{dx} = I - O \tag{7.35}$$

During the month of July, an aquifer is being pumped to irrigate a corn crop at $20\,m^3\,sec^{-1}$ and is being recharged with rainwater at a net rate of $5\,m^3\,sec^{-1}$; what is the change in volume during July?

Solving for dS:

$$dS = (I - O)\,dt$$

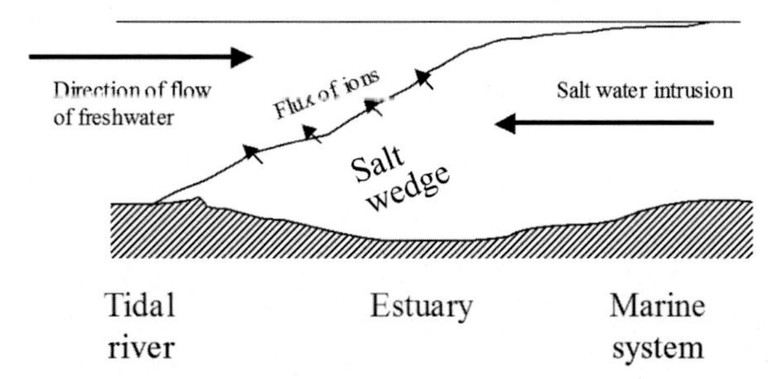

FIG. 7.7 Saltwater intrusion into a freshwater system. This denser salt water submerges under the lighter freshwater system. The same phenomenon can occur in coastal aquifers.

TABLE 7.7 Important general ionic composition classifications of fresh waters and marine waters

Composition	River water	Salt water
pH	6–8	8
Ca^{2+}	$4 \times 10^{-5} M$	$1 \times 10^{-2} M$
Cl^{-}	$2 \times 10^{-4} M$	$6 \times 10^{-1} M$
HCO_3^{-}	$1 \times 10^{-4} M$	$2 \times 10^{-3} M$
K^{+}	$6 \times 10^{-5} M$	$1 \times 10^{-2} M$
Mg^{2+}	$2 \times 10^{-4} M$	$5 \times 10^{-2} M$
Na^{+}	$4 \times 10^{-4} M$	$5 \times 10^{-1} M$
SO_4^{2-}	$1 \times 10^{-4} M$	$3 \times 10^{-2} M$

M = molarity (moles per liter solution).
Sources of data: P.M. Gschwend, Environmental Organic Chemistry. John Wiley & Sons, 2016; J.P. Kim, K.A. Hunter, M.R. Reid, Factors influencing the inorganic speciation of trace metal cations in fresh waters, Mar. Freshw. Res. 50 (4) (1999) 367–372; R.P. Schwarzenbach, P.M. Gschwend, D.M. Imboden, Organic acids and bases: acidity constant and partitioning behavior, in: Environmental Organic Chemistry, 1993, pp. 245–274; D. Vallero, Environmental Biotechnology: A Biosystems Approach, Elsevier Science, 2015.

$$dS = (5 - 20)\,\text{m}^3\,\text{sec}^{-1} \times \left(31\,\text{days} \times 24\,\text{h}\,\text{day}^{-1} \times 60\,\text{min}\,\text{hr.}^{-1} \times 60\right)\,\text{sec}$$

$$dS \approx -4 \times 10^7\,\text{m}^3$$

Thus, the aquifer is losing 40 million cubic meters of water (about ten and a half billion gallons) in July.

The continuity equation can be further described as.

$$\frac{dS}{dx} = P - R - E - T - I \tag{7.36}$$

where P is precipitation, R is surface runoff (e.g., to streams), E is evaporation, T is transpiration, and I is infiltration (i.e., the movement of water through the troposphere-soil interface).

During August and September, Townsville (area (A) = 50 km^2) received 250 mm of rain. The evaporation rate was 100 mm, transpiration was 50 mm, and infiltration was 30 mm. What was the volume of surface runoff during this period?

Solving for R:

$$R = P - E - T - I$$

$$R = (250 - 100 - 50 - 30)\,\text{mm}$$

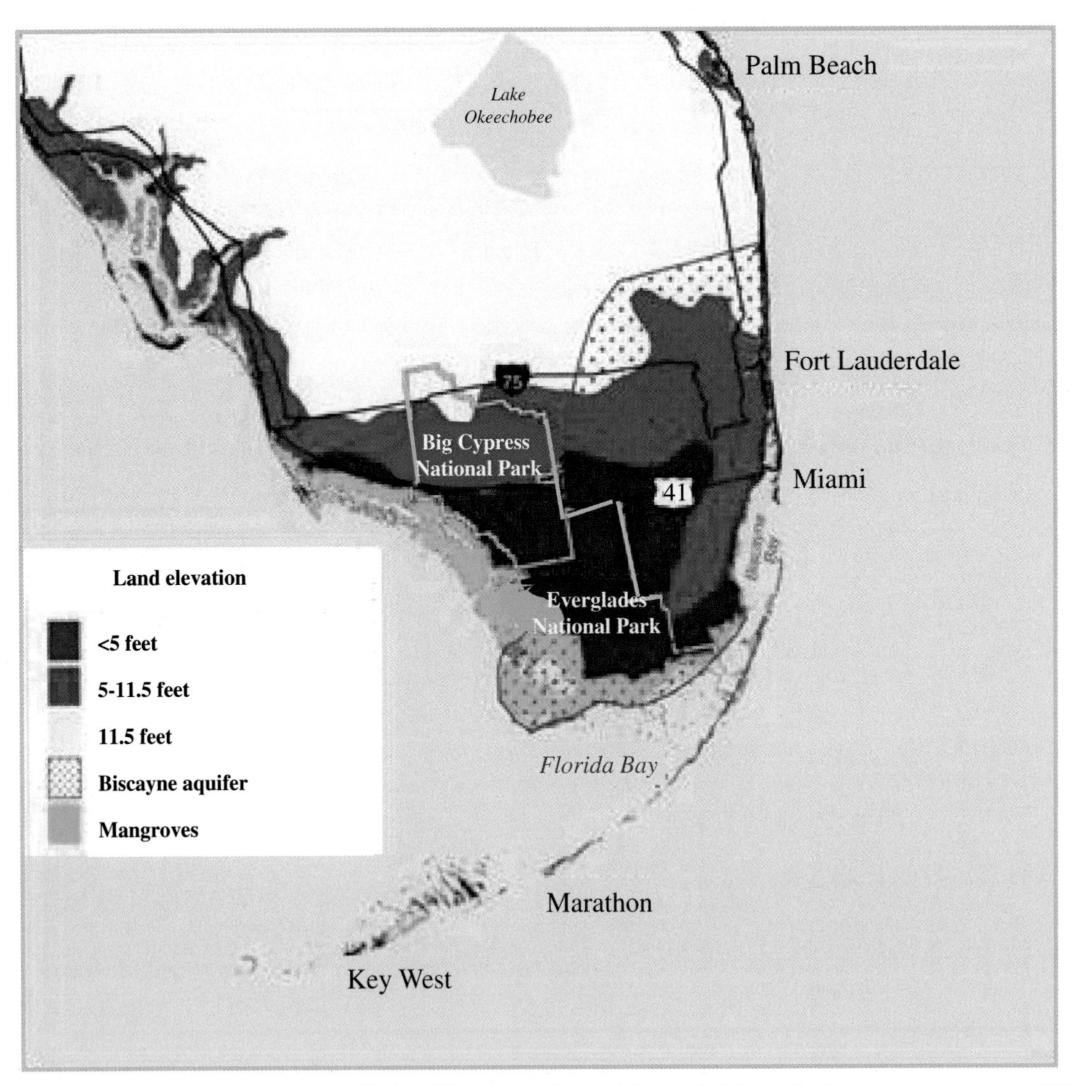

FIG. 7.8 Elevation and aquifer locations in southern Florida. Although a small part of the aquifer is beneath salty mangrove area, most of it is recharged by the freshwater Everglades, rendering the area vulnerable to saltwater intrusion and increased salinity of both surface and groundwater sources of drinking water. *(Source: U.S. Environmental Protection Agency. Saving Florida's Vanishing Shores. (2002). 1 November 2013. Available: http://www.epa.gov/climatechange/Downloads/impacts-adaptation/saving_FL.pdf.)*

$$R = 70\,\text{mm} = 0.07\,\text{m}$$

Volume of surface runoff (V) is $R \times A$, so

$$V = 0.07\,\text{m} \times 50 \times 10^6\,\text{m}^2$$
$$V = 3.5 \times 10^6\,\text{m}^2$$

Nearby city, Villetown, is also 50 km^2 in area. They have instituted a number of drainage improvements, including tree plantings and permeable surfaces in parking lots. During the same August and September, Villetown also received 250 mm of rain. However, the evaporation rate was 60 mm, transpiration was 60 mm, and infiltration was 100 mm. What was the volume of surface runoff for Villetown for this time period?

Solving for R:

$$R = P - E - T - I$$

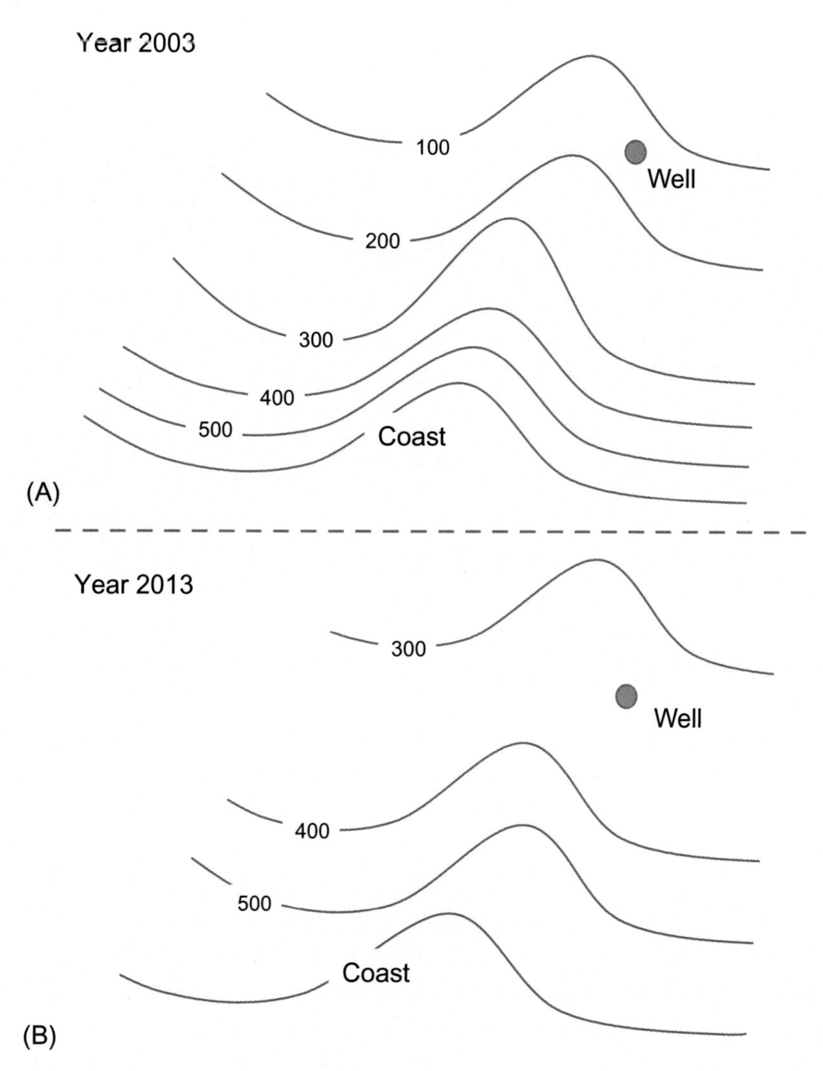

FIG. 7.9 Hypothetical isopleths of total dissolved solid ($mg L^{-1}$) concentrations in groundwater (e.g., 30m depth). The increases over the decade, that is, increased concentrations in 2013 (A) versus 2003 (B), indicate saltwater intrusion and potential contamination at the drinking water well site. *(Source: D. Vallero, Environmental Biotechnology: A Biosystems Approach, Elsevier Science, 2015.)*

$$R = (250 - 60 - 60 - 100)\,\text{mm}$$

$$R = 30\,\text{mm} = 0.03\,\text{m}$$

Volume of surface runoff (V) is $R \times A$, so.

$$V = 0.03\,\text{m} \times 50 \times 10^6\,\text{m}^2$$

$$V = 1.5 \times 10^6\,\text{m}^2$$

Therefore, Villetown's surface runoff was 2 million cubic meters less than Townville's during this 2-month period, in part at least to the increased infiltration and decreased evaporation rates. The transpiration rate increased due to the greater number of trees. Thus, the tree planting increased E, T, and I, by holding water in the plant tissue, releasing water to the atmosphere, and moving water to the soil via the root systems.

Note that all the variables on the right side of Eq. (7.10) are fluxes, adhering to the diffusion, advective, and turbulence principles discussed in Chapter 6.

The mass of water vapor and air is of the same units, usually kilograms. If water vapor in the parcel has the same temperature and pressure as the air, then.

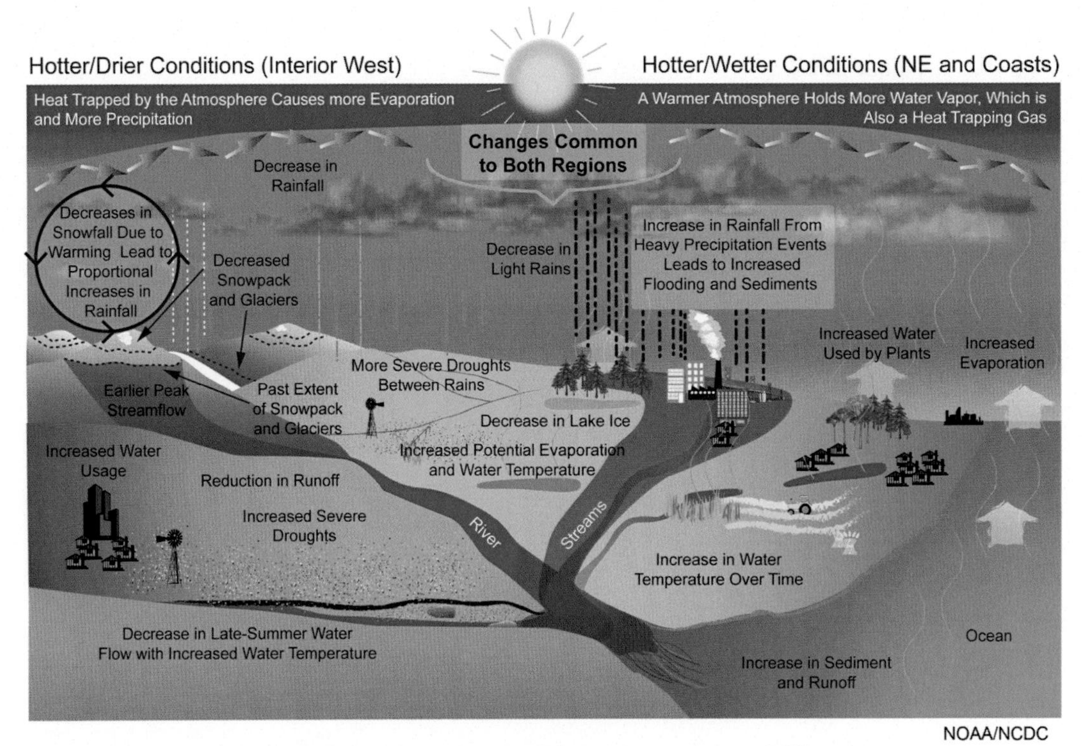

FIG. 7.10 Potential impacts of climate change of the hydrologic cycle. *(Source: U.S. Environmental Protection Agency. Climate Impacts on Water Resources. (2016). 1 March 2018. Available: https://19january2017snapshot.epa.gov/climate-impacts/climate-impacts-water-resources_.html. Based on information from T.R. Karl, J. Melillo, T. Peterson, Global climate change impacts in the United States, Global Climate Change Impacts in the United States, 2009.)*

$$\omega = 0.62198 \cdot \frac{p_w}{p_{at} - p_w} \tag{7.37}$$

where p_w is the partial pressure of the water vapor in the moist air parcel and p_{at} is the atmospheric pressure.

The degree of saturation (μ) is the ratio of ω of moist air to humidity ratio of saturated moist air (ω_{sat}) at a specific temperature and pressure:

$$\mu = \frac{\omega}{\omega_{sat}} \tag{7.38}$$

7.7 WATER, OXYGEN AND ENERGY RELATIONSHIPS

Matter and energy cycling takes place at all levels, each of which is important to air pollution management. Energy cycling can begin with the release of heated plumes not only from incinerators but also from composting facilities and landfills. Exchanges between the hydrosphere, atmosphere, and lithosphere, for example, soil and sediment, create sinks and sources of energy. Heat reservoirs in terrestrial and aquatic systems receive added heat, which when released can alter habitats (e.g., changes to freeze-thaw cycles, seasonal variations, and selectivity of certain soil bacteria genera).

Mass and energy cycling commonly occurs within the hydrologic cycle. The first law of thermodynamics dictates that these cycles be both mass and energy balanced. Some of these water-energy relationships are obvious and direct; for example, the thermal inversions that lead to urban air pollution are the result of differences in heat energy in water vapor at various layers in the troposphere. Some of the water-energy relationships are incremental and indirect, such as the transfer of energy between trophic states in an ecosystem, which relies in part on energy and water exchanges among biotic tissue and abiotic substrates (e.g., aqueous-phase phosphorous (K) compounds that transfer energy during photosynthesis).

Water-energy cycling is also important in anthropogenic systems and is part of a design of pollution control equipment selection and application. For example, at the facility scale, water and energy are often addressed together, such as the management of water and energy at a factory or power plant. Managing water is a simultaneous process with managing energy. Indeed, mismanagement of water systems at the facility level can involve trade-offs between types of pollution. Again, the first law requires, for example, that allowing heated water to be released in any amount, even the permitted level, would increase the overall temperature of the receiving waters.

The heat from boilers and other industry-scale operations is going to be exchanged. This is a consideration of not only air pollution and water pollution control decisions but also air pollution management decisions. Facility design determines in large part where the energy goes (see Fig. 7.11). In fact, a pollution control and heat control design can directly affect dissolved oxygen (DO) content of the receiving water temperature since temperature is directly proportional to DO content.

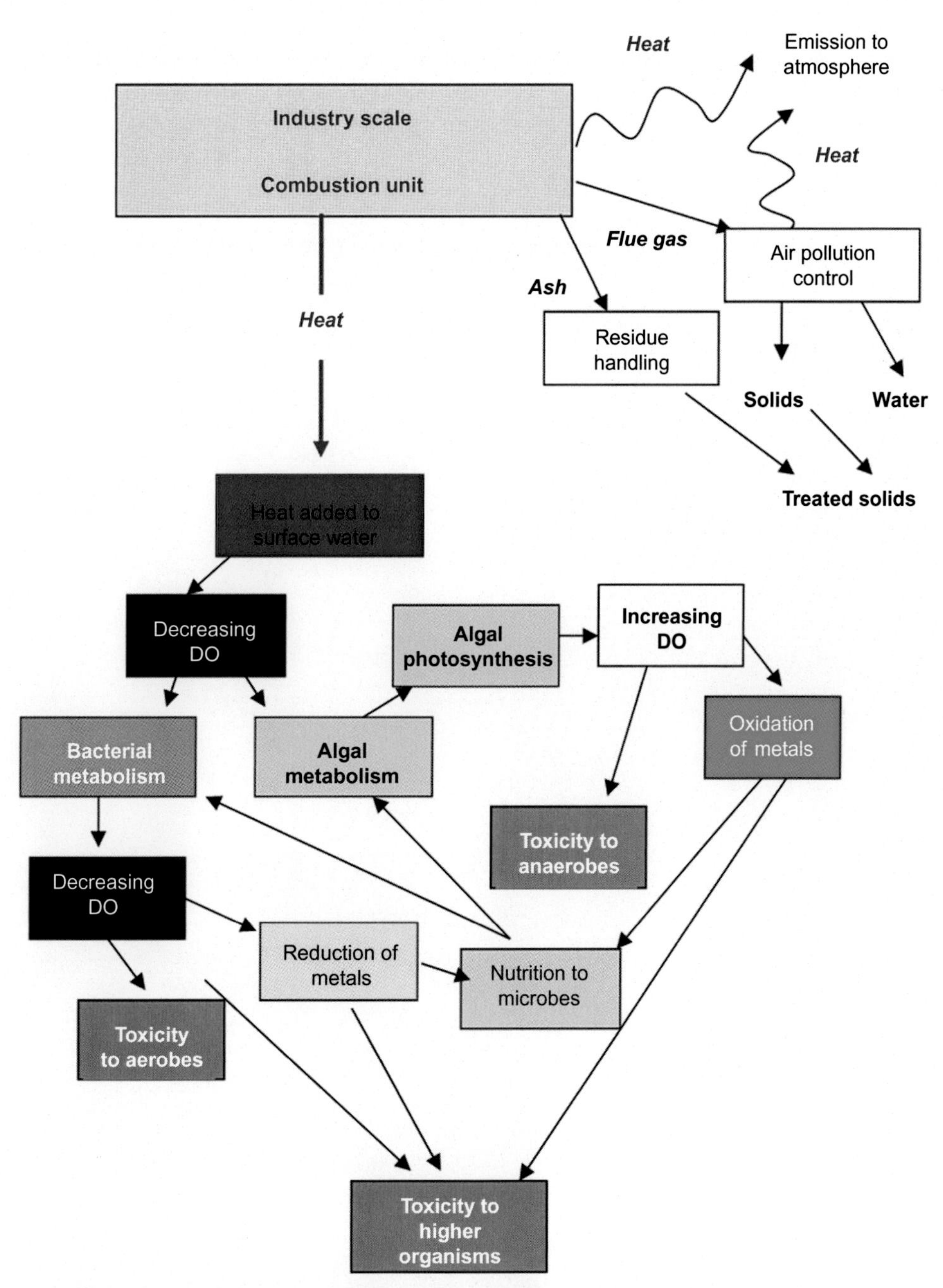

FIG. 7.11 Physical, chemical, and biological linkages in environmental management. In this example, both energy and matter must be considered in a management decision, for example, to incinerate industrial or municipal wastes. The control of matter leaving the combustion unit must be optimized with the transfer of energy. The added heat results in an abiotic response (i.e., decreased dissolved oxygen (DO) concentrations in the water), which leads to biotic processes that either increase or decrease DO and increase the toxic responses (e.g., low DO and increase in metal bioavailability). *(Source: D.A. Vallero, Fundamentals of Air Pollution, 5th ed., Elsevier Academic Press, Waltham, MA, 2014, p. 999 pages cm.)*

The DO is a limiting factor of the type of fish communities that can be supported by a water body (see Tables 7.8 and 7.9). The resulting net increase in heat may directly stress the biotic integrity of a surface water ecosystem; for example, fish species vary in their ability to tolerate higher temperatures, meaning that the less tolerant, higher-value fish will be inordinately threatened.

The increased temperature can also increase the aqueous solubility of toxic substances to organisms. For example, greater concentrations of mercury and other toxic metals will occur with elevated temperatures. The lower DO concentrations will lead to a reduced environment where the metals and compounds will form sulfides and other compounds that can be toxic to the aquatic life. Thus, a cascade if harm begins with the change in temperature, followed by the decrease in DO and increasing metal concentrations. The synergistic impact of combining the hypoxic water and reduced metal compounds result in fish kills and other adverse effects on the stream's ecosystems. The cascade may continue beyond the stream ecosystem to cause human health problems, such as when mercury and other toxic metals migrate into the food chain and enter the human diet.

TABLE 7.8 Relationship between water temperature and maximum dissolved oxygen (DO) concentration in water (at 1 atm (atm))

Temperature (°C)	Dissolved oxygen (mg L^{-1})	Temperature (°C)	Dissolved oxygen (mg L^{-1})
0	14.60	23	8.56
1	14.19	24	8.40
2	13.81	25	8.24
3	13.44	26	8.09
4	13.09	27	7.95
5	12.75	28	7.81
6	12.43	29	7.67
7	12.12	30	7.54
8	11.83	31	7.41
9	11.55	32	7.28
10	11.27	33	7.16
11	11.01	34	7.16
12	10.76	35	6.93
13	10.52	36	6.82
14	10.29	37	6.71
15	10.07	38	6.61
16	9.85	39	6.51
17	9.65	40	6.41
18	9.45	41	6.41
19	9.26	42	6.22
20	9.07	43	6.13
21	8.90	44	6.04
22	8.72	45	5.95

Sources: D. Vallero, Environmental Biotechnology: A Biosystems Approach, Academic Press, 2010; E. Dohner, A. Markowitz, M. Barbour, J. Simpson, J. Byrne, G. Dates, 5.2. Monitoring and assessing water quality, in: Volunteer Stream Monitoring: A Methods Manual, 1997.

TABLE 7.9 Normal temperature tolerances of aquatic organisms

Organism	Taxonomy	Range in temperature tolerance (°C)	Minimum dissolved oxygen (mg L^{-1})
Trout	*Salma, Oncorhynchus,* and *Salvelinus* spp.	5–20	6.5
Smallmouth bass	*Micropterus dolomieu*	5–28	6.5
Caddisfly larvae	*Brachycentrus* spp.	10–25	4.0
Mayfly larvae	*Ephemerella invaria*	10–25	4.0
Stonefly larvae	*Pteronarcys* spp.	10–25	4.0
Catfish	Order Siluriformes	20–25	2.5
Carp	*Cyprinus* spp.	10–25	2.0
Water boatmen	*Notonecta* spp.	10–25	2.0
Mosquito larvae	Family Culicidae	10–25	1.0

Source: D. Vallero, Environmental Biotechnology: A Biosystems Approach, Academic Press, 2010. Data acquired from V. Corporation. Computer 19: Dissolved oxygen in water. (2018). 5 March 2018. Available: http://www2.vernier.com/sample_labs/BWV-19-COMP-dissolved_oxygen.pdf.

Biota also plays a role in the heat-initiated effect. Combined abiotic and biotic responses occur. Notably, the growth and metabolism of the bacteria result in even more rapidly decreasing DO levels. Algae both consume DO for metabolism and produce DO by photosynthesis. The increase in temperature increases their aqueous solubility, and the decrease in DO is accompanied by redox changes, for example, the formation of reduced metal species, such as metal sulfides. This is also being mediated by the bacteria, some of which will begin reducing the metals as the oxygen levels drop (reduced conditions in the water and sediment). However, the opposite is true in the more oxidized regions, that is, the metals are forming oxides. The increase in the metal compounds combined with the reduced DO and combined with the increased temperatures can act synergistically to make the conditions toxic for higher animals, for example, a fish kill.

Although DO is reported frequently as mg L^{-1}, it is also expressed as percent of air saturation. The mass per volume and percent air saturation comparisons vary not only by temperature and pressure but also by dissolved solid concentration of the water, that is, salinity. As discussed in Chapter 1, water vapor content in the troposphere ranges from 0 to about 3%.

The solubility of O_2 varies with temperature, that is, the solubility in water decreases with increasing temperature (see Table 7.8). Along with overall atmospheric pressure, the partial pressure exerted by O_2 decreases with altitude. The solubility of O_2 decreases 1.4% per 100 m of altitude increase. Salinity also decreases O_2; for example, salt water can hold approximately 20% less O_2 when saturated than fresh water; for example, at 0°C, salt water is saturated at 12.0 mg L^{-1}, but fresh water is saturated at 14.2 mg L^{-1}.

Dissolved oxygen presents an example of using psychrometry, that is, the science of moist air, to calculate water-air exchanges according to the principles of Henry's law. That is, the mass of a slightly soluble gas like O_2 that dissolves in a mass of a liquid at a given temperature is almost directly proportional to the partial pressure of that gas. So, the point when O_2 in the air and O_2 in the surface water are in equilibrium defines the DO saturation concentration.

Most of the O_2 added to surface waters comes from the absorption from air in the atmosphere, that is, reaeration, and from photosynthesis of aquatic plants and algae. The humidity ratio (ω) and degree of water vapor saturation (μ) are indicators of evaporation and the rate of reaeration. According to Fick's law, the water vapor concentration gradient increases with decreasing ω and μ. Thus, more water vapor moves from the surface water to the air for dry air than for moist air. The rate of reaeration is increased with turbulence of both the air layer at the interface and in the water.

Surface waters lose O_2 by the oxidation of organic matter in the water and underlying sediment, the oxidation of reduced chemical species, especially ammonia (NH_3^-) in the water and sediment, and respiration by aquatic biota. Increasing water temperature also leads to drops in DO.

Incidentally, the sediment underlying the water column is also dynamic in terms of DO. For example, the sediment is not always quiescent and, in fact, can vary between advection-dominated and diffusion-dominated, depending on the turbulence of the water at the sediment-water interface (see Fig. 7.12). Thus, just as turbulence at the atmosphere-water interface

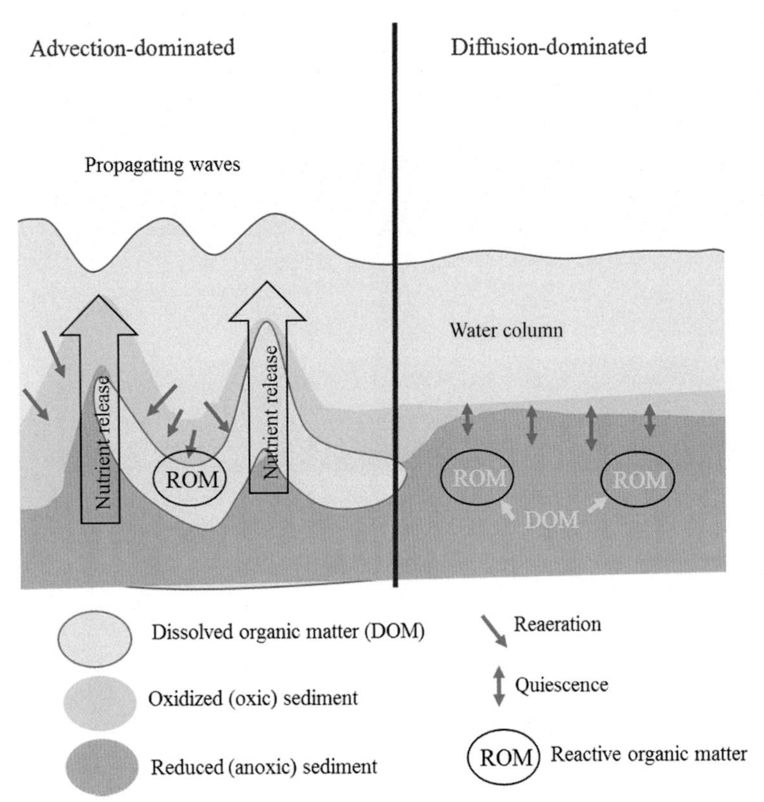

FIG. 7.12 Advection-dominated *(left)* and diffusion-dominated *(right)* sediment reaeration. Microbially mediated aerobic degradation, including the conversion of dissolved organic matter into more reactive organic matter (ROM), is enhanced by advection, which combines reaeration with the stirring of sediment matter. *(Modified from U. Franke, L. Polerecky, E. Precht, M. Huettel, Wave tank study of particulate organic matter degradation in permeable sediments, Limnol. Oceanogr. 51 (2) (2006) 1084–1096.)*

increases surface water reaeration rates, turbulence at the sediment-water interface increases sediment reaeration rates [4]. This increase in sediment DO increases the rate of aerobic degradation of organic matter that demonstrates the relationship between physical transport mechanisms, in this case advection, and chemical transformation mechanisms, in this case aerobic degradation.

The interface between the water and air surfaces is affected by the extent to which the air is saturated with water vapor. The mass per volume and percent air saturation comparisons vary not only by temperature and pressure but also by dissolved solid concentration of the water, that is, salinity. As discussed in Chapter 1, water vapor content in the troposphere ranges from 0 to about 3%.

The initiating abiotic effect, that is, increased temperature, causes or plays a role in other effects, for example, increased microbial population. The associated decline in DO, however, is a net decline between photosynthetic and nonphotosynthetic microbes. The growth and metabolism of the bacteria result in decreasing DO levels. Algae are different than bacteria. Algae both consume DO for metabolism and produce DO by photosynthesis. Thus, the relationship between temperature and microbial growth represents a stress that leads to an effect. The effect then is itself a stressor for a subsequent effect. This can occur multiple times, cascading in an ecosystem. Let us consider two simplified scenarios:

Scenario 1

Stressor 1: Deposition of airborne NH_4^- directly to a lake → Effect 1: Lower DO/reduced conditions.
Stressor 2: Anaerobic bacteria growth → Effect 2: Release of reduced compounds, for example, H_2S.
Stressor 3: Low pH/increased metal solubility in water → Effect 3: Lower reproduction and survival rates for sensitive fish species.
Stressor 4: Bioavailable, toxic metals → Effect 4: Increased metal uptake by aquatic life, especially benthic organisms.
Stressor 5: Mercury, lead, and other toxic metals increase in food chain → Effects 5 and 6: Humans exposed to toxic metals and fish kills.

Scenario 2

Stressor 1: Deposition nitrates NO_3^- → Effect 1: Increased growth of algae.
Effect 1: Photosynthesis produces less O_2 than is used by algae and bacteria → Lower DO/reduced conditions.
Stressor 2: Algal blooms block light penetration → **Effect 2:** Increasingly less photosynthesis and O2 production.
Stressor 3: Anaerobic bacteria growth → Effect 3: Release of reduced compounds, for example, CN^-.
Stressor 4: Low pH/increased metal solubility in water → Effect 4: Lower reproduction and survival rates for sensitive fish species.
Stressor 5: Bioavailable, toxic metals → Effect 5: Increased metal uptake by aquatic life, especially benthic organisms.
Stressor 6: Mercury, lead, and other toxic metals increase in food chain → Effects 6 and 7: Humans exposed to toxic metals and growth of opportunistic species (e.g., *Pfiesteria* spp.).
Stressor 7: Toxins released from opportunistic species → Effect 8: Fish kill.

Thus, a sequence of abiotic and biotic events ultimately leads to increased concentrations of bioavailable metals and toxins. For the purposes of illustration, these stressor-response relationships have been simplified. For example, Stressor 5 in the Scenario 2 is composed of numerous stressor-response relationships, including feedbacks between the water column and sediment (e.g., diffusion rates and fluxes [4]), microbial growth (e.g., changes in species abundance of algae, bacteria, and fungi), water quality (both improved and degraded), variations in types and rates of ion exchanges, changes in pH, and damage to tissue in biota, which changes rates of absorption, metabolism, distribution, and elimination of chemical species by organisms. For example, the increase in temperature increases the aqueous solubility of most metals. The effect of acidity is more complicated. Metal solubility increases with acidity, that is, increases with decreasing pH, but will decrease after some threshold pH that varies among metals (see Fig. 7.13), which decreases DO concentrations and modifies reduction/oxidation (redox) condition. Reduction results in the formation of reduced metal species, such as metal sulfides. These changes can increase the exposure of human populations to toxic substances.

Thus, the increasing aqueous solubility of most metals brought on by the increase in temperature shown in Fig. 7.13 also initiates the release of metals from suspended or settled solids in the water. The process is further complicated in that it is also being mediated by the bacteria, some of which will begin reducing the metals as the oxygen levels drop (reduced conditions in the water and sediment). However, the opposite is true in the more oxidized regions, where the metals are forming oxides.

Microbes use redox reactions for metabolism, that is, to build cells (anabolism) and to break down organic matter to survive (catabolism). Those that require O_2 as the energy source are aerobes; those that use compounds other than O_2 as the energy source (e.g., NO_3^-, SO_4^{2-}, and CH_2O) are anaerobes [4]. Those that can survive under aerobic or anaerobic conditions are classified as facultative microbes. Facultative aerobes, for example, can produce the adenosine triphosphate (ATP) needed for energy transfer by aerobic respiration when O_2 is available, but if not, the microbe can switch to anaerobic respiration, that is, fermentation, when O_2 is unavailable. ATP is a transporter of energy when one of its phosphate groups breaks off, which converts the ATP to adenosine diphosphate (ADP). When a phosphate breaks off, energy is released which the plant uses for cellular processes. In microbiology, microbes are distinguished as either obligate or facultative. Obligate aerobes can only survive with a sufficient DO concentration, that is, they cannot produceADP if O_2 is absent. Obligate anaerobes cannot survive if O_2 is present [18, 19].

The increase in the metal compound concentrations combined with the reduced DO, in the presence of increased temperatures, can act synergistically to make the conditions toxic for higher animals, for example, a fish kill [20]. Predicting the likelihood of ecosystem change and adverse events like fish kills can be quite complicated, with many factors that either mitigate or exacerbate the outcome (see Fig. 7.14). The increase in metal concentrations in water increases not only the potential exposure to and risks from all metals but also the transformation of metals into chemical species. If these new compounds have higher bioavailability, then exposure, dose, and body burden of the metals will also likely increase (e.g., methylmercury in fish ingested by humans).

7.8 REACTIONS WITH WATER

On their way to becoming air pollutants, water is not only their residence, for example, droplets, but also the source of chemical transformation. As mentioned, the redox conditions in water, whether a surface water, groundwater, or droplet,

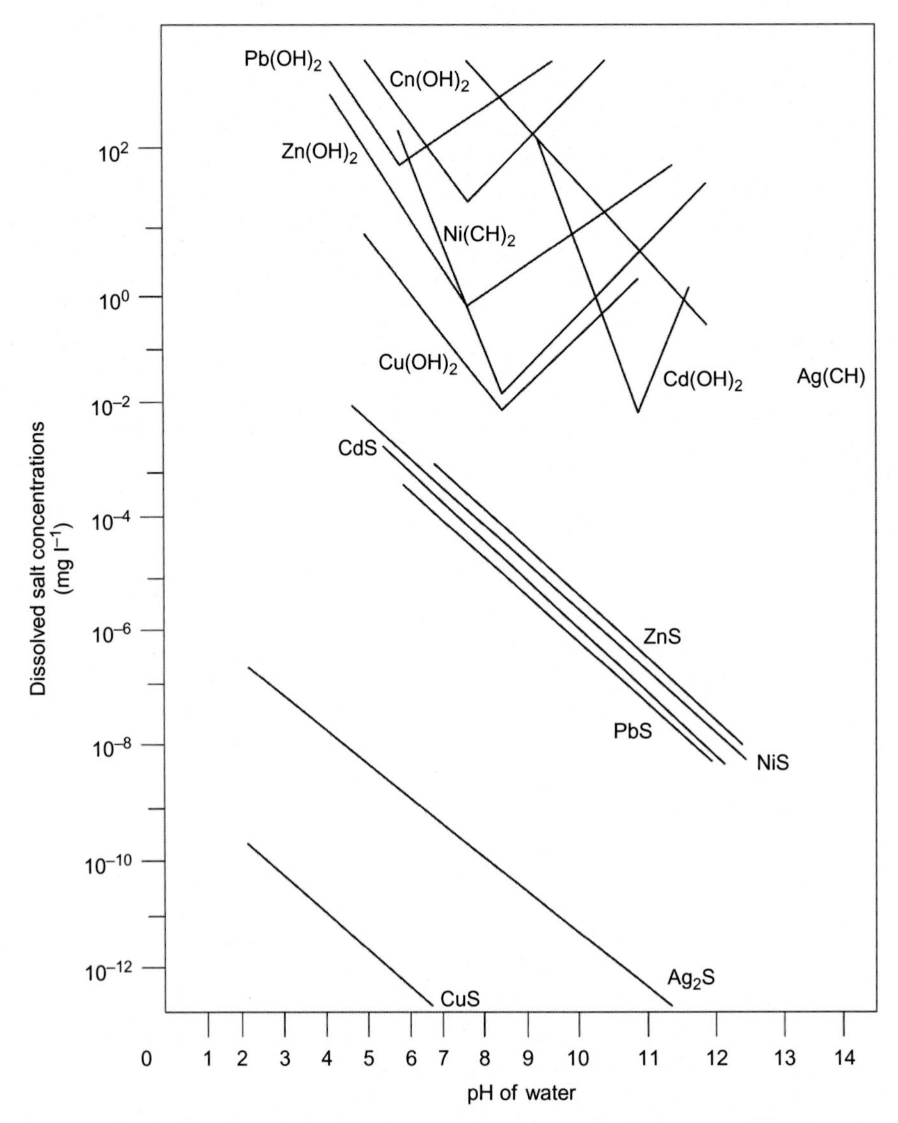

FIG. 7.13 Effect of pH on aqueous solubility of metal salts. The temperature is not reported in the data source [17]. *(Reproduced with permission D.A. Vallero, Fundamentals of Air Pollution, 5th ed., Elsevier Academic Press, Waltham, MA, 2014, p. 999 pages cm; adapted from EPA/625/8-80-003, Summary Report: Control and Treatment Technology for the Metal Finishing Industry – Sulfide Precipitation, 1999.)*

determine the ionization and speciation of chemicals. Other chemical processes occur, but no discussion of water's role in air pollution would be complete without discussing hydrolysis, that is, the reaction with water. The hydrolyzed compound can degrade into compounds that are less toxic, such as what occurs in Phase 1 metabolism within an organism. Conversely, hydrolysis can produce more highly toxic metabolites, a process known as bioactivation. Fig. 7.15 illustrates the forms that the pesticide methyl parathion can take after it is released into the environment, including pathways that produce compounds that may be more toxic than the parent compound. This is an example of an air pollutant that is transformed in water and by reacting with water during and after deposition, that is, there at least three pathways dominated by hydrolysis.

Almost any industrial, commercial, or residential activity will release substances, including contaminants, into the environment. After release, many processes and exchanges occur among and within the environmental compartments (see Fig. 7.16) and those at the interface between the organism and the environment. Water is part of every process shown.

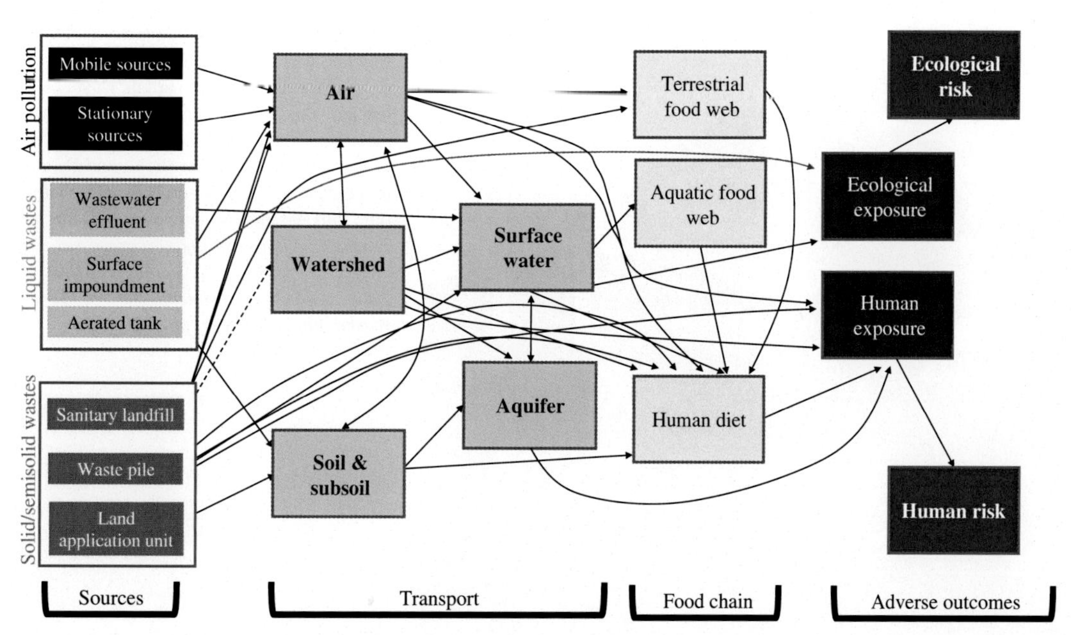

FIG. 7.14 Environmental transport pathways can be affected by net heat gain. Compounds (nutrients and contaminants), microbes, and energy (e.g., heat) follow the path through the environment indicated by *arrows*. The residence time within any of the boxes is affected by abiotic conditions, including temperature. *(Modified from D. Vallero, Translating Diverse Environmental Data into Reliable Information: How to Coordinate Evidence from Different Sources. Academic Press, 2017; and D.A. Vallero, K.H. Reckhow, A.D. Gronewold, Application of Multimedia Models for Human and Ecological Exposure Analysis. International Conference on Environmental Epidemiology and Exposure, Durham, NC. International Society of Exposure Analysis, 2007.)*

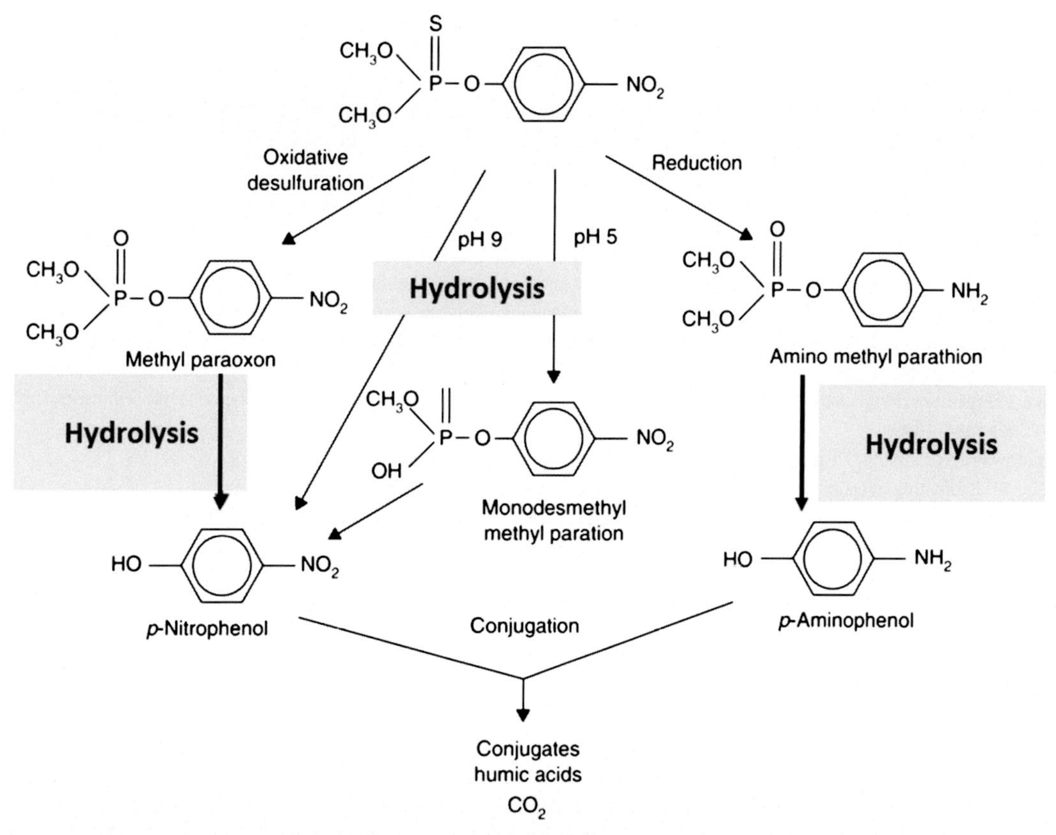

FIG. 7.15 Pathways of methyl parathion in water under varying redox and acidity conditions. Pathways are affected by the pH of water and sediment, available oxygen and water, and numerous other environmental conditions. *(Reproduced with permission D. Vallero, Environmental Biotechnology: A Biosystems Approach. Elsevier Science, 2015. Information and graphic sources: Methyl parathion. Mol. Nutri. Food Res. 38 (1993) 350. doi:10.1002/food.19940380333; A. Bourquin, R. Garnas, P. Pritchard, F. Wilkes, C. Cripe, N. Rubinstein, Interdependent microcosms for the assessment of pollutants in the marine environment, Int. J. Environ. Stud. 13 (2) (1979) 131–140; R. Wilmes, Parathion-Methyl: Hydrolysis Studies, 1987.)*

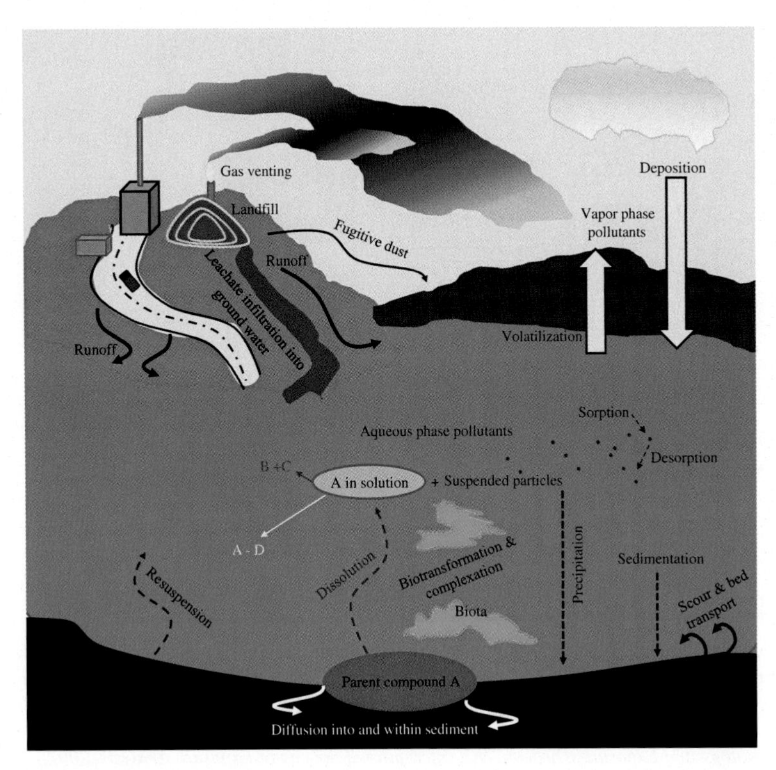

FIG. 7.16 Transport and transformation of chemicals in a water system. Contaminants from a human activity may reach a water system in various ways. Once the substance reaches the water, numerous transformation processes, including dissociation and degradation to form metabolites and degradation products (B, C, and D), are brought about by both abiotic (e.g., hydrolysis and photolysis) and biotic (i.e., biodegradation) processes. *(Data from W. Lyman, Transport and transformation processes, Rand GM Fundamentals of Aquatic Toxicology. Effects, Environmental Fate, and Risk Assessment, 2nd ed., Taylor and Francis, Washington, DC, 1995, pp. 449–492.)*

The equilibrium and speciation of elements in the atmosphere, hydrosphere, and biosphere determine their importance to air pollution.

Thus, environmental engineers and air quality managers need a modicum of knowledge about the biogeochemical cycling, especially for carbon and nutrients. This is considered in detail in the next chapter.

REFERENCES

[1] D.A. Vallero, Fundamentals of Air Pollution, fifth ed., Elsevier Academic Press, Waltham, MA, 2014. p. 999 pages cm.

[2] T. Chouhan, I. Jaising, C.A. Alvares, Bhopal, the inside story: carbide workers speak out on the world's worst industrial disaster, Apex Pr, 1994.

[3] U.S. Geological Survey, The Water Cycle: Water Storage in the Atmosphere, 10 July 2018. Available: https://water.usgs.gov/edu/watercycleatmosphere.html, 2018.

[4] U. Franke, L. Polerecky, E. Precht, M. Huettel, Wave tank study of particulate organic matter degradation in permeable sediments, Limnol. Oceanogr. 51 (2) (2006) 1084–1096.

[5] W.J. Shuttleworth, Chapter 4: Evaporation, in: Handbook of Hydrology, McGraw-Hill, USA, 1993.

[6] H.L. Penman, Estimating evaporation, Eos Trans. Am. Geophys. Union 37 (1) (1956) 43–50.

[7] J. Monteith, M. Unsworth, Principles of Environmental Physics: Plants, Animals, and the Atmosphere, Academic Press, 2013.

[8] D.A. Vallero, Excerpted and Revised From: Fundamentals of Air Pollution, fifth ed., Elsevier Academic Press, Waltham, MA, 2014. p. 999 pages cm.

[9] USEPA, Dermal Exposure Assessment: Principles and Applications (EPA/600/8–91/011B), US Environmental Protection Agency, 1992.

[10] M. Sutton, C.E. Pitcairn, D. Fowler, The exchange of ammonia between the atmosphere and plant communities, in: Advances in Ecological Research, vol. 24, Elsevier, 1993, pp. 301–393.

[11] W.H. Schlesinger, Biogeochemistry, Elsevier, 2005.

[12] J. Prospero, R. Arimoto, Atmospheric Transport and Deposition of Particulate Material to the Oceans, 2008.

[13] U.S. Environmental Protection Agency, Climate Impacts on Water Resources, 1 March 2018. Available: https://19january2017snapshot.epa.gov/climate-impacts/climate-impacts-water-resources_.html, 2016.

[14] T.R. Karl, J. Melillo, T. Peterson, Global climate change impacts in the United States, in: Global Climate Change Impacts in the United States, 2009.

[15] H. Sarptas, N. Alpaslan, D. Dolgen, GIS supported solid waste management in coastal areas, Water Sci. Technol. 51 (11) (2005) 213–220.

[16] J.P. Riley, R. Chester, Chemical Oceanography, Elsevier, 2016.

[17] J. E. S. a. A. J. Freedman, Modern alkaline cooling water treatment, Ind. Water Eng. January–February (1973) 31.

[18] B.E. Rittmann, Environmental biotechnology in water and wastewater treatment, J. Environ. Eng. 136 (4) (2009) 348–353.
[19] B.E. Rittmann, P.L. McCarty, Environmental Biotechnology: Principles and Applications, Tata McGraw-Hill Education, 2012.
[20] D.A. Vallero, K.H. Reckhow, A.D. Gronewold, Application of Multimedia Models for Human and Ecological Exposure Analysis, in: International Conference on Environmental Epidemiology and Exposure, Durham, NC, International Society of Exposure Analysis, 2007.

FURTHER READING

[21] P.H. Gleick, Water in Crisis: A Guide to the Worlds Fresh Water Resources, 1993.
[22] U.S. Geological Survey, The World's Water, 28 February 2018. Available: https://water.usgs.gov/edu/earthwherewater.html, 2016.
[23] Q. Cronk, Vapour Pressure Deficit Calculation. University of British Columbia, Center for Plant Research, 2018. Available online: http://cronklab.wikidot.com/calculation-of-vapour-pressure-deficit.
[24] U.S. Geological Survey, Summary of the Water Cycle, 28 February 2018. Available: https://water.usgs.gov/edu/watercyclesummary.html, 2016.
[25] V. Evangelou, Environmental Soil and Water Chemistry: Principles and Applications, John Wiley Sons, Inc, Canada, 1998.
[26] EM 1110-1-4012, Engineering and Design Precipitation/Coagulation/Flocculation, 2001.
[27] J.N. Spencer, G.M. Bodner, L.H. Rickard, Chemistry: Structure and Dynamics, John Wiley & Sons, 2010.
[28] M. Olia, A.S. Casparian, How to Prepare for the Fundamentals of Engineering, FE/EIT Exam, Barron's Educational Series, 2000.
[29] Illinois State Water Survey. University of Illinois at Urbana-Champaign, NADP Data Report 2013-01, National Atmospheric Deposition Program: 2012 Annual Summary, 2013.
[30] U.S. Environmental Protection Agency, Saving Florida's Vanishing Shores, 1 November 2013. Available: http://www.epa.gov/climatechange/Downloads/impacts-adaptation/saving_FL.pdf, 2002.
[31] P.M. Gschwend, Environmental Organic Chemistry, John Wiley & Sons, 2016.
[32] J.P. Kim, K.A. Hunter, M.R. Reid, Factors influencing the inorganic speciation of trace metal cations in fresh waters, Mar. Freshw. Res. 50 (4) (1999) 367–372.
[33] R.P. Schwarzenbach, P.M. Gschwend, D.M. Imboden, Organic acids and bases: acidity constant and partitioning behavior, in: Environmental Organic Chemistry, 1993, pp. 245–274.
[34] D. Vallero, Environmental Biotechnology: A Biosystems Approach, Elsevier Science, 2015.
[35] D. Vallero, Environmental Biotechnology: A Biosystems Approach, Academic Press, 2010.
[36] E. Dohner, A. Markowitz, M. Barbour, J. Simpson, J. Byrne, G. Dates, 5.2. Monitoring and assessing water quality, in: Volunteer Stream Monitoring: A Methods Manual, 1997.
[37] V. Corporation, Computer 19: Dissolved oxygen in water, 5 March 2018. Available: http://www2.vernier.com/sample_labs/BWV-19-COMP-dissolved_oxygen.pdf, 2018.
[38] EPA/625/8-80-003, Summary Report: Control and Treatment Technology for the Metal Finishing Industry – Sulfide Precipitation, (1999).
[39] D. Vallero, Translating Diverse Environmental Data into Reliable Information: How to Coordinate Evidence from Different Sources, Academic Press, 2017.
[40] Methyl parathion. Mol. Nutri. Food Res. 38 (1993) 350, https://doi.org/10.1002/food.19940380333.
[41] A. Bourquin, R. Garnas, P. Pritchard, F. Wilkes, C. Cripe, N. Rubinstein, Interdependent microcosms for the assessment of pollutants in the marine environment, Int. J. Environ. Stud. 13 (2) (1979) 131–140.
[42] R. Wilmes, Parathion-Methyl: Hydrolysis Studies, (1987).
[43] W. Lyman, Transport and transformation processes, in: Rand GM Fundamentals of Aquatic Toxicology. Effects, Environmental Fate, and Risk Assessment, 2nd ed., Taylor and Francis, Washington, DC, 1995, pp. 449–492.

Chapter 8

Air pollution biogeochemistry

8.1 INTRODUCTION

Air pollutants exist in the earth's dynamic systems of matter and energy. A chemical species' amount, chemical form, and location in the earth's systems determine whether and to what extent the substance is essential or detrimental. The formation, transformation, transport, and fate of a substance are determined by biogeochemistry, that is, the relationships among biological ("bio"), geophysical ("geo"), and chemical ("chemistry") mechanisms, processes, and changes in the environment, especially the cycling of matter and energy within and among the earth's atmosphere, hydrosphere, lithosphere, and biosphere.

In the previous chapter, we discussed the water cycle's role in the transport and transformation of chemical species, which is a key aspect of biogeochemistry. However, the biogeochemistry of a handful of elements is particularly important to air pollution and environmental science in terms of how they are affected by and how they affect the earth's other cycles.

Consider nitrogen (N), which is an indispensable nutrient for most life-forms. Reading the label on a bag of fertilizer, one may see something akin to "12-12-12" or "20-10-10." The first of these numbers is total nitrogen, with phosphorus (e.g., available phosphate, P_2O_2) and potassium (e.g., soluble potash, K_2O). The nitrogen may be of one chemical species, for example, ammoniacal N. It may seem that since N_2 is the most abundant gas in the troposphere that plants would not be N-limited, so why would anyone need to add N to help plants grow? The answer is that plants can only use certain bioavailable N chemical species. Thus, there need to be symbiotic relationships with microbes in roots and soil to transform the abundant N_2 into bioavailable forms. Under many conditions, such as excessive concentrations, transformations by biota into toxic chemical forms, or moving into drinking water supplies, this same N can become air, water, and soil pollutants. Air pollutants are formed, transformed, and transported by these and myriad other biogeochemical cycles.

Matter and energy cycle through the environment by physical, chemical, and biological mechanisms. The extent and likelihood that a substance will move and where it may move and end up in the environment, that is, its fate, can be estimated by quantifying these mechanisms and matching them with the inherent properties of the substance. The physical and chemical mechanisms and properties will also determine whether the substance and its transformation products are likely to become more or less toxic and more or less mobile.

The biogeochemical cycle is a means of depicting how matter and energy move through trophic states within biological organizations and how the abiotic, nonliving components of a system interact with the biotic, living components. It is a means of keeping track of the mass and energy balances, both thermodynamically and fluid dynamically [1].

This chapter specifically addresses the cycles of carbon; environmental nutrients, especially N and sulfur (S); and metals. The cycles of other substances are also important to air pollution management. Energy is discussed in this chapter as part of the biogeochemical cycles, for example, light in photosynthesis and heat's role in solubility and volatilization and thermal reactions. These cycles depend on the water cycle and processes that are discussed in Chapter 7.

8.2 SPHERES AND CYCLES

The lithosphere affects air pollution in many ways. The geomorphology of an area can influence and even dictate where people live; for example, most cities are in coastal plains and in valleys near waterways. The terrain also determines wind patterns and air circulation; for example, inversions in urban valleys can greatly exacerbate air pollution episodes, and land-sea breezes can improve or degrade air quality depending on the direction.

An example of a physical mechanism occurs when the molecular configuration engenders its substantial expansion by volume as it freezes, that is, change of state from liquid to solid, or when it sublimates, that is, change of state from gas to solid. These changes in state initiate frost heaving, one of the mechanisms needed to break down rock formations. The resulting unconsolidated materials provide surfaces for chemical and biological processes that eventually create soil [2].

Chemically, this molecular configuration, especially polarity, induces the water to dissolve many chemical compounds, including acids that react with compounds in solid rock, for example, within fissures and crack, changing the geomorphology

Air Pollution Calculations. https://doi.org/10.1016/B978-0-12-814934-8.00008-9

of the terrain. The alignment of a water molecule's oxygen and two hydrogen atoms gives a slightly negative charge at the oxygen end and a slightly positive charge at the hydrogen ends. Given that "like dissolves like," polar substances have an affinity to become dissolved in water, and nonpolar substances resist being dissolved in water. The hydrogen atoms form an angle of 105 degrees with the oxygen atom. The asymmetry of the water molecule leads to a dipole moment in the symmetry plane toward the more positive hydrogen atoms.

Biologically, these and countless other physical chemical processes make the original rock material suitable for microbial and root growth. Lichen and other simple plant systems further break down the rock and its unconsolidated materials, setting the stage for larger plant growth. As a result, the terrain hardly resembles the original rock formation.

8.3 CARBON EQUILIBRIUM AND CYCLING

Many of carbon-based molecules exist in equilibrium with one another. For example, Fig. 8.1 demonstrates the equilibrium among carbonate (CO_3^{2-}) compounds, bicarbonate (HCO_3^-) compounds, carbonic acid (H_2CO_3), carbon dioxide (CO_2), and organic compounds (e.g., formed by microbial degradation).

Thus far, we have differentiated organic from inorganic compounds, but have not defined these terms. Organic compounds contain carbon, so do a few inorganic compounds, like CO and CO_2. Generally, organic compounds are those that contain at least one carbon-carbon or carbon-hydrogen bond. Furthermore, an organic molecule is one that fills its outermost valence shell by having its carbon atom share electrons with other atoms, that is, forming a covalent bond. Each atom contributes an electron in what has been described anthropomorphically as a "mutual deception," whereby the two atoms are tricking one another into acting like the shell is filled [3]. Since CO and CO_2 are not covalently bonded, they are considered to be inorganic compounds.

There is some debate over other compounds like the cyanides (CN −), which from the foregoing discussion would be inorganic. The same goes for the various physical forms, that is, allotropes, of carbon, for example, diamonds and fullerenes, for example, C-60 buckyballs. Diamonds are seldom an air pollution concern, but fullerenes and other nanoparticles can be pollutants and can be used to remove and treat pollutants.

For the purposes of air pollution discussion, the principal concern in distinguishing organic from inorganic molecules lies in the harm they cause, how they behave in a reactor or in the atmosphere, and how they can be removed from the air.

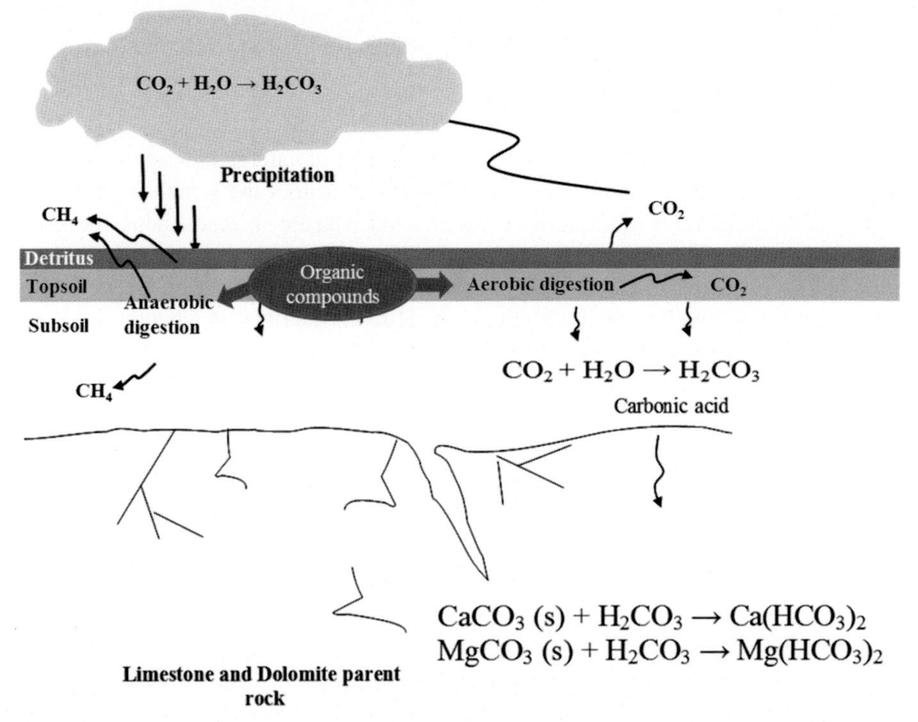

FIG. 8.1 Biogeochemistry of carbon equilibrium. The processes that release carbonates are responsible for much of the buffering capacity of natural soils against the effects of acid rain. *(Reproduced with permission from D.A. Vallero, Fundamentals of Air Pollution, 5th ed., Elsevier Academic Press, Waltham, MA, 2014, p. 999 pages cm.)*

Thus, the covalent C—C and C—H definition fits most air pollution scenarios, that is, these molecules tend to share much in common, for example, polarity, solubility, and other inherent properties compared with inorganic molecules.

Carbon dioxide, with water, is the ultimate product of aerobic microbial respiration, as it is for complete combustion. However, it is also an important greenhouse gas. It also contributes to the droplet acidity. On a global scale, uncontaminated rain's mean pH is often reported to be about 5.6, owing mainly to its dissolution CO_2 and the formation of H_2CO_3 (see Chapter 7 for an explanation of acidity and alkalinity and pH). As the water droplets fall through the air, the CO_2 in the atmosphere becomes dissolved in the water, setting up an equilibrium condition:

$$CO_2\ (\text{gas in air}) \leftrightarrow CO_2\ (\text{dissolved in the water}) \tag{8.1}$$

The CO_2 in the water reacts to produce hydrogen ions:

$$CO_2 + H_2O \leftrightarrow H_2CO_3 \leftrightarrow H^+ + HCO_3^- \tag{8.2}$$

$$HCO_3^- \leftrightarrow 2H^+ + CO_3^{2-} \tag{8.3}$$

We can begin to find the pH of water in equilibrium by applying the mean partial pressure (p_a) of CO_2 in the air mixture, which is about 4.0×10^{-4} atm. Henry's law states that the concentration of a dissolved gas is directly proportional to the partial pressure of that gas above the solution:

$$p_a = K_H[c] \tag{8.4}$$

where K_H = Henry's law constant and $[c]$ = molar concentration of the gas.

In water, the relationship is

$$p_a = K_H C_W \tag{8.5}$$

where C_W is the concentration of gas in water.

When CO_2 dissolves in water and since Henry's law is a function of a substance's solubility in water and its vapor pressure, the equilibrium can be expressed the proportionality between the CO_2 concentration in water and CO_2's partial pressure in the open atmosphere. Thus, as discussed in Chapters 4 and 5, the CO_2 concentration of the water droplet at equilibrium with air is obtained from the partial pressure of Henry's law constant:

$$p_{co_2} = K_H[CO_2]_{aq} \tag{8.6}$$

The change from CO_2 in the atmosphere to carbonate ions in water droplets follows a sequence of equilibrium reactions:

$$CO_{2(g)} \overset{K_H}{\leftrightarrow} CO_{2(aq)} \overset{K_r}{\leftrightarrow} H_2CO_{3(aq)} \overset{Ka1}{\leftrightarrow} HCO_{3(aq)}^- \overset{Ka2}{\leftrightarrow} CO_{3(aq)}^{2-} \tag{8.7}$$

The processes that release carbonates increase the buffering capacity of natural soils against the effects of acidic water ($pH < 5$). The ionic strength of the receiving soil or surface waters determines the actual change in pH. For example, the carbonate-rich soils like those in central North America can withstand even elevated acid deposition compared with the thin soil areas, such as those in the Canadian Shield, the New York Finger Lakes region, and much of Scandinavia.

The concentration of CO_2 is constant, since the CO_2 in solution is in equilibrium with the air that has a constant partial pressure of CO_2. The equilibrium involves the hydronium ion (H_3O^+), described in Chapter 7.

The two reactions and ionization constants for carbonic acid are

$$H_2CO_3 + H_2O \leftrightarrow HCO_3^- + H_3O^+ \quad K_{a1} = 4.3 \times 10^{-7} \tag{8.8}$$

$$HCO_3^- + H_2O \leftrightarrow CO_3^{-2} + H_3O^+ \quad K_{a2} = 4.7 \times 10^{-11} \tag{8.9}$$

Since K_{a1} is four orders of magnitude greater than K_{a2}, the second reaction can be ignored for the purposes of C equilibrium. The solubility of gases in liquids can be described quantitatively by Henry's law. Thus, CO_2 in the atmosphere at 25°C, the Henry's law constant, and the partial pressure can be applied to find the equilibrium.

What are the mean molar concentration of carbon dioxide and carbonic acid and the pH of water droplets in the atmosphere at equilibrium?

This calculation begins with Henry's law. The K_H for $CO_2 = 3.4 \times 10^{-2}\ \text{mol L}^{-1}\ \text{atm}^{-1}$. The partial pressure of CO_2 is found by calculating the fraction of CO_2 in the atmosphere. On 16 July 2018, the Mauna Loa, Hawaii, Observatory measured the atmospheric CO_2 to be 408.85 ppm. Assuming a mean concentration of CO_2 in the earth's troposphere to be the same as the Mauna Loa measurement, the fraction of CO_2 must be 408.85 divided by 1000,000 or 4.0885×10^{-4} atm. Thus, the carbon dioxide and carbonic acid molar concentration can now be found:

$$[CO_2] = [H_2CO_3] = 3.4 \times 10^{-2} \text{mol L}^{-1} \text{atm}^{-1} \times 4.0885 \times 10^{-4} \text{atm} \approx 1.4 \times 10^{-5} M$$

The equilibrium is $[H_3O^+] = [HCO^-]$. Taking this and the previously calculated CO_2 molar concentration gives

$$K_{a1} = 4.3 \times 10^{-7} = \frac{[H_3O^+]^2}{[CO_2(aq)]} = \frac{[H_3O^+]^2}{1.4 \times 10^{-5}}$$

$$[H_3O^+]^2 = 6.0 \times 10^{-12}$$

$$[H_3O^+] = 3.0 \times 10^{-6} M$$

The log of 3.0×10^{-6} is -5.5, so the negative log is 5.5. Or the droplet pH is about 5.5.

Note that this was calculated in 2008 in the fourth edition of *Fundamentals of Air Pollution* [4], with the assumption that the atmospheric CO_2 concentration was 350 ppm, and the found to be 5.6. Thus, the increase in CO_2 in the recent decade has been accompanied by increased acidity (0.1 decrease in pH and 4.0×10^{-7} increase in $[H_3O^+]$ moles).

Thus, only accounting for CO_2-H_2CO_3 equilibriums, "normal" rainfall is more acidic than what has often been reported, that is, pH = 5.6. Droplet acidity would be expected to increase when we consider the contribution of other acids from the other biogeochemical cycles, especially nitrogen and sulfur.

What would the pH of precipitation be if carbon dioxide concentrations increase to an average of 500 ppm?

From the preceding discussion, a global increase in CO_2 concentrations must also change the mean acidity of precipitation (see Fig. 8.2).

Many models expect a rather constant increase in tropospheric CO_2 concentrations. For example, the increase from the present concentration of about 410–500 ppm tropospheric CO_2 concentrations would be accompanied by a proportional decrease in precipitation pH. The molar concentration can be adjusted using the previous equations:

$$3.4 \times 10^{-2} \text{ mol L}^{-1} \text{ atm}^{-1} \times 5.0 \times 10^{-4} \text{ atm} = 1.7 \times 10^{-5} \text{ M, so } 4.3 \times 10^{-7} = \frac{[H_3O^+]^2}{[CO_2(aq)]} = \frac{[H_3O^+]^2}{1.7 \times 10^{-5}} \text{ and}$$

$$[H_3O^+]^2 = 7.3 \times 10^{-12} \text{ and } [H_3O^+] = 3.7 \times 10^{-6} M$$

Thus, average water droplet pH would be decreased to about 5.4.

This means that the incremental increase in atmospheric CO_2 can be expected to contribute to greater acidity in natural rainfall. The precipitation rates themselves would also be affected if greenhouse gas concentrations continue to increase, so any changes in atmospheric precipitation rates would also, on average, be expected to be more acidic. This is an interesting example of how the earth is actually a very large bioreactor. Changing one variable can profoundly change the entire system; in this instance, the release of one gas changes numerous physical (e.g., temperature) and chemical (e.g., precipitation pH) factors, which in turn evoke a biological response (biome and ecosystem diversity).

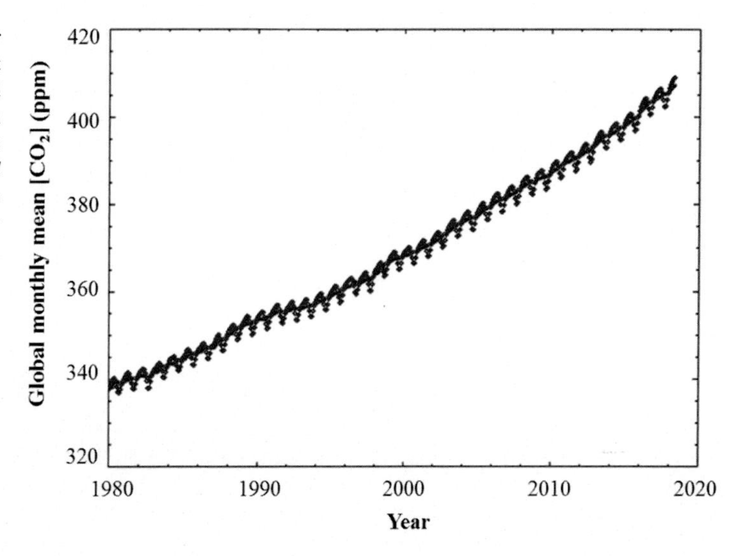

FIG. 8.2 Monthly mean carbon dioxide globally averaged over marine surface sites. A global average is constructed by first fitting a smoothed curve as a function of time to each site, and then, the smoothed value for each site is plotted as a function of latitude for 48 equal time steps per year. A global average is calculated from the latitude plot at each time step [5]. *(Modified from National Oceanic & Atmospheric Administration, Trends in Atmospheric Carbon Dioxide. (2018). 17 July 2017. Available: https://www.esrl.noaa.gov/gmd/ccgg/trends/gl_full.html.)*

8.3.1 The carbon cycle and greenhouse effect

Carbon dioxide is a prominent greenhouse gas, so it is logical to expect the increases in its atmospheric concentrations to be associated with changes in climate. Modeling climate change is difficult given the many drivers and constraints involved (see Fig. 8.3). The changes can affect biological systems. Ecological structure and function can be particularly sensitive. The effects can snowball or diminish as a result of the interactions of these drivers and constraints. For example, wetland structures may change if anaerobic microbial decomposition increases, which would result in increasing releases of CH_4, which would mean increasing global temperatures, all other factors being held constant. However, if greater biological activity and increased photosynthesis is triggered by the increase in CO_2 and wetland depth is decreased, CH_4 global concentrations would fall, leading to less global temperature rise. Conversely, if this increased biological activity and photosynthesis leads to a decrease in forest floor detritus mass, then less anaerobic activity may lead to lower releases of CH_4. There will be increases and decreases at various scales, so the net effects on a complex, planetary system are highly uncertain.

Another important carbon-based greenhouse gas is methane (CH_4), which is the product of anaerobic decomposition from myriad natural and human activities and facilities, such as from landfills. Methane also is emitted during the combustion of fossil fuels and cutting and clearing of forests. The concentration of CH_4 in the atmosphere has been steady at about 0.75 for over a thousand years and then increased to 0.85 ppm in 1900. Since then, CH_4 concentrations have doubled to about 1.7 ppm.

The main mechanism for removing CH_4 from the atmosphere is by reaction with free radicals, an atom, molecule, or ion that has an unpaired electron. The most important of these reactions is with the hydroxyl radical (•OH):

$$CH_4 + \bullet OH + 9O_2 \rightarrow CO_2 + 0.5H_2 + 2H_2O + 5O_3 \quad (8.10)$$

This indicates that the reaction creates carbon dioxide, water vapor, and ozone, all of which are greenhouse gases. Thus, CH_4 molecule, itself a greenhouse gas, is the source of others.

The difference in gas concentrations and the exchange coefficients between the atmosphere and surface waters determines how quickly a molecule of gas can move across the ocean–atmosphere boundary. It takes about 1 year to equilibrate

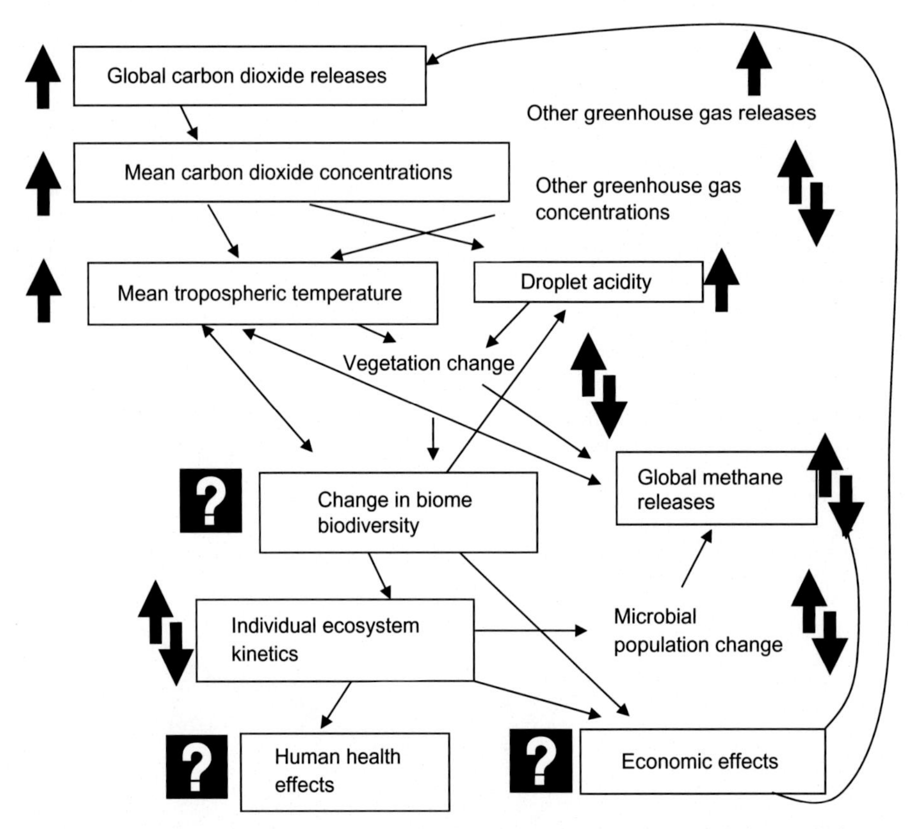

FIG. 8.3 Systematic view of changes in tropospheric carbon dioxide. *Thick arrows* indicate whether this factor will increase *(up arrow)*, decrease *(down arrow)*, or vary depending on the specifics (e.g., some greenhouse gas releases have decreased, e.g., the chlorofluorocarbons, and some gases can cool the atmosphere, e.g., sulfate aerosols). Question mark indicates that the type and direction of change are unknown or mixed. *Thin arrows* connect the factors as drivers toward downstream effects. *(Source: D. Vallero, Environmental Biotechnology: A Biosystems Approach. Elsevier Science, 2015.)*

CO_2 in the surface ocean with atmospheric CO_2; thus, large atmosphere-ocean differences in CO_2 concentrations are common. Biota and ocean circulation account for most of the difference. The oceans contain vast C reservoirs, with which the atmosphere exchanges since CO_2 reacts with water to form carbonic acid and its dissociation products. With the increased atmospheric CO_2 concentrations, the interaction with the ocean surface alters the chemistry of the seawater resulting in ocean acidification [6, 7].

Ocean uptake of anthropogenic CO_2 is primarily a physical response to increasing atmospheric CO_2 concentrations. Increasing the partial pressure of a gas in the atmosphere directly above the body of water causes the gas to diffuse into that water until the partial pressures across the air-water interface are equilibrated. The effects are complex; for example, increasing CO_2 also modifies the climate that in turn may change ocean circulation, which changes the rate of ocean CO_2 uptake. Marine ecosystem changes also alter the uptake [7].

Halocarbons, that is, those that contain halogen and carbon atoms, are the same classes of compounds involved in the destruction of atmospheric ozone that are also at work in promoting global warming. The most effective global warming gases are the chlorofluorocarbons (CFCs), for example, CFC-11 and CFC-12, both of which are no longer manufactured, and the banning of these substances has shown a leveling off in the stratosphere.

Many, but not all, greenhouse gases are C-based. Notably, nitrous oxide (N_2O; also known as dinitrogen monoxide) is the third most emitted to the atmosphere predominantly from human activities that also release C-based greenhouse gases, especially the cutting and clearing of tropical forests, which are vital sources of atmospheric oxygen. The greatest problem with nitrous oxide is that there appear to be no natural removal processes for this gas, and so, its residence time in the stratosphere is quite long. This is prime example of the interrelationships and interdependencies between the C cycle and the N cycle discussed later in this chapter.

8.3.2 Global warming potential

Greenhouse gases warm the earth by absorbing energy and decreasing the rate at which the energy escapes the atmosphere. These gases differ in their ability to absorb energy, that is, they have various radiative efficiencies. They also differ in their atmospheric residence times. Each gas has a specific global warming potential (GWP), which allows comparisons of the amount of energy the emissions of 1 ton of a gas will absorb over a given time period, usually a 100-year averaging time, compared with the emissions of 1 ton of CO_2.

Because CO_2 has a very long residence time in the atmosphere, its emissions cause increases in atmospheric concentrations of CO_2 that will last thousands of years [8]. Methane's average atmospheric residence time is about a decade. However, its capacity to absorb substantially more energy than CO_2 gives it a GWP ranging from 28 to 36. The GWP also accounts for some indirect effects; for example, CH_4 is a precursor to another greenhouse gas, ozone.

The GWP of N_2O is 265–298 times that of CO2, with an average residence time of 100 years. The CFCs, hydrofluorocarbons (HFCs), hydrochlorofluorocarbons (HCFCs), perfluorocarbons (PFCs), and sulfur hexafluoride (SF_6) are referred to as high-GWP gases, that is, GWP can be up to tens of thousands. For a given amount of mass, they hold substantially more heat than does CO_2.

What happens to the methane GWP if a 20-year averaging time is used?

A 20-year GWP is sometimes used as an alternative to the 100-year GWP. The 20-year GWP is based on the energy absorbed over 20 years, which prioritizes gases with shorter lifetimes, since it ignores any impacts that occur after 20 years from the emission. The GWPs are calculated relative to CO_2, so the GWPs are based on an 80% shorter time frame that will be larger for gases with atmospheric residence times shorter than that of CO_2 and smaller for gases with residence times greater than CO_2.

Since CH_4 has a shorter atmospheric residence time than CO_2, *the 100-year GWP is much less than the 20-year GWP.* The CH_4 20-year GWP has been estimated [8] to be 84–87, compared with the 100-year GWP of 28–36.

What happens to the GWP for the simplest fluorocarbon, tetrafluoromethane (CF_4), if a 20-year averaging time is used?

The estimated atmospheric residence time for CF_4 is 50,000 years [8]. The 100-year GWP is estimated to range from 6630 to 7350, which is larger than the 20-year GWP of 4880–4950. Therefore, the shorter averaging time gives a *100-year GWP that is much larger than the 20-year GWP.*

An alternate metric is the global temperature potential (GTP). Unlike the GWP, which is a measure of the heat absorbed over a given time period due to emissions of a gas, the GTP is a measure of the temperature change at the end of that time period relative to CO_2. The calculation of the GTP is more complicated than that for the GWP, based on models of response of the climate system to increased greenhouse gas concentrations, that is, climate sensitivity, and the amount of time it takes the system to respond, especially on the oceans' capacity to absorb the heat [8].

Carbon dioxide equivalency (CDE) is one way to estimate how much CO_2 would be needed for a mixture of emissions to have the same GWP, if measured over a defined period of time, most often 100 years. CDE, then, is the time-integrated radiative forcing of a quantity or rate of gas emissions to the troposphere. It is not an instantaneous value of the radiative forcing by a concentration of greenhouse gases in the atmosphere. The CDE is the product of the greenhouse gas mass (in million metric tonnes) and GWP. A company is proposing to produce 500 kg per year of a refrigerant gas mixture containing 10% sulfur hexafluoride (SF_6), 40% perfluorohexane (C_6F_{14}), and 50% diflurομethane (CH_2F_2; also known as hydrofluorocarbon-32). What is the CDE for this mixture?

Using data from the Intergovernmental Panel on Climate Change (IPCC) [9], the 100-year GWP values for these gases are

SF_6: 22,800
C_6F_{14}: 9300
CH_2F_2: 679

Calculate the gas mixture GWP for the mixture ratio:

$$(0.1 \times 22800) + (0.4 \times 9300) + (0.5 \times 675) = 2280 + 3720 + 337.5 = 6337.5$$

Multiply the mixture GWP times the mass to obtain the CDE. For industrial emissions, CO_2 equivalents are often expressed as "million metric tonnes of carbon dioxide equivalents" (MMTCDE):

$$500\,\text{kg} = 5 \times 10^{-7}\ \text{million tonnes}$$

$$\text{MMTCDE} = 6337.5 \times \left(5 \times 10^{-7}\right) = 3.17 \times 10^{-3}$$

Since this is equilibrated to CO2, that is, the 100-year GWP = 1, this means that the annual addition of this newly formulated gas mixture to the atmosphere is equivalent to 0.0317 million metric tonnes of CO_2.

8.3.3 Combustion and carbon

The cycling of C indicates the complexity of possible outcomes of air quality management and engineering. For example, pollution control efficiency and success have often been based on CO_2 production rates. Thus, ironically, the greater the amount of CO_2 emitted from a source, the more efficiently the pollution control equipment is operating because complete combustion results in the production of only CO_2 and H_2O. Likewise, the production of CH_4 has been an indication of complete anaerobic digestion of organic compounds within the landfill or during fermentation. Both of these indicators of pollution control efficiency also happen to be greenhouse gases.

The combustion of organic compounds is how most energy is generated. These carbon-based coupons undergo reactions that differ in rates and products, depending on temperature, mixtures with other substances, oxygen concentrations, and numerous other factors. As a result, the types and amounts of product released from combustion vary but include some of the most toxic and otherwise harmful pollutants. For this reason, the next chapter's focus on combustion and other thermal reactions is also an extension of the carbon cycle.

Carbon is essential throughout the earth's biogeochemical cycles. These include natural and synthetic chemical species, which can be used to provide needed goods and services but which can harm human health and the environment.

8.4 NUTRIENT CYCLING

Nitrogen (N), sulfur (S), phosphorous (P), and potassium (K) are macronutrients on which microbial and plant life depend for survival and growth. However, from air pollution perspective, N and S compounds deserve special attention, given that many sources generate and release substantial amounts of N and S compounds into the environment.

The same nutrient compounds that are essential are also potential pollutants. Nitrate and nitrite compounds and ammonia and amines are important water pollutants. Two of the six National Ambient Air Quality Standard (NAAQS) pollutants are oxides of these elements, that is, sulfur dioxide (SO_4) and nitrogen dioxide (NO_2). In addition, the oxides of nitrogen (NO_x), made up of NO_2 and nitric oxide (NO), are major precursors of another NAAQS pollutant, that is ozone. However, some of these same and other nutrient compounds are essential for crops, woodlands, and other ecosystems.

In addition to N and S, the cycling of other nonmetals is also crucial to a complete understanding of most air pollutants. In fact, P molecules are essential parts of energy transfer during photosynthesis. When P compounds deposit onto surface waters, the increased P concentrations accelerate eutrophication and, along with increasing N concentrations, contribute to the so-called dead zones in large water bodies, such as the Chesapeake Bay. In fact, P and K are intricately woven into the N and S cycles. P is a component of atmospheric deposition, along with N and S compounds. The P component is usually higher when wet deposition is greater than dry deposition, since P is usually bound to aerosols. Thus, in areas where large amounts of P are applied as a fertilizer to land or in arid regions with strong winds to transport soils, airborne P can be substantial, albeit still less than airborne N [10]. Not only most of the total P in the atmosphere is in mineral form as aerosols, but also there are large global sources of P from biomass burning and even less from energy and industrial combustion sources [11].

Phosphorous is also indirectly responsible for pollution, since the production of P fertilizer releases fluoride compounds. The movement of fluoride through the atmosphere and into a food chain illustrates an air-water interaction at the local scale (<100 km) [12]. Other industrial sources of fluoride include aluminum processing and glass manufacturing plants. Domestic livestock in the vicinity of substantial fluoride sources are exposed to fluoride by ingestion of forage crops. Fluoride released into the air by industry is deposited and accumulated in vegetation. Its concentration is sufficient to cause damage to the teeth and bone structure of the animals that consume the crops. Potassium is also part of the biogeochemistry of N and S. For example, when base cations like K deposited on soil and other surfaces, they increase surface's pH. This alkalinity buffers the acidity generated by N and S acidic compounds (e.g., H_2, NO_3, and H_2SO_4, respectively).

8.4.1 The nitrogen cycle [13]

Nitrogen is ubiquitous throughout the earth, involved in countless abiotic and biochemical reactions. The element exists in myriad chemical forms that move and change at varying rates within the nitrogen cycle. Like carbon, N is stored in reservoirs such as the atmosphere, living organisms, soils, and oceans [14]. From these sinks, N chemical species cycle between the atmosphere and the biosphere, moving into and out of ecosystems its organisms, including humans, within these ecosystems, continuously transforming into new compounds.

Most of the Earth's mass of N resides in the atmosphere. Approximately 79% of the molecules in Earth's atmosphere are molecular nitrogen (N_2). All organisms must metabolize N into larger molecules, notably amino acids, proteins, and deoxyribonucleic acid (DNA). Amino acids contain the functional group amine (NH_2) and are the building blocks of proteins. DNA consists of adenine, cytosine, guanine, and thymine (T) and consists of rings of carbon and nitrogen atoms, with various side chains.

The N_2 is not directly available to most organisms, with N-fixing bacteria in legume roots being the notable exception. The atmospheric N_2 is converted to bioavailable N species (e.g., nitrates, NO_3^-) abiotically (e.g., lightning strikes or fires) and biotically (e.g., by the N-fixing bacteria known as diazotrophs).

Many ecosystems are N-limited as a result of N-limits of the flora and microorganisms living within these systems. Indeed, along with P, the amount of N is the major limiting factor on the rate of net primary production of biomass in most ecosystems. That is, an increase in bioavailable N is associated with an increase in the production of biomass of these organisms in an ecosystem. For example, increasing the amount of carbon in a lake system will have little effect on the production of biomass, since most systems have ample mass of bioavailable C. Conversely, increasing bioavailable N in most lake systems increases total biomass. Thus, N is a limiting factor for plant growth.

The N cycle mirrors the trophic state and energy cycles in ecosystems. Producers convert mineral forms of N to N-containing molecules, for example, amino acids and DNA. Consumers then receive N by consuming producers and forming even larger N-containing molecules, that is, proteins (built from amino acids). Higher trophic state consumers receive N from consuming lower-level organisms, that is, both consumers and producers. Decomposers convert N in the opposite direction, that is, mineralization, converting the N-containing, organic molecules of dead organisms to inorganic compounds, for example, ammonium (NH_4^+) salts. The NH_4^+ salts have high aqueous solubility and readily absorb to soil particles. At these sites, bacteria and other microbes can oxidize the NH_4^+ into nitrite (NO_2^-) compounds and in turn to nitrate (NO_3^-) compounds. This oxidation process is known as nitrification. The NO_3^- compounds are quite bioavailable to most plant life (i.e., macrophytes). These dissolved oxidized forms of N are reduced (i.e., denitrified) by bacteria known as denitrifiers. This is a major process by which N is returned to the atmosphere.

Ionization is an important part of N and S cycling, as it is for all nutrient cycling. This is due to the configuration of electrons in an atom. The arrangement of the electrons in the atom's outermost shell, that is, valence, determines the ultimate chemical behavior of the atom. The outer electrons become involved in transfer to and sharing with shells in other atoms, that is, forming new compounds and ions. An atom will gain or lose valence electrons to form a stable ion that have the same number of electrons as the noble gas nearest the atom's atomic number. As mentioned, the N cycle includes three principal forms that are soluble in water under environmental conditions: the cation (positively charged ion) ammonium (NH_4^+) and the anions (negatively charged ions) nitrate (NO_3^-) and nitrite (NO_2^-). Nitrates and nitrites combine with various organic and inorganic compounds. Once taken into the body, NO_3^- is converted to NO_2^-. Since NO_3^- is soluble and readily available as a nitrogen source for plants (e.g., to form plant tissue such as amino acids and proteins), farmers are the biggest users of NO_3^- compounds in commercial fertilizers (although even manure can contain high levels of NO_3^-).

Nitrogen in several forms finds its way from the atmosphere to the soil through abiotic and biotic processes. As mentioned, NO_3^- compounds are formed from lightning and by nitrogen-fixing bacteria in legumes' root nodules. The rhizobia (genera *Rhizobium*, *Sinorhizobium*, and *Bradyrhizobium*) are gram-negative soil bacteria motile, rod-shaped, aerobic bacteria that infect legume roots (see Fig. 8.4) in a symbiotic relationship with legumes, for example, *Medicago*, *Melilotus*, and *Trigonella*. The symbiotic relationship results from the bacteria fixing atmospheric nitrogen, providing ammonium for protein production in the plant. In exchange, the bacteria obtain energy from the plant [15].

Fungi and bacteria also degrade nitrate compounds in nature. With moisture, aerobes and anaerobes in detritus on the forest floor induce numerous simultaneous chemical reactions along with abiotic reactions, making for a balance among various chemical forms of N, as well as those of S, P, K, C, and other nutrients. The chemical reactions in the N cycle, as in any nutrient cycle, require various energy sources, especially light, heat, and metabolic. As mentioned, some biochemical processes with an organism fix N_2 from the atmosphere, to form simple N compounds (e.g., diazotrophs in root nodules), which in turn form amino acids in the tissues of plants and animals.

The mineralization and denitrification occur by numerous processes, in addition to microbial degradation, including photolysis, hydrolysis, and reduction or oxidation. The result is a wide array of conversions of nitrogen-containing organic compounds (e.g., proteins and amino acids) to inorganic (mineral) forms, such as ammonia, ammonium hydroxide, nitrite, and nitrate. Note that the gases at the top of Fig. 8.5 include those that are important in air pollution. For example, NO is one of the compounds involved in the photochemistry that leads to the formation of the pollutant ozone (O_3) in the troposphere. Note also that macrophytes are central in the figure. Much of the chemistry occurs on the floor in the detritus where microbes degrade complex molecules. Nutrients in the soil are transported by the root capillary action to plant cells. Gases are transpired through leaves back to the atmosphere.

The atmospheric speciation at the top of Fig. 8.5 is an oversimplification, with many competing processes. For example, soil NO_2^- compounds can release nitrous acid (HONO) directly to the atmosphere. When soil contains elevated amounts of nitrates due to fertilization, HONO is released to the atmosphere. The more acidic the soil, the greater will be the release of HONO. In the troposphere, HONO leads to the formation of hydroxyl radicals, which both degrade and increase the deposition of air pollutants. Large volumes of acids are released from soil continuously to the atmosphere. Soils with high N concentrations form the acids from NO_2^- ions. These anions are first released into the soil by microbes that have transformed ammonium and nitrate ions into nitrite ions. Increasing soil acidity produces high nitrite concentrations, leading to greater

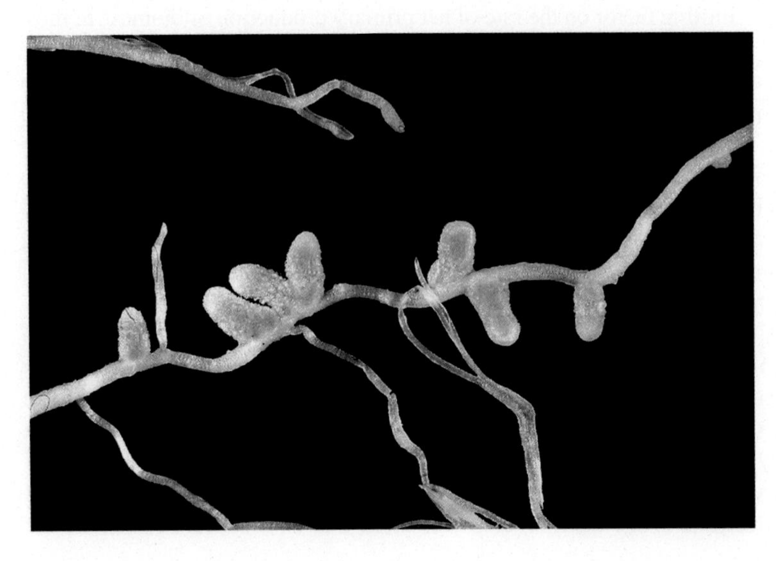

FIG. 8.4 The root nodules of a 4-week-old *M. italica* inoculated with *S. meliloti*. *(Source: Wikipedia photo; https://upload.wikimedia.org/wikipedia/commons/b/b3/Medicago_italica_root_nodules_2.JPG (Accessed 3 October 2009).)*

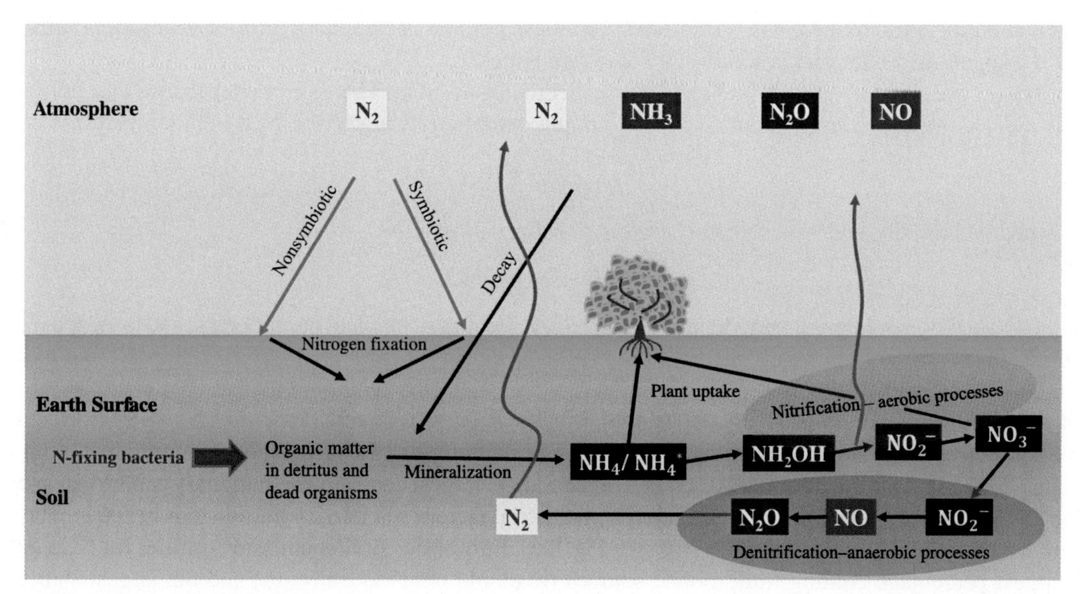

FIG. 8.5 Nitrogen cycling in the troposphere. *(Modified from D.A. Vallero, Environmental Contaminants: Assessment and Control. Elsevier Academic Press, Amsterdam; Boston, 2004, pp. xxxix, 801 p.)*

concentrations of HONO emitted to the troposphere [16]. About 30% of the primary OH radical production is attributed to the photolysis of HONO.

Thus, what may at first appear as a soil-water interaction (fertilizer added to increase crop yield) is indeed a soil-water-biota (microbial) process, followed by an air-water interaction (chemical transformation of air pollutants, followed by precipitation). In this instance, the air pollutants are decreased by increasing concentrations of water pollutants (nitrates and ammonium).

Fig. 8.5 also shows that when organic compounds are degraded by microbes, especially nitrifying bacteria, oxides of nitrogen (NO_x) are released to the atmosphere. Thus, the flux of gases nitric oxide (NO) and nitrogen dioxide (NO_2) from the soil to the lower troposphere is inversely related to the rate of degradation of organic compounds in the soil.

Ion exchange is a type of sorption, that is, movement of a chemical species from the liquid or gas phase to the solid phase. Plants grow as a function of available nutrients and other cycles within the forest ecosystem [17]. With this growth, compounds of N and other nutrients find their way to the atmosphere.

Nitrogen has three common oxidation states, +3, +4, and +5. Like carbon, N can form many compounds, even with a single, other element. For example, it can combine with oxygen to form N_2O, NO, NO_2, or N_2O_5, all important air pollutants or precursors to air pollutants.

A report shows that 2.054 kg of the gas NO_2 was measured at STP. However, it does not give the volume. Calculate this volume.

$$\text{Mole of } NO_2 = 1\,\text{M N and } 2\,\text{M O}_2 = 14.0\,\text{g} + 2(16.00)\,\text{g} = 46.01\,\text{g}$$

$$\frac{2054\,\text{g}}{1} \times \frac{1\,\text{mol}}{46.01} \times \frac{22.4\,\text{L}}{\text{mol}} \approx 1000\,\text{L} = 1\,\text{m}^3$$

Nitrite is an intermediate compound in both nitrification and denitrification. The N in the nitrate anion (NO_3^-) has an oxidation state of +5 (i.e., each of three O atoms with −2 charge + nitrogen's 5 = the nitrate charge of −1). The nitrite anion has a trivalent N, that is, N^{3+} (i.e., two O atoms each with −2 charge plus 3 yields the nitrite charge of −1). The major mechanism for producing N^{3+} in soil is by biological nitrification and denitrification processes. Nitrifying microbes produce nitrite ions from ammonium. Denitrifying microbes produce nitrite from nitrate [16].

The N compounds enter the troposphere by several mechanisms. In addition to the ordinary concentrations of molecular N, various nitrogen compounds are formed from reactions ranging from very fast (especially combustion) to quite slow, multistage (microbial) reaction rates. The two principal air pollutants regulated throughout the world are nitric oxide and nitrogen dioxide. Nitric oxide (NO) is a colorless, odorless gas and is essentially insoluble in water. Nitrogen dioxide (NO_2) has a pungent acid odor and is somewhat soluble in water.

Air pollution experts now refer to NO and NO_2 collectively as NO_x. This is in part because the NO lifetime after emission is quite short (> minute < hour) [18]. The simple oxidation of molecular N at high temperature is

$$N_2 + O_2 \xrightarrow{\Delta} 2NO \tag{8.11}$$

The emitted NO rapidly undergoes photochemical transformation to NO_2:

$$2NO + O_2 \rightarrow 2NO_2 \tag{8.12}$$

Indeed, NO an NO_2 can interconvert, depending on redox and photochemical conditions in the troposphere and stratosphere, For example, in sunlight, ultraviolet radiation splits NO_2 into NO and O. So, it is often preferred to consider the dynamic relationship of the two molecules rather than to ascribe a single speciation, especially for characterizing the atmosphere. However, the two compounds must be distinguished in risk assessments given their very different health effects.

Mobile sources, that is, cars, trucks, trains, boats, and aircraft, comprise the largest contributor to NO_x emissions in the United States (59%; 12.7 million tons in 2002). The United States' largest contributing stationary source category is electric generation (22%; 4.6 million tons in 2002), followed by industrial factories (11%; 2.4 million tons in 2002). The eastern and midwestern US regions had the highest emissions (see Fig. 8.6). Europe has similar emission profiles for NOx (see Fig. 8.7), with the largest percentage coming from mobile sources (road and nonroad transport) and the largest stationary source being energy production and distribution.

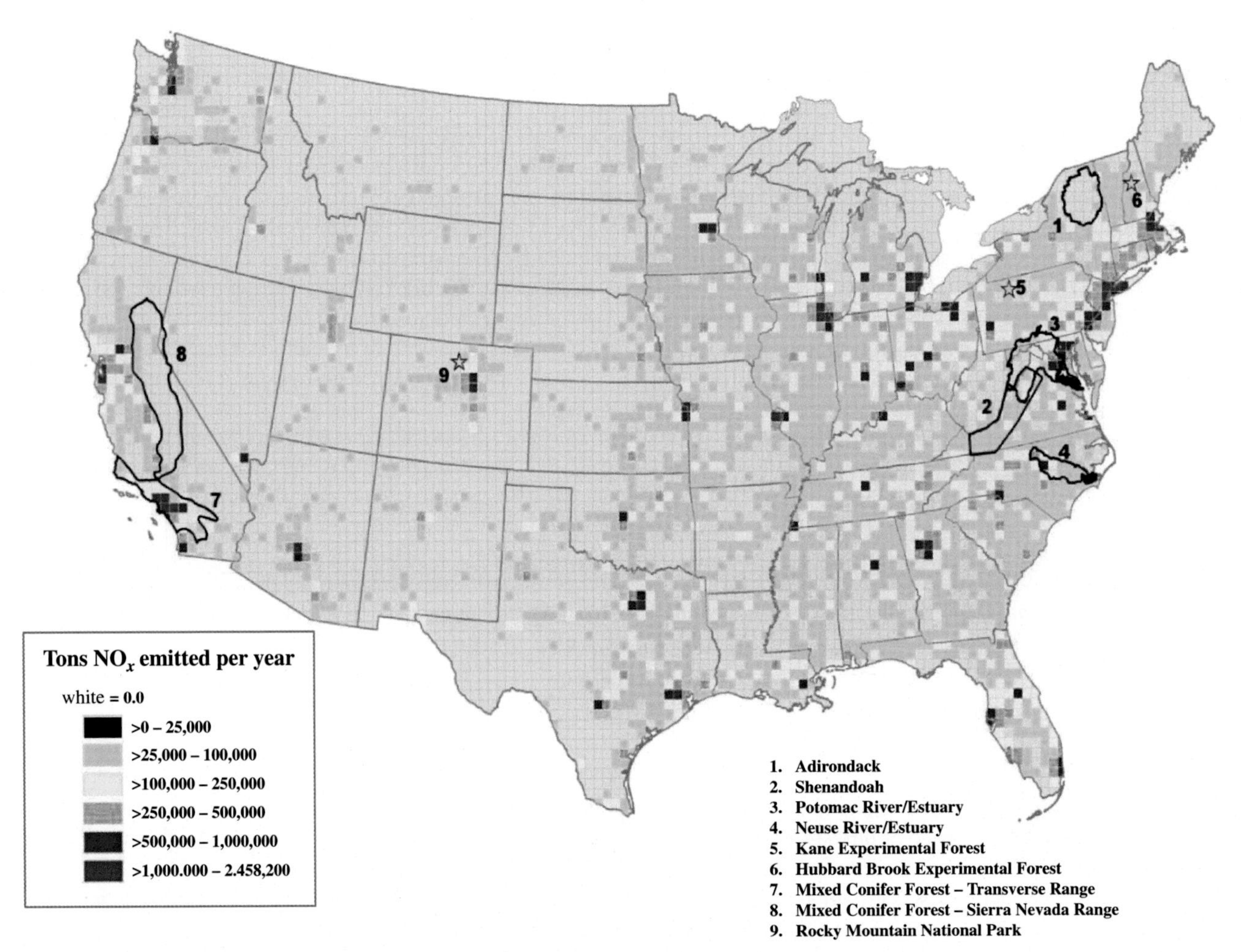

FIG. 8.6 Total oxides of nitrogen (NO_x) emissions (tons yr^{-1}) in 2002. *(Source: US Environmental Protection Agency, Policy Assessment for the Review of the Secondary National Ambient Air Quality Standards for Oxides of Nitrogen and Oxides of Sulfur. EPA-452/R-11-005a. Research Triangle Park, NC, 2011.)*

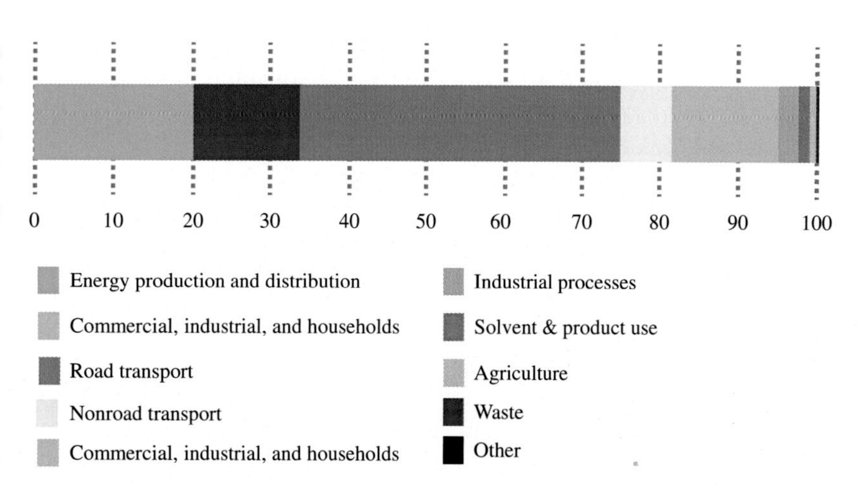

FIG. 8.7 European Union nations' percentage contribution to nitrogen (NO_x) emissions by industrial sector in 2010. *(Source: European Environmental Agency, Air pollution statistics. European Commission Eurostat, 2013. http://epp.eurostat.ec.europa.eu/statistics_explained/index.php/Air_pollution_statistics (Accessed 12 November 2013).)*

Air pollutants can react with one another when they are forming (e.g., in a reactor) in the atmosphere or in any volume, for example, within a canister awaiting chemical analysis. The presence of a catalyst will increase the reaction rate by lowering the demand for activation energy (see Chapter 9).

Although both NO and NO_2 can be harmful to humans and ecosystems, NO is also actually essential to metabolism so is harmful at much higher concentrations than that of NO_2. On the other hand, NO_2 is inherently toxic, causing respiratory problems. Both compounds are precursors to tropospheric ozone formation so present indirect health problems when the smog and O_3 are inhaled.

Nitrogen species important to air quality include many compounds in addition to NO_x. Atmospheric N reactions can be more complicated than may be inferred from the formation of NO and NO_2. For example, the NO_x reaction from NO to NO_2 likely involves water and an intermediate, that is, nitrous acid (HONO). The larger suite of N air pollution compounds is known as total reactive nitrogen, which is denoted as NO_y. This suite includes NO and NO_2, plus their oxidation products [19]. The major vapor-phase and particulate constituents of NO_y include NO, NO_2, nitric acid (HNO_3), peroxyacetyl nitrate (PAN), HONO, organic nitrates, and particulate nitrates.

Every NO_y compound is derived directly from NO_x emissions or through transformations in the troposphere [20]. Most of the compounds originate from combustion processes (see Chapter 9). As noted, the atmosphere itself is the source of much of the nitrogen leading to the formation of nitrogen compounds. Molecular nitrogen (N_2) makes up the largest share of gaseous content of the earth's atmosphere (79% by volume). Because N_2 is relatively nonreactive under most atmospheric conditions, it seldom enters into chemical reactions, but under pressure and at very high temperatures, such as in an internal combustion engine or industrial boiler, the molecular N will react with O_2, that is, Reaction (8.11).

This is known as "thermal NO_x" since the oxides form at high temperatures, such as near burner flames in combustion chambers. Approximately 90%–95% of the nitrogen oxides generated in combustion processes are in the form of NO. NO_2 and other nitrogen oxides can also form at high heat and pressure. Motor vehicles contribute largely to the tropospheric concentrations of NO_x. Another source of N is the fuel (see Chapter 9).

Cyanide (CN^-) is another N anion that is important to air pollution. Cyanide in the bloodstream impairs oxidative phosphorylation, a process by which oxygen is taken up for the production of essential cellular energy in the form of adenosine triphosphate (ATP). This process transfers electrons from nicotinamide adenine dinucleotide (NADH) to form water from H^+ and O_2, through a series of reactions catalyzed by enzymes. With less O_2 available for the cytochrome *c* to react with and hence complete the electron transport process, ATP production is diminished. The high binding affinity of CN^- to the ferric ion in hemoglobin is responsible for the decrease in O_2 carried in the blood. Thus, CN^- binds preferentially to hemoglobin and prohibits oxygen binding.

Methemoglobinemia is a particularly troublesome outcome resulting from endogenous production of the CN^- anion that results from the exposure of infants to nitrates. Ingesting high concentrations of nitrates, for example, in drinking water, can cause serious short-term illness and even death in infants 6 months or younger. The serious illness in infants is due to the conversion of microbial NO_3^- to NO_2^- in the gastrointestinal tract. Small children's lower stomach acidity (greater pH) allows for bacterial growth than do adult stomachs. Also, NO_3^- is more easily converted to NO_2^- than in adult hemoglobin, and circulatory system is too mature in small children to return to normal hemoglobin. Small children are also susceptible given their greater fluid intake to body weight ratio compared with adults and lower enzyme levels needed to convert methemoglobin to hemoglobin [21]. As a result, the increased NO_2^- concentrations interfere with the oxygen-carrying capacity of the blood. Especially in small children, when nitrates compete successfully against molecular oxygen, the blood carries methemoglobin (as opposed to healthy hemoglobin), giving rise to clinical symptoms. This acute condition can deteriorate a child's health rapidly over a period of days, especially if the water source continues to be used. Long-term, elevated exposures of nitrates and nitrites can cause an increase in the kidneys' production of urine (diuresis) and increased starchy deposits and hemorrhaging of the spleen [22].

A few animal studies suggest that elevated NO_3^- and NO_2^- in exposures may also elicit other effects, for example, increased miscarriage rates and anencephaly. Nitrate exposure may also cause hypothyroidism (i.e., by mimicking and blocking iodide from reaching the thyroid) [23].

8.4.2 The sulfur cycle

Compounds of sulfur (S), as those of N, exist at atmospheric concentrations well in excess of what would be expected from equilibrium geochemistry in an atmosphere with 21% O_2 [24]. Sulfur is released to the atmosphere as either reduced forms, for example, hydrogen sulfide, or oxidized forms, for example, sulfur dioxide (see Fig. 8.8). Both forms include air pollutants. Hydrogen sulfide is oxidized to sulfur dioxide in a three-step process. Note that the hydroxyl radical initiates the transformation from hydrogen sulfide to sulfur dioxide:

$$H_2S + HO^\bullet \rightarrow HS^\bullet + H_2O \quad (8.13)$$

$$HS^\bullet + O_2 \rightarrow HO^\bullet + SO^\bullet \quad (8.14)$$

$$SO^\bullet + O_2 \rightarrow SO_2 + O^\bullet \quad (8.15)$$

The atmospheric reactions of SO_2 are very complex and proceed through three different pathways to the sulfate ion (SO_4^{2-}). Sulfur dioxide can react with the hydroxyl radical to form the HSO_3 radical, which then can react with another hydroxyl radical to form water and SO_3 or H_2SO_4. Sulfur dioxide has sufficient aqueous solubility to dissolve in water droplets where it can react with oxygen gas to form SO_4^{2-}. The third pathway to sulfate occurs when sulfur dioxide reacts with hydrogen peroxide to form sulfuric acid:

$$HO^\bullet + SO_2 \rightarrow HOSO_2^\bullet \quad (8.16)$$

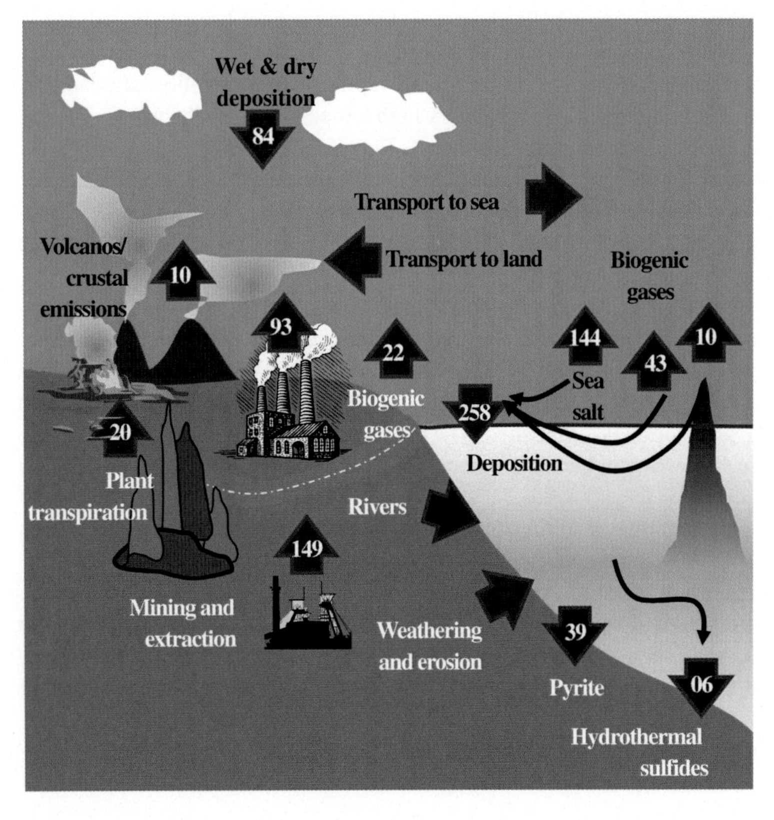

FIG. 8.8 Fluxes of sulfur to and from the atmosphere (teragrams of S in *arrows*). *(Source: R. Foust, Atmospheric reactions of sulfur and nitrogen. (2004). Available: http://jan.ucc.nau.edu/~doetqp-p/courses/env440/env440_2/lectures/lec37/lec37.htm.)*

$$HOSO_2^\bullet + HO^\bullet \rightarrow H_2SO_4 \tag{8.17}$$

$$SO_{2(\text{aqueous})} + [O]_{(\text{aqueous})} \rightarrow H_2SO_4 \tag{8.18}$$

$$SO_2 + H_2O_2 \rightarrow H_2SO_4 \tag{8.19}$$

With sufficient residence time in the atmosphere, S will be oxidized to the sulfate ion, usually as sulfuric acid (H_2SO_4). Ammonia (NH_3), the most common base in the atmosphere, reacts with H_2SO_4 to form ammonium bisulfate (NH_4HSO_4) and ammonium sulfate ($(NH_4)_2SO_4$). Sulfuric acid, NH_4HSO_4 and $(NH_4)_2SO_4$, all are hydroscopic substances, that is, readily dissolved in water. Thus, they wash out of the atmosphere during precipitation events [25].

A reduced form of sulfur that is highly toxic and an important pollutant is hydrogen sulfide (H_2S). Certain microbes, especially bacteria, reduce nitrogen and sulfur, using the N or S as energy sources through the acceptance of electrons. For example, sulfur-reducing bacteria can produce hydrogen sulfide (H_2S), by chemically changing oxidized forms of sulfur, especially sulfates (SO_4). To do so, the bacteria must have access to the sulfur, that is, it must be in the water, which can be in surface or groundwater or the water in soil and sediment. These sulfur reducers are often anaerobes, that is, bacteria that live in water where concentrations of molecular oxygen (O_2) are deficient. The bacteria remove the O_2 molecule from the sulfate leaving only the S, which in turn combines with hydrogen (H) to form gaseous H_2S.

In groundwater, sediment, and soil water, H_2S is formed from the anaerobic or nearly anaerobic decomposition of deposits of organic matter, for example, plant residues. Thus, redox principles can be used to treat H_2S contamination, that is, the compound can be oxidized using a number of different oxidants (see Table 8.1). Strong oxidizers, like molecular oxygen and hydrogen peroxide, most effectively oxidize the reduced forms of S, N, or any reduced compound.

8.4.3 Interactions between sulfur and nitrogen

From the standpoint of air pollution, the biogeochemical cycles must not be limited to single elements. As shown in Fig. 8.9, after a compound is emitted, it reacts with numerous other compounds and changes physically and chemically. It is transported advectively, is dispersed, and is finally deposited to the earth's surface, where it continues to undergo physical, chemical, and biological changes. This is demonstrated by the interactions of N and S in acid deposition and by the integration of myriad atmospheric chemical species that influence and are influenced by N and S [20]. This includes both the formation of conventional NAAQS pollutants, like ozone and particulate matter, and the transformation and changes in bioavailability of toxic air pollutants, for example, mercury (Hg) and SVOCs like dioxins and organochlorine pesticides.

The atmospheric movement of N and S compounds and other pollutants from sources to receptors is only one form of translocation. A second one involves the attempt to control air pollutants at the source. For example, the control of SO_2 and particulate matter by wet or dry scrubbing techniques yields large quantities of waste materials—often toxic—which are subsequently stored onsite at the facility or taken to landfills and other long-term disposal sites. If these wastes are not properly stored, they can be released to soil or water systems, for example, from runoff of acids. The prime examples involve the disposal of toxic materials in dump sites or landfills.

TABLE 8.1 Theoretical amounts of various agents required to oxidize 1 mg L^{-1} of sulfide ion

Oxidizing agent	Amount (mg L^{-1}) needed to oxidize 1 mg L^{-1} of S^{2-} based on practical observations	Theoretical stoichiometry (mg L^{-1})
Chlorine (Cl_2)	2.0–3.0	2.2
Chlorine dioxide (ClO_2)	7.2–10.8	4.2
Hydrogen peroxide (H_2O_2)	1.0–1.5	1.1
Potassium permanganate ($KMnO_4$)	4.0–6.0	3.3
Oxygen (O_2)	2.8–3.6	0.5
Ozone (O_3)	2.2–3.6	1.5

Source: Water Quality Association, Ozone Task Force Report "Ozone for POU, POE & Small Water System Applications," Lisle, IL, 1999.

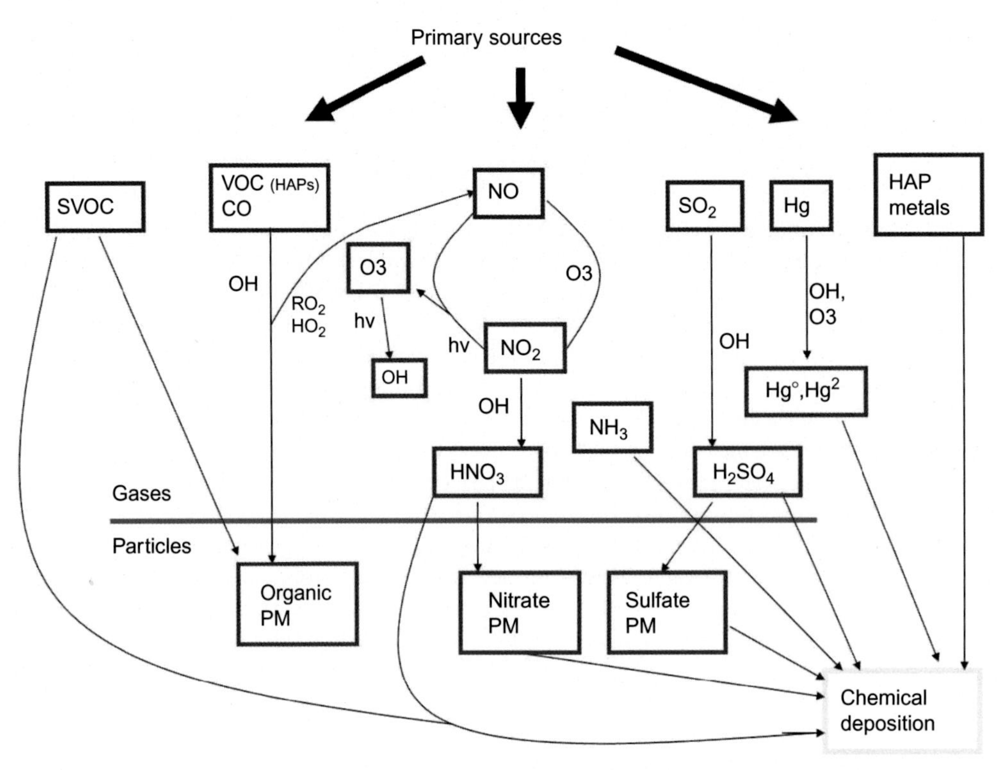

FIG. 8.9 Interactions between nitrogen and sulfur compounds and among S and N compounds and other air pollutants. Notes: *SVOC*, semivolatile organic compound; *VOC*, volatile organic compound; RO_2, radical consisting of a chain of organic compounds with H substituted with O_2 (e.g., propane, C_3H_8, reacts with OH to form the radical, $C_3H_7O_2\cdot$); *Hg*, mercury. *(Source: US Environmental Protection Agency, Policy Assessment for the Review of the Secondary National Ambient Air Quality Standards for Oxides of Nitrogen and Oxides of Sulfur. EPA-452/R-11-005a. Research Triangle Park, NC, 2011.)*

The oxidized chemical species of sulfur and nitrogen (e.g., sulfur dioxide (SO_2) and nitrogen dioxide (NO_2)) form acids when they react with water. This can occur in any media, for example, the atmosphere (i.e., acid deposition) and in mining waste and ash runoff to surface waters and groundwater. The lowered pH is responsible for numerous environmental problems. Many compounds contain both N and S along with the typical organic elements (C, H, and O). The reaction for the combustion of such compounds, in general form, is

$$C_aH_bO_cN_dS_e + 4a + b - 2c) \rightarrow aCO_2 + \left(\frac{b}{2}\right) H_2O + \left(\frac{d}{2}\right) N_2 + eS \tag{8.20}$$

Reaction (8.20) demonstrates the incremental complexity as additional elements enter the reaction. In the real world, pure reactions are rare. The environment is filled with mixtures and heterogeneous reactions. Reactions can occur in sequence, parallel, or both. For example, a feedstock to a municipal incinerator contains myriad types of wastes, from garbage, to household chemicals, to commercial wastes, to even small (and sometimes) large industrial wastes that may be illegally dumped. The N content of typical cow manure is about 5 kg per metric ton (about 0.5%). If the fuel used to burn the waste also contains S along with the organic matter, then the five elements will react according to the stoichiometry of Reaction (8.15).

As mentioned, oxidation and reduction of N and S occur in ecosystems as represented by trophic state and energy levels. The formation of sulfur dioxide (SO_2) and nitric oxide (NO) by acidifying molecular sulfur is a redox reaction:

$$S(s) + NO_3^-(aq) \rightarrow SO_2(g) + NO(g) \tag{8.21}$$

The designations in parentheses give the physical phase of each reactant and product: "s" for solid, "aq" for aqueous, and "g" for gas.

The oxidation half-reactions for this reaction are

$$S \rightarrow SO_2 \tag{8.22}$$

$$S + 2H_2O \rightarrow SO_2 + 4H^+ + 4e^- \tag{8.23}$$

The reduction half-reactions for this reaction are

$$NO_3^- \rightarrow NO \tag{8.24}$$

$$NO_3^- + 4H^+ + 3e^- \rightarrow NO + 2H_2O \tag{8.25}$$

Therefore, the balanced oxidation-reduction reactions are

$$4NO_3^- + 3S + 16H^+ + 6H_2O \rightarrow 3SO_2 + 16H^+ + 4NO + 8H_2O \tag{8.26}$$

$$4NO_3^- + 3S + 4H^+ \rightarrow 3SO_2 + 4NO + 2H_2 \tag{8.27}$$

8.4.4 Acid deposition

As mentioned during the discussion of the carbon cycle, carbonic acid contributes to droplet acidity and is the major reason for unpolluted rainwater being acidic. However, N and S compounds, notably those resulting from the reactions mentioned in the previous sections, are the major contributors to acid deposition, commonly known as acid rain. Acid deposition may include precipitation with a pH of <5.6 but typically has a pH of 4.4 or less.

The atmosphere near a smelter contains 10 ppb SO_2. At 20°C and 1 atm pressure, assuming that all of the SO_2 reacts with water to form sulfuric acid, how many moles of sulfuric acid would be in a cubic meter of air?

Find moles of air in 1 m^3 air:

$$1\ \text{atm} = 101325\ \text{Pa}$$

$$20^\circ\text{C} = 293^\circ\text{K}$$

$$PV = nRT$$

$$n = \frac{PV}{RT} = \frac{(101325\ \text{Pa})(1\ \text{m}^3)}{\left(8.314\ \text{J mol}^{-1}\ \text{K}^{-1}\right)(293\text{K})} = 41.6\ \frac{\text{Pa} \cdot \text{m}^3 \cdot \text{mol}}{\text{J}}$$

Since J = Pa · m^3, the volume contains 41.6 mol

$$1\ \text{ppb} = \frac{\text{mols } H_2SO_4}{\text{mols air}} \times 10^9$$

$$10\ H_2SO_4\ \text{ppb} = \frac{\text{mols } H_2SO_4}{41.6\ \text{mols air}} \cdot (10^9)$$

$$\text{Moles of } H_2SO_4 = \frac{(10\ H_2SO_4\ \text{ppb})(41.6\ \text{mol air})}{(10^9)} = 4.16 \times 10^{-7}$$

An aircraft measures the amount of water in the liquid phase in this cubic meter of air and finds it to be 2 mL. What is the pH of this plume's suspended water if all of the sulfuric acid is dissolved in the plume water?

Moles of H^+ produced by H_2SO_4 ionization in the plume water:

$$4.16 \times 10^{-7}\ \text{mol } H_2SO_4 \times \frac{2\ \text{mol 7 mol } H^+}{\text{mol } H_2SO_4} = 8.32 \times 10^{-7}\ \text{mol l } H^+$$

Calculate the H^+ moles in the volume of water, that is, 2 mL:

$$\frac{8.32 \times 10^{-7}\ \text{mol } H^+}{2\ \text{mL}} = \frac{8.32 \times 10^{-7}\ \text{mol } H^+}{2 \times 10^{-3}\ \text{L}} = 4.16 \times 10^{-4}\ \text{mol } H^+$$

Since this is for a volume of only 2 mL, the H^+ concentration in mol L^{-1} is 500 times than this, or 0.208 mol H^+ L^{-1}.

log 0.208 = −0.68. Since pH is the negative log of hydrogen ion concentration, the droplets in this plume volume have a pH of 0.68. If it were to reach to ground at this pH, it would clearly be extremely strong acid rain. However, this assumes that all of the acid in the air is in the water droplets and has not speciated into vapor and particulate species (e.g., SO_2 and particulate sulfates). Also, before the water is precipitated, it may be buffered by basic compounds. In any event, this is a very strong acid and is likely near the stack.

Two hours later during a rain shower, a pH meter is used to sample the rainwater that reads 4.4. What is the H^+ concentration of this sample?

$$pH = -\log_{10}[H^+]$$
$$4.4 = -\log_{10}[H^+]$$
$$10^{-4.4} = [H^+]$$

Finding the antilog:

$$[H^+] = 3.98 \times 10^{-5}\ \text{mol L}^{-1}$$

N and S compounds are ubiquitous in the atmosphere and the environment. They are constituents of both essential and detrimental compounds. As such, their complete biogeochemical cycles must be understood to explain the formation and fate of air pollutants.

8.5 METAL AND METALLOID CYCLES

The atmosphere is an important compartment in the cycling of toxic metals and metalloids and their compounds. Several metals have received much attention as air pollutants. For example, the United States lists nine metals and two metalloids and their compounds as hazardous air pollutants subject to US regulations under Section 112 of the Clean Air Act Amendments:

- Beryllium
- Cadmium
- Chromium
- Cobalt
- Lead
- Manganese
- Mercury
- Nickel
- Selenium
- Antimony (metalloid)
- Arsenic (metalloid)

One of the key adverse outcomes of certain metals is damage to the central and peripheral nervous systems, especially lead (Pb) and mercury (Hg). Other metals, such as manganese (Mn), have also been associated with chronic, neurological disorders. Metals and their inorganic and organometallic compounds are well known for major contamination events and exposures to large numbers of susceptible populations, such as small children. Mercury is particularly difficult to address since its mobility in the environment and its toxicity to humans and animals are determined by its chemical form. For example, dimethylmercury compounds are highly toxic, accumulate in the food chain, and have high affinity for organic tissues, but elemental mercury is much less toxic and is slower to bioaccumulate.

All elements exist in the environment in one or more oxidation or valence states. Compounds can be made less harmful by changing these states and by removing elements to which they are combined, for example, halogens like chlorine. This is true for both organic and inorganic compounds. For example, chromium (Cr) is highly toxic and carcinogenic in compounds in which it has a hexavalent oxidation state (Cr^{+6}), for example, chromium trioxide (CrO_3). Conversely, in its trivalent form (Cr^{+3}), the United States considers it to be essential for metabolism, whereas the European Food Safety Authority has concluded that there is no evidence of Cr^{+3}'s beneficial effects in healthy persons [26]. However, the trivalent forms are not considered to be toxic.

Organic compounds can be treated by destruction conversion to simpler compounds, finally becoming carbon dioxide and water. Metals are often hazardous even in elemental (zerovalent) forms. Thus, unlike organic materials, they are not "destroyed," but merely transformed to different oxidation states. For example, at high temperatures, the organic groups in an organometallic compound, like other organics, may decompose to CO_2 and water, but the metal may react with O_2 and become a metal oxide, or it may simply become a zerovalent metal. These inorganic substances may end up in and on particulate matter that is released to the atmosphere or in the bottom ash as a solid and likely hazardous waste.

Other elements, in addition to or instead of organic functional groups can also change the speciation of metals. For example, in a recent reactor study [27], sulfur and lead (Pb) react with one another and speciate differently with increasing temperature and the presence of oxides. With lime (CaO), solid-phase lead sulfate ($PbSO_4$) formed below 730°C, whereas solid-phase calcium sulfate ($CaSO_4$) formed between 730 and 900°C. Pb and CaO did not react. When silicon dioxide (SiO_2) was present, Pb predominately speciated to solid-phase $PbSO_4$ up to about 950°C and decomposed to form gas-phase lead oxide (PbO) or reacted with SiO_2 to form solid-phase lead (II) silicate (Pb_2SiO_4). Since metals and S are constituents of coal, for example, the S content can change metal speciation. It also seems to show that air pollution control equipment may be effective at metal capture by adding certain sorbents, for example, the reaction of PbO in the presence of SiO_2.

Some metal compounds may be sufficiently volatile, even under standard temperature and pressure, to enter the atmosphere in the gas phase, especially mercury (Hg). This can be in the form of gaseous metal or gaseous salt, for example, metal-halogen salts like $HgCl_2$.

Each form of metal has its own toxicity and dictates its fate in the environment. Chromium, for example, in its trivalent (Cr^{+3}) is an essential form of the metal. Although toxic at higher level, Cr^{+3} is much less toxic than the hexavalent Cr^{+6}, which is highly toxic to aquatic fauna and is a suspected human carcinogen. Toxicity, persistence, and fate are also determined by the metal's equilibrium chemistry, especially the amount in ionic forms and the amount that forms salts with nonmetals.

Atmospheric deposition is an important source of many metals, including cadmium, copper, lead, and zinc, but less important for other metals, for example, iron, aluminum, and manganese [28]. Dust is a major source of metal-laden particles that can be transferred to numerous compartments in the biosphere and hydrosphere. The actual metal loads vary seasonally, depending on rainfall and prevailing winds. For example, particulate matter transport to oceans is highest in waters downwind of deserts [29]. Other metal sinks include melt from glaciers and icebergs, seasonal sea ice melting, island wakes, volcanism, and hydrothermal activity [30].

Metal cycling is influenced by other cycles. For example, acidification of the oceans and acid deposition onto terrestrial systems will change the rates and types of metal cycling. Metal cycling is intricately linked to organisms in the biosphere. The concentrations, transformation, and cycling are driven by the growth rates, biomass sequestration, species diversity, and trophic interactions of biota. Indeed, many metals are micronutrients, without which plants and microbes suffer deficiencies. Above these levels, organisms suffer toxic effects. In forests, metals may become stable and immobile on the floor due to adsorption onto detritus. On the other hand, organic acids in the vegetation (e.g., leaf leachate) and bacterial decomposition in soil organic matter (i.e., mineralization) may increase solubility and otherwise mobilize the metals [28].

Metal mobility in the biosphere and bioavailability vary among genera of organisms and among metals. Generally, metals can be assimilated into environmental compartments by two processes (see Fig. 8.10). One mechanism depends on ionic species forming surface complexes with carrier proteins during active transport, allowing metal cations to be transported across cellular membranes. The second mechanism is nonionic. Metal species may be transported across biological membranes by concentration gradients, that is, molecular or passive diffusion [28].

A ligand is a functional group (either a molecule or ion) that binds to a central metal atom to form a coordination complex. The bonding between the metal and ligand generally involves formal donation of one or more of the ligand's electron pairs. The metal-ligand bonding can be covalent or ionic. Metals associate with a range of ligands to form the complexes shown in the center of Fig. 8.10. These include metal aquo complexes, that is, coordination compounds containing metal (Me) ions with water serving as the only ligand; general stoichiometry is $[Me(H_2O)_n]^{z+}$. The aquo complexes are the predominant metal species in aqueous solutions of many metal salts, for example, metal nitrates, sulfates, and perchlorates. Metal complexes also include those with the OH^- functional group (i.e., hydroxo complexes), other inorganic ligands (e.g., Cl^-, SO_4^{2-} and HCO_3^-), and organic ligands [28].

Trace metals can be mobilized and immobilized from the solid phase, for example, by sorption and desorption. The solid phase may include suspended particles, soil, and sediments. Solid-solution interactions of trace metals occur through adsorption/desorption reactions on organic and inorganic functional groups. They may also occur directly as precipitation/dissolution reactions. In the solid phase, metals associate with surface functional groups, hydroxide, carbonate or sulfide minerals (in amorphous or crystalline forms), or organic matter. Finally, biota may alter the cycling of trace metals through uptake by assimilation or surface [28].

The ocean is a vast sink for metals. For example, zinc (Zn), cadmium (Cd), nickel (Ni), and copper (Cu) have residence times ranging from 3000 to 100,000 years, given mean deep ocean ventilation time of about 1000 years. Generally, metal concentrations increase with depth (see Fig. 8.11).

Most metals exist at low concentrations in the ocean, ranging from 0.002 nanomolar (nM) for cadmium (Cd) and manganese (Mn) to 0.1 nM for iron (Fe) and zinc (Zn). Metals can be taken up incidentally with essential metals; for example, Cd is a nutrient analog for Zn in surface waters [31]. Metal concentrations tend to increase by orders of magnitude from

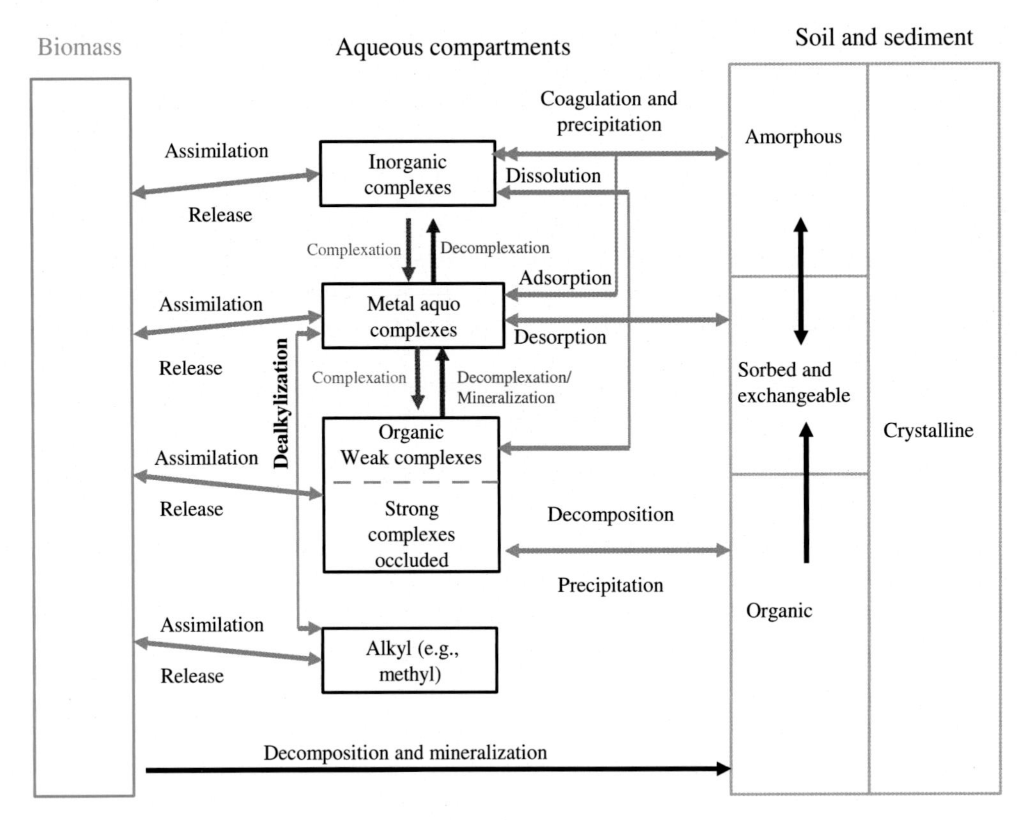

FIG. 8.10 Sources, sinks, and mobilization of metals in the biosphere. Individual behavior and pathways vary for each metal. *(Based on C.T. Driscoll, J.K. Otton, A. Iverfeldt, Trace metals speciation and cycling, in: B. Moldan, J. Cerny (Eds.), Biogeochemistry of Small Catchments: A Tool for Environmental Research, John Wiley & Sons, Hoboken, New Jersey, 1994.)*

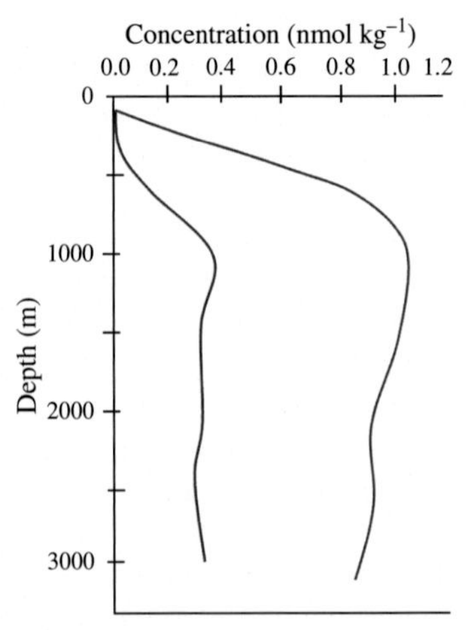

FIG. 8.11 Depth profiles of cadmium in the central North Pacific (32.7°N, 145.0°W, September 1977) and North Atlantic (34.1°N, 66.1°W, July 1979). Manganese concentrations in the Pacific were analyzed in acidified, unfiltered seawater samples. *(Sources: K.W. Bruland, R.P. Franks, Mn, Ni, Cu, Zn and Cd in the western north Atlantic, in: C.S. Wong, E. Boyle, K.W. Bruland, J.D. Burton, E.D. Goldberg (Eds.),* Trace Metals in Sea Water. *Plenum Press, New York, NY, 1983. pp. 395–414; W.G. Sunda, Feedback interactions between trace metals and phytoplankton in the Ocean. Front. Microbiol. 3 (2012) 204. https://doi.org/10.3389/fmicb.2012.00204.)*

open ocean water to coastal and estuarine systems. Inland streams, groundwater, dust, and sediments contribute metal loads to the ocean [32]. The heavy metal concentration gradients between coastal and open water are consistent [33, 34]. Iron concentrations, for example, may reach 20 micromolar (μM) in high humic rivers, which is 4–6 orders of magnitude higher than Fe concentrations in surface ocean waters. Most iron in freshwater streams exists as colloidal suspensions (0.02–0.4 μm diameter). These colloids degrade rapidly when they contact high salinity, which induces coagulation and settling. Reported gradient concentrations for cobalt (Co) in the coastal waters of Baja California ranged from 0.93 to as much as 3.6 nM [35]. That is about four times more concentrated at the coast, near the source than in more oceanic waters. In other parts of the world, a similar pattern has been identified. Therefore, the freshwater-saltwater boundary effectively serves to precipitate many metals to the sediment [36].

Conversely, anoxic and reduced environments, such as in high-organic-content sediments, may remobilize a portion of the metals that have settled, via reduction of particulate metals to more soluble species [37]. Metals in the lithosphere reach the atmosphere and other environmental compartments by physical and chemical mechanisms. Weathering breaks down the rocks into smaller materials such as clays and silts. With the increase in the surface area as the particle size decreases, the materials become capable of trapping larger loads of charged molecules and ions and incorporating lipophilic organic molecules. Due to their characteristic hydrophobicity and their tendency to be excluded from the water phase (i.e., decreased polarity), these organic molecules sorb to these particles as well. Thus, the receiving areas will often contain larger quantities of fine sediments and, with attendant, higher metal and organic content [38].

Another important factor in the metal cycle is temperature. For example, greater concentrations of metals will occur in water with increasing temperatures. Temperature change also changes biomass and metal exchanges with organisms. As mentioned, increasing temperature decreases dissolved oxygen (DO) concentrations, which in turn leads to a reduced environment where the metals and compounds will form sulfides and other compounds that can be toxic to the fish. Thus, the change in temperature, the resulting decrease in DO and increasing metal concentrations, and the synergistic impact of combining the hypoxic water and reduced metal compounds are a cascade of harm to the stream's ecosystems.

Combined abiotic and biotic processes are affected by and affect metal geochemical cycling. Notably, the growth and metabolism of the bacteria result in even more rapidly decreasing DO levels. Algae both consume DO for metabolism and produce DO by photosynthesis. The increase in temperature increases their aqueous solubility, and the decrease in DO is accompanied by redox changes, for example, the formation of reduced metal species, such as metal sulfides. This is also being mediated by the bacteria, some of which will begin to reduce the metals as the molecular oxygen levels decline, leading to overall reduced conditions in the water and sediment. The opposite process occurs with increasing oxidation. In the more oxidized regions, the metals form metal oxides. The increase in the metal compounds combined with the changing concentrations of DO, combined with the increased temperatures, acts synergistically [1].

The biota-metal cycling relationship is not entirely linear. For example, the initiating abiotic effect mentioned above, that is, increased temperature, results in an increased microbial population. Recall from Chapter 6 that microbes vary in their intake and release of O_2. Bacteria's metabolism decreases the amount of O_2 in the water, whereas algae both consume O_2 for metabolism and produce DO during photosynthesis. Thus, algae behave similarly to bacteria in the dark, but similarly to plants in light. Meanwhile, a combined abiotic and biotic response occurs with the metals. At higher temperatures, the aqueous solubility of metals increases. The decrease in DO concentrations is accompanied by redox changes, for example, the formation of reduced metal species, such as metal sulfides. However, the metal speciation is mediated by the bacteria, some of which will begin reducing the metals as the oxygen levels drop with the concomitant reduced conditions in the water and sediment. The opposite is true in the more oxidized regions, where the metals are forming oxides. The net increase or decrease in the metal concentrations; the speciation of metal compounds; and the changes in redox, combined with the increased temperatures, complicating any predictions of whether these factors will act synergistically or antagonistically to make the conditions toxic for higher animals, for example, a fish kill [39].

Metal-induced or metal-mediated ecosystem damage depends on species diversity, for example, the balance of algae and bacteria, as well as on the types of metallic compounds formed and any catalysts that may increase metal chemistry. Of course, metals are not the only cause of fish kills and other ecosystem effects. Indeed, they are usually merely contributing causes unless the metal concentrations are extremely high. Indeed, the problems may be mediated by metals. Given that a number of metals are micronutrients, if metal concentrations are decreased or the chemical forms are completely not bioavailable, this could be a contributing factor in ecosystem damage (i.e., a metal-deficient species could adversely affect predator-prey, biomass production, etc.).

Human activities change the geochemistry of metals. After extraction, the processes used to refine and manufacture products from the ores expose populations to metal concentrations well above undisturbed conditions. In addition, metal compounds end up in products or otherwise come into contact with people and the environment.

Extraction of metals is an example of how humans have changed the biogeochemical cycling throughout the hydrosphere, atmosphere, and biosphere. Human activities can greatly accelerate the mobilization of metals, severalfold over natural releases from weathering and natural fluxes [40], that is, degassing from the Earth's crust and mantle, which is the sum of volcanic emissions to the atmosphere and net hydrothermal flux to the sea (see Table 8.2).

For both metals and nonmetals, the activities by humans greatly increase the amount of potentially toxic substances released to the environment. For example, the human-to-natural weathering ratio for copper (Cu) is 14.2, meaning that

TABLE 8.2 Estimates of the global flux in the biogeochemical cycles of certain elements (10^{12} g yr^{-1})

Element	Juvenile flux[a] (1)	Chemical weathering (2)	Natural cycle[b] (3)	Biospheric recycling ratio 3/(1+2)	Human mobilization[c] (4)	Human-to-natural ratio 4/(1+2)
Boron	0.02	0.19	8.8	42	0.58	2.8
Carbon	30	210	107,000	446	8700	36.3
Nitrogen	5	20	9200	368	221	8.8
Phosphorous	~0	2	1000	500	25	12.5
Sulfur	10	70	450	5.6	130	1.6
Chlorine	2	260	120	0.46	170	0.65
Calcium	120	500	2300	3.7	65	0.10
Iron	6	1.5	40	5.3	1.1	0.14
Copper	0.05	0.056	2.5	23.6	1.5	14.2
Mercury	0.0005	0.0002	0.003	4.3	0.0023	3.3

[a] *Emission from the Earth's crust and mantle, that is, sum of volcanic emissions to the atmosphere and net hydrothermal flux to the sea.*
[b] *Annual biogeochemical cycle to/from the Earth's biota on land and in the oceans, in the absence of humans.*
[c] *Direct and indirect mobilization by extraction and mining from the Earth's crust or (for N) industrial fixation.*

Source: E.S. Bernhardt, W. Schlesinger, Biogeochemistry: An Analysis of Global Change, Elsevier Inc, Waltham, MA, USA and Oxford, UK, 2013.

all of the weathering processes around the globe account for 14 times less than the extraction and processing of Cu by humans. Put another way, mercury's (Hg) ratio of 3.3 means that without human extractions and manufacturing, only one-third of the Hg would be released into the environment. Thus, since the total Hg loading to the environment drives the amount of toxic forms (e.g., methylmercury) reaching and accumulating in the food chain, this means that human activities are substantially increasing the amount of toxic Hg in the food being ingested and air be inhaled by humans.

Considering that weathering takes place over most of the earth's surface, including land and water, this indicates the extent of anthropogenic activities in mobilizing potential air pollutants. As such, the composition of the atmosphere has been changing substantially and relatively rapidly in recent centuries and decades compared with the previous millennia.

So-called heavy metals dominate the list of "toxic metals." The designation of "heavy" is any metal with the atomic mass greater than or equal to that of iron. These include the transition metals, lanthanides, and actinides [41].

Most toxic metals exist naturally in very small concentrations, but for some, the toxic levels can be just above the background concentration. In many cases, small amounts are essential. The dose-response curve for these metals would be U-shaped, with deficiencies at low doses, toxicity at high doses, and an optimal dose range between deficiency and toxicity.

Many texts refer to this group of metals and metalloids as trace metals. However, "trace" is not so much a descriptive term as an adjective for low concentrations. In most parts of the world, metal toxicity is a relatively uncommon medical condition. Poisonings, for example, lead and arsenic poisoning, are often meant to be acute effects, including death, at high doses. Such doses are unusual in general air pollution scenarios, except in and near specialized industrial environments. Generally, atmospheric concentrations are much lower than those that induce acute effects, so the principal air pollution concern about metals is longer-term, low-concentration exposures.

8.5.1 Metals and heat

As is the case for carbon and nutrient cycling, combustion is also a key means of releasing and transforming metals in the environment. Many of these substances can be completely combusted at much lower temperatures than larger, more recalcitrant compounds like the dioxins. For example, high-temperature incineration may not be needed to treat wastes contaminated with many volatile organic compounds (VOCs). In addition, high-temperature incineration of wastes with heavy metals will likely increase the volatilization of some of these metals into the combustion flue gas (see Tables 8.3 and 8.4). High concentrations of volatile trace metal compounds in the flue gas poses increased challenges to air pollution control. Thus, other thermal processes, that is, thermal desorption and pyrolysis, can provide an effective alternative to incineration.

TABLE 8.3 Conservative estimates of heavy metals and metalloids partitioning to flue gas as a function of solid temperature (t) and chlorine content. Note that the remaining percentage of metal is contained in the bottom ash. Partitioning for liquids is estimated at 100% for all metals. The combustion gas temperature is expected to be 40–500°C higher than the solid temperature

	T = 871°C		*T* = 1093°C	
Metal or metalloid	**Cl = 0%**	**Cl = 1%**	**Cl = 0%**	**Cl = 1%**
Antimony	100%	100%	100%	100%
Arsenic	100%	100%	100%	100%
Barium	50%	30%	100%	100%
Beryllium	5%	5%	5%	5%
Cadmium	100%	100%	100%	100%
Chromium	5%	5%	5%	5%
Lead	100%	100%	100%	100%
Mercury	100%	100%	100%	100%
Silver	8%	100%	100%	100%
Thallium	100%	100%	100%	100%

Source: U.S. Environmental Protection Agency, Handbook: Guidance on Setting Permit Conditions and Reporting Trial Burn Results, EPA 625/6-89-019, 1989.

TABLE 8.4 Metal and metalloid volatilization temperatures

	Without chlorine		With 10% chlorine	
Metal or metalloid	**Volatility temperature (°C)**	**Principal species**	**Volatility temperature (°C)**	**Principal species**
Chromium	1613	CrO_2/CrO_3	1611	CrO_2/CrO_3
Nickel	1210	$Ni(OH)_2$	693	$NiCl_2$
Beryllium	1054	$Be(OH)_2$	1054	$Be(OH)_2$
Silver	904	Ag	627	AgCl
Barium	841	$Ba(OH)_2$	904	$BaCl_2$
Thallium	721	Tl_2O_3	138	TlOH
Antimony	660	Sb_2O_3	660	Sb_2O_3
Lead	627	Pb	−15	$PbCl_4$
Selenium	318	SeO_2	318	SeO_2
Cadmium	214	Cd	214	Cd
Arsenic	32	As_2O_3	32	As_2O_3
Mercury	14	Hg	14	Hg

Source: B.C. Willis, M.M. Howie, R.C. Williams, Public Health Reviews of Hazardous Waste Thermal Treatment Technologies: A Guidance Manual for Public Health Assessors. US Department of Health and Human Services, Agency for Toxic Substances and Disease Registry, Division of Health Assessment and Consultation, 2002.

Fossil fuels, especially coal, contain metals. The engineer and plant manager need a chemical analysis of a fuel's metal content, especially toxic heavy metals, along with sulfur, volatile matter, ash, and heat value. Combustion of the fuel will release some of the metals to the atmosphere as vapors and particulates, that is, fly ash, and the rest will reside in the bottom ash. Thus, the metals can find their way to surface and groundwater, air, and soil.

For the plant engineer, fuels are easier to control than more heterogeneous combustible matter, for example, solid waste. These wastes vary considerably not only in heat value but also in metal content. Thus, it is best to take steps to remove metals before combustion, for example, by desorption. Thermally desorbing contaminants from the solid matrix has the additional benefits of lower fuel consumption, no formation of slag, less volatilization of metal compounds, and less complicated air pollution control demands. So, beyond monetary costs and ease of operation, a less energy (heat)-intensive system can be more advantageous in terms of actual pollutant removal efficiency.

8.5.2 Mercury biogeochemistry

The metal mercury (Hg) has been of interest to scientists and the general public of late. For example, there has been much concern as to whether a preservative used in vaccines might be part of the cause for the apparent increase in autism in recent decades. The metal has also been central to the debate about the safety of seafood and trade-offs of hazards to unborn and nursing babies and the benefits of breastfeeding. Actually, mercury is a natural component of the earth, with an average abundance of approximately 0.05 $mg\,kg^{-1}$ in the earth's crust, with significant local variations.

It is an unusual metal in that it is a liquid at room temperature (it melts at 38.9°C and boils at 303°C) and is heavy, with a density of 13.5 $g\,cm^{-3}$ at 25°C. Studies have indicated that Hg can be a toxic substance and that one of its compounds, methylmercury chloride, is a highly toxic substance. Mercury not only persists in the environment but also is readily transformed. When reduced and alkylated, Hg becomes more toxic; for example, methylmercury biomagnifies up the food chain and reaches a wide variety of species and ecosystems. Thus, both human and ecosystems may be exposed to excessive levels of mercury in the environment [42]. Not only is the exposure to total Hg problematic, but also the conditions of many systems make for the most toxic forms of Hg; for example, monomethyl and dimethyl Hg compounds are produced in reduced- and low-oxygen environments and metabolic processes within organisms.

Mercury ores that are mined generally contain about 1% mercury, but at the Almaden mine in Spain, they are mining an ore that contains about 12% mercury. The mining of mercury continues in a number of countries. In the early 1980s, the annual production from mining HgS was 7000 t but is now about 2000 t [43]. Mercury is also obtained as a by-product of mining or refining of other metals (such as zinc, gold, and silver) or minerals and even from natural gas, recycling spent products, and reprocessing mine tailings containing Hg. In the United States, this recovery of mercury as a by-product of gold and other metal mining, from recycling and from the decommissioning of chlor-alkali plants, amounts to about 250 t per year.

Concentrations of total mercury in the atmosphere of the northern hemisphere have recently been estimated at 2 $ng\,m^{-3}$, those in the southern hemisphere being half this value. Values in urban areas are usually higher (e.g., 10 $ng\,m^{-3}$). Mercury has an appreciable vapor pressure (2×10^{-3} mmHg or 0.3 Pa at 25°C), which is the reason for it being in the atmosphere [44, 45]. It has been estimated that each year, 2000–3000 t [46] enter the atmosphere from natural and anthropogenic sources. Natural sources, such as volcanoes, are responsible for approximately half of atmospheric mercury emissions.

Mercury in stack emissions from coal-fired power plants is the major source of Hg deposition in North America. Thus, atmospheric transport and deposition of Hg contributes to the buildup of Hg in the food chain. As result, airborne Hg is a substantial source of Hg exposure, not only by inhalation but also by ingestion. The exchanges with the atmosphere occur between both land and surface waters. Indeed, most of the exchanges occur between water and air, with sediments as a large global Hg sink. The natural sources include volcanoes and biogenic (especially flora) emissions.

The mercury cycle includes numerous compartments and cycling at various scales. The chemical speciation of Hg is quite complex, consisting of numerous chemical and physical forms with very different properties. However, airborne Hg mercury takes three fundamental physicochemical forms:

1. Elemental or zerovalent mercury (Hg^0), which is the uncombined chemical element
2. Divalent mercury ($Hg2^+$), which is chemically reactive and under most environmental conditions is found to be combined with other substances into mercury salts (e.g., $HgCl_2$)
3. Particulate-phase mercury [Hg(p)], most of which is actually the chemical species Hg(II) but is only slightly reactive because it is mixed with solid-phase material in the atmosphere

These three Hg species vary in their affinities for tissue, their persistence, and their bioavailability. Mercury compounds may remain in the atmosphere for extended periods depending on the oxidation state. For example, Hg^0 has an atmospheric residence time of about 1 year but Hg^{2+} for just a few weeks [47, 48]. This is in part because the Hg^0 is oxidized to Hg^{2+}, the major chemical species in both wet and dry Hg deposition to the earth's surface [49].

During biogeochemical cycling, Hg is readily interconverted among Hg^0, Hg^{2+}, and Hg(p). After Hg in oxidized states is deposited, it may reach reduced environments and reenter the atmosphere as vapor-phase Hg^0 [50]. Under reduced conditions, Hg will also become alkylated, which may be accelerated and mediated by microbial metabolism.

The remaining nonvolatilized Hg fraction accumulates in soil organic matter [51] or binds to organic compounds suspended in surface waters [52].

In oceans, part of the Hg^{2+} that is deposited onto the surface is reduced photochemically to Hg^0. This Hg^0 fraction is then reemitted to the atmosphere advectively by winds and by partitioning as a dissolved gas to the atmosphere [53] (see discussion of Henry's law and air-water equilibrium constants in Chapter 5). Mercury accumulates with increasing latitude. In polar regions, oxidation to Hg^{2+} is likely mediated by bromine (Br) in the atmosphere [40].

The transformation chemistry of Hg, therefore, determines not only how Hg will speciate in the atmosphere and other parts of the environment but also the likely exposure scenarios that will result. The food chain characteristics of Hg are very important. For example, Hg increases in bioavailability when it resides in sediment, that is, decreased dissolved O_2 allows for reduction and methylation of other forms of Hg. Thus, the form of Hg in atmospheric deposition is likely to be changed as it moves among compartments and as it is taken up and metabolized by organisms. As mentioned, this leads to alkylation. For example, sulfate-reducing bacteria convert oxidized forms of Hg to methylmercury (CH_3Hg) compounds, which are much more bioavailable and toxic than the oxidized precursors.

Even identical amounts of atmospheric Hg deposition onto a surface water body may yield dramatically different tissue concentrations in fish than in another water body nearby, for example, if the two systems have varying depths of organic sediment. The system with deeper organic sediments is likely to have fish with higher CH_3Hg concentrations due to the bacterial sulfate reduction. There also appears to be a direct relationship between acid deposition and methylation of Hg, if the acid is H_2SO_4 [40].

Mercury is an air pollutant in several forms, including as a divalent cation in water vapor and droplets. The Hg is in precipitation that reaches the earth's surface. However, Hg found in the environment is often alkylated, for example, an organometallic with a chain of carbon atoms. How does mercury deposited onto surface waters transform to monomethylmercury compounds?
This can happen by numerous mechanisms, including ion complexes discussed in Chapter 7:

$$\text{Cation} + \text{Anion (ligand)} \rightarrow \text{Complex}$$

$$\text{Association reaction: } Hg^{2+} + Cl^- \leftrightarrow HgCl^- \quad K_{eq} = 10^{7.2}$$

$$\text{Dissociation reaction: } Hg(SH)_{2^\circ} \leftrightarrow Hg^{2+} + 2HS^- \quad \log K = -36.6$$

$$K_{diss} = (Hg^{2+})(HS^-)^2 / \left(Hg(SH)_2{}^0\right)$$

$$\text{Sulfide complexes: } Hg^{2+} + HS^- \rightarrow Hg(SH)_{2^\circ}$$

$$Hg^{2+} + HS^- \rightarrow HgS(SH)^-$$

$$Hg^{2+} + HS^- \rightarrow Hg(S_n)SH$$

Ultimately, the product is $(CH_3)HgCl$, $(CH_3)HgOH$, or another monomethylmercury compound. Therefore, sulfate-reducing bacteria commonly take these complexes or other forms of Hg and reduce them to obtain energy.

Bottom line: The environmental factors that control bacterial methylation of mercury are not well understood, but it is obvious that after atmospheric Hg is deposited, it is transformed to methylmercury, which in turn is bioaccumulated and biomagnified in aquatic food webs. At low concentrations ($\sim 3\,mg\,L^{-1}$), natural dissolved organic matter strongly complexes with ionic Hg^{2+} and CH_3Hg^+, which enhances biological uptake and methylation of Hg in aquatic environments [54].

8.5.3 Biogeochemistry of other heavy metals

Besides mercury, the mass of most heavy metals reaches the environment as aerosols or as compounds, not in the zerovalent state. Therefore, the inherent properties of the metal compounds dictate the fate of the metals.

The emissions of heavy metals are predominantly from human activities. The emission rates from human and natural sources can be described by the mobilization factor (MF) [55]:

$$MF = \frac{E_h}{E_n} \tag{8.28}$$

where E_h is the emission rate of a metal from human sources and E_n is the emission rate of the same metal from natural sources, for example, volcanoes. The MF calculations for selected metals and metalloids are given in Table 8.5. Only cobalt (Co) and manganese (Mn) show greater emissions from natural sources, that is, $MF < 1$.

TABLE 8.5 Mobilization factors (MF) calculated from annual emission rates

	Emissions ($10^9\,g\,yr^{-1}$)		
Element	**Natural sources**	**Human activities**	**Mobilization Factor**
Antimony (Sb)	0.98	38.00	38.8
Arsenic (As)	2.80	78.00	27.9
Cadmium (Cd)	0.29	5.50	19.0
Chromium (Cr)	58.00	94.00	1.6
Cobalt (Co)	7.00	4.40	0.6
Copper (Cu)	19.00	260.00	13.7
Lead (Pb)	5.90	2000.00	339.0
Manganese (Mn)	610.00	320.00	0.5
Mercury (Hg)	0.04	11.00	275.0
Molybdenum (Mo)	1.10	51.00	46.4
Nickel (Ni)	28.00	98.00	3.5
Selenium (Se)	0.41	14.00	34.1
Silver (Ag)	0.06	5.00	83.3
Tin (Sn)	5.20	43.00	8.3
Vanadium (V)	65.00	210.00	3.2
Zinc (Zn)	36.00	840.00	23.3

Data from J.N. Galloway, J.D. Thornton, S.A. Norton, H.L. Volchok, R.A. McLean, Trace metals in atmospheric deposition: a review and assessment, Atmos. Environ. (1967) 16 (7) (1982) 1677–1700.

Another technique for comparing natural with anthropogenic metal emissions is the enrichment factor (EF), which is the ratio of two ratios, that is, the ratio of the concentration of a metal in the earth's crust and the concentration of a reference metal, often iron or aluminum, divided by the ratio of the concentration of the metal in the atmosphere and the atmospheric concentration of the reference metal. For example, the EF of cadmium is

$$EF = \frac{\left(\frac{[Cd]}{[Al]}\right)_{air}}{\left(\frac{[Cd]}{[Al]}\right)_{crust}} \tag{8.29}$$

If $EF > 1$, the metal is being added, that is, the atmosphere is being enriched, which is the case for Cd, Cu, Pb, Se, Sb, and Zn at a global scale. However, these and other metals can be quite high, for example, >10, in urbanized and industrialized areas. In the United States, for example, Se and Pb are the most enriched [55].

The US Agency for Toxic Substances and Disease Registry (ATSDR) in 1999 ranked lead (Pb) second in the list of prioritized hazardous substances. The noxious effects of lead have been known for a long time—we are aware of the Romans inadvertently poisoning themselves by drinking and eating out of lead vessels. The World Health Organization has guidelines for lead in the air and in water of $0.5\,\mu g\,m^{-3}$ and $0.1\,mg\,L^{-1}$, respectively. The Centers for Disease Control and Prevention in the United States states that a blood level of $10\,\mu g\,dL^{-1}$ ($100\,\mu g\,L^{-1}$ or $0.48\,\mu mol\,L^{-1}$) in children is a cause for concern. Lead poisoning occurs through ingestion, inhalation, and dermal contact, with inhalation perhaps the most common exposure route. Factory workers dealing with lead and lead products and people living near such factories are the most likely to be affected. In countries where lead tetraethyl is still added to gasoline, the general public is exposed to significant lead levels through vehicle exhausts and even lead-containing dust in the air [56].

Generally, exposures to lead can result in anemia; nervous system dysfunction; impaired cognitive, motor, and language skills; learning difficulties; nervousness and emotional instability; insomnia; nausea; lethargy; weakness; hypertension; kidney problems; decreased fertility and increased level of miscarriages; and low birth weight and premature deliveries. Children are at particular risk to Pb toxicity, including neurological impairment, growth retardation, delayed sexual maturation, and impaired vitamin D metabolism. Children are very much more susceptible to damage from lead than adults, and neurological impairment can occur in children with blood lead levels $<10\,\mu g\ dL^{-1}$ [57].

The biogeochemical cycling of Pb appears to be similar to that of other divalent metals. As evidence, uptake by plant life increases with increasing concentrations in soil and water. For example, studies have shown the most influential factor in biotic Pb concentrations (also Zn, Cu, and Ni) is the distance from the ocean shore. For example, in arctic willow growing a few meters from the tide line, the Pb content was higher than that of the same plant species growing about 1 km from the coastline and sheltered from the sea [58].

A major transport mechanism for Pb is by aerosols. In continental climates with dry and hot summer and severe winter, strong winds can transport particulates for short and long distance, depending on the consistency of wind speed and inversely proportional to aerosol size. The sources are often anthropogenic. For example, the atmospheric transport of soil particles with sorbed organic compounds, heavy metals, and metalloids (e.g., Ni, Cd, Co, Zn, Pb, and As) has origins from both industrial emissions to the atmosphere and waste landfill sites [58].

Lead and other metals vary by ecosystem type, depending on deposition rates, uptake by organisms, and sequestration within abiotic and biotic sinks. For example, averaged Pb fluxes in biological cycling of boreal and subboreal forest ecosystems vary considerably among types. The flux in subboreal northern taiga coniferous forests had a turnover flux of 1.9 g Pb per hectare per year (g Pb $ha^{-1}\ yr^{-1}$), subboreal coniferous and small-leaved forests had 4.2 g Pb $ha^{-1}\ yr^{-1}$, subboreal broad-leaved forests had 7.9 Pb $ha^{-1}\ yr^{-1}$, and sphagnum forest swamps had 12.0 Pb $ha^{-1}\ yr^{-1}$ [58].

Cadmium (Cd) occurs in the earth's crust at a concentration of 0.1–0.5 ppm and is commonly associated with other metal ores (e.g., zinc, lead, and copper). Ocean water Cd concentrations range between <5 and $110\,ng\,L^{-1}$, with higher levels reported near coastal areas and in marine phosphates and phosphorites. Surface waters and groundwater generally contain $<1\,\mu g\,L^{-1}$ Cd. Soil Cd concentrations are dependent on several factors, for example, the specific Cd compound's mobility, natural geochemistry, and Cd loading from agricultural chemicals and atmospheric deposition. Volcanoes, forest fires, sea-salt aerosols, and other natural phenomena also load Cd into the atmosphere [59].

In recent years, cadmium (Cd) is most prominent as an element in nickel-cadmium (NiCad) batteries, but Cd has been used extensively in metal plating and other industries. Exposure to Cd has been associated with kidney damage (proteinuria) and respiratory effects, for example, reduction in forced vital capacity (FVC) and reduction in peak expiratory flow rate [60].

The molecular weight of Cd is $112.41\,g\,mol^{-1}$. Other physicochemical properties are as follows: density, $8.642\,g\,cm^{-3}$ at 20°C; boiling point, 765°C; melting point, 320.9°C; and vapor pressure, 0.13 kPa at 394°C. Thus, under most environmental conditions, Cd^0 is nonvolatile and would likely need to be sorbed to aerosols to reach the atmosphere. However, Cd exists in various chemical species. Two that are particularly toxic are cadmium chloride ($CdCl_2$) and cadmium oxide (CdO). The aqueous solubility of $CdCl_2$ is rather high (about $100\,mg\,mL^{-1}$ at 20°C). Its vapor pressure is higher than Cd^0, although it is nonvolatile under environmental conditions (0 kPa at 20°C). The aqueous solubility of CdO is low ($<1\,mg\,mL^{-1}$). CdO is nonvolatile under environmental conditions (0 kPa at 20°C). Cadmium carbonate ($CdCO_3$) is also insoluble in water. Cadmium sulfate ($CdSO_4$) and cadmium sulfide (CdS) are soluble in water (e.g., CdS aqueous solubility $=1.3\,mg\,L^{-1}$ at 18°C).

Given the low solubilities and vapor pressures of most Cd compounds, dissolution and volatilization must be assisted by other mechanisms to mobilize Cd. For example, when Cd is ground to powder, it burns and vaporizes. Thus, a major mechanism for entry into the atmosphere is combustion. Concentrations of Cd in fossil fuels, for example, coal and crude oil, and incineration of municipal wastes (e.g., batteries in the waste stream) are the largest sources of airborne Cd.

The oxidation state of Cd not only is most often +2 but also occurs as +1. The +1 state is reached in a manner similar to that of Hg. That is, Cd must first exist as a divalent cation, before conversion to Cd^+. For example, dissolving elemental Cd in a mixture of cadmium chloride and aluminum chloride (wherein Al has a +3 oxidation state) yields a compound wherein Cd has +1 oxidation state:

$$Cd + CdCl_2 + 2AlCl_3 \rightarrow 2AlCl_3 + Cd_2(AlCl_4)_2 \qquad (8.30)$$

This reaction is unlikely to occur often in nature, so the +1 oxidation state is more likely to be found in industrial and laboratory settings.

8.5.4 Metalloid cycling

Arsenic (As) and antimony (Sb) are the two metalloids of principal concern in air pollution. Both As and Sb belong to Group 15 of the periodic table.

Arsenic has been known for centuries, and its compounds have been used in all sorts of preparations, including cosmetics; paints; artist's colors and dyes; alloys of lead, copper, and brass; pesticides; wood preservatives; tonics; glass production (it removes the green tint from the iron impurities); and even medicines. There is a theory that the once-popular arsenic-based wallpaper pigment, Paris green, (copper(II) acetoarsenite) was the cause of Napoleon's mysterious death in 1821. In the late 19th century, arsenic compounds were linked to the deaths of >1000 children through inhaling arsenic vapors from moldy green wallpaper—probably the Paris green type. About the same time, arsenic-based medications for syphilis, asthma, and psoriasis were considered the cause of skin cancer. Paris green (copper(II) acetoarsenite) is a highly toxic emerald-green crystalline powder that has been used as a rodenticide and insecticide and also as a pigment, despite its toxicity. It is also used as a blue colorant in fireworks. The color is apparently vivid, blue green when very finely ground and a deeper true green when coarsely ground [61, 62].

In the United States, the main use of arsenic is wood preservatives for outdoor and industrial environments. The use of arsenic compounds has dropped dramatically in recent decades. In 1999, the United States imported 23,400t of arsenic metal and compounds, over 80% from China [63]. In 2016, the United States import total fell to 6120t, with 86% from China [64]. Almost 90% of the imports are being used in chromated copper arsenate (CCA) wood preservative for "pressure-treated" decking, landscaping, walkways, and industrial usage. The rest is used in semiconductors, specialized metal, a few remaining pesticides, and treatments for acute leukemia and other cancers.

Arsenic (As) ores (arsenic sulfides, orpiment, and realgar) have been mined and used since ancient times. It is probable that its toxic nature has been known since ancient times [65]. Arsenic and many As compounds are very toxic to humans. Arsenic disrupts the ATP reactions in the body and is very toxic to all animal life. Arsenic and its compounds, notably As^{3+}, have in the past been used in the production of pesticides (e.g., in treated wood products), herbicides, and insecticides. In 1980, the US National Toxicology Program listed inorganic arsenic compounds as known human carcinogen. Subsequently, in 1981, inorganic arsenic-based pesticides were banned, and by 1985, the United States had stopped producing arsenic. The applications of arsenic have declined considerably since that time, but in many parts of the world, arsenic compounds are used in wood preservatives for industrial applications. Developed nations and the United States continue to import As for this and other specialized purposes.

The main route for human poisoning by arsenic is from drinking groundwater that contained high concentrations of arsenic. A 2007 study found that over 137 million people in >70 countries are probably affected by arsenic poisoning of drinking water [66]. Arsenic is toxic to aquatic life, birds, and land animals and in soils; where arsenic content is high, plant growth and crop yields may be poor. Aquatic life is particularly sensitive to arsenic compounds and moderately toxic to birds and land animals.

Arsenic provides an illustration of the interactions among abiotic, geologic systems and biotic, especially bacterial, systems in the cycling of metals and metalloids. In most of its forms, As compounds are very persistent in the environment but are transformed within and outside of microbial populations. As such, As can bioaccumulate in fish and shellfish. In spite of its inherent toxicity, some microorganisms have evolved to tolerate relatively high concentrations of As. Some even thrive on it, using As for an energy source. The energy generation is based on the redox chemistry of arsenic. Arsenic has four main oxidation states: −3, 0, +3, and +5 with the predominant inorganic forms being arsenates (+5) and arsenites (+3). Under aerobic conditions, some microorganisms obtain energy by the oxidation of As^{3+} to As^{5+}, with the resultant electrons being transferred to electron acceptors such as nitrate ions or oxygen and the ATP pump [67]. Under anaerobic conditions, energy for growth is obtained by the microbial reduction of As^{5+} to As^{3+} coupled to the oxidation of organic matter or inorganic electron donors such as hydrogen or sulfide ions. These microbial processes together with inorganic and physical processes make up the global arsenic cycle [67].

The trivalent form is the most common and is the basis for the arsenates and most organoarsenic compounds. Both As_2S_3 (orpiment) and As_4S_4 (realgar) are composed of trivalent arsenic. The As_4S_4 species is made possible by As—As bonds in the molecule [68]. Arsenic shares a column of the periodic table with phosphorus, so its compounds resemble those of phosphorus (K). Indeed, most organisms metabolize As via the K-transport system and eliminate As species by reducing them to arsenite [69].

An advocacy nongovernmental organization in 2001 in the United States, wanting to ban arsenic in all consumer products, found that pressure-treated wood leached >1 mg of arsenic onto a moistened hand wipe the size of a 4-year-old's hand. This is 100 times the US Environmental Protection Agency's $10\,\mu g\,L^{-1}$ "allowable daily exposure level" for drinking water, assuming the child has a liter a day. Likely due in part to these findings, the wood processing industry agreed to stop using arsenic-based wood preservatives for household use as of December 2003.

Herbicides containing organometallic As compounds, which are not as inherently toxic as their inorganic compounds, remain in the United States and other country's agriculture and landscaping. For example, monosodium methanearsonate (MSMA) is currently registered as a broad-spectrum herbicide to control grasses and broad-leaved weeds. The registration is limited to cotton, sod farms, golf courses, and highway rights-of-way and cannot be used in Florida, except for cotton in specified counties [70].

Europe restricts the marketing and use of arsenic, including banning of chromated copper arsenate (CCA) wood treatment [71]. Treated CCA wood is not permitted to be used in residential or domestic constructions but is permitted for use in various industrial and public works, such as bridges, highway safety fencing, electric power transmission, and telecommunication poles. This directive is not unlike the ruling in the United States.

In many other parts of the world, the same type of ruling exists: In Australia, the use of CCA preservative for treatment of timber was restricted to certain applications from March 2006, and CCA may no longer be used to treat wood used in "intimate human contact" applications and that includes children's play equipment, furniture, residential decking, and hand railing. Use for low-contact residential, commercial, and industrial applications remains unrestricted, as does its use in all other situations. Similarly, to the US EPA regulations, the Australian authorities did not recommend dismantling or the removal of existing CCA-treated wood structures. In many parts of the world (parts of SE Asia, Chile, and Argentina and parts of Western United States, Taiwan, Thailand, Mainland China, Bangladesh, and West Bengal), the arsenic concentration in groundwater is high enough to cause serious arsenic poisoning to people drinking the water. However, nowhere in the world is the problem greater than in the Ganges delta of Bangladesh and to a lesser extent in West Bengal [72, 73].

Bangladesh, in the 1970s, had one of the highest infant mortality rates in the world. This was largely due to ineffective water purification and sewage systems. Most drinking water used to be collected from open dug wells and ponds with little or no arsenic but contaminated with water transmitting diseases such as diarrhea, dysentery, typhoid, cholera, and hepatitis. At that time, the UNICEF and the World Bank aid workers worked out a solution; they decided to drill tube wells (boreholes) deep underground to tap into the groundwater [74].

As a result, 8 million tube wells (boreholes) were constructed to provide "safe" drinking water, and in a short time, the infant mortality and diarrheal illness were reduced by 50%. Most of the wells were augured to depths between 20 and 100 m. Unfortunately, one in five of these wells was contaminated with arsenic, which was above the government's drinking water standard set at $0.05\,mg\,L^{-1}$, this being five times the World Health Organization's acceptable maximum concentrations of arsenic in safe drinking water, which was $0.01\,mg\,L^{-1}$. The crisis came to the attention of international organizations in 1995, and the resultant study involved the analysis of thousands of water samples and hair, nail, and urine samples [75–77]. Indeed, 900 villages in 61 out of 64 provinces in Bangladesh and in West Bengal 17 provinces had water supplies with arsenic water levels above the government limit. Results showed that many of wells that had been drilled at depths $>20\,m$ and $<100\,m$ were contaminated, whereas arsenic was not detected in groundwater and well water collected from depths $>100\,m$. The explanation was that groundwater closer to the surface had spent a shorter time in the ground and therefore had not dissolved significant amounts of arsenic from the soil, and furthermore, the water from the wells deeper than 100 m was exposed to older sediments that have already been depleted of arsenic [78]. The total number of people using arsenic-rich water (levels $>0.1\,mg\,L^{-1}$) was 1.4 million in West Bengal and a staggering 46–57 million in Bangladesh [79–81]. The crisis is still raging as many of the remedial programs have not worked.

As mentioned, As biogeochemistry is like that of the heavy metals. Microbes play a key role in As cycling and toxicity. Anaerobic bacteria can reduce As^{5+} during cometabolism. When an As undergoes cometabolism, the reactions are catalyzed by an enzyme that is fortuitously produced by the organisms for other purposes. The microbe does not directly benefit from the degradation of the compound. The biotransformation of the As compound could actually be harmful or can inhibit growth and metabolism on microbe. In addition, aerobic microorganisms can oxidize As^{3+}. Thus, both aerobic and anaerobic microorganisms can cycle and transform As in the environment.

Microbes degrade soilborne contaminants microbially, but these mechanisms are enhanced by the biochemodynamics of the root zone (the rhizosphere). This process is slower usually than phytodegradation. This process can enhance As cometabolism because plant roots release natural substances, for example, sugars, alcohols, and acid, that are food for soil microorganisms. The fixation and release of nutrients are also a natural type of biostimulation for the microbes. In addition, the roots provide conduits and physical loosening of soil, improving microbial contact with oxygen and nutrients.

Antimony (Sb) exists in fossil fuels, especially in coal. Fossil fuel combustion is likely the largest source of anthropogenic Sb emissions to the atmosphere. Abundant in sulfide minerals and co-occurring with heavy metals, Sb emissions likely result from the mining and metallurgy of nonferrous metals, especially Pb, Cu, and Zn.

Their identical outer orbital electron configuration dictates that Sb and As have the same range of oxidation states in environmental systems (i.e., from −3 to +5). The major chemical forms of both are the oxides, hydroxides, or oxoanions, in the +5 state in oxidized environments (i.e., antimonates and arsenates) or in the +3 state in more reduce environments (i.e., antimonites and arsenites). Like As, the toxicity of Sb depends upon speciation [82]. The organoantimonials (e.g., methylated Sb species) are less toxic than the antimonates (Sb^{5+}), which are less toxic than the antimonites (Sb^{3+}).

Metalloids seem to behave similarly to heavy metals regarding biogeochemistry. However, this cannot be assumed in all environments, given the lack of relevant and specific data. These data are particularly lacking for Sb [82].

8.6 BIOGEOCHEMICAL CYCLES AND DECISION MAKING

Air quality management decisions are often made among competing interests and perspectives. These involve trade-offs. For example, remediating a contaminated site may call for excavation of soil, which is then incinerated. The complete incineration will result in generating and emitting greenhouse gases. Determining whether this approach is successful and appropriate depends on the extent and quality of options from which these emissions occur. For example, if dioxin-laden soil is incinerated, this may have been the only viable approach to detoxify a very toxic and persistent compound. Releasing CO_2 in this case is truly a measure of success.

One of the biggest engineering challenges is how to select and operate control technologies in a manner fully cognizant and deferential of the biogeochemical cycles. This view can support work to reduce the impact of global climate change debate, in light of the seeming paucity of ways to deal with the problem. The National Academy of Engineering has identified the most important challenges to the future of engineering. Both the nitrogen and carbon biogeochemical cycles are explicitly identified among the most pressing engineering needs.

The biogeochemical cycle that extracts nitrogen from the air for its incorporation into plants—and hence food—has become altered by human activity. With widespread use of fertilizers and high-temperature industrial combustion, humans have doubled the rate at which nitrogen is removed from the air relative to preindustrial times, contributing to smog and acid rain, polluting drinking water, and even worsening global warming. Nitrogen and other nutrients present a challenge to scientists, engineers, and decision makers who strive for efficient agricultural production in a sustainable and environmentally acceptable way [83–85]. Food consists predominantly of proteins, carbohydrates, fats, and other organic compounds. Therefore, the C cycle and nutrient cycles are inextricably linked. Like C, as discussed in the next chapter, chemical species of N and the other nutrient elements are essential and toxic, depending on its dose and form. The Academy articulates this challenge:

The biogeochemical cycle that extracts nitrogen from the air for its incorporation into plants—and hence food—has become altered by human activity. With widespread use of fertilizers and high-temperature industrial combustion, humans have doubled the rate at which nitrogen is removed from the air relative to preindustrial times, contributing to smog and acid rain, polluting drinking water, and even worsening global warming. Engineers must design countermeasures for nitrogen cycle problems while maintaining the ability of agriculture to produce adequate food supplies [86].

Engineers can expect to be increasingly asked to recommend improvements to the biogeochemical cycling, such as enhancements to the food life cycles (e.g., animal feeding operations, farmlands, rangelands, and groceries). How can engineering innovation improve the efficiency of various human activities related to nitrogen, from making fertilizer to recycling food wastes? Currently, less than half of the fixed nitrogen generated by farming practices actually ends up in harvested crops. And less than half of the nitrogen in those crops actually ends up in the foods that humans consume. In other words, fixed nitrogen leaks out of the system at various stages in the process—from the farm field, to the feedlot, to the sewage treatment plant. Engineers not only need to identify the leakage points and devise systems to plug them, that is, the structural and mechanical solutions, but also must engage biological solutions, such as understanding the processes that lead to increased C and nutrient emissions and applying this understanding to modify the processes accordingly [87].

REFERENCES

[1] D. Vallero, Environmental Biotechnology: A Biosystems Approach, Elsevier Science, 2015.
[2] D.A. Vallero, Fundamentals of Air Pollution, fifth ed., Elsevier Academic Press, Waltham, MA, 2014. p. 999 pages cm.
[3] P.M. Gschwend, Environmental Organic Chemistry, John Wiley & Sons, 2016.
[4] D.A. Vallero, Fundamentals of Air Pollution, fourth ed., Elsevier, Amsterdam; Boston, 2008. pp. xxiii, 942 p.
[5] K.A. Masarie, P.P. Tans, Extension and integration of atmospheric carbon dioxide data into a globally consistent measurement record, J. Geophys. Res. Atmos. 100 (D6) (1995) 11593–11610.
[6] D.A. Vallero, Engineering aspects of climate Change, in: Climate Change, second ed., Elsevier, 2015, pp. 547–568.

[7] National Oceanic and Atmospheric Administration, Carbon Education Tools, 2 March 2018. Available: https://www.pmel.noaa.gov/co2/file/Carbon+Cycle+Graphics, 2018.
[8] U.S. Environmental Protection Agency, Understanding Global Warming Potentials, 9 July 2018. Available: https://www.epa.gov/ghgemissions/understanding-global-warming-potentials, 2018.
[9] I. P. O. C. Change, IPCC, in: Climate Change, 2014.
[10] R.A. Duce, The impact of atmospheric nitrogen, phosphorus, and iron species on marine biological productivity, in: The Role of Air-Sea Exchange in Geochemical Cycling, Springer, 1986, pp. 497–529.
[11] N. Mahowald, et al., Global distribution of atmospheric phosphorus sources, concentrations and deposition rates, and anthropogenic impacts, Glob. Biogeochem. Cycles 22 (4) (2008).
[12] J.L. Shupe, Fluorides: Effects on Vegetation, Animals and Humans, Paragon Press, Inc, 1983.
[13] D.A. Vallero, Excerpted and Revised From: Fundamentals of Air Pollution, fifth ed., Elsevier Academic Press, Waltham, MA, 2014. p. 999 pages cm.
[14] National Center for Atmospheric Research, Biogeochemical Cycles, 4 March 2015. Available: http://scied.ucar.edu/longcontent/biogeochemical-cycles, 2015.
[15] EPA/630/R-94/003 A review of ecological assessment case studies from a risk assessment perspective, vol. II, (1994).
[16] H. Su, et al., Soil nitrite as a source of atmospheric HONO and OH radicals, Science 333 (6049) (2011) 1616–1618.
[17] T. Green, G. Brown, L. Bingham, Environmental impacts of conversion of cropland to biomass production, Tennessee Valley Authority Southeastern Regional Biomass Energy Program, Muscle Shoals, AL, United States, 1996.
[18] UK Air Pollution Information System, Nitrogen Oxides (NOx), 2 March 2018. Available: http://www.apis.ac.uk/overview/pollutants/overview_NOx.htm, 2012.
[19] B. Finalyson-Pitts, J. Pitts, Chemistry of the Upper and Lower Atmosphere: Theory, Experiments, and Application, Academic Press, San Diego, CA, 1999.
[20] EPA-452/R-11-005a, Policy Assessment for the Review of the Secondary National Ambient Air Quality Standards for Oxides of Nitrogen and Oxides of Sulfur, 2011.
[21] L. Fewtrell, Drinking-water nitrate, methemoglobinemia, and global burden of disease: a discussion, Environ. Health Perspect. 112 (14) (2004) 1371.
[22] U.S. Environmental Protection Agency, National Primary Drinking Water Regulations, 27 February 2018. Available: https://www.epa.gov/ground-water-and-drinking-water/national-primary-drinking-water-regulations, 2017.
[23] S.M.A. Adelana, Nitrate health effects, in: Water Encyclopedia, 2005.
[24] W.H. Schlesinger, Biogeochemistry, Elsevier, 2005.
[25] R. Foust, Atmospheric reactions of sulfur and nitrogen, Available: http://jan.ucc.nau.edu/~doetqp-p/courses/env440/env440_2/lectures/lec37/lec37.htm, 2004.
[26] N. EFSA Panel on Dietetic Products and Allergies, Scientific opinion on dietary reference values for chromium, EFSA J. 12 (10) (2014) 3845.
[27] T.C. Ho, Control of trace metal emissions during coal combustion, Federal Energy Technology Center (FETC), Morgantown, WV, and Pittsburgh, PA, 1998.
[28] C.T. Driscoll, J.K. Otton, A. Iverfeldt, Trace metals speciation and cycling, in: B. Moldan, J. Cerny (Eds.), Biogeochemistry of Small Catchments: A Tool for Environmental Research, John Wiley & Sons, Hoboken, New Jersey, 1994.
[29] S. Vink, Seasonal variations in the distribution of Fe and Al in the surface waters of the Arabian Sea, Deep-Sea Res. II Top. Stud. Oceanogr. 46 (8–9) (1999) 1597–1622.
[30] P. Boyd, M. Ellwood, The biogeochemical cycle of iron in the ocean, Nat. Geosci. 3 (10) (2010) 675.
[31] K.W. Bruland, Oceanographic distributions of cadmium, zinc, nickel, and copper in the North Pacific, Earth Planet. Sci. Lett. 47 (2) (1980) 176–198.
[32] K.W. Bruland, R.P. Franks, Mn, Ni, Cu, Zn and Cd in the western North Atlantic, in: Trace Metals in Sea Water, Springer, 1983, pp. 395–414.
[33] D.C. Girvin, A.T. Hodgson, M.E. Tatro, R.N. Anaclerio, Spatial and Seasonal Variations of Silver, Cadmium, Copper, Nickel, Lead, and Zinc in South San Francisco Bay Water during Two Consecutive Drought Years, 1978.
[34] J. Stauber, S. Andrade, M. Ramirez, M. Adams, J. Correa, Copper bioavailability in a coastal environment of Northern Chile: comparison of bioassay and analytical speciation approaches, Mar. Pollut. Bull. 50 (11) (2005) 1363–1372.
[35] S.A. Sanudo-Wilhelmy, A.R. Flegal, Anthropogenic silver in the Southern California Bight: a new tracer of sewage in coastal waters, Environ. Sci. Technol. 26 (11) (1992) 2147–2151.
[36] E. Boyle, J. Edmond, E. Sholkovitz, The mechanism of iron removal in estuaries, Geochim. Cosmochim. Acta 41 (9) (1977) 1313–1324.
[37] J. Moore, O. Braucher, Sedimentary and mineral dust sources of dissolved iron to the world ocean, Biogeosciences 5 (3) (2008) 631–656.
[38] J.V. Macías-Zamora, Ocean pollution, in: Waste, Elsevier, 2011, pp. 265–279.
[39] D.A. Vallero, K.H. Reckhow, A.D. Gronewold, Application of multimedia models for human and ecological exposure analysis, in: International Conference on Environmental Epidemiology and Exposure, International Society of Exposure Analysis, Durham, NC, 2007.
[40] E.S. Bernhardt, W. Schlesinger, Biogeochemistry: An Analysis of Global Change, Elsevier Inc, Waltham, MA, USA and Oxford, UK, 2013.
[41] J.H. Duffus, "Heavy metals"—a meaningless term?(IUPAC technical report), Pure Appl. Chem. 74 (5) (2002) 793–807.
[42] F. Senese, Why is mercury a liquid at STP? in: General Chemistry, vol. 1, Frostburg State University, 2007.
[43] Background paper for stakeholder panel to address options for managing U.S. non-federal supplies of commodity-grade mercury, Available: http://www.epa.gov/mercury/stocks/backgroundpaper.pdf, 2007.
[44] Northeast Waste Management Officials' Association, Indoor Air Mercury, 26 July 2003. Available: http://www.newmoa.org/prevention/mercury/MercuryIndoor.pdf, 2018.

[45] O. Lindqvist, et al., Mercury in the Swedish environment—recent research on causes, consequences and corrective methods, Water Air Soil Pollut. 55 (1–2) (1991) xi–261.

[46] G. Rice, R. Ambrose, O. Bullock, J. Swartout, Mercury Study Report to Congress. Vol. 3. Fate and Transport of Mercury in the Environment, Environmental Protection Agency, Research Triangle Park, NC, United States, 1997. Office of Air Quality Planning and Standards.

[47] M.A. Engle, et al., Comparison of atmospheric mercury speciation and deposition at nine sites across central and eastern North America, J. Geophys. Res. Atmos. 115 (D18) (2010).

[48] E.M. Prestbo, D.A. Gay, Wet deposition of mercury in the US and Canada, 1996–2005: Results and analysis of the NADP mercury deposition network (MDN), Atmos. Environ. 43 (27) (2009) 4223–4233.

[49] C.-J. Lin, S.O. Pehkonen, The chemistry of atmospheric mercury: a review, Atmos. Environ. 33 (13) (1999) 2067–2079.

[50] J.A. Graydon, et al., The role of terrestrial vegetation in atmospheric Hg deposition: pools and fluxes of spike and ambient Hg from the METAALICUS experiment, Glob. Biogeochem. Cycles 26 (1) (2012).

[51] J.D. Demers, C.T. Driscoll, T.J. Fahey, J.B. Yavitt, Mercury cycling in litter and soil in different forest types in the Adirondack region, New York, USA, Ecol. Appl. 17 (5) (2007) 1341–1351.

[52] J.A. Dittman, et al., Mercury dynamics in relation to dissolved organic carbon concentration and quality during high flow events in three northeastern US streams, Water Resour. Res. 46 (7) (2010).

[53] J. Kuss, C. Zülicke, C. Pohl, B. Schneider, Atlantic mercury emission determined from continuous analysis of the elemental mercury sea-air concentration difference within transects between 50° N and 50° S, Glob. Biogeochem. Cycles 25 (3) (2011).

[54] W. Dong, L. Liang, S. Brooks, G. Southworth, B. Gu, Roles of dissolved organic matter in the speciation of mercury and methylmercury in a contaminated ecosystem in Oak Ridge, Tennessee, Environ. Chem. 7 (1) (2010) 94–102.

[55] J.N. Galloway, J.D. Thornton, S.A. Norton, H.L. Volchok, R.A. McLean, Trace metals in atmospheric deposition: a review and assessment, Atmos. Environ. (1967) 16 (7) (1982) 1677–1700.

[56] National Institute for Occupational Safety and Health, Lead compounds (as Pb), 26 July 1994. Available: https://www.cdc.gov/niosh/idlh/7439921.html, 2018.

[57] V. Vella, E. O'Brien, E. Idris, Health impacts of lead poisoning a preliminary listing of the health effects & symptoms of lead poisoning, LEAD (Lead Education and Abatement Design) Action News 6 (2010).

[58] V.N. Bashkin, R.W. Howarth, Modern Biogeochemistry, Springer Science & Business Media, 2002.

[59] Toxicological Profile for Cadmium, Available: https://www.atsdr.cdc.gov/toxprofiles/tp.asp?id=48&tid=15, 2012.

[60] Canadian Office of Environmental Health Hazard Assessment, Chronic toxicity summary: Cadmium and cadmium compounds, Available: http://oehha.ca.gov/air/chronic_rels/pdf/7440439.pdf, 2000. Accessed 15 November 2013.

[61] C. o. T. Environmental Management Division, Arizona, Health & Safety in the Arts: Painting & Drawing Pigments, 9 February 2012. Available: http://www.tucsonaz.gov/arthazards/paint1.html, 2012.

[62] New Jersey Department of Health and Senior Services, Hazardous Substance Fact Sheet: Copper Acetoarsenate, 26 July 2018. Available: https://nj.gov/health/eoh/rtkweb/documents/fs/0529.pdf, 2018.

[63] R.G. Reese Jr., Arsenic, in: U.S. Geological Survey Minerals Yearbook – 1999, 1999. pp. 7.0–7.4.

[64] M.W. Geroge, Arsenic (advance release), in: U.S. Geological Survey Minerals Yearbook – 2016, 2018. pp. 7.0–7.4.

[65] A. Vahidnia, G. Van der Voet, F. De Wolff, Arsenic neurotoxicity—a review, Hum. Exp. Toxicol. 26 (10) (2007) 823–832.

[66] P.L. Smedley, D. Kinniburgh, A review of the source, behaviour and distribution of arsenic in natural waters, Appl. Geochem. 17 (5) (2002) 517–568.

[67] J.R. Lloyd, R.S. Oremland, Microbial transformations of arsenic in the environment: from soda lakes to aquifers, Elements 2 (2) (2006) 85–90.

[68] J. Tossell, Theoretical studies on arsenic oxide and hydroxide species in minerals and in aqueous solution, Geochim. Cosmochim. Acta 61 (8) (1997) 1613–1623.

[69] W.R. Cullen, K.J. Reimer, Arsenic speciation in the environment, Chem. Rev. 89 (4) (1989) 713–764.

[70] Organic Arsenicals; Amendments to Terminate Uses; Amendment to Existing Stocks Provisions, U. S. E. P. Agency, 2013.

[71] WRITTEN QUESTION E-0582/03 by Erik Meijer (GUE/NGL) to the Commission. Phasing out of carcinogenic impregnated wood following the ban on new use and labelling of used material as toxic waste, E. U. Commission, 2003.

[72] A. Mukherjee, et al., Arsenic contamination in groundwater: a global perspective with emphasis on the Asian scenario, J. Health Popul. Nutr. (2006) 142–163.

[73] U.K. Chowdhury, et al., Groundwater arsenic contamination in Bangladesh and West Bengal, India, Environ. Health Perspect. 108 (5) (2000) 393–397.

[74] WaterAid, Silent Killer, 10 February 2012. Available: www.wateraid.org/uk/what_we_do/sustainable_technologies/technology_notes/243.asp, 2012.

[75] S. Yadav, A. Gaur, A. Srivastava, S. Yadav, Monitoring of Arsenic contents in water bodies of Bundelkhand region (India), Int. J. Environ. Sci. 7 (1) (2016) 62–69.

[76] A. Chatterjee, D. Das, B.K. Mandal, T.R. Chowdhury, G. Samanta, D. Chakraborti, Arsenic in ground water in six districts of West Bengal, India: the biggest arsenic calamity in the world. Part I. Arsenic species in drinking water and urine of the affected people, 1995.

[77] D. Das, A. Chatterjee, B.K. Mandal, G. Samanta, D. Chakraborti, B. Chanda, Arsenic in ground water in six districts of West Bengal, India: the biggest arsenic calamity in the world. Part 2. Arsenic concentration in drinking water, hair, nails, urine, skin-scale and liver tissue (biopsy) of the affected people, Analyst 120 (3) (1995) 917–924.

[78] A. Singh, Chemistry of arsenic in groundwater of Ganges–Brahmaputra river basin, Curr. Sci. (2006) 599–606.

[79] J. Rehwaldt, B. Ness, C. Nes, S. Monda, M. Abdul, A. Khan, Arsenic Mitigation in Bangladesh, 2014.
[80] W. H. Organization, Arsenic in drinking water, http://www.who.int/mediacentre/factsheets/fs210/en/index.html, 2001.
[81] D.A. Vallero, T.M. Letcher, Unraveling Environmental Disasters, Newnes, 2012.
[82] S.C. Wilson, P.V. Lockwood, P.M. Ashley, M. Tighe, The chemistry and behaviour of antimony in the soil environment with comparisons to arsenic: a critical review, Environ. Pollut. 158 (5) (2010) 1169–1181.
[83] C. Mote Jr., D.A. Dowling, J. Zhou, The power of an idea: the international impacts of the grand challenges for engineering, Engineering 2 (1) (2016) 4–7.
[84] W. El Maraghy, Future trends in engineering education and research, in: Advances in Sustainable Manufacturing, Springer, 2011, pp. 11–16.
[85] J.-Y. Yoon, M.R. Riley, Grand challenges for biological engineering, J. Biol. Eng. 3 (1) (2009) 16.
[86] National Academy of Engineering, Introduction to the Grand Challenges for Engineering, Available: http://www.engineeringchallenges.org/challenges/16091.aspx, 2018.
[87] J. Galloway, et al., Human Alteration of the Nitrogen Cycle–Threats, Benefits, and Opportunities, UNESCO-SCOPE Policy Briefs, 2007.

FURTHER READING

[88] National Oceanic & Atmospheric Administration, Trends in Atmospheric Carbon Dioxide, 17 July 2017. Available: https://www.esrl.noaa.gov/gmd/ccgg/trends/gl_full.html, 2018.
[89] D.A. Vallero, Environmental Contaminants: Assessment and Control, Elsevier Academic Press, Amsterdam; Boston, 2004. pp. xxxix, 801 p.
[90] U.S. Environmental Protection Agency, Handbook: Guidance on Setting Permit Conditions and Reporting Trial Burn Results, EPA 625/6-89-019, 1989.
[91] B.C. Willis, M.M. Howie, R.C. Williams, Public Health Reviews of Hazardous Waste Thermal Treatment Technologies: A Guidance Manual for Public Health Assessors. US Department of Health and Human Services, Agency for Toxic Substances and Disease Registry, Division of Health Assessment and Consultation, 2002.

Chapter 9

Thermal reactions

9.1 INTRODUCTION

Arguably, the burning of substances produces the most toxic and largest quantity of air pollutants worldwide. No matter if the burning is intentional, for example, for energy production and cookstoves, or unintentional, for example, wildfires, air pollution at every scale is affected.

The biogeochemical cycles considered in the previous chapter are greatly affected by combustion, especially the release of compounds of carbon, nitrogen, sulfur, and metals into these cycles.

In particular, carbon compounds are included in metrics of pollution control efficiency and success, especially CO_2, CO, and other products of incomplete combustion (PICs).

As mentioned in Chapter 8, following the discussion of CO_2's role in global warming, it is somewhat ironic that the greater the amount of CO_2 emitted from a source, the more efficiently the pollution control equipment is operating. Indeed, complete combustion results in the production of only CO_2 and H_2O. Likewise, the production of CH_4 has been an indication of complete anaerobic digestion of organic compounds within the landfill or during fermentation. Both of these indicators of pollution control efficiency also happen to be greenhouse gases.

The combustion of organic compounds is how most energy is generated. These carbon-based coupons undergo reactions that differ in rates and products, depending on temperature, mixtures with other substances, oxygen concentrations, and numerous other factors. As a result, the types and amounts of product released from combustion vary but include some of the most toxic and otherwise harmful pollutants. For this reason, the next chapter's focus on combustion is also an extension of the carbon cycle.

9.2 AIR AND COMBUSTION

The theoretical, sufficient concentration of O_2 to achieve complete combustion is that needed to react with the total C in the combustible material, that is, fuel. The air needed to achieve this is known as "theoretical air" or "stoichiometric air," which depends on the chemical makeup of the fuel and the fuel feed rate. The feed rate is expressed as a volume or mass per time, for example, $L\,hr^{-1}$ and $kg\,hr^{-1}$, respectively.

The ideal combustion process, that is, stoichiometric combustion, is that that burns the fuel completely. The deficit between stoichiometric combustion and incomplete combustion determines the percentage of combustion inefficiency (see Fig. 3.8 in Chapter 3). Therefore, at or above theoretical air, the process is 100% efficient. To avoid products of incomplete combustion (PICs), especially carbon monoxide (CO), excess air is usually added.

The excess air or excess fuel for a combustion system is based on the stoichiometric air-fuel ratio, the precise, ideal fuel ratio in which chemical mixing proportion is reached. For safety reasons, prevention of explosive conditions from fouling and generation of high temperatures, and because fuel composition and conditions even in well-controlled reactors vary, combustors are designed to achieve "on-ratio" combustion, that is, requiring a known amount of excess air, often 10%–20% above the expect stoichiometric air value [1] (see Table 9.1). Fuel-lean mixtures have air content greater than the stoichiometric ratio; fuel-rich mixtures have air content less than the stoichiometric ratio.

For fuels (C_mH_n) in the gas phase, the stoichiometric combustion reaction is [2]

$$\underbrace{C_mH_n}_{fuel} + \left(m+\frac{n}{4}\right)\underbrace{(O_2+3.76N_2)}_{air} \rightarrow mCO_2+\frac{n}{2}H_2O+3.76\left(m+\frac{n}{4}\right)N_2 \tag{9.1}$$

Air Pollution Calculations. https://doi.org/10.1016/B978-0-12-814934-8.00009-0

TABLE 9.1 Approximate O_2 and CO_2 in flue gas at various excess air conditions

	Percent volume of CO_2 in flue gas					
Percent excess air	Natural gas	Propane butane	Fuel oil	Bituminous coal	Anthracite coal	Percent O_2 in flue gas for all fuels
0	12	14	15.5	18	20	0
20	10.5	12	13.5	15.5	16.5	3
40	9	10	12	13.5	14	5
60	8	9	10	12	12.5	7.5
80	7	8	9	11	11.5	9
100	6	6	8	9.5	10	10

Data from Engineering Toobox, Stoichiometric Combustion and Excess of Air, January 31, 2019, Available from: https://www.engineeringtoolbox.com/stoichiometric-combustion-d_399.html.

For solid and liquid fuels ($C_aH_bO_cN_dS_e$), the stoichiometric combustion reaction is [2]

$$\underbrace{C_aH_bO_cN_dS_e}_{fuel} + \underbrace{x(O_2 + 3.76N_2)}_{air} \rightarrow aCO_2 + \frac{b}{2}H_2O + \left(\frac{d}{2} + 3.76x\right)N_2 + eSO_2 \tag{9.2}$$

where $x = a + \frac{b}{4} - \frac{c}{2} + e$

Note that the $C_aH_bO_cN_dS_e$ empirical formula of solid or liquid fuel is assumed to account of 100% of the mass and is calculated assuming the fuel is a dry, ash-free elemental composition.

Combustion efficiency increases with excess air until heat loss in the amount of excess air reaches the heat loss to a point where the net efficiency begins to drop (see Fig. 9.1).

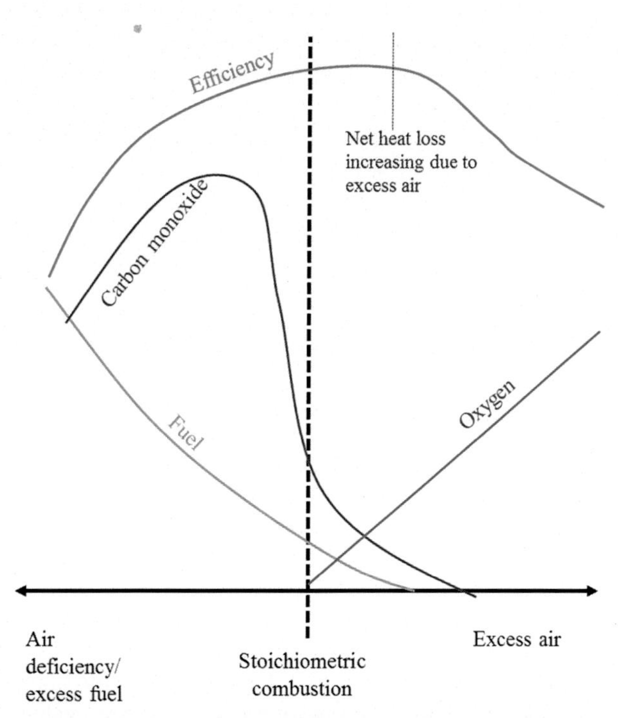

FIG. 9.1 Air-fuel ratio's effect on combustion efficiency and generation of carbon monoxide. *(Data from Engineering Toobox, Stoichiometric Combustion and Excess of Air. (2003). July 16, 2018. Available: https://www.engineeringtoolbox.com/stoichiometric-combustion-d_399.html).*

What is the reaction of methane burned at 25% excess air?

Since air is about 79% N_2 and 21% O_2, the N_2-to-O_2 mole ratio is 3.76. Thus, the stoichiometric combustion of CH_4 is

$$CH_4 + 2(O_2 + 3.76\,N_2) \rightarrow CO_2 + 2\,H_2O + 7.52\,N_2$$

Therefore, CH_4 combusted with 25% excess air can be expressed as

$$CH_4 + 1.25 \times 2(O_2 + 3.76\,N_2) \rightarrow CO_2 + 2\,H_2O + 0.5\,O_2 + 9.4\,N_2$$

Note that the amounts of carbon dioxide and water remain the same after the theoretical air concentration is achieved.

Propane is being combusted at 80% theoretical air, that is, 20% air deficiency. What is the combustion efficiency of this reaction?

$$C_3H_8 + 5 \cdot (0.80) \cdot (O_2 + 3.76N_2) \rightarrow 2CO + CO_2 + 4H_2O + 15.04N_2$$

Calculate the volume:

Volume of product species is proportional to the coefficient of each species,

$$\text{CE of } C_3H_8 \text{ gas} = \frac{\text{coefficient of } CO_2}{\text{coefficient of } CO_2 + \text{coefficient of } CO} = \frac{1}{1+2} = 0.33 \text{ or } \underline{\mathbf{33\%\ by\ volume}}.$$

Calculate the mass:

From the stoichiometry, there is one mole of CO_2 and 2 mol of CO in this reaction, thus,

$$CE = \frac{\text{mass } CO_2}{\text{mass } CO_2 + \text{mass } CO} = \frac{12+32}{(12+32)+2(12+16)} = 0.44 \text{ or } \underline{\mathbf{44\%\ by\ mass}}$$

9.3 VOLATILITY OF COMBUSTION PRODUCTS

These combustion principles must be applied to air pollution control technologies. For example, one of the cross media concerns that tie solid waste to air quality is the thermal treatment of waste. This is commonly done not only with excess O_2, that is, incineration, but also when O_2 is absent or at very low concentrations, that is, pyrolysis. Depending on the fuel, that is, waste stock, numerous volatile organic compounds (VOCs) and semivolatile organic compounds (SVOCs), notably polycyclic aromatic hydrocarbons (PAHs), dioxins and furan, are generated under high temperatures in waste combustors and in various industrial reactors. As evidence, Figs. 9.2 and 9.3 indicate that the highly varying optimal temperature ranges to form VOCs and SVOCS. The aliphatic (carbon-chain) compounds in this fire, notably 1-dodecene, 9-nonadecane, and 1-hexacosene, are generated in higher concentrations at lower temperatures (about 800°C), whereas the aromatic (carbon-ring) compounds require higher temperatures. The total mass and the total concentrations of these thermally generated compounds continue to increase with temperature, but the chemical speciation changes [3, 4]. At some point, the temperature, pressure, sorbent, and oxygen conditions will supplant these compounds to form others that may be even more toxic, for example, the halogenated dioxins and furans. The engineer and plant operator seek the optimal combination of factors in the thermal process to completely convert as much of the organic mass to nontoxic products [5].

9.4 FLUE GAS

Combustion is used primarily for heat by changing the potential chemical energy of the fuel to thermal energy. This occurs in a fossil fuel-fired power plant, a home furnace, or an automobile engine. Combustion is also used as a means of destruction for our unwanted materials. This decreases the volume of a solid waste by burning the combustibles in an incinerator. The combustion reactor, for example, kiln, accepts combustible solids, liquids, and gases, with undesirable properties such as odors, to a high temperature in an afterburner system to convert them to less objectionable gases. Note that incineration is often used to denote any thermal destruction of waste, but the process covers a large O_2 range. As O_2 concentrations approach zero, pyrolysis dominates, but combustion occurs at higher available O_2 concentrations.

As a general rule, a fuel is assumed to produce dry, stoichiometric gas emissions at a rate of $0.25\ N\ m^3\ M^{-1}\ J^{-1}$, deducting for the water vapor that is formed during combustion [6]. The caloric value (H_{inf}) of the fuel is expressed in energy per mass (e.g., megajoules per kg, MJ kg^{-1}). Since air contains 21% O_2, the dry flue gas at y% O_2 is expressed as

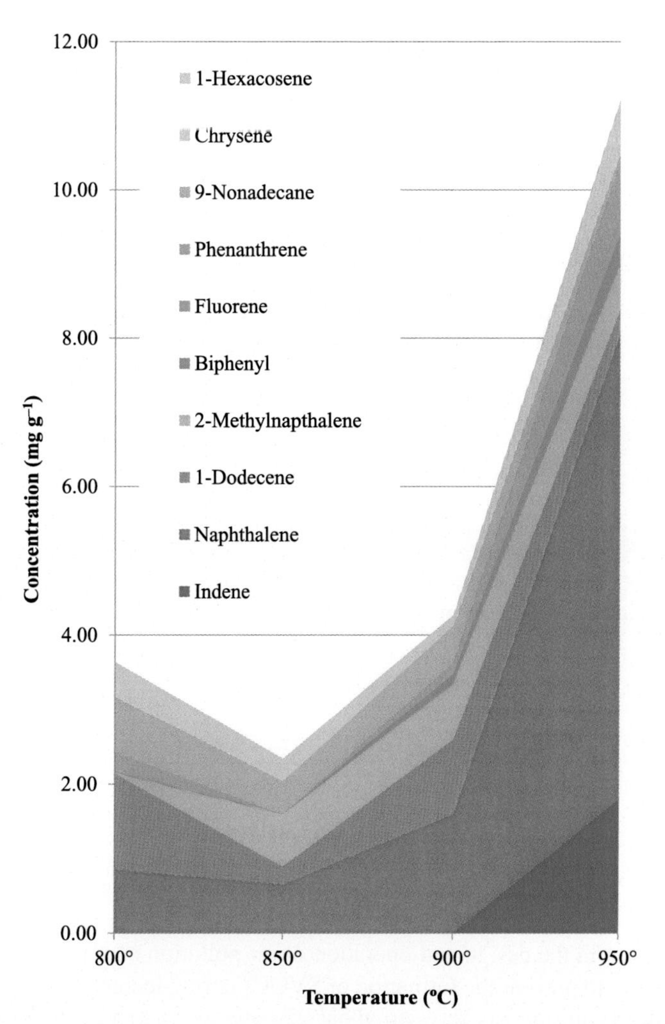

FIG. 9.2 Selected hydrocarbon compounds generated in a low-density polyethylene pyrolysis; low-oxygen conditions, in four temperature regions. *(Source: D.A. Vallero, Fundamentals of Air Pollution, 5th ed., Elsevier Academic Press, Waltham, MA, 2014, p. 999 pages cm; Data from R. Hawley-Fedder, M. Parsons, F. Karasek, Products obtained during combustion of polymers under simulated incinerator conditions: I. Polyethylene, J. Chromatogr. A 314 (1984) 263–273).*

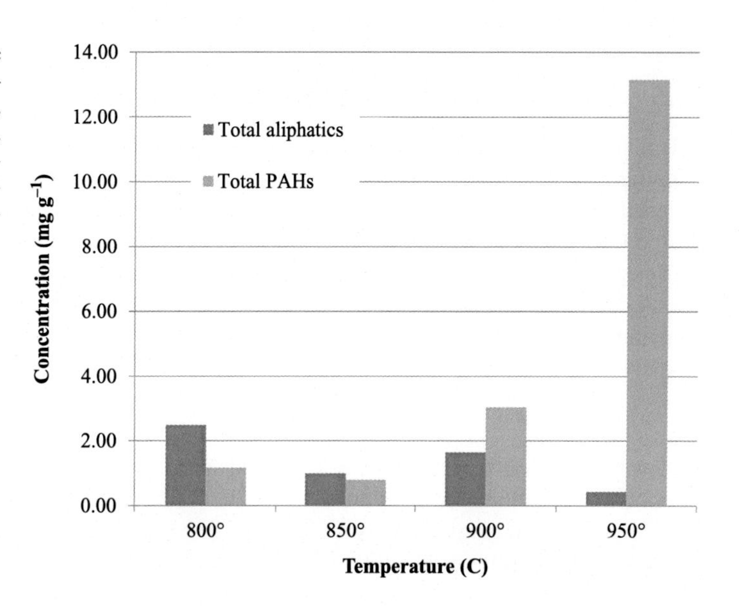

FIG. 9.3 Total aliphatic (chain) hydrocarbons versus polycyclic aromatic hydrocarbons (PAHs) generated in a low-density polyethylene pyrolysis in four temperature regions. *Source: D.A. Vallero, Fundamentals of Air Pollution, 5th ed., Elsevier Academic Press, Waltham, MA, 2014, p. 999 pages cm; Data from R. Hawley-Fedder, M. Parsons, F. Karasek, Products obtained during combustion of polymers under simulated incinerator conditions: I. Polyethylene, J. Chromatogr. A 314 (1984) 263–273.*

$$\text{Dry flue gas in Nm}^3\,\text{kg}^{-1} = \text{H}_{\text{inf}} \times 0.25 \times \frac{21}{21-y} \tag{9.3}$$

The wet flue gas at z% H_2O is compared with saturation and is expressed as

$$\text{Wet flue gas in Nm}^3\,\text{kg}^{-1} = \text{H}_{\text{inf}} \times 5.25 \times \frac{21}{21-y} \times \frac{100}{100-z} \tag{9.4}$$

Actual flue gas volume (V_{AFG}) at T (°C) is expressed as

$$\text{V}_{\text{AFG}}\text{ in m}^3\,\text{kg}^{-1} = \text{H}_{\text{inf}} \times 525 \times \frac{21-y}{100-z} \times \frac{273+\text{T}}{273} \tag{9.5}$$

which is simplified to

$$\text{V}_{\text{AFG}} = 2\text{H}_{\text{inf}} \times \frac{273+\text{T}}{(21-y)\times(100-z)}$$

A waste with an average caloric value of 5 MJ kg^{-1} is burned at 100°C in a reactor with 14% oxygen content and 15% water vapor content. What is the actual flue gas volume for this system?

$$\text{V}_{\text{AFG}} = 2\text{H}_{\text{inf}} \times \frac{273+\text{T}}{(21-y)\times(100-z)}$$

$$\text{V}_{\text{AFG}} = 2\times 5\frac{273+100}{(21-14)\times(100-15)}$$

$$\text{V}_{\text{AFG}} = 6.27\,\text{m}^3\,\text{kg}^{-1}$$

Flue gas volume is important in the design and operation of air pollution control equipment. For example, it is often preferable to lower temperatures to prevent the formation of SVOCs during incineration; for example, chlorinated dioxins and furans can form catalytically in flue gas between about 250 and 350°C, so cooling to <230°C would substantially inhibit the formation of these toxic compounds [7]. There are a number of ways to cool flue gases, including evaporative cooling (humidification, for example, using water sprays), which increases the area of heat transfer. Addition of cooler air to the flue gas and heat recovery are also options. Addition of cooler air, however, would increase V_{AFG}. Therefore, if a large volume of air is added, the combustion efficiency would be hampered.

9.5 THERMAL POLLUTANT DESTRUCTION

Plant managers, chemists, engineers, and regulators want to know the efficiency of combustion. This is important for determining the extent to which a process is functioning as designed. For example, is the fuel being burned as efficiently as needed to boil water to turn an electric turbine to produce a sufficient amount of energy? If the combustion efficiency is too low, the steam generation is insufficient to move the turbine, and fewer calories of heat are produced. Combustion efficiency is also crucial to air pollution, since incomplete combustion generates toxic substances.

A common calculation of combustion efficiency (*CE*) is the ratio of the actual combustion relative to complete combustion of a fuel by comparing the CO concentration with combined CO and CO_2 concentrations:

$$CE = \left[\frac{C_{\text{CO2}}}{C_{\text{CO2}}+C_{\text{CO}}}\right] \times 100 \tag{9.6}$$

where C_{CO2} = concentration of carbon dioxide and C_{CO} = concentration of carbon monoxide. Note that the concentrations can be in any volume or mass units, since the *CE* is a unitless value between 0 and 100%.

After a certain temperature is reached, *CE* increases with increasing chamber temperature. Other factors that affect *CE* include mixing and available O_2.

The efficiency of destroying toxic organic compounds by combustion is similar to *CE*, but rather than being a function of CO and CO_2, waste incinerator performance is based on the destruction and removal efficiency (DRE), that is, the percentage of the number of molecules of an organic compound emitted to the atmosphere compared with the number of molecules entering the incinerator [8]. Commonly, for hazardous wastes, a minimum 99.9999% DRE is required. This means that one molecule of an organic compound is released to the air for every million molecules of the organic compound entering the incinerator. The *DRE* calculation is

$$DRE = \left[\frac{W_{in} - W_{out}}{W_{in}}\right] \times 100 \tag{9.7}$$

where W_{in} is the mass feed rate of organic compound in the waste stream feeding the incinerator and W_{out} is the mass emission rate of the organic compound present in exhaust emissions prior to release to the atmosphere.

A regulatory agency requires that the combustion of 20 kg polychlorinated biphenyl (PCB) be incinerated in a hazardous waste incinerator at 99.9% efficiency and that the DRE be 99.9999%. The burn took 1 h. The exhaust rate was 1.5 L sec^{-1}. The emission rate was measured to be 950 μg m^{-3} PCBs, 5 ppm CO, and 1% CO_2. Does the incinerator meet the regulated standards?

Since 1% = 10,000 CO_2 is 1%, combustion efficiency is as follows:

$$CE = \left[\frac{10000}{10000+5}\right] \times 100 = 99.95\%$$

The total volume of exhaust for the hour is 1.5 L sec^{-1} × 3600 s $hr.^{-1}$ = 5400 L = 5.4 m^3.
Thus, at 950 μg m^{-3}, the total mass of PCBs exiting the stack is 950 μg m^{-3} × 5.4 m^3 = 5130 μg PCB.
Since 20 kg of PCB was fed into the incinerator, the *DRE* is

$$\left[\frac{20\,\text{kg}\left(10^9\,\mu\text{g}\,\text{kg}^{-1}\right) - 5130\,\mu\text{g}}{20\,\text{kg}\left(10^9\,\mu\text{g}\,\text{kg}^{-1}\right)}\right] \times 100 = 99.99997\%$$

Therefore, this incinerator appears to be meeting both *CE* and *DRE* standards.

Indeed, the number of nines in the *DRE* is an indicator of the toxicity of a compound. For example, a less toxic compound may only require four nines, that is, ≥99.99%, whereas a highly toxic compound like a PCB or dioxin would require more nines, for example, the six nines *DRE* (≥99.9999%).

Combustion efficiency is defined in other ways, such as a measure of energy conservation from the fuel into useful energy, for example, steam generation [9]. This may be found by subtracting the heat content of the exhaust gases, expressed as a percentage of the fuel's heating value, from the total fuel-heat potential, that is, 100%:

$$CE = 100\% - \left[\frac{\text{stack heat losses}}{\text{fuel heating value}} \times 100\right]$$

Stack *heat losses* are calculated from measurements of gas concentration and temperature measurements, along with the fuel's unique composition and heat content. The heat losses are primarily from the heated dry exhaust gases (CO_2, N_2, and O_2) and from water vapor formed from the reaction of hydrogen in the fuel with O_2 in the air. When water goes through a phase change from liquid to vapor, it absorbs a tremendous amount of heat energy in the process. This heat of vaporization, or latent heat, is usually lost and not recovered. This is evident from the white cloud that exits a stack on a cool day, that is, predominantly from condensing water vapor losing its latent heat to the atmosphere [9]. Table 9.2 gives combustion efficiency for fuel oil under varying conditions of temperature, CO_2, and O_2.

9.6 ACTIVATION ENERGY

Air pollutants can react with one another when they are forming (e.g., in a reactor), in the atmosphere or in any volume, for example, within a canister awaiting chemical analysis. Activation energy is needed to amount of energy needed for the reaction to take place, that is, the energy needed to bring all molecules in one mole of a substance to their reactive state at a given temperature. An exothermic reaction releases energy, whereas an endothermic reaction requires energy. If a catalyst is present, the reaction rate increases, that is, the catalyst decreases the demand for activation energy (see Fig. 9.4).

TABLE 9.2 Combustion efficiency of a fuel oil derived from net temperature (°C), that is, the difference of stack temperature (T_{stack}) and supply temperature (T_{supply}), relative to carbon dioxide and oxygen percent and temperature

		Net temperature ($T_{stack} - T_{supply}$)												
%CO_2	**%O_2**	**149**	**160**	**171**	**182**	**193**	**204**	**216**	**227**	**238**	**249**	**260**	**288**	**316**
15.6	0.0	31	31	31	30	30	30	30	30	29	29	29	28	28
14.1	2.0	31	31	30	30	30	30	29	29	29	29	28	28	27
13.4	3.0	31	30	30	30	30	29	29	29	29	28	28	28	27
12.6	4.0	31	30	30	30	29	29	29	29	28	28	28	27	27
11.9	5.0	30	30	30	29	29	29	29	28	28	28	28	27	26
11.1	6.0	30	30	29	29	29	29	28	28	28	27	27	26	26
10.4	7.0	30	29	29	29	29	28	28	28	27	27	27	26	25
9.6	8.0	29	29	29	28	28	28	27	27	27	26	26	25	24

Source of data: I. TSI, Combustion Analysis Basics: An Overview of Measurements, Methods and Calculations Used in Combustion Analysis. (2004). Available: http://www.tsi.com/uploadedFiles/_Site_Root/Products/Literature/Handbooks/CA-basic-2980175.pdf (Accessed 23 July 2018).

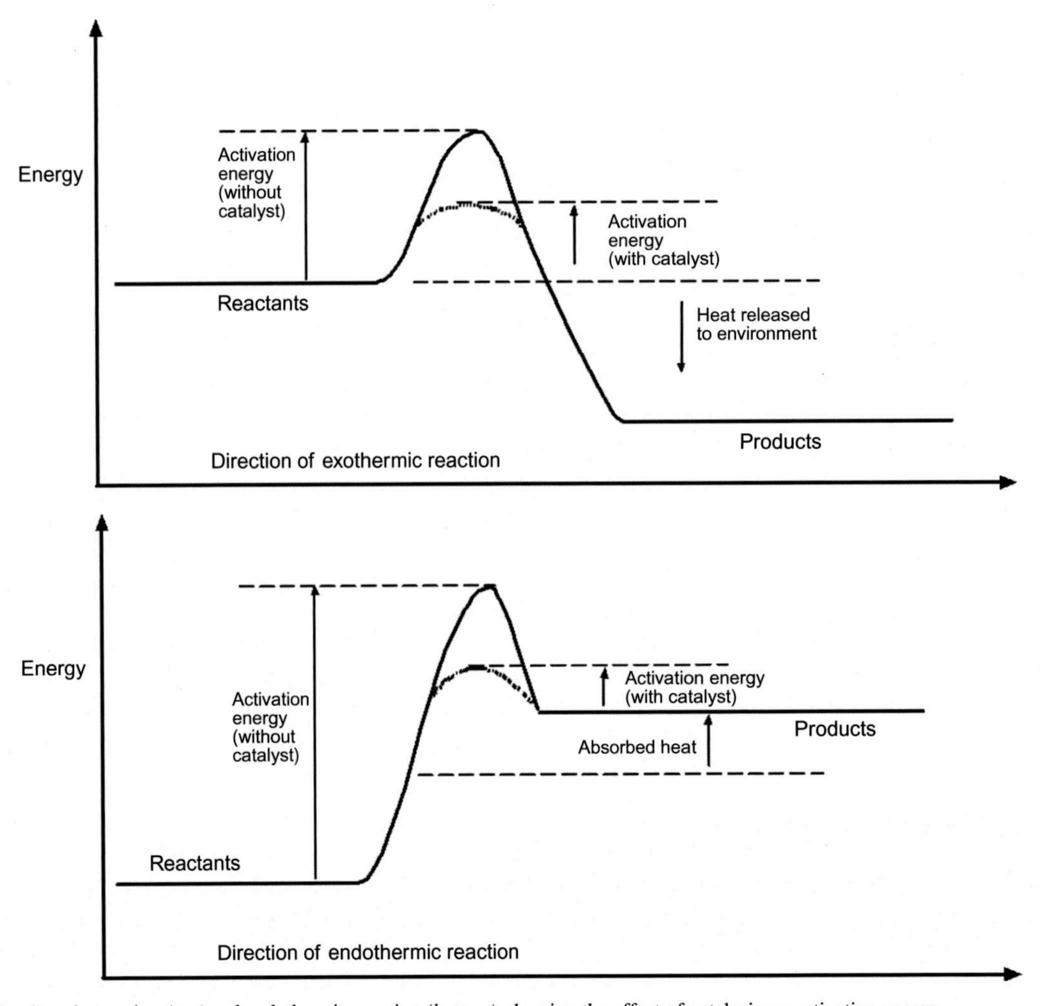

FIG. 9.4 Exothermic reaction *(top)* and endothermic reaction *(bottom)*, showing the effect of catalysis on activation energy.

This energy requirement is a reminder that reaction rates are temperature-dependent. Indeed, the rate constant k reflects this in the Arrhenius equation:

$$k = A \cdot e^{\frac{E_a}{RT}} \tag{9.8}$$

where A is the preexponential factor (constant), E_a is the Arrhenius activation energy, R is the universal gas constant, and T is the absolute temperature.

Nitrogen dioxide and carbon monoxide are reacting in a chamber at 70. The measured rate constant for this reaction is at 701°K is 2.57 $M^{-1}s^{-1}$ and at 895°K is 567 $M^{-1}s^{-1}$. What is the activation energy?

Convert the Arrhenius equation to natural log form:

$$\ln k = \frac{E_a}{RT} + \ln A \tag{9.9}$$

Set up a "two-point" form of the Arrhenius equation:

$$\ln k_1 = \frac{E_a}{RT_1} + \ln A \tag{9.10}$$

$$\ln k_2 = \frac{E_a}{RT_2} + \ln A \tag{9.11}$$

The difference of Eq. (8.18) from Eq. (8.17):

$$\ln \frac{k_1}{k_2} = \frac{E_a}{R}\left(\frac{1}{T_2} - \frac{1}{T_1}\right) \tag{9.12}$$

$$\ln \frac{567\,M^{-1}s^{-1}}{2.57\,M^{-1}s^{-1}} = \frac{E_a}{R}\left(\frac{1}{701\,K} - \frac{1}{895\,K}\right)$$

$$5.40 = \frac{E_a}{R}\left(\frac{3.09\times10^{-1}}{K}\right)$$

$$E_a = 5.40\left(\frac{K}{3.09\times10^{-1}}\right)R$$

$$E_a = 5.4 = \left(\frac{K}{3.09\times10^{-1}}\right)8.314\frac{J}{mol\cdot K} = 1.45\times10^5\,J\,mol^{-1} = 145\,kJ\,mol^{-1}$$

We could have also solved for E_a at the outset:

$$E_a = \frac{RT_1T_2}{(T_1 - T_2)}\ln\frac{k_1}{k_2} \tag{9.13}$$

9.7 NITROGEN AND SULFUR

Nitrogen species important to air quality include both oxidized and reduced forms. Atmospheric N reactions can be more complicated than may be inferred from Reactions (8.11) and (8.12). For example, the NO_x reaction from NO to NO_2 likely involves water and an intermediate, that is, nitrous acid (HONO):

$$4NO + O_2 + 2H_2O \rightarrow 4HONO \tag{9.14}$$

Combustion processes, both stationary (e.g., power plants) and mobile (e.g., automobiles), entail reactions that oxidize both atmospheric N_2 and nitrogen in the fuel. As discussed in Chapter 8, the term "oxides of nitrogen" (NO_x) does not include all oxidized species, just NO and NO_2. However, the larger suite of N air pollution compounds, that is, total reactive nitrogen (NO_y), includes not only NO and NO_2 but also their oxidation products [10]. Combustion generates other vapor-phase and aerosol species in NO_y including HONO, nitric acid (HNO_3), peroxyacetyl nitrate (PAN), organic nitrates, and particulate nitrates:

$$NO_y = NO_2 + NO + HNO_3 + PAN + 2N_2O_5 + HONO + NO_3 + NO_3^-\ \text{compounds} + NO_3^-\ \text{aerosols} \tag{9.15}$$

The NO_y compounds are products of reactions with NO_x emissions or from transformations in the troposphere [11]. Most of the compounds originate from combustion processes. The N source is either molecular nitrogen in the troposphere, that is, in the air intake, or the N-content of the fuel. The tropospheric N_2 makes up the largest share of gaseous content in the combustion air (79% by volume).

When pressure and temperatures are sufficiently high, for example, in an internal combustion engine or industrial boiler, the plentiful N_2 will react with O_2. NO makes up about 90%–95% of this thermal NO_x. Other nitrogen oxides can also form at high heat and pressure, especially NO_2. Most motor vehicles around the world still employ high-temperature/high-pressure internal combustion engines. Thus, such mobile sources contribute largely to the tropospheric concentrations of NO_x, making it a major mobile source air pollutant in terms of human health directly (e.g., respiratory toxicity of NO_2) and indirectly (i.e., NO_x as the key components in tropospheric ozone production). These conditions of high temperature and pressure can also exist in boilers such as those in power plants, so NO_x is also commonly found in high concentrations leaving fossil fuel power generating stations.

In addition to the atmospheric molecular nitrogen as a precursor of nitrogen air pollutants of combustion, fossil fuels themselves contain varying concentrations of N. Nitrogen oxides that form from the fuel or feedstock are called "fuel NO_x." Unlike the sulfur compounds, which mainly exit stationary source stacks as vapor-phase compounds (e.g., SO_2 and other oxides of sulfur), a significant fraction of the fuel nitrogen burned in power plants and other stationary sources remains in the bottom ash or in unburned aerosols in the gases leaving the combustion chamber, that is, the fly ash. Nitrogen oxides can also be released from nitric acid plants and other types of industrial processes involving the generation and/or use of nitric acid (HNO_3).

At temperatures far below combustion, such as those often present in the ambient atmosphere, NO_2 can form the molecule NO_2-O_2N or simply N_2O_4 that consists of two identical simpler NO_2 molecules. This molecular configuration is known as a dimer. The dimer N_2O_4 is distinctly reddish brown and contributes to the brown haze that is often associated with photochemical smog incidents.

In addition to the health effects associated with NO_2 exposure, much of the concern for regulating emissions of nitrogen compounds is to suppress the reactions in the atmosphere that generate the highly reactive molecule ozone (O_3). Nitrogen oxides play key roles in O_3 formation. Ozone forms photochemically (i.e., the reaction is caused or accelerated by light energy) in the lowest level of the atmosphere, known as the troposphere, where people live. Nitrogen dioxide is the principal gas responsible for absorbing sunlight needed for these photochemical reactions. So, in the presence of sunlight, the NO_2 that forms from the NO incrementally stimulates the photochemical smog-forming reactions because nitrogen dioxide is very efficient at absorbing sunlight in the ultraviolet portion of its spectrum. This is why ozone episodes are more common during summer and in areas with ample sunlight. Other chemical ingredients, that is, ozone precursors, in O_3 formation include volatile organic compounds (VOCs) and carbon monoxide (CO). Governments regulate the emissions of precursor compounds to diminish the rate at which O_3 forms.

Sulfur, like nitrogen, is oxidized during combustion but is also an important component of particulate matter (PM; see Table 9.3). For example, diesel particulate matter is formed by a number of simultaneous physical processes during cooling and dilution of exhaust, that is, nucleation, coagulation, condensation, and adsorption. The core of the particles is formed by nucleation and coagulation from primary spherical particles consisting of solid carbonaceous matter (known as elemental carbon, EC) and ash (metals and other elements). By coagulation, adsorption, and condensation, various organic and S compounds (e.g., sulfates) are added and combined with other condensed material [12].

The small diameter of diesel PM ($<0.5\,\mu m$) makes for very large surface areas. <0.5 mm, these particles have a very large surface area per gram of mass, which allows them to adsorb large quantities of ash, organic compounds, and sulfate (see Fig. 9.5). The specific surface area of the EC core is approximately 30–50 $m^2\,g^{-1}$. The organic constituents originate from unburned fuel, engine lubrication oil, and small quantities of partial combustion and pyrolysis products. During combustion, S compounds in the fuel are oxidized to sulfur dioxide (SO_2). From 1% to 4% of fuel, S is oxidized to form H_2SO_4. Upon cooling, this sulfuric acid and water condense into an aerosol that is nonvolatile under ambient conditions [12].

The oxidized chemical species of sulfur and nitrogen (e.g., sulfur dioxide (SO_2) and nitrogen dioxide (NO_2)) form acids when they react with water. This can occur in any media, for example, the atmosphere (i.e., acid deposition) and in mining waste and ash runoff to surface waters and groundwater. The lowered pH is responsible for numerous environmental problems. Many compounds contain both N and S along with the typical organic elements (C, H, and O). The reaction for the combustion of such compounds, in general form, is

$$C_aH_bO_cN_dS_e + 4a + b - 2c) \rightarrow aCO_2 + \left(\frac{b}{2}\right) H_2O + \left(\frac{d}{2}\right) N_2 + eS \tag{9.16}$$

TABLE 9.3 Typical chemical composition (percent) of fine particulate matter ($PM_{2.5}$)

	Eastern United States	Western United States	Diesel $PM_{2.5}$
Elemental carbon	4	15	75
Organic carbon	21	39	19
Sulfate, nitrate, and ammonium	48	35	1
Minerals	4	15	2
Unknown	23	–	3

Source: US Environmental Protection Agency. Air quality criteria for particulate matter. External review draft, October 1999.

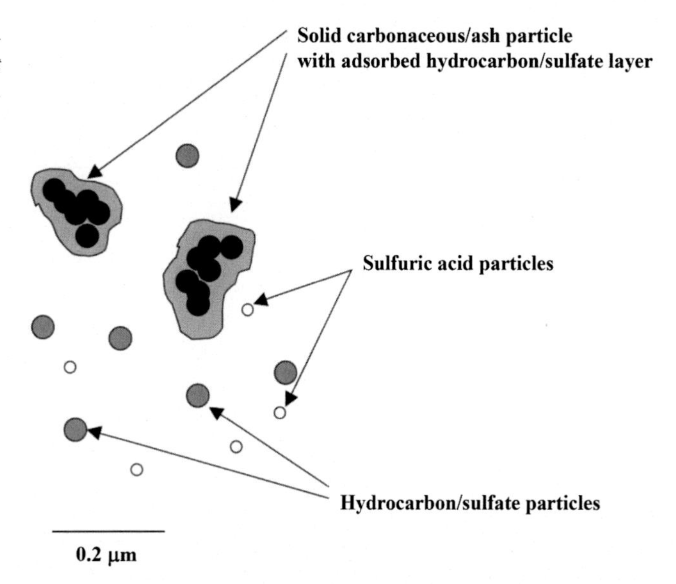

FIG. 9.5 Diesel particulate matter. *Source: US Environmental Protection Agency, Health Assessment for Diesel Engine Exhaust. Report No. EPA/600/8-90/057F. National Center for Environmental Assessment. Washington, DC, 2002.*

Reaction (9.16) demonstrates the incremental complexity as additional elements enter the reaction. In the real world, pure reactions are rare. The environment is filled with mixtures and heterogeneous reactions. Reactions can occur in sequence, parallel, or both. For example, a feedstock to a municipal incinerator contains myriad types of wastes, from garbage to household chemicals to commercial wastes, and even small and sometimes large industrial wastes that may be illegally dumped. The N content of typical cow manure is about 5 kg per metric ton (about 0.5%). If the fuel used to burn the waste also contains S along with the organic matter, then the five elements will react according to the stoichiometry of Reaction (8.15).

Redox occurs in certain regions of a combustion chamber, stack, or other locations inside the combustor or in the ambient air. Recall from Chapter 8 that the formation of sulfur dioxide (SO_2) and nitric oxide (NO) by acidifying molecular sulfur, for example, in coal, includes the redox reaction:

$$S(s) + NO_3^-(aq) \rightarrow SO_2(g) + NO(g) \tag{9.17}$$

The oxidation half-reactions for this reaction are

$$S \rightarrow SO_2 \tag{9.18}$$

$$S + 2H_2O \rightarrow SO_2 + 4H^+ + 4e^- \tag{9.19}$$

The reduction half-reactions for this reaction are

$$NO_3^- \rightarrow NO \tag{9.20}$$

$$NO_3^- + 4H^+ + 3e^- \rightarrow NO + 2H_2O \tag{9.21}$$

Therefore, the balanced oxidation-reduction reactions are

$$4NO_3^- + 3S + 16H^+ + 6H_2O \rightarrow 3SO_2 + 16H^+ + 4NO + 8H_2O \quad (9.22)$$

$$4NO_3^- + 3S + 4H^+ \rightarrow 3SO_2 + 4NO + 2H_2 \quad (9.23)$$

This indicates the importance of the sulfur content of a fuel in estimating the amount of SO_2 that would be expected to be emitted during combustion.

A company receives weekly deliveries of coal and takes a sample of each week, with the analyses below. Calculate the weighted 28-day average that can be used to predict sulfur dioxide emissions.

Week	Percent sulfur	$kJ\,kg^{-1}$	kg SO_2 per gigajoule ($kg\,GJ^{-1}$)	Coal burned (tonnes)
1	2.9	26,229	2.3	23
2	2.7	26,693	2.2	34
3	3.2	25,068	2.7	47
4	2.4	25,997	2.0	27

The emission rates will be weighted according to the amount of fuel burned each week.

Calculate the heat input during the period represented by each sample for weeks 1 through 4:

Week 1: $23\,t \times 26{,}229\,kJ\,kg^{-1} \times 1000\,kg\,tonne^{-1} = 6.03 \times 10^8\,kJ$
Week 2: $34\,t \times 26{,}693\,kJ\,kg^{-1} \times 1000\,kg\,tonne^{-1} = 9.08 \times 10^8\,kJ$
Week 3: $47\,t \times 25{,}068\,kJ\,kg^{-1} \times 1000\,kg\,tonne^{-1} = 1.18 \times 10^9\,kJ$
Week 4: $27\,t \times 25{,}997\,kJ\,kg^{-1} \times 1000\,kg\,tonne^{-1} = 7.01 \times 10^8\,kJ$

Add the product of the heat input times the emission rate for each sample and divide by the total heat input for the 30-day period. Weighted 30-day average SO_2 emission rate:

$$\left[\left(6.03 \times 10^8\,kJ\,hr^{-1}\right)\left(2.3\,kg\,GJ^{-1}\right) + \left(9.08 \times 10^8\,kJ\,hr^{-1}\right)\left(2.2\,kg\,GJ^{-1}\right) + \left(1.18 \times 10^9\,kJ\,hr^{-1}\right)\left(2.7\,kg\,GJ^{-1}\right)\right.$$
$$\left. + \left(7.01 \times 10^8\,kJ\,hr^{-1}\right)\left(2.0\,kg\,GJ^{-1}\right)\right]$$
$$\div \left(6.03 \times 10^8 + 9.08 \times 10^8 + 1.18 \times 10^9 + 7.01 \times 10^8\right)\,kJ \approx 2.4\,\textit{kg SO}_2\textit{ per gigajoule}$$

Thus, this plant would be expected to have emitted, on average, a 2.4 kg of sulfur dioxide over this 4 – week period, based on their fuel supply.

These types of calculations are needed to ensure compliance with air quality standards. In the United States, most states conduct these rolling average calculations in units of BTU per pound, pounds of SO_2 per million BTUs (mmBtu), and tons.

The air quality regulatory agency requests a 30-day rolling weighted average in addition to the previously reported 28-day weighted average. The company has purchased lower-quality coal the most recent 2 weeks. Calculate the rolling weighted average SO_2 emission rates.

Week	Percent sulfur	$kJ\,kg^{-1}$	kg SO_2 per gigajoule ($kg\,GJ^{-1}$)	Coal burned (tonnes)
1	2.9	26,229	2.3	23
2	2.7	26,693	2.2	34
3	3.2	25,068	2.7	47
4	2.4	25,997	2.0	27
5	3.5	24,590	2.9	48
6	3.3	25,875	2.8	52

There are four samples for the latest 30 days, so these will be used in the rolling average. The emission rates will be weighted according to the amount of fuel burned each week.

Week 5: $49\,t \times 24{,}590\,kJ\,kg^{-1} \times 1000\,kg\,tonne^{-1} = 1.20 \times 10^9\,kJ$

Week 6: $52\,t \times 25{,}875\,kJ\,kg^{-1} \times 1000\,kg\,tonne^{-1} = 1.35 \times 10^9\,kJ$

The rolling, weighted 30-day average SO_2 emission rate for weeks 2–5 is

$$\left[\left(9.08 \times 10^8\,kJ\,hr^{-1}\right)\left(2.2\,kg\,GJ^{-1}\right) + \left(1.18 \times 10^9\,kJ\,hr^{-1}\right)\left(2.7\,kg\,GJ^{-1}\right) + \left(7.01 \times 10^8\,kJ\,hr^{-1}\right)\left(2.0\,kg\,GJ^{-1}\right) + \left(1.20 \times 10^9\,kJ\,hr^{-1}\right)\left(2.9\,kg\,GJ^{-1}\right)\right]$$

$$\div \left(9.08 \times 10^8 + 1.18 \times 10^9 + 7.01 \times 108 + 1.20 \times 10^9\right)\,kJ \approx 2.5\,\textit{kg } SO_2 \textit{ per gigajoule}$$

The rolling, weighted 30-day average SO_2 emission rate for weeks 3–6 is

$$\left[\left(1.18 \times 10^9\,kJ\,hr^{-1}\right)\left(2.7\,kg\,GJ^{-1}\right) + \left(7.01 \times 10^8\,kJ\,hr^{-1}\right)\left(2.0\,kg\,GJ^{-1}\right) + \left(1.20 \times 10^9\,kJ\,hr^{-1}\right)\left(2.9\,kg\,GJ^{-1}\right) + \left(1.35 \times 10^9\,kJ\,hr^{-1}\right)\left(2.8\,kg\,GJ^{-1}\right)\right]$$

$$\div \left(9.08 \times 10^8 + 1.18 \times 10^9 + 7.01 \times 108 + 1.35 \times 10^9\right)\,kJ \approx 2.6\,\textit{kg } SO_2 \textit{ per gigajoule}$$

This is problematic in that the SO_2 emission rates are increasing, which means that the rolling, weighted average emission rates increase by 0.1 kg GJ^{-1} each week on record.

The emitted nitrogen and sulfur compound can cause direct and indirect damage to health and ecosystems, for example, acid deposition, and to materials and structures (see Chapter 8).

9.8 METALS IN THERMAL REACTIONS

Metals like mercury (Hg) and lead (Pb) are emitted into the atmosphere from combustion and other thermal reactions. The metals are usually trace constituents of the fuel but may also be intentionally added as catalysts and for other purposes during the thermal processes, for example, organometallic compounds in gasoline or diesel fuel. Metals may also scavenge from the walls or components of the chamber or reactor. The reactions can be a single step, for example, the direct formation of a metal oxide from a reaction with the O_2 in the air, or lead to various metal compounds through a series or branching of reactions depending on the conditions of the flue gas. The flue gas can have widely varying regions of flow rate (e.g., change in contact time), temperature, O_2 concentration, particulate matter (PM) concentration, availability of sorbents, redox, and reactants (e.g., oxide gases and acids). These and other conditions can lead to different reactions, even if the fuel does not change.

REFERENCES

[1] Engineering Toobox, Stoichiometric Combustion and Excess of Air, July 16, 2018. Available: https://www.engineeringtoolbox.com/stoichiometric-combustion-d_399.html, 2003.

[2] I.V. Ion, Energy Saving Technologies, Universitatea "Dunărea de Jos" din Galaţi, Romania, 2013, Available from: https://scholar.google.com/scholar?cluster=4618665913383406658&hl=en&as_sdt=1,34 (Accessed on 31 January 2019).

[3] D.A. Vallero, Fundamentals of Air Pollution, fifth ed., Elsevier Academic Press, Waltham, MA, 2014. p. 999 pages cm.

[4] R. Hawley-Fedder, M. Parsons, F. Karasek, Products obtained during combustion of polymers under simulated incinerator conditions: I. Polyethylene, J. Chromatogr. A 314 (1984) 263–273.

[5] EPA/540/R-92/074 B, Guide for Conducting Treatability Studies under CERCLA: Thermal Desorption, 1992.

[6] W. Bank, Municipal Solid Waste Incineration: World Bank Technical Guidance Report, 1999.

[7] EPA-45013-89-27e Municipal Waste Combustors: Background Information for Proposed Guidelines for Existing Facilities, 1989.

[8] M.D. Erickson, P.G. Gorman, D.T. Heggem, Relationship of destruction parameters to the destruction/removal efficiency of PCBs, J. Air Pollut. Control Assoc. 35 (6) (1985) 663–665.

[9] I. TSI, Combustion Analysis Basics: An Overview of Measurements, Methods and Calculations Used in Combustion Analysis, Available: http://www.tsi.com/uploadedFiles/_Site_Root/Products/Literature/Handbooks/CA-basic-2980175.pdf, 2004. Accessed 23 July 2018.

[10] B. Finalyson-Pitts, J. Pitts, Chemistry of the Upper and Lower Atmosphere: Theory, Experiments, and Application, Academic Press, San Diego, CA, 1999.

[11] EPA-452/R-11-005a, Policy Assessment for the Review of the Secondary National Ambient Air Quality Standards for Oxides of Nitrogen and Oxides of Sulfur, 2011.

[12] US Environmental Protection Agency, Health Assessment Document for Diesel Engine Exhaust, National Centre for Environmental Assessment, Washington, DC, 2002.

Chapter 10

Air pollution phases and flows

10.1 INTRODUCTION

Air pollutants may be classified in many ways. Often, regulatory agencies distinguish gas-phase and aerosol pollutants according to their sources, that is, they are emitted from two primary source categories, mobile and stationary. The source of gas-phase emissions from a mobile source, for example, an automobile, is mostly from the exhaust of the internal combustion engine, with some from noncombustion ancillary systems, for example, fuel tank and injection systems [1]. Combustion is also the major source of aerosol emissions from mobile sources, but particulate matter (PM) can also result from reentrained road dust, tire wear remnants, and PM generated by other mechanical processes [2–4].

Stationary sources, for example, electric-generating power plants, also produce most of their gas-phase and PM emissions from combustion. However, depending on the industrial or commercial operations, gas-phase compounds may also be released from more diffuse operations like chemically treated lawns, fields, and forests and from waste and storage sites like pits, ponds, lagoons, and piles. Aerosols can also be generated by these noncombustion sources [5], for example, blowing off coal, gob, and ash piles, and bioaerosols, for example, pollen, from fields and forests. Air pollutants may also be categorized according to not only the harm they may cause, usually on human health and ecology, but also damage to materials and structures.

The primary distinction of types of air pollution scientists and engineers, however, lies in the pollutants' inherent properties, beginning with their physical attributes. These are often the first clues as to how a substance will move and change in the atmosphere and the damage it may ultimately cause. As mentioned, air pollutants are first separated into two-"phase" categories: aerosols and gas phase. Phase is in quotations because air pollution scientists and engineers differ a bit from physicists, who differentiate matter into three basic phases based on thermodynamics, that is, gases, liquids, and solids. As indicated in Fig. 10.1, any compound can take the form of these three phases, depending on the temperature and pressure.

Classifying air pollutant by only two phases is logical within the context of equilibrium. As discussed in Chapter 4, matter exists in equilibrium between phases. The equilibrium is controlled by environmental and atmospheric conditions, especially pressure and temperature. In the laboratory, these conditions are usually kept at or adjusted to standard temperature and pressure (STP), that is, 0°C and 1 bar (0.986923 atm).

Under most tropospheric conditions, the equilibrium of most importance is between the gas phase and another phase. That is, the substance will become airborne by either volatilization or aerosolization. Volatilization depends on partial pressures to reach the atmosphere, whereas aerosolization depends on mechanical and kinetic processes, for example, condensation, advection, and settling. Another way to look at this is that a gas is in solution in the air and an aerosol is a liquid, solid, or both that is in suspension in the air. Indeed, the latter is the textbook definition of an aerosol [6–8]. In this sense, aerosols find their way to the atmosphere very differently than do gases.

Outside of the confines of the laboratory, STP is not usually the best way to adjust, for example, when taking measurements of emissions or ambient concentrations of airborne substances. In the lower troposphere, a compound's phases fall within a relatively tight range of temperatures and atmospheric pressures. Thus, even slight changes in atmospheric conditions can substantially change the thermodynamics, kinetics, and equilibria discussed in previous chapters.

Preferably, test results must be adjusted to conditions to be comparable with other measurements. This is often near what is known as normal temperature and pressure (NTP), which is 20°C and 1 atm. Room temperature and pressure (RTP), that is, 20°C and 1 atm, conditions are also sometimes used. For atmospheric measurement and modeling purposes, the US Environmental Protection Agency (EPA) defines a "standard condition" (SC) to be 25°C and 1 atm [9]. Ambient air and emission measurements are often adjusted to SC, so measurements are often required to be adjusted to these conditions for regulatory and permitting purposes. As such, SC-adjusted values appear in governmental reports and in measurement methods published in handbooks, manuals, and textbooks [10].

Measurements are part of the system of determining whether the air is fit to breathe and, if not, identifying the pollutants and their concentrations. Gas-phase pollutants include carbon monoxide (CO); oxides of nitrogen (NO_x); oxides of sulfur,

Air Pollution Calculations. https://doi.org/10.1016/B978-0-12-814934-8.00010-7

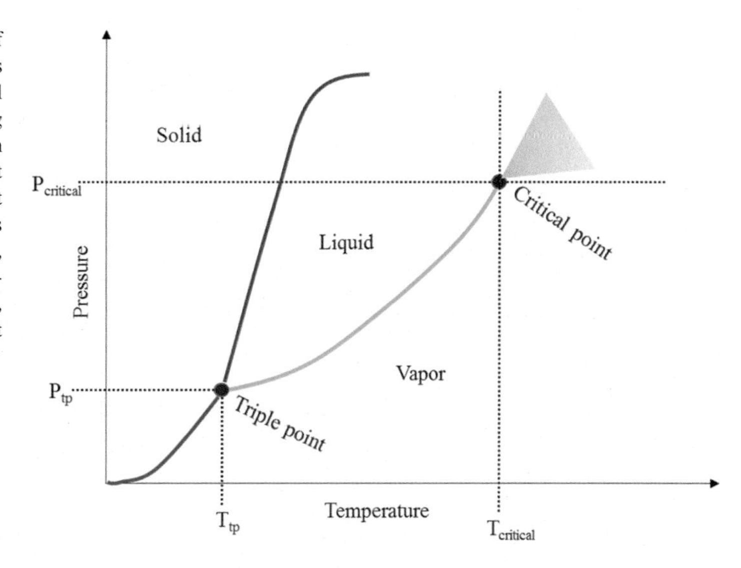

FIG. 10.1 Phase diagram for a hypothetical substance. All of the phase regions shown in the diagram will exist, but their shapes and slopes of the polygon sides and points of phase changes will differ. The vapor-liquid boundary line temperature is the boiling point, and the pressure is the vapor pressure. The line between the vapor and solid phases is known as the sublimation point (i.e., a phase change from solid to gas or gas to solid, without first becoming a liquid). The liquid-solid boundary line temperature is freezing and melting point. T_{tp} and P_{tp} define the triple point, that is, where the three phases of pure substance can exist in equilibrium. $T_{critical}$ and $P_{critical}$ are the pressure and temperature, respectively, of a fluid at its critical point, that is, the point at which a gas cannot be liquefied by an increase of pressure.

especially sulfur dioxide (SO_2); and ozone (O_3). Many of the hazardous air pollutants (HAPs) are organic vapors, for example, the volatile organic compounds (VOCs) like benzene, formaldehyde, and vinyl chloride, which have sufficiently high vapor pressures so that at NTP, RTP and SC will send many molecules into the atmosphere compared with less volatile compounds, for example, semivolatile compounds (SVOCs) like the polychlorinated biphenyls (PCBs), which send fewer compounds to the atmosphere as vapors and often many more via aerosolization, for example, sorbed to particulate matter. However, even these few molecules, for example, a few parts per billion (ppb), can be hazardous given the toxicity of many SVOCs. The HAPs also include inorganic vapors, for example, mercury (Hg) and its compounds. Metals and their compounds can be gas phase and aerosol phase. Lead (Pb), for example, can reach the atmosphere as gaseous oxides, as aerosols, and as organometallic vapors.

The phases of these compounds must be known to estimate the concentrations in the air and to predict what concentrations may be given various emissions. For example, the EPA uses modeling to estimate air pollution emissions from mobile sources. The modeled results are used to determine the mobile source contribution to concentrations of pollutants in an area, that is, a nonattainment or attainment area. Ambient temperature affects the emissions from an automobile's exhaust, that is, the emissions of CO, NO_x, and VOCs during normal operations will be higher in colder temperatures than in an "allowable" test range, that is, 68–75°F. This is because when an engine starts in cold weather, the engine and emission control equipment require warm-up time to operate within performance specifications. With colder temperatures, CO and VOC and to a lesser extent NO_x emissions increase because the fuel, for example, gasoline, condenses on the colder surfaces and because catalytic converters are less efficient [11]. Thus, any exhaust test measurements taken at higher temperatures, that is, within the allowable range, would need to be adjusted. That is, the concentrations during cold starts would need to be increased for gas-phase emissions during colder months to predict ambient concentrations and to determine whether an area will meet the standards. Thus, air pollution control actions like effective vehicle emission testing must account for the phases of pollutants under various temperatures and atmospheric conditions.

10.2 GAS-PHASE POLLUTANTS

In air pollution literature, gases and vapors are grouped together into "gas-phase" substances. Gases and vapors behave similarly in the atmosphere and require similar types of treatment technologies, for example, absorption, that differ from those for aerosols, for example, filtration. After being emitted, the gases and vapors will remain in the atmosphere until they are either transformed or transported. The transformation products not only may be in the gas phase but also may be liquid or solid, whereupon they become aerosols.

The concentration of a substance in the air, including an air pollutant, is reported as either mass or volume per volume. Gas-phase pollutants are most often reported in units of volume of the pollutant within a volume of air, for example, parts per million or billion (ppm and ppb, respectively). Thus, if the pollutant mass is known, it must be converted to pollutant volume.

10.2.1 The gas laws

The gas laws underpin air pollution science by explaining the interrelationships among pressure, temperature, and volume. Air is a mixture of gases, with some liquids and solids present in varying amounts. Collectively, liquids and solids are aerosols. Of the states of matter, only gases can be compressed.

Thus, temperature, pressure, and density are interdependent. Robert Boyle (1627–91) first identified the inverse relationship between volume and pressure at a given temperature:

$$V \propto P^{-1} \tag{10.1}$$

Proportionalities can be converted to equalities if a constant (R) is applied:

$$PV = R \tag{10.2}$$

Later, Jacques Charles (1746–1823) introduced the temperature-volume law, which states that a gas volume held at a constant pressure is directly proportional to temperature:

$$V \propto T \tag{10.3}$$

Next, Joseph Gay-Lussac (1778–1850) introduced that pressure-temperature law, which states that a gas held at constant volume is directly proportional to temperature:

$$P \propto T \tag{10.4}$$

Another important gas law introduced by Amedeo Avogadro (1776–1856) states that the number of moles (n) of gas increases with volume:

$$V \propto n \tag{10.5}$$

Combining these laws gives

$$V \propto \frac{T}{P} \tag{10.6}$$

Putting the laws together,

$$P = \rho RT \tag{10.7}$$

where P is pressure (atm), R is the universal gas constant, T is temperature in K, and ρ is density (kg m^{-3}, discussed in Section 10.2.5).

The units of the universal gas constant R vary but are commonly expressed as

$$R = \frac{0.08212\,\mathrm{L \cdot atm}}{\mathrm{mol \cdot K}} \text{ or } \frac{8.314\,\mathrm{joules}}{\mathrm{mol \cdot K}} \tag{10.8}$$

where K is the temperature expressed in degrees Kelvin. The value of the gas constant for a specific gas (R_{gas}) depends on its molecular weight (m_{gas}).

Some commonly used units of R are shown in Table 10.1.

Thus, the specific gas constant (R_{gas}) is distinguished from the universal gas constant (R) by the molecular weight (m) of the specific gas $\left(\frac{R}{m_{gas}}\right)$:

$$R_{gas} = \frac{R}{m_{gas}} \tag{10.9}$$

R_{gas} can be expressed in units of energy per temperature per mass, for example, J K^{-1} kg^{-1}.

What is the gas constant for dry tropospheric air?

Ignoring the minor gases, find m for 79/21% mixture of molecular nitrogen and oxygen:

$$R_{air} \approx \frac{8.314}{0.79 m_{N2} + 0.21 m_{O2}} = \frac{8.314}{28.8} \approx 289\,\mathrm{J\,K^{-1}\,kg^{-1}}.$$

TABLE 10.1 Commonly applied units of the universal gas constant (R)

Values of R	Units
8.3144598	$kg\,m^2\,s^{-2}\,K^{-1}\,mol^{-1}$
8.3144598	$J\,K^{-1}\,mol^{-1}$
8.3144598×10^{-3}	$kJ\,K^{-1}\,mol^{-1}$
8.3144598×10^{7}	$erg\,K^{-1}\,mol^{-1}$
8.3144598×10^{-3}	$amu\,(km/s)^2\,K^{-1}$
8.3144598	$m^3\,Pa\,K^{-1}\,mol^{-1}$
8.3144598×10^{6}	$cm^3\,Pa\,K^{-1}\,mol^{-1}$
8.3144598	$L\,kPa\,K^{-1}\,mol^{-1}$
8.3144598×10^{3}	$cm^3\,kPa\,K^{-1}\,mol^{-1}$
8.3144598×10^{-6}	$m^3\,MPa\,K^{-1}\,mol^{-1}$
8.3144598	$cm^3\,MPa\,K^{-1}\,mol^{-1}$
8.3144598×10^{-5}	$m^3\,bar\,K^{-1}\,mol^{-1}$
8.3144598×10^{-2}	$L\,bar\,K^{-1}\,mol^{-1}$
83.144598	$cm^3\,bar\,K^{-1}\,mol^{-1}$
62.363577	$L\,Torr\,K^{-1}\,mol^{-1}$
1.9872036×10^{-3}	$kcal\,K^{-1}\,mol^{-1}$
8.2057338×10^{-5}	$m^3\,atm\,K^{-1}\,mol^{-1}$
0.082057338	$L\,atm\,K^{-1}\,mol^{-1}$
82.057338	$cm^3\,atm\,K^{-1}\,mol^{-1}$
0.082057338	$atm\,gmol^{-1}\,K^{-1}$
62.363577	$L\,mmHg\,gmol^{-1}\,K^{-1}$
1545.3	$ft\text{-}lbf\,lbmol^{-1}\,°R^{-1}$
10,73	$psia\text{-}ft^3\,lbmol^{-1}\,°R^{-1}$

Sources of data: US National Institute of Standards and Technology, The NIST reference on constants, units, and uncertainty. (2018). August 11, 2018. Available: https://physics.nist.gov/cgi-bin/cuu/Value?r; G.D. Wight, Fundamentals of Air Sampling, CRC Press, 1994.

Calculate the gas constant water vapor:

$$R_{H2O} \approx \frac{8.314}{m_{H2O}} = \frac{8.314}{18.02} \approx 461\,J\,K^{-1}\,kg^{-1}.$$

What is the density of water vapor that water vapor exerts 900 Pa pressure at 20°C?

$$P = \rho RT$$

$$R_{H2O} \approx \frac{8.314}{m_{H2O}} = \frac{8.314}{18.02} \approx 461\,J\,K^{-1}\,kg^{-1}.$$

$$T = 273.15 + 20 = 293.15\,K$$

$$\rho = \frac{P}{RT} = \frac{900}{461 \times 293.15} = 0.0067\,kg\,m^{-3}.$$

10.2.2 Gas characteristics

Determining molecular speed from pressure calculations is a way to distinguish gases from other states of matter. Consider a gas molecule with its molar mass = m, which is confined to cubic box with perfectly elastic walls (see Fig. 10.2), that is, there is no loss of kinetic energy in the collision of a gas molecule with the wall. Assume that the molecule is moving at speed u, which can be resolved to the three dimensions of the box edges, x, y, and z, that is, u_x, u_y, and u_z. The assumption of perfect elasticity means that after the molecule hits the wall, it will rebound, but not lose speed, remaining at u, and will move in the opposite direction, normal to the wall. Thus, the change in the molecule's momentum is [12]

$$\Delta mu = (\Delta mu_x)_f - (\Delta mu_x)_i = -(mu_x) - (mu_x) = -2\,mu_x \tag{10.10}$$

where f and i subscripts represent final and initial states, respectively.

Since the total momentum in this system is conserved, the molecule will transfer $2mu$ momentum onto surface A. If after rebounding the molecule avoids all other gas molecules and reaches surface A′, it would take $l\,u^{-1}$ time to cross the cube, and the round-trip back to A would take a total of $2l\,u^{-1}$, meaning that the gas molecule's rate of momentum transferred onto the surface A is

$$\frac{\Delta mu}{\Delta t} = 2mu_x \frac{u_x}{2l} = \frac{mu_x^2}{l} \tag{10.11}$$

This rate of momentum transfer onto surface A is the force exerted on the surface according to Newton's law of momentum. The transfer rate for all gas molecules onto A can be found by summing the $\frac{mu_x^2}{l}$ values (n):

$$F_{total} = \frac{m}{l}(u_{x1} + u_{x2} + \cdots u_{xn}) \tag{10.12}$$

Since pressure is force divided by area, we can find pressure (P):

$$P = \frac{F_{total}}{A} = \frac{m}{l^3}(u_{x1} + u_{x2} + \cdots u_{xn}) = \frac{mN}{V}\left(\frac{u_{x1} + u_{x2} + \cdots u_{xn}}{N}\right) \tag{10.13}$$

where N is the total number of gas molecules in the box and V is box volume. mN is the total gas mass in the box, so $\frac{mN}{V}$ is the gas density (ρ), that is, mass per unit volume. Also, $\frac{u_{x1} + u_{x2} + \cdots u_{xn}}{N}$ is the average u_x^2 for all gas molecules in the box, that is, $\overline{u_x^2}$. Therefore, Eq. (10.3) can be rewritten as

$$P = \rho\overline{u_x^2} \tag{10.14}$$

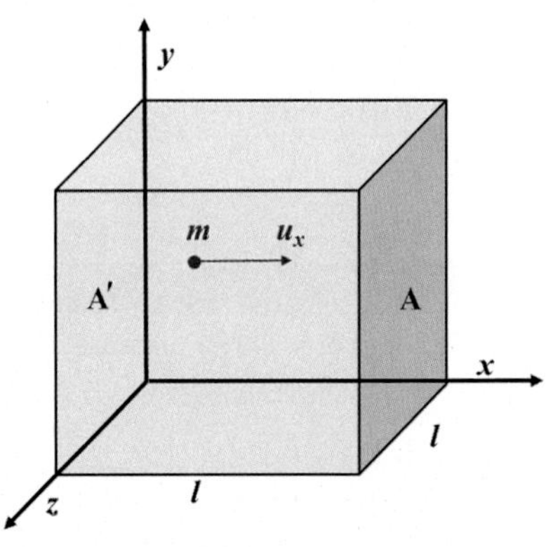

FIG. 10.2 Movement of an ideal-gas molecule along the x-axis at the speed of u_x in cubic box, with all sides having length l, with perfectly elastic sides (i.e., walls having area = l^2). *(Based on Y. Zhang, Indoor Air Quality Engineering CRC Press, Boca Raton, FL, 2004.)*

Since $u^2 = u_x^2 + u_y^2 + u_z^2$ for any molecule and the large number of molecules is moving randomly, the average values of u_x^2, u_y^2, and u_z^2 are equal. Thus, each is one-third of u^2. The movement has no preference in direction, so the pressure exerted in the wall is

$$P = \rho\overline{u_x^2} = \frac{\rho\overline{u^2}}{3} \tag{10.15}$$

The square root of $\overline{u_x^2}$ is known as the gas molecules' root-mean-square (rms) speed, that is, u_{rms}. This is the average molecular speed, expressed in m sec^{-1}. Eq. (10.14) would be true if it accounts for collisions among the molecules, providing that these collisions are elastic and occur between identical molecules and that the time of the collision is negligible compared with time between collisions [12]. Thus, we can use the previous equations to establish the relationship between macroscopic pressure and microscopic rms speed:

$$u_{rms} = \left(\overline{u^2}\right)^{1/2} = \left(\frac{3P}{\rho}\right)^{1/2} \tag{10.16}$$

Thus, if the gas pressure and density are known, the root-mean-square speed of the gas molecules can be calculated. Within the time and density constraints of most air pollution scenarios, Eq. (10.7) is a reasonable approximation of molecular speed. We can further simplify the equations to find molecular speed with the universal gas constant (R) and time (T).

These calculations show that u_{rms} is directly proportional to the square root of temperature in K. Thus, increasing temperature by a factor of 9 means that the molecular speed (velocity) will increase by a factor of 3, increasing T by a factor 16 results in a gas velocity increase by a factor of 4, and so on.

Increasing the molar mass of the gas decreases the gas velocity proportionately. Thus, increasing mass by a factor of 9 means that the molecular speed (velocity) will decrease by a factor of 3, increasing m by a factor 16 results in a gas velocity decrease by a factor of 4, and so on.

We now need to consider energy to complete our understanding on gas velocity. The kinetic energy of q single gas molecule (E_g) is.

$$E_g = \frac{mu^2}{2} \tag{10.17}$$

However, given the large number of molecules and collisions, the average kinetic energy of a gas $\left(\overline{E}_g\right)$ is entirely dependent on temperature:

$$\overline{E}_g = \frac{3RT}{2} \tag{10.18}$$

Since kinetic energy is expressed as joules, m is expressed in kg, and velocity in m sec^{-1}, the two energy equations need to be converted to be expressed in moles by introducing the inverse of number of molecules $\left(\frac{1}{n}\right)$ to KE:

$$\overline{E}_g = E_g \cdot \frac{1}{n} \tag{10.19}$$

$$\frac{3RT}{2} = \frac{mu^2}{2} \cdot \frac{1}{n} \tag{10.20}$$

$$3RT = \frac{mu^2}{n} \tag{10.21}$$

Rearranging gives.

$$\mathrm{u} = \sqrt{\frac{3RT}{m}} \tag{10.22}$$

What is the molecular speed of molecular oxygen at room temperature? Use this to demonstrate how gas velocity is indeed expressed in meters per second

These calculations must use temperature in degrees Kelvin.

Assuming room temperature is 25°C, this is 298.15 K.

Molar mass of O_2 is 32 g mol^{-1} or 0.032 kg mol^{-1}.

R contains joules (J), that is, the energy unit equal to Newton (N) times meters (m). N is kg·m·sec^{-2}, and J is derived as kg·m^2·sec^{-2}. Thus, we must convert the molar mass to kg mol^{-1}.

Thus, replacing joules with its derivation units,

$$u_{rms} = \sqrt{\frac{3\left(8.3145\,\frac{\text{kg m}^2}{\text{mol}\cdot\text{K}\cdot\text{sec}^2}\right)298.15\text{K}}{0.032\,\frac{\text{kg}}{\text{mol}}}} = \sqrt{\frac{3\left(8.3145\,\frac{\cancel{\text{kg}}\,\text{m}^2}{\cancel{\text{mol}}\cdot\cancel{\text{K}}\cdot\text{sec}^2}\right)298.15\cancel{\text{K}}}{0.032\,\frac{\cancel{\text{kg}}}{\cancel{\text{mol}}}}} = \sqrt{2.24\times10^5\,\frac{\text{m}^2}{\text{sec}^2}} \approx 473\,\text{m sec}^{-1}$$

The square root of m^2 sec^{-2} is m sec^{-1}; this shows that gas velocity is indeed expressed as length per time, for example, m sec^{-1}.

Thus, at room temperature, the velocity of molecular oxygen gas is about 130 m sec^{-1} faster than the speed of sound, which is 343 m sec^{-1}.

Assume an insulated reactor window contains argon (Ar). What is the u_{rms} at 300 K?

R= 8.3145 J mol^{-1} K^{-1}.

Argon has molar amount of 40 g mol^{-1}. Since we are using joules, we need to convert g to kg: (40 g mol^{-1})(kg 1000 g^{-1})= 0.04 kg mol^{-1}

$$u = \sqrt{\frac{3(8.3145)(300)}{0{,}04}} = 432.5231\text{ m sec}^{-1}$$

What is the u_{rms} at 900 K?

$$u = \sqrt{\frac{3(8.3145)(900)}{0{,}04}} = 749.1520\,\text{m sec}^{-1}.$$

Increasing the temperature by a factor of 3 results in an increase in u_{rms} of square root of 3.

The ideal-gas laws underpin sampling and analysis of gas-phase pollutant measurements. Also, air pollution models include algorithms that adhere to these laws. An "ideal gas" obeys these laws completely, so models assume ideal conditions or allow for user-defined or default assumptions when a gas deviates from them. The interrelationships among the pressure, temperature, and volume of ideal gases mean that collisions between atoms or molecules are perfectly elastic and in which intermolecular attractive forces are absent. As such, an ideal gas consists of a collection of perfectly hard spheres that collide but otherwise do not interact with one another. All internal energy is kinetic energy, which means that a change in internal energy concordantly changes temperature. Thus, an ideal gas has three state variables: absolute pressure (P), volume (V), and absolute temperature (T). The ideal-gas law states that V is directly proportional to T and inversely proportional to P.

Even though "real gases" often deviate to varying degrees from the ideal-gas assumptions, depending on this deviation, models must recognize how a particular gas will behave in the atmosphere. Generally, the real gases encountered in most air pollution scenarios deviate very little from ideal-gas behavior [13, 14].

10.2.3 Standard conditions

As mentioned, for most air pollution, calculations involving the real gases and vapors deviate very little from ideal-gas behavior [13, 14].

At STP (223.15 K and 1 bar, i.e., 0.986923 atm), 1 mol of gas occupies 22.4 L, that is, the molar volume is 22.4 mol L^{-1}. At NTP (293.15 K and 1 atm), 1 mol of gas occupies a volume of 22.7 L. At SC (298.15 K and 1 atm), the molar volume is 24.5 mol L^{-1}. At SC (298.15 K and 1 atm), the higher temperature further increases the molar volume, that is, 1 mol of gas occupies a volume of 24.47 L (24.47 mol L^{-1}).

In the United States, suppose the annual ambient concentrations of carbon monoxide (CO) exceed 9 ppmv averaged over 8 h or 35 ppmv averaged over 1 h. The city has installed a device at a busy intersection in town. The device reported one 8-h period that averaged 10,000 μg m^{-3} and a month later a 1 h average of 41,450 μg m^{-3}. The values were adjusted for SC. Did either of these readings exceed the CO standard?

$$x\,\mu g\,m^{-3} = \text{ppmv} \times M \times \frac{10^3}{24.47} = \text{ppmv} \times M \times 40.9$$

CO: molecular weight = 28.

8 h average:

$$10{,}000\,\mu g\,m^{-3} = x\,\text{ppmv} \times 28 \times 40.9$$

$$x\,\text{ppmv CO} = \frac{10000\,\mu g\,m^{-3}}{28 \times 40.9}$$

$$x \approx 8.73\,\text{ppmv}$$

1 h average:

$$41{,}450\,\mu g\,m^{-3} = x\,\text{ppmv} \times 28 \times 40.9$$

$$x\,\text{ppmv CO} = \frac{41450\,\mu g\,m^{-3}}{28 \times 40.9}$$

$$x \approx 36.2\,\text{ppmv}$$

Therefore, the 8-h reading meets the standard, but the 1-h reading exceeds the CO standard by 1.2 ppmv.

A term similar to STP that is used in air pollution and atmospheric sciences, which was borrowed from aeronautics, is the "standard atmosphere." A standard atmosphere is a vertical depiction of temperature, pressure, and density that is representative of atmosphere and used to compare and calibrate observed conditions. A standard atmosphere has 1 atm pressure, temperature equal to 15°C, and ρ equal to 1.225 g m^{-3}, at sea level [15]. The variations of temperature and density of air are shown in Table 10.2.

TABLE 10.2 Effect of temperature on the density of air

Temperature T (°C)	Density of air ρ (kg m^{-3})
35	1.1455
30	1.1644
25	1.1839
20	1.2041
15	1.2250
10	1.2466
5	1.2690
0	1.2922
−5	1.3163
−10	1.3413
−15	1.3673
−20	1.3943
−25	1.4224

Source: U. Standard Atmosphere, US Standard Atmosphere 1976, US Government Printing Office, Washington, DC, 1976.

Another standard condition concept is the virtual temperature (T_v), which is a way to compare and calibrate various moisture conditions in the air. Under common tropospheric conditions, air moisture ranges from almost completely dry, that is, 0% water vapor, to a few percent water vapor. Thus, the gas constant, *R* is larger for a mass of moist air than *R* for the same mass of dry air. The difference increases with increasing moisture. Thus, to facilitate calculations, the *R* for dry can be used by assuming a fictitious temperature, that is, T_v, in the ideal-gas equation. The value of T_v is the temperature that the dry air must be to have the same density of moist air at the same pressure. Each molecule of lighter water vapor (atomic weight of 16) displaces a heavier molecule of air (atomic weight of 29), so moist air is less dense than dry air. Therefore, T_v is always larger than actual temperatures.

T_v in degrees Kelvin can be found by

$$T_v = \frac{T}{1-\left(\frac{e}{P}\right)(1-\varepsilon)} \tag{10.23}$$

where *T* is the actual temperature (Kelvin), *e* is the partial pressure of water vapor, *P* is the total pressure (partial pressure of dry air + *e*), and ε is the ratio of gas constants for dry and moist air, that is, $\frac{R_{dry}}{R_{H2O}}$, which is 0.622.

A moist air sample collected at 25°C and 1.1 atm shows the water vapor partial pressure to be 0.01 atm. The next day, a sample was collected at 22°C and 0.9 atm and water vapor pressure measured at 0.02 atm. On day 3, the temperature was 25°C and 0.9 atm and water vapor pressure measured at 0.02 atm. What are the virtual temperatures for these conditions?

$$\text{Day 1}: T_v = \frac{298.15\text{K}}{1-\left(\frac{0.01}{1.1}\right)(1-0.622)} = 299.17\,\text{K} = 26.02^{\circ}\text{C}$$

$$\text{Day 2}: T_v = \frac{295.15\text{K}}{1-\left(\frac{0.02}{0.9}\right)(1-0.622)} = 297.65.\text{K} = 24.5^{\circ}\text{C}$$

$$\text{Day 2}: T_v = \frac{298.15\text{K}}{1-\left(\frac{0.02}{0.9}\right)(1-0.622)} = 300.68.\text{K} = 27.5^{\circ}\text{C}$$

So, day 3 had the highest virtual temperature; even the actual temperatures for days 1 and 3 were the same.

Incidentally, these conditions not only are limited to gas-phase pollutants but also apply to aerosol-phase measurements. Often, a variety of pollutants are measured together, simultaneously at single site (see Fig. 10.3). All measurements are adjusted to SC or are otherwise standardized when reported.

10.2.4 Gases versus vapors

To this point, we have used vapor to describe water's gas phase but have referred to molecular oxygen and nitrogen as gases. For air pollution purposes, a gas is matter that has a single thermodynamic state within these typical atmospheric ranges, for example, NTP, RTP, or SC, whereas a vapor is one that exists in more than one phase in these ranges. For example, molecular oxygen (O_2) is a gas because every O_2 molecule is in the gas phase at and near SC. Indeed, the boiling point, the temperature at which O_2 changes from a gas to a liquid, at 1 atm is −183°C. Conversely, gas-phase benzene is a vapor because at SC, some molecules are behaving as a gas and the rest are behaving as a liquid. The phase diagram in Fig. 10.1 near SC for O_2 would only show the gas phase, whereas for benzene, it would show the liquid and gas phases, but not the solid phase.

The principal example of phase change for gases in air pollution studies is the use of liquidized gases, especially N_2, in tanks and cylinders. For example, air pollution studies make much use of liquid N_2. Cryogenic liquids are liquefied gases that have an STP boiling point less than −90°C. At STP, N_2 has a boiling point equal to −196°C. A typical gas liquifying system has a cryogenic storage tank, one or more vaporizers, and a pressure-temperature control system. The system is designed to repel heat by enclosing the liquid in a separate compartment within an inner vessel. If the N_2 is needed in its liquid state, this is all that is needed. Otherwise, the vaporizers convert the liquid to its gaseous state. A pressure control manifold mediates the pressure during the feed process. This is dramatic evidence that gases can be compressed and can be expanded to fill their containers uniformly.

FIG. 10.3 Monitoring site for several near-road mobile source air pollutants in Las Vegas, Nevada. The pollutants included benzene, 1,3-butadiene, carbon monoxide, oxides of nitrogen, and particulate matter. *Top photo*: rooftop inlets for various monitors and microphones to determine traffic volume and types. *Bottom photo*: author during the installation of monitors.

At STP, NTP, RTP, and SC, a gas will remain in the gaseous state notwithstanding concentration, that is, it will not condense. Molecular nitrogen (N_2) will not condense at STP, even at 100%. Vapors at STP will exist as both gases and aerosols, but at SC, many of the vapor-phase pollutants rarely reach saturation. Thus, vapors and gases are often treated synonymously as air pollutants, that is, they differ from aerosols in the manner in which they are measured, removed, and treated [13, 14].

Of course, the distinctions between gases and vapors become less clear outside of the lower troposphere. For example, within reactors and combustors and even within stacks, temperatures and/or pressures are highly elevated, so substances that are pure liquids and solids under SC will volatilize. However, within the atmosphere, the two major phases apply.

10.2.5 Gas density

The gas laws imply that density is a function of pressure and temperature, since density is indirectly proportional to gas volume. The density of air is important to air pollution, especially because air density and the density of the pollutant gas are factors in transport. Air density is the mass of air in a given volume. Recall that density varies with temperature and pressure per the ideal-gas law, that is, $P=\rho RT$. P is pressure in atm, the universal gas constant (R) equals $\frac{0.08212\ \text{L}\cdot\text{atm}}{\text{mol}\cdot\text{K}}$, T is temperature in K, and ρ is density.

Thus, we can find density by rearranging the gas law equation:

$$\rho=\frac{P}{RT} \tag{10.24}$$

Typically, air density is about 1.2 kg m^{-3} at sea level.

What is the density of air at 1 atm and 20°C? How about in the Nevada desert in summer, if the atmospheric pressure is 0.92 atm and $T=40$°C?

The MW of air$=28.9$ g gmol^{-1}.

Another way to state the gas law is

$$PV=nRT \tag{10.25}$$

where V=volume and n is the number of moles.

$$\text{The standard air's gas density}=\rho_{\text{air}}=\frac{P\cdot MW}{RT}=\frac{1\cdot 28.9}{0.0812\cdot 293.15}=1.21\ kg\ m^{-3}$$

What is the density of air in the Nevada desert in summer, if the atmospheric pressure is 0.92 atm and $T=40$°C?

$$\text{Air's gas density}=\rho_{\text{air}}=\frac{P\cdot MW}{RT}=\frac{0.92\cdot 28.9}{0.0812\cdot 313.15}=1.05\ kg\ m^{-3}$$

Note that air density is substantially lower under the desert conditions than under standard atmospheric conditions.

If the number density of benzene in a garage is 2.4×10^{18} molecules per m^3, what is the mass density in μmol? In μg m^{-3}?

$$2.4\times10^{18}\ C_6H_6\ \text{molecules per m}^3]\div 6.022\times10^{23}\ \text{molecules}\approx 4\times10^{-6}\ \text{mol m}^{-3}$$

2.4×10^{12} C_6H_6 molecules per cm^3 is the same as the mass density 4×10^{-6} mol m^{-3} or 4 μmol (μmol) m^{-3}. The atomic mass of carbon (C) is 12 and of hydrogen (H) is 1, so the atomic mass of C_6H_6 is 78. Thus, 1 mol C_6H_6 $=78$ g$=7.8\times10^{-5}$ μg.

Since 1 μmol$=10^{-6}$ mol, then 2 μmol benzene$=4\times78\times10^{-6}$ g$=1.92\times10^{-4}$ g$=$ *192 μg*.

10.2.6 Using the gas laws to calculate pollutant classes

Gas laws can be used to calculate some, but not all, pollutant classes. For example, the units for the air pollutant, nitrogen dioxide (NO_2), can be converted using Eq. (10.2). However, when addressing smog and ozone in the troposphere, the oxides of nitrogen, not just NO_2, are calculated. The concentrations of NO_2 and nitric oxide (NO) are commonly expressed in ppm (vol.) and expressed as a sum, known as "NO_x." Thus, the conversion from ppm (vol.) to mg m^{-3} must be carried out separately for NO_2 and for NO, after which the two-chemical species are added to ascertain NO_x.

This is the case for all chemical classes, that is, each species is calculated using Eq. (10.2), and then, the species are added together. For example, if you were going to convert polycyclic aromatic hydrocarbons from ppb to nanograms per cubic meter (nm m^{-3}), you would first convert each of the species:

1. Acenaphthene
2. Acenaphthylene
3. Anthracene
4. Benz(a)anthracene
5. Benzo(a)pyrene
6. Benzo(e)pyrene
7. Benzo(b)fluoranthene
8. Benzo(ghi)perylene
9. Benzo(j)fluoranthene
10. Benzo(k)fluoranthene
11. Chrysene
12. Dibenz(a,h)anthracene
13. Fluoranthene
14. Fluorene
15. Indeno(1,2,3-cd)pyrene
16. Phenanthrene
17. Pyrene

Then, you would add the conversions together to show the total PAH concentration in ng m^{-3}.

What may cause the above calculations of benzene and sulfuric acid to be incorrect?
Ozone is a gas, so the assumption that it will follow the ideal-gas law is valid. However, both benzene and sulfuric acid are vapors, so the same assumption may not be true. Thus, they may exist in particulate and gas phases in the atmosphere. The particulate phase would also be expressed as μg m^{-3}, so the two phases would be added together. In addition, H_2SO_4 in the atmosphere involves a dynamic system among numerous chemical species, including sulfate-bound aerosols, dissolved H_2SO_4 in water droplets and gas-phase SO_2 and other oxides of sulfur in the atmosphere.

10.3 FLUID DYNAMICS

Dynamics combines the properties of the fluid and how it moves. This means that the continuum fluid mechanics vary by whether the fluid is viscous or inviscid and compressible or incompressible and whether flow is laminar or turbulent. For example, the properties of the two principal environmental fluids, that is, water in an aquifer and an air mass in the troposphere, are shown in Table 10.3. Thus, in air pollution mechanics, turbulent systems are quite common. However, within smaller systems, such as control technologies, laminar conditions can be prominent.

Dynamics is divided into kinematics and kinetics. Kinematics is concerned with the study of a body in motion independent of forces acting on the body. That is, kinematics is the branch of mechanics concerned with motion of bodies with reference to force or mass. This is accomplished by studying the geometry of motion irrespective of what is causing the motion. Therefore, kinematics relates position, velocity, acceleration, and time. Aerodynamics and hydrodynamics are the important branches of environmental mechanics. Both are concerned with deformable bodies and with the motion of fluids. Therefore, they provide an important underlying aspect of pollutant transport and movements of fluids and

TABLE 10.3 Contrasts between plume in groundwater and atmosphere

	Groundwater plume	Air mass plume
General flow type	Laminar	Turbulent
Compressibility	Incompressible	Compressible
Viscosity	Low viscosity (1×10^{-3} kg m^{-1} s^{-1} at 288 K)	Very low viscosity (1.781×10^{-5} kg m^{-1} s^{-1} at 288 K)

consider fluid properties, for example, compressibility and viscosity. These are crucial properties needed to understand the movement of contaminants within plumes, flows in vents and pipes, and design of air pollution control systems.

Kinetics is the study of motion and the forces that cause motion. This includes analyzing force and mass as they relate to translational motion. Kinetics also considers the relationship between torque and moment of inertia for rotational motion.[1] A key concept for environmental dynamics within the context of physical motion is that of linear momentum, the product of mass and velocity. A body's momentum is conserved unless an external force acts upon a body. Kinetics is based on Newton's first law of motion, which states that a body will remain in a state of rest or will continue to move with constant velocity unless an unbalanced external force acts on it. Stated as the law of conservation of momentum, linear momentum is unchanged if no unbalanced forces act on a body. Or, if the resultant external force acting on a body is zero, the linear momentum of the body is constant.

Kinetics is also based upon Newton's *second law of motion*, which states that the acceleration of a body is directly proportional to the force acting upon that body and inversely proportional to the body's mass. The direction of acceleration is the same as the force of direction. The equation for the second law is

$$\mathbf{F} = \frac{d\mathbf{p}}{dt} \tag{10.26}$$

where

$\mathbf{p}$ = momentum.

Newton's third law of motion states that for every acting force between two bodies, there is an equal but opposite reacting force on the same line of action:

$$\mathbf{F}_{reacting} = -\mathbf{F}_{acting} \tag{10.27}$$

Another force that is important in air pollution is friction, which is a force that always resists motion or an impending motion. Friction acts parallel to the contacting surfaces. When bodies encounter one another, friction acts in the direction opposite to that that is bringing the objects into contact.

Air pollutants may move within one environmental compartment, such as a source within a home. Most often, however, pollutants move among numerous compartments, such as when a contaminant moves from the source to the atmosphere, until it is deposited to the soil and surface waters, where it is taken up by plants and eaten by animals. Science is by necessity compartmentalized among physics, chemistry, and biology. However, complex topics like pollutant transport require that all three of the sciences be considered. The dynamics, both thermos and fluid, of gases are important for two basic reasons. First, many precursors and pollutants are themselves gases and vapors, which require an understanding of their mass continuity and energy conservation and changes in momentum. Second, all pollutants are carried by gases, notwithstanding their phase. Thus, aerosols move with the gas flow, that is, advectively. In the atmosphere, the carrier gas is the mixture of gases we call air. In the gas chromatograph, the carrier gas may be helium (He).

In thermodynamics, mass is conserved when a fluid is in motion, mass continuity must equal the rate of mass that enters and exits a system, plus any accumulation of mass. Accumulation is simply input minus output. This mass balance is the difference between the deposited mass and the withdrawn mass. The system can be any control volume, for example, the entire troposphere, a parcel of atmosphere, a home, a room, a pipe, or a cell within an organism.

10.3.1 Dimensions of flow and mass continuity

The dimension of flow refers to the number of space coordinates. The velocity of flow depends on position, as defined by coordinates:

$$u = f(x, y, z) \tag{10.28}$$

A real flow is three-dimensional, that is, the flow is described by the x, y, and z axes. However, even simple, three-dimensional flows are very difficult to calculate. For example, source emissions often assume a one-dimensional flow, that is, in the x direction only. In this case, the velocity is constant over entire section, that is, the fluid (i.e., gas or liquid) moves along parallel flow lines (see Fig. 10.4).

1. Note that up to this point, we have used the term in its chemical connotation, that is, kinetic reactions have yet to reach equilibrium, which means there is an inequality between both sides of the reaction. Thus, kinetic reactions employ a one-direction arrow (→), whereas equilibria show a two-direction arrow (↔).

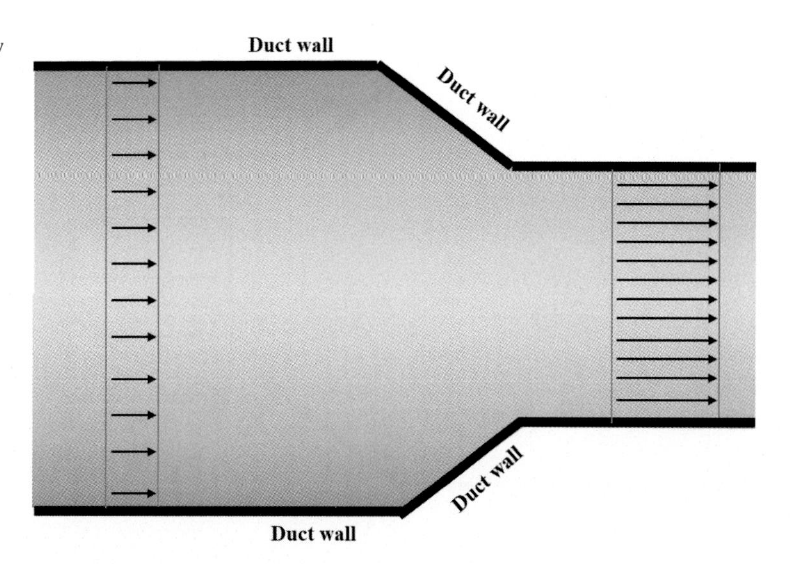

FIG. 10.4 One-dimensional flow in a duct. Velocity vectors are shown in two sections, that is, velocity fields.

TABLE 10.4 Comparison of velocity (*u*) flow dimensions and types by position coordinates and time (*t*)

	Flow type	
Flow dimension	**Steady**	**Unsteady**
One-dimensional	$u=f(x)$	$u=f(x,t)$
Two-dimensional	$u=f(x,y)$	$u=f(x,y,t)$
Three-dimensional	$u=f(x,y,z)$	$u=f(x,y,z,t)$

Thus, the distance (x) and time (t) in which the mass (m) moves are interchangeable:

$$\frac{dm}{dt}=\dot{m}=\rho\times \mathrm{A}\times u \tag{10.29}$$

where A is the cross-sectional area perpendicular to the flow lines, for example, a duct, and u is the gas velocity.

The expressions of flow are shown in Table 10.4. Note that all unsteady flow is a function of both position and time.

Flow can be laminar or turbulent. A laminar flow may be steady or unsteady. A steady flow is one whose flow field at any instant of time is the same as at any other instant of time. A turbulent flow is always unsteady. However, given randomness, that is, if a turbulent flow regresses to its mean, we may assume it to be steady at the mean. This is known as stationary turbulent flow. Laminar flow occurs at low Reynolds numbers (Re or N_R), whereas turbulent flow occurs at high Reynolds numbers.

If there are no losses (e.g., leaks) or gains (e.g., connecting sources of the gas) within the duct or other conduit, the flow rate is steady. That is, the parameters in Eq. (10.25) do not change. Thus, as shown in Fig. 10.5, at the upflow location (subscript = 1) and downflow location (subscript = 2), the mass continuity at steady state equals a constant:

$$\frac{dm}{dt}=\rho_1\times \mathrm{A}_1\times u_1=\rho_2\times \mathrm{A}_2\times u_2=\text{constant} \tag{10.30}$$

This is the expression of the law of the conservation of mass, which is the basis for calculating emission rates and other gas flows.

A two-dimensional flow occurs when the flow velocity at every point is parallel to a fixed plane. At any point normal to the plane, the velocity is constant (see Fig. 10.5).

The mass flow rate ($\dot{m}$), logically, is mass per time. The density units are mass (M) per volume (L^3), and velocity units are length (L) per time (T) and units of area (L^2):

$$\dot{m}=\rho\times \mathrm{A}\times u=\frac{\mathrm{M}}{\mathrm{L}^3}\times\frac{\mathrm{L}^2}{\square}\times\frac{\mathrm{L}}{\mathrm{T}}=\frac{\mathrm{M}}{\mathrm{T}} \tag{10.31}$$

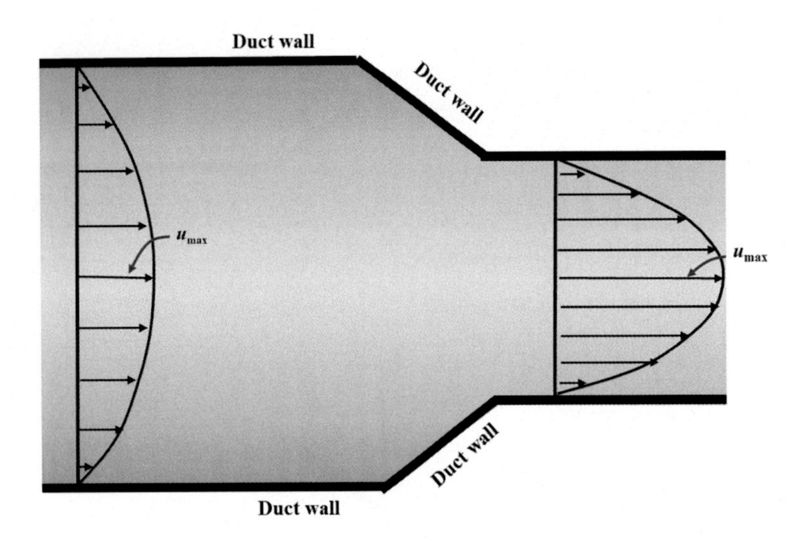

FIG. 10.5 Continuity of mass: two-dimensional flow. Maximum velocity is u_{max}. The subscripts denote the values of these variables at locations 1 and 2, respectively. *(Based on information from G.D. Wight, Fundamentals of Air Sampling, CRC Press, 1994.)*

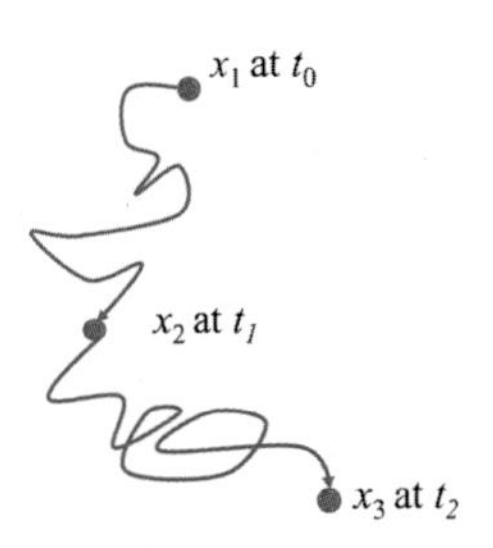

FIG. 10.6 Random walk representing the movement of a particle, that is, a hypothetical point that is moving in random paths from position x_1 to position x_2 during time interval $t_0 - t_1$ and from x_2 to x_3 during time interval $t_1 - t_2$. This is the theoretical basis for Lagrangian gas flow, which is used in atmospheric plume models. *(Reproduced with permission from D.A. Vallero, Fundamentals of Air Pollution, 5th ed., Elsevier Academic Press, Waltham, MA, 2014, p. 999 pages cm.)*

Thus, in a one-dimensional flow, if one parameter changes, the others must also change proportionately. For example, if the gas density remains the same, but the cross-sectional area doubles in size, the velocity must decrease by 50% to preserve continuity.

Figs. 10.4 and 10.5 indicate that, at constant density, a decrease in the cross-sectional area would result in the gas flow velocity increasing and a decrease in density across a constant area would also result in a velocity increase. Thus, since an increase in pressure would result in increased density, so velocity would likewise decrease with increasing pressure (see Fig. 10.6).

A technician at an air monitoring station is using a compressor. The compressor's air enters a 36 cm^2 inlet. The entering air has a density of 1.2 kg m^{-3} and a mean velocity of 4 m sec^{-1}. The air exits the compressor at 3 m sec^{-1} through a 5 cm-diameter duct. Calculate the mass flow rate and density at the nozzle

The area at the inlet is

$$A_1 = 36\,\text{cm}^2 = 0.0036\,\text{m}^2$$

The area of a circle, that is, the nozzle, is

$$A_2 = \pi r^2 = \pi\left(\frac{D}{2}\right)^2 = \frac{\pi D^2}{4} \tag{10.32}$$

$$A_2 = \frac{3.142 \times (0.05)^2}{4} = 0.00196375 \approx 1.96 \times 10^{-3}\,\text{m}^2$$

Mass flow rate:

$$\dot{m} = \rho \times A \times u = 1.2 \times 0.0036 \times 4 = 1.73 \times 10^{-2}\,\text{kg sec}^{-1}$$

Recall that the conservation of mass dictates that $\rho_1 \times A_1 \times u_1 = \rho_2 \times A_2 \times u_2$.

Therefore, the density of the air at the nozzle must be

$$\rho_2 = \frac{\dot{m}}{A_2 \times u_2} = \frac{1.73 \times 10^{-2}}{1.96 \times 10^{-3} \times 3} \approx 2.94\,kg\,m^{-3}$$

The air exiting the compressor is more than twice as dense as the air entering given the smaller area lower exit velocity compared with the air intake.

Air pollution generally focuses on gas movement, that is, either the gas-phase substance or the gas or mixture of gases, especially air, that contain and carry substances. However, air pollution involves all fluids, including liquids and gases. Fluids are generally divided into two types: *ideal* and *real*. The former has zero viscosity, which means there is no resistance to shear. An ideal fluid is also incompressible and flows with uniform velocity distributions. It has no friction between moving layers and no turbulence (i.e., eddy currents). On the contrary, a real fluid has finite viscosity, has nonuniform velocity distributions, is compressible, and experiences friction and turbulence. Real fluids are further subdivided according to their viscosities. A Newtonian fluid is one that has a constant viscosity at all shear rates at a constant temperature and pressure. Water and most solvents are Newtonian fluids. Non-Newtonian fluids are those with viscosities that are not constant at all shear rates. Although many of the liquid-phase pollutants are behave has Newtonian fluids, sediment, soil and other environmental compartments can become contaminated with Non-Newtonian fluids, for example, drilling fluids and oil residues.

In physics, the particle is a theoretical point that has mass and location, but no geometric extension. This particle can be visualized moving within the fluid as a representation of where that portion of the fluid is going and at what velocity. The thermodynamic control volume is an arbitrary region in space that is defined by boundaries that are either stationary or moving. As mentioned, the control volume is a depiction of conservation of mass and energy. The amount of mass or energy entering a control volume must equal the amount exiting the control volume minus what remains in the control volume. Thus, the control volume is useful in measuring and modeling the amount of an air pollutant entering, remaining, and exiting a parcel of air, water, and soil or even a cell or parcel of tissue in an organism.

The forces acting on air, water, or any fluid are either body forces or surface forces. The former are forces that act on every particle within the fluid, occurring without making physical contact, for example, gravitation force. The latter are forces that are applied directly to the fluid's surface by physical contact.

Stress represents the total force per unit area acting on a fluid at any point within the fluid volume. So, stress at any point *P* is

$$\sigma(P) = \lim_{\delta A \to 0} \frac{\delta \mathbf{F}}{\delta \mathbf{A}} \tag{10.33}$$

where

$\sigma(P)$ = vector stress at point *P*
δA = infinitesimal area at point *P*
δF = force acting on δA

Fluid properties are the fluid's characteristics that are used to predict how the fluid will react when subjected to applied forces. A fluid is in continuum if it is infinitely divisible, that is, it is made up of many molecules that are constantly in motion and colliding with one another. A fluid in continuum has no holes or voids, meaning its properties are continuous, that is, temperature, volume, and pressure fields are continuous. The assumption of continuum allows the fluid's properties to be functions of position and time.

Fluid properties are important at every scale of air pollution, such as the cascades of Newtonian and non-Newtonian fluids in bronchial tubules in the lung and the difference in flow from turbulent at the entry of air by inhalation to laminar at the blood-air exchange sites. Fluid properties must also be considered at the cellular scale, for example, flow through cell membranes.

These fluid properties can be represented by the density field:

$$\rho = \rho(x, y, z, t) \tag{10.34}$$

where

ρ = density of the fluid
x, y, z = coordinates in space
t = time

The fluid properties are also represented by the velocity field:

$$\vec{v}=\vec{v}\,(x, y, z, t) \tag{10.35}$$

The density field and the velocity field imply that the distribution of density and velocity within a control volume is a function of space (x, y, and z) and time (t). Thus, if the fluid properties and the flow characteristics at each position do not vary with time, the fluid is said to be at steady flow:

$$\rho=\rho(x, y, z) \text{ or } \frac{\partial \rho}{\partial t}=0 \tag{10.36}$$

and

$$\vec{v}=\vec{v}\,(x, y, z) \text{ or } \frac{\partial \vec{v}}{\partial t}=0 \tag{10.37}$$

Conversely, a time-dependent flow is considered to be an unsteady flow. Any flow with unchanging magnitude and direction of the velocity vector $\vec{v}$ is considered to be a uniform flow.

Fluids, then, can be classified according to observable physical characteristics of flow fields. Laminar flow is in layers, while turbulent flow has random movements of fluid particles in all directions. In incompressible flow, the variations in density are assumed to be constant, while the compressible flow has density variations, which must be included in flow calculations. Viscous flows must account for viscosity, while inviscid flows assume viscosity is zero.

The velocity field provides a way to characterize the motion of fluid particles and provides the means for computing these motions. Eqs. (10.31) and (10.32) are representations of the Euler approach, that is, fluid properties are written as functions of space and time, so the flow is determined by analysis of the behavior of functions. This differs from the Lagrange approach, which characterizes the fluid by tagging a piece of the fluid, for example, a parcel, and determining the flow properties by tracking the movement of these properties in time (see Fig. 10.6). This random walk of the particle provides what is known as the Lagrangian viewpoint, which is expressed mathematically as

$$\vec{v}=[x(t), y(t), z(t)] \tag{10.38}$$

A Lagrangian plume model characterizes the plume by calculating the air dispersion from statistics of the trajectories of many parcels, that is, sufficient to represent the whole plume. Thus, statistics, that is, standard deviations, are employed to analyze the expected movement of a gas flow or a plume in the atmosphere with time in all three dimensions, that is, the Gaussian model.

The major means by which gas-phase and aerosol substances move in the atmosphere is advection, that is, they move with the flow of the fluid where they reside, that is, the air. This change in position in space is the fluid velocity (u). This is a vector field quantity. Speed (v) is the magnitude of the vector velocity u at some given point in the fluid, and average speed ($\overline{\text{v}}$) is the mean fluid speed through a control volume's surface. Therefore, velocity is a vector quantity (magnitude and direction), while speed is a scalar quantity (magnitude only). The standard units of velocity and speed are meter per second (m sec^{-1}).

Gas velocity[2] must be calculated for air quality measurements and to determine mixing rates within various control volumes, for example, in an air quality monitor, in a room in a building, or in a plume in the atmosphere.

Engineers and scientists often assume a mean velocity for an entire control volume, for example, a well-mixed chamber, to simplify calculations of a substance's concentration compared with an unmixed chamber in which the substances and carrier gases move at multiple velocities. Unfortunately, this assumption results in calculations of average concentrations within the entire chamber or plume. This is an example of the difficulty in stepping up laboratory or mesoscale findings to the real world. For example, an air pollutant plume is never homogeneous. It is certainly likely to have higher concentrations near the source, but it will also have pockets of higher and lower concentrations given variability of terrain, structures, albedo, clouds, temperature, and moisture. Even a small compartment, for example, a room in a house, is never completely mixed, with regions of varying pollutant concentrations, given that gas velocity approaches zero near the walls.

Pressure (p) is a force per unit area:

$$p=\frac{F}{A} \tag{10.39}$$

2. The distinction between velocity and speed is seldom made in air pollution.

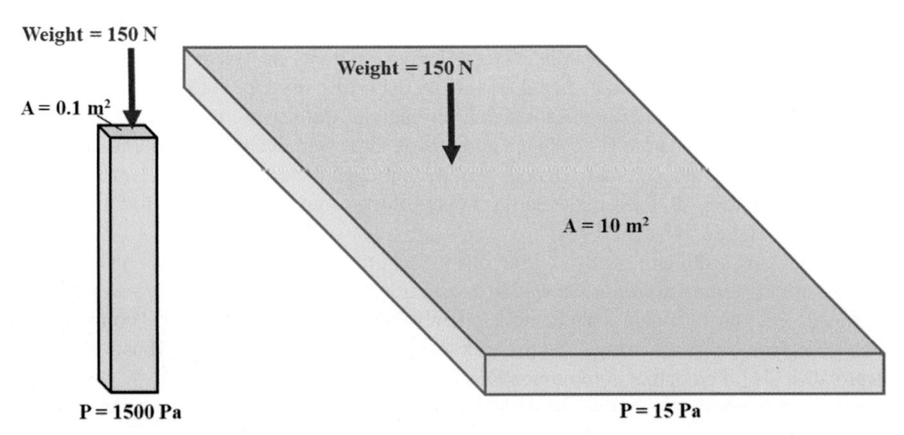

FIG. 10.7 Difference in pressure with same weight over different-sized areas.

Thus, p is a type of stress that is exerted uniformly in all directions. This is arguably the most important type of stress in air pollution engineering, where it is common to use the term pressure instead of force to describe the factors that influence the behavior of fluids. The standard unit of p is the pascal (P), which is equal to 1 N m^{-2}. For vapor pressure, partial pressures, and partitioning coefficients discussed in the next chapters, the most common units are kilopascal (kP) and millimeters of mercury (mm Hg). Pressure varies with area, as shown in Fig. 10.7. In this example, the same weight (force) over different areas leads to different pressures, resulting in much higher pressure when the same force is distributed over a much smaller area.

For a liquid at rest, the medium is considered to be a continuous distribution of matter. However, when considering p for a gas, it is an average of the forces against the vessel walls, that is, gas pressure. Fluid pressure is a measure of energy per unit volume per the Bernoulli's principle, which states that the static pressure in the flow plus one-half of the density times the velocity squared is equal to a constant throughout the flow, referred to as the total pressure of the flow:

$$p + \tfrac{1}{2}p\mathrm{V}^2 + pgh = \text{constant} \tag{10.40}$$

where

p = pressure
V = fluid velocity
h = elevation (head)
g = gravitational acceleration

Symbolic Ambiguity

Mea culpa! I often tell anyone listening that one of the key rules of science is that it must strive to eliminate ambiguity. This is even more important for the technical professions. A physician must take care that a script can only be interpreted by fellow professionals in a singular way. "I thought the good doctor meant arsenic, but it turns out it was supposed to be aspirin…." The same goes for the engineer. "Was that supposed to be a factor of safety applied to tank volume? I thought it was gas velocity!" Context is key, but always be certain of the meaning of the symbols and abbreviations.

The engineering and scientific literature applies several symbols for velocity, including V, *v*, and *u*. Some physics texts even use *c* for linear velocity. Angular velocity is commonly expressed as ω, which is also used to denote the humidity ratio (see Chapter 7) and in chemical kinetics to denote the effectiveness factor, that is, the actual rate of reaction divided by the rate of reaction unaffected by diffusional resistance. Both uppercase *P* and lowercase *p* denote pressure; physicists also use *P* to denote polarization.

The symbol V is used here since this is how the Bernoulli's principle is often expressed, whereas *u* is often used to express velocity in mass flow equations, and *v* is commonly used in air pollution models, for example, to express Euler and Lagrange viewpoints. However, this and other texts use V to denote volume. In this chapter, we distinguish volume from velocity by italicizing the *V* when denoting volume. This text also uses the lower case, nonitalicized v to denote speed, that is, the scalar quantity of the magnitude of an object's velocity. However, in air pollution science, the term speed is usually considered to be synonymous with velocity, the vector quantity of distance traveled per unit time.

The letter K is also ambiguous, with various meanings depending on whether it is lower or uppercase and/or italicized. In its uppercase form, it is used to denote Kelvin temperature and equilibrium constants, as well as being the symbol for potassium. In this text, the lowercase k denotes a rate constant in a chemical reaction but is also used for the Boltzmann's constant (see Chapter 3).

Some symbols are used almost uniformly and consistently throughout the scientific and engineering disciplines, including ρ (density), A (area), λ (wavelength), dynamic viscosity (η), and flux (J). However, even ρ is sometimes also used by electrical engineers to denote specific resistance. And η can also denote entropy (although this text prefers S). Unfortunately, symbols and units vary with the air pollution literature, so, even in the same chapter of the same text, one must to take care to distinguish among the symbols.

This elimination of ambiguity is difficult enough within one's scientific or engineering discipline but increases with interdisciplinary communications among technical groups. However, it increases substantially when the receiver of the information is a more general audience. Indeed, symbolic language is often undecipherable for nonscientists. Technical communication can be seen as a critical path, where the engineer sends a message and the audience receives it (see Fig. 10.8). Thus, communication can be either perceptual or interpretative [16]. Perceptual communications are directed toward the senses.

Whereas the communication methods toward the right-hand side of Fig. 10.8 may be appropriate for the fellow technologists within the same field, this may be unclear to others.

Interpretive communications encode messages that require intellectual effort by the receiver to understand the sender's meanings. This type of communication can be either verbal or symbolic. Scientists and engineers draw heavily on symbolic information when communicating among themselves. Attending a meeting in which experts are discussing scientific subject matter unfamiliar to you and in which unrecognizable symbols are used is an example of symbolic miscommunication.

I have coined the term "techno-polysemy problem." Homonyms are terms with more than one meaning. Polysemes are words that also have one meaning, but at least two of the meanings overlap. Let us assume that in the meeting you attended, the experts are using words and symbols that are used in your area of expertise but with very different meanings. For example, psychologists speak of "conditioning" [17] with a very different meaning than that of an engineer [18], which is different from that of a statistician or mathematician [19, 20], and altogether different from that of sludge or compost handler [21, 22]. Perhaps, most people in a waste-related public meeting or hearing or even a meeting with decision-makers in a jurisdiction have *no* concept of the term, except when applied to air conditioning or a swimming pool. The waste manager may have all of these in the audience and must be certain that both clearly understand such homonyms. A symbol, a word, or a phrase *will mean* what it *can mean* [23].

10.3.2 Energy and the Bernoulli principle

Three forms of energy are at play at any point in the flow:

1. Kinetic energy results from the motion of the fluid at that point.
2. Potential energy is due to the positional elevation (h) above the point.
3. Pressure energy results from the absolute pressure of the fluid at that point.

Per the first law of thermodynamics, the total energy must remain constant, but it can be transformed within these three forms. Between two states, the net work on a fluid is equal to the gain in kinetic energy plus the gain in potential energy. Thus, from the perspective of potential energy, that is, head, flow work can be described as $\frac{p}{\rho g}$ and velocity head (i.e., kinetic energy) as $\frac{V^2}{2g}$, where g is the graviational acceleration. This leads to the relationship of flow at two points, x = 1 and x = 2, at z height above each point:

$$\left(\frac{p}{\rho g}+\frac{V^2}{2g}+z\right)_1=\left(\frac{p}{\rho g}+\frac{V^2}{2g}+z\right)_2 \tag{10.41}$$

This is the Bernoulli's equation.

A wet scrubber is used to remove SO_2 from a power plant's stack. At 1 atm, the scrubber has a vertical nozzle that produces an upward 10 m sec^{-1} jet of water that is 20 mm in diameter. If the jet retains a circular cross section, will its diameter be large enough to traverse a 40 mm-diameter opening 3 m above the nozzle, assuming the density of water is 1000 kg m^{-3}?

$$\left(\frac{p}{\rho g}+\frac{V^2}{2g}+z\right)_1=\left(\frac{p}{\rho g}+\frac{V^2}{2g}+z\right)_2$$

$z_1 = 0$

$z_2 = 4.5$ m

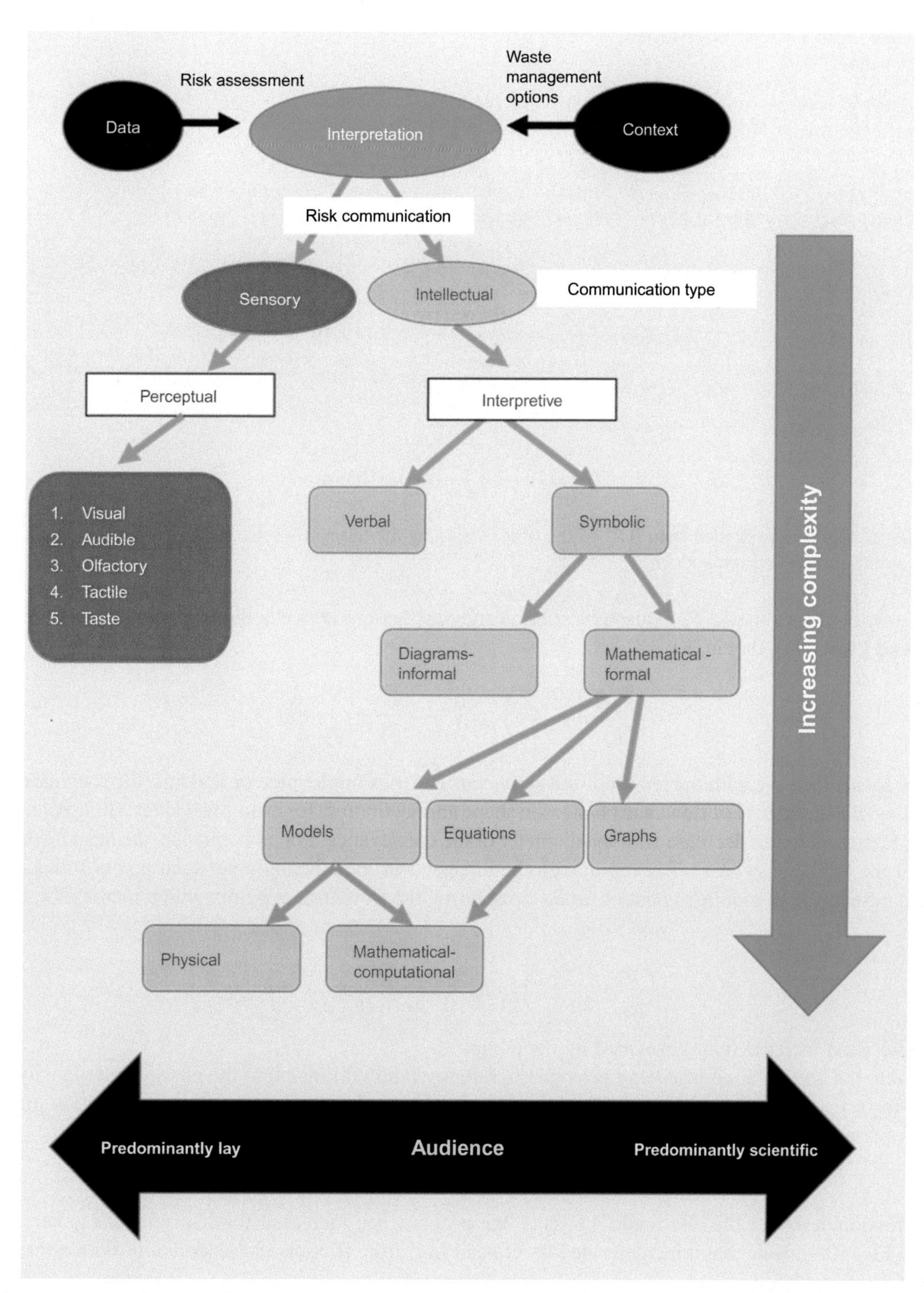

FIG. 10.8 Communication techniques. All humans use perceptual communication, such as observing body language of an engineer or smelling an animal feeding operation. The right side of the figure is the domain of technical communication. Thus, nonscientists or even technically savvy audiences from different disciplines may be overwhelmed by perceptive cues or may not understand the symbolic, interpretive language being used by a bioengineer and others in the risk communication process. Thus, the type of communication in a scientific briefing is quite different from that of a public meeting or a briefing for a neighborhood group potentially affected by a risk management decision. *(Data from M. Myers, A. Kaposi, A First Systems Book: Technology and Management, Imperial College Press, 2004; T.R. Green, Cognitive dimensions of notations, in: People and Computers V, 1989, pp. 443–460.)*

p_1 and p_2 are both 1 atm, so no change.
Energy equation:

$$\frac{V_2^2}{2g} = \frac{V_1^2}{2g} + z_1 - z_2$$

g=9/8 m sec^{-1}
Therefore, $V_2 = \sqrt{V_1^2 - 2gh} = \sqrt{10^2 - 2(9.8)(4.5)} = 3.44$ m sec^{-1}
The diameter can be found by using the continuity equation:

$$r = \frac{D}{2} = \frac{0.02}{2} = 0.01$$

$$A_2 = \frac{A_1 \times V_1}{V_2} = \frac{\pi(0.01^2)(10)}{3.44} = 9.13 \times 10^{-4}\,\text{m}^2$$

$$A_2 = \frac{\pi \times D_2^2}{4}$$

$$D_2 = \sqrt{\frac{4 \times 9.13 \times 10^{-4}}{\pi}} = 0.034\,\text{m}$$

Thus, the fountain has expanded from 0.02 to 0.034 m. Assuming the fountain keeps its circular cross section, the 400 mm (0.04 m) opening *should be large enough.*

In the real world, the flow will be reduced by friction and obstructions between the two points, expressed as the loss of elevation (head loss $= \Delta h$), that is, h_i:

$$\left(\frac{p}{\rho g} + \frac{V^2}{2g} + z\right)_1 = \left(\frac{p}{\rho g} + \frac{V^2}{2g} + z\right)_2 + h_i \tag{10.42}$$

Frictional losses increase with the length of the conveyance, for example, pipe or conduit; the roughness of walls; the velocity of flow; the turbulence of flow; and changes in shape and sectioning, for example, valves, fittings, and bends, of the conveyance. Frictional losses decrease with the diameter of the conveyance. For convenience, the head losses are included as fraction of the total energy at $x = 2$, even though the losses occur incrementally between points 1 and 2.

Engineers restate the Bernoulli's equation when considering the flow from a pump, which increases z, to a downflow point as

$$\left(\frac{p}{\rho g} + \frac{V^2}{2g} + z\right)_1 + h_p = \left(\frac{p}{\rho g} + \frac{V^2}{2g} + z\right)_2 + h_i \tag{10.43}$$

where h_p is the head increase (gain) provided by the pump.

To adhere to the conservation of energy principle, a flowing fluid will maintain the energy, but velocity and pressure can change. In fact, velocity and pressure will compensate for each other to adhere to the conservation principle, so the Bernoulli's equation can be states as

$$p_1 + \tfrac{1}{2} pV_1^2 + pgh_1 = p_2 + \tfrac{1}{2} pV_2^2 + pgh_2 \tag{10.44}$$

This expression describes the "Bernoulli's effect" that occurs when increased fluid speed leads to decreased internal pressure (see Fig. 10.9). Note that it includes the loss of head (Δh) from friction and obstructions within the flow between $x = 1$ (i.e., h_1) and $x = 2$ (i.e., h_2).

10.3.3 Measurement

Air pollution data are based on measurements. Sometimes, only one or a few measurements are input into models or are otherwise used as the basis for describing the air quality for an entire area. Among the most common reasons for measurements is to whether the air quality meets standards imposed by various levels of government. As such, the measurements must meet minimum quality standards for precision and accuracy, which will be discussed in detail in the next chapter.

Gas measurement devices fall into three basic classes: (1) direct volume measurement, (2) rate of gas flow, and (3) gas velocity.

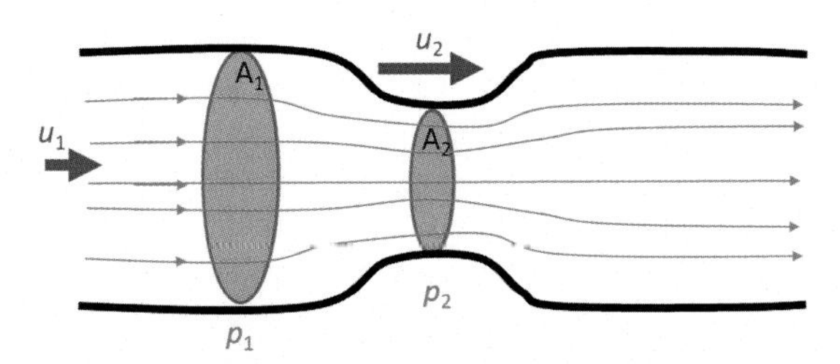

FIG. 10.9 The Bernoulli's principle and the effect of relationship between pressure energy, area, and velocity. As the cross-sectional area of flow decreases, the velocity increases, and the pressure decreases. At the center of the diagram, the increase in pressure results in decreasing ideal-gas velocity. *Note*: $p_2 < p_1$, $u_2 > u_1$, and $A_2 < A_1$.

Volume measurements are based on an average volumetric flow rate (Q), that is, the volume in motion per time:

$$Q = V t^{-1} \tag{10.45}$$

Commonly used devices for direct volume measurement include the spirometer, displacement bottle, bubble meter, mercury-sealed piston, wet test meter, dry gas meter, and positive displacement meter [13]. Gas flow measurement devices include the orifice meter, venture meter, capillary tube, rotameter, and mass flow meter. Gas velocity measurement is commonly conducted with variations of the pitot tube and electronic devices. These are each discussed in Chapter 11.

10.4 AEROSOLS

As mentioned, air pollution engineers and scientists generally consider substances to be in either the gas phase or the aerosol phase. Thus, any substance that is not a gas or vapor must be an aerosol. There are both scientific and everyday categories of aerosols. Dust is formed by disintegration of solid matter and dispersion of powder. Dust can include living and nonliving constituents, for example, pollen, house dust, roadway matter, and fly ash. Fog is vapor concentration, which appears as clouds, no matter its altitude, but in everyday use, fog is restricted to the earth's surface. Mists and sprays are produced by atomization and condensation, for example, pesticide sprays or industrial mists, for example, sulfuric acid. Smoke is often induced by thermal processes, for example, soot, diesel exhaust, and tobacco smoke. Smog is a combination of gases, especially ozone, and aerosols that form in the presence of sunlight. These and other classes of aerosols are used in scientific literature and in newspapers, newscasts, and everyday communications.

The key aspect of aerosols is that they exist in a dispersed system, that is, a two-phase system of a dispersed phase with a continuous medium [24]. Aerosols are solids and liquids whose continuous medium is a gas or mixture of gases, especially air [25]. Other dispersed systems in hydrosols, which are solids or liquids whose continuous medium is a liquid, for example, milk, are butter fat suspended in water, and latex paint is latex suspended in water. A foam is a gas suspended in a solid or liquid; for example, polyurethane foam's dispersed phase is air within a solid-phase plastic.

Many air pollutants exist in liquid and solid forms suspended in air, that is, aerosols. The aerosol itself is a pollutant, that is, particulate matter (PM), which varies in damage and harm that it causes by physical features, especially its diameter and shape. Aerosols may also be the vehicles that hold and carry air pollutants, that is, solid, liquid, and gaseous forms of harmful compounds that are sorbed[3] to or encapsulated within the aerosol.

The textbook definition of an aerosol is a suspension of solid or liquid particles in a gas. Of course, for air pollutants, this gas is usually air. Sometimes, the pollutant is solution in suspension or a suspension within suspension; for example, the pollutant is dissolved or suspended in a water droplet that is itself suspended in air. Also, as the pollutant is being formed, it may exist as a liquid and solid that is suspended in gases other than air in reactors, stacks, and flues. However, air quality agencies and regulators usually consider a substance to be merely a "potential air pollutant" until it reaches the air. It is the amount exiting the stack, vent, or flue that is regulated as stack emissions.

From an ambient air perspective, the amount of an air pollutant is measured in the air by pollution control agencies at the local, state, provincial, and federal agencies, which use various sampling methods. The method of choice is often determined first by the phase of the pollutant. For example, the mass of aerosols of a given size suspended in the air is sampled by pumping air through devices with inlets of a specific size to allow only particles of that size range to pass through and

3. Environmental engineers often use the term "sorbed," which is inclusive of all types of sorption. A substance is adsorbed if it is on the aerosol's surface, whereas the substance is absorbed if it is integrated into the aerosol. Other types of sorption include chemisorption and ion exchange. See Chapter 4 for a detailed discussion of sorption.

to collect on filters [13, 26]. The filters are weighed before and after the sampling period, for example, 1 or 24 h, to represent the amount of PM in the ambient air. These and other samples form the basis for actions needed, for example, in the United States, those delineated in the state implementation plans [8, 27]. A pollutant must first enter the atmosphere to be an air pollutant. Most nations and jurisdictions regulate a handful of substances under ambient air quality standards. Almost universally, the class of air pollutants known as particulate matter (PM) is one of these ambient air pollutants, which is an expression of all liquid and solid matter of a certain size that is suspended in air. In addition, PM is an important vector for pollutant transport of harmful chemical and biological agents, which are moved by advection along with the aerosol, for example, dissolved in or sorbed.

Certainly, regulations are oriented toward emissions and ambient concentrations of pollutants, but recent green engineering and pollution prevention are asking engineers, scientists, and managers to take a more expansive view of pollutants. For instance, optimizing conditions by substituting or varying concentrations of reagents in a reactor, changing pressure, temperature and other operational conditions in industrial processes, and deploying other mechanisms can produce fewer aerosols and gas-phase pollutants (see Fig. 10.10). In addition to eliminating pollutants, these system approaches can lead to more efficient air pollution controls, for example, producing PM of a size and shape that is more easily collected and treated compared with the PM that would have be generated by conventional processes [28, 29].

Compared with gases and vapors that exist as individual molecules in random motion, with each gas or vapor exerting its proportionate partial pressure, aerosols are aggregates of many molecules, sometimes of similar molecules, often of dissimilar ones. Aerosols age in the air by several processes. Some particles serve as nuclei upon which vapors condense. Some particles react chemically with atmospheric gases or vapors to form different compounds. When two particles collide in the air, they tend to adhere to each other because of attractive surface forces, thereby forming progressively larger and larger particles by agglomeration. The larger a particle becomes, the greater its weight and the greater its likelihood of falling to the ground rather than remaining airborne. The process by which particles fall out of the air to the ground is called *sedimentation*. Washout of particles by snowflakes, rain, hail, sleet, mist, or fog is a common form of agglomeration and sedimentation. Still, other particles leave the air by impaction onto and retention by the solid surfaces of vegetation, soil, and buildings. The particulate mix in the atmosphere is dynamic, with continual injection into the air from sources of small particles; creation of particles in the air by vapor condensation or chemical reaction among gases and vapors; and removal of particles from the air by agglomeration, sedimentation, or impaction.

Air pollution is generally associated with industry, transportation, energy production, and other human activities, that is, anthropogenesis. However, many harmful substances are natural in origin. For example, particles are emitted into the atmosphere from both human activities and natural sources. Particulate forms emitted naturally include condensed water vapor; the condensed and reacted forms of natural organic vapors; salt particles resulting from the evaporation of

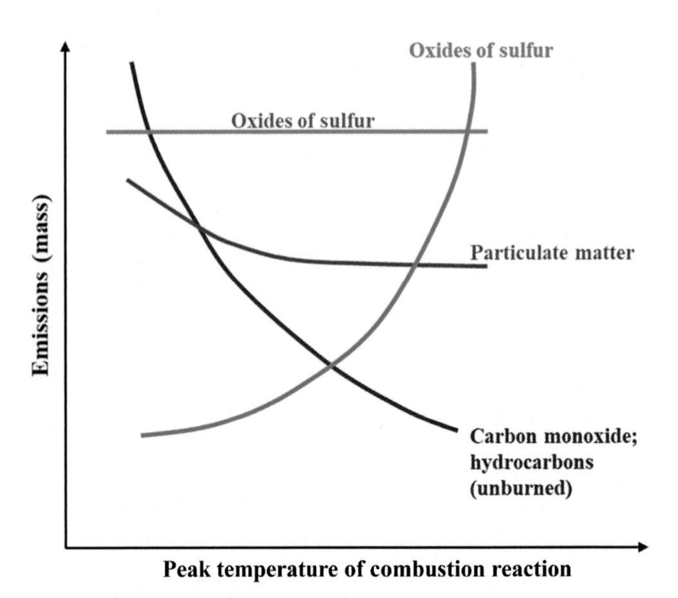

FIG. 10.10 Operating conditions change the mix of gas phase and particulate matter during combustion, leading to emissions. In this case, the emissions of oxides of sulfur remain unchanged in the peak temperature regime. The formation of oxides of nitrogen increases, and carbon monoxide decreases with increasing temperature. Particulates decrease up to a certain temperature, then reach steady state. *(Data from D.A. Vallero, Fundamentals of Air Pollution, 5th ed., Elsevier Academic Press, Waltham, MA, 2014, p. 999 pages cm.)*

water from sea spray; wind-borne pollen, fungi, molds, algae, yeasts, rusts, bacteria, and debris from live and decaying plant and animal life; particles eroded by the wind from beaches, desert, soil, and rock; particles from volcanic and other geothermal eruption and from forest fires started by lightning; and particles entering the troposphere from outer space.

Particulates are measured on a dry basis, thereby eliminating from the measurement not only water droplets and snowflakes but also all vapors, both aqueous and organic, that evaporate or are desiccated from the PM during the drying process. Since different investigators and investigative processes employ different drying procedures and definitions of dryness, it is important to know the procedures and definition employed when comparing data.

In contrast with gas-phase air pollutants that are unique chemical species, particles are first classified according to their physical properties. Indeed, the physical structure of the aerosol can play a large part in the damage it causes. For example, asbestos and other fibers cause damage to lung tissue predominately owing to their shape, for example, length, width, and angularity. Such physical properties also apply other types of particles found in the air, such as dust, dirt, soot, smoke, and liquid droplets. In contrast, the other ambient criteria pollutants, for example, O_3, CO, SO_2, and NO_2 in the United States, only require a concentration of the chemical species.

The other national ambient air pollutant in the United States is lead (Pb). Like other metals, Pb concentrations are total, that is, all gas-phase and aerosol-Pb species. However, since under standard atmospheric conditions, most Pb is in the aerosol phase, ambient Pb measurements are usually extractions of Pb from total suspended particulates (TSP) or PM_{10} [30]. The PM was previously only collected on glass fiber filters but recently allowed it also to be collected on polytetrafluoroethylene (PTFE) or quartz filter media. After extraction, for example, by heated ultrasonic methods, the extract is analyzed using inductively coupled plasma mass spectrometry (ICP-MS), a commonly accepted technique for metal analysis.

In addition to the physical properties, characterizing the hazards posed by aerosols depends on both the PM concentration and the concentrations of various chemical species that comprise the PM. An aerosol is usually not a specific chemical entity but is a mixture of particles from different sources and of different sizes, compositions, and properties. A particulate can be a single particle, for example, a crystal with density about the same as the parent mineral, like quartz. A particulate may also be an aggregate, that is, a group of particles held together by strong, for example, molecular, forces. An example is cement. A particulate may also be an agglomerate, that is, an adhering or cohering group of particles. An example is soot or other low-density structures. Finally, a flocculate is the weakest grouping that can be easily dispersed mechanically [31].

The chemical composition of PM is highly variable. In addition to its importance in determining the potential harm of PM, knowing the chemical composition of a particle can indicate its source; for example, receptor models use chemical composition and morphology of particles to trace pollutants back to the source [32].

The chemical composition of tropospheric particles includes inorganic ions, metallic compounds, elemental carbon, organic compounds, and crustal (e.g., carbonates and compounds of alkali and rare earth elementals) substances. The organic fraction can be particularly difficult to characterize, since it often contains thousands of organic compounds. The composition varies by source types, that is, stationary sources differ depending on the type of manufacturing, energy production, metal refining, and other factors, and mobile sources vary by fuel type, for example, diesel differs substantially from gasoline in the chemical composition of emitted PM [33]. In addition, activities such as biomass burning, wildfires, and volcanoes contribute PM of various compositions. Particulates also differ in composition by seasons and other temporal factors [34].

The size of a particle is determined by how the particle is formed. For example, combustion can generate very small particles, while coarse particles are often formed by mechanical processes (see Fig. 10.11). If particles are sufficiently small and of low mass, they can be suspended in the air for long periods of time. Larger particles (e.g., > 2.5 but <10 μm aerodynamic diameter) are found in smoke or soot (see Fig. 10.12), while very small particles (2.5 μm) may be apparent only indirectly, such as the way they diffuse, diffract, absorb, and reflect light (see Fig. 10.13).

Aerosol characteristics, including diameter and shape, vary by type of emissions. Air pollutants and precursors may be emitted directly to the air from stationary sources, such as factories and power plants, and from mobile sources, like cars and trucks. Aerosols can form directly with emissions, that is, immediately exiting the system such as exhaust from internal combustion engines or from stacks from power and industrial boilers and furnaces. Direct emissions almost always include PM. The PM emission can also be indirect, as when these and other particles are reentrained due to the movement of vehicles, for example, in a "near-road" situation. Area or nonpoint sources of particles include construction, agricultural activities such as plowing and tilling, mining, and forest fires.

Particulates may also form from gases that have been previously emitted, such as when gases released from burning fuels react with sunlight and water vapor. A common production of such "secondary particles" occurs when gases undergo chemical reactions in the atmosphere involving O_2 and water vapor (H_2O). Photochemistry can be an important step in secondary particle formation, resulting when chemical species like ozone (O_3) are involved in step reactions with radicals,

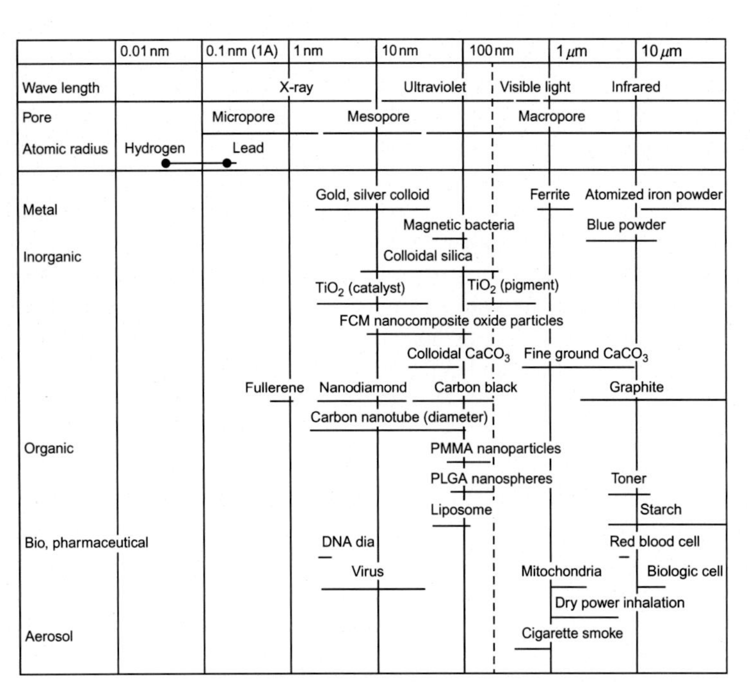

FIG. 10.11 Characteristics of particles according to diameter. *(Reprinted with permission from M. Hosokawa, T. Yokoyama, M. Naito, K. Nogi, T. Yokoyama, Nanoparticle Technology Handbook, Elsevier, 2007, https://doi.org/10.1016/S1748-0132(07)70119-6.)*

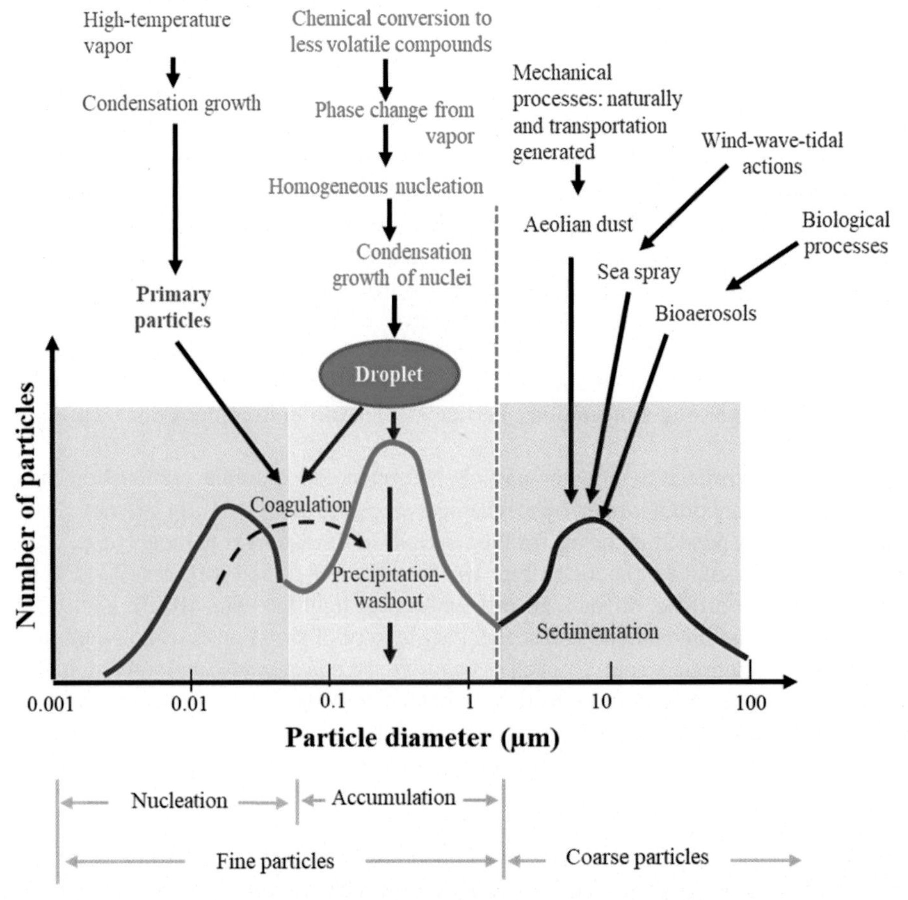

FIG. 10.12 Prototypical size distribution of tropospheric particles with selected sources and pathways of how the particles are formed. *Dashed line* is approximately 2.5 mm diameter. *(Based on information from K. Whitby, B. Cantrell, Atmospheric aerosols-characteristics and measurement, in: International Conference on Environmental Sensing and Assessment, Las Vegas, NV, 1976, p. 1; K. Whitby, K. Willeke, Single particle optical counters: principles and field use, in: Aerosol Measurement, 1979, p. 145; B.J. Finlayson-Pitts, J.N. Pitts Jr., Chemistry of the Upper and Lower Atmosphere: Theory, Experiments, and Applications, Elsevier, 1999.)*

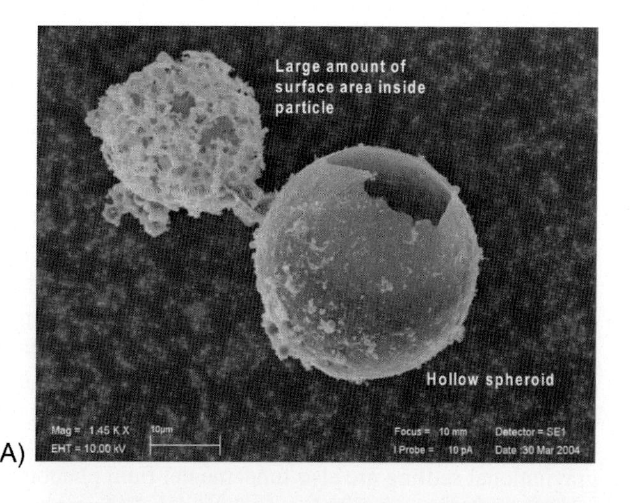

(A)

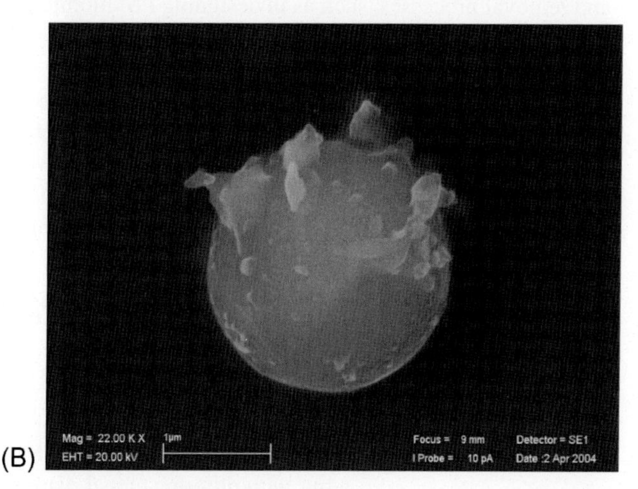
(B)

FIG. 10.13 Scanning electron micrograph of coarse aerosols emitted from an oil-fired plant (A) and a fine, spherical aluminosilicate aerosol emitted as fly ash (B). *(Photograph courtesy of US Environmental Protection Agency.)*

for example, the hydroxyl (OH) and nitrate (NO_3) radicals. Organic compounds also commonly form during these processes, that is, secondary organic aerosols [35, 36]. In addition, nucleation of particles from low-vapor-pressure gases emitted from sources or formed in the atmosphere, condensation of low-vapor-pressure gases on aerosols already present in the atmosphere, and coagulation of aerosols can contribute to the formation of particles. The chemical composition, transport, and fate of particles are directly associated with the characteristics of the surrounding gas.

The difference between a secondary aerosol and an indirect emission is that secondary aerosols were not aerosols until they subsequently formed from gas and vapor precursors after the emission. Conversely, an indirect emission consists of particulate matter that already existed but that was reemitted after a disturbance.

10.4.1 Particulate matter size and mass

Very small particles may remain suspended for some time. Their buoyancy allows them to travel longer distances than larger, heavier particles. Smaller particles are also challenging because they are associated with numerous health effects, mainly because they can penetrate more deeply into the respiratory system than larger particles.

Presently, the mass of PM falling in two major size categories:

- $PM_{2.5}$: <2.5 μm diameter
- PM_{10}: >2.5 μm and <10 μm diameter

Both size ranges are important air pollutants and differ in some of the harm they cause to human health and ecosystems. Thus, the measurements $PM_{2.5}$ and PM_{10} are commonly taken with colocated monitors but more commonly by dichotomous samplers, that is, instruments with inlets that are designed for size exclusions, that is, to segregate the mass of each size fraction [37]. Particles with diameters >10 μm are generally of less concern, since these particles rarely travel long distances. However, exposure to larger PM can occur near a large particulate-emitting source, for example, roadside or a nearby mine, especially if wind speeds are sufficiently high.

$PM_{2.5}$ and PM_{10} are defined by "diameters," which may imply that the particle is spherical. However, this is seldom the case. Instead, PM diameters are often described using an equivalent diameter, that is, the diameter of a sphere that would have the same fluid properties (see Table 10.5).

Aerosol diameter has been expressed in various ways. Microscopists characterize a particle geometrically according to its Feret's diameter (D_f) or Martin's diameter (D_m) (see Fig. 10.14). D_f is defined as the mean distance between two tangents drawn perpendicular to each other on opposite sides of the particle. This gives the longest dimension, either horizontal or vertical. D_m is the mean length of a chord that divides the particle into two equal areas. The projected area diameter is the diameter of circle with the same projected area as the particle [38, 39].

Mass can be determined for a nearly spherical particle by microscopy, either optical or electron; by light scattering and Mie theory; by the particle's electric mobility; or by its aerodynamic behavior. Another term, optical diameter, is the diameter of a spherical particle that has an identical refractive index as the particle. Optical diameters are used to calibrate the optical particle-sizing instruments, which scatter the same amount of light into the solid angle measured. Diffusion and gravitational settling are also fundamental fluid phenomena used to estimate the efficiencies of PM transport, collection, and removal processes, such as in designing PM monitoring equipment and ascertaining the rates and mechanisms of how particles infiltrate and deposit in the respiratory tract.

Only for extremely small-diameter particles, for example, nanoparticles and ultrafines, with diameters around 0.001 μm, is diffusion sufficiently important that the Stokes diameter is often used. The Stokes diameter for a particle is the diameter of a sphere with the same density and settling velocity as the particle. The Stokes diameter is derived from the aerodynamic drag force caused by the difference in velocity of the particle and the surrounding fluid. Thus, for smooth, spherical particles, the Stokes diameter is identical to the physical or actual diameter. The aerodynamic diameter (D_{pa}) for all particles >0.5 μm can be approximated[5] as the product of the Stokes particle diameter (D_{ps}) and the square root of the particle density (ρ_p):

$$D_{pa} = D_{ps}\sqrt{\rho_p} \tag{10.46}$$

TABLE 10.5 The particles with the same aerodynamic diameter (d_p) but with differing actual diameters (d) and density (r_p)

	Shape	Density (ρ_p) and diameter (d)	Aerodynamic diameter (d_p)
	Solid sphere	$\rho_p = 2.0\ g\ cm^{-3}$ $d = 1.4\ \mu m$	$d_p = 2.0\ \mu m$
	Hollow sphere	$\rho_p = 0.5\ g\ cm^{-3}$ $d = 2.8\ \mu m$	
	Irregular	$\rho_p = 2.3\ g\ cm^{-3}$ $d = 1.3\ \mu m$	

Source: D.A. Vallero, Fundamentals of Air Pollution, 5th ed., Elsevier Academic Press, Waltham, MA, 2014, p. 999 pages cm; adapted from U.S. Environmental Protection Agency, Chapter 4: particle collection mechanisms, in: J.R. Richards (Ed.), APTI Course 413: Control of Particulate Matter Emissions, 3rd ed., ICES Ltd, Research Triangle Park, NC, 2000. [Online]. Available: http://www.4cleanair.org/APTI/413Combined.pdf.

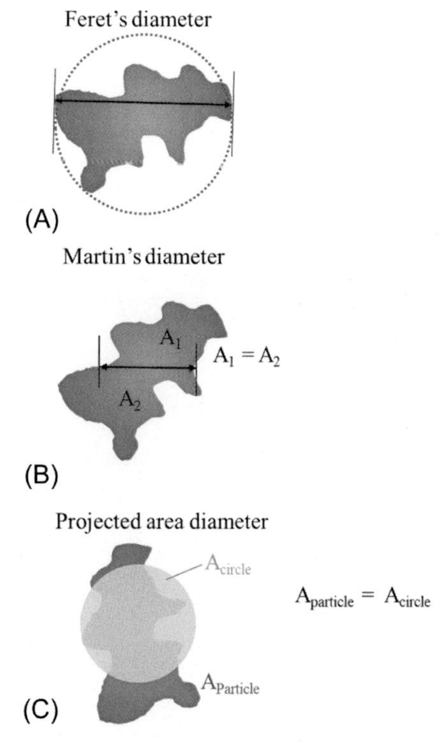

FIG. 10.14 Aerosol sizing techniques for same particle: (A) Feret's diameter, (B) Martin's diameter, and (C) projected area diameter. Note that for Martin's diameter, the particle orientation is random.

10.4.2 Particulate matter and respiratory fluid mechanics

In the United States, the Clean Air Act established the ambient standard (NAAQS) for particulate matter (PM), with notable changes in requirements over the decades since. In 1971, the US EPA required measurements of total suspended particulates (TSP) as measured by a high-volume sampler, that is, a device that collected a large range of sizes of particles (aerodynamic diameters up to 50 μm). Smaller particles are more likely to be inhaled than larger particles, so in 1987, the US EPA changed the standard for PM from TSP to PM_{10}, that is, particle matter $\leq$10 μm diameters. The diameter most often used for airborne particle measurements is the "aerodynamic diameter."

The NAAQS for PM_{10} became a 24 h average of 150 μg m^{-3} (not to exceed this level more than once per year) and an annual average of 50 μg m^{-3} arithmetic mean. However, subsequent research showed the need to protect people breathing even smaller PM in air, since most of the particles that penetrate deeply into the air-blood exchange regions of the lung are very small. Thus, in 1997, the US EPA added a new fine particle (diameters $\leq$2.5), known as $PM_{2.5}$.

Whereas gases flow with the air, aerosols deposit differentially within the airway. The main mechanisms are the same as those for any filter, removing or missing an inhaled aerosol.

The motion of air and gases carrying the aerosols within the respiratory system follow fundamental fluid dynamics theory [40]. The motion of these fluids is governed by the conservation of mass (continuity) equation and conservation of momentum (Navier–Stokes) equation. Under most conditions, the flow of air in the respiratory airways is assumed to be incompressible flow of a Newtonian fluid. For incompressible flow, the continuity equation is expressed as

$$\nabla \cdot \boldsymbol{u} = 0 \tag{10.47}$$

where ∇ is a gradient operator, that is, the vector differential operator "del," and $\boldsymbol{u}$ is the velocity vector field.[4] The gradient operator of the scalar function $f(x_1, x_2, \ldots x_n)$ is expressed as ∇f.

4. Bold type indicates a vector quantity.

In addition to assuming the fluid is incompressible, the continuity assumes that the velocity of the fluid at the position $\mathbf{r}_0(x, y, z)$ at time t is $\boldsymbol{u}(x, y, z, t)$ and that the velocity at the same time at a neighboring position $\mathbf{r}_0 + \mathbf{r}$ is $\boldsymbol{u} + \delta \boldsymbol{u}$. In matrix form, this is

$$\begin{pmatrix} \delta u \\ \delta v \\ \delta w \end{pmatrix} = \begin{pmatrix} \frac{\partial u}{\partial x} & \frac{\partial u}{\partial y} & \frac{\partial u}{\partial z} \\ \frac{\partial v}{\partial x} & \frac{\partial v}{\partial y} & \frac{\partial v}{\partial z} \\ \frac{\partial w}{\partial x} & \frac{\partial w}{\partial y} & \frac{\partial w}{\partial z} \end{pmatrix} \begin{pmatrix} x \\ y \\ z \end{pmatrix} \tag{10.48}$$

The vector form of this matrix is

$$\Delta \boldsymbol{u} = \mathbf{D} \cdot \mathbf{r} \tag{10.49}$$

where **D** is the deformation tensor, that is, the expression of how a nonuniform motion field is deforming the fluid element. Stated in three-dimensional coordinates,

$$\nabla f = \frac{\partial f}{\partial x}\mathbf{i} + \frac{\partial f}{\partial y}\mathbf{j} + \frac{\partial f}{\partial z}\mathbf{k} \tag{10.50}$$

where **i**, **j**, and **k** are standard unit vectors.

The Navier-Stokes equation is

$$\rho\left[\frac{\partial \boldsymbol{u}}{\partial t} + (u \cdot \nabla)u\right] = -\nabla p + \eta \nabla^2 u + \mathrm{f} \tag{10.51}$$

where ρ is the fluid density, p is the fluid pressure, η is the dynamic viscosity of the fluid, ∇^2 is the Laplacian operator $[\nabla^2 u = \partial^2_{xx} u + \partial^2_{yy} u + \partial^2_{zz} u]$, and **f** denotes the body forces per unit volume. The solution to the Navier-Stokes equation is a flow field or a velocity field, that is, the description of the velocity of the fluid at a specific point in space and time.

For cylindrical profiles like bronchi, the gradient operator ∇ can be expressed in cylindrical coordinates:

$$\frac{\partial}{\partial r} + \frac{1}{r}\frac{\partial}{\partial \theta_\theta} + \frac{\partial}{\partial z} \tag{10.52}$$

Thus, the continuity equation can also be expressed cylindrically:

$$\frac{1}{r}\frac{\partial}{\partial r}(ru_r) + \frac{1}{r}\frac{\partial}{\partial \theta}u_\theta + \frac{1}{r}\frac{\partial}{\partial z}u_z = 0 \tag{10.53}$$

where ∇_r, ∇_θ and ∇_z are the components of the fluid velocity, which are depicted in Fig. 10.15, that is, radial (r), circumferential (θ), and axial (z) directions, respectively.[5]

Airway velocities are complicated by numerous factors including lung and other tissue morphologies and the airway generations, that is, the branching levels through which the air is flowing. Equations can be tailored to these conditions or idealized velocity profiles can be assumed for the cascade of generations. These include parabolic flow (laminar fully developed), plug flow (laminar undeveloped), and turbulent flow [40]. For example, the upper tracheobronchial airways may be assumed to be turbulent, but in the pulmonary region, plug and parabolic profiles may be assumed [6].

The right and left lung are connected via their primary bronchi to the trachea and upper airway of the nose and mouth. From there, the bronchi, that is, airways, subdivide into a branching network of many levels. Each level, called a generation, is designated with an integer. The trachea are generation $n = 0$, the primary bronchi are generation $n = 1$, and so forth. Thus, theoretically, there are $2n$ airway tubes at generation n. For generations $0 \leq n \leq 16$, known as the conducting zone, gas flow is restricted to entry and exit in the airway [41]. In this zone, the air is moving, but there is no air-blood gas exchange of O_2 and CO_2.

The generations $n > 16$ comprise the respiratory zone, where air exchange takes place. Generations $17 \leq n \leq 19$ are where air sacs (alveoli), which range from 75 to 300 μm in diameter, become present on the airway walls. Alveoli are thin-walled and owing to the rich capillary blood supply are designed for gas exchange. These are the respiratory bronchioles.

5. For the derivation of the momentum equations mentioned here, see [14] D.A. Vallero, *Fundamentals of air pollution*, Fifth edition. ed. Waltham, MA: Elsevier Academic Press, 2014, p. 999 pages cm.

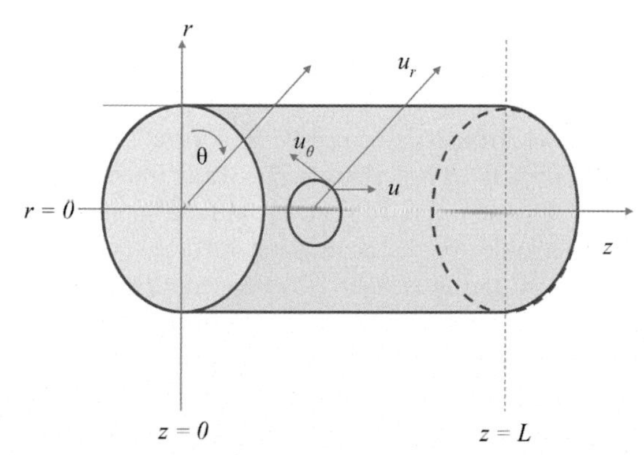

FIG. 10.15 Coordinate system for an ideal cylindrical airway, depicting velocity component at an arbitrary point. *(Data from K. Isaacs, J. Rosati, T. Martonen, L. Ruzer, N. Harley, Modeling Deposition of Inhaled Particles, Aerosols Handbook, CRC Press, Boca Raton, FL, 2012, pp. 83–128.)*

For generations $20 \leq n \leq 22$, the walls of the tubes or ducts consist entirely of alveoli. At generation $n = 23$, terminal alveolar sacs are made up of clusters of alveoli.

Two principal factors that are relevant to gas exchange are the airway volume (V_{aw}) and airway surface area (A_{aw}), which are proportional to the size of the person. Air exchange increases in proportion to A_{aw}. The V_{aw} (mL) for children is proportional to height and is approximated as [42]

$$V_{\text{aw}} = 1.018 \times \text{Height (cm)} - 76.2 \tag{10.54}$$

What is the airway volume of a 5 ft-tall child?

5 ft. = 1.524 m = 152.4 cm

$$V_{\text{aw}} = 1.018 \times 1.52.4 - 76.2 = 78.9\,\text{mL}$$

For adults, V_{aw} (mL) can be estimated by adding the ideal body weight (pounds) plus age in years [43].

What is the airway volume for 35-year-old adult whose ideal body weight is 75 kg and 60 year-old adult whose ideal body weight is 90 kg?

Adult 1:

75 kg = 165.3 lb.

$$V_{aw} = 165.3 + 35 = 200.3\,mL$$

Adult 2:

90 kg = 198.4 lb.

$$V_{aw} = 198.4 + 60 = 258.4\,mL$$

Thus, the combination of weight and age gives a much higher airway volume for the *second* adult [44].

The average lung has from 300 to 500 million alveoli. An average adult lung has about 70 m^2 alveolar surface area. This large A_{aw} allows for not only efficient gas exchange to supply O_2 for normal respiration but also large increases in gas exchange needed when a person is stressed (e.g., during exercise, injury, or illness).

The lung consists of fluids of various types, both Newtonian and non-Newtonian. The Reynolds number varies by generation (very high in the trachea but low in the alveoli) [41]. Airways have liquid lining, with two layers in the first generations (up to about $n = 15$). A watery (Newtonian), serous layer is next to the airway wall. In this layer, the cilia pulsate toward the mouth. A mucosa cover over this layer has several non-Newtonian fluid properties, for example, viscoelasticity, shear thinning, and a yield stress.

Alveolar cells produce surfactants that orient at the air-liquid interface and reduce the surface tension significantly. Air pollutants can adversely affect the surfactant chemistry, which can make the lungs overly rigid, thus hindering inflation [41].

A pulmonary surfactant is a surface-active lipoprotein complex (phospholipoprotein) produced by type II pneumocytes, which are also called alveolar type II cells. These pneumocytes are granular and comprise 60% of the alveolar lining cells. Their morphology allows them to cover smaller surface areas than type I pneumocytes. Type I cells are highly attenuated, very thin (25 nm) cells that line the alveolar surfaces and cover 97% of the alveolar surface. Surfactant molecules have both a hydrophilic head and a lipophilic tail. Surfactants adsorb to the air-water interface of the alveoli with the hydrophilic head that collects in the water, while the hydrophobic tail is directed toward the air. The principal lipid component of surfactant, dipalmitoylphosphatidylcholine, is responsible for the decreasing surface tension. The actual surface tension decreases depend on the surfactant's concentration on the interface. This concentration's saturation limit depends on temperature and the presence of other compounds in the interface. Surface area of the lung varies during compliance (i.e., lung and thorax expansion and contraction) during ventilation. Thus, the surfactant's interface concentration is seldom at the level of saturation. During lung expansion (inspiration), the surface increases, opening space for new surfactant molecules to join the interface mixture. During expiration, lung surface area decreases, compressing the surfactant and increasing the density of surfactant molecules, thus further decreasing the surface tension. Therefore, surface tension varies with air volume in the lungs, which protects the lungs from atelectasis at low air volume and form tissue damage at high air volume [45–47].

The mechanisms in Fig. 10.15 are affected by aerosol size. The size of the particle is indirectly proportional to its depth of lung penetration. This aerosol measurement is an expression of the mass of a particle size. Prior to 1997, the primary (health) standard for PM_{10} was a 24 h average of 150 μg m^{-3} (not to exceed this level more than once per year) and an annual average of 50 μg m^{-3} arithmetic mean. However, this revised standard did not provide sufficient protection for people breathing PM-contaminated air, since most of the particles that penetrate deeply into the air-blood exchange regions of the lung are quite small (see Fig. 10.16). The smallest particles, known as the respirable fraction, required additional attention. As mentioned, the US EPA added the $PM_{2.5}$ standard in 1997 [48].

As mentioned, lung filtration is like any filter, which consists of four mechanical processes: (1) diffusion, (2) interception, (3) inertial impaction, and (4) electrostatics (see Fig. 10.17). These will be revisited along with other mechanisms, especially gravitational settling, turbulent mixing, and phoresis, in Chapter 12, since they are also at work in aerosol control technologies. However, they deserve present attention given their importance in lung fluid dynamics.

Diffusion is important only for very small particles ($\leq$0.1 μm diameter) because the Brownian motion allows them to move away in a "random walk" away from the airstream. Diffusion is important only for very small particles ($\leq$0.1 μm diameter) because the Brownian motion allows them to move in a random walk away from the airstream.

Interception works mainly for particles with diameters between 0.1 and 1 μm. The particle does not leave the airstream but encounters matter (e.g., lung tissue).

Inertial impaction results from a particle's momentum, that is, mass times acceleration. If the momentum is large enough, the particle is not able to follow a change in gas flow direction, for example, in the upper airway or at bifurcations in the airway [49], eventually contacting airway walls. Inertial impaction, whether in the lung, an air monitor, or a pollution control device, collects particles that are sufficiently large to leave the airstream by inertia (see Fig. 10.18). This can include particles with diameters $\geq$1 μm [50] but mainly particles with diameters $\geq$5 μm [49]. Note that drag forces and inertial forces are competing, that is, drag acts in the opposite direction of the particle's motion. Thus, the particle's deviation from the air streamline increases with particle mass and flow rate, which can be expressed as a function of the dimensionless

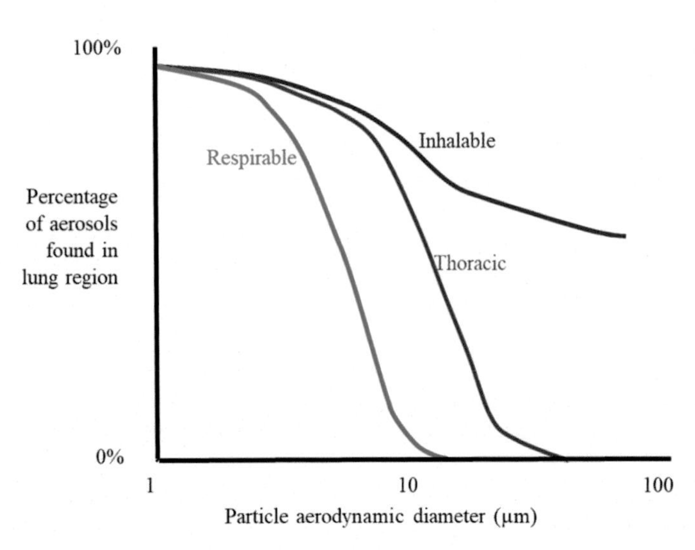

FIG. 10.16 Three regions of the respiratory system where particulate matter is deposited. The inhalable fraction remains in the mouth and head area. The thoracic fraction is the mass that penetrates the airways of the respiratory system, while the smallest fractions, that is, the respirable particulates, are those that can infiltrate most deeply into the alveolar region. *(Reproduced with permission from D.A. Vallero, Fundamentals of Air Pollution, 5th ed., Elsevier Academic Press, Waltham, MA, 2014, p. 999 pages cm.)*

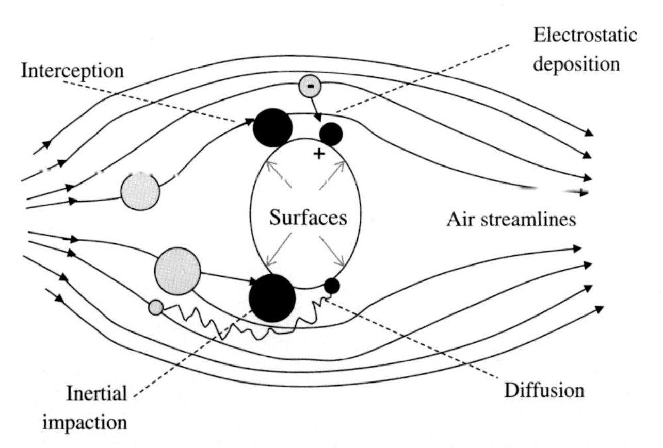

FIG. 10.17 Mechanical processes involved the deposition of particulate matter on a filter or in the lung. *(Source: D.A. Vallero, Fundamentals of Air Pollution, 4th ed., Elsevier, Amsterdam; Boston, 2008, pp. xxiii, 942 p; Reproduced with permission from K.L. Rubow, Filtration: fundamentals and applications, in: Aerosol and Particle Measurement, University of Minnesota, Minneapolis, MN, 2004.)*

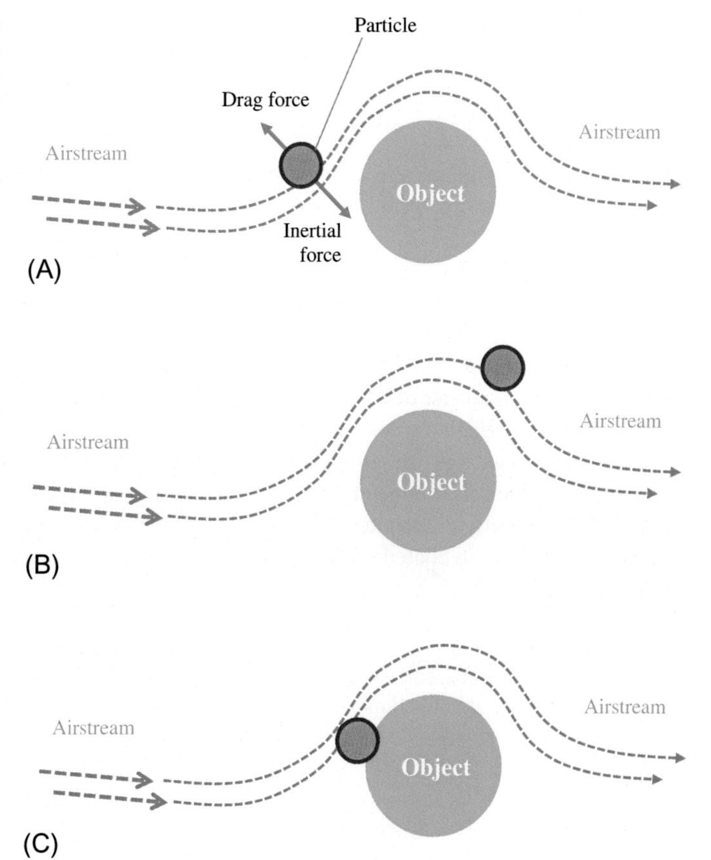

FIG. 10.18 Inertial impaction. The airstream flows around the object (A). If the momentum around the airstream and around the object and drag force are higher than the inertial force, the particle will miss the object (B). However, if the inertial force sufficiently exceeds the drag force and the particle velocity is high enough, the particle will contact the object and be captured (C). *(Source: D.A. Vallero, Excerpted and Revised From: Fundamentals of Air Pollution, 5th ed., Elsevier Academic Press, Waltham, MA, 2014, p. 999 pages cm.)*

Stokes number, that is, the ratio of the characteristic time of the particle to the characteristic time of the flow or of an obstacle:

$$\text{Stk} = \frac{t_0 u_0}{l_0} \tag{10.55}$$

where t_0 is the particle's relaxation time, that is, the time constant of exponential decay of velocity due to drag; u_0 is the fluid velocity not affected by the obstacle; and l_0 is the characteristic dimension of the obstacle, especially diameter.

When $Re < 1$, that is, a Stokes flow, the drag coefficient of the particle is inversely proportional to the Re. Under Stokes flow conditions, the relaxation time is

$$t_0 = \frac{\rho_p d_p^2}{18\mu_g} \tag{10.56}$$

where ρ_p is the particle density; d_p is the particle diameter; and μ_g is the dynamic viscosity of the carrier gas or gas mixture, that is, in the case of respiratory fluid dynamics, the air.

Therefore, the Stokes number can be restated as [49]

$$\text{Stk} = \frac{\rho_p d_p^2 \dot{u}}{18\mu_g d} \tag{10.57}$$

where $\dot{u}$ is the mean velocity of the air and d is a characteristic length equal to the diameter of the airway.

The smaller the Stk, the more likely the particle will follow the streamline. Conversely, a particle with a large Stk is dominated by its inertia and is more likely to continue along its trajectory.

The inertial force competing with the drag force is similar to the competition between gravitational and centrifugal forces. Indeed, impaction efficiencies may be evaluated with the same approaches used for gravitational and centrifugal forces. For particles with $Re < 1$, impaction effectiveness can relate to a dimensionless inertial impaction parameter (Ψ), calculated as

$$\Psi = \frac{C_C d_p^2 3 y_p \rho_p}{18\mu_g D_C} \tag{10.58}$$

where C_C = Cunningham slip correction factor (dimensionless), d_p = physical particle diameter (cm), y_p = difference in velocity and collection object (cm sec^{-1}), D_C = diameter of the collection object (cm), ρ_p = density of particle, and μ_g = gas velocity (g $cm^{-1}s^{-1}$).

With increasing Ψ values, particles are more likely to move radially toward a collection surface. Conversely, smaller Ψ values mean that the particle is likely to remain in the airstream and miss the target collection surface. Therefore, larger diameter, faster moving, and denser particles are more likely to be captured by inertial impaction. Also, smaller collection targets (e.g., fine wires or filter fibers) increase the likelihood of inertial capture.

Electrostatics consist of electric interactions between the atoms in the filter and those in the particle at the point of contact (van der Waals forces) and electrostatic attraction (charge differences between particle and filter medium).

Other important factors affecting filtration efficiencies include the thickness and pore diameter or the filter; the uniformity of particle diameters and pore sizes; the solid volume fraction; the rate of particle loading onto the filter (e.g., affecting particle "bounce"); the particle phase (liquid or solid); capillarity and surface tension (if either the particle or the filter media are coated with a liquid); and characteristics of air or other carrier gases, such as velocity, temperature, pressure, and viscosity.

What is the inertial impaction of a particle with an aerodynamic diameter equal to 20 μm and a density of 2000 kg m^{-3} inhaled by an adult, if the velocities are [51] the following?

Airway	Airway diameter (mm)	Airway length (mm)	Mean air velocity (m sec^{-1})	Residence time (sec)
Mouth	20	70	3.2	0.022
Pharynx	30	30	1.4	0.021
Trachea	18	120	4.4	0.027
Left main bronchus	13	37	3.7	0.01
Right main bronchus	13	37	3.7	0.01

Assume that the air temperature is 25°C, that is, the dynamic viscosity is about 1.84×10^{-5} kg $m^{-1}s^{-1}$.

$$\text{Stk} = \frac{\rho_p d_p^2 \dot{u}}{18\mu_g d} = \frac{2000 \times (20 \times 10^{-6})^2 \times \dot{u}}{18 \times 1.84 \times 10^{-5} \times d} = 2.415 \times 10^{-3} \frac{\dot{u}}{d}$$

$$\text{Mouth}: \text{Stk} = 2.415 \times 10^{-3} \times \frac{3.2}{20 \times 10^{-3}} = 0.386$$

$$\text{Pharynx}: \text{Stk} = 2.415 \times 10^{-3} \times \frac{1.4}{30 \times 10^{-3}} = 0.113$$

$$\text{Trachea}: \text{Stk} = 2.415 \times 10^{-3} \times \frac{4.4}{18 \times 10^{-3}} = 0.590$$

$$\text{Left main bronchus}: \text{Stk} = 2.415 \times 10^{-3} \times \frac{3.7}{13 \times 10^{-3}} = 0.687$$

$$\text{Right main bronchus}: \text{Stk} = 2.415 \times 10^{-3} \times \frac{3.7}{13 \times 10^{-3}} = 0.687$$

Since the greater the Stk is above 1, the more likely the particles will impact with the airway walls and deposit, and conversely, the further the Stk is below 1, the likelihood increases for the particle to follow the air streamline. Thus, a substantial number of particles would be expected to deposit on the walls of the trachea and bronchi, since their Stk values are relatively close to unity.

How would the inertial impaction change for a 2.5 μm-diameter particle of the same density as above?

$$\text{Stk} = \frac{\rho_p d_p^2 \dot{u}}{18\mu_g d} = \frac{2000 \times (2.5 \times 10^{-6})^2 \times \dot{u}}{18 \times 1.84 \times 10^{-5} \times d} = 3.774 \times 10^{-5} \frac{\dot{u}}{d}$$

$$\text{Mouth}: \text{Stk} = 3.774 \times 10^{-5} \times \frac{3.2}{20 \times 10^{-3}} = 0.006$$

$$\text{Pharynx}: \text{Stk} = 3.774 \times 10^{-5} \times \frac{1.4}{30 \times 10^{-3}} = 0.002$$

$$\text{Trachea}: \text{Stk} = 3.774 \times 10^{-5} \times \frac{4.4}{18 \times 10^{-3}} = 0.009$$

$$\text{Left main bronchus}: \text{Stk} = 3.774 \times 10^{-5} \times \frac{3.7}{13 \times 10^{-3}} = 0.011$$

$$\text{Right main bronchus}: \text{Stk} = 3.774 \times 10^{-5} \times \frac{3.7}{13 \times 10^{-3}} = 0.011$$

Thus, even in the bronchus, impaction is less likely, and smaller particles will continue past these airway walls and will depend on other mechanisms and decreases in velocity to deposit. This is one of the reasons for the deposition patterns depicted in Fig. 10.16, wherein smaller diameter particles penetrate more deeply into the respiratory system than do coarser particles.

10.5 CONCENTRATION

Whether in the gas or aerosol phase, air pollutants and other air constituents are expressed as concentrations in air. Air pollution literature always specifies the concentration of the chemical of concern. In fact, the term was used over 40 times in Chapter 1 and several times already in this chapter. Indeed, many of the calculations conducted in air quality work involve concentrations of the reactants and the products in chemical reactions in which air pollutants are formed. For example, the engineer needs to know the concentration of reactants in an industrial process so that the proper air pollution control devices are installed. The regulator needs to know the concentration of a pollutant that can be emitted from a stack in combination with all other sources so that the air quality is protected. This protection is an expression of mass (e.g., so many tons released) and concentration; for example, these tons would be expected to lead to ambient concentrations of the pollutant in the air that people breathe.

As mentioned in the previous answer, smaller air constituents, including all air pollutants, can be expressed in numerous ways. Fundamentally, chemical concentration is of one of two types. It is either a number density, that is, molecules per unit volume, such as molecules cm^{-3}, or a mass density, such as micrograms per cubic meter ($\mu g\,m^{-3}$).

10.5.1 Volume and mass conversions

When a concentration is expressed simply as ppm or ppb, it is unclear whether a volume or weight basis is intended. To avoid confusion caused by different units, air pollutant concentrations in this book are generally expressed as mass per volume, for example, micrograms per cubic meter of air ($mg\,m^{-3}$), at a given temperature (e.g., 20 or 25°C) and a given atmospheric, for example, 1 atm (atm), which is 760 mm mercury (mm Hg).

Note that some of the units used in air pollution engineering are numbers, that is, numbers of "particles."[6] Table 10.1 lists gram-mole (gmol) and pound-mole (lbmol). Often, air pollution calculations use kilogram-moles, that is, kgmoles (kgmol). A mole is an expression of the number of particles. The International System of Units, that is, the SI system, describes the mole as having about 6.023×10^{23} particles. The kgmol equals 1000 gmol, so it has 6.023×10^{26} particles. A lbmol is equal to 453.59237 mol. This is demonstrated by dividing the mass (m) in grams by molecular weight (M):

$$1\ \text{gmol} = \frac{m\,[\text{in grams}]}{M} \tag{10.59}$$

Similarly,

$$1\text{lbmol} = \frac{m\,[\text{in pounds}]}{M} \tag{10.60}$$

Thus, gmol, lbmol, and kgmol are expressions and units of n, that is, the number of moles.

The molar volume of an ideal gas at 273 K and 1 atm, that is, STP, will occupy $22.414\,m^3$.

Air pollution is expressed as concentrations, especially chemical concentrations. However, concentration is expressed in different ways. The vapor phase of formaldehyde provides example of a number density concentration of air pollutant. For instance, we may measure the air and find 1.2×10^{12} formaldehyde (CH_2O) molecules per $cm^3 = 1.2 \times 10^{18}$ molecules per m^3. Note that this is a unitless density because both the chemical and the air are expressed as the number of molecules. As we stated previously, a unitless density requires that the units of the substance must be the same as the units of the medium.

Air pollution literature commonly uses the unspecified term "parts" rather than specific terms, for example, molecules. Gas concentrations are commonly reported as parts per million (ppm). For hazardous air pollutants like formaldehyde, they are usually reported in parts per billion (ppb) or parts per trillion (ppt). These all assume this is parts of formaldehyde per parts air. However, the parts of contaminant per parts fluid would differ. If this was a water pollution text, the parts would be parts of formaldehyde per parts water. If this was a pharmaceutical or biomedical engineering text, the parts may be parts of formaldehyde per parts blood, fat, or serum.

Often, air pollutant concentration is expressed as a mass density (e.g., 12 micrograms per cubic meter ($\mu g\,m^{-3}$) of air). A number density can be converted to mass density using the gram molecular weight (mole) per volume. Avogadro's constant (N_A) is the number of constituent particles (molecules or atoms). Given that one mole (mol) of any particle contains 6.022×10^{23} molecules, we can use the number density of formaldehyde to find its mass density from its number of moles per volume [52]:

$$\left[1.2 \times 10^{18}\,CH_2O\,\text{molecules per}\,m^3\right] \div 6.022 \times 10^{23}\,\text{molecules} \approx 2 \times 10^{-6}\,\text{mol}\,m^{-3}$$

In this case, the number density of 1.2×10^{12} formaldehyde (CH_2O) molecules per cm^3 is the same as the mass density $2 \times 10^{-6}\,\text{mol}\,m^{-3}$ or 2 μmol (μmol).

6. "Particles" here are the physics connotation of the smallest, elementary particle, for example, atoms and molecules, and not the air pollution shorthand meaning for particulate matter or aerosols.

At 25°C and 760 mmHg, express 15 ppm benzene, 180 ppb ozone. and 500 ppm of sulfuric acid in μg m^{-3}.
To convert from units of ppm (vol.) to mg m^{-3}, we assume that the ideal-gas law is accurate under ambient conditions. Although benzene and sulfuric acid are vapors and ozone is a gas, we are treating all three as ideal gases. The error introduced from doing so is often sufficiently small and ignored in tropospheric calculations.

This allows for a generalized formula for the conversion at 25°C and 760 mmHg:

$$1\,\text{ppm (vol) pollutant} = \frac{1\,\text{L pollutant}}{10^6\,\text{L air}} = \frac{\left(\frac{1\,\text{L}}{22.4}\right) \times \text{MW} \times 10^6 \times \mu\text{g} \times \text{g} \times \text{m}^{-1}}{10^6\text{L} \times 298\text{K}^{\circ} \times 10^{-3}\text{m}^3 \times \text{L}^{-1}} \quad (10.61)$$

$$= 40.9 \times \text{MW}\,\mu\text{g}\,\text{m}^{-3}$$

where MW is the molecular weight in grams per mole (g mol^{-1}).

Thus, to answer this question, we apply the general relationship:

Gas-phase contaminant concentration (μg m^{-3}) = contaminant concentration (ppm) × molecular weight of the contaminant × 40.9.

Thus, x μg m^{-3} of $C_6H_6 = 15 \times 78.11 \times 40.9 = 47{,}920\mu g m^{-3}$

y μg m^{-3} $O_3 = 0.180 \times 48 \times 40.9 = 353\mu g m^{-3}$

z μg m^{-3} $H_2SO_4 = 500 \times 98 \times 40.9 = 2{,}004{,}100\mu g m^{-3}$ (this is a very high concentration of an air pollutant, that is, 2 g in cubic meter of air!)

10.5.2 Concentration gradients

In previous chapters, the concept of the concentration gradient was introduced and explained, like numerous other air pollution concepts. Recall that it is the change in concentration with distance. Among other applications related to air pollution, concentration gradient calculations are needed to aid in linking sources to exposed populations.

What is the concentration of 2 μmol of formaldehyde if expressed in units of micrograms per cubic meter?
The 2 μmol formaldehyde per m^3 of air can be converted to the more common units of μg m^{-3}. The atomic mass of carbon (C) is 12, hydrogen (H) is 1, and oxygen (O) is 16, so the atomic mass of CH_2O is 30. Thus, 1 mol CH_2O = 30 g = 3×10^{-5} μg.

Since 1 μmol = 10^{-6} mol, then 2 μmol formaldehyde = $2 \times 30 \times 10^{-6}$ g = 6×10^{-5} g = 60 μg. Expressed as a mass density, 2 μmol per m^3 CH_2O in air is the same as a chemical concentration of 60 μg m^{-3}. Mass density is more commonly used than number density to express air pollutant concentrations. However, the expression of mass density is highly variable, especially when applied by standards and regulations. Even in the same standards, different units are used (e.g., ppb and μg m^{-3}), depending on the jurisdiction.

The World Health Organization (WHO) is an excellent source of more information about the physical, chemical, and biological characteristics and health effects from formaldehyde [53] and other air toxics [54]. The WHO, for example, includes conversion factors for gaseous compounds. For formaldehyde, at pressure (P), these are the following:

At 20°C, 1 ppm = 1.249 mg m^{-3}, and 1 mg/m^3 = 0.801 ppm.

At 25°C, 1 ppm = 1.228 mg/m^3, and 1 mg/m^3 = 0.814 ppm.

The EPA [55], Agency for Toxic Substances and Disease Registry (ATSDR) [56, 57], National Institute of Occupation Safety and Health (NIOSH) [58], and the Consumer Products Safety Commission [59] also provide information on air toxics, including formaldehyde. The companion to this book, *Fundamentals of Air Pollution*, considers formaldehyde and other hazardous air pollutants from various perspectives, for example, exposure, risk, transport, and fate [14].

In the formaldehyde example discussed above, we said that it would be preferable to take measurements at the potential source and at incremental distances from the source. Put in simplest terms, a concentration gradient is the change in concentration of chemical A with distance x:

$$\frac{d[A]}{dx} \quad (10.62)$$

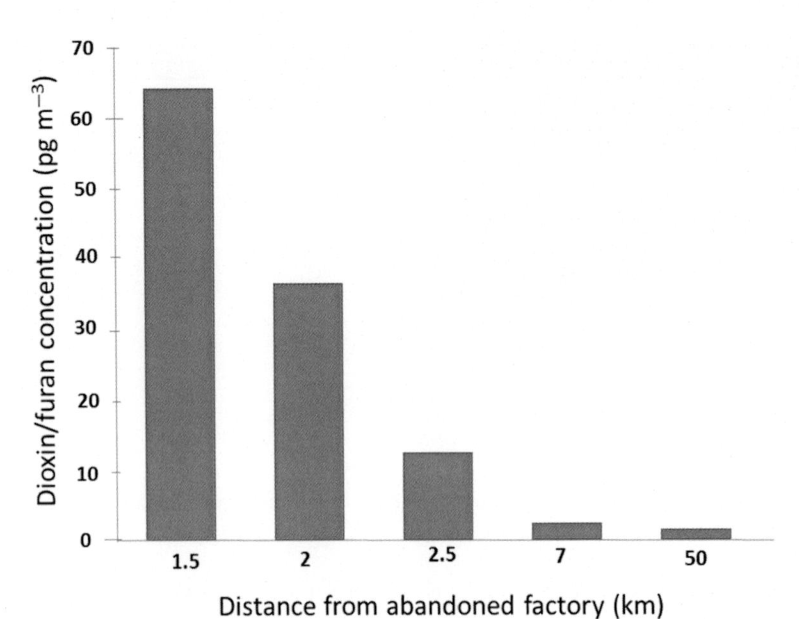

FIG. 10.19 Total concentrations in picograms per cubic meter ($pg\,m^{-3} = 10^{-12}\,g\,m^{-3}$) of octachlorodibenzodioxin, heptachlorodibenzo-p-dioxin, and octachlorodibenzofuran in ambient air at distances away from an abandoned pentachlorophenol plant factory in China. *(Data from C. Li et al., Long-term persistence of polychlorinated dibenzo-p-dioxins and dibenzofurans in air, soil and sediment around an abandoned pentachlorophenol factory in China, Environ. Pollut. 162 (2012) 138–143.)*

In air pollution, the concept of gradient becomes much more complicated because the air moves, and chemical reactions take place after a substance is released. Some chemical compounds are highly persistent, so that they do not change substantially, for example, break down or form new compounds, after release.

An example of such compounds is the chlorinated dioxins and furans (see Fig. 10.19). Three compounds measured at distances away from the potential source, an abandoned pentachlorophenol (PCP) factory [60]. All three compounds are very persistent, so it is likely they did not form after being released. Note that the concentrations decrease with distance from the PCP plant. This is a concentration gradient. This is a negative concentration gradient, that is, the concentration is decreasing with distance from the source. The slope of the gradient appears almost linear near the plant and begins to flatten at 7 km. However, even at 50 km away, however, the compounds are still detected. Thus, at 50 km, the gradient has not reached steady state, that is, zero gradient. Once steady state is reached, this could be the background concentration for these compounds, that is, the concentration not directly attributable to this source.

How is the concentration gradient used in air pollution models?

The concentration gradient is a thermodynamic phenomenon. The zone of influence for an air pollution model is denoted by a control volume. As mentioned, air pollution concentration gradients are more complex than the simple concept described in the previous answer. In addition to thermodynamics, the air pollution gradient must also incorporate the laws of motion. Thus, we must consider not only distance (x) but also time (t):

$$\frac{d[A]}{dt} = -v \cdot \frac{d[A]}{dx} + \frac{d}{dx}\left(D\frac{d[A]}{dx}\right) + r \qquad (10.63)$$

where [A] is the concentration of chemical A, u is the velocity of the air, D is a diffusion coefficient, and r represents the net contribution or decrease in concentration resulting from reactions within the control volume. Note that the right-hand side of the equation is negative. This means that the concentration decreases in time and distance from the source of chemical A. If u is known, then t can be found. An obvious example is if $u = 0.5\,m\,sec^{-1}$, after 1000 s, then $x = 500\,m$. Of course, u is always variable in the real world, so measures of central tendency must be defined, for example, mean or median t and x.

Environmental and other regulatory agencies frequently require that the air in which a pollutant exists, for example, an emission, be converted to dry gas values. Indeed, in the United States, some agencies require reporting in units of standard cubic foot of dry air (SCFD). Other countries require similar corrections. For example, licenses issued by the Irish Environmental Protection Agency often require that emission data from combustion sources be reported on a dry gas basis, that is, any pollutant sampling by stack tests must include simultaneous readings of stack moisture [61].

A measurement of nitrogen dioxide (NO_2) released from a power plant stack is 180 ppm by volume (ppmv). The air contains 10% H_2O by volume. What is the dry basis for this sample?

Wet basis of a gas can be converted to dry basis by

$$C_{dry} = \frac{C_{wet}}{1 - w} \tag{10.64}$$

where C_{dry} is the dry basis concentration; C_{wet} is the dry basis concentration; and w is the fraction, by volume, of the gas that is water vapor. Thus, we can find the dry basis of the NO_2:

$$C_{dry} = \frac{180}{1 - 0.10} = 200\ ppmv$$

The next day, the process was changed. NO_2 released from a power plant stack was found to be 150 ppmv, but the air's humidity increased to 30% H_2O by volume. The plant manager claims that he improved the emissions by 50 ppmv. You are not convinced and calculate the dry basis for this sample. Are you correct?

The dry basis for the Day 2 NO_2 measurement is

$$C_{dry} = \frac{150}{1 - 0.3} = 214\ \text{ppmv}$$

Thus, even though the concentration is 50 ppmv less than NO_2 on a wet basis, tripling the moisture made the NO_2 concentration higher by 14 ppmv using the new process. *So, indeed, you were correct.*

The state pollution control agency's emission permit limit for particulate matter (PM) emissions is not to exceed 750 mg m^{-3}. They inform you that oxygen content can also vary in air pollutant emission measurement. This is important since O_2 and CO_2 concentrations interrelate. This applies to both gases and aerosols. After correcting to a dry gas with a specified CO_2 content of 10%, PM concentrations in the emission are found to be 400 mg m^{-3} in a dry gas with 5% CO_2. A junior engineer is pleased that the emission limits are being met. Correcting for oxygen, do you agree?

The corrected concentration of a dry gas with a specified reference volume percentage of CO_2 (C_r) is

$$C_r = C_m \cdot \frac{(referencevolume\%CO_2)}{(measuredvolume\%CO_2)} \tag{10.65}$$

where C_m is the measured concentration of a dry gas with a measured volume percentage of CO_2. Thus, C_r is

$C_r = 400 \cdot \frac{(10)}{(5)} = 800 mg m^{-3}$. The corrected PM emission concentration is twice as high as the uncorrected emission concentration. This exceeds the permit limit of 750 mg m^{-3}. *So, you do not agree and must inform the junior engineer that the existing processes and controls must be changed.*

Increasingly, air pollutants are monitored on a continuous basis, that is, using continuous emission monitoring systems (CEMS). These systems may be needed to ensure ongoing compliance with a specific pollutant emission limit or to monitor compliance with source and pollution control equipment operating limits [62]. Like other monitors, however, corrections are needed for CEMS. The regulatory agency requests that your CEMS measurements be corrected to a standard diluent oxygen concentration. Your continuous readings give a mean 24 h CO concentration of 200 ppmv, 5% O2 and 12% CO_2. Correct the CO concentration to a dry concentration, 3% oxygen and 12% carbon dioxide

Using Eq. (10.4), you find that the dry basis CO is 250 ppmv. To find the wet basis for O_2, use and solve wet basis [62]:

$$\frac{C_{O2wet}}{1 - w} \tag{10.66}$$

$$= \frac{5}{1 - 0.8} = 6.25\%, \text{dry}$$

And, for the dry basis for CO_2, use

$$\frac{C_{CO2dry}}{1-w} \tag{10.67}$$

$$=\frac{12}{1-0.8}=15\%,\text{dry}$$

Since O_2 represents 20.9% of air, the following equation can be used:

$$C_{corr}=C_{meas}\cdot\frac{20.9-X}{20.9-Y} \tag{10.68}$$

where C_{corrO2} is the concentration corrected for O_2, C_{meas} is the CEMS reading, X is the target corrected percent volumetric O_2, and Y is the measured average percent volumetric O_2. Thus, the oxygen-corrected concentration for the target 3% O_2 is

$$C_{CO,3\%02}=C_{CO,dry}\frac{20.9-3}{20.9-6.25}=250\cdot\frac{20.9-3}{20.9-6.25}=305.5\,ppmv,dry\ at\ 3\%O_2$$

And the oxygen correction for target 12% CO_2 is

$$C_{CO,3\%02}=C_{CO,dry}\frac{20.9-3}{20.9-6.25}=250\cdot\frac{12}{15}=200\,ppmv,dry\ at\ 12\%CO_2$$

As mentioned, the physical state of a substance is a function of pressure and temperature. Since air pollution is generally a tropospheric phenomenon, a substance is generally considered to be a solid, liquid, or gas under "normal" atmospheric conditions, for example, 1 atm at 20 °C. Thus, molecular oxygen and nitrogen, O_2 and N_2, are considered gases, even though we sometimes have N_2 in the liquid state in the field or laboratory, under high pressure and low temperature in a steel tank.

Give some possible reasons for the differences in formaldehyde concentrations shown in Fig. 10.20 and benzene concentrations in Fig. 10.21.

Each of the sites must be considered for their unique emission profiles. Some of the sites may have sources of formaldehyde, for example, wood processing plants, which may have seasonally variable emissions. Formaldehyde is emitted from natural and industrial sources. Combustion processes are the largest sources. Some oxygenated fuels lead to higher formaldehyde concentrations than ordinary gasoline. These fuels are used most frequently during the winter. In addition, hydrocarbons like isoprene will undergo photochemical reactions in places that have smog (e.g., Los Angeles). The key is that formaldehyde can form numerous ways.

Like formaldehyde, benzene is produced from natural and anthropogenic sources. Natural sources include crude oil seeps, forest fires, and transpiration and volatilization from flora. Major anthropogenic sources to the ambient air include vehicle exhaust and refueling. The concentration of benzene tends to increase with lower ambient temperatures and with poor atmospheric mixing. Also, there may be sources that vary seasonally. Thus, it is important to look for possible sources near the sites, including petrochemical releases, hazardous waste sites, and leaking tanks (e.g., active or abandoned gasoline stations).

Why might formaldehyde concentrations be higher in manufactured homes versus conventional homes? How might time since construction affect these data?

Formaldehyde is a common by-product of wood processing, especially the adhesive component of composites, meaning that plywood, fiberboard, paneling, flooring, and other construction materials commonly found in manufactured housing have sufficient concentrations that volatilize into the indoor air. In addition, the substance is off-gassed from furniture and other materials, for example, drapes, textiles, and glues. Although these are also common in other housing types, the manufactured homes may be more confined (e.g., more formaldehyde-containing products per volume of indoor air). If the air exchange rates are lower, this would also increase concentrations. In addition, the population of persons living in mobile homes has an overall lower mean socioeconomic status (SES) than the general US population. Lower SES is associated with higher likelihood and amount of tobacco smoking (generating the so-called environmental tobacco smoke or ETS) and the use of ancillary heating devices (e.g., kerosene heaters and gas stoves), which are also sources of formaldehyde.

The discussion of the two main phases of air pollutants sets the stage for the next two chapters that address measurement and control technologies, respectively.

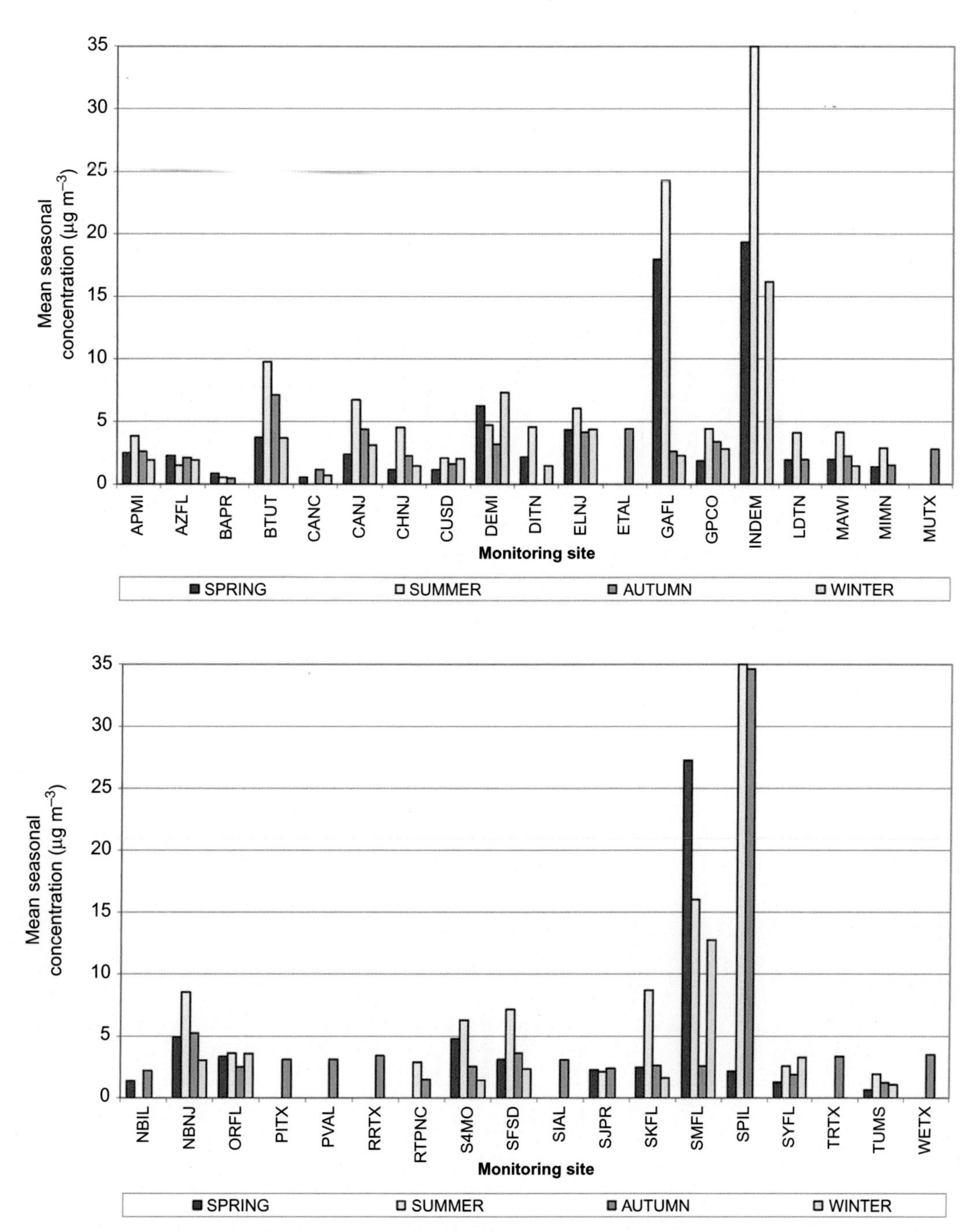

FIG. 10.20 Measured concentrations of formaldehyde in several urban areas during four seasons of the year 2005. *APMI*, Allen Park, Detroit, MI; *AZFL*, Azalea Park, *St*. Petersburg, FL; *BAPR*, Barcelona, Puerto Rico; *BTUT*, Bountiful, UT; *CANC*, Candor, NC; *CANJ*, Camden, NJ; *CHNJ*, Chester, NJ; *CUSD*, Custer, SD; *DITN*, Dickson, TN; *ELNJ*, Elizabeth, NJ; *ETAL*, East Thomas, Birmingham, AL; *FLFL*, Davie, FL; *GAFL*, Gandy in Tampa, FL; *GRMS*, Grenada, MS; *INDEM*, Gary, IN; *ITCMI*, Sault Sainte Marie, MI; *LDTN*, Loudon, TN; *MAWI*, Madison, WI; *MIMN*, Minneapolis, MN; *MUTX*, Murchison Middle School in Austin, TX; *NBAL*, North Birmingham, AL; *NBNJ*, New Brunswick, NJ; *ORFL*, Winter Park, FL; *PCOK*, Site 1 in Ponca City, OK; *PGMS*, Pascagoula, MS; *PITX*, Pickle Research Center, Austin, TX; *POOK*, Site 2 in Ponca City, OK; *PVAL*, Providence, RI; *RRTX*, Round Rock, TX; *RTPNC*, Research Triangle Park, NC; *SFSD*, Sioux Falls, SD; *SIAL*, Sloss Industries, Birmingham, AL; *SJPR*, San Juan, Puerto Rico; *SMFL*, Simmons Park in Tampa, FL; *SPIL*, Schiller Park in Chicago, IL; *TRTX*, Travis High School in Austin, TX; *TUMS*, Tupelo, MS; *WETX*, Webberville Road in Austin, TX. *(Source: EPA-454/R-07-001, Urban Air Toxics Monitoring Program (UATMP), 2005.)*

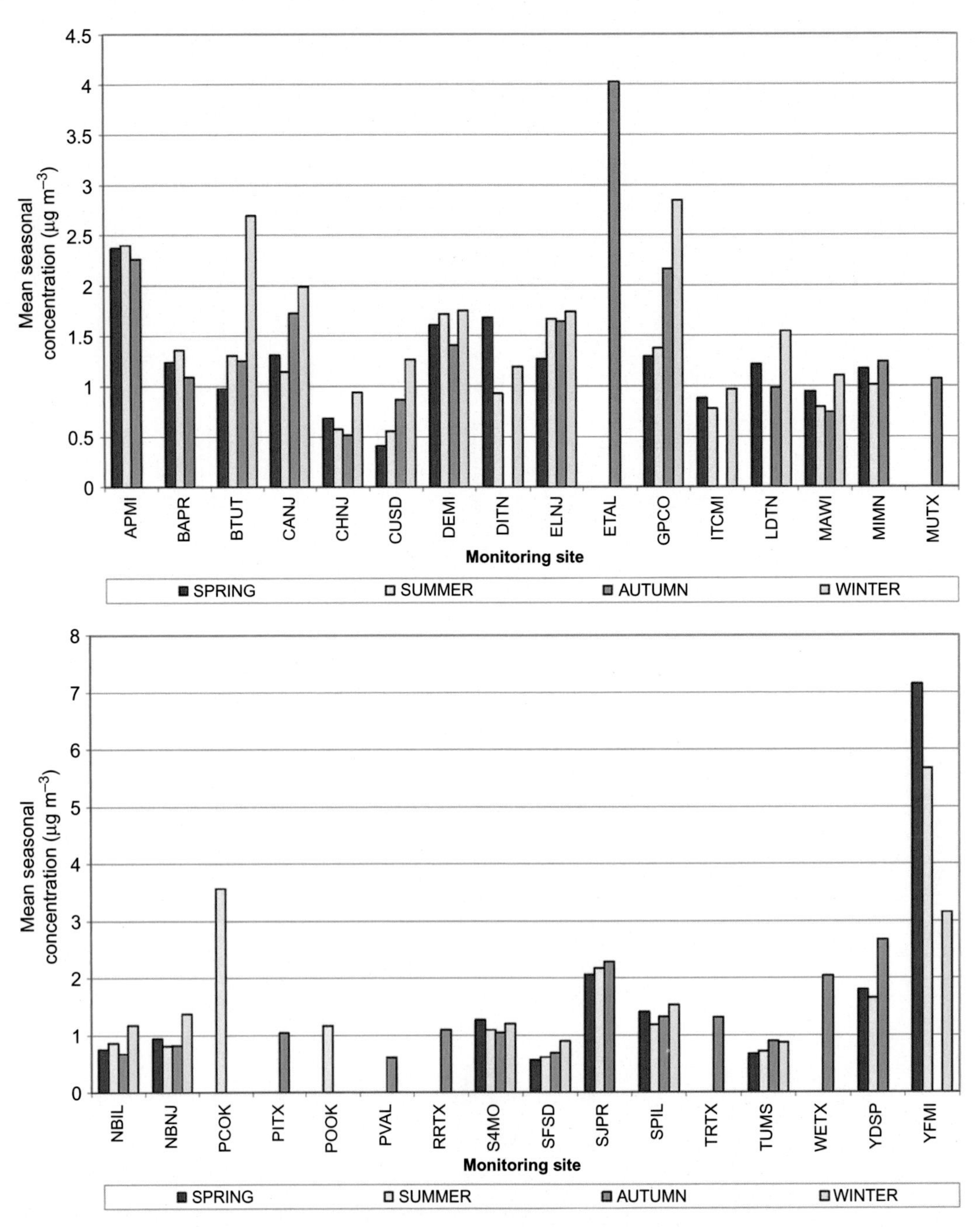

FIG. 10.21 Measured concentrations of benzene in several urban areas during four seasons of the year 2005. *APMI*, Allen Park, Detroit, MI; *AZFL*, Azalea Park, *St.* Petersburg, FL; *BAPR*, Barcelona, Puerto Rico; *BTUT*, Bountiful, UT; *CANC*, Candor, NC; *CANJ*, Camden, NJ; *CHNJ*, Chester, NJ; *CUSD*, Custer, SD; *DITN*, Dickson, TN; *ELNJ*, Elizabeth, NJ; *ETAL*, East Thomas, Birmingham, AL; *FLFL*, Davie, FL; *GAFL*, Gandy in Tampa, FL; *GRMS*, Grenada, MS; *INDEM*, Gary, IN; *ITCMI*, Sault Sainte Marie, MI; *LDTN*, Loudon, TN; *MAWI*, Madison, WI; *MIMN*, Minneapolis, MN; *MUTX*, Murchison Middle School in Austin, TX; *NBAL*, North Birmingham, AL; *NBNJ*, New Brunswick, NJ; *ORFL*, Winter Park, FL; *PCOK*, Site 1 in Ponca City, OK; *PGMS*, Pascagoula, MS; *PITX*, Pickle Research Center, Austin, TX; *POOK*, Site 2 in Ponca City, OK; *PVAL*, Providence, RI; *RRTX*, Round Rock, TX; *RTPNC*, Research Triangle Park, NC; *SFSD*, Sioux Falls, SD; *SIAL*, Sloss Industries, Birmingham, AL; *SJPR*, San Juan, Puerto Rico; *SMFL*, Simmons Park in Tampa, FL; *SPIL*, Schiller Park in Chicago, IL; *TRTX*, Travis High School in Austin, TX; *TUMS*, Tupelo, MS; *WETX*, Webberville Road in Austin, TX; *YDSP*, El Paso, TX; *YFMI*, Yellow Freight in Detroit, MI. *(Source: EPA-454/R-07-001, Urban Air Toxics Monitoring Program (UATMP), 2005.)*

REFERENCES

[1] H. Huo, Y. Wu, M. Wang, Total versus urban: well-to-wheels assessment of criteria pollutant emissions from various vehicle/fuel systems, Atmos. Environ. 43 (10) (2009) 1796–1804.

[2] A. Thorpe, R.M. Harrison, Sources and properties of non-exhaust particulate matter from road traffic: a review, Sci. Total Environ. 400 (1–3) (2008) 270–282.

[3] J. Bukingham, R. Macdonald, Dictionary of Organic Chemistry, Chapman and Hall, New York, 1982.

[4] F. Laden, L.M. Neas, D.W. Dockery, J. Schwartz, Association of fine particulate matter from different sources with daily mortality in six US cities, Environ. Health Perspect. 108 (10) (2000) 941.

[5] P. Kassomenos, et al., Study of PM10 and PM2. 5 levels in three European cities: Analysis of intra and inter urban variations, Atmos. Environ. 87 (2014) 153–163.

[6] K. Isaacs, J.A. Rosati, T.B. Martonen, Modeling deposition of inhaled particles, in: Aerosols Handbook: Measurement, Dosimetry, and Health Effects, CRC Press, 2012.

[7] C. Liu, S. Shi, C. Weschler, B. Zhao, Y. Zhang, Analysis of the dynamic interaction between SVOCs and airborne particles, Aerosol Sci. Technol. 47 (2) (2013) 125–136.

[8] D.A. Vallero, Air pollution, in: Kirk-Othmer Encyclopedia of Chemical Technology, John Wiley & Sons, Inc, 2000.

[9] EPA/625/R-96/010b, Compendium Method IO-2.4, Calculations for Standard Volume, 1999.

[10] C. Lee, G.L. Huffman, Thermodynamics used in environmental engineering, in: C. Lee, S.D. Lin (Eds.), Handbook of Environmental Engineering Calculations, McGraw Hill, 2007.

[11] T. Jobson, Y. Huangfu, G. VanderSchelden, S. Chung, Impact of Cold Climates on Vehicle Emissions: The Cold Start Air Toxics Pulse, 2016.

[12] Y. Zhang, Indoor Air Quality Engineering, CRC Press, Boca Raton, FL, 2004.

[13] G.D. Wight, Fundamentals of Air Sampling, CRC Press, 1994.

[14] D.A. Vallero, Fundamentals of Air Pollution, fifth ed., Elsevier Academic Press, Waltham, MA, 2014. p. 999 pages cm.

[15] H.F. Hemond, E.J. Fechner, Chemical Fate and Transport in the Environment, Elsevier, 2014.

[16] C. Mitcham, R.S. Duval, Engineering Ethics, 2000.

[17] R. Lachman, J.L. Lachman, E.C. Butterfield, Cognitive Psychology and Information Processing: An Introduction, Psychology Press, 2015.

[18] D.N. Bangala, N. Abatzoglou, J.-P. Martin, E. Chornet, Catalytic gas conditioning: application to biomass and waste gasification, Ind. Eng. Chem. Res. 36 (10) (1997) 4184–4192.

[19] O. Schabenberger, C.A. Gotway, Statistical Methods for Spatial Data Analysis, CRC Press, 2017.

[20] J. Paris, A Note on*-Conditioning, Unpublished note, 2017, pp. 1–3.

[21] Z. Yan, B. ÖRMECI, J. Zhang, Effect of sludge conditioning temperature on the thickening and dewatering performance of polymers, J. Residuals Sci. Technol. 13 (3) (2016).

[22] R.T. Haug, The Practical Handbook of Compost Engineering, CRC Press, 1993.

[23] D.A. Vallero (Ed.), Technical Writing Course, in: Instructor said that "a word will mean what it can mean" and "a phrase will mean what it can mean", U. S. Environmental Protection Agency, Kansas City, Missouri, 1978. Region VII, circa.

[24] B. Jowett, The Republic of Plato, Clarendon Press, 1888.

[25] E. Mesza´ros (Ed.), Chapter 4. The atmospheric aerosol, In: Studies in Environmental Science, vol. 11, Elsevier, 1981, pp. 91–132.

[26] J.C. Chow, Measurement methods to determine compliance with ambient air quality standards for suspended particles, J. Air Waste Manage. Assoc. 45 (5) (1995) 320–382.

[27] J. Cao, J.C. Chow, F.S. Lee, J.G. Watson, Evolution of PM2. 5 measurements and standards in the US and future perspectives for China, Aerosol Air Qual. Res. 13 (4) (2013) 1197–1211.

[28] D.A. Vallero, C. Brasier, Sustainable Design: The Science of Sustainability and Green Engineering, John Wiley & Sons, 2008.

[29] D.A. Vallero, Air pollution monitoring changes to accompany the transition from a control to a systems focus, Sustainability 8 (12) (2016) 1216.

[30] J.M. Harrington, C.M. Nelson, F.X. Weber, K.D. Bradham, K.E. Levine, J. Rice, Evaluation of methods for analysis of lead in air particulates: an intra-laboratory and inter-laboratory comparison, Environ. Sci. Process. Impacts 16 (2) (2014) 256–261.

[31] D.Y. Pui, B.Y. Liu, Advances in instrumentation for atmospheric aerosol measurement, Phys. Scr. 37 (2) (1988) 252.

[32] Z. Lu, Q. Liu, Y. Xiong, F. Huang, J. Zhou, J.J. Schauer, A hybrid source apportionment strategy using positive matrix factorization (PMF) and molecular marker chemical mass balance (MM-CMB) models, Environ. Pollut. 238 (2018) 39–51.

[33] D.A. Vallero, Fundamentals of Air Pollution, fourth ed., Elsevier, Amsterdam; Boston, 2008, pp. xxiii, 942 p.

[34] S. Police, S.K. Sahu, G.G. Pandit, Chemical characterization of atmospheric particulate matter and their source apportionment at an emerging industrial coastal city, Visakhapatnam, India, Atmospheric Pollution Research 7 (4) (2016) 725–733.

[35] M. Claeys, et al., Formation of secondary organic aerosols through photooxidation of isoprene, Science 303 (5661) (2004) 1173–1176.

[36] J.H. Kroll, J.H. Seinfeld, Chemistry of secondary organic aerosol: formation and evolution of low-volatility organics in the atmosphere, Atmos. Environ. 42 (16) (2008) 3593–3624.

[37] W. Boubel, D. Fox, D. Turner, A. Stern, Sources of air pollution, in: Fundamentals of Air Pollution, 1994, pp. 72–96.

[38] H.G. Merkus, Particle Size Measurements: Fundamentals, Practice, Quality, Springer Science & Business Media, 2009.

[39] A.C. Stern, Air pollution. Vol. II. Analysis, monitoring, and surveying, in: Air Pollution. Vol. II. Analysis, Monitoring, and Surveying, 1968.

[40] K. Isaacs, J. Rosati, T. Martonen, L. Ruzer, N. Harley, Modeling Deposition of Inhaled Particles, Aerosols Handbook, CRC Press, Boca Raton, FL, 2012, pp. 83–128.

[41] J.B. Grotberg, Respiratory fluid mechanics, Phys. Fluids 23 (2) (2011) 021301.
[42] A. Kerr, Dead space ventilation in normal children and children with obstructive airways disease, Thorax 31 (1) (1976) 63–69.
[43] A. Bouhuys, Respiratory dead space, in: Handbook of Physiology. Section III, vol. 1, 1964, pp. 699–714.
[44] S.C. George, M.P. Hlastala, Airway gas exchange and exhaled biomarkers, in: Comprehensive Physiology, 2011.
[45] S. Schurch, M. Lee, P. Gehr, Pulmonary surfactant: Surface properties and function of alveolar and airway surfactant, Pure Appl. Chem. 64 (11) (1992) 1745–1750.
[46] F. Possmayer, N. K, K. Rodrigueza, R. Qanbarb, S. Schürch, Surface activity in situ, in vivo, and in the captive bubble surfactometer, Comp. Biochem. Physiol. A Mol. Integr. Physiol. 129 (1) (2001) 209–220.
[47] S. Schürch, H. Bachofen, F. Possmayer, Surface activity in situ, in vivo, and in the captive bubble surfactometer, Comp. Biochem. Physiol. A Mol. Integr. Physiol. 129 (1) (2001) 195–207.
[48] U.S. Environmental Protection Agency, Air quality Criteria for Particulate Matter (Final Report, 1996), U.S. Environmental Protection Agency, Washington, DC, 1996.
[49] C. Darquenne, Aerosol deposition in health and disease, J. Aerosol Med. Pulm. Drug Deliv. 25 (3) (2012) 140–147.
[50] K.L. Rubow, Filtration: fundamentals and applications, in: Aerosol and Particle Measurement, University of Minnesota, Minneapolis, MN, 2004.
[51] M.J. Rhodes, M. Rhodes, Introduction to Particle Technology, John Wiley & Sons, 2008.
[52] D.R. Bullard, et al., APTI Course SI:409: Basic Air Pollution Meteorology: Student Guidebook, U.S. Environmental Protection Agency, Air Pollution Training Institute, Research Triangle Park, NC, 1982.
[53] D.A. Kaden, C. Mandin, G.D. Nielsen, P. Wolkoff, Formaldehyde, 2010.
[54] World Health Organization, WHO Guidelines for Indoor Air Quality: Selected Pollutants, 2010.
[55] U.S. Environmental Protection Agency, Formaldehyde, April 25, 2018. Available: https://www.epa.gov/formaldehyde, 2018.
[56] Toxicological Profile for Formaldehyde (Draft), 1997.
[57] Agency for Toxic Substances and Disease Registry, Formaldehyde and Your Health, April 15, 2018. Available: https://www.atsdr.cdc.gov/formaldehyde/index.html, 2018.
[58] The National Institute for Occupational Safety and Health, Formaldehyde, April 25, 2018. Available: https://www.cdc.gov/niosh/topics/formaldehyde/, 2018.
[59] An Update on Formaldehyde: 2013 Revision, Available: https://www.cpsc.gov/s3fs-public/pdfs/AN%2520UPDATE%2520ON%2520FORMALDEHYDE%2520final%25200113.pdf, 2013.
[60] C. Li, et al., Long-term persistence of polychlorinated dibenzo-p-dioxins and dibenzofurans in air, soil and sediment around an abandoned pentachlorophenol factory in China, Environ. Pollut. 162 (2012) 138–143.
[61] Emissions Monitoring Guidance (Note AG2), Available: http://www.epa.ie/pubs/advice/air/emissions/AG2%20Air%20Emissions%20Monitoring%20Guidance%20Note_rev3.pdf, 2015.
[62] Continuous Emission Manual, Available: https://digital.osl.state.or.us/islandora/object/osl%3A20999/datastream/OBJ/view, 1992.

FURTHER READING

[63] US National Institute of Standards and Technology, The NIST reference on constants, units, and uncertainty, August 11, 2018. Available: https://physics.nist.gov/cgi-bin/cuu/Value?r, 2018.
[64] U. Standard Atmosphere, US Standard Atmosphere 1976, US Government Printing Office, Washington, DC, 1976.
[65] M. Myers, A. Kaposi, A First Systems Book: Technology and Management, Imperial College Press, 2004.
[66] T.R. Green, Cognitive dimensions of notations, in: People and Computers V, 1989, pp. 443–460.
[67] M. Hosokawa, T. Yokoyama, M. Naito, K. Nogi, T. Yokoyama, Nanoparticle Technology Handbook. Elsevier, 2007. https://doi.org/10.1016/S1748-0132(07)70119-6.
[68] K. Whitby, B. Cantrell, Atmospheric aerosols-characteristics and measurement, in: International Conference on Environmental Sensing and Assessment, Las Vegas, NV, 1976, p. 1.
[69] K. Whitby, K. Willeke, Single particle optical counters: principles and field use, in: Aerosol Measurement, 1979, p. 145.
[70] B.J. Finlayson-Pitts, J.N. Pitts Jr., Chemistry of the Upper and Lower Atmosphere: Theory, Experiments, and Applications, Elsevier, 1999.
[71] U.S. Environmental Protection Agency, Chapter 4: particle collection mechanisms, in: J.R. Richards (Ed.), APTI Course 413: Control of Particulate Matter Emissions, third ed., ICES Ltd, Research Triangle Park, NC, 2000 [Online]. Available: http://www.4cleanair.org/APTI/413Combined.pdf.
[72] D.A. Vallero, Excerpted and Revised From: Fundamentals of Air Pollution, fifth ed., Elsevier Academic Press, Waltham, MA, 2014. p. 999 pages cm.
[73] EPA-454/R-07-001, Urban Air Toxics Monitoring Program (UATMP), 2005.

Chapter 11

Sampling and analysis

11.1 INTRODUCTION

The first step in assessing and managing the risks posed by air pollution is to measure the amount of contaminants in the air and in any environmental medium that serves as a source of an air pollutant or its precursor [1]. Such estimates must be based on reliable data and information, beginning by characterizing the release of a substance, for example, an emission from a stack, the substance's movement and transformation in the environment, and its concentrations near or within an organism, that is, the receptor (see Fig. 11.1) [2].

Air pollution extent and characteristics must be documented to the progress made by control and prevention in improving air quality over time and space. Ultimately, these air quality changes must be compared with improvements in human health, ecosystem condition, and public welfare, for example, less material destruction from air pollutants.

Air quality measurement is an encompassing term that includes developing methods, applying those methods, deploying monitoring technologies, and interpreting the results from these technologies. An air quality assessment may be part of a larger assessment, for example, an environmental assessment to determine whether a project requires and environmental impact statement (EIS), a site assessment (e.g., of an abandoned hazardous waste site) or a state implementation plan (SIP) to determine where and to what extent an area is out of compliance with ambient air quality standards. Conversely, a measurement may be confined in time and space, for example, a single stack test to see what is being emitted during a facility's operations. These measurements are usually focused on chemical concentrations but will also include physical measurements, for example, temperature, and sometimes biological factors, for example, leaf damage associated with sulfate deposition [3].

Air quality measurements underpin health, exposure, and risk assessments (see Chapter 12). The focus is often limited to essential measurements tailored to a specific decision-making objective. For example, exposure assessments require information about the pollutant concentrations in the locations where specific human activities occur. A measurement of a pollutant at a central monitoring site, therefore, is not an exposure measurement, per se, since it does not reflect the concentration where the activity takes place. A personal monitor worn during a day would be a more precise and accurate measurement of exposure for that day if it were matched with the person's activities, for example, using a diary. However, even the central site data can be used to estimate exposures if the measurement results are extended with models.

Since measurements are only relevant to the time and place where they are taken, air pollution models are needed to interpolate between measurements and extrapolate from samples to characterize the extent and severity of air pollution and to predict future air pollution scenarios. The models can be as simple as averaging concentrations from measurements, or they may be very complex.

Air pollution modeling is discussed in detail in Chapter 13, but it is important to keep in mind that measurements and models are complementary. Indeed, measurements are used to evaluate and to ground-truth models, that is, evaluating the model's predictions against a well-designed and controlled set of predictions. Such "knowns" are compared with model outcomes, that is, "predicted" results. The deviation of a modeled outcome and measurement is an indication of model uncertainty, that is, the measurement is what the value should be, but the modeled outcome deviates from this value. Under tightly controlled conditions, laboratory measurements are used to build and to validate models, but under real-world conditions, models are seldom fully validated; rather, they are merely evaluated (see Fig. 11.2). For example, the models are "benchmarked" to evaluate their performance and find ways to improve them.

11.1.1 Direct and indirect sampling

Air pollution measurements may be direct or indirect. Direct measurements are those in which the substance of concern is what is collected and analyzed. For example, a measurement of particulate matter (PM) would be directly measured by pumping air through a PM monitor and collecting particles on a filter. The particles would then be measured, for example, weighed, sized, and chemically analyzed. Direct measurements can be in situ, that is, taken in the environment or ex situ collected and taken elsewhere for measurement.

Air Pollution Calculations. https://doi.org/10.1016/B978-0-12-814934-8.00011-9

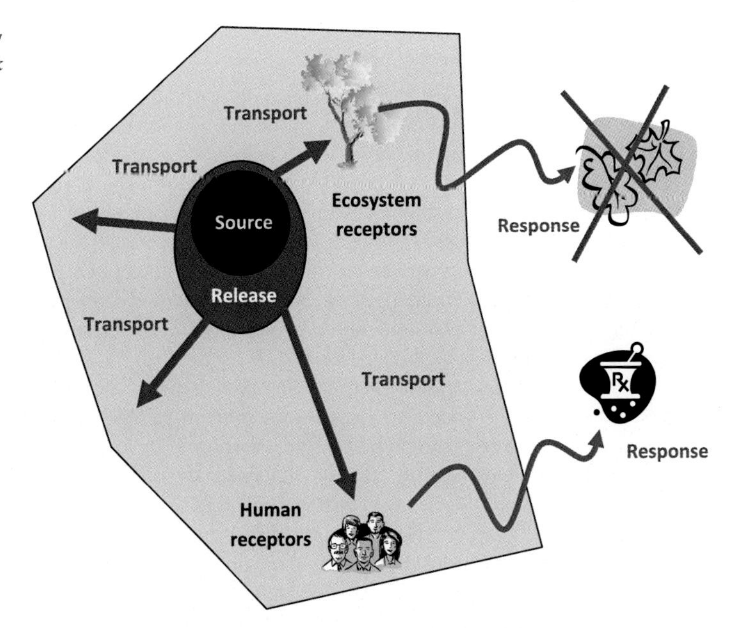

FIG. 11.1 Sites of measurements needed to support air quality assessments. *(Source: T.M. Letcher, D. Vallero, Waste: A Handbook for Management. Academic Press, 2011.)*

An indirect measurement is one where the substance itself is not collected but would be characterized by an indicator. For example, $PM_{2.5}$ concentrations can be measured directly by collecting particles on a filter, as mentioned above, but may also be quantified by measuring the backscatter from a beam of light through air sampled in a chamber within a nephelometer [4, 5]. The amount and type of scattering would indicate the quantity and size of particles. The advantage of the indirect method is that it is in real time, that is, the light scattering is immediately translated to a $PM_{2.5}$ concentration and appears on the screen or readout. Another advantage is that sampling density and duration can be much higher than direct sampling using a portable device (see Discussion Box "Using Light to Measure Air Pollution").

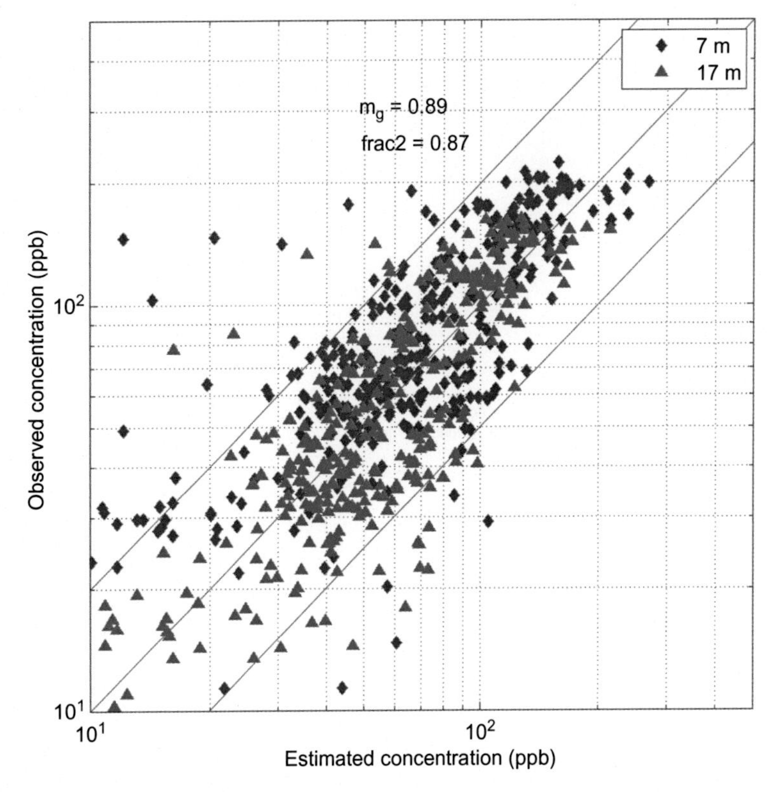

FIG. 11.2 Model estimates compared with the nitric oxide (NO) measurements collected at 7 and 17 m from a roadway shoulder at a height of 2 m. The observed concentrations are consistent with model estimates: 87% of the model estimates are within a factor of two of the observations, and the under-prediction bias is about 10%. *(Sources: D.A. Vallero, Fundamentals of Air Pollution, 5th ed., Elsevier Academic Press, Waltham, MA, 2014, p. 999 pages cm; T. Barzyk, Nitric Oxide Model, in: D.A. Vallero, U.S. Environmental Protection Agency, Research Triangle Park, NC, 2013; A. Venkatram, M. Snyder, V. Isakov, S. Kimbrough, Impact of wind direction on near-road pollutant concentrations, Atmos. Environ. 80 (2013) 248–258.)*

Using Light to Measure Air Pollution

The interplay of energy and matter has been discussed throughout this book. Changes in energy are accompanied by responses in matter and vice versa. The concentrations of aerosols and gases in the atmosphere affect the intensity and wavelengths of electromagnetic energy. Thus, for example, what can be seen is affected by the chemical and particle composition of the air. Indeed, one of the earliest observations of air pollution was the impairment of visibility, that is, degradation of the psychophysical, that is, eye-brain, linkage to perceive objects through the atmosphere (see Fig. 11.3). Several factors influence our concept of visibility, including the characteristics of the source, the human observer, the object, and pollution in the atmosphere [6]. Measuring visibility at a site can take two approaches: human observations and optical measurements. In pristine locations such as national parks, use of human observers has permitted us to gain an understanding of the public's concept of visibility impairment. Although it is difficult to quantify the elements of human observations, this type of research, when coupled with optical measurements, provides a better measure of visibility at a given location [7, 8].

To see an object, an observer must be able to detect the contrast between the object and its surroundings. If this contrast decreases, the object becomes increasingly difficult to observe. In the atmosphere, visibility decreases with distance from the object, the sun's angle, and pollution concentrations. The lowest limit of contrast for human observers is called the *threshold contrast* and is important because this value influences the maximum distance at which we can see various objects. Threshold contrast is illustrated in Fig. 11.4. I is the intensity of light received by the eye from the object, and $I + \Delta I$ represents the intensity coming from the surroundings. The threshold contrast can be as low as 0.018–0.03, and the object can still be perceptible. Other factors, such as the physical size of the visual image on the retina of the eye and the brain's response to the color of the object, influence the perception of contrast.

Light scattering results from the interaction of light with gases or particles in such a manner that the direction or frequency of the light is altered. Absorption of the light results when the electromagnetic radiation is transferred into the gas or particle. Light scattering by gaseous molecules is wavelength-dependent, for example, scattering of sunlight into the blue wavelength range during the day. Blue scattering is dominant in atmospheres that are relatively free of aerosols or light-absorbing gases. Light scattering by particles is the most important cause of visibility reduction and is aerosol size-dependent.

Light absorption by gases in the lower troposphere is limited to the absorption characteristics of air pollutant, nitrogen dioxide (NO_2), which absorbs the shorter or blue wavelengths of visible light, causing the eye to observe the red wavelengths. A yellow to reddish-brown tint results in atmospheres containing quantities of NO_2. Light absorption by aerosols is related principally to carbonaceous or black soot in the atmosphere. Other types of fine particles such as sulfates, although not good light absorbers, are very efficient at scattering light.

The interaction of light in the atmosphere is expressed as.

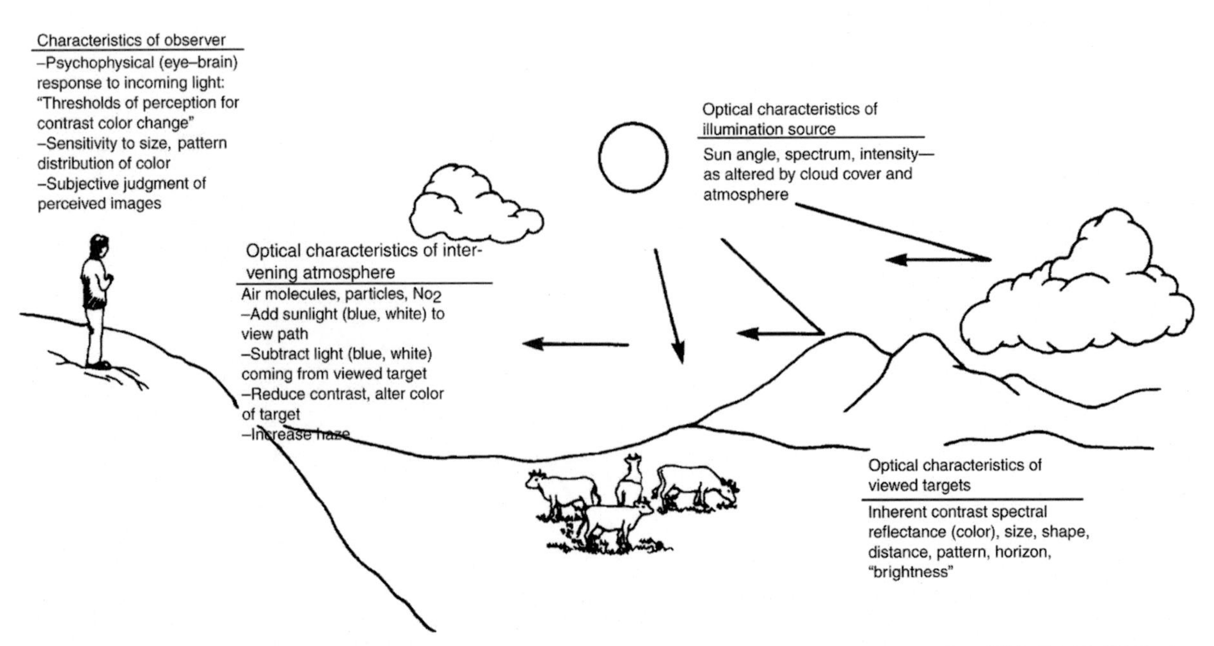

FIG. 11.3 Factors determining visibility in the atmosphere. *(Source: P. Visibility, E. An, Report to Congress, EPA-450/5-79-008. US Environmental Protection Agency, 1979.)*

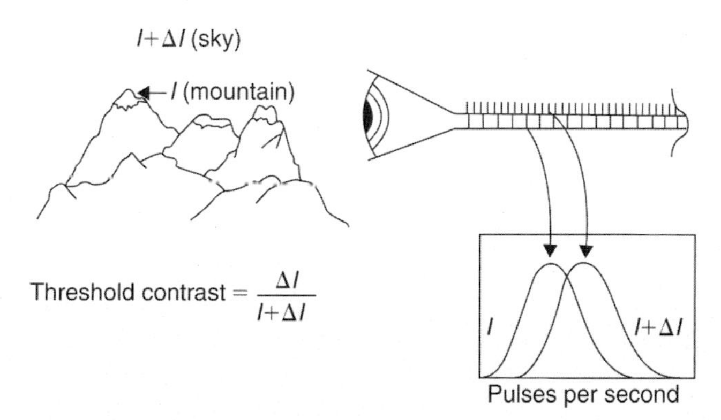

FIG. 11.4 Threshold contrast in distinguishing an object from its surroundings. The eye responds to an increment in light intensity by increasing the number of signals (pulses) sent to the brain. The detection of threshold contrast involves the ability to discriminate between the target (I) and the brighter background ($I+\Delta I$). *(Source: D.A. Vallero, Fundamentals of Air Pollution, 4th ed., Elsevier, Amsterdam; Boston, 2008, pp. xxiii, 942 p; reproduced with permission from K. James, R. Gregory, Eye and Brain: The Psychology of Seeing, 1976.)*

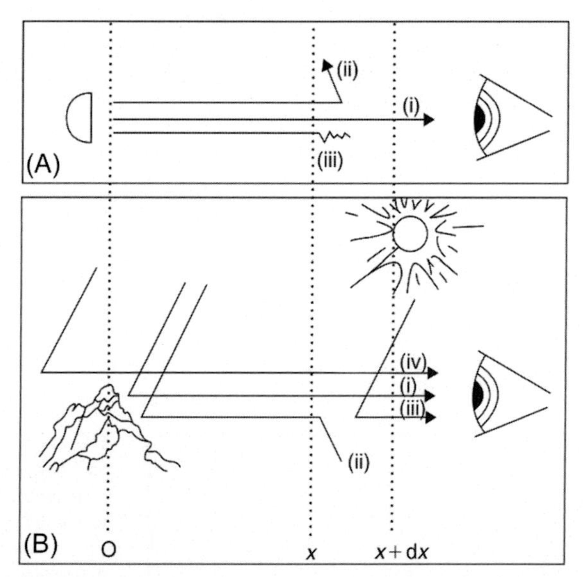

FIG. 11.5 (A) A diagram of extinction of light from a source such as an electric light in a reflector, illustrating (i) transmitted, (ii) scattered, and (iii) absorbed light. (B) A diagram of daylight visibility, illustrating (i) residual light from a target reaching an observer, (ii) light from a target scattered out of an observer's line of sight, (iii) air light from the intervening atmosphere, and (iv) air light constituting horizon sky. *(Source: D.A. Vallero, Fundamentals of Air Pollution, 4th ed., Elsevier, Amsterdam; Boston, 2008, pp. xxiii, 942 p; reproduced with permission from P. Visibility, E. An, Report to Congress, EPA-450/5-79-008. US Environmental Protection Agency, 1979.)*

$$-dI = b_{ext} I \cdot dx \tag{11.1}$$

where $-dI$ is the decrease in intensity, b_{ext} is the extinction coefficient, I is the original intensity of the beam of light, and dx is the length of the path traveled by the beam of light.

Fig. 11.5A shows a beam of light transmitted through the atmosphere. The intensity of the beam $I(x)$ decreases with the distance from the illumination source as the light is absorbed or scattered out of the beam. For a short period, this decrease is proportional to the intensity of the beam and the length of the interval at that point. Here, b_{ext} is the extinction or attenuation coefficient and is a function of the degree of scattering and absorption of the particles and gases that are present in the beam path.

Fig. 11.5B illustrates a slightly more complicated case but one more applicable to atmospheric visibility. The observer still depends on the ability to perceive light rays emanating from the target object and on the scattering and absorption of those rays out of the beam but must contend with additional light scattered into the line of sight from other angles. Such extraneous light is sometimes called *air light.* Eq. (11.1) is modified to account for this phenomenon by adding a term to represent this background intensity:

$$-dI = dI(\text{extinction}) + dI(\text{air light}) \tag{11.2}$$

This air light term contributes to the reduced visibility, known as atmospheric haze.

A simplified relationship [9] that relates the visual range and the extinction coefficient is.

$$L_v = \frac{392}{b_{ext}} \tag{11.3}$$

where L_v is the distance at which a black object is just barely visible [10]. Eq. (11.3) assumes the following:

1. The background behind the target is uniform.
2. The object is black.
3. An observer can detect a contrast of 0.02.
4. The ratio of air light to extinction is constant over the path of sight.

While this is useful as a first approximation for determining visual range, many situations exist in which the results are qualitative. The extinction coefficient b_{ext} depends on the presence of gases and molecules that scatter and absorb light in the atmosphere. The extinction coefficient is the sum of the air and pollutant scattering and absorption interactions:

$$b_{ext} = b_{rg} + b_{ag} + b_{scat} + b_{ap} \tag{11.4}$$

where b_{rg} is scattering by gaseous molecules, that is, Rayleigh scattering; b_{ag} is absorption by NO_2 gas; b_{scat} is scattering by particles; and b_{ap} is absorption by particles. These various extinction components are a function of wavelength. As extinction increases, visibility decreases.

The Rayleigh scattering extinction coefficient for particle-free air is 0.012 km^{-1} for green light ($\gamma = 0.05\,\mu m$) at sea level [11]. This permits a visual range of ~320 km. The aerosol-free or Rayleigh scattering case represents the best visibility possible with the current atmosphere on earth.

The absorption spectrum of NO_2 shows significant absorption in the visible region [12] (see Fig. 11.6). As a strong absorber in the blue region, NO_2 can color plumes red, brown, or yellow. Fig. 11.7 shows a comparison of extinction coefficients of 0.1 ppm

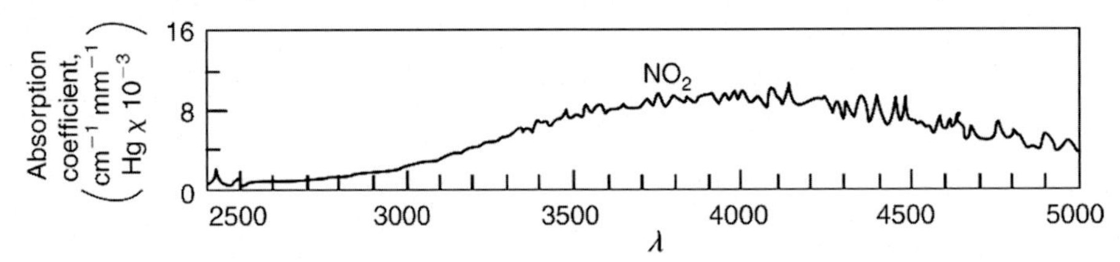

FIG. 11.6 Absorption spectrum of nitrogen dioxide. *(Source: T. Hall Jr., F. Blacet, Separation of the absorption spectra of NO2 and N2O4 in the range of 2400–5000A, J. Chem. Phys. 20 (11) (1952) 1745–1749.)*

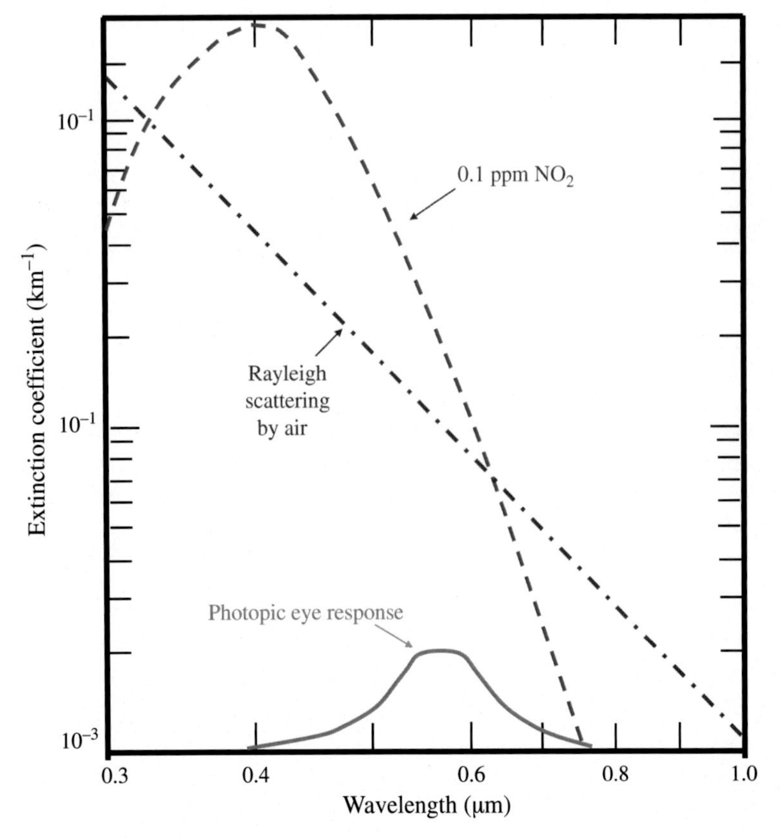

FIG. 11.7 Comparison of b_{ext} for 0.1 ppm NO_2 and Rayleigh scattering by air. The photopic eye response represents the range of wavelengths over which the eye detects light. *(Source: D.A. Vallero, Fundamentals of Air Pollution, 5th ed., Elsevier Academic Press, Waltham, MA, 2014, p. 999 pages cm.)*

NO_2 and Rayleigh scattering by air [13]. In urban areas, some discoloration can be due to area-wide NO_2 pollution. In rural areas, the biggest problem with NO_2 is that in coherent plumes from power plants, it contributes to the discoloration of the plume.

Removing suspended particulate matter (PM) is the most important factor in visibility reduction. In most instances, the visual quality of air is controlled by particle scattering and is characterized by the extinction coefficient b_{scat}. An aerosol's diameter is crucial in its interaction with light. Other factors are the refractive index and shape of the aerosol, although these are less understood than PM size. Alternatively, the extinction coefficient associated with an aerosol can be measured directly.

Aerosol chemical composition determines the amount of light absorbed and scattered. The darkness of pollutant plumes is affected by incoming light and the chemical makeup of the particles in the plume. In Fig. 11.8, the two plumes on the left contain large amounts of PM, whereas the two plumes on the right consist primarily of water droplets. The plume on the far right, which is illuminated by direct sunlight, appears to be white. The second identical water droplet plume, which is shaded, is much darker.

Light and aerosols interact in the four basic ways: refraction, diffraction, phase shift, and absorption (see Fig. 11.9). For particles with a diameter of 0.1–1.0 μm, scattering and absorption can be calculated by using the Mie equations [14]. Fig. 11.10 shows the

FIG. 11.8 Effect of aerosol chemical composition on plume color. During June 7, 2018 an eruption of Kilauea in Hawaii, lava enters the ocean. This results in the emission a white plume called "laze," which drifts downwind. Laze is composed of steam, hydrochloric acid, and very small diameter volcanic glass aerosols. The plume color changes with altitude as the laze becomes diluted. *(Source and photo credit: U.S. Geological Survey. Available at: https://www.usgs.gov/news/k-lauea-volcano-erupts.)*

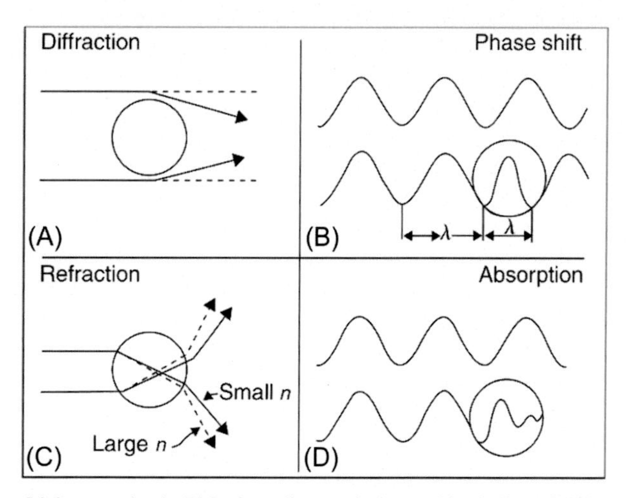

FIG. 11.9 Aerosol-light interaction. Light scattering by PM >2 μm diameter is the combined effect of diffraction and refraction. (A) Diffraction is an edge effect whereby the light is bent to fill in the shadow behind the particle. (B) The speed of a wavefront entering a particle with refractive index $n > 1$ (for water, $n = 1.33$) is reduced. (C) Refraction produces a lens effect. The angular dispersion resulting from bending incoming rays increases with n. (D) For absorbing media, the refracted wave intensity decays within the particle. When the particle size is comparable to the wavelength of light (0.1–1.0 μm), these interactions (A)–(D) are complex and enhanced. *(Source: D.A. Vallero, Fundamentals of Air Pollution, fourth ed., Elsevier, Amsterdam; Boston, 2008, pp. xxiii, 942 p.)*

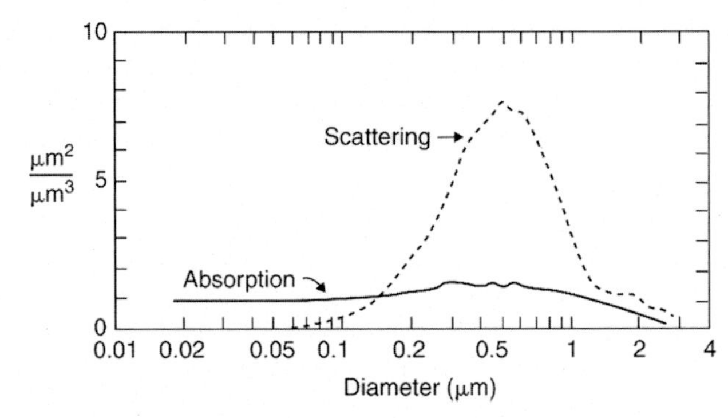

FIG. 11.10 Scattering and absorption cross section per unit volume as a function of particle diameter. *(Reproduced with permission from D.A. Vallero, Fundamentals of Air Pollution, 4th ed., Elsevier, Amsterdam; Boston, 2008, pp. xxiii, 942 p; data from R.J. Charlson, A.P. Waggoner, J. Thiekle, Visibility protection for class 1 areas: the technical basis. Final report, Council on Environmental Quality, Washington, DC, USA, 1978.)*

relative scattering and absorption efficiency per unit volume of particle for a typical aerosol containing some light-absorbing soot [15]. This clearly shows the importance of atmospheric particles in the diameter range of 0.1–1.0 μm as efficient light-scattering centers. With particles of larger and smaller diameters, scattering decreases. Absorption generally contributes less to the extinction coefficient than does the scattering processes. Atmospheric particles of different chemical composition have different refractive indexes, resulting in different scattering efficiencies. Fig. 11.11 shows the scattering-to-mass ratio for four different materials. Clearly, carbon or soot aerosols and aerosols of the same diameter with water content scatter with different efficiencies at the same diameter.

Visibility is also affected by alteration of particle size due to hydroscopic particle growth, which is a function of relative humidity. In Los Angeles, California, the air, principally of marine origin, has numerous sea-salt particles. Visibility is noticeably reduced when humidity exceeds about 67%. Relative humidity and origin of air (maritime or continental) also affect visibility, presumably due to numerous hygroscopic aerosols from air pollution sources. Some materials, such as sulfuric acid mist, exhibit hygroscopic growth at humidity as low as 30%.

The ratio of the incident solar transmissivity to the extraterrestrial solar intensity can be as high as 0.5 in clean atmospheres but can drop to 0.2–0.3 in polluted areas, indicating a decrease of 50% in ground-level solar intensity. The turbidity coefficient can also be derived from these measurements and used to approximate the aerosol loading of the atmosphere. By assuming a particle size distribution in the size range of 0.1–10.0 μm and a particle density, the total number of particles can be estimated. The mass loading per cubic meter can also be approximated. Because of the reasonable cost and simplicity of the sun photometer, comparative measurements can now be made around the world.

As mentioned, atmospheric haze is the condition of reduced visibility caused by the presence of fine aerosols and/or NO_2 in the atmosphere. The haze can be uniform or layered (see Fig. 11.12). The particles must be 0.1–1.0 μm in diameter, the size range in which light scattering occurs. The source of these particles may be natural or anthropogenic. Atmospheric haze has been observed in both the western and eastern portions of the United States. Typical visual ranges in the East are <15 miles and in the Southwest >50 miles. The desire to protect visual air quality in the United States is focused on the national parks in the West. The ability to see

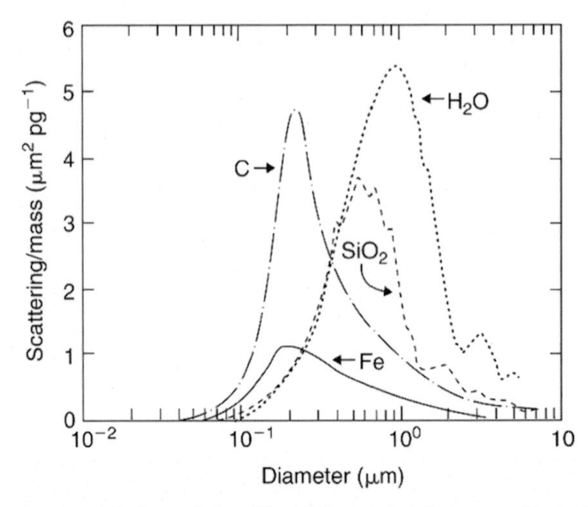

FIG. 11.11 Single-particle scattering-to-mass ratio for particles of four different chemical compositions. Carbon particles are also very efficient absorbers of light. *(Source: D.A. Vallero, Fundamentals of Air Pollution, 4th ed., Amsterdam; Boston: Elsevier, 2008, pp. xxiii, 942 p; reproduced with permission from R.J. Charlson, A.P. Waggoner, J. Thiekle, Visibility protection for class 1 areas: the technical basis. Final report, Council on Environmental Quality, Washington, DC, USA, 1978.)*

FIG. 11.12 Example of uniform (A) and layered (B) haze. (A) Uniform haze: Pinnacles National Monument, California. *Top photo* has about 200 mile visibility, and the *bottom photo* with uniform haze has about 15 mile visibility. (B) Layered haze: Bryce Canyon, Navaho Mountain. *Top photo*, lower altitude layer of haze obscures bottom of Navaho Mountain, whereas higher altitude haze in the *bottom photo* obscures the top. *(Source: Interagency Monitoring of Protected Visual Environments. (2013), 22 November 2013. Available: http://vista.cira.colostate.edu/improve/Education/Glossary/Haze.htm.)*

vistas over 50–100 km in these locations makes them particularly vulnerable to atmospheric haze. This phenomenon is generally associated with diffuse or widespread atmospheric degradation as opposed to individual plumes.

The major component of atmospheric haze is sulfate particulate matter, particularly ammonium sulfate, along with varying amounts of nitrate particulate matter, which in some areas can equal the sulfate. Other components include graphitic material, fine fly ash, and organic aerosols. The haze-inducing PM can be primary, that is, directly injected into the atmosphere, or secondary, formed in the atmosphere by gas-to-particle conversion processes.

Optical measurements permit the quantification of visibility degradation under different conditions. Several instruments can measure visual air quality, for example, cameras, photometers, telephotometers, transmissometers, and scattering instruments.

Photography can provide a permanent record of visibility conditions at a particular place and time. This type of record can preserve a scene in a photograph in a form similar to the way it is seen. Photometers measure light intensity by converting brightness to representative electric signals with a photodetector. Different lenses and filters may be used to determine color and other optical properties. When used in combination with long-range lenses, photometers become telephotometers. This type of instrument may view distant objects with a much smaller viewing angle. The output of the photodetector is closely related to the perceived optical properties of distant targets. Telephotometers are often used to measure the contrast between a distant object and its surroundings, a measurement much closer to the human observer's perception of objects.

A transmissometer is similar to a telephotometer except that the target is a known light source. If we know the characteristics of the source, the average extinction coefficient over the path of the beam may be calculated. Transmissometers are not very portable in terms of looking at a scene from several directions. They are also very sensitive to atmospheric turbulence, which limits the length of the light beam.

Scattering instruments are also used to measure visibility degradation. The most common instrument is the integrating nephelometer, which measures the light scattered over a range of angles. The physical design of the instrument, as shown in Fig. 11.13, permits a point determination of the scattering coefficient of extinction, b_{ext} [16]. In clean areas, b_{ext} is dominated by scattering, so that the integrating nephelometer yields a measure of the extinction coefficient. The b_{ext} can be related to visual range as described in Eq. (11.3).

Other measurements important to visual air quality are pollutant-related, that is, the size distribution, mass concentration, and number concentration of airborne particles and their chemical composition. From the size distribution, the Mie theory of light scattering can be used to calculate the scattering coefficient. Table 11.1 summarizes the different types of visual monitoring methods.

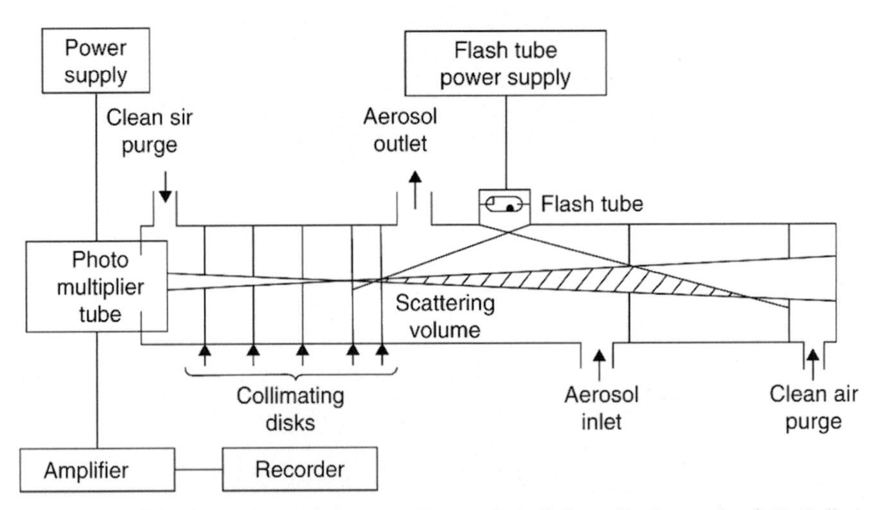

FIG. 11.13 Schematic diagram of the integrating nephelometer. *(Source: D.A. Vallero, Fundamentals of Air Pollution, 4th ed., Elsevier, Amsterdam; Boston, 2008, pp. xxiii, 942 p; reproduced with permission from N.C. Ahlquist, R.J. Charlson, A new instrument for evaluating the visual quality of air, J. Air Pollut. Control Assoc. 17 (7) (1967) 467–469; A. Waggoner, N. Ahliquist, R. Charlson, Measurement of the aerosol total scatter–backscatter ratio, Appl. Opt. 11 (12) (1972) 2886–2889.)*

TABLE 11.1 Visibility monitoring methods

Method	Parameters measured	Advantages	Limitations	Preferred use
Human observer	Perceived visual quality, atmospheric color, plume blight, and visual range	Flexibility, judgment; large existing database (airport visual range)	Labor-intensive; variability in observer perception; suitable targets for visual range not generally available	Complement to instrumental observations; areas with frequent plume blight, discoloration; visual ranges with available target distances
Integrating nephelometer	Scattering coefficient (b_{scat}) at site	Continuous readings; unaffected by clouds, night; b_{scat} directly relatable to fine aerosol concentration at a point; semiportable; used in several previous studies; sensitive models available; automated	Point measurement, requires assumption of homogeneous distribution of particles; neglects extinction from absorption, coarse particles (>3–10 μm); must consider humidity effects at high relative humidity	Areas experiencing periodic, well-mixed general haze; medium to short viewing distances; small absorption coefficient (b_{abs}); relating to point composition measurements

Continued

TABLE 11.1 Visibility monitoring methods—cont'd

Method	Parameters measured	Advantages	Limitations	Preferred use
Multiwavelength telephotometer	Sky and/or target radiance, contrast at various wavelengths	Measurement over long view path (up to 100 km) with suitable illumination and target, contrast transmittance, total extinction, and chromaticity over sight path can be determined; includes scattering and absorption from all sources; can detect plume blight; automated	Sensitive to illumination conditions; useful only in daylight; relationship to extinction, aerosol relationship possible only under cloudless skies; requires large, uniform targets	Areas experiencing mixed or inhomogeneous haze, significant fugitive dust; medium to long viewing distances (one-fourth of visual range); areas with frequent discoloration; horizontal sight path
Transmissometer	Long path extinction coefficient (b_{ext})	Measurement over medium view path (10–25 km); measures total extinction, scattering and absorption; unaffected by clouds, night	Calibration problems; single wavelength; equivalent to point measurement in areas with long view paths (50–100 km); limited applications to date still under development	Areas experiencing periodic mixed general haze, medium to short viewing distance areas with significant absorption (b_{abs})
Photography	Visual quality, plume blight, color, contrast (limited)	Related to perception of visual quality; documentation of vista conditions	Sensitive to lighting conditions; degradation in storage; contrast measurement from film subject to significant errors	Complement to human observation, instrumental methods; areas with frequent plume blight, discoloration
Particle samplers	Particles	Permit evaluation of causes of impairment	Not always relatable to visual air quality; point measurement	Complement to visibility measurements
Hi vol.	TSP	Large database, amenable to chemical analysis; coarse particle analysis	Does not separate sizes; sampling artifacts for nitrate, sulfate; not automated	Not useful for visibility sites
Cascade impactor	Size-segregated particles (more than two stages)	Detailed chemical, size evaluation	Particle bounce, wall losses; labor-intensive	Detailed studies of scattering by particles,2 μm
Dichotomous and fine particle samplers (several fundamentally different types)	Fine particles (2.5 μm) coarse particles (2.5–15 μm) inhalable particles (0–15 μm)	Size cut enhances resolution, optically important aerosol analysis, low artifact potential, particle bounce; amenable to automated compositional analysis; automated versions available; large networks under development	Some large-particle penetration; 24 h or longer sample required in clean areas for mass measurement; automated version relatively untested in remote locations	Complement to visibility measurement, source assessment for general haze, ground-level plumes

Source: P. Visibility, E. An, Report to Congress, EPA-450/5-79-008. US Environmental Protection Agency, 1979. Source: US Environmental Protection Agency. *Office of Air Quality Planning and Standards.* Protecting Visibility, Research Triangle Park, North Carolina, 1979. Report No. EPA-450/5-79-008.

11.2 QUALITY ASSURANCE

An air pollution monitoring plan must be in place before samples are collected and arrive at the laboratory. The plan includes quality assurance provisions and describes the procedures to be employed. There are advantages to combining a project plan with a quality assurance plan, that is, aptly known as quality assurance project plan (QAPP) [17, 18]. The advantage of a QAPP over two separate documents is the assurance that quality is "built in" and that the project's objectives will be met. That is, it would be unfortunate and irresponsible to come to the end of an air pollution study only to find that the data are not sufficient to support an assessment. Perhaps, the monitors were in an area that is not representative of a target population's exposure to a pollutant. Or there were too few samples to draw the necessary conclusions about the air quality. Even high-quality measurements, that is, highly precise and accurate, are useless or inadequate if they are made in the wrong place, are too few, or are measuring the wrong things.

A device used for measurement must be reliable. The reliability of an air monitoring technology is similar to that of any item, as discussed in Chapter 16. The first step in ensuring measurement reliability is calibration, that is, the comparison of the device response to a known response. Calibration is the means of finding the deviation of the measurement from what is known to be precise and accurate, so it is worthwhile to consider what these terms mean.

Environmental decisions must be based upon reliable information, which begins with measurements of the constituents of the air in various places. For information to be reliable, it must be of a quality matched to the needs of the decision-maker. For example, a farmer needs to know the air's relative humidity (RH) within ±10% during growing season. A county governmental official may need to know the same air's hourly concentration of one or two toxic compounds to five significant figures (see Chapter 1), because the county has a factory that has released these compounds in the past. These two instances are both demanding that sampling and analysis be conducted, but the precision and accuracy of the results are very different. The farmer needs much less precision but is hoping that within these defined ranges of moisture that the results will be accurate, that is, close enough to the actual RH. The county needs highly precise information that must also be sufficiently accurate.

Although often taken together, the terms precision and accuracy mean two very different statistical terms. *Precision* describes how refined and repeatable an operation can be performed, such as the exactness in the instruments and methods used to obtain a result [19]. It is an indication of the uniformity or reproducibility of a result. This can be likened to shooting arrows,[1] with each arrow representing a data point. Targets A and B in Fig. 11.14 are equally precise. If the center of the target, that is, the bull's-eye, represents the "true value," then data set B is more accurate than data set A. Consistently missing the bull's-eye in the same direction at the same distance is an example of bias or systematic error. Measurement bias can be due to an instrument's inherent flaws, for example, always assuming pressure is X% greater than the true value. The key here is knowing the deviation. If it is known, the equipment can be calibrated and adjusted according to the correct,

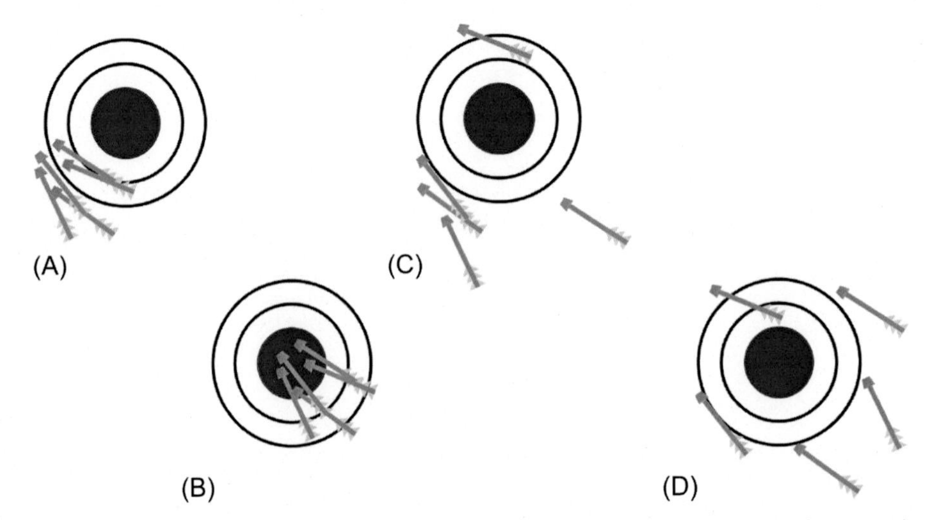

FIG. 11.14 Precision and accuracy. The bull's-eye represents the true value. Targets A and B demonstrate data sets that are precise, targets B and D data sets that are accurate, and targets C and D data sets that are imprecise. Target B is the ideal data set, which is precise and accurate.

1. My apologies to the originator of this analogy, who deserves much credit for this teaching device. The target is a widely used way to describe precision and accuracy.

known value (e.g., the deviation is subtracted from the measurement). Keeping with the archery analogy, the archer would move her sight up and to the right.

Thus, *accuracy* is an expression of how well a study conforms to some defined standard, that is, the true value. Thus, accuracy expresses the quality of what is found, whereas precision expresses the quality of the operation by which we obtained our finding. Thus, the four possibilities are that our data are precise but inaccurate (Target A), precise and accurate (Target B), imprecise and inaccurate (Target C), and imprecise and accurate (Target D).

11.2.1 Measurement precision

The two precise groupings in Fig. 11.14, that is, Targets A and B, appear to indicate that the instrument (bow and arrow) gives repeatable results. Indeed, the two targets demonstrate bias. The known grouping (A) is obtained under controlled conditions; for example, a precise amount of a substance is put in an instrument, and the instrument is calibrated to read that amount. More than one concentration is needed to calibrate an instrument. A sufficient number of knowns must be put into the instrument to construct a calibration curve. In addition to laboratory-generated calibration curves, field instruments must also account for real-world atmospheric conditions; for example, several concentration curves under various RH conditions may need to be constructed to see how an instrument performs, for example, the extent to which concentration curves differ at varying RH values (see Fig. 11.15).

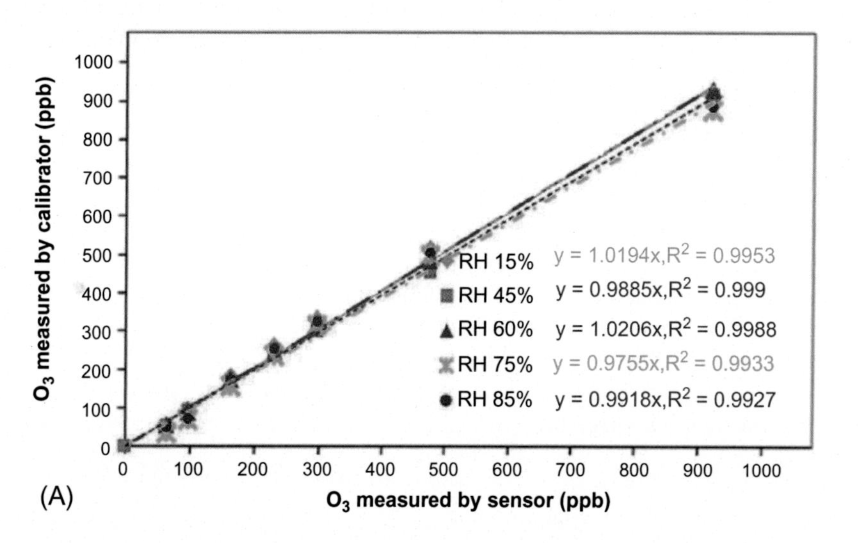

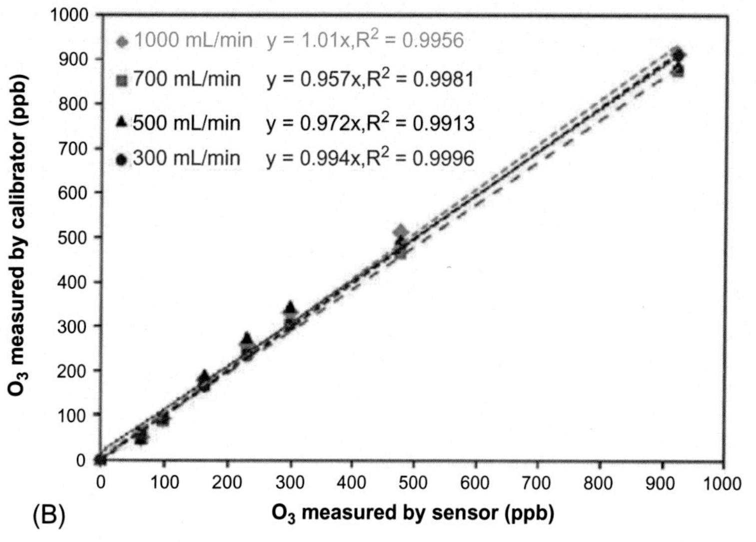

FIG. 11.15 Known (y-axis) to measured (x-axis) readings from a sensor: (A) Ozone (O_3) sensor calibrations under different relative humidity (RH). Experimental conditions: flow rate of sample gas, 500 mL minute^{-1}; temperature, 20°C. (B) Sensor calibrations under various gas flow rates. Experimental conditions: RH, 60±5%; temperature, 20°C. The deviation between higher and lower appears to increase with O_3 concentration. Since the sensor shows lower O_3 concentrations than the knowns at 15% and 75% RH, this is a negative bias. At other RH conditions, they are positive biases. The accuracy deviations are highest at 75%. However, given the consistently small deviations, this could be simply inherent variability of the sensors. *(Source: X. Pang, M.D. Shaw, A.C. Lewis, L.J. Carpenter, T. Batchellier, Electrochemical ozone sensors: a miniaturised alternative for ozone measurements in laboratory experiments and air-quality monitoring, Sens. Actuators B: Chem. 240 (2017) 829–837.)*

The imprecise yet accurate scenario represented by Target D may appear to be counterintuitive. However, it does occur in atmospheric measurements. Indeed, Targets B and D only differ in amount. Target D simply has greater "spread," that is, the variance and standard deviation of D is larger than that of B. However, the values of one of their measures of central tendency, that is, the means, are nearly the same. So, both data sets are giving us the right answer, but almost all the measurements in B are near the true value. None of the data points in D are near the true value, yet the mean is close to the center of the target.

Numerous factors can lead to a loss of precision in air measurements. Uncertainties in environmental measurement are introduced through human error, equipment malfunctions, and experimental bias [20]. Measurements are useful to the extent they produce valid data, beginning with the assurance that each figure or digit in the numerical expression of a measurement should be significant. A significant figure is a number that is correct within a specified or implied limit of error. Thus, if the height of a woman, expressed in significant figures, is written as 5.18 ft, it is assumed that only the *last figure* (i.e., the 8) may be in error, but the next to the last figure is assumed to be correct. Uncertainty in the first or second figure would obviate the significance from the last figure (e.g., the number of feet must first be known before worrying about inches). If the last figure is uncertain by a specified amount, the woman's height can be given as 5.18 ± 0.01 ft.

As mentioned in Chapter 1, the significant figures of a number can be found by reading the number from the left to the right and counting all digits starting with the first digit that is not zero. The decimal point is ignored since it is determined by the units, not by the precision of the operation. So, the measurements 33.35 cm and 333.5 mm are equivalent, and both have four significant figures.

When adding and subtracting, any figure in the answer is significant only if each number in the problem contributes a significant figure at that decimal level (i.e., the level of greatest magnitude will determine how many significant figures should be carried in the answer):

507.7812
0.00034
20.31
528.09

When "rounding off" (discarding nonsignificant figures), the last significant figure is unchanged if the next figure is <5 and is increased by 1 if the next figure is 5 or more:

4.7349 rounded to 4.735 (four significant figures)
4.7349 rounded to 4.73 (three significant figures)
2.8150 rounded to 2.82 (three significant figures)

In multiplication and division, the number of significant figures in the answer is the same as that in the quantity with the fewest significant figures:

$$(2.0 \times 5199)/0.0834 = 5.4 \times 10^7$$

In a calculation with multiple steps, first, find the number of significant figures in the answer as discussed above and then round off each number with excess significant figures to one or more significant figures than necessary. Then, round off the answer to the correct significant figures. This procedure preserves significance without too much work. For example, $x = 4.3 \times (311.8/273.1) \times [760/(784 - 2)]$.

There are two significant figures in the number 4.3; therefore, the answer will have two significant figures. Round off according to Rules 1 and 2, to one extra significant figure. Note that the presence of only one significant figure in the number 2 does not mean that there is only one significant figure in the answer because $784 - 2 = 782$, which has three significant figures. So,

$$x = 4.3 \times (312/273) \times (760/782)$$

Solve and round off to two significant figures

$$x = 4.8$$

One means of showing the level of precision of laboratory and field measurements is to compare duplicate measurements. This is expressed by relative percent difference (RPD):

$$\text{RPD} = \frac{(C_1 - C_2)}{(C_1 + C_2)/2} \times 100\% \tag{11.5}$$

Relative percent difference example 1.

An environmental field team has just completed a stack test for the toxic gas hydrogen sulfide (H_2S) for a manufacturing plant. This resulted in duplicate samples that, upon analysis, the H_2S stack emission concentrations were found to be 111 and 102 $\mu g\,m^{-3}$. The QA plan for the project calls for an RPD $\leq$ 10.00%. Do these duplicates meet the QA performance requirements for the H_2S stack test?

Solution

Calculate the RPD:

$$\text{RPD} = \frac{(111 - 102)}{(111 + 102)/2} \times 100\% = 8.45\%$$

Since the RPD is <10.00%, the results are sufficiently precise to meet the QA requirements.

In the same study mentioned above, a new QA requirement of an RPD $\leq$ 3.00% was added to improve precision. To address this need, the team purchased new equipment that yielded H_2S concentrations of 111 and 107 $\mu g\,m^{-3}$. Does the new equipment allow the team to meet the new precision requirements?

$$\text{RPD} = \frac{(111 - 107)}{(111 + 107)/2} \times 100\% = 3.67\%$$

Even though the precision improved by almost 5%, the RPD is still greater than the 3.00% QA requirement, so the results are too imprecise to be used.

When more than two replicates are available, the relative standard deviation (RSD) is used:

$$\text{RSD} = \frac{s}{\bar{Y}} \times 100\% \quad (11.6)$$

where s = the standard deviation of the replicates and $\bar{Y}$ = mean of the replicate measurements.[2]

The standard deviation, s, is calculated as

$$s = \sqrt{\left[\sum_{n=1}^{n}\left[\frac{(Y_1 - \bar{Y})^2}{n-1}\right]\right]} \quad (11.7)$$

What is the RSD for benzene at the concentrations of 23, 18, 17, 20, and 23 $\mu g\,m^{-3}$ from a near-road air pollution study?

First, calculate the sample mean: $\bar{Y}$ = 101 ÷ 5 = 20.

Next, find s:

$$s = \sqrt{\left[\sum_{n=1}^{n}\left[\frac{(Y_1 - \bar{Y})^2}{n-1}\right]\right]} = \sqrt{\frac{3^2 + 2^2 + 3^2 + 0^2 + 3^2}{4}} \quad (11.8)$$

$$= 7.8$$

Finally, calculate the RSD:

$$\text{RSD} = \frac{s}{\bar{Y}} \times 100\% = 7.8/20 \times 100\% = 39\%$$

2. For many environmental situations, the random error, that is, the difference between the value that is measured and the actual value, is assumed to be distributed normally. This is the prototypical "bell curve" that is also known as a Gaussian distribution. The mean of the entire population has the symbol μ, and the variance (i.e., the "spread" of values as indicated by how far the tails extend from the mean on the bell curve) is represented by σ^2. The population standard deviation is square root of the variance so is simply σ. The mean of the sample of the population is the "Y bar" ($\bar{Y}$) shown here, and the sample standard deviation is represented by s.

TABLE 11.2 Hypothetical concentrations of benzo(a)anthracene in sample extracts

1	2	3	4	5
Concentration of spike sample	Concentration of sample only	Recovery (1–2)	Known concentration added	Deviation from expected (3–4)
1.13	1.15	−0.02	0.05	−0.07
1.10	1.15	−0.05	0.05	−0.10
2.11	2.03	0.08	0.05	0.03
3.09	3.03	0.06	0.05	0.01
9.67	9.07	0.60	0.05	0.55
5.57	5.69	−0.12	0.05	−0.17
9.90	9.77	0.13	0.05	0.08
9.54	9.52	0.02	0.05	−0.03
8.90	8.85	0.05	0.05	0.00
1.11	0.99	0.12	0.05	0.07
			$\sum$ Deviation =	0.37

A known amount of the chemical being tested is often added to samples as a QA check. This is known as a "spiked" sample. The analysis of the unspiked samples is first determined, followed by an analysis after the samples are spiked.

You are measuring the amount of benzo(a)anthracene in the air near a power plant. The extracts from the filters that collected the air samples are shown in Table 11.2. The table also shows that the lab technician has added 0.05 mg L^{-1} benzo(a)anthracene to each sample. Calculate the sample precision for the 10 samples collected and spiked.

Solution

$$s=\sqrt{\frac{\left(\sum \text{deviation}^2\right)}{n-1}} \tag{11.9}$$

$$=\sqrt{\frac{(0.37)^2}{10-1}}=0.123$$

11.2.2 Measurement accuracy

Accuracy combines bias and precision to reflect how close the measured data are to the true value. Spiking is a valuable analytic tool to assure accuracy. A known amount of the substance of interest (i.e., the analyte) is added to the sample. The percent recovery (%R) for the spiked sample is found as follows:

$$\%R=\frac{S-U}{C}\times 100\% \tag{11.10}$$

where

S = the measured concentration in the spiked sample
U = the measure concentration of the unspiked sample
C = actual concentration of known addition to the sample

A tire recycling facility has emitted polycyclic aromatic hydrocarbons (PAHs) into the air for several decades. Soils downwind from the facility were sampled. What is the percent recovery of a sample to measure benzo(a)anthracene in soil that has a measured concentration in the spiked aliquot of 5.25 ng kg^{-1} and a measured concentration of 3.85 ng kg^{-1} in the unspiked aliquot? The soil has been spiked with a known amount equal to 2.00 ng kg^{-1} of benzo(a)anthracene.

$$\%R=\frac{S-U}{C}\times 100\%=[(5.25-3.85)/2.00]\times 100\%=70\%$$

A standard reference material (SRM) is often used to determine accuracy. An SRM contains a known analyte concentration that is certified by an outside source. An aliquot of the standard reference material is processed as a sample and processed through the complete analytic procedure used for all environmental samples. When standard reference material (SRM) is available, the percent recovery can be calculated as

$$\%R = \frac{C_M}{C_{SRM}} \times 100\% \quad (11.11)$$

where C_M is the concentration of the SRM that is measured and C_{SRM} is the actual concentration of the SRM. So, if a certified laboratory provides us with an aliquot that they state has a concentration of 3.95 ng kg^{-1} benzo(a)anthracene, but our analysis shows 3.85 ng kg^{-1}, then the $\%R = (3.85/3.95) \times 100\% = 97.47\%$.

As the numerator in the %R equation approaches the value of the denominator, the sample accuracy improves. Values deviate from 100% recovery because of bias and analytic imprecision, including operator and equipment deficiencies.

The overall accuracy of all samples in a study can be evaluated by summing their individual percent recoveries and dividing by the number of samples. This gives a mean percent recovery for the study. The variance and standard deviation of the overall recoveries can also be calculated.

11.3 DATA QUALITY OBJECTIVES

Even highly precise and accurate data are less useful if they are incomplete and do not properly represent the objectives of an investigation. One measure of completeness is to determine the fraction of measurements that are valid against the total number of measurements taken. In this way, percent completeness also represents operational efficiency. Representativeness is also a consideration of data, particularly if decisions are going to be made. For example, do the data represent an area beyond where the measurements were taken. This involves judgment as to whether information may be extrapolated from the measurements. Models are sometimes used to do these. Conversely, representativeness may also include interpolation, for example, deciding how to model the areas between measurements.

Air pollutant monitoring must strictly adhere to a well-designed QAPP to investigate environmental conditions. This enables measurements to be statistically representative, comparable and useful beyond the specific study [21]. The plan describes in detail the sampling apparatus (e.g., real-time probes, sample bags, bottles, and soil cores), the number of samples needed, sample handling, and transportation [2]. The quality and quantity or samples are determined by data quality objectives (DQOs), which are defined by the objectives of the overall contaminant assessment plan. DQOs are also useful in communications within and among different interest groups. For example, the air pollution expert may receive a general mandate from policy makers and managers but must convert these into sound science. The DQOs are qualitative and quantitative statements that translate nontechnical project goals into scientific and engineering outputs needed to answer technical questions [22].

A quantitative DQO defines a required level of scientific and data certainty, whereas a qualitative DQO expresses decision goals without specifying those goals in a quantitative manner. Even when expressed in technical terms, DQOs must specify the decision that the data will ultimately support, but not the way those data will be collected. DQOs guide the determination of the data quality that is needed in both the sampling and analytic efforts. The US Environmental Protection Agency (EPA) has listed three examples of the range of detail of quantitative and qualitative DQOs [23, 24]:

1. *Example of a less detailed, quantitative DQO:* Determine with >95% confidence that contaminated surface soil will not pose a human exposure hazard.
2. *Example of a more detailed, quantitative DQO*: Determine to a 90% degree of statistical certainty whether the concentration of mercury in each bin of soil is <96 ppm.
3. *Example of a detailed, qualitative DQO:* Determine the proper disposition of each bin of soil in real time using a dynamic work plan and a field method able to turn around lead (Pb) results on the soil samples within 2 h of sample collection.

This means that if the conditions in question are tightly defined, for example, the seasonal change in pH in rainfall downwind from a power plant, a small number of samples using simple pH probes would be defined as the DQO. Conversely, if the environmental assessment is more complex and larger in scale, for example, the characterization of year-round concentrations of a suite of air pollutants downwind from a power plant and chemical manufacturing facility, the sampling plan's DQO may dictate that numerous samples at various points be continuously sampled for gases and PM, as well as certain toxic air pollutants, depending on the stack tests and other evidence of what is and has been emitted by these and other source. This is even more complicated for biotic systems, including humans, which may also require microbiological monitoring, for example, biomarkers of exposure to the pollutants.

As mentioned, an air pollution sampling plan may also include other environmental media, for example, soil, water, and biota, for complete characterization of the exposure and risk. The sampling and analysis plan should explicitly point out which methods will be used. The EPA, for example, if toxic chemicals are being monitored, specifies specific sampling and analysis methods [25, 26].

An air pollution sampling plan must lead to results that are spatially representative. The geographic area where samples are to be collected not only is more than simply representing two-dimensional area but also includes distinctive physical features of a metropolitan, for example, high rises in the central business district. It must also account for land-sea linkages, valleys, and other elements of topography.

Air pollution measurements and models are complementary. The sampling plan with a QAPP must reflect the modeling that characterizes the likely extent of a pollutant plume (see Fig. 6.2 in Chapter 6). Air pollution plumes vary substantially in size and extent, with models ranging from highly localized, for example, one or two buildings (see Fig. 11.16) to urban scale (see Figs. 11.17 and 11.18) to continental (see Fig. 11.19).

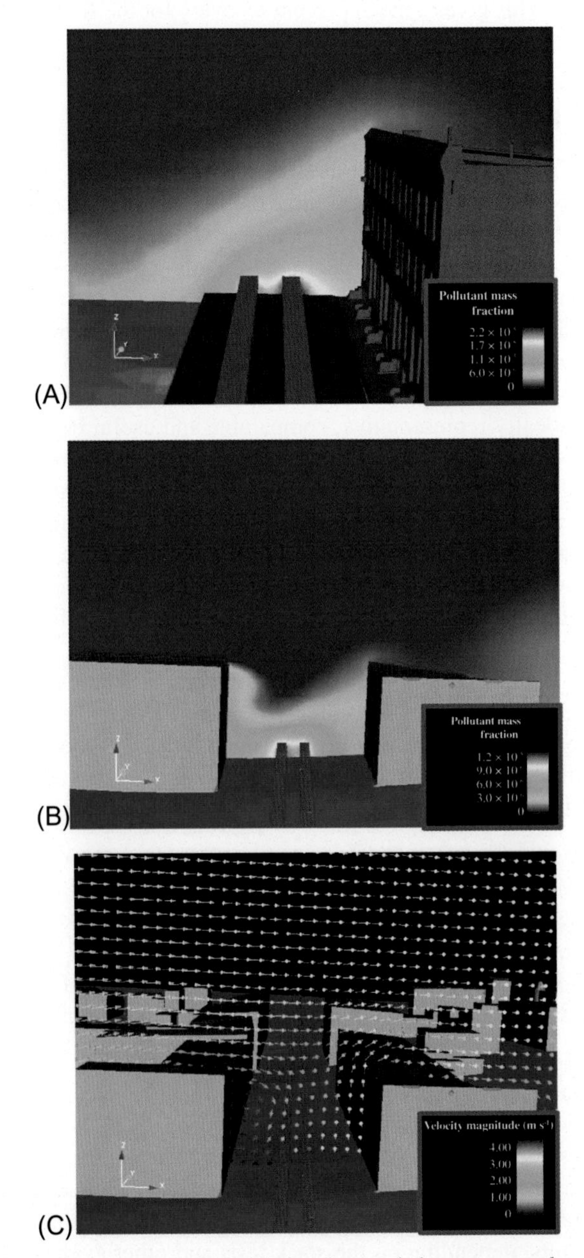

FIG. 11.16 Vertical slice view of a computation fluid dynamic model of roadway emissions represented as a source box along the roadway: (A) concentrations for street bounded by building on one side, (B) concentrations for street canyon, and (C) wind vectors for street canyon. *(Modified from U.S. Environmental Protection Agency, Research Triangle Park, NC.)*

FIG. 11.17 Simulated plume of neutrally buoyant smoke released from the World Trade Center, using a physical wind-tunnel model, showing flow from left to right is displayed; natural light is illuminating the smoke, and a vertical laser sheet is additionally illuminating the plume near the centerline of source. Vertical laser sheet is additionally illuminating the plume near the centerline of source. *(Source: D.A. Vallero, T.M. Letcher, Unraveling Environmental Disasters. Newnes, 2012. Photo used with permission from S. Perry, D. Heath. Fluid Modeling Facility, U.S. Environmental Protection Agency, Research Triangle Park, North Carolina, 2006.)*

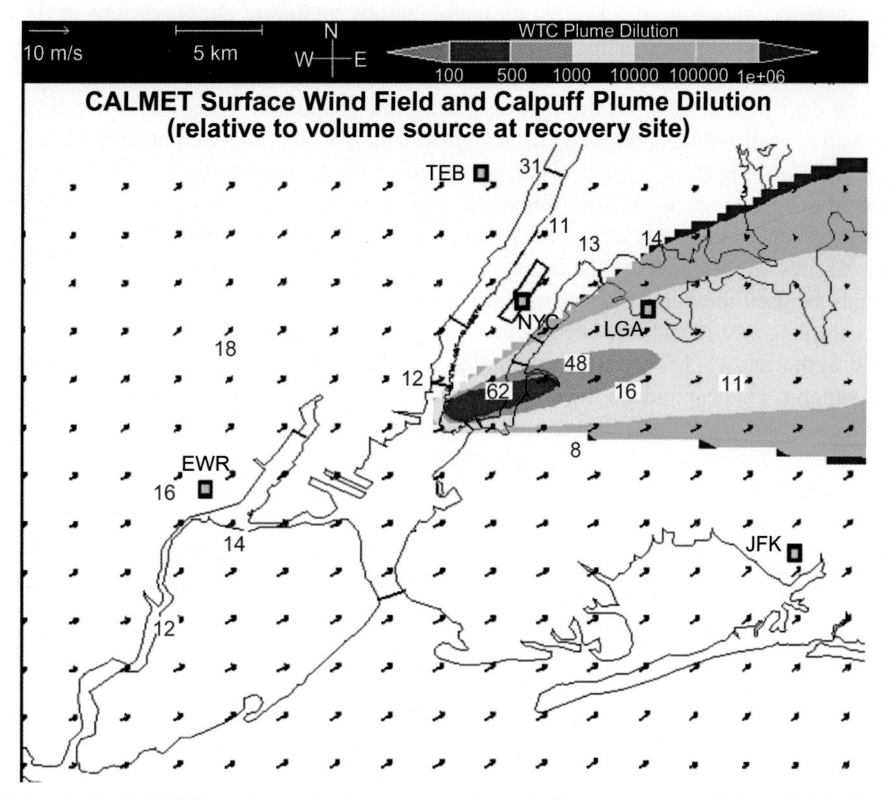

FIG. 11.18 Plume simulated with CALPUFF model showing the average hourly particulate concentration $\leq 2.4\,\mu m$ ($PM_{2.5}$) dilution of a volume source at the World Trade Center, New York City. *(Photo used with permission from U.S. Environmental Protection Agency.)*

FIG. 11.19 Dust plume from the Rub' al Khali swept across the southeastern part of the Arabian Peninsula and over the Arabian Sea between 28 July and 31 July 2018. Photo from Moderate Resolution Imaging Spectroradiometer (MODIS) on board NASA's Aqua satellite. The plume is the result of hot and dry air and low-level nearly constant northwesterly winds. *(Source: J. Schmaltz, MODIS: Moderate Resolution Imaging Spectroradiometer, National Aeronautics and Space Administration, 2018.)*

Airborne substances can be transported long distances in winds aloft. This includes bioaerosols, which are particles that consist of microorganisms, that is, viruses, bacteria, and fungi, or particles released by organisms, for example, pollen, spores, and cysts [1]. For example, some of the invasive bacteria that threaten coral reef habitats may originate from Africa in the form of Saharan dust. Deserts commonly contain gravel and bedrock, along with some sand. The Sahara is the exception, with sand covering 20% of the spatial extent of the desert. This means that the Sahara often loses large amounts of dust by winds that advectively transport particles in plumes that can travel across the Atlantic Ocean, depositing dust along the way. Continental scale windblown dust is common in Africa and the Arabian Peninsula (see Fig. 11.20).

The extent and shape of plumes are determined by many factors, including the conditions of the atmosphere and the characteristics of the substances that are being carried. Assuming the simplest scenario (never the case in environmental science), the distribution of the substances will be random. Gaussian dispersion algorithms (see Fig. 6.7 in Chapter 6) are often used to estimate drift over simple terrain, that is, having low roughness index and low relief. This means that the substances are assumed to be distributed randomly according to wind vectors. That is, standard deviations of the substances in three dimensions, that is, along the x, y, and z axes, are calculated to determine the location of the plume carrying these particles. As discussed in Chapter 6, actual atmospheric dispersion methods may be applied to more complex ecosystems characterized by vertical venting in forest areas, channeling down canyons, and both horizontal and vertical recirculations that may occur at local sites. In fact, some of the recent computational models that have been developed for air pollution in complex airsheds may be used in ecological risk assessments for disasters.

Boundaries of a sampling area must be defined carefully. For example, Fig. 11.21 shows a sampling grid, with a sample taken from each cell in the grid [17]. The target population may be divided into relatively homogeneous subpopulations within each area or subunit. This can reduce the number of samples needed to meet the tolerable limits on decision errors and to improve efficiency. For air pollution studies of a stationary source, at least one upwind site may be needed, along with a sufficient number of downwind sites to characterize the plume. A complicating factor is the contribution of other sources both upwind and downwind.

A complicating factor is the contribution of other sources both upwind and downwind. For example, the top of Fig. 11.22 depicts a hypothetical plume of an air pollutant released from a stationary source that may be expected in simple terrain. In this case, there are low concentration of the pollutant upwind and a concentration gradient downwind. Thus, if the wind continues in this direction and velocity during the investigation, one upwind monitor and two or three downwind monitors may be sufficient [27] to characterize the plume and to provide measurement data for models. However, if there is another source of the pollutant, it may contort the plume by adding concentration at the intersection of the two plumes (Fig. 11.22B). Incidentally, the other source may be a stationary source or mobile source. For example, in a recent study to investigate the contribution of an interstate highway to air pollution in Las Vegas [28, 29], the downwind monitors periodically showed elevated concentrations of some air pollutants when a train passed by the downwind sites. This was expected by the investigators, since the selection of the monitoring sites were otherwise optimal, that is, electric services were available, agreement by owners to use the right-of-way, and the ability to operate the sites continuously along a

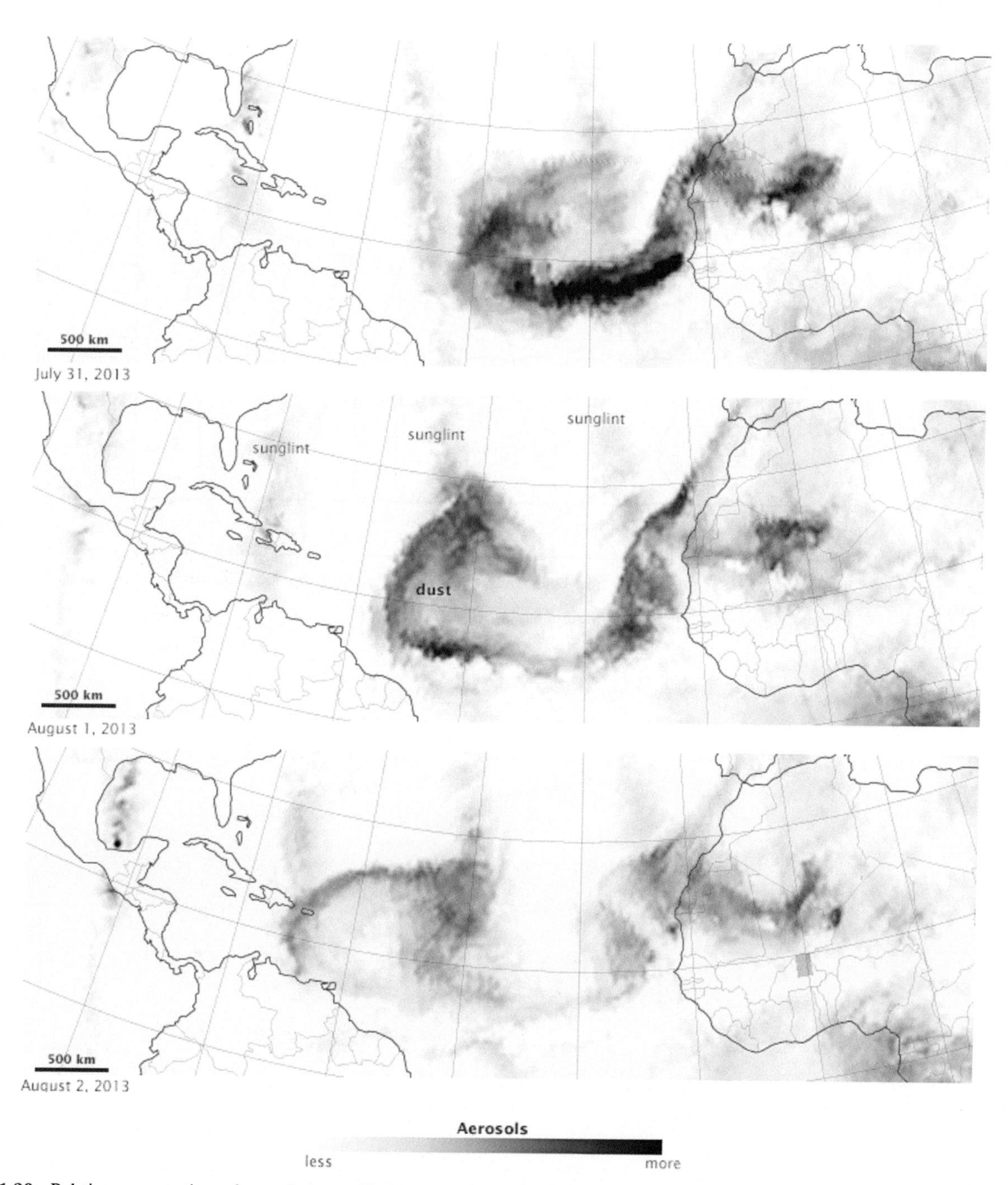

FIG. 11.20 Relative concentrations of aerosols detected by NASA's Ozone Mapping Profiling Suite (OMPS) on 31 July through 2 August 2013. Lower concentrations appear in yellow, and greater concentrations appear in orange-brown. Each map includes approximately six satellite passes. Note that sunlight also causes some vertical banding in these images. *(Source: C. Seftor, Tracking Dust Across the Atlantic, E. O. National Aeronautics and Space Administration, 2013.)*

transect expected based on Las Vegas's meteorology, especially its wind roses [28, 30]. In air pollution monitoring, siting is almost never ideal, so the site is usually chosen based on the sensitivity of the key variables and objectives. In this case, the periodic increase in pollutant concentrations was known and quantified. Known variability is preferable to unknown variability.

Time is another essential parameter that determines the type and extent of monitoring needed. Conditions vary over the course of a study due to changes in weather conditions, seasons, operation of equipment, and human activities. These include seasonal changes in groundwater levels, seasonal differences in farming practices, daily or hourly changes in

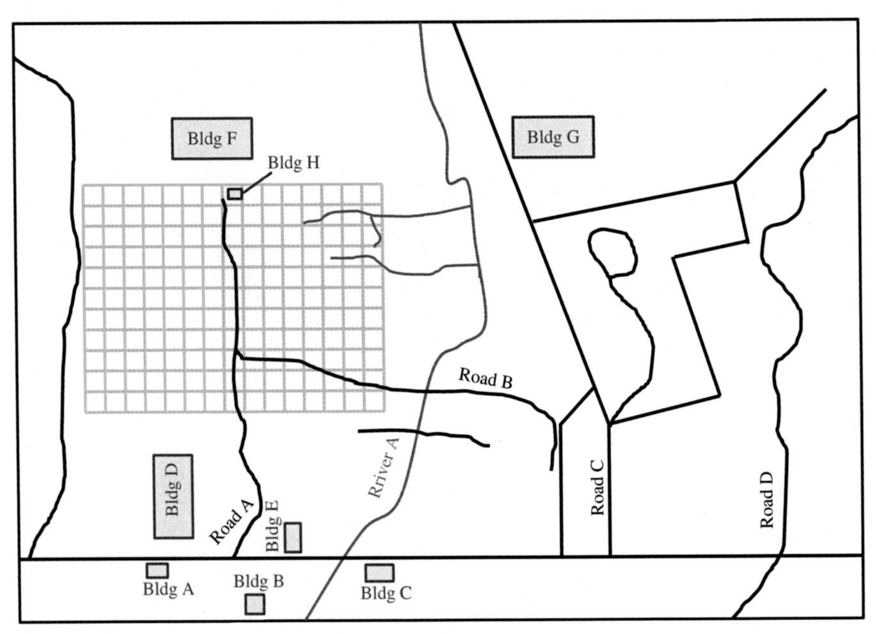

FIG. 11.21 Environmental assessment area delineated by map boundaries. *(Based on EPA/600/R-96/055, Guidance for the data quality objectives process, EPA QA/G-4 (2002).)*

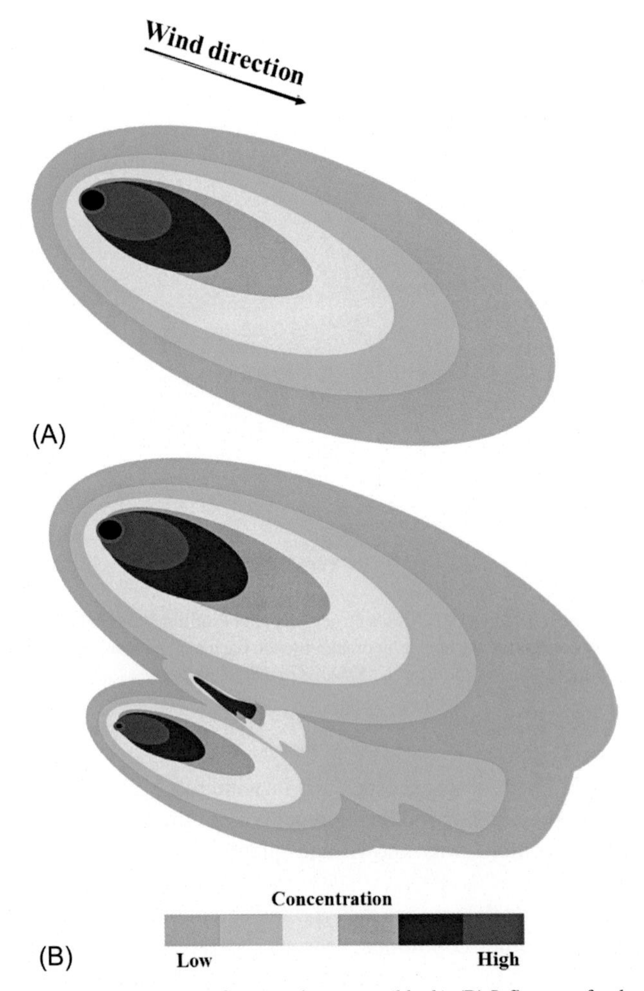

FIG. 11.22 Hypothetical plume (A) of an emitted pollutant from a point source *(black)*. (B) Influence of a downwind source.

airborne contaminant levels, and intermittent pollutant discharges from industrial sources. Such variations must be considered during data collection and in the interpretation of results. Some examples of environmental time sensitivity are as follows:

- Concentrations of lead in dust on windowsills may show higher concentrations during the summer when windows are raised, and paint/dust accumulates on the windowsill.
- Terrestrial background radiation levels may change due to shielding effects related to soil dampness.
- Amount of pesticides on surfaces may show greater variations in the summer because of higher temperatures and volatilization.
- Instruments that may not give accurate measurements when temperatures are colder.
- Airborne PM measurements that may not be accurate if the sampling is conducted in the wetter winter months rather than the drier summer months.

Feasibility should also be considered. This includes gaining legal and physical access to the properties, equipment acquisition and operation, environmental conditions, times, and conditions when sampling is prohibited (e.g., freezing temperatures, high humidity, and noise).

11.4 SAMPLING AIR POLLUTANT EMISSIONS AND RELEASES

The measurements at the point of a pollutant's release into the environment usually represent the highest concentrations of an air pollutant during its life cycle, excluding measurements within the reactor. A reactor may be specifically designed to produce a compound, for example, a pesticide or household chemical, with releases to the environment through stacks and vents. The next steps in the chemical's life cycle involve dilution of concentrations until the substance is again concentrated by organisms (see Fig. 11.23).

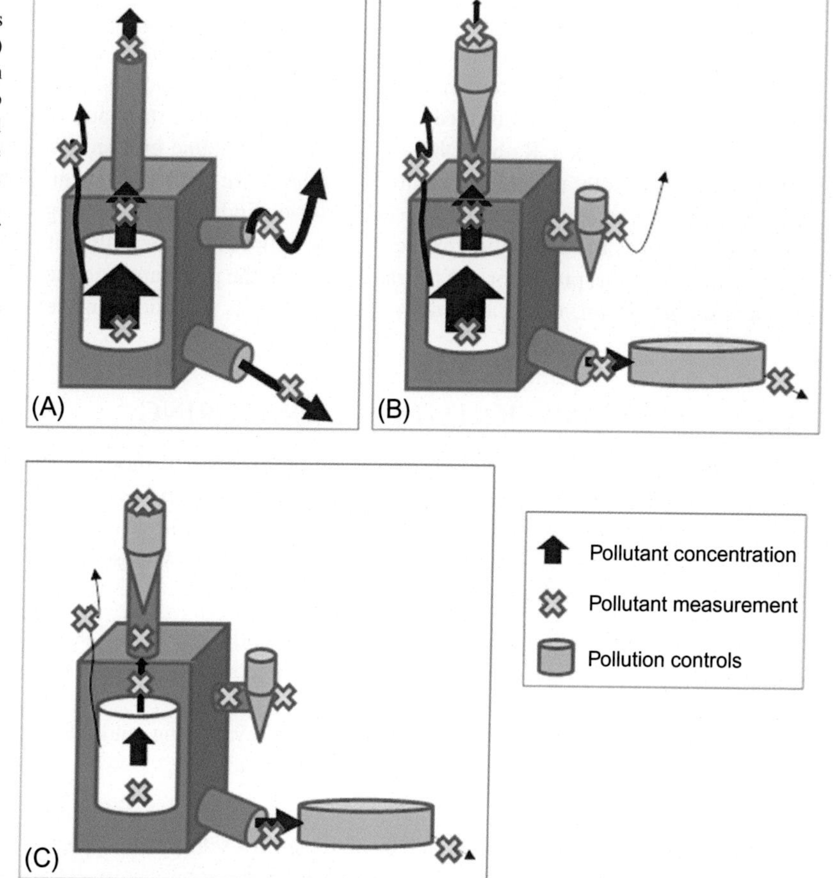

FIG. 11.23 Potential sites for measuring hypothetical chemical compound. The width of arrows represents relative pollutant concentration. (A) Sampling locations prior to installation of pollution control equipment. (B) Sampling locations to determine effectiveness of pollution control equipment. Ideally, sampling is needed at the entrance and exit of each pollution control device to determine pollution control efficiency. (C) Sampling locations to determine effectiveness of changes to reactor, for example, green chemistry and substitution of reagents to decrease or eliminate pollution, such as the substitution of a chlorinated compound with a nonchlorinated solvent, thereby eliminating organochlorine emissions. The scenario shown in C has completed eliminated stack and vent emissions and decreased fugitive emissions from the reactor (see Fig. 11.24), as well as decreased the pollutant concentrations in the effluent being released into surface and groundwater.

FIG. 11.24 Types of fugitive emissions. *(Source: D.K. Mikel, R. Merrill, J. Colby, EPA Handbook: Optical Remote Sensing for Measurement and Monitoring of Emissions Flux, US Environmental Protection Agency, North Carolina, 2011.)*

The measurement locations in Fig. 11.24 are at the emission level for a facility. Ambient concentrations, discussed in the next section, result from this facility and all other sources. Thus, regulatory agencies compare emissions from each source with ambient measurements that represent a type of mixed average concentration. For example, emission measurements indicate that Facility X emits one ton of Chemical A annually. Given mixing, meteorology, and other factors, models indicate that the ambient concentrations would require a total of 100 tons of Chemical A to be emitted annually to reach the current annual average air concentration of $10\,\mu g\,m^{-3}$. Thus, Facility X accounts for 1% of the mass of Chemical A in the ambient air. If the models show that the ambient air concentration must decrease to $5\,\mu g\,m^{-3}$, assuming all other factors remain the same, the total amount of Chemical A allowed to be emitted would be 50 tons. If all facilities must reduce their emissions of Chemical A proportionately, then Facility X must reduce its emission to 0.5 tons per year.

Of course, this example is very crude, and linkages are limited in time and space, for example, the concentrations of benzene or carbon monoxide for several city blocks downtown from an industrial district of road intersection. Most such source apportionments are more complicated. For example, tropospheric ozone (O_3) results from the presence of hydrocarbons, NO_x, and sunlight. Thus, measuring O_3 from a facility's stack and vents would be meaningless in linking the source to ambient concentrations. Rather, NO_x and organic compound emission measurements are needed, along with measurements of incoming solar radiation measured over the entire airshed, since ground-level ozone is a regional pollutant for the entire region. Source apportionment also depends on winds, terrain, mixing, dispersion, and chemical transformation of a pollutant, as well as the population expected to be exposed to that pollutant. Transformation works for or against ambient concentrations, that is, precursor can be transformed to the pollutant, and the pollutant can be degraded. The emission measurement method, then, varies according to the accuracy, precision, and representativeness needed.

11.5 AMBIENT AIR POLLUTION MONITORING

The complexity of factors needed for proper citing of ambient monitors can be illustrated by a recent plan to measure mobile source air toxic (MSAT) concentrations and variations in concentrations as a function of distance from the highway and to establish relationships between MSAT concentrations as related to highway traffic flows including traffic count, vehicle types and speeds, and meteorologic conditions such as wind speed and wind direction. Specifically, the monitoring plan has the following goals [30]:

1. Identify the existence and extent of elevated air pollutants near roads.
2. Determine how vehicle operations and local meteorology influence near-road air quality for criteria and toxic air pollutants.
3. Collect data that will be useful in ground truthing, evaluating, and refining models to determine the emissions and dispersion of motor-vehicle-related pollutants near roadways.

A complex monitoring effort requires management and technical staff with a diversity of skills that can be brought to bear on the implementation of this project. This diverse skill set includes program management, contract administration, field monitoring experience, laboratory expertise, and quality assurance oversight.

The purpose of any site selection process is to gather and analyze sufficient data that would lead one to draw informed conclusions regarding the selection of the most appropriate site for the monitoring at a specific location. Moreover, the site selection process needs to include programmatic issues to ensure an informed decision is reached.

11.5.1 Selection of a monitoring site

Selecting a monitoring site must be based on scientific and feasibility factors, as shown in Table 11.3 and Fig. 11.25. Each step has varying degrees of complexity due to "real-world" issues. The first step was to determine site selection criteria (see Table 11.4). The follow-on steps include (1) developing list of candidate sites and supporting information, (2) applying site selection filter ("coarse" and "fine"; filter composition, e.g., quartz, glass, polycarbonate, mixed cellulose ester membrane), (3) site visit, (4) selecting candidate site(s) via team discussion, (5) obtaining site access permission(s), and (6) implementing site logistics. These and other feasibility steps must be tailored to the type of study. For example, quartz filters may not be used for asbestos measurements since both the filter and asbestos are comprised of silicates and because transmission electron microscopy often requires destruction of the filter, making mixed cellulose ester membrane preferable.

Following the development and application of site selection criteria, a list of candidate sites based on these criteria can then be developed. Geographic information system (GIS) data, tools and techniques, and on-site visits would be used to compare various sites that meet these criteria.

After applying site selection criteria as a set of "filters," candidate sites are incrementally eliminated. For example, the first filter would be sites with low traffic counts; the next filter, the presence of extensive sound barriers, eliminates additional sites; and other filters, for example, complex geometric design or the lack of available traffic volume data, eliminate additional sites. Next, feasibility considerations would eliminate additional candidate sites.

Even a well-designed environmental monitoring plan will need to be adjusted during the implementation phase. For example, investigators may discover barriers or differing conditions from what was observed in the planning phase (e.g., different daily traffic counts or new road construction).

An important component of "ground truthing" or site visit is to obtain information from local sources. Local businesses and residents can provide important information needed in a decision process, such as types of chemicals stored previously at a site, changes in vegetation, or even ownership histories.

Spatial tools are very useful in making and explaining environmental decisions [31, 32]. Until recently, the use of GIS and other spatial tools in decision processes has required the acquisition of large amounts of the data. In addition, the software had not been user-friendly. GIS data have now become more readily available in both quantity and quality, and GIS exists in common operating system environments. In fact, environmental regulatory agencies increasingly use data layers to assess and describe environmental conditions. A prominent example is the EnviroAtlas, which the US

TABLE 11.3 Example of steps in selecting an air quality monitoring site applied in a recent near-road air pollution study.

Step	Site selection steps	Method	Comment
1	Determine site selection criteria	Monitoring protocol	
2	Develop list of candidate sites	Geographic information system (GIS) data; on-site visit(s)	Additional sites added as information is developed
3	Apply coarse site selection filter	Team discussions and management input	Eliminate sites below acceptable minimums
4	Site visit	Field trip	Application of fine site selection filter
5	Select candidate site(s)	Team discussions and management input	
6	Obtain site access permissions	Contact property owners	If property owners do not grant permission, then the site is dropped from further consideration
7	Site logistics (i.e., physical access and utilities—electric and communications)	Site visit(s) and contact utility companies	

Source: S. Kimbrough, D. Vallero, R. Shores, A. Vette, K. Black, V. Martinez, Multi-criteria decision analysis for the selection of a near road ambient air monitoring site for the measurement of mobile source air toxics, Transp. Res. Part D: Transp. Environ. 13 (8) (2008) 505–515.

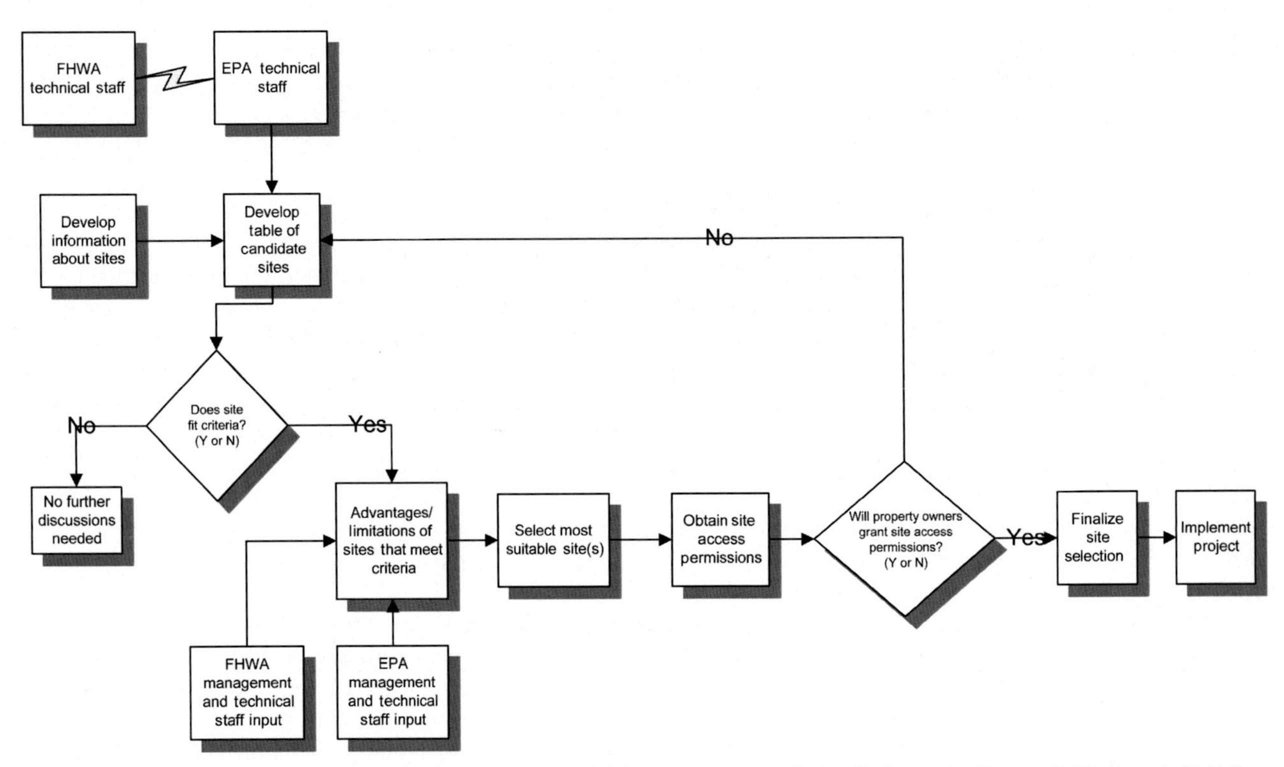

FIG. 11.25 Monitoring location selection decision flow chart applied in a recent near-road air pollution study. *(Source: S. Kimbrough, D. Vallero, R. Shores, A. Vette, K. Black, V. Martinez, Multi-criteria decision analysis for the selection of a near road ambient air monitoring site for the measurement of mobile source air toxics, Transp. Res. Part D: Transp. Environ. 13 (8) (2008) 505–515.)*

TABLE 11.4 Monitoring site selection considerations and criteria applied in a recent near-road air pollution study

Selection considerations	Monitoring protocol criteria
Essential criteria for this monitoring study:	
AADT (> 150,000)	Only sites with >150,000 annual average daily traffic (AADT) are considered as candidates
Geometric design	The geometric design of the facility, including the layout of ramps, interchanges, and similar facilities, will be considered. Where geometric design impedes effective data collection on MSATs and $PM_{2.5}$, those sites will be excluded from further consideration. All sites have a "clean" geometric design
Topology (i.e., sound barriers and road elevation)	Sites located in terrain making measurement of MSAT concentrations difficult or that raise questions of interpretation of any results will not be considered. For example, sharply sloping terrain away from a roadway could result in under representation of MSAT and $PM_{2.5}$ concentration levels on monitors near the roadway simply because the plume misses the monitor as it disperses
Geographic location	Criteria applicable to representing geographic diversity within the United States as opposed to within any given city
Availability of data (traffic volume data)	Any location where data, including automated traffic monitoring data, meteorologic data, or MSAT concentration data, are not readily available or instrumentation cannot be brought in to collect such data
Meteorology	Sites will be selected based on their local climates to assess the impact of climate on dispersion of emissions and atmospheric processes that affect chemical reactions and phase changes in the ambient air
Desirable, but not essential criteria:	
Downwind sampling	Any location where proper siting of downwind sampling sites is restricted due to topology, existing structures, meteorology, etc. may exclude otherwise suitable sites for consideration and inclusion in this study
Potentially confounding air pollutant sources	The presence of confounding emission sources may exclude otherwise suitable sites for consideration and inclusion in this study
Site access (admin/physical)	Any location where site access is restricted or prohibited either due to administrative or physical issues will not be considered for inclusion in the study

Source: S. Kimbrough, D. Vallero, R. Shores, A. Vette, K. Black, V. Martinez, Multi-criteria decision analysis for the selection of a near road ambient air monitoring site for the measurement of mobile source air toxics, Transp. Res. Part D: Transp. Environ. 13 (8) (2008) 505–515.

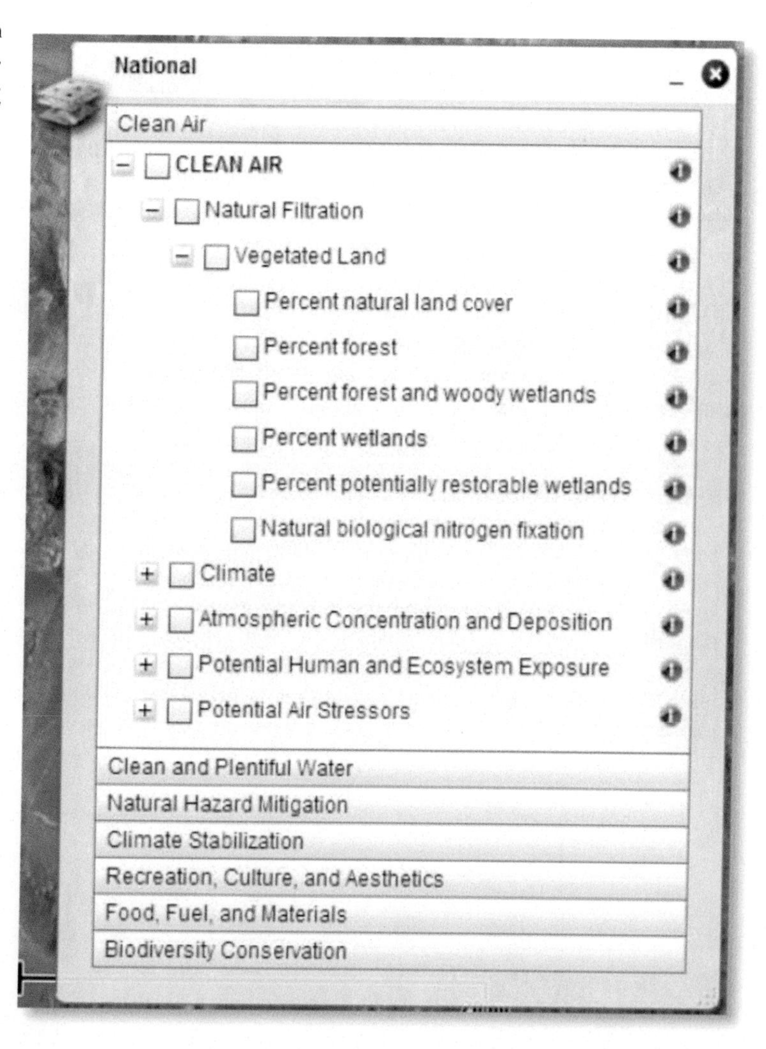

FIG. 11.26 Menu of map layers related to air quality in EnviroAtlas. *(Source: U.S. Environmental Protection Agency, EnviroAtlas Interactive Map: User's Guide, in EnviroAtlas, (2016), July 2016. Available: http://www.epa.gov/enviroatlas/enviroatlas-data-layer-matrix.)*

EPA has developed to provide a system of interactive tools to support and to document "ecosystem goods and services," that is, ecological benefits to humans from nature, including food supply, water supply, flood control, security, public health, and economy [33]. Fig. 11.26 shows some of the map layers that underpin the EnviroAtlas [34]. More specific to air quality data, the US EPA maintains interactive mapping of monitors across the nation, with several map layers [35]:

- Monitors for all criteria pollutants (CO, Pb, NO_2, ozone, PM_{10}, $PM_{2.5}$, and SO_2)
- $PM_{2.5}$ Chemical Speciation Network monitors
- Interagency Monitoring of PROtected Visual Environments (IMPROVE) monitors
- National Air Toxics Trends Stations (NATTS)
- Multipollutant Monitoring Network (NCORE)
- Nonattainment areas for all criteria pollutants
- Tribal areas
- Federal Class I areas (national parks and wilderness areas)

The GIS data layers that are commonly needed in air quality assessments include the location of suitable soils, wells, surface water sources, residential areas, schools, airports, and roads. From these data, layer queries are formulated to provide the most suitable sites (e.g., depth to water table may help identify sources of pollution). Typically, quantitative weighting criteria are associated with the siting criteria and elements of the data layers, for example, certain types of soils would be more suitable than others and thus would have applicable quantitative values [36].

11.5.2 Air sampling approaches

Ambient air sampling varies a spatially, but the sampling resolution generally ranges from neighborhood (~1–5 km scale) to regional (up to 1000 km scale) [37–39], with special, more focused sampling efforts, for example, near-road studies and industrial, fence-line assessments at finer scales [29, 40, 41]. Indoor and personal sampling are even finer scales.

As mentioned, the grab sample is simply a measurement at a site at a single point in time. Composite sampling physically combines and mixes multiple grab samples (from different locations or times) to allow for physical, instead of mathematical, averaging. The acceptable composite provides a single value of contaminant concentration measurement that can be used in statistical calculations.

Multiple composite samples can provide improved sampling precision and reduce the total number of analyses required compared with noncomposite sampling [42], for example, "grab" or integrated soil sample of x mass or y volume, the number of samples needed (e.g., for statistical significance), the minimum acceptable quality as defined by the quality assurance (QA) plan and sampling standard operating procedures (SOPs), and sample handling after collection.

Consider an evenly distributed grid of homes used to sample neighborhood-scale air quality. The sampling needs to represent the residents' potential exposure to a contaminant, as shown in Fig. 11.27. If an assessment downwind from a chemical manufacturing facility reported values of are 3, 1, 2, 12, and 2 μg m^{-3} of benzene, the mean contamination concentration is only 4 mg m^{-3}. Does this meet the local health standard of 5 μg m^{-3}?

If benzene concentrations must remain below the threshold of 5 μg m^{-3}, the mean concentration would indicate the area does not exceed local health standards and would be reported below the threshold level. However, the fourth home is well above the health level. This is an example of a false negative effect that can occur with composite sampling.

A false positive effect, conversely, occurs when the sampling results indicate conditions that are worse than actual condition.

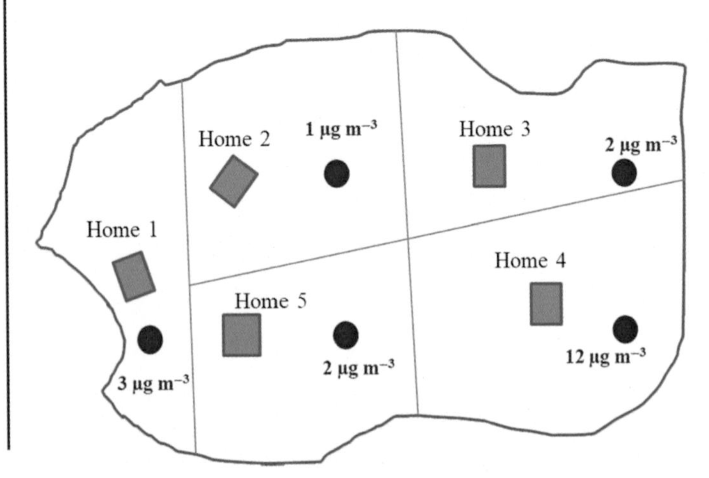

FIG. 11.27 Hypothetical composite sampling grid in neighborhood.

How might the values in Fig. 11.28 lead to a false positive interpretation?

Simply calculating a mean value from the results from the benzene monitors that are located over the extensive area shown in Fig. 11.28 would indicate an annual mean benzene air concentration averaged for five sites were 6 μg m^{-3}. If this mean value is used, the whole area appears to violate the benzene threshold, when only the site downwind form the industrial site is in violation. Indeed, there appears to be a steep concentration gradient from the industrial district moving southwest that within 3 km downwind returns to background concentrations, that is, approximately 3 μg m^{-3}.

Selecting appropriate sampling methods and considerations on their use are key parts of any study design and environmental assessment, since the results will be the basis for exposure models, risk assessments, feasibility studies, land use and zoning maps, and other information used by fellow engineers, clients, and regulators [43]. As indicated in Table 11.5, measurement errors and uncertainties will accompany the results and even be compounded as the data are translated into information [44], so it is important to include all necessary metadata to ensure others may deconstruct, quality assure, and ensure appropriate applications.

FIG. 11.28 Hypothetical annual mean benzene air concentrations at five widely dispersed sampling sites.

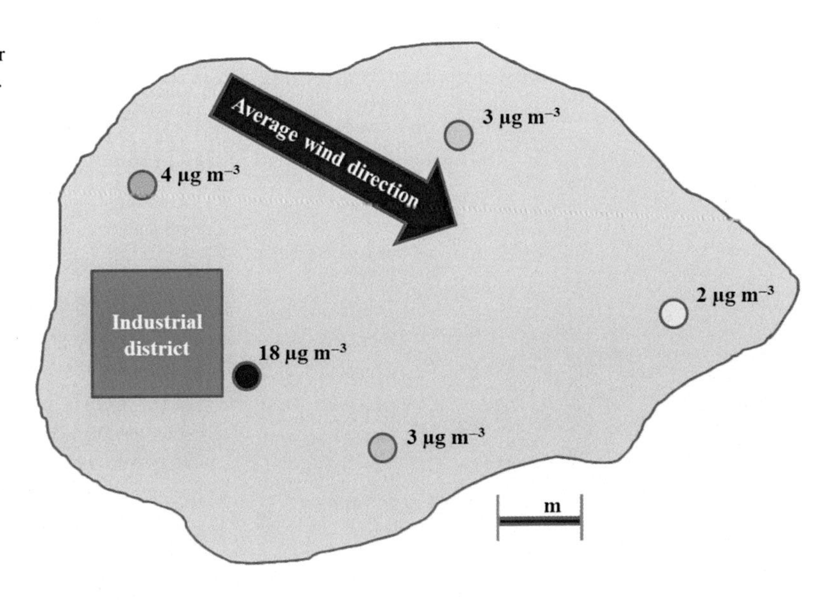

TABLE 11.5 Types of uncertainty and contributing errors in environmental engineering

Type of uncertainty	Type of error causing uncertainty	Description or example
Emissions	Misclassification and miscalculation	Reliance on third-party and other sources of information with little or no metadata regarding quality. Confusing actual emission measurements with reported estimates
Transport and transformation	Incorrect model application	Applying a model for the wrong chemistry, atmospheric, terrain, and other conditions, for example, using a simple dispersion model in a complex terrain. Applying a quantitative structural activity relationship (QSAR) to inappropriate compounds, for example, for metals when the QSAR is only for organic compounds or for semivolatile compounds when the QSAR is only for volatile organic species
Exposure scenario	Misclassification	Failure to adequately identify exposure routes, exposure media, and exposed population
Sampling or measurement (parameter uncertainty)	Measurement: random	Random errors in analytic devices (e.g., imprecision of continuous monitors that measure stack emissions)
	Measurement: systemic	Systemic bias (e.g., estimating inhalation from indoor ambient air without considering the effect of volatilization of contaminants from hot water during showers)
	Surrogate data	Use of alternate data for a parameter instead of direct analysis of exposure (e.g., use of population figures as a surrogate for population exposure)
	Misclassification	Incorrect assignment of exposures of subjects in historical epidemiological studies resulting from faulty or ambiguous information
	Random sampling error	Use of a small sample of individuals to estimate risk to exposed workers
	Nonrepresentativeness	Developing exposure estimates for a population in a rural area based on exposure estimates for a population in a city

TABLE 11.5 Types of uncertainty and contributing errors in environmental engineering—cont'd

Type of uncertainty	Type of error causing uncertainty	Description or example
Observational or modeling	Relationship errors	Incorrectly inferring the basis of correlations between environmental concentrations and urinary output
	Oversimplification	Misrepresentations of reality (e.g., representing a three-dimensional aquifer with a two-dimensional mathematical model)
	Incompleteness	Exclusion of one or more relevant variables (e.g., relating a biomarker of exposure measured in a biological matrix without considering the presence of the metabolite in the environment)
	Surrogate variables	Use of alternate variables for ones that cannot be measured (e.g., using wind speed at the nearest airport as a proxy for wind speed at the facility site)
	Failure to account for correlations	Not accounting for correlations that cause seemingly unrelated events to occur more frequently than expected by chance (e.g., two separate components of a nuclear plant are missing a washer because the same newly hired assembler put them together)
	Model disaggregation	Extent of (dis)aggregation used in the model (e.g., separately considering subcutaneous and abdominal fat in the fat compartment of a physiologically based pharmacokinetic (PBPK) model)
	Biological plausibility	Applying a PBPK model to chemicals for which it has not been coded, for example, a one-compartment model for a compound with known metabolites

Sources: D.A. Vallero, Fundamentals of Air Pollution, 5th ed., Elsevier Academic Press, Waltham, MA, 2014, p. 999 pages cm; EPA/100/B-04/001, U.S. Environmental Protection Agency. An examination of EPA risk assessment principles and practices. (2004). Available: http://www.epa.gov/OSA/pdfs/ratf-final.pdf.

11.5.3 Sampling approaches

Random sampling assigns the location of sites based on statistical randomization techniques. The advantage of this approach is that provides an opportunity for statistical significance. However, significance requires a sufficient number of samples for the defined confidence levels, for example, x samples needed for 95% confidence. A possible drawback is that large areas of the site may be missed for sampling due to chance distribution of results. Stratified random sampling may address this weakness, that is, by dividing the site into areas and randomly sampling within each area. This helps to avoid the omission problems of random sampling alone.

Random sampling may also neglect deterministic factors, for example, complex terrain, sea breezes, valley effects, and other factors that work against randomization. In these cases, random sampling may need to be replaced or augmented by Bayesian techniques.

Stratified sampling targets contaminants or other parameters. The sampling area is subdivided, and sampling patterns and densities are adjusted and vary in different areas. Stratified sampling can be used for complex and large sites, such as mining.

As mentioned, grid or systematic sampling covers the entire site. Sampling locations are readily identifiable, which is valuable for follow-on sampling, if necessary. The grid does not have to be rectilinear. In fact, rectangles are not the best polygon to use in the value that is to be representative of a cell. Circles provide equidistant representation but overlap. Hexagons are sometimes used as a close approximation to the circle. The US Environmental Monitoring and Assessment Program (EMAP) has used a hexagonal grid pattern, for example.

Judgment samples are collected base upon knowledge of the site. This overcomes the problem of ignoring sources or sensitive areas but is vulnerable to bias of both inclusion and exclusion. Obviously, this would not be used for spatial representation, but for pollutant transport; plume characterization; or monitoring near a sensitive site, for example, a school.

At each stage of monitoring from sample collection through analysis and archiving, only qualified and authorized persons should be in possession of the samples. This is usually assured by requiring chain-of-custody manifests. Sample handling includes specifications on the temperature range needed to preserve the sample, the maximum amount of time the sample can be held before analysis, special storage provisions (e.g., some samples need to be stored in certain solvents), and chain-of-custody provisions (only certain, authorized persons should be in possession of samples after collection).

Each person in possession of the samples must require that recipient sign and date the chain-of-custody form before transferring the samples. This is because samples have evidentiary and forensic content, so any compromising of the sample integrity must be avoided.

11.6 AMBIENT AIR POLLUTANT SAMPLING

Criteria air pollutants and hazardous air pollutants differ in terms of the types of sampling needed. For example, criteria pollutants are most often monitored to determine compliance with the air pollution laws and regulations. The resulting data from these monitors can be downloaded for specific study areas and time periods for assessments of air quality, attainment designations, plans to address nonattainment, modeling, and reports, for example, to the US Congress or other legislative bodies [39, 40, 45]. Generally, gas-phase pollutant concentrations are recorded continuously and calculated as hourly averages, and PM mass concentrations vary, often reported as hourly or 24-h averages [40].

An ambient air sampling system must be able to collect a sample that is representative of the atmosphere at a particular place and time and that can be evaluated as a mass or volume concentration [46]. Any credible sampling system is as non-intrusive as possible, that is, the sampling process must avoid introducing error, and sampling apparatus should not alter the chemical or physical characteristics of the sample in an undesirable manner.

Air sampling systems are configured in various ways, with four of the most common configuration shown in Fig. 11.29. Fig. 11.29A depicts a configuration typical of many extractive sampling techniques in practice, that is, those in which the liquid- or solid-phase collection medium is extracted using wet chemistry. Examples are SO_2 in liquid sorbents and poly-nuclear aromatic hydrocarbons on solid sorbents. Fig. 11.29B is used for "open-face" filter collection, in which the filter is directly exposed to the atmosphere being sampled. Fig. 11.29C is an evacuated container used to collect an aliquot of air or gas to be transported to the laboratory for chemical analysis; for example, polished stainless steel canisters are used to collect ambient hydrocarbons for air toxic analysis. Fig. 11.29D is the basis for many of the automated continuous analyzers, which combine the sampling and analytic processes in one piece of equipment, for example, continuous ambient air monitors for SO_2, O_3, and NO_x.

For any of any sampling system configuration or the specific material sampled, several characteristics are important for all ambient air sampling systems. In particular, a system's soundness is based on four criteria [4]:

1. Collection efficiency
2. Sample stability
3. Analyte recovery
4. Minimal interference

Ideally, the first three would be 100%, and there would be no interference or change in the material when collected.

One example of these factors is demonstrated when sampling for sulfur dioxide (SO_2). Liquid sorbents for SO_2 depend on the solubility of SO_2 in the liquid collection medium. Certain liquids at the correct pH can remove ambient concentrations of SO_2 with 100% efficiency until the characteristics of the solution are altered so that no more SO_2 may be dissolved in the volume of liquid provided. Under these circumstances, sampling is 100% efficient for a limited total mass of SO_2 transferred to the solution, and the technique is acceptable as long as sampling does not continue beyond the time that the sampling solution is saturated [46]. A second example is the use of solid sorbents such as Tenax for volatile hydrocarbons

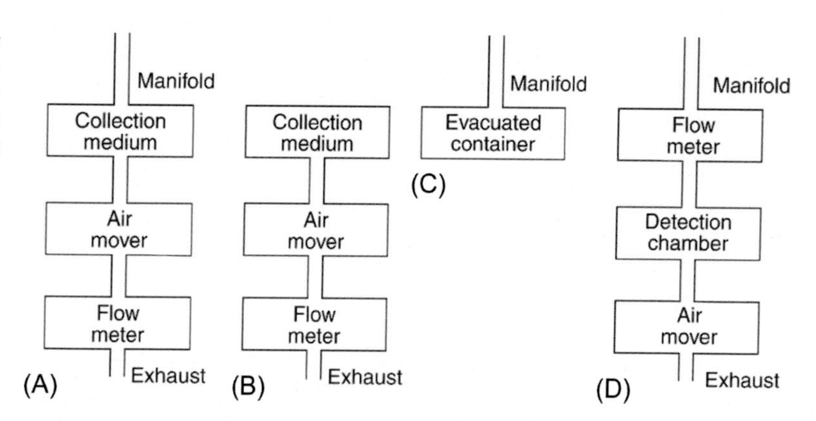

FIG. 11.29 Schematic diagram of various types of air sampling systems. (A) Sampler using liquid- or solid-phase collection medium; (B) "open-face" filter collection configuration; (C) gas or liquid aliquot collected in an evacuated container; and (D) real-time sampling and analytic configuration.

by the physical adsorption of the individual hydrocarbon molecules on active sites of the sorbent [47]. Collection efficiency drops drastically when the active sites become saturated.

Sample stability becomes increasingly important with increasing length of time between sampling and analysis. Effects of temperature, trace contaminants, and chemical reactions can cause the collected species to be lost from the collection medium or to undergo a transformation that will prevent its recovery. Consistent recovery rates near 100% are also required because a variable recovery rate will prevent quantification of the analysis. Interference should be minimal and, if present, well understood.

11.6.1 Passive sampling

Passive sampling uses the free flow of air carrying pollutants, that is, no pump. The concentration is usually derived using Fick's law of diffusion (see Chapter 6), so electric power is not needed, and the sampler can be placed in almost any location [48, 49]. The lack of pumps also means no noise or lessened safety issues, so the devices are very useful for personal and indoor measurements.

Passive monitors may be either a diffusive sampler, that is, the pollutant freely diffuses through a gas layer (see Fig. 6.8 in Chapter 6) or a penetration sampler, that is, the pollutant moves into a semipermeable membrane. Passive systems are usually the most effective approach for estimating spatial gradients around compliance monitors; surveying additional pollutants; identifying areas of unusually high concentrations, that is, hot spots; and estimating individual exposures to gas-phase pollutants [40]. Passive samplers usually must be colocated and with calibrated active sampling systems, for example, at central monitoring sites, to establish equivalence and comparability [40].

The time-weighted mean concentration ($\overline{C}$) of a substance in the air over sampling time (t) is calculated as

$$\overline{C} = \frac{M}{Q \times t} \tag{11.12}$$

where M is the mass of diffusing substance trapped and retained in the sampler's medium and Q is the uptake or sampling rate, which is the diffusion coefficient discussed in Chapter 6.

A passive sampler is placed on a column on a chemical manufacturing floor. The sampling rate is 16.7 mL min^{-1}. After 2 days, the mass of benzene was found to be 5 mg. What is the time-weighted mean benzene concentration? Is this a safe level of exposure?

$$1\ \text{mL} = 10^{-6}\ \text{m}^3$$

1 day = 1440 min; minutes cancel in the denominator.

$$\overline{C} = \frac{5\ \text{mg}}{16.7 \times 10^{-6} \times 2 \times 1440} = 104\ \text{mg m}^{-3}\ \text{benzene}$$

Benzene's molecular weight is 78.11 g mol^{-1}.

$$\text{At 1 atm and } 25^\circ\text{C}, x\ \text{ppm} = \frac{104 \times 24.45}{78.11} = 32.6\ \text{ppm}$$

This violates the Occupational Safety and Health Administration standards for benzene. The 8-h time-weighted average (TWA) must not exceed 1 ppm, and the short-term exposure limit (STEL) cannot exceed 5 ppm over any 15-min time period [50]. This means that the air needs to be cleaned or that the floor can only be occupied by persons using personal protective equipment (PPE) that allows them to breathe air below the TWA and STEL concentrations.

When an uptake rate is determined for a compound under established conditions of temperature and pressure, the airborne concentrations can be estimated for a desired sampling period. These can be input in exposure equations (see Chapter 12). Generally, passive monitors work best for exposure durations ranging from 15 min to 8 h. Extending the exposure time leads to a lower reportable concentration limit, up to about 30 days. Of course, this is only advantageous for compounds that are not substantially reactive; for example, if the compound has a half-life of 15 days, half of the mass will break down in the sorbing media.

The example calculation highlights one of the problems of passive monitors, that is, their resolution and detection limits are typically worse than those of active sampling devices. This is because the air flow into the device is much lower and inconsistent in passive monitoring, that is, their less air volume from which to calculate concentrations and more uncertainty as to the actual sampling rate.

Incidentally, workplace microenvironments may contain levels far higher than atmospheric means and even higher than those found in highly polluted urban air; especially in air pollution, control devices are not deployed. Thus, the allowable concentration of an air pollutant will vary according to the protection needed in ambient air and various personal and workplace microenvironments (see Discussion Box "Workplace Exposure").

Workplace Exposures

Higher concentrations of air pollutants are often permitted in workplace environments than in the general environment. This is because workplace environments often require personal protection equipment (PPE), which is enforced by occupational regulatory agencies. In addition, exposures in workplace environments are generally much shorter in duration and frequency, for example, 8 h per day, 5 days per week for 30 years, compared with general exposures, for example, 24 h per day, 7 days per week for a lifetime. The minimum risk level (MRL) is an estimate of the daily human exposure to a hazardous substance that is likely to be without appreciable risk of adverse noncancer health effects over a specified duration of exposure. Thus, both the cancer and MRL values for formaldehyde are much lower than the occupational levels on the right-hand side of Fig. 2.1 in Chapter 2 [51].

The recommended or regulated amount of exposure to air pollutants varies by scenario and activity. In the United States, the Occupational Safety and Health Administration's (OSHA) workplace exposure standard is the permissible exposure limit (PEL) for each air pollutant. The PEL is set to protect workers day in and day out for their entire career, that is, a long-term exposure.

The example discussed two of the three types of PELs:

1. The 8 h TWA is very commonly used. It is based on measurements taken near or on the work, that is, personal exposure. The measured level may exceed the TWA value at times during the 8 h time period, but the 8 h average concentration must be below the PEL.
2. The ceiling limit (PEL-C) is the maximum allowable concentration, not an average. Thus, any exceedance is considered a violation of the PEL, even for a very short time.
3. As mentioned, the STEL is the concentration that must not be exceeded when averaged over a specified short period of time (usually 15 min). When there is an STEL for a substance, exposure still must never exceed the PEL-C or the 8 h TWA.

Most air pollutants with PELs have a published TWA value. Some air pollutants also have ceiling or short-term exposure limits (STELs). Others have ceilings and STELs instead of TWA values.

The recommended levels of workplace exposure are based on research; for example, the National Institute of Occupational Safety and Health (NIOSH) publishes air quality limits for substances to provide guidance on safe worker exposures. The recommendations can also come from professional and scientific associations. For example, the American Conference of Governmental Industrial Hygienists publishes occupational exposure guidelines, that is, the threshold limit values (TLVs). Both the TLV and REL are advisory, but the PEL is a legal requirement that carries penalties when exceeded.

NIOSH publishes four types of exposure recommendations, structured similarly to the PELs:

1. The REL-TWA: 8 h time-weighted average concentration
2. The REL-TWA: 15 min time-weighted average concentration
3. REL-C: Ceiling concentration—not to be exceeded
4. IDHL: 30 min concentration that is immediately dangerous to life and health

Some industrial operations may require work in higher workplace exposures. This is permitted only if the workers wear appropriate personal protection equipment (PPE), especially respirators. Assigned protection factors (APFs) define the level of respiratory protection expected to be provided by a respirator or class of respirators. [52] The APFs range from 5 to 10,000. Similarly, the maximum use concentration (MUC) is the highest atmospheric concentration of a hazardous substance from which a worker would be expected to be protected when wearing the respirator. The relationship between the MUC and APF informs the decision on type of respirator needed:

$$\text{MUC} = \text{PEL} \times \text{APF} \tag{11.13}$$

An employee wears a half-mask air purifying respirator with an APF equal to 10 in an atmosphere known to have detectable concentrations of SO_2. What is the maximum use concentration?

The PEL for SO_2 gas is 5 ppm:

$$\text{MUC} = 10 \times 5 = 50\,\text{ppm}$$

Incidentally, the IDHL for SO_2 is 100 ppm, so the half-mask respirator appears to maintain exposure at acceptable levels [52].

Passive samplers can detect long-term averages to very low concentrations. The actual levels of detection vary with environmental conditions [48, 49, 53, 54]. Passive monitoring is often used for semiquantitative purposes, for example, to determine the presence or absence of a compound and screening [55], that is, unknown precision and accuracy, prior to quantitative studies using active monitors in which air volume is more precisely known so that concentrations can be calculated.

The choice of sorbent material in a diffusion tube or badge passive sampler, that is, the material and its impregnants, varies according to the compound being measured. Indeed, more than one sorbent may be needed to trap volatile compounds [56]. For both passive and active trapping, a combination is Tenax, and various types of activated carbon have been deployed for volatile organic compounds (VOCs). Care must be taken, however, since carbon-based sorbents are somewhat chemically reactive, causing labile compounds, for example, sulfur compounds, to become degraded while trapped [57, 58].

11.6.2 Active sampling

As mentioned, the major difference between passive and active sampling is that active systems deploy an air mover, for example, pump, to draw air through the sorbing material. In addition to the pump, the major components of most sampling systems are an inlet manifold, a collection medium, and a flow measurement device. The inlet manifold transports material from the ambient atmosphere to the collection medium or analytic device, preferably in an unaltered condition. The inlet opening may be designed for a specific purpose. All inlets for ambient sampling must be rainproof. Inlet manifolds are made of glass, Teflon, stainless steel, or other relatively unreactive materials and permit the remaining components of the system to be located at a distance from the sample manifold inlet. The air mover provides the force to create a vacuum or lower pressure at the end of the sampling system. The collection medium for a sampling system may be a liquid or solid sorbent for dissolving gases, a filter surface for collecting particles, or a chamber to contain an aliquot of air for analysis. The flow device measures the volume of air associated with the sampling system. Examples of flow devices are mass flow meters, rotameters, and critical orifices.

Thereafter, the air is pumped into a container or directed into an in situ sensor. Thus, a consistent, nonvarying pumping rate is needed for accurate and precise quantitation of air volume, flow rates, and sampling time [39, 59]. In addition, active systems have the advantage of not requiring diffusion, which allows for measurements of particles and other low diffusion substances, including airborne microbes and bioaerosols [60].

Many microsensors now use battery-powered miniature pumps that allow for air sampling in microenvironments or worn by individuals, that is, personal sampling [60–62]. These measurements are valuable in risk assessments, since humans are exposed to most chemical compounds in near-field scenarios, for example, consumer product use [63–68] and from indoor air pollutants [69]. One would suspect that such near-field exposures would also be highest for bioaerosols, given experience with contagious diseases, but there is a scarcity of reliable studies to support this hypothesis at present [70].

Detection limits of optical light scattering and absorption systems and electrochemical gas sensors continue to improve [40]. Particle scattering by nephelometer with a $PM_{2.5}$ inlet and a "smart" heater to remove moisture under high relative humidity (e.g., $> 65\%$) can provide a surrogate for $PM_{2.5}$ mass [71, 72].

11.6.3 Remote sensing

Remote sensing of many air pollutants employs indirect methods, such as calibrating energy levels to the pollutant's mass. For example, air pollutant mass can be represented by the amount of scattering and absorption of infrared, visible, and ultraviolet radiation at different wavelengths along a sight path. Path lengths may range from a few meters, for example, in-plume monitoring, to thousands of kilometers, for example, geostationary satellites [40, 73–75]. Satellite remote sensing estimates for PM, NO_2, SO_2, and some other pollutants often correspond to several kilometers, for example, urban and industrial areas, with resolution currently limited to about 10km [40].

11.7 GAS AND VAPOR PHASE MEASUREMENT METHODS

Gaseous pollutants are generally collected using methods different from aerosols. The gas-phase sampling systems shown in Fig. 11.29 must be able to move the gas from the manifold inlet to the collection medium in an unaltered state. The inlet and other components that contact the air must be made of material that will not react with the gas-phase pollutant of interest. Tests of material for manifold construction can be made for specific gases to be sampled. For most gases, glass or Teflon is sufficiently nonreactive. No condensation should be allowed to occur in the sampling manifold.

The volume of the manifold and the sampling flow rate determine the time required for the gas to move from the inlet to the collection medium. This residence time can be minimized to decrease the loss of reactive species in the manifold by keeping the manifold as short as possible.

The liquid or gas collection medium for gases may be held in an evacuated flask or in a cryogenic trap. Liquid collection systems take the form of bubblers that are designed to maximize the gas-liquid interface. Each design is an attempt to optimize gas flow rate and collection efficiency. Higher flow rates permit shorter sampling times, but increasing flow rates can begin to cause the collection efficiency to drop below 100%.

11.7.1 Extractive sampling

When bubbler systems are used for collection, the gaseous species generally undergoes hydration or reaction with water to form anions or cations. For example, when SO_2 and NH_3 are absorbed in bubblers, they form HSO_3^- and NHO_4^+; the analytic techniques measure these ions, rather than the gases. Table 11.6 gives examples of gases that may be sampled with bubbler systems.

Bubblers are often used when the number samples are small and when sampling is frequent. The advantages of these types of sampling systems are low-cost and portability. However, this technique calls training given the need for skill and care required. Solid sorbents such as Tenax, XAD, and activated carbon (charcoal) trap gases on the active sorption sites of the surface of the material. Fig. 11.30 illustrates the loading of active sites with increasing sample time. It is critical that the breakthrough sampling volume, the amount of air passing through the tube that saturates its absorptive capacity, not be exceeded. The breakthrough volume is dependent on the concentration of the gas being sampled and the absorptive capacity of the sorbent. This means that the user must have an estimate of the upper limit of concentration for the gas being sampled.

After sampling is completed, the tube is sealed and transported to the analytic laboratory. To recover the sorbed gas, two techniques may be used. The tube may be heated while an inert gas is flowing through it. At a sufficiently high temperature, the absorbed molecules are desorbed and carried out of the tube with the inert gas stream. The gas stream may then be passed through a preconcentration trap for injection into a gas chromatograph for chemical analysis. The second technique is liquid extraction of the sorbent and subsequent liquid chromatography. Sometimes, a derivatization step is necessary to convert the collected material chemically into compounds that will pass through the column more easily, for example, conversion of carboxylic acids to methyl esters. Solid sorbents have increased our ability to measure hydrocarbon species under a variety of field conditions. Care must be taken to minimize problems of contamination of the collection medium, sample instability on the sorbent, and incomplete recovery of the sorbed gases [4].

Although most texts, including this one, discuss the measurement of gases and particulate matter separately, there are sampling systems that collect both phases. Sampling systems are available to remove gas-phase molecules from a moving airstream by diffusion to a coated surface, while permitting the passage of particulate matter downstream for collection on a filter or other trapping medium. Such diffusion denuders can be used to sample for SO_2 or acid gases in the presence of particulate matter [76]. This type of sampling has been developed to minimize the interference of gases in particulate sampling and vice versa.

The configuration shown in Fig. 11.30C is used to collect an aliquot of air in its gaseous state for transport back to the analytic laboratory. Use of a preevacuated flask permits the collection of a gas sample in a specially polished stainless steel container. Pressure-volume relationships allow for the removal of a known volume from the tank for subsequent chemical analysis. Another means of collecting gaseous samples is the collapsible bag. Bags made of polymer films can be used for collection and transport of samples. The air may be pumped into the bag by an inert pump such as one using flexible metal bellows, or the air may be sucked into the bag by placing the bag in an airtight container that is then evacuated. This forces the bag to expand, drawing in the ambient air sample.

Most air pollutants are sampled by obtaining a known amount of air in a container. Canisters or bags are used for many gas-phase compounds. Depending on how long it takes to fill the container, this technique provides a prolonged snapshot of an air pollutant's concentration. The sample represents a specific time interval.

At 5:00 p.m., you begin collecting air samples in a 6-L stainless steel canister with a valve that allows air to enter on a 6-l canister at a rate of 0.5 L min^{-1}. When will the canister be full and what are your next steps if you are interested in 1,3-butadiene?

The canister will be full in 12 min at a flow rate. Thus, if the sample begins at 5:00 p.m. on Monday, it will be full at 5:12 p.m. that day. If only a few cans are available but a longer time is needed, for example, 1 h during the highest traffic, the valve may be adjusted to a lower flow rate, so that the canister fills five times more slowly, that is, 0.1 L min^{-1}, and the concentration is integrated over the entire hour.

TABLE 11.6 Collection of gases by absorption

Gas	Sampler	Sorption medium	Airflow ($l\ m^{-1}$)	Minimum sample (l)	Collection efficiency	Analysis	Interferences
Ammonia	Midget impinger	25 mL 0.1 N sulfuric acid	1–3	10		Nessler reagent	–
	Petri bubbler	10 mL of above	1–3	10	+95	Nessler reagent	–
Benzene	Glass bead column	5 mL nitrating acid	0.25	3–5	+95	Butanone method	Other aromatic hydrocarbons
Carbon dioxide	Fritted bubbler	10 mL 0.1 N barium hydroxide	1	10–15	60–80	Titration with 0.05 N oxalic acid	Other acids
Ethyl benzene	Fritted bubbler or midget impinger	15 mL spectrograde isooctane	1	20	+90	Alcohol extraction, ultraviolet analysis	Other aromatic hydrocarbons
Formaldehyde	Fritted bubbler	10 mL 1% sodium bisulfite	1–3	25	+95	Liberated sulfite titrated, 0.01 N iodine	Methyl ketones
Hydrochloric acid	Fritted bubbler	0.005 N sodium hydroxide	10	100	+95	Titration with 0.01 N silver nitrate	Other chlorides
Hydrogen sulfide	Midget impinger	15 mL 5% cadmium sulfate	1–2	20	195	Add 0.05 N iodine, 6 N sulfuric acid, back-titrate 0.01 N sodium thiosulfate	Mercaptans, carbon disulfide, and organic sulfur compounds
Lead, tetraethyl, and tetramethyl	Dreschel-type scrubber	100 mL 0.1 M iodine monochloride in 0.3 N	1.8–2.9	50–75	100	Dithizone	Bismuth, thallium, and stannous tin
Mercury, diethyl, and dimethyl	Midget impinger	15 mL of above	1.9	50–75	91–95	Same as above	Same as above
	Midget impinger	10 mL 0.1 M iodine monochloride in 0.3 N hydrochloric acid	1–1.5	100	91–100	Dithizone	Copper
Nickel carbonyl	Midget impinger	15 mL 3% hydrochloric acid	2.8	50–90	190	Complex with alpha-Furil dioxime	–
Nitrogen dioxide	Fritted bubbler (60–70 μm pore size)	20–30 mL Saltzman reagent[a]	0.4	Sample until color appears; probably 10 mL of air	94–99	Reacts with absorbing solution	Ozone in fivefold excess peroxyacyl nitrate
Ozone	Midget impinger	1% potassium iodide in 1 N potassium hydroxide	1	25	+95	Measures color of iodine liberated	Other oxidizing agents
Phosphine	Fritted bubbler	15 mL 0.5% silver diethyl dithiocarbamate in pyridine	0.5	5	86	Complexes with absorbing solution	Arsine, stibine, and hydrogen sulfide
Styrene	Fritted midget impinger	15 mL spectrograde isooctane	1	20	+90	Ultraviolet analysis	Other aromatic hydrocarbons
Sulfur dioxide	Midget impinger, fritted rubber	10 mL sodium tetrachloromercurate	2–3	2	99	Reaction of dichlorosulfito-mercurate and formaldehyde-depararosaniline	Nitrogen dioxide,[b] hydrogen sulfide[c]
Toluene diisocyanate	Midget impinger	15 mL Marcali solution	1	25	95	Diazotization and coupling reaction	Materials containing reactive hydrogen attached to oxygen (phenol); certain other diamines
Vinyl acetate	Fritted midget impinger and simple midget impinger in series	Toluene	1.5	15	+99 (84 with fritted bubbler only)	Gas chromatography	Other substances with same retention time on column

[a] *5 g sulfanilic, 140 mL glacial acetic acid, and 20 mL 0.1% aqueous* N-(1-naphthyl) ethylene diamine.
[b] *Add sulfamic acid after sampling.*
[c] *Filter or centrifuge any precipitate.*

Source: D.A. Vallero, Fundamentals of Air Pollution, 5th ed., Elsevier Academic Press, Waltham, MA, 2014, p. 999 pages cm.

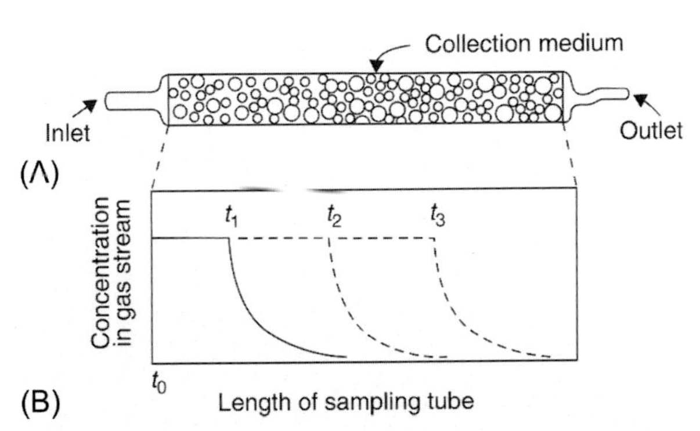

FIG. 11.30 Solid sorbent collection tube. (A) The tube is packed with a granular medium. (B) As the hydrocarbon-containing air is passed through the collection tube at t_1, t_2, and t_3, the collection medium becomes saturated at increasing lengths along the tube. *(Source: D.A. Vallero, Fundamentals of Air Pollution, 5th ed., Elsevier Academic Press, Waltham, MA, 2014, p. 999 pages cm.)*

After you complete the analysis of the contents of all of the canisters, you find that the concentrations in the 12 min canister are 10 mg m^{-3}, but the 1 h canister is found to be 25 mg m^{-3} for the same compound at the same location. Why might this have happened?

These results could indicate that traffic is higher after 5:12 p.m. or that a higher emission sources, for example, a series of poorly maintained trucks leaving a construction site after work hours. Then again, this could be an anomalous, episodic event. Perhaps, one large source happened to pass by the sampler. 1,3-Butadiene is commonly emitted in motor vehicle exhaust and is particularly high in exhaust from vehicles with malfunctioning pollution control equipment [77]. For example, if three other similar sites find little difference between the 12 min and 1 h integration times, the site with the difference would probably need a few more comparisons. Finding the optimal sampling interval and the appropriate sampling locations depends on comparisons such as these. This is why investigations of traffic and other factors need to be conducted before deciding when, where, and how long to sample air pollutants.

Canisters and bags can be filled two ways. In the examples above, the air in the canister had been removed and the empty canister's vacuum pulled in the air, since fluids flow from higher to lower pressure. The canister or bag can then be returned to the laboratory whereupon the analyst removes small amounts of the contents into detectors.

Air pollution measurements can be conducted using evacuated stainless steel canisters with electropolished inner surfaces, known as Summa canisters (see Fig. 11.31). The electropolishing and chemical deactivation yields an internal surface with very low chemical reactivity. These canisters are employed to sample for vapors, especially volatile organic compounds (VOCs). The VOCs sampled with Summa canisters consist of both aliphatic and aromatic hydrocarbons, including halogenated forms. Several of these are shown in Table 11.7. This technique has also been applied to a variety of practical applications, such as indoor air quality problems. Canisters are cleaned and evacuated, so that the lower pressure inside the canister allows air to enter without the need for a sampling pump [4].

Many air pollutants are chemically reactive and may combine with chemicals on the surfaces inside of collection systems. Thus, stainless steel canisters are now often coated with relatively inert substances, especially fused silica, which allows them to be stored longer prior to chemical analysis. The same considerations need to be made for other air pollutant container, for example, selecting a collection bag must consider the rate of degradation induced by the compounds that will be stored in the bag. When the canister is delivered to the laboratory, it is pressurized with nitrogen, and the contents are analyzed by gas chromatography/mass spectrometry (GC/MS).

Several canisters can be used to provide a map of air pollution. The number of canisters per unit area is known as the sampling density. Highly reactive contaminants with short half-lives may require greater sampling density than less reactive pollutants, since the former may breakdown in relatively short distances from their sources.

The amount of time that a sample can be stored in a container varies by the type of compounds and the container. For example, the US Environmental Protection Agency recommends that VOCs not be held in a Summa canister for $>$14 days from collection and 12 days from the receipt at the laboratory. In Tedlar bags, the holding times are much shorter, that is, 40 h from collection and 36 h from receipt by the laboratory. Also, the conditions of storage are usually specified. For VOCs, both bags and canisters can be preserved at ambient temperatures and near atmospheric pressure. However, for some highly reactive compounds, refrigeration and other preservation techniques will be required.

FIG. 11.31 Six-liter Summa canister. *(Source: A.C. Stern, Fundamentals of Air Pollution, Elsevier, 2014.)*

TABLE 11.7 Select organic compounds that can be collected using Summa canisters and Tedlar bags

Analyte	CRQL(ppbv)
Acetone	5
Acetonitrile	5
Acetonitrile	5
Acrolein	5
Acrylonitrile	5
Benzene	2
Benzyl chloride	5
Bromodichloromethane	2
Bromomethane	5
1,3-Butadiene	5
2-Butanone	5
Chlorobenzene	2
Chlorodifluoromethane	2
Chloroethane	2
Chloroform	2
1,2-Dichlorobenzene	2
trans-1,2-Dichloroethene	2
1,2-Dichloropropane	2

Continued

TABLE 11.7 Select organic compounds that can be collected using Summa canisters and Tedlar bags—cont'd

Analyte	CRQL(ppbv)
Dichlorofluoromethane	2
t-1,2-Dichloropropene	2
cis-1,2-Dichloropropene	5
1,2-Dichloro-1,1,2,2-tetra-fluoroethane	2
n-Pentane	2
Propylene	5
Styrene	5
1,2,4-Trichlorobenzene	2
1,1,1-Trichloroethane	5
1,1,2-Trichloroethane	5
1,1,2,2-Tetrachloroethane	2
Tetrachloroethene	5
Tetrachloromethane (carbon tetrachloride)	2
Toluene	5
Trichloroethene	2
Trichlorofluoromethane	2
1,1,2-Trichloro-1,2,2-trifluoroethane	2
Xylenes (m- and p-)	5
Xylene (o-)	5

CRQL, contract required quantitation limit; *ppbv*, parts per billion by volume. The CRQL is the lowest concentration that must be detected by a US EPA superfund contract laboratory.
Source: E. P. A. U.S., Managing the Quality of Environmental Data in EPA Region 9. (2013), 13 August 2013. Available: http://www.epa.gov/region9/qa/pdfs/aircrf.pdf.

11.7.2 Continuous emission sampling and analysis

A real-time analytic technique can be incorporated into a continuous monitoring instrument placed at the sampling location. Most often, the monitoring equipment is located inside a shelter such as a trailer or a small building, with the ambient air drawn to the monitor through a sampling manifold. The monitor then extracts a small fraction of air from the manifold for analysis by an automated technique, which may be continuous or discrete. Instrument manufacturers have developed automated in situ monitors for several air pollutants, including SO_2, NO, NO_2, O_3, and CO. This approach is also being increasingly applied for organic pollutants. For example, real-time gas chromatograph-mass spectrometers (GC-MS) are being used to monitoring a wide range of organic compounds, including air toxics [29].

Continuous emission monitoring (CEM) is an important type of combined real-time sampling. Indeed, regulatory agencies are increasingly requiring enhanced and periodic measurements to ensure compliance with emission standards for specific pollutants, which often makes CEM to method of choice. In the United States, these are known as the "compliance assurance monitoring (CAM) rule." The rule is designed to ensure proper operation of air pollution control equipment. The CAM rule has two general means of compliance: direct and indirect. Direct compliance assurance can be accomplished using CEM. Indirect compliance assurance is based on measurements of key parameters related to how well the equipment is operating, for example, temperature, flow, pressure drop, and voltage changes [78]. The Clean Air Act requires CEM for large sources and sources whose emissions must be continuously monitored, that is, under the provisions of new source review [79, 80].

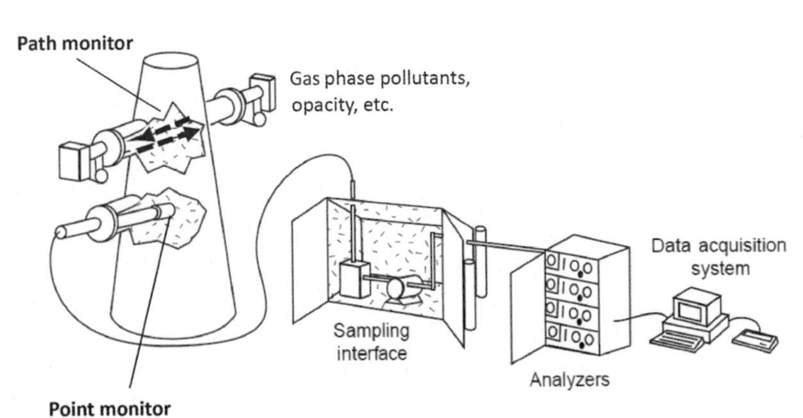

FIG. 11.32 Components and flow of a continuous emission monitoring (CEM) system for more than one gas-phase pollutant. *(Reproduced with permission from D.A. Vallero, Fundamentals of Air Pollution, 5th ed., Elsevier Academic Press, Waltham, MA, 2014, p. 999 pages cm; based on D.C. Mussatti, R. Srivastava, P. M. Hemmer, R. Strait, EPA air pollution control cost manual, Air Quality Strategies and Standards Division of the Office of Air Quality Planning and Standards, vol. 27711, US Environmental Protection Agency, Research Triangle Park, NC, 2002.)*

The CEM is an integrated system that collects samples directly from the stack or other air pollutant conveyance to the atmosphere. The system consists of the equipment needed to determine an air pollutant's concentration or emission rate. The system components shown in Fig. 11.32 are as follows:

1. The sampling and conditioning system.
2. Chemical analyzers.
3. The data acquisition system (DAS).
4. The controller system.

In addition to measuring various gas and particulate phase pollutant, CEM also is used for opacity and volumetric flow rates.

A CEM system may be either extractive or in situ. The monitor and analyzers of an in situ CEM are located within the stack. Extractive CEM systems capture a sample from within a stack, condition the sample (e.g., remove moisture and impurities), and move the sample to the analyzer. As shown in Fig. 11.42, point measurements measure pollutant concentrations at a single precise point where the sampling cell is located, whereas path measurements are taken across a given path in the emissions stream. Path readings are often from a signal across the stack and reflecting it back to a detector near the source of the signal. The measured concentrations crossing that path are averaged over a given time period [4].

Numerous gaseous compounds are being sampled with CEM, including NO_x, SO_2, CO, CO_2, O_2, total hydrocarbons, and hydrochloric acid (HCl).

11.7.3 Open-path technologies for gas-phase air pollutants [81]

The CEM concept also applies to measurements of ambient air, albeit by different names. For example, several open-path, continuous monitoring methods are being used successfully for a wide range of pollutants in the ambient air. Open-path technologies are being used to detect and quantify air pollution in the ambient air. These include tunable lasers, differential absorption techniques, and Fourier transform approaches.

The tunable diode laser (TDL) provides real-time concentration measurements for several air pollutants (see Table 11.8). The TDL requires only minimal maintenance, needs no consumables, and is user-friendly, for example, not sensitive to common vibration levels, minimal drift, frequent calibration unnecessary, immune to ambient relative humidity and temperature changes and laser intensity fluctuations, no sample preconditioning or treatment required before analysis, simple field deployment and installation, and able to use low-power optical sources [4]. The system can cover long open paths and provides high temporal resolution. It includes a stable multipass optical cell, which is highly sensitive. Temperature and pressure controls are internal. The drawbacks of TDL are that it can only detect a single compound per laser and it is limited in the number of detectable compounds. Indeed, sensitivity is limited to those compounds with overtone absorbencies in the near- to mid-infrared range. Particulate matter and airborne objects interfere with the path [82].

Differential absorption lidar (DIAL) is used to detect concentrations of gases within a transect in the atmosphere. Lidar is an acronym for light detection and ranging, and the technique is used for measuring physical characteristics of the atmosphere [83]. DIAL measures air pollutants by probing the atmosphere with pulsed laser radiation at two wavelengths. One

TABLE 11.8 Detection limits for airborne compounds quantified using tunable diode laser technologies

Gaseous compounds measured by OP-TDL systems	Approximate λ (nm)	Reported detection limit (ppm)
Ammonia	760, 1500	0.5–5.0
Carbon monoxide	1570	40–1000
Carbon dioxide	1570	40–1000
Hydrogen chloride	1790	0.15–1
Hydrogen cyanide	1540	1.0
Hydrogen fluoride	1310	0.1–0.2
Hydrogen sulfide	1570	20
Methane	1650	0.5–1
Nitric oxide	1800	30
Nitrogen dioxide	680	0.2
Oxygen	760	50
Water	970, 1200, 1450	0.2–1.0
Acetylene	1520	*
Ethylene	1693	*
Formaldehyde	1930	*
Hydrogen bromide	1960	*
Hydrogen iodide	1540	*
Nitrous oxide	2260	*
Phosphine	2150	*
Propane	1400, 1500, 1700	*

Notes: λ = tuned wavelength; * = no established detection limit since these are not commonly measured.
Source: C. Secrest, D. Hoyt, New technology applications, in: Presented at the National Multimedia Inspection/Enforcement Workshop, Dallas, Texas, May 15 and 16, 2012, 2012.

wavelength is efficiently absorbed by the trace gas, and the other wavelength is less efficiently absorbed. The radiation source projects packets of energy through the atmosphere, which interact with the trace gas. The optical receiver collects radiation backscattered from the target. By controlling the timing of source pulses and processing of the optical receiver signal, one can determine the concentration of the trace gas over various distances from the analyzer. This capability permits three-dimensional mapping of pollutant concentrations. Applications are plume dispersion patterns and three-dimensional gaseous pollutant profiles in urban areas.

The DIAL system provides spatially resolved pollutant concentration in two dimensions, in a relatively short period of time. They are deployable in many different applications and configurations and are quite portable. The path lengths are generally between 1 and 3 km but can be of varying lengths. For example, near a facility's fence line, the paths may be shorter to determine differences in concentrations at various directions from the source and under different wind directions.

Numerous air pollutants are detected with DIAL. SO_2 and O_3 are detected by an ultraviolet DIAL system operating at wavelengths near 300 nm.[3] Tunable infrared CO_2 lasers are used in applications of IR-DIAL systems that can measure several gases [84], including SO_2, CO, hydrochloric acid (HCl), methane (CH_4), CO_2, H_2O, nitrous oxide (N_2O), ammonia (NH_3), and hydrogen sulfide (H_2S). The number of air toxics measured by DIAL is increasing [82] (see Table 11.9). The laser source is switched between the low-absorption and high-absorption frequencies for the trace gas to be detected [85]. The system is pointed toward a target, and focusing lenses are used to collect the returning signal. The beam splitter diverts a

3. Code of Federal Regulations, Title 40, Part 58, *Ambient Air Quality Surveillance, Appendix D—Network Design for State and Local Air Monitoring Stations (SLAMS)*, pp. 158–172. US Government Printing Office, Washington, DC, July 1992.

TABLE 11.9 Air toxics detected using differential absorption lidar (DIAL)

Species measured by DIAL	Concentration sensitivity	Maximum range
Benzene	10 ppb	800 m
Sulfur dioxide	10 ppb	3 km
Toluene	10 ppb	800 m
Ethane	20 ppb	800 m
Ethylene	10 ppb	800 m
Methane	50 ppb	1 km
General hydrocarbons	40 ppb	800 m
Hydrogen chloride	20 ppb	1 km
Methanol	200 ppb	500 m

Notes: Concentration sensitivities for measurements of a 50 m-wide plume at a range of 200 m, under typical meteorologic conditions (from National Physical Laboratory (NPL), Middlesex, the United Kingdom). The range value represents the typical working maximum range for the NPL DIAL system.
Source: C. Secrest, D. Hoyt, New technology applications, in: Presented at the National Multimedia Inspection/Enforcement Workshop, Dallas, Texas, May 15 and 16, 2012, 2012.

portion of the transmitted beam to a detector. The backscattered and transmitted pulses are integrated to yield direct current electric signals. Examples of lidar readings for toluene and SO_2 are shown in Figs. 11.33 and 11.34, respectively.

The DIAL system is limited to those chemical species with the unique chemical properties required to be detected. For example, only a few wavelengths are measured. Spectral artifacts often appear at the same energy spectrum as the pollutant of concern, but the specific chemical causing the interference cannot be identified [82]. This could lead to positive bias, that is, the peak will be higher than the peak that would have been present if only the pollutant of concern were present. The reported concentration is the maximum concentration detected in a cell in the measurement plane. The path is one-dimensional, but the measurement plane is two-dimensional. Thus, the resolution of the planes can be calculated from the DIAL system resolution. For example, if the path is 3.75 m, each cell is 3.75 m square. Fig. 11.35 indicates the method for calculating an emission rate from DIAL readings. The concentration assigned to each cell is multiplied by the perpendicular wind field determined for that cell. The individual emission rates are summed to give the total emission rate through the plane. This figure shows two hypothetical plume calculations, that is, the one to the left has a small plume cross section and therefore a small integrated emission rate, and the one to the right has a larger plume cross section and therefore represents a larger emissions rate. Note that the peak concentration in both is similar and indeed may even be higher in the small plume than the large plume but the angle of the path dilutes the concentration.

Other widely used open-path technologies are the open-path infrared (IR) and differential optical absorption spectroscopy (DOAS). Open-path technologies take advantage of energy-molecule relationships. For example, the vibrational frequencies of all the IR absorbing molecules in the IR beam path are captured in the IR spectrum. Likewise, vibrational frequencies of all the ultraviolet (UV)-absorbing molecules in the UV beam path are captured in the UV spectrum and so on. When a molecule absorbs light, the energy of the molecule is increased, and the molecule is promoted from its lowest energy state (i.e., ground state) to an excited state. Light energy in a specific wavelength region of the electromagnetic spectrum stimulates molecular vibrations. Molecular species display their own characteristic vibrational structure when stimulated by that specific wavelength region of radiation. Fig. 11.36 shows an example of the IR absorption spectra for nitrous oxide, CO2, CO, NO, NO_2, and NH_3. Vibrational frequency is represented by wave number, and the wave number and vibrational structure identify a molecule.

The open-path Fourier transform infrared (OP-FTIR) spectroscopy is quite versatile in that it can quantify many chemicals in air simultaneously, whereas most open-path systems must be tuned to a specific compound (see Fig. 11.37). OP-FTIR detects numerous compounds in the low ppb range. FTIR provides mean concentrations over designated path lengths. Tunable diode lasers have similar detection limits as FTIR but are an example of a compound-specific open-path approach or at least are restricted to detecting chemicals that respond to the small frequency range they operate in. Like FTIR, open-path Raman spectroscopy can quantify many chemicals and may achieve low detection limits; however, its weak signal diminishes resolution. Relative humidity is a major interference of FTIR but not for Raman, so Raman is likely preferred in high moisture conditions [86].

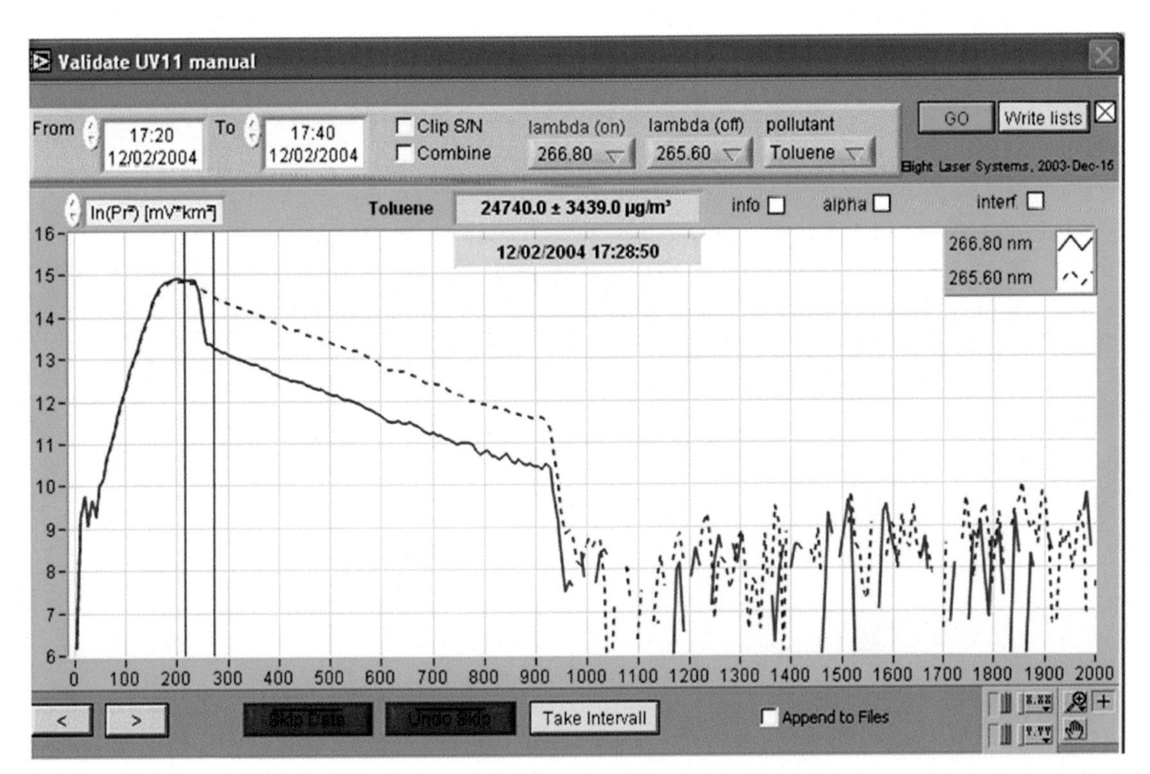

FIG. 11.33 Screenshot from lidar system, showing the backscatter signal of a pair of wavelengths used to measure toluene. A known amount of toluene was released in a chamber and measured 275 m from the lidar. The *blue line* represents the λ_{on}. The *red line* is λ_{off}, and it is unaffected by the presence of toluene. *(Source: J. Moreno, S. Moreno-Grau, A. Garcıa-Sánchez, Lidar probes air pollution (n.d.).)*

Absorption of energy is often calibrated with chemical concentration in spectrographic analysis, that is, the greater the amount of energy absorbed, the greater the concentration. Indeed, the Beer-Lambert law states that for a constant path length, the intensity of the incident, that is, direct, light energy traversing an absorbing medium diminishes exponentially with concentration, that is, energy absorption is proportional to chemical concentration:

$$A = eb[C] \tag{11.14}$$

where A is the absorbency of the molecule, e is the molar absorptivity (proportionality constant for the molecule), b is the light's path length, and $[C]$ is the chemical concentration of the molecule. Thus, the concentration of the chemical can be ascertained by measuring the light absorbed. The FTIR instrument measures intensity in the IR spectrum and applies the intensity signature of the returning incident spectrum, along with Beer's law, to identify and quantify compounds.

Like other open-path systems, FTIR measurements can be made actively or passively, respectively, that is, providing its own energy or receiving it externally. In the active mode, an instrument focuses a light beam before passing it through an interferometer, which converts the light beam into a modulated signal as a function of optical path difference. When the beams are recombined, certain wavelengths recombine constructively and some destructively, which creates an interference pattern, which is called an interferogram [73]. The recombined IR beam then passes from the beam splitter into the open path where a portion of the IR energy is absorbed by the gaseous compounds to be measured. The resulting IR beam reaches the IR detector where the interference pattern is detected, digitized, and transformed mathematically into a standard single beam infrared frequency spectrum using a mathematical algorithm known as a Fourier transform. The encoded light beam passes from the sending optics across the transect being measured for contaminants (i.e., the open path), then back to receiving optics and onto a detector. The detector then records the signal and mathematically derives the concentration using a Fourier transform technique. This produces a spectrum, such as the one depicted in Fig. 11.37, which can identify specific contaminants and their concentrations.

The passive mode works in the same manner as the active mode but uses no sending unit. Rather, the instrument uses an external energy source (e.g., the sun or combustion gases) to provide the infrared light. Receiving optics focus the light into the interferometer that encodes them into an interferogram format. This is then directed onto a detector for recording. As with the active approach, the detector then records the signal and mathematically derives the concentration using a Fourier transform technique to quantify the concentrations of the pollutants.

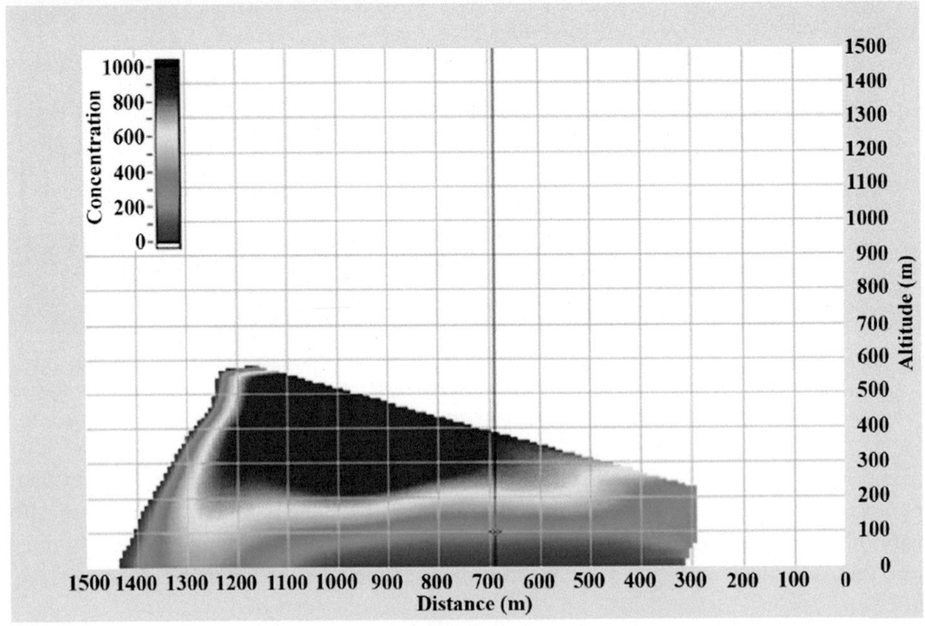

FIG. 11.34 Lidar measurement of concentrations of sulfur dioxide ($\mu g\,m^{-3}$) downwind of an electric power generating facility. The lidar was located beside the chimneys. Photo was taken during the lidar measurement. *(Source and photo credit: J. Moreno, S. Moreno-Grau, A. Garcıa-Sánchez, Lidar probes air pollution (n.d.).)*

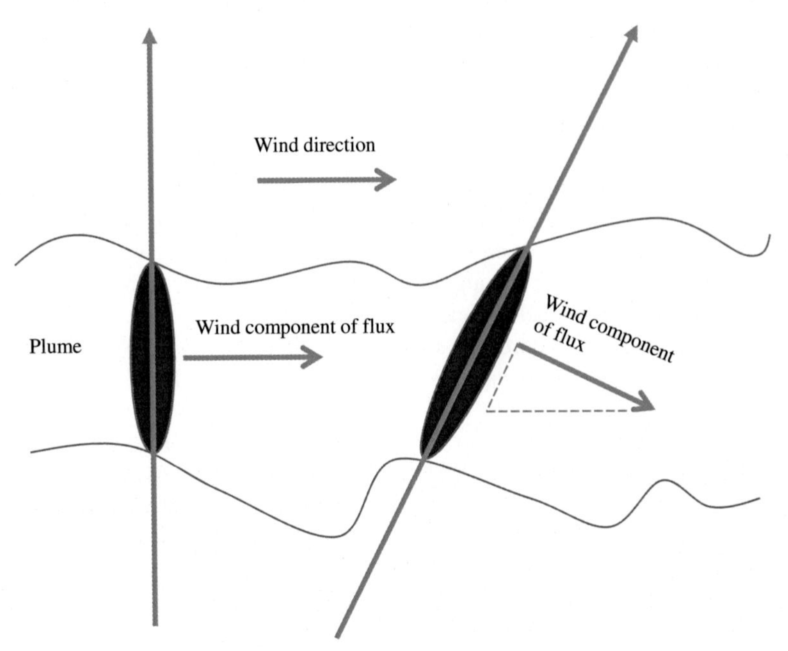

FIG. 11.35 Two hypothetical plume cross sections used for concentration calculations, that is, the one to the left has a small plume cross section and therefore a small integrated emission rate, and the one to the right has a larger plume cross section and therefore represents a larger emissions rate. *(Reproduced with permission from D.A. Vallero, Fundamentals of Air Pollution, 5th ed., Elsevier Academic Press, Waltham, MA, 2014, p. 999 pages cm; based on D.K. Mikel, R. Merrill, J. Colby, EPA Handbook: Optical Remote Sensing for Measurement and Monitoring of Emissions Flux, US Environmental Protection Agency, North Carolina, 2011.)*

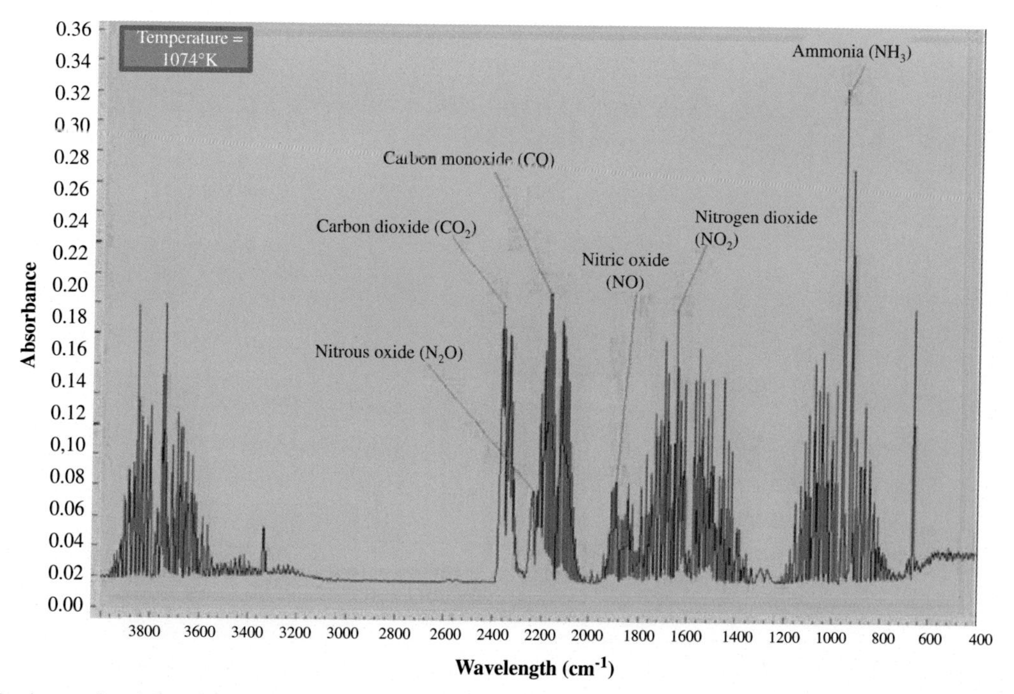

FIG. 11.36 Fourier transform infrared absorption spectrum recorded at 1075 K. *(Modified from D.K. Mikel, R. Merrill, J. Colby, EPA Handbook: Optical Remote Sensing for Measurement and Monitoring of Emissions Flux, US Environmental Protection Agency, North Carolina, 2011.)*

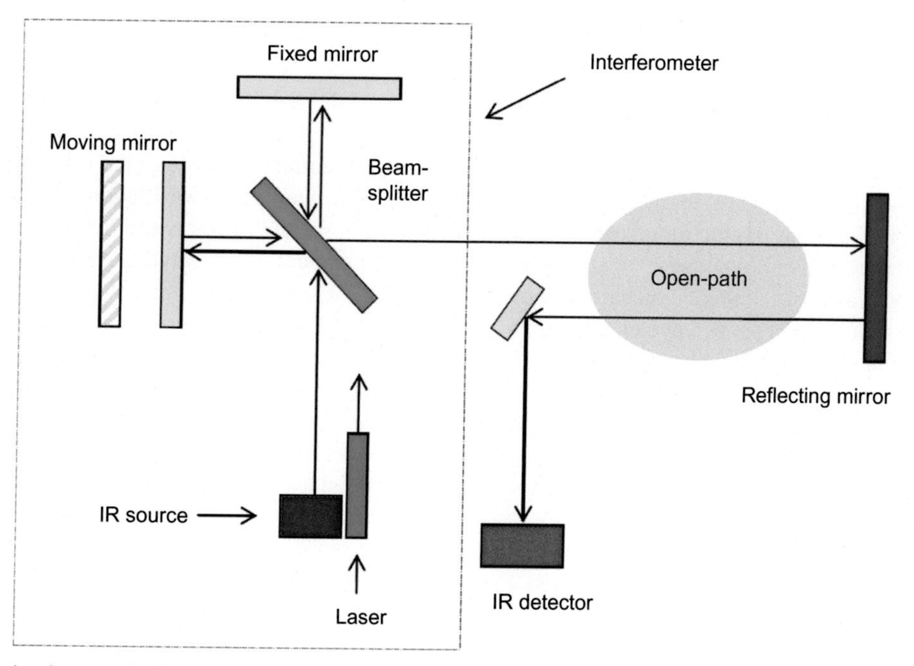

FIG. 11.37 Schematic of open-path Fourier transform infrared (OP-FTIR) spectroscopy system. *(Source: D. K. Mikel, R. Merrill, J. Colby, EPA Handbook: Optical Remote Sensing for Measurement and Monitoring of Emissions Flux, US Environmental Protection Agency, North Carolina, 2011.)*

The DOAS system (see Fig. 11.38) quantifies concentrations of gaseous compounds by measuring the absorption of light by chemical compounds in the air, again, by applying the Beer-Lambert law. However, the wavelength is in the higher frequency wavelengths, that is, the visible, and often in the ultraviolet (UV) range, that is, UV-DOAS.

A positive attribute of the UV-DOAS is its extremely long path-length capability—typically 500 m with some research applications up to 10 km [73]. The UV-DOAS is most frequently used to measure or monitor criteria and smog-related air

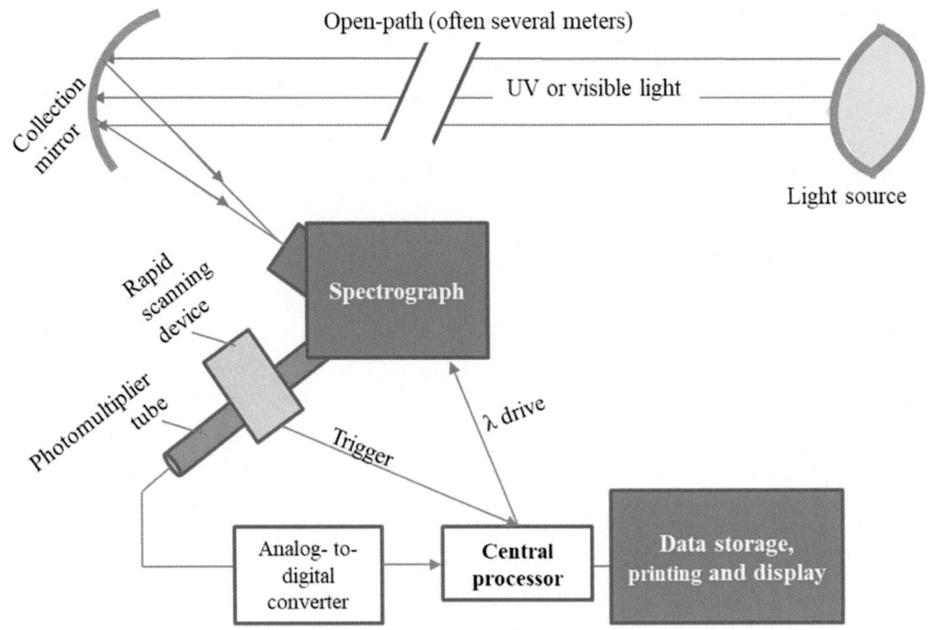

FIG. 11.38 Schematic of an open-path air measurement device: differential optical adsorption spectrometer (DOAS), which measures the absorption through an atmospheric path (typically 0.5–1.5 km) of two closely spaced wavelengths of light from an artificial source.

pollutants and for pollutants that do not produce ideal IR absorption bands. However, because the absorption bands for UV-DOAS are very wide, its specificity is limited, that is, many compounds cannot be accurately quantified by UV-DOAS. In addition, N_2 and O_2 molecules in the air cause broad spectral scattering and interfere with many of the compounds that can be measured. The UV-DOAS has detection limits in the low ppb range and has reached parts per trillion (ppt) in some research applications when used with optimum measurement path lengths [73].

In DOAS, one wavelength is chosen to match an absorption line of the specific compound of interest, and the other is near that line to account for atmospheric effects [87]. The term absorption line is used to mean a wavelength that a given atom or molecule absorbs more than it does other wavelengths. For example, if ozone is the compound of interest, the absorption line would be the wavelength at which ozone absorbs more than other wavelengths. As in other open-path systems (e.g., FTIR), a beam of electromagnetic radiation (in this case, visible or ultraviolet light) containing this wavelength is directed at a segment (open path) of the atmosphere, and the amount of the wavelength absorbed would be measured and calibrated against a known concentration of the compound of interest [87].

At the UV, visible, and near-IR wavelengths (approximately 180–780 nm), changes occur in the energy between the bonding electrons in molecules that absorb the light. The wavelength ranges produced by UV-DOAS include the rotational and vibrational transitions caused by near-IR light, so the typical application of UV-DOAS restricts the UV light to a wavelength range of 245–380 nm. Due to the range of excitations measured, molecular absorption bands tend to be far broader than that of the FTIR systems. Compounds that can be accurately detected and measured with the UV-DOAS must have specific chemical structure characteristics that allow for unique absorption bands. This limits the number of compounds that can be monitored.

Generally, the approaches discussed in this section have been and continue to be used effectively to measure numerous gas-phase pollutants. However, the state of the science in measurement continues to grow. Sensors, open-path detectors, and other real-time detection systems, for example, transform infrared, differential optical absorption spectroscopy, and tuned lasers, are being used with much success. Also, new materials, particularly nanomaterials, are being applied to environmental sampling and analysis.

11.7.4 Mobile measurement technologies

Remote sensing involves monitoring in which the analyzer is physically removed from the volume of air being analyzed. Satellites have been used to monitor light-scattering aerosol over large areas [73, 88, 89]. Emissions from large point sources such as volcanic activity and wildfires can be tracked by aircraft and satellite [89].

Air pollution instrumentation installed on mobile platforms allows for measurements that provide data to help understand the formation and transport of photochemical smog, acidic deposition, and the dispersion of air pollutants from sources. Mobile monitoring platforms may also be moved to "hot spots," that is, areas suspected of having high concentrations of specific air pollutants. These areas may be nearby locations downwind of a large source or a particular location that is an unfavorable receptor due to meteorologic conditions. Vehicular and aircraft monitoring systems can also be moved to locations where hazardous chemical spills, nuclear and chemical plant accidents, or volcanoes or earthquakes have occurred.

The major advantage of a mobile monitoring system is its ability to obtain air quality information in the intermediate region between source monitors and stationary fixed monitors. The major disadvantage is the paucity of suitable instrumentation that operates properly in the mobile platform environment. Limitations of existing instrumentation for the use on movable platforms are inadequate temperature and pressure compensation; incompatible power, size, and weight requirements; and excessive response time. Most movable platforms are helicopters, airplanes, trucks, or vans. These platforms do not provide the relatively constant-temperature environment required by most air quality instrumentation. Equipment mounted in aircraft is subject to large pressure variations with changing altitude. Most instrumentation is designed to operate with alternating current electric power, whereas relatively low amounts of direct current power are available in aircraft or vans. Space is at a premium, and response times are often too slow to permit observation of rapid changes in concentration as aircraft move in and out of a plume.

Despite these limitations, mobile monitoring systems have been used for decades to obtain useful information, such as the verification and tracking of a well-defined urban plume spreading downwind. These samples have been collected by a combination of instrumented aircraft and mobile vans. Cross-sectional paths were flown by the aircraft at increasing distances downwind. Meteorologic conditions of low wind speed in the same direction helped to maintain this urban plume in a well-defined condition for several hundred kilometers downwind. These systems can also link precursors to air pollutants; for example, the presence of large point sources of NO and SO_2 was observed by the changes in the O_3 and b_{scat} profiles. Sharp decreases in ozone concentration occurred when large amounts of NO were present, and rapid increases in b_{scat} were caused by the primary and secondary particulate matter from power plant plumes embedded in the larger urban plume. The overall increasing level of ozone and b_{scat} at greater downwind distances was caused by the photochemical reactions as the urban plume was transported farther away from St. Louis. This type of plume mapping can be accomplished only by mobile monitoring systems.

Recently, the mobile technologies have improved. For example, the trace ambient gas analyzer (TAGA) includes a portable gas chromatography system. The system has been used to measure various organic compounds in the air, for example, chlorine dioxide (ClO_2) in the ambient air near residences adjacent to a congressional building in Washington, DC [90]. The ClO_2 detection limits for this system were 900 ppqv (parts per quadrillion by volume) in real time. Incidentally, such low detection limits are needed since the action levels for ClO_2 are also quite low, that is, 25 ppbv for 15 min at a location. Obviously, action limits must be higher than detection limits to ensure safety, for example, whether to evacuate a building.

11.8 METHODS FOR SPECIFIC POLLUTANTS

The methods for measuring air pollutants are many and varied. Regulatory agencies require specific methods to ensure that ambient air quality standards are being met. At national levels, rules and regulations often require a specific method. In the United States, this is known as the federal reference method (FRM), but other methods may also be allowed under specified conditions, that is, a federal equivalent method (FEM). Scientific evidence must underpin an FRM or FEM. This section introduces a few of these methods, but this is merely a snapshot of an ever-changing list. Before deciding which method to use, contact the appropriate jurisdiction, given the ongoing advances in technologies and higher standards. For example, the 1 h SO_2 NAAQS was changed in 2010, with the intention of increased protection of public health by reducing exposure to high, short-term SO_2 concentrations. This will increase the need for exploratory monitoring conducted with highly time-resolved data, that is, frequency in minutes to hours. A good way to stay abreast of such changes is to read the US EPA's list of active FRMs and FEMs for National Ambient Air Quality Standard (NAAQS) pollutants [91], which is updated twice a year: https://cfpub.epa.gov/si/si_public_record_report.cfm?dirEntryId=321491 (accessed 21 September 2018).

The discussions here describe how the methods work. It is also important to note that FRMs and FEMs may apply to other compounds besides those discussed, for example, open-path and other continuous systems are also useful for air pollutant detection for criteria pollutants and air toxics [25]. As mentioned, continuous methods can show short-term peak SO_2 concentrations because of the high time resolution. These peaks would be missed by longer time averaging. Unfortunately, continuous methods may not be economically or logistically feasible for many situations. However, like many other aspects of society, for example, smart phones, recent technological advances in sensor, and sampling technology are providing more compact and less costly continuous methods. Although most sensors are not yet reference or equivalent methods, many will likely be useful in other ways, for example, screening and increasing sampling density in time and space, including exploratory monitoring. Most new sensors use proprietary methods, based on electrochemical principles. The accuracy and precision of these sensors must be evaluated, likely by colocation and comparison with reference and equivalent methods in the field [92]. In addition, passive methods commonly used in the past for saturation studies are commercially available, easy to use, and are relatively inexpensive. However, passive methods do not resolve the temporal need to see the peaks, since they are integrated samples collected over days to weeks. They can certainly improve the sampling spatial density, for example, to determine long-term concentration trends, but they do not offer the same insight into where the short-term peak concentrations may be occurring in an area around or otherwise impacted by an air pollutant source. Thus, passive methods are not the recommended method of choice for conducting exploratory monitoring to aid in determining where to place source-oriented monitors and should be considered as only a complementary or backup strategy to other methods and approaches to identify criteria and air toxic monitoring sites. The length of any exploratory monitoring is determined by the source (e.g., does the facility have peak operating periods, or does it change operations seasonally?), meteorology, and other intraannual variability.

Hundreds of chemical species can be found in emissions and polluted air. The gas-phase air pollutants most commonly monitored are CO, O_3, NO_2, SO_2, and nonmethane volatile organic compounds (NMVOCs). All of these are NAAQS criteria pollutants, although NMVOCs are not directly regulated and are measured as an indication of the hydrocarbons as a component of ground-level O_3. Indeed, the measurement of specific hydrocarbon compounds has become routine in the United States for two reasons: (1) their potential role as air toxics and (2) the need for detailed hydrocarbon data for control of urban ozone concentrations. Hydrochloric acid (HCl), ammonia (NH_3), and hydrogen fluoride (HF) are measured less often, depending on the sources in an airshed. Calibration standards and procedures are available for all of these analytic techniques, ensuring the quality of the analytic results [4]. See Table 11.10 for a summary of emission limits for one type of hazardous air pollutant source, incinerators.

11.8.1 Carbon monoxide sampling

Like all other criteria pollutants, measurement apparatus to detect and quantify carbon monoxide (CO) concentrations consists of three basic systems: pneumatic, analytic, and electronic. The pneumatic system consists of pumps, valves, lines, conduits, filters, flow meter, and other components to bring the air to the analytic system, which detects physical response (e.g., IR absorbance). The electronic system converts this basic physical and chemical information, for example, absorbance of light energy, to concentrations for a particular pollutant.

The primary reference method used for measuring carbon monoxide in the United States is based on nondispersive infrared (NDIR) photometry [93]. The principle involved is the preferential absorption of infrared radiation by carbon monoxide. Fig. 11.39 is a schematic representation of an NDIR analyzer. Like the FTIR methods addressed previously, NDIR is based on IR absorption and concentration relationships according to the Beer-Lambert law. The analyzer has a hot filament source of infrared radiation, a chopper, a sample cell, reference cell, and a detector. The reference cell is filled with a non-infrared-absorbing gas, and the sample cell is continuously flushed with ambient air containing an unknown amount of CO. The detector cell is divided into two compartments by a flexible membrane, with each compartment filled with CO. Movement of the membrane causes a change in electric capacitance in a control circuit whose signal is processed and fed to a recorder.

The chopper intermittently exposes the two cells to infrared radiation. The reference cell is exposed to a constant amount of infrared energy that is transmitted to one compartment of the detector cell. The sample cell, which contains varying amounts of infrared-absorbing CO, transmits to the detector cell a reduced amount of infrared energy that is inversely proportional to the CO concentration in the air sample. The unequal amounts of energy received by the two compartments in the detector cell cause the membrane to move, producing an alternating current (AC) electric signal whose frequency is established by the chopper spacing and the speed of chopper rotation. The signal is sent to a data acquisition system (DAS). The CO instrument needs to be calibrated often, that is, injecting air with known concentrations of CO into the system and comparing readings.

Water vapor absorbs IR, so it is a serious interfering substance in this technique. A moisture trap such as a drying agent or a water vapor condenser is required to remove water vapor from the air to be analyzed. Instruments based on other techniques are available that meet the performance specifications, such as those listed in Table 11.10. Equivalent methods for CO and the other criteria pollutants are published online and updated as new methods are approved [91].

TABLE 11.10 Sampling required to demonstrate compliance with emission limits for the new source performance standards (NSPS) for hospital/medical/infectious waste incinerators (HMIWI), pursuant to the US Court of Appeals for the District of Columbia Circuit Ruling of 2 March 1999, remanding the rule to the US Environmental Protection Agency for further explanation of the agency's reasoning in determining the minimum regulatory "floors" for new and existing HMIWI

Pollutant (units)	Unit size*	Proposed remand limit for existing HMIWI[§]	Proposed remand limit for new HMIWI[§]
HCl (ppmV)	L, M, S	78 or 93% reduction[¶]	15[¶] or 99% reduction[¶]
	SR	3100[¶]	N/A[‖]
CO (ppmV)	L, M, S	40[¶]	32
	SR	40[¶]	N/A[‖]
Pb (mg/dscm)	L, M	0.78 or 71% reduction	0.060 or 98% reduction[¶]
	S	0.78 or 71% reduction	0.78 or 71% reduction
	SR	8.9	N/A[‖]
Cd (mg/dscm)	L, M	0.11 or 66% reduction[¶]	0.030 or 93% reduction
	S	0.11 or 66% reduction[¶]	0.11 or 66% reduction[¶]
	SR	4[¶]	N/A[‖]
Hg (mg/dscm)	L, M	0.55[¶] or 87% reduction	0.45 or 87% reduction
	S	0.55[¶] or 87% reduction	0.47 or 87% reduction
	SR	6.6	N/A[‖]
PM (gr/dscf)	L	0.015[¶]	0.009
	M	0.030[¶]	0.009
	S	0.050[¶]	0.018
	SR	0.086[¶]	N/A[‖]
CDD/CDF, total (ng $dscm^{-1}$)	L, M	115	20
	S	115	111
	SR	800[¶]	N/A[‖]
CDD/CDF, TEQ (ng $dscm^{-1}$)	L, M	2.2	0.53
	S	2.2	2.1
	SR	15[¶]	N/A[‖]
NO_x (ppmV)	L, M, S	250[¶]	225
	SR	250[¶]	N/A[‖]
SO_2 (ppmV)	L, M, S	55[¶]	46
	SR	55[¶]	N/A[‖]

**L: large; M: medium; S: small; SR: small rural.*
[§]All emission limits are measured at 7% oxygen.
[¶]No change proposed.
[‖]Not applicable.

Source: D.A. Vallero, Fundamentals of Air Pollution, 5th ed., Elsevier Academic Press, Waltham, MA, 2014, p. 999 pages cm.

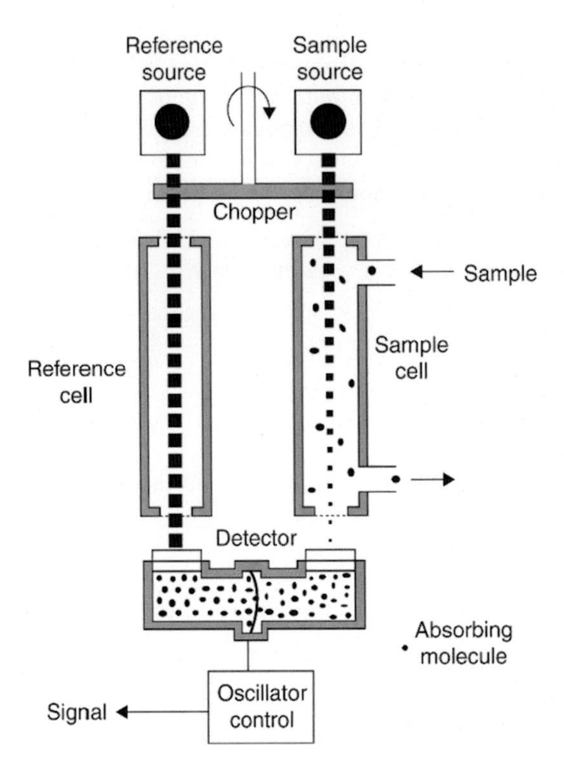

FIG. 11.39 Nondispersive infrared (NDIR) analyzer. *(Source: R.J. Bryan, Ambient air quality surveillance, Air Pollution, 3 (1976) 343–392.)*

11.8.2 Ozone sampling

The principal method used for measuring ozone is based on chemiluminescence [94]. When ozone and ethylene react chemically, products are formed that are in an excited electronic state. These products fluoresce, releasing light. The principal components are a constant source of ethylene, an inlet sample line for ambient air, a reaction chamber, a photomultiplier tube, and signal-processing circuitry. The rate at which light is received by the photomultiplier tube is dependent on the concentrations of O_3 and ethylene. If the concentration of ethylene is made much higher than the ozone concentration to be measured, the light emitted is proportional only to the ozone concentration.

Instruments based on this principle may be calibrated by a two-step process [95] shown in Fig. 11.40. A test atmosphere with a known source of ozone is produced by an ozone generator, a device capable of generating stable levels of O_3. Step 1 involves establishing the concentration of ozone in the test atmosphere by ultraviolet (UV) photometry. This is followed by step 2, calibration of the instrument's response to the known concentration of ozone in the test atmosphere.

The O_3 method is presently being investigated for possible improved detection, that is, the nitric oxide (NO)-chemiluminescence ozone analyzer. This method is based on the dry, gas-phase reaction between NO and O_3 that generates nitrogen dioxide in an activated state (NO_2^*) and molecular oxygen (O_2). As the unstable, activated nitrogen dioxide (NO_2^*) returns to a lower energy state (NO_2), it emits a photon ($h\nu$), while it luminesces in a broadband spectrum from visible light to infrared light (approximately 590 nm–2800 nm). The two-step gas-phase reaction proceeds as follows:

$$NO + O_3 \rightarrow NO_2{}^* + O_2 \tag{11.15}$$

Then, the activated nitrogen dioxide breaks down to nitrogen dioxide and a photon:

$$NO_2{}^* \rightarrow NO_2 + h\nu \tag{11.16}$$

The number of photons emitted during the reaction is directly proportional to the O concentration in the sample. As the gas-phase reaction proceeds in the analyzer mixing chamber/reaction cell, the emitted photons are counted by a photomultiplier tube (PMT). The electronics of the analyzer converts the $h\nu$ count to O concentration, using a mathematical algorithm.

In the United States, the NO-chemiluminescence method is designated as an FEM for ozone, that is, it has met all of the same performance specifications as the existing FRM and is comparable with the existing FRM from a measurement perspective. As of 2013, the NO-chemiluminescence devices are becoming increasingly available from vendors.

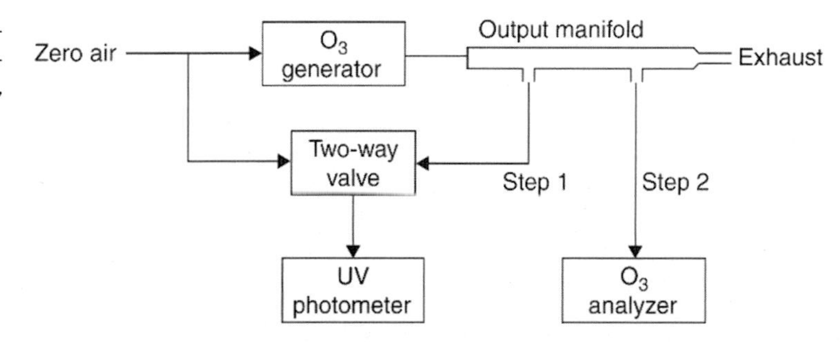

FIG. 11.40 Two-step calibration apparatus for an ultraviolet (UV) ozone analyzer. *(Source: D.A. Vallero, Fundamentals of Air Pollution, 5th ed., Elsevier Academic Press, Waltham, MA, 2014, p. 999 pages cm.)*

The NO-chemiluminescence method behaves in the same manner as the ethylene (C_2H_4) chemiluminescence FRM when exposed to the same interferents[4] found in ambient air that may have an effect on the response of the widely used ultraviolet-spectrophotometric ozone methods, Care must be taken with this method in that NO is toxic in high concentrations, including the concentrations of NO found in an NO cylinder source (asphyxiant).

This method is very similar to the NO_2 method discussed below.

11.8.3 Nitrogen dioxide sampling

The principal method used for measuring NO_2 is also based on chemiluminescence (Fig. 11.41). NO_2 concentrations are determined indirectly from the difference between the NO and NO_x (NO + NO_2) concentrations in the atmosphere. These concentrations are determined by measuring the light emitted from the chemiluminescent reaction of NO with O_3 (similar to the reaction of O_3 with ethylene noted for the measurement of O_3), except that O_3 is supplied at a high constant concentration, and the light output is proportional to the concentration of NO present in the ambient air stream.

Fig. 11.42 shows the monitoring system based on these principles. To determine the NO_2 concentration, the NO and NO_x (NO + NO_2) concentrations are measured. The block diagram shows a dual pathway through the instrument, one to measure NO and the other to measure NO_x. The NO pathway has an ambient airstream containing NO (as well as NO_2), an ozone stream from the ozone generator, a reaction chamber, a photomultiplier tube, and signal-processing circuitry. The NO_x pathway has the same components, plus a converter for quantitatively reducing NO_2 to NO. The instrument can also electronically subtract the NO from NO_x and yield as output the resultant NO_2.

Air passing through the NO pathway enters the reaction chamber, where the NO present reacts with the O_3. The photons produced are measured by the photomultiplier tube and converted to an NO concentration. The NO_2 in the airstream in this pathway is unchanged. In the NO_x pathway, the NO- and NO_2-laden air enters the converter, where the NO_2 is reduced to form NO; all of the NO_x exits the converter as NO and enters the reaction chamber. The NO reacts with O_3, and the output signal is the total NO_x concentration. The NO_2 concentration in the original airstream is the difference between NO_x and NO. Calibration techniques use gas-phase titration of an NO standard with O_3 or an NO_2 permeation device.

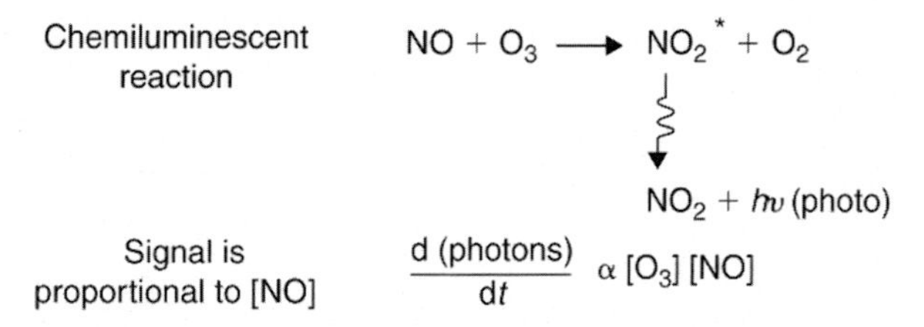

FIG. 11.41 NO_2 chemiluminescent detection chemistry and physics based on the reaction of NO and O_2. *(Source: D.A. Vallero, Fundamentals of Air Pollution, 5th ed., Elsevier Academic Press, Waltham, MA, 2014, p. 999 pages cm.)*

4. These interferents are listed, in 40 CFR Part 53, Table B-3 (e.g., H_2S, CO_2, and H_2O).

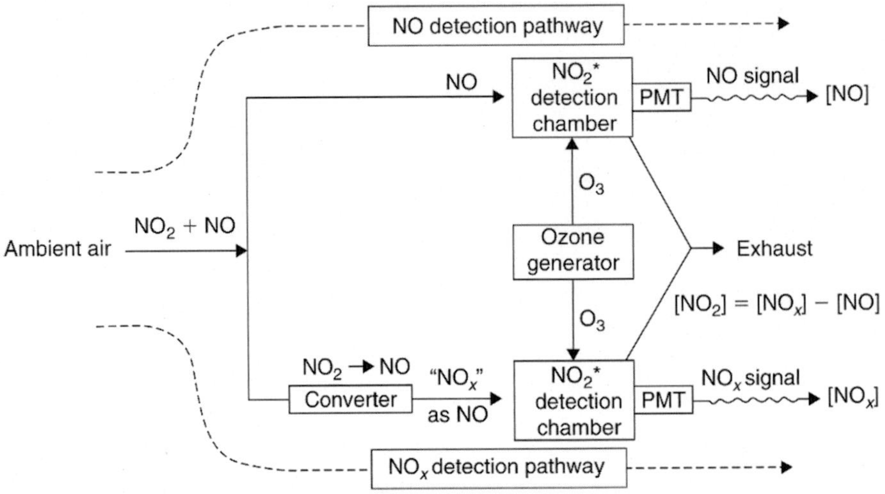

FIG. 11.42 Schematic diagram of chemiluminescent detector for NO_2 and NO. PMT: photomultiplier tube. *(Source: D.A. Vallero, Fundamentals of Air Pollution, 5th ed., Elsevier Academic Press, Waltham, MA, 2014, p. 999 pages cm.)*

11.8.4 Sulfur dioxide sampling

Several manual and continuous analytic techniques are used to measure SO_2 in the atmosphere. Methods for determining levels in the air include ion chromatography, titration, calorimetry, mass spectrometry, conductometry, amperometric detection, flame photometric detection, and turbidimetry. Of these methods, ion chromatography generally has the lowest detection limit for SO_2 [96, 97].

The manual techniques involve two-stage sample collection and measurement. Samples are collected by bubbling a known volume of gas through a liquid collection medium. Collection efficiency is dependent on the gas-liquid contact time, bubble size, SO_2 concentration, and SO_2 solubility in the collection medium. The liquid medium contains chemicals that stabilize SO_2 in solution by either complexation or oxidation to a more stable form. Field samples must be handled carefully to prevent losses from exposure to high temperatures. Samples are analyzed at a central laboratory by an appropriate method.

The West and Gaeke manual method, which has been modified since its inception, had been the basis for the US EPA reference method for measurement of SO_2 [98, 99]. The method uses the colorimetric principle; that is, the amount of SO_2 collected is proportional to the amount of light absorbed by a solution. The collection medium is an aqueous solution of sodium or potassium tetrachloromercurate (TCM). Absorbed SO_2 forms a stable complex with TCM. This enhanced stability permits the collection, transport, and short-term storage of samples at a central laboratory. The analysis proceeds by adding bleached pararosaniline dye and formaldehyde to form red-purple pararosaniline methylsulfonic acid. Optical absorption at 548 nm is linearly proportional to the SO_2 concentration. Procedures are followed to minimize interference by O_3, oxides of nitrogen, and heavy metals.

In the United States, the current SO_2 reference method employs ultraviolet fluorescence (UVF). The UVF system depicted in Fig. 11.43 includes a measurement cell, a UV light source of appropriate wavelength, a UV detector system with appropriate wave length sensitivity, a pump and flow control system for sampling the ambient air and moving it into the measurement cell, filters, and other conditioning components needed to minimize measurement interferences, suitable control, and measurement processing capability. The SO_2 concentrations are derived from an automated measurement of the intensity of fluorescence released by SO_2 in an ambient air sample contained in a measurement cell of an analyzer when the air sample is irradiated by UV light that passed through the cell. The fluorescent light released by the SO_2 is also in the ultraviolet region but at longer wavelengths than the excitation light. Typically, optimum instrumental measurement of SO_2 concentrations is obtained with an excitation wavelength in a band between approximately 190–230 nm and measurement of the SO_2 fluorescence in a broad band around 320 nm. Interferents include aromatic hydrocarbon species and possibly other compounds [100].

The continuous methods combine sample collection and the measurement technique in one automated process. The measurement methods used for continuous analyzers include conductometric, colorimetric, coulometric, and amperometric techniques for the determination of SO_2 collected in a liquid medium [101]. Other continuous methods utilize physico-chemical techniques for detection of SO_2 in a gas stream. These include flame photometric detection (described earlier) and fluorescence spectroscopy [102]. Instruments based on these principles are available that meet standard performance specifications.

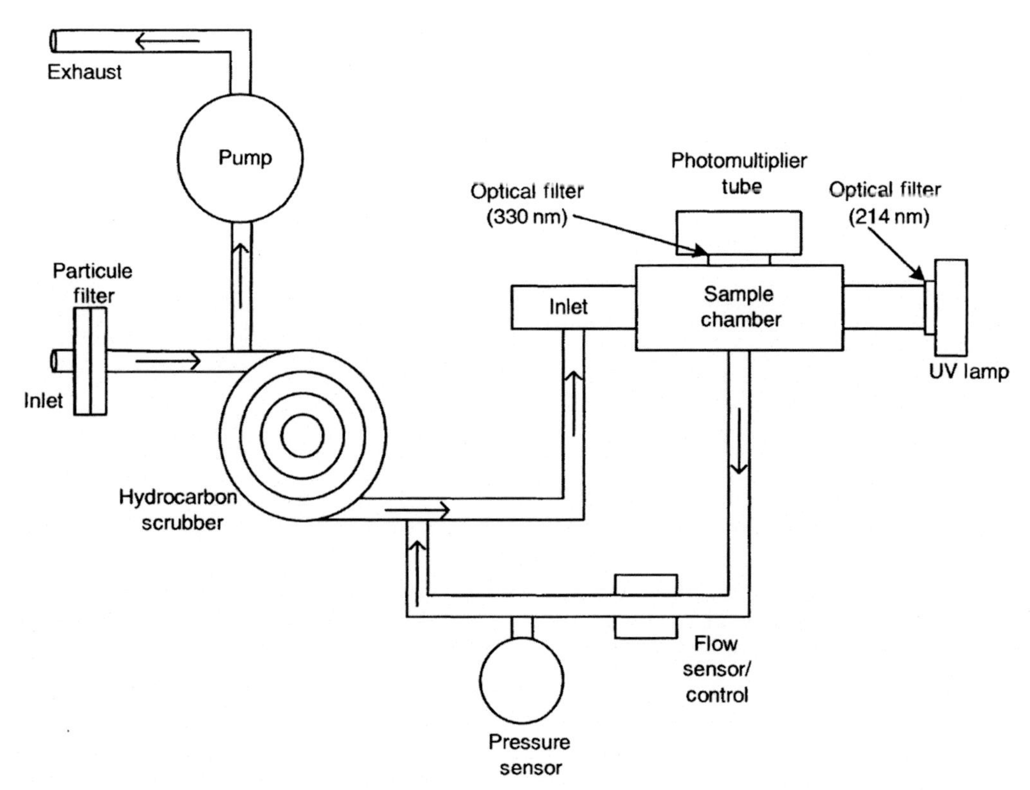

FIG. 11.43 Schematic of ultraviolet fluorescence (UVF) federal reference method for analyzing sulfur dioxide in ambient air. *(Source: Primary National Ambient Air Quality Standard for Sulfur Dioxide; Final Rule, 2010.)*

11.8.5 Nonmethane volatile organic compounds

The large number and species of hydrocarbons in the atmosphere render ambient air monitoring difficult. The ambient atmosphere contains a ubiquitous concentration of methane (CH_4) at approximately 1.6 ppm worldwide [6]. The concentration of all other hydrocarbons in ambient air can range from 100 times less to 10 times greater than the methane concentration for a rural versus an urban location. The terminology of the concentration of hydrocarbon compounds is potentially confusing. Hydrocarbon concentrations are referred to by two units—parts per million by volume (ppmV) and parts per million by carbon (ppmC). Thus, 1 μL of gas in 1 L of air is 1 ppmV, so the following is true [4]:

Mixing ratio	ppmv	ppmC
$\frac{1\ \mu L\ of\ O_3}{1\ L\ of\ air} =$	1 ppm ozone	–
$\frac{1\ \mu L\ of\ SO_2}{1\ L\ of\ air} =$	1 ppmv SO_2	–
$\frac{1\ \mu L\ of\ CH_4}{1\ L\ of\ air} =$	1 ppmv CH_4	1 ppmC CH_4
$\frac{1\ \mu L\ of\ C_2H_6}{1\ L\ of\ air} =$	1 ppmv C_2H_6	2 ppmC C_2H_6

The unit ppmC takes into account the number of carbon atoms contained in a specific hydrocarbon and is the generally accepted way to report ambient hydrocarbons. This unit is used for three reasons: (1) The number of carbon atoms is a very crude indicator of the total reactivity of a group of hydrocarbon compounds; (2) historically, analytic techniques have expressed results in this unit; and (3) considerable information has been developed on the role of hydrocarbons in the atmosphere in terms of concentrations determined as ppmC. With decreasing detection limits, the units are also expressed as parts per billion by volume (ppbV) and ppbC.

Measurement of hydrocarbons in the ambient air fall into two classes: methane (CH_4) and all other NMVOCs. Analyzing hydrocarbons in the atmosphere involves a three-step process: collection, separation, and quantification. Collection involves obtaining an aliquot of air, for example, with an evacuated canister. The principal separation process is gas

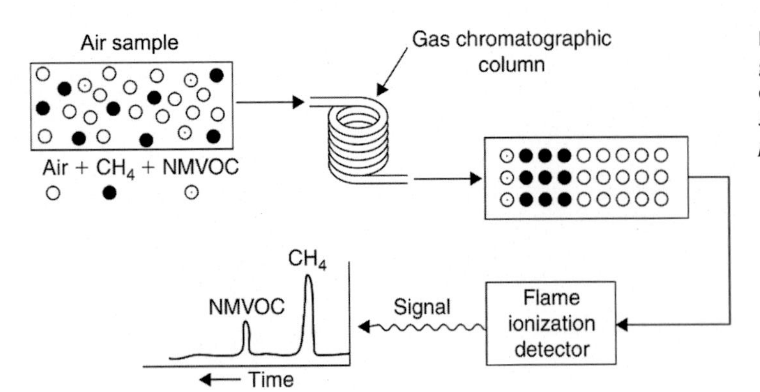

FIG. 11.44 Schematic diagram of hydrocarbon detection by gas chromatography. NMVOC = nonmethane volatile organic carbon. *(Source: D.A. Vallero, Fundamentals of Air Pollution, 5th ed., Elsevier Academic Press, Waltham, MA, 2014, p. 999 pages cm.)*

chromatography (GC), and the principal quantification technique is with a calibrated flame ionization detector (FID). Mass spectroscopy (MS) is used along with GC to identify individual hydrocarbon compounds. Since the NMVOCs are in the gas phase, they are evacuated from the container directly to the GC.

A simple schematic diagram of the GC/FID principle is shown in Fig. 11.44. Air containing CH_4 and other hydrocarbons classified as NMVOCs passes through a GC column, and the air, CH_4, and NMVOC molecules are clustered into groups because of different absorption/desorption rates. As CH_4 and NMVOC groups exit the column, they are "counted" by the FID. The signal output of the detector is proportional to the two groups and may be quantified when compared with standard concentrations of gases. This simplified procedure has been used extensively to collect hydrocarbon concentration data for the ambient atmosphere. A major disadvantage of this technique is the grouping of all hydrocarbons other than CH_4 into one class. Hydrocarbon compounds with similar structures are detected by an FID in a proportional manner, but for compounds with significantly different structures, the response may be different. This difference in sensitivity results in errors in measurements of NMVOC mixtures.

More sophisticated GC columns and techniques perform more detailed separations of mixtures of hydrocarbons into discrete groups.

Other types of detectors include the flame photometric detector (FPD) and the electron capture detector (ECD). The FID is composed of a H_2 flame through which the hydrocarbon gases are burned, forming charged carbon atoms, and an electrometer grid that generates a signal current proportional to the number of carbon atoms in the flame. The example of 1 ppmV methane (CH_4) and 1 ppmV (but 2 ppmC) ethane (C_2H_6) is related to this detection principle. One ppmV of CH_4 and 1 ppmV of C_2H_6 in air have the same number of molecules of hydrocarbon in a given volume of air, but if an aliquot of each mixture was run through an FID, the signal for ethane would be nearly twice the methane signal: 2 ppmC ethane compared with 1 ppmC methane.

The FPD is also used to measure sulfur-containing compounds and therefore is useful for measurement of sulfur-containing hydrocarbons such as dimethylsulfide or furan. The FPD has an H_2 flame in which sulfur-containing gases are burned. In the combustion process, electronically excited S_2^* is formed. A photomultiplier tube detects light emitted from the excited sulfur at ~395 nm. The ECD is preferred for measuring nitrogen-containing compounds such as peroxyacetyl nitrate (PAN) and other oxidized species of nitrogen. The ECD contains a radioactive source that establishes a stable ion field. Nitrogen-containing compounds capture electrons in passing through the field. Alterations in the electronic signal are related to the concentration of the nitrogen species.

11.9 PARTICULATE MATTER

Particulate matter (PM) includes any airborne substances that are not in the gas phase, either liquid or solid. They are generally classified according to the particle's diameter; for example, total suspended particulates (TSP) include all aerosols in the atmosphere, but smaller particles are of most concern because they remain suspended and they cause the most health effects. These are usually classified as the coarse fraction, that is, aerodynamic diameter, that is, 2.5–10 μm diameters (PM_{10}) and the fine fraction ≤2.5 μm diameters ($PM_{2.5}$). The three major characteristics of particulate pollutants in the ambient atmosphere are total mass concentration, size distribution, and chemical composition. Most nations' standards are now based on atmospheric concentrations of fine PM, that is, $PM_{2.5}$, to protect human health from effects caused by inhalation of suspended particulate matter. However, there remains a strong interest in the course fraction (PM_{10})

because it may be linked with certain diseases (e.g., asthma) and because it often has toxic components (e.g., sorbed metals and semivolatile organic compounds like dioxin). The size distribution of particulate pollutants is very important in understanding the transport and removal of particles in the atmosphere and their deposition behavior in the human respiratory system. Their chemical composition may determine the type of effects caused by particulate matter on humans, vegetation, and materials.

Recently, there has been a growing interest in ultrafines, that is, PM with diameters of about 0.1 μm, that is, 100 nm or less. Coincidentally, this is the same size as that of nanoparticles. However, a distinction between nanoparticles and ultrafines are that nanoparticles are "engineered," that is, made intentionally, whereas ultrafines are unintentional by-products from other activities (e.g., diesel combustion). Another distinction is that the 100 nm length is a single dimension, that is, a nanoparticle has at least one dimension that is $\leq$100 nm, whereas an ultrafine particle has an aerodynamic diameter of 100 nm. Thus, a long chain of carbon-60 spheres would still be considered a nanoparticle because its width is $<$100 nm, but an aggregate of ultrafine particle may be considered larger than an ultrafine particle because its Stokes diameter is $>$100 nm. Recall from Chapter 5 that the Stokes particle diameter (D_{ps}) is the diameter of a sphere with the same density and settling velocity as the particle. The diameter is derived from the aerodynamic drag force caused by the difference in velocity of the particle and the surrounding fluid. For spherical particles, D_{ps} is identical to the physical or actual diameter, but for aggregates and other nonspherical aerosols, they can differ. A hollow particle is obviously different from a solid particle, so density also comes into play. For example, for particles with dimensions $>$500 nm, the aerodynamic diameter is approximated[5] as the product of the D_{ps} and the square root of the particle density.

Particles can be directly emitted into the atmosphere (primary aerosols) or form in the atmosphere as a result of reactions and physical processes, that is, from gas-phase constituents become aerosols after emission (secondary aerosol). The secondary source of fine particles in the atmosphere is gas-to-particle conversion processes, considered to be the more important source of particles contributing to atmospheric haze (see Discussion Box "Using Light to Measure Air Pollution"). In gas-to-particle conversion, gaseous molecules become transformed to liquid or solid particles. This phase transformation can occur by three processes: absorption, nucleation, and condensation. Absorption is the process by which a gas goes into solution in a liquid phase. Absorption of a specific gas is dependent on the solubility of the gas in a particular liquid, for example, SO_2 in liquid H_2O droplets. Nucleation and condensation are terms associated with aerosol dynamics.

Nucleation is the growth of clusters of molecules that become a thermodynamically stable nucleus. This process is dependent on the vapor pressure of the condensable species. The molecular clusters undergo growth when the saturation ratio, S, is $>$1, where saturation ratio is defined as the actual pressure of the gas divided by its equilibrium vapor pressure. The condition when $S > 1$ is referred to as a supersaturated condition.

The size at which a cluster may be thermodynamically stable is influenced by the Kelvin effect. The equilibrium vapor pressure of a component increases as the droplet size decreases. Vapor pressure is determined by the energy necessary to separate a single molecule from the surrounding molecules in the liquid. As the curvature of the droplet's surface increases, fewer neighboring molecules will be able to bind a particular molecule to the liquid phase, thus increasing the probability of a molecule escaping the liquid's surface. Thus, smaller droplets will have a higher equilibrium vapor pressure. This would affect the minimum size necessary for a thermodynamically stable cluster, suggesting that components with lower equilibrium saturation vapor pressures will form stable clusters at smaller diameters.

Condensation is the result of collisions between a gaseous molecule and an existing aerosol droplet when supersaturation exists. Condensation occurs at much lower values of supersaturation than nucleation. Thus, when particles already exist in sufficient quantities, condensation will be the dominant process occurring to relieve the supersaturated condition of the vapor-phase material.

A simple model for the formation and growth of an aerosol at ambient conditions involves the formation of a gas product by the appropriate chemical oxidation reactions in the gas phase. This product must have a sufficiently low vapor pressure for the gas-phase concentration of the oxidized product to exceed its saturation vapor pressure. When this condition occurs, nucleation and condensation may proceed, relieving supersaturation. These processes result in the transfer of mass to the condensed phase. Aerosol growth in size occurs, while condensation is proceeding.

Coagulation is the process by which discrete particles come into contact with each other in the air and remain joined together by surface forces. Coagulation represents another way in which aerosol diameter will increase. However, it does not alter the mass of material in the coagulated particle.

The clearest example of this working model of homogeneous gas-to-particle conversion is sulfuric acid aerosol formation. Sulfuric acid (H_2SO_4) has an extremely low saturation vapor pressure. Oxidation of relatively small amounts of sulfur dioxide (SO_2) can result in a gas-phase concentration of H_2SO_4 that exceeds its equilibrium vapor pressure in the ambient atmosphere, with the subsequent formation of sulfuric acid aerosol. In contrast, nitric acid (HNO_3) has a much higher saturation vapor pressure. Therefore, the gas-phase concentration of HNO_3 is not high enough to permit nucleation of nitric acid aerosol in typical atmospheric systems.

11.9.1 Sampling particulates

The complexity and diversity of particulate composition, size, mass, and concentrations in the atmosphere dictate the need for accurate and representative PM monitoring. The major purpose of ambient particulate sampling is to obtain mass concentration and chemical composition data, preferably as a function of particle diameter. This information is valuable for a variety of problems: effects on human health, identification of particulate matter sources, understanding of atmospheric haze, and particle removal processes.

The primary approach is to separate the particles from a known volume of air and subject them to weight determination and chemical analysis. The principal methods for extracting particles from an airstream are filtration and impaction. All sampling techniques must be concerned with the behavior of particles in a moving airstream. The difference between sampling for gases and sampling for particles begins at the inlet of the sampling manifold and is due to the discrete mass associated with individual particles.

Particulate matter is collected using equipment that separates out the size fraction of concern. As discussed in Chapter 5, PM filtration consists of four mechanical processes: (1) diffusion, (2) interception, (3) inertial impaction, and (4) electrostatics [4, 103] (see Fig. 10.17 in Chapter 10).

Diffusion applies to only the extremely small particles ($\leq 0.1\,\mu m$ diameter), that is, those that experience Brownian motion and undergo a "random walk" away from the airstream. Interception captures PM with diameters between 0.1 and 1 μm. The particle remains in the airstream but contacts the filter medium. Inertial impaction collects PM with large enough diameters and mass to exit the airstream via inertia (diameters $\geq 1\,\mu m$). Electrostatics consist of electric interactions between the atoms in the filter and those in the particle at the point of contact (van der Waal's forces) and electrostatic attraction (charge differences between particle and filter medium). Other important factors include the thickness and pore diameter or the filter, the uniformity of particle diameters and pore sizes, the solid volume fraction, the rate of particle loading onto the filter (e.g., affecting particle "bounce"), the particle phase (liquid or solid), capillarity and surface tension (if either the particle or the filter media are coated with a liquid), and atmospheric and stack conditions, for example, velocity, temperature, pressure, and viscosity.

Mass concentration units for ambient measurements are mass (μg) per unit volume (m^3). Size classification involves the use of specially designed inlet configurations, for example, $PM_{2.5}$ sampling. To determine mass concentration, all the particles are removed from a known volume of air, and their total mass is measured. This removal is accomplished by two techniques, filtration and impaction. Mass measurements are made by pre- and postweighing of filters or impaction surfaces. To account for the absorption of water vapor, the filters are generally equilibrated at standard or other conditions specified in the PM method (e.g., $T = 20°C$ and 50% relative humidity).

Filters vary in pore size, strand diameter, chemical composition, and quality. Vendors document the important physical properties of the commonly used filter materials and the expected filter efficiencies (see Table 11.11). In airborne PM purposes, most fibrous filters consist of glass fibers, cellulose, quartz, polystyrene, and polycarbonate materials. Membrane filters are made of cellulose esters, polytetrafluoroethylene (PTFE), polyethylene, polycarbonate, silver, and other materials. If physical properties dictate the filter choice, then glass fiber filters would be used. However, these inorganic silicate materials can become highly radioactive during certain analyzes and are therefore not useful during a nondestructive activation technique is applied. Similarly, carbon-based filters would not generally be used in a destructive technique, since the carbon in cellulose, PFTE, etc. would show up in the analysis and produce a positive C bias (e.g., for an organic C-to-elemental C ratio calculation). If the analysis will include fibers, especially asbestos, mixed cellulose would be preferred for transmissive electron microscopy (TEM), since the filter material can be destroyed during etching (i.e., the cellulose is dissolved leaving the fibers more visible to TEM). However, the pore size would vary according to the type of investigation (e.g., if longer fibers are of major interest, an 8 μm pore size would be used, whereas if fibers of interest are shorter, the 4.5 μm pore size filter would be the best choice) [104]. Thus, subsequent elemental analysis after collection is an important selection criterion for PM filters [105, 106].

Size distributions are determined by classifying airborne particles by aerodynamic diameter, electric mobility, or light-scattering properties. The most common technique is the use of multistage impactors, each stage of which removes particles of progressively smaller diameter. Fig. 11.45 shows a four-stage impactor. The particulate matter collected on each stage is weighed to yield a mass size distribution or is subjected to chemical analysis to obtain data on its chemical size distribution. Impactors are used to determine size distributions for particle diameters of 0.1 μm and larger.

Electric mobility is utilized to obtain size distribution information in the 0.01–1.0 μm diameter range. This measurement method requires unipolar charging of particles and their separation by passage through an electric field [107]. By incrementally increasing the electric field strength progressively, larger charged particles may be removed from a flowing airstream. The change in the amount of charge collected by an electrometer grid is then related to the number of particles present in a particular size increment. Instruments based on this principle yield a number size distribution.

TABLE 11.11 Published impurities (ng cm^{-2}) for blank values for selected fibrous and membrane filters

Element	W 41	W 541	MFHA + W 41	MFAA + W 41	Nucl.	QM-A
Na	150	100	800	740	45	58,000
Mg	<80	<60	<400	<370	<5	–
Al	12	11	30	17	4	–
Cl	100	400	1200	1400	–	2900
K	15	12	145	62	2	6700
Ca	140	35	810	560	3	–
Sc	<0.005	0.004	0.006	0.008	0.0003	0.12
Ti	10	2	25	<30	1.4	–
V	<0.03	0.03	<0.10	<0.21	0.01	–
Cr	3	2	25	36	1	24
Mn	0.5	0.5	8	2.1	<0.2	3.9
Fe	40	30	80	100	12	580
Co	0.1	0.04	0.3	0.25	0.01	9.3
Cu	<4	2	24	17	1.2	29
Zn	<25	3	50	<180	0.7	880
As	<0.1	0.04	<0.4	0.13	0.04	–
Se	<0.3	0.2	–	1.0	6.2	–
Br	5	2	9	7.7	2	–
Sb	0.15	0.07	0.8	0.23	0.006	4.1
La	<0.2	0.04	<0.6	<0.5	<0.003	–

W 41, Whatman 41, cellulose fiber; *W 541*, Whatman 541, cellulose fiber; *MFHA*, millipore cellulose ester membrane (0.45 μm pore size); *MFAA*, Millipore cellulose ester membrane (0.8 μm pore size); *Nucl.*, Nuclepore membrane (pore size 0.4 μm); *QM-A*, Whatman QM-A, quartz microfiber filter.

Source: R. Dams, A critical review of nuclear activation techniques for the determination of trace elements in atmospheric aerosols, particulates and sludge samples (Technical Report), in: Pure and Applied Chemistry vol. 64, 1992, p. 991.

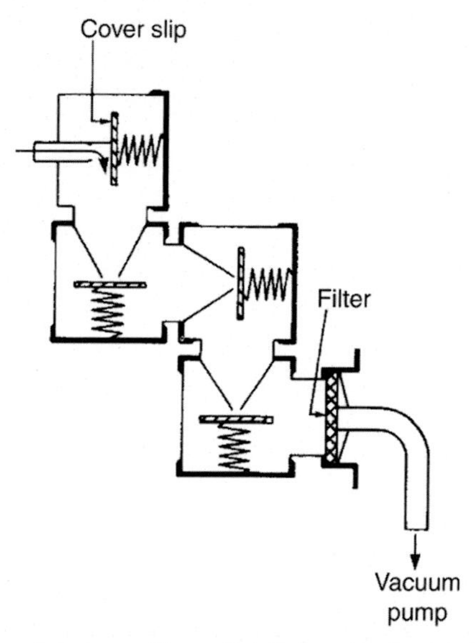

FIG. 11.45 Schematic diagram of a four-stage cascade impactor. *(Source D.A. Vallero, Fundamentals of Air Pollution, 5th ed., Elsevier Academic Press, Waltham, MA, 2014, p. 999 pages cm; reproduced with permission from P.M. Giever, Particulate matter sampling and sizing, in: Air Pollution: Measuring, Monitoring, and Surveillance of Air Pollution, vol. 3, p. 1, 1976.)*

Light-scattering properties of particles are also utilized to determine a number size distribution [5, 108]. Individual particles interact with a light beam and scatter light at an angle to the original direction of the beam. The intensity of the scattered light is a function of the diameter and the refractive index of the particle. Inlet systems are designed to dilute a particle-laden airstream sufficiently to permit only one particle in the beam at a time. The intensity of the scattered light, as measured by a photomultiplier tube, is proportional to particle size. The number of electric pulses of each magnitude is accumulated in a multichannel analyzer. By sampling at a known flow rate, the number of particles of different diameters is counted with this type of instrument. This principle is also used in nephelometers, that is, portable PM detectors.

The chemical composition of particulate pollutants is determined in two forms: specific elements or specific compounds or ions. Knowledge of their chemical composition is useful in determining the sources of airborne particles and in understanding the fate of particles in the atmosphere. Elemental analysis is used to describe the individual elements present in a sample, for example, the percentage of iron, manganese, and sulfur of a PM, but not the various chemical species. From elemental analysis techniques, we do not obtain direct information about the chemical form of S in a sample such as sulfate or sulfide. Two nondestructive techniques used for direct elemental analysis of particulate samples are X-ray fluorescence (XRF) spectroscopy and neutron activation analysis (NAA).

XRF is a technique in which a sample is bombarded by X-rays [109]. Inner shell electrons are excited to higher energy levels. As these excited electrons return to their original state, energy with wavelengths characteristic of each element present in the sample is emitted. These high-energy photons are detected and analyzed to give the type and quantity of the elements present in the sample. The technique is applicable to all elements with an atomic number of 11, that is, sodium (Na) or higher. In principle, complex mixtures may be analyzed with this technique. Difficulties arise from a matrix effect, so that care must be taken to use appropriate standards containing a similar matrix of elements. This technique requires relatively sophisticated equipment and highly trained personnel.

With NAA, the PM sample is bombarded with neutrons, which interact with the sample to form different isotopes of the elements in the air sample [105]. Many of these isotopes are radioactive and may be identified by comparing their radioactivity with standards. The NAA technique is not quite as versatile as XRF and requires a neutron source.

Pretreatment of the collected PM may be required for chemical analysis. Pretreatment generally involves extraction of the particulate matter into a liquid. The solution may be further treated to transform the material into a form suitable for analysis. Trace metals may be determined by AA spectroscopy, emission spectroscopy, polarography, and anodic stripping voltammetry. Analysis of anions is possible by colorimetric techniques and ion chromatography. Sulfate (SO_4^{2-}), sulfite (SO_3^{2-}), nitrate (NO_3^-), chloride (Cl^-), and fluoride (F^-) may be determined by ion chromatography [110].

11.9.2 PM sampler design considerations

An inlet of the $PM_{2.5}$ sampler (see Fig. 11.46 and top of Fig. 10.3 in Chapter 10) is often designed to extract ambient aerosols from the surrounding airstream, remove particles with aerodynamic diameters $>10\,\mu m$, and move the remaining smaller particles to the next stage. An impactor and filter assembly can remove particles with diameters $<10\,\mu m$ and $>2.5\,\mu m$ but will allow particles of $2.5\,\mu m$ in diameter to pass and be collected on a filter surface. Particles $<10\,\mu m$ but $<2.5\,\mu m$ are removed downstream from the inlet by a single-stage, single-flow, single-jet impactor assembly. Aerosols are collected on filters that are weighed before and after sampling, for example, 37 mm diameter glass filters immersed in low-volatility, low-viscosity diffusion oil. The oil is added to reduce the impact of "bounce," that is, particles hit the filter and are not reliably collected.

Sampling errors may occur at the inlet, and particles may be lost in the sampling manifold while being transported to the collection surface. Fig. 11.47 depicts possible flow patterns around a sampling inlet in a uniform flow field. Fig. 11.47A shows that when no air is permitted to flow into the inlet, the streamline flow moves around the edges of the inlet. As the flow rate through the inlet increases, more and more of the streamlines are attracted to the inlet. Fig. 11.47B is known as the isokinetic condition, that is, the sampling flow rate is equal to the flow field rate. An example of an isokinetic condition is an inlet with its opening into the wind pulling air at the wind speed. When one is sampling for gases, this is not a serious constraint because the composition of the gas will be the same under all inlet flow rates; that is, there is no fractionation of the air sample by different gaseous molecules [111].

Fig. 11.47B, the isokinetic case, is the ideal case. Often, the isokinetic condition is assumed for gas-phase pollutants, but not for particle-containing airstreams. The ideal sample inlet would always face into the wind and sample at the same rate as the instantaneous wind velocity, which is improbable. Under isokinetic sampling conditions, parallel airstreams flow into the sample inlet, carrying with them particles of all diameters capable of being carried by the stream flow. When the sampling rate is lower than the flow field, as depicted in Fig. 11.47C, the streamlines start to diverge around the edges of the inlet, and the larger particles with more inertia are unable to follow the streamlines and are captured by the sampling inlet.

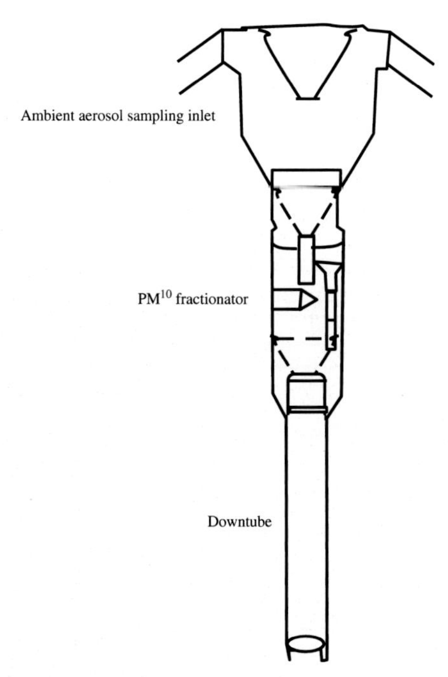

FIG. 11.46 Schematic of the inlet in a particulate matter sampler.

The opposite effect occurs when the sampling rate exceeds the flow field, as depicted in Fig. 11.47D. The inlet captures more streamlines, but the larger particles near the edges of the inlet may be unable to follow the streamline flow and escape collection by the inlet. The inlet may be designed for particle size fractionation; for example, a $PM_{2.5}$ inlet will exclude particles larger than 2.5 μm aerodynamic diameter (see Figs. 10.11 and 10.12 in Chapter 10).

These inertial effects become less important for particles with diameters <5 μm and for low wind velocities, but for samplers attempting to collect particles above 5 μm, the inlet design and flow rates become important parameters. In addition, the wind speed has a much greater impact on sampling errors associated with particles >5 μm in diameter.

After passing the inlet, care must be taken to move the particles to the collection medium in an unaltered form. Potential problems arise from too long or too twisted manifold systems. Gravitational settling in the manifold will remove a fraction of the very large particles. Larger particles are also subject to loss by impaction on walls at bends in a manifold. Particles may also be subject to electrostatic forces that will cause them to migrate to the walls of nonconducting manifolds. Other problems include condensation or agglomeration during transit time in the manifold. These constraints require sampling manifolds for particles to be as short and have as few bends as possible.

The collection technique involves the removal of particles from the airstream. The two principal methods are filtration and impaction. Filtration consists of collecting particles on a filter surface by three processes: direct interception, inertial impaction, and diffusion [112]. Filtration is used to remove a very high percentage of the mass and number of particles by these three processes. Any size classification is done by a preclassifier, such as an impactor, before the particle stream reaches the surface of the filter.

11.9.3 Indirect PM measurements

In addition to collecting, weighing and analyzing PM by mass and size, various indirect methods are employed. As mentioned, PM concentrations by size can be estimated by light scattering. The amount of scattering is calibrated against known PM concentrations, for example, collocating one of the direct measurement instruments with the nephelometer and comparing results.

The concentration of particulates is directly proportionate to the amount of scattering detected, so an optical device can provide an indirect measure of $PM_{2.5}$ concentration that can be determined by a ratio of scattering to $PM_{2.5}$ concentrations compared at several sites and over time. Of course, since this is an indirect method, interferences can be substantial, including changes in environmental conditions, especially relative humidity. Since nephelometers do not collect PM, no analysis of particles for chemical composition is possible. In addition, extremely small particles in the air may be missed by an optical microscope, individually weighing so little that their presence is masked in gravimetric analysis by the presence of a few large particles.

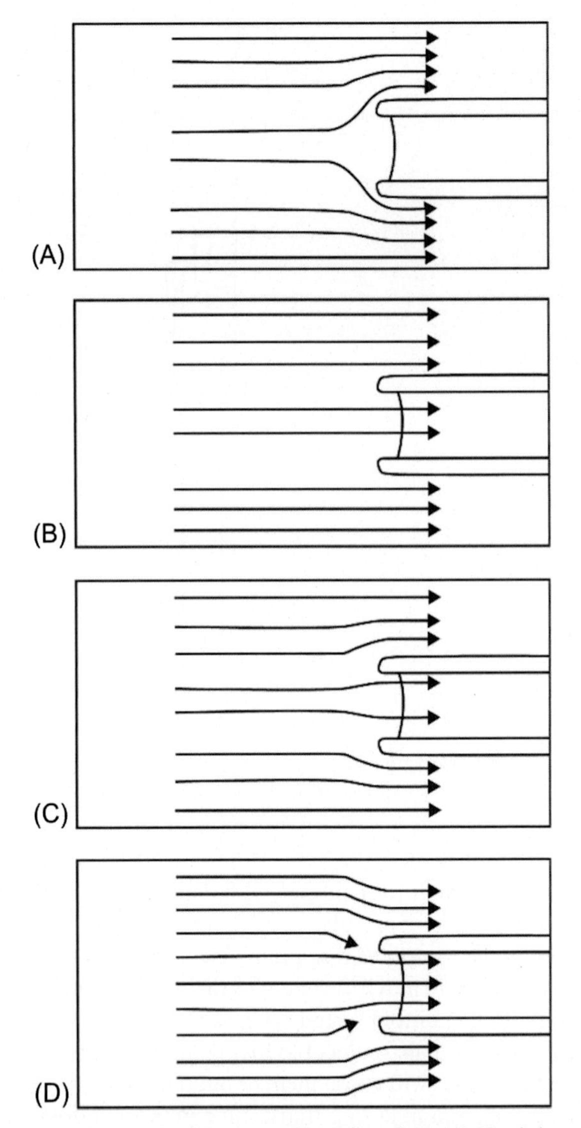

FIG. 11.47 The streamline flow patterns around a sampling inlet in a uniform flow field. (A) No air is permitted to flow into the inlet, so that the streamline flow moves around the edges of the inlet; (B) isokinetic condition, where the sampling flow rate is equal to the flow field rate; (C) sampling rate is lower than the flow field; and (D) sampling rate exceeds the flow field. *(Source D.A. Vallero, Fundamentals of Air Pollution, 5th ed., Elsevier Academic Press, Waltham, MA, 2014, p. 999 pages cm.)*

The open-path technologies discussed earlier for gas-phase pollutants can be used for PM. Given the relationship between PM size and plume light extinction, each air pollution plume has unique features in specific regions of the open-path extinction spectra. The extinction measurements a PM plume can be combined with algorithms to retrieve an estimate of mean mass distribution. Generally, larger PM (1–20 mm diameter) has interferences in IR extinction spectra, whereas smaller PM will affect the visible and UV wavelength region [113].

The DIAL, OP-FTIR, and other open-path systems are being evaluated to address aerosols. This is challenging in that aerosol cloud is much more complex than a gaseous cloud given the many physical properties of the specific material, meaning that radiation interactions between particles and the atmosphere add variability beyond that of gas-phase pollutants. This will include new algorithms for analyzing the open-path signals [89, 114].

11.10 MEASURING GAS AND PARTICULATE PHASES TOGETHER

Ambient air always contains particulate matter, which can interfere with the measurement of gases. For example, if one is trying to determine the amount of NO_2 in the air, any solid-phase nitrate particles and nitric acid vapor will interfere with the results. Thus, there are two basic reasons to know the phase distribution of an air sample, that is, to know and decrease interference and to account for the total amount of a contaminant in both phases.

11.10.1 Denuders

Diffusion denuders can be used for particle-gas-phase separation. In a denuder, air is pulled through a hollow tube that is coated with sorbing media. The flow is kept laminar by modulating the flow rate to the tube diameter. This increases diffusion several orders of magnitude greater than Brownian molecular diffusivity of the particulate matter, allowing the gas to move into the sorbing media in the interior walls of the tubes much faster than under normal, turbulent conditions. The tube length is chosen to ensure that all or at least a known and consistent amount of the gas moves into the walls before the air exits the tube. However, most of the particles remain suspended in the air flow and can be captured, for example, with filters.

Annual denuders are a specific type that employs two concentric tubes, with inner surface of the outer tube and the outer surface of the inner tube coated with the sorbing media. This doubles the surface area into which the gas will move, thus improving collection efficiency, shortening overall tube length, shortening sampling time, and/or allowing for higher flow rates. These and other denuders are often used in sampling airborne semivolatile organic compounds (SVOCs). These substances are often found in both vapor and particulate phases, so the separation of the denuder is ideal.

11.10.2 Analysis and measurement of acidic deposition

The two components of acidic deposition are wet deposition and dry deposition. The collection and subsequent analysis of wet deposition are intuitively straightforward. A sample collector opens to collect rainwater at the beginning of a rainstorm and closes when the rain stops. The water is then analyzed for pH, anions (negative ions), and cations (positive ions). The situation for dry deposition is much more difficult [115]. Collection of particles settling from the air is very dependent on the surface material and configuration. The surfaces of trees, plants, and grasses are considerably different from that of the round, open-top canister often used to collect dry deposited particles. After collection, the material must be suspended or dissolved in pure water for subsequent analysis.

The National Atmospheric Deposition Program has established the nationwide sampling network of ~100 stations in the United States. The sampler is shown in Fig. 11.48, including a wet deposition collection container. The wet collection bucket is covered with a lid when it is not raining. A sensor for rain moves the lid to open the wet collector bucket and cover the dry bucket at the beginning of a rainstorm. This process is reversed when the precipitation ends.

The primary constituents to be measured are the pH of precipitation, sulfates, nitrates, ammonia, chloride ions, metal ions, phosphates, and specific conductivity. The pH measurements help to establish reliable long-term trends in patterns of acidic precipitation. The sulfate and nitrate information is related to anthropogenic sources where possible. The measurements of chloride ions, metal ions, and phosphates are related to sea spray and windblown dust sources. Specific conductivity is related to the level of dissolved salts in precipitation.

The flowchart for analysis of wet and dry precipitation shown in Fig. 11.48 includes the weight determinations, followed by pH and conductivity measurements and finally chemical analysis for anions and cations. The pH measurements are made with a well-calibrated pH meter, with extreme care taken to avoid contaminating the sample. The metal ions Ca^{21}, Mg^{21}, Na^{1}, and K^{1} are determined by flame photometry, which involves absorption of radiation by metal ions in a hot flame. Ammonia and the anions Cl^-, SO_4^{2-}, NO_3^-, and PO_4^{3-} are measured by automated colorimetric techniques.

11.10.3 Measuring hazardous air pollutants

In addition to the criteria pollutants, there are numerous hazardous air pollutants that must be measured. The characteristics of these pollutants vary considerably in inherent properties, including vapor pressure, ranging from gas phase to highly volatile to semivolatile to nonvolatile (e.g., sorbed to aerosols). The methods for the gas-phase fraction mirror those of the gas-phase pollutants if they possess the key physical and chemical similarities, for example, absorbance and fluorescence by UV, visible, or IR wavelengths. The compounds bound to particles may be analyzed by PM chemical constituent determinations, such as X-ray fluorescence. However, such methods do not speciate, but merely give elemental composition. To ascertain specific compounds of concern will require wet chemistry and other more laborious approaches.

Given the diversity of air toxics, individual methods of collection, handling, storage, extraction, separation, and detection must be tailored to the specific characteristics of these compounds. To this end, air quality regulation agencies have developed compendia of methods. For example, the US Environmental Protection Agency maintains a set of 17 methods in a standardized format with a variety of applicable sampling methods and several analytic techniques, for specific classes of organic pollutants, as appropriate to the specific pollutant compound, its level, and potential interferences.

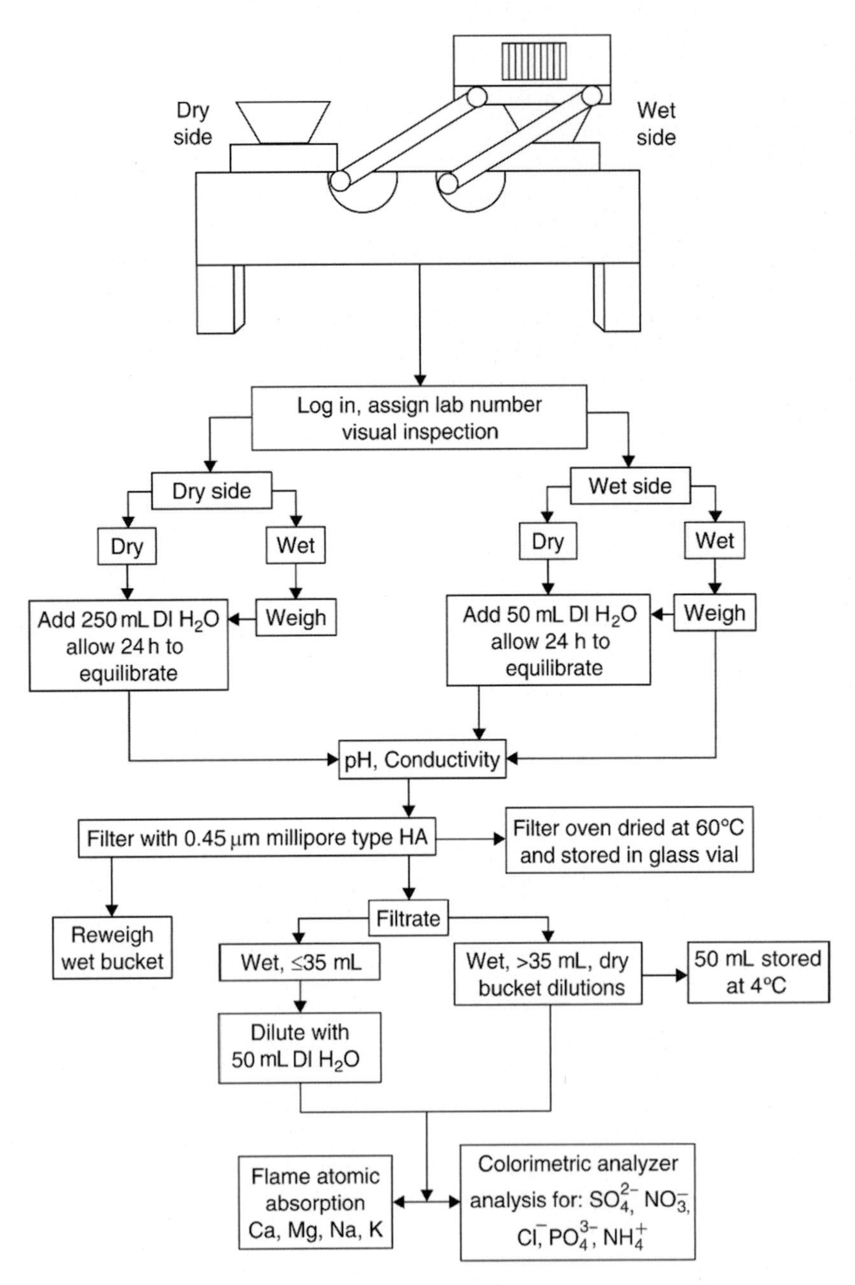

FIG. 11.48 Schematic diagram of an integrated precipitation collector *(top)* and flow chart of the steps needed for sample analysis (DI H_2O= distilled water). *(Source: G.J. Stensland et al., NADP Quality Assurance Report: Central Analytical Laboratory, 1980.)*

Consequently, this treatment allows the user flexibility in selecting alternatives to complement his or her background and laboratory capability. These methods may be modified from time to time as advancements are made [25].

The visual guide to the organization of the compendium for specific methods covering a variety of organic compounds is shown in Fig. 11.49.

The methods assigned the "A" notation are methods that were published in the first edition (Compendium Methods TO-1 through TO-14) of the compendium and have now been updated due to technological advances in either the sampling or analysis methodology. In addition, three new methods (Compendium Methods TO-15 through TO-17) have been added to make the complete second edition of the compendium. These methods were added due to their advanced technology application involving specially treated canisters (Compendium Method TO-15), long-path (open-path) Fourier transform infrared spectroscopy (Compendium Method TO-16), and multibed sorbent techniques (Compendium Method TO-17).

Given that the methods for hazardous air pollutants are diverse and changing, the reader is advised to visit the compendium site (Link https://www3.epa.gov/ttnamti1/airtox.html. Accessed 21 September 2018) and browse for the most

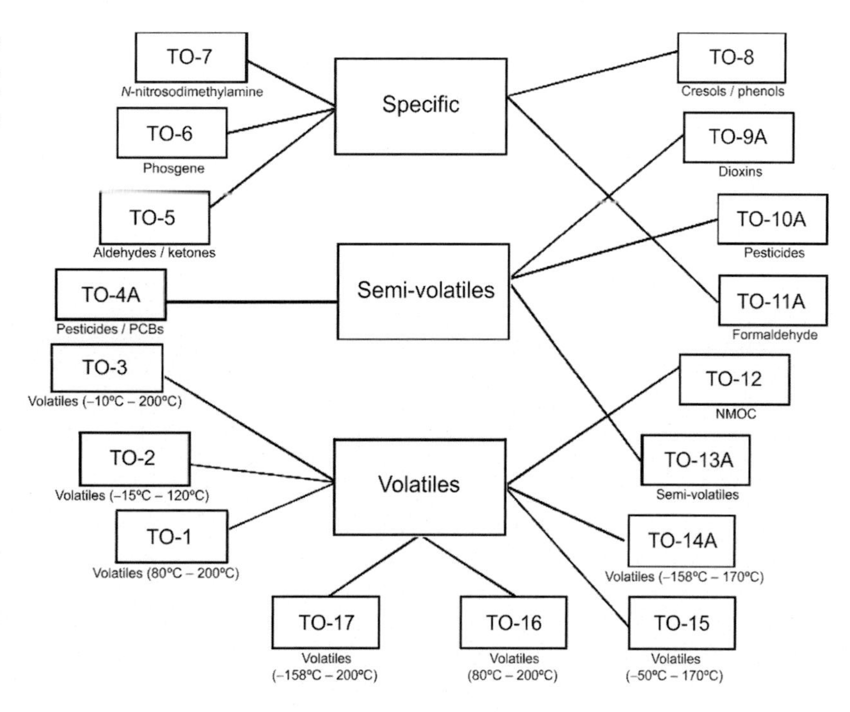

FIG. 11.49 Visual guide to the air toxic methods compendium. *(Source: EPA-625/SR-96/010b, Project summary: Compendium of Methods for the Determination of Toxic Organic Compounds in Ambient Air, 2nd ed. (1999).)*

appropriate method for a given pollutant (see Table 11.12). For example, if the compound of concern is a semivolatile compound, there are four possible methods, according to the visual guide. If the compound is a dioxin, then Method TO-9A should be used. However, if the compound is a polycyclic aromatic hydrocarbon, browsing the likely methods shows that TO-13A is most applicable. The method provides every step from sample selection through laboratory analyses, including quality control and assurance.

11.11 LABORATORY ANALYSIS

Real-time analysis of air quality is becoming more commonplace, yet for many air pollutants, most environmental samples must be brought to a laboratory analyzed after collection. The steps that must be taken to interpret the concentration of a chemical in the sample are known as "wet chemistry."

11.11.1 Extraction

When an air sample that is trapped on a filter or other media arrives at the laboratory, the next step is to remove it from the media for analysis, that is, extraction. Extraction is needed for two reasons. Numerous toxic chemicals have low vapor pressures and may not be readily dissolved in water. Thus, they may be found in various media, for example, sorbed to particles, in the gas phase, or in the water column suspended to colloids (and very small amounts dissolved in the water itself). To collect such chemicals in the gas phase, a common method calls for trapping it on polyurethane foam (PUF). Thus, to analyze dioxins in the air, the PUF and particle matter must first be extracted, and to analyze dioxins in soil and sediment, those particles must also be extracted.

Extraction makes use of physics and chemistry. For example, many compounds can be simply extracted with solvents, usually at elevated temperatures. A common solvent extraction is the Soxhlet extractor, named after the German food chemist, Franz Soxhlet (1848–1913). The Soxhlet extractor (the US EPA Method 3540) removes sorbed chemicals by passing a boiling solvent through the media. Cooling water condenses the heated solvent, and the extract is collected over an extended period, usually several hours. Other automated techniques apply some of the same principals as solvent extraction but allow for more precise and consistent extraction, especially when large volumes of samples are involved. For example, supercritical fluid extraction (SFE) brings a solvent, usually carbon dioxide to the pressure and temperature near its critical point of the solvent, where the solvent's properties are rapidly altered with very slight variations of pressure [116]. Solid-phase extraction (SPE) uses a solid and a liquid phase to isolate a chemical from a solution and is often used to

TABLE 11.12 Listing of compounds classes in *Compendium of Methods for the Determination of Toxic Organic Compounds in Ambient Air*—Second Edition

Compendium method no.	Type of compounds determined	Sample collection device	Analytic methodology[a]
TO-1[b]	Volatile organic compounds	Tenax solid sorbent	GC/MS
TO-2[b]	Volatile organic compounds	Molecular sieve sorbent	GC/MS
TO-3[b]	Volatile organic compounds	Cryotrap	GC/FID
TO-4A	Pesticides/PCBs	Polyurethane foam	GC/MD
TO-5[b]	Aldehydes/Ketones	Impinger	HPLC
TO-6[b]	Phosgene	Impinger	HPLC
TO-7[b]	Anilines	Adsorbent	GC/MS
TO-8[b]	Phenols	Impinger	HPLC
TO-9A	Dioxins	Polyurethane foam	HRGC/HRMS
TO-10A	Pesticides/PCBs	Polyurethane foam	GC/MD
TO-11A	Aldehydes/ketones	Adsorbent	HPLC
TO-12[b]	Nonmethane organic compounds (NMOC)	Canister or on-line	FID
TO-13A	Polycyclic aromatic hydrocarbons	Polyurethane foam	GC/MS
TO-14A	Volatile organic compounds (nonpolar)	Specially treated canister	GC/MS and GC/MD
TO-15	Volatile organic compounds (polar/nonpolar)	Specially treated canister	GC/MS
TO-16	Volatile organic compounds	Open-path monitoring	FTIR
TO-17	Volatile organic compounds	Single/multibed adsorbent	GC/MS, FID, etc.

[a]*GC/MS, gas chromatography/mass spectrometry; GC/FID, gas chromatography/flame ionization detector; HPLC, high-performance liquid chromatography; GC/MD, gas chromatography/multidetector; GC/IT, gas chromatography/ion trap detector; FTIR, Fourier transform infrared spectroscopy; HRGC/HRMS, high resolution gas chromatography/high resolution mass spectrometry.*
[b]*Methods denoted by "2" have not been changed since their publication in the first edition of the compendium, so the full content of these methods is not repeated in the second edition. Therefore, the full content of these methods must be obtained from the original compendium (EPA 600/4-89-017).*
Source: EPA-625/SR-96/010b, Project summary: Compendium of Methods for the Determination of Toxic Organic Compounds in Ambient Air, 2nd ed. (1999).

clean up a sample before analysis. Combinations of various extraction methods can enhance the extraction efficiencies, depending upon the chemical and the media in which it is found. Ultrasonic and microwave extractions may be used alone or in combination with solvent extraction. For example, the US EPA Method 3546 provides a procedure for extracting hydrophobic (i.e., not soluble in water) or slightly water-soluble organic compounds from particles such as soils, sediments, sludges, and solid wastes. In this method, microwave energy elevates the temperature and pressure conditions (i.e., 100–115°C and 50–175 psi) in a closed extraction vessel containing the sample and solvent(s). This combination can improve recoveries of chemical analytes and can reduce the time needed compared than the Soxhlet procedure alone.

11.11.2 Sample analysis

Air monitoring using canisters and bags allows the air to flow directly into the analyzer. Water samples may also be directly injected. Surface methods, such as fluorescence, sputtering, and atomic absorption, require only that the sample be mounted on specific media (e.g., filters). Also, continuous monitors like the chemiluminescent system mentioned previously provide ongoing, direct measurements. These and the solutions holding the extracts mentioned in the previous section must be analyzed.

Separation science embodies the techniques for separating complex mixtures of analytes, which is the first stage of chromatography. The second is detection. Separation makes use of the chemicals' different affinities for certain surfaces under various temperature and pressure conditions. The first step, injection, introduces the extract to a "column." The term column is derived from the time when columns were packed with sorbents of varying characteristics, sometimes meters in length, and the extract was poured down the packed column to separate the various analytes. Today, columns are of two major types, gas and liquid. Gas chromatography (GC) makes use of hollow tubes ("columns") coated inside with compounds that hold organic chemicals. The columns are in an oven, so that after the extract is injected into the column, the temperature is increased, as well as the pressure, and the various organic compounds in the extract are released from the column surface differentially, whereupon they are collected by a carrier gas (e.g., helium) and transported to the detector. Generally, the more volatile compounds are released first (they have the shortest retention times), followed by the semi-volatile organic compounds. So, boiling point is often a very useful indicator as to when a compound will come off a column. This is not always the case, since other characteristics such as polarity can greatly influence a compound's resistance to be freed from the column surface. For this reason, numerous GC columns are available to the chromatographer (different coatings, interior diameters, and lengths). Rather than coated columns, liquid chromatography (LC) makes use of columns packed with different sorbing materials with differing affinities for compounds. Also, instead of a carrier gas, LC uses a solvent or blend of solvents to carry the compounds to the detector. In the high-performance LC (HPLC), pressures are also varied.

Detection is the final step for quantifying the chemicals in a sample. Several detection approaches are also available. Arguably, the most common is absorption. Chemical compounds absorb energy at various levels, depending upon their size, shape, bonds, and other structural characteristics. Chemicals also vary in whether they will absorb light or how much light they can absorb depending upon wavelength. Some absorb very well in the ultraviolet (UV) range, while others do not. Diode arrays help to identify compounds by giving a number of absorption ranges in the same scan. Some molecules can be excited and will fluoresce. The Beer-Lambert law (Eq. 11.14) states that energy absorption is proportional to concentration of chemical species. Using this relationship, in each separation-detection analytic system, solutions with several known concentrations of the chemical species of interest are injected, resulting in a calibration curve (see Fig. 11.50). The theoretical Beer-Lambert relationship is the idealized (green) curve of the expect absorbance response, whereas there will be a deviation (red curve), response curve, due to inherent instrumentation error, imprecise known solutions, stray energy sources (e.g., light), unaccounted chemical reactions within the solvent, and other factors. As such, the most reliable part of the curve is its most linearity response range. Generally, at least five standards and one blank (i.e., same pure solvent absent the target chemical species) are needed to fit the standard curve. The owner's manual and websites for the instruments must be consulted, especially to see if a manual curvature correction function is available for the instrument.

The type of detector needed depends upon the kinds of pollutants of interest. Detection gives the "peaks" that are used to identify compounds (Fig. 11.51). For example, if hydrocarbons are of concern, GC with flame ionization detection (FID) may be used. GC-FID gives a count of the number of carbon atoms, so, for example, long chains can be distinguished from short chains. The short chains come off the column first and have peaks that appear before the long-chain peaks. However, if pesticides or other halogenated compounds are of concern, electron capture detection (ECD) is a better choice.

One of the most popular detection methods for environmental pollutants is mass spectrometry (MS), which can be used with either GC or LC separation. The MS detection is highly sensitive for organic compounds and works by using a stream of electrons to consistently break apart compounds into fragments. The positive ions resulting from the fragmentation are separated according to their masses. This is referred to as the "mass-to-charge ratio" or m/z. No matter which detection device is used, software is used to decipher the peaks and to perform the quantitation of the amount of each contaminant in the sample.

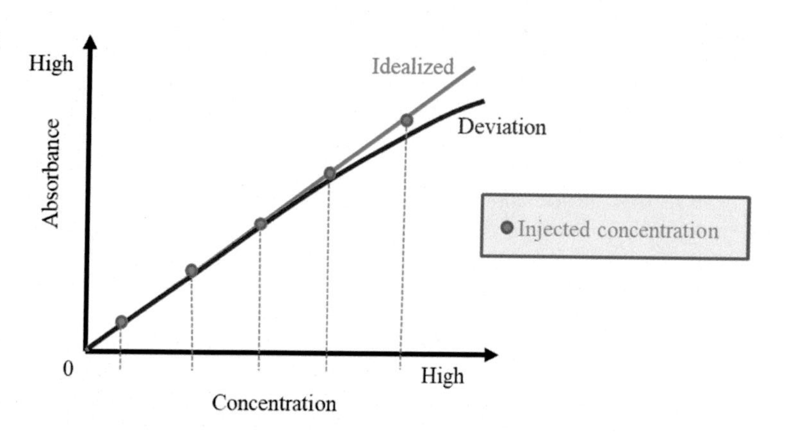

FIG. 11.50 Calibration curve for a hypothetical chemical species.

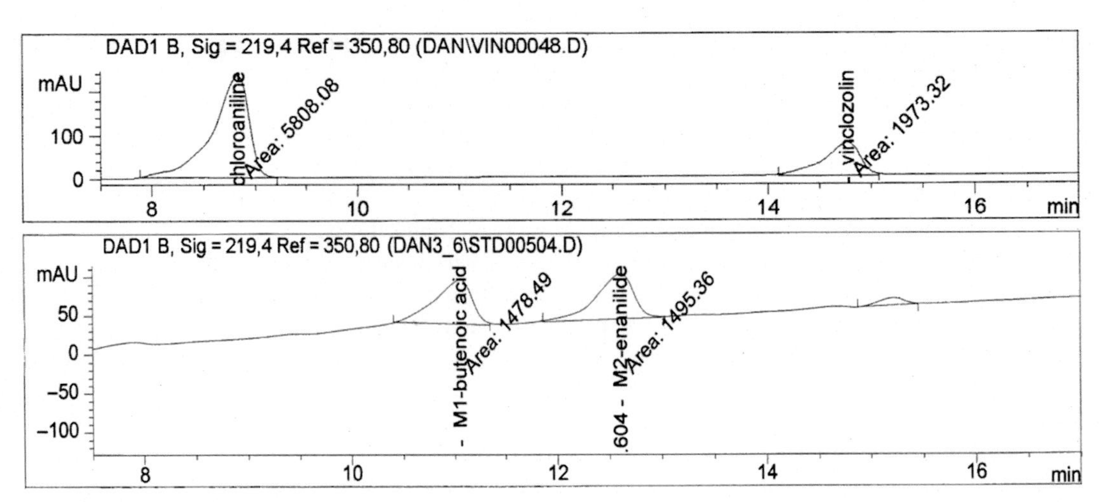

FIG. 11.51 Chromatogram of compounds trapped on polyurethane foam, extracted by supercritical fluid extraction, separated by gas chromatography, and detected by diode-array ultraviolet detection (DAD). *(Source: D.A. Vallero, J.J. Peirce, Engineering the Risks of Hazardous Wastes, Butterworth-Heinemann, Burlington, MA, 2003, pp. xxi, 306 p.)*

Consider an analytic laboratory that has generated the chromatogram and table shown in Fig. 11.52, using an HPLC/UV at 254 nm using a 5 μm, C_{18}, 4.6 × 250 mm column from an extract.

What are the retention times of compound A and B? Which compound is present in a larger amount? Which compound has the higher boiling point? What would happen to the retention times of compounds A and B if the column temperature were raised? What if compound B were suspected to be benzo(a)pyrene based on company records and the type of industries nearby?

The retention time of compound A is 21.31 min, shown above of the peak and in the table's retention time column. The retention time of compound B is 24.53 min. You cannot tell from this table or chromatogram which compound is present in a larger amount, since the only way to do so is to have calibration curve from known concentrations of compound A and compound B (at least three but preferably five). For example, the HPLC could be run successively with injections of pure solutions of 0.01, 0.1, 1, 10, and 100 μg L^{-1} concentrations of compound A and again with pure solutions of the same concentrations of compound B. These concentrations would give peak areas associated with each known concentration. Then, the HPLC software can construct the calibration curve.

Thus, for example, if a peak with an area of 200 is associated with 1 μg L^{-1} of compound A and a peak with an area of 2000 is associated with 10 μg L^{-1} of compound A, that is, a linear calibration curve, at 21.31 min after the aliquot is injected into the HPLC, then when you run your unknown sample and a peak at 21.31 min with an area of 1000 would mean you have about 5 μg L^{-1} concentration of compound A in the sample. The same procedure would be followed to draw a calibration curve for compound B at a retention time of 24.53 min.

The reason it is not sufficient to look at the percent area is that each compound is physically and chemically different. Per Beer-Lambert law, the amount of energy absorbed (in this case, the UV light) is what gives the peak. If a molecule of compound A absorbs UV at this wavelength (i.e., 254 nm) at only 25% as that of compound B, compound A's concentration would be higher than that of compound B (because even though compound B has twice the percent area, its absorbance is four times that of compound B).

Compound A has the lower boiling point since it comes off the column first. Of course, this is only true if other factors, especially polarity, are about the same. For example, if compound B has about the same polarity as the column being used, but compound A has a very different polarity, compound A will have a greater tendency to leave the column. Generally, however, retention time is a good indicator of boiling point; that is, lower retention times mean lower boiling points.

If the column temperature were raised, both compounds A and B would come off the column in a shorter time. Thus, the retention times of both compounds A and B would be shorter than before the temperature was raised.

To determine whether the peak at 24.53 min is benzo(a)pyrene, you must first obtain a true sample of pure benzo(a)pyrene to place in a standard solution. This is the same process as you used to develop the calibration curve above. That is, you would inject this standard of known benzo(a)pyrene into the same HPLC and the same volume of injection. If the standard gives a peak at a retention time at about 25 min, there is a good chance it is benzo(a)pyrene. As it turns out, benzo(a)pyrene absorbs UV at 254 nm and does come off an HPLC column at about 25 min.

The column type also affects retention time and peak area. The one used by the laboratory is commonly used for polycyclic aromatic hydrocarbons, including benzo(a)pyrene. However, numerous columns can be used for semivolatile organic compounds, so both the retention time and peak area will vary somewhat. Another concern is coelution, that is, two distinct compounds that have nearly the same retention times. One means of reducing the likelihood of coelution is to target the wavelength of the UV

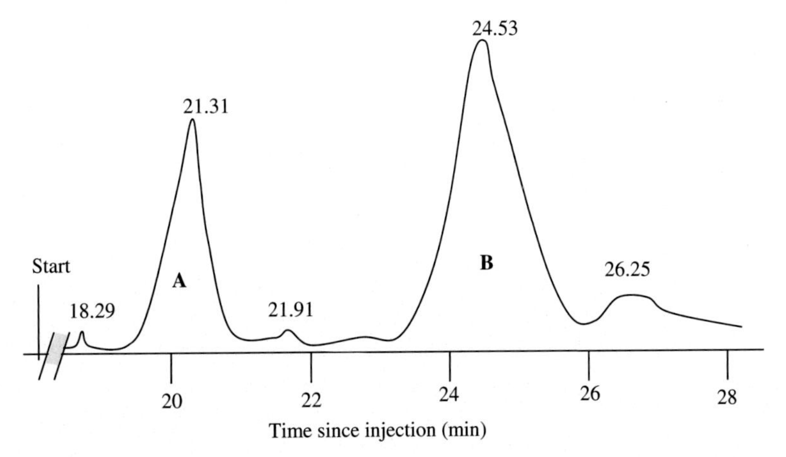

FIG. 11.52 Hypothetical HPLC-UV curve and data table.

Retention time	Area	Type	Area/height	Area %
18.29	NA	NA	NA	0.1
21.31	NA	NA	NA	31.4
21.91	NA	NA	NA	0.2
24.53	NA	NA	NA	67.2
26.25	NA	NA	NA	1.1

detector. For example, the recommended wavelength for benzo(a)pyrene is 254 nm, but 295 nm is preferred by environmental chromatographers because the interference peak in the benzo(a)pyrene window is decreased at 295 nm. Another way to improve detection is to use a diode-array detection system with the UV detector. This gives a number of different chromatograms simultaneously at various wavelengths. Finally, there are times when certain detection systems are completely unable to detect a specific molecule. For example, if a molecule does not absorb UV light (i.e., it lacks a group of atoms in a molecule responsible for absorbing the UV radiation, known as chromophores), there is no way to use any UV detector. In this case, another detector, for example, mass spectrometry, must be used.

For inorganic substances and metals, the additional extraction step may not be necessary. The actual measured media (e.g., collected airborne particles) may be measured by surface techniques like atomic absorption (AA), X-ray fluorescence (XRF), inductively coupled plasma (ICP), or sputtering. As for organic compounds, the detection approaches can vary. For example, ICP may be used with absorption or MS. If all one needs to know is elemental information, for example, to determine total lead or nickel in a sample, AA or XRF may be sufficient. Unlike organic compounds, these metal analytic methods do not speciate. For speciation, for example, finding the concentrations of a specific organometallic species of Ni, sample preparation is more involved, including a process known as "derivatization." Derivatizing a sample is performed by adding a chemical agent that transforms the compound in question into one that can be recognized by the detector. This is done for both organic and inorganic compounds, for example, when the compound in question is too polar to be recognized by MS.

The physical and chemical characteristics of the compounds being analyzed must be considered before visiting the field and throughout all the steps in the laboratory. Although it is beyond the scope of this book to go into detail, it is worth mentioning that the quality of results generated about contamination depends upon the sensitivity and selectivity of the analytic equipment. Table 11.13 defines some of the most important analytical chemistry metrics.

TABLE 11.13 Types of analytic metrics used in air pollution studies

Type of limit	Description
Limit of detection (LoD)	Lowest concentration or mass that can be differentiated from a blank with statistical confidence. This is a function of sample handling and preparation, sample extraction efficiencies, chemical separation efficiencies, and capacity and specifications of all analytic equipment being used (see IDL below)
Instrument detection limit (IDL)	The minimum signal greater than noise detectable by an instrument. The IDL is an expression of the piece of equipment, not the chemical of concern. It is expressed as a signal-to-noise (S-N) ratio. This is mainly important to the analytical chemists, but the engineer should be aware of the different IDLs for various instruments measuring the same compounds, so as to provide professional judgment in contracting or selecting laboratories and deciding on procuring for appropriate instrumentation for all phases of remediation
Limit of quantitation (LoQ)	The concentration or mass above which the amount can be quantified with statistical confidence. This is an important limit because it goes beyond the "presence-absence" of the LoD and allows for calculating chemical concentration or mass gradients in the environmental media (air, water, soil, sediment, and biota)
Practical quantitation limit (PQL)	The combination of LoQ and the precision and accuracy limits of a specific laboratory, as expressed in the laboratory's quality assurance/quality control (QA/QC) plans and standard operating procedures (SOPs) for routine runs. The PQL is the concentration or mass that the engineer can consistently expect to have reported reliably

Source: D.A. Vallero, J.J. Peirce, Engineering the Risks of Hazardous Wastes, Butterworth-Heinemann, Burlington, MA, 2003, pp. xxi, 306 p.

Air pollution measurements are usually an early part of a comprehensive assessment needed to support a decision and actions. Often, they are input into models that extend the measurement results in space and time (see Chapter 13). These and other measurements are needed for ground truthing and evaluating these models. As such, air pollution measurement and modeling are mutually dependent.

REFERENCES

[1] D. Vallero, Environmental Biotechnology: A Biosystems Approach, Elsevier Science, 2015.
[2] D.A. Vallero, Measurements in environmental engineering, in: M. Kutz (Ed.), Handbook of Measurement in Science and Engineering, 2013.
[3] U.S. Environmental Protection Agency, Environmental Measurement, 26 January 2016. Available: http://www.epa.gov/measurements, 2015.
[4] D.A. Vallero, Fundamentals of Air Pollution, fifth ed., Elsevier Academic Press, Waltham, MA, 2014, p. 999 pages cm.
[5] K. Whitby, K. Willeke, Single particle optical counters: principles and field use, Aerosol Meas. (1979) 145.
[6] S. Friedlander, Dust, Smoke and Haze, Wiley, New York, 1977.
[7] W.C. Malm, K.K. Leiker, J.V. Molenar, Human perception of visual air quality, J. Air Pollut. Control Assoc. 30 (2) (1980) 122–131.
[8] T.C. Daniel, Measuring the quality of the natural environment: a psychophysical approach, Am. Psychol. 45 (5) (1990) 633.
[9] H. Koschmieder, Theory of the horizontal visual range, Beitr Phys Freien Atm 12 (1924) 171–181.
[10] T.M. Letcher, D. Vallero, Waste: A Handbook for Management, Academic Press, 2011.
[11] H.C. van de Hulst, Scattering in the atmospheres of the earth and planets, in: The Atmospheres of the Earth and Planets, 1949, p. 49.
[12] T. Hall Jr., F. Blacet, Separation of the absorption spectra of NO2 and N2O4 in the range of 2400–5000A, J. Chem. Phys. 20 (11) (1952) 1745–1749.
[13] R.B. Husar, J.B. Elkins, W.E. Wilson, US visibility trends, 1960–1992, regional and national, in: A and WMA Annual Meeting, vol. 1, Air & Waste Management Association, 1994. pp. 94-MP3. 05.
[14] S. Twomey, Atmospheric Aerosols, 1977.
[15] R.J. Charlson, A.P. Waggoner, J. Thiekle, Visibility protection for class 1 areas: the technical basis. Final report, Council on Environmental Quality, Washington, DC, USA, 1978.
[16] N.C. Ahlquist, R.J. Charlson, A new instrument for evaluating the visual quality of air, J. Air Pollut. Control Assoc. 17 (7) (1967) 467–469.
[17] EPA/600/R-96/055, Guidance for the data quality objectives process, EPA QA/G-4, 2002.
[18] Elements of a Quality Assurance Project Plan (QAPP) For Collecting, Identifying and Evaluating Existing Scientific Data/Information, Available: https://www.epa.gov/osa/elements-quality-assurance-project-plan-qapp-collecting-identifying-and-evaluating-existing, 2017.
[19] D. Vallero, Paradigms Lost: Learning from Environmental Mistakes, Mishaps and Misdeeds, Butterworth-Heinemann, 2005.
[20] Public Health Assessment Guidance Manual, Available: https://www.atsdr.cdc.gov/hac/PHAManual/toc.html, 2005.
[21] G.M. Lovett, et al., Who needs environmental monitoring? Front. Ecol. Environ. 5 (5) (2007) 253–260.
[22] EPA/240/B-06/001, Data quality objectives guidance, 2006.
[23] EPA 542-R-01-014, Clarifying DQO Terminology Usage to Support Modernization of Site Cleanup Practice, 2001.
[24] Method 1613: Tetra-through octa-chlorinated dioxins and furans by isotope dilution HRGC/HRMS (Rev. B), 1994.

[25] EPA/625/R-96/010b, Method TO-9A in Compendium of Methods for the Determination of Toxic Organic Compounds in Ambient Air, second ed., 1999.
[26] SW846 Method 8290A-1: Polychlorinated dibenzodioxins (PCDDs) and polychlorinated dibenzofurans (PCDFs) by high resolution gas chromatograph/high resolution mass spectrometry (HRGC/HRMS), Available: http://www3.epa.gov/epawaste/hazard/testmethods/sw846/pdfs/8290a.pdf, 2007.
[27] D.V.S. Kimbrough, R. Shores, A. Vette, K.T. Black, V. Martinez, Multi-criteria decision analysis for the selection of a near road ambient air monitoring site for the measurement of mobile source air toxics, Transp. Res. Part D: Transp. Environ. 13 (8) (2008) 505–515.
[28] S. Kimbrough, et al., Long-term continuous measurement of near-road air pollution in Las Vegas: seasonal variability in traffic emissions impact on local air quality, Air Qual. Atmos. Health 6 (1) (2013) 295–305.
[29] D. Vallero, FHWA and EPA National Near-Road Study, Final ReportU.S. Environmental Protection Agency and U.S. Department of Transportation, Federal Highway Administration, 2010.
[30] S. Kimbrough, D. Vallero, R. Shores, A. Vette, K. Black, V. Martinez, Multi-criteria decision analysis for the selection of a near road ambient air monitoring site for the measurement of mobile source air toxics, Transp. Res. Part D: Transp. Environ. 13 (8) (2008) 505–515.
[31] J. Malczewski, GIS and Multicriteria Decision Analysis, John Wiley & Sons, 1999.
[32] V. Sumathi, U. Natesan, C. Sarkar, GIS-based approach for optimized siting of municipal solid waste landfill, Waste Manag. 28 (11) (2008) 2146–2160.
[33] U.S. Environmental Protection Agency, Ecosystem Services in EnviroAtlas, 29 April 2017. Available: http://www.epa.gov/enviroatlas/ecosystem-services-enviroatlas, 2015.
[34] U.S. Environmental Protection Agency, EnviroAtlas Data Layer Matrix, 1 February 2016. Available: http://www.epa.gov/enviroatlas/enviroatlas-data-layer-matrix, 2015.
[35] U.S. Environmental Protection Agency, Interactive Map of Air Quality Monitors, 18 September 2018. Available: https://www.epa.gov/outdoor-air-quality-data/interactive-map-air-quality-monitors, 2018.
[36] Environmental Systems Research Institute, Understanding GIS: The ARC/INFO Method: Self Study Workbook: Version 7 for UNIX and OpenVMS, Esri Press, 1995.
[37] J.G. Watson, J.C. Chow, D. DuBois, M. Green, N. Frank, Guidance for the network design and optimum site exposure for PM2. 5 and PM10, Environmental Protection Agency, Office of Air Quality Planning and Standards, Research Triangle Park, NC, United States, 1997. Nevada Univ. System, Desert Research Inst., Reno, NV, United States; National Oceanic and Atmospheric Administration, Las Vegas, NV, United States.
[38] U. EPA, Locating and Estimating Air Emissions from Sources of Polycyclic Organic Matter, United States Environmental Protection Agency, Research Triangle Park, NC, USA, 1998.
[39] J.C. Chow, J.P. Engelbrecht, J.G. Watson, W.E. Wilson, N.H. Frank, T. Zhu, Designing monitoring networks to represent outdoor human exposure, Chemosphere 49 (9) (2002) 961–978.
[40] I. W. G. o. t. E. o. C. R. t. Humans, Outdoor air pollution, in: IARC monographs on the evaluation of carcinogenic risks to humans, 109, 2016, p. 9.
[41] S. Kimbrough, R.C. Shores, D.A. Whitaker, FHWA and EPA National near-Road Study, Federal Highway Administration. Office of Natural Environment, Detroit, MI, USA, 2013.
[42] U.S. Environmental Protection Agency, Test Methods: Frequent Questions, 1 February 2016. Available: http://www3.epa.gov/epawaste/hazard/testmethods/faq/, 2015.
[43] Environmental Health Australia, Environmental Health Risk Assessment: Guidelines for Assessing Human Health Risks from Environmental Hazards, Commonwealth of Australia, Canberra, ACT, 2012.
[44] EPA/100/B-04/001, U.S. Environmental Protection Agency. An examination of EPA risk assessment principles and practices, Available: http://www.epa.gov/OSA/pdfs/ratf-final.pdf, 2004.
[45] U.S. Environmental Protection Agency, Air Quality System (AQS), 18 September 2018. Available: https://www.epa.gov/aqs, 2018.
[46] L.D. Pagnotto, R. Keenan, Sampling and Analysis of Gases and Vapors, The Industrial Environment-Its Evaluation and Control, National Institute for Occupational Safety and Health, US Public Health Service, Washington, DC, 1973.
[47] T. Tanaka, Chromatographic characterization of porous polymer adsorbents in a trapping column for trace organic vapor pollutants in air, J. Chromatogr. A 153 (1) (1978) 7–13.
[48] A. Kot-Wasik, B. Zabiegała, M. Urbanowicz, E. Dominiak, A. Wasik, J. Namieśnik, Advances in passive sampling in environmental studies, Anal. Chim. Acta 602 (2) (2007) 141–163.
[49] B. Zabiegała, A. Kot-Wasik, M. Urbanowicz, J. Namieśnik, Passive sampling as a tool for obtaining reliable analytical information in environmental quality monitoring, Anal. Bioanal. Chem. 396 (1) (2010) 273–296.
[50] Z. Subpart, D. Subpart, Toxic and Hazardous Substances, Code of Federal Regulations, Occupational Safety and Health Administration, US Department of Labor, 1910. Available at https://www.osha.gov/pls/oshaweb/owadisp.show_document.
[51] Toxicological Profile for Formaldehyde (Draft), 1997.
[52] Permissible Exposure Limits (PELs), 2013.
[53] J.M. Parnis, D. Mackay, T. Harner, Modelling PUF passive air samplers: temperature dependence of polyurethane foam–air partitioning of PAHs and their O-, N-, and S-derivatives computed by COSMO-RS, Atmos. Pollut. Res. 7 (1) (2016) 155–161.
[54] G.D. Wight, Fundamentals of Air Sampling, CRC Press, 1994.
[55] C. Chaemfa, et al., Screening of atmospheric short-and medium-chain chlorinated paraffins in India and Pakistan using polyurethane foam based passive air sampler, Environ. Sci. Technol. 48 (9) (2014) 4799–4808.

[56] M.R. Ras, F. Borrull, R.M. Marcé, Sampling and preconcentration techniques for determination of volatile organic compounds in air samples, TrAC Trends Anal. Chem. 28 (3) (2009) 347–361.
[57] A. Ribes, G. Carrera, E. Gallego, X. Roca, M.J. Berenguer, X. Guardino, Development and validation of a method for air-quality and nuisance odors monitoring of volatile organic compounds using multi-sorbent adsorption and gas chromatography/mass spectrometry thermal desorption system, J. Chromatogr. A 1140 (1–2) (2007) 44–55.
[58] Ö.O. Kuntasal, D. Karman, D. Wang, S.G. Tuncel, G. Tuncel, Determination of volatile organic compounds in different microenvironments by multibed adsorption and short-path thermal desorption followed by gas chromatographic–mass spectrometric analysis, J. Chromatogr. A 1099 (1–2) (2005) 43–54.
[59] J.G. Watson, J.C. Chow, R.J. Tropp, X.L. Wang, S.D. Kohl, L.A. Chen, Standards and traceability for air quality measurements: flow rates and gaseous pollutants, Mapan 28 (3) (2013) 167–179.
[60] H.-S. Moon, J.-H. Lee, K. Kwon, H.-I. Jung, Review of recent progress in micro-systems for the detection and analysis of airborne microorganisms, Anal. Lett. 45 (2–3) (2012) 113–129.
[61] M. Marć, B. Zabiegała, J. Namieśnik, Mobile systems (portable, handheld, transportable) for monitoring air pollution, Crit. Rev. Anal. Chem. 42 (1) (2012) 2–15.
[62] S. Steinle, et al., Personal exposure monitoring of PM2. 5 in indoor and outdoor microenvironments, Sci. Total Environ. 508 (2015) 383–394.
[63] K.L. Dionisio, et al., Exploring consumer exposure pathways and patterns of use for chemicals in the environment, Toxicol. Rep. 2 (2015) 228–237.
[64] P.P. Egeghy, et al., Computational exposure science: an emerging discipline to support 21st-century risk assessment, Environ. Health Perspect. 124 (6) (2016) 697.
[65] M. Goldsmith, et al., Ex Priori: Exposure-based Prioritization across Chemical Space, in: International Society of Exposure Science Annual Conference, Cincinnati, Ohio, 2014.
[66] J. Mitchell, et al., Comparison of modeling approaches to prioritize chemicals based on estimates of exposure and exposure potential, Sci. Total Environ. 458 (2013) 555–567.
[67] J.F. Wambaugh, et al., High-throughput models for exposure-based chemical prioritization in the ExpoCast project, Environ. Sci. Technol. 47 (15) (2013) 8479–8488.
[68] J.F. Wambaugh, et al., High throughput heuristics for prioritizing human exposure to environmental chemicals, Environ. Sci. Technol. 48 (21) (2014) 12760–12767.
[69] J.W. Schaeffer, et al., Size, composition, and source profiles of inhalable bioaerosols from Colorado dairies, Environ. Sci. Technol. 51 (11) (2017) 6430–6440.
[70] S.M. Walser, et al., Evaluation of exposure–response relationships for health effects of microbial bioaerosols – a systematic review, Int. J. Hyg. Environ. Health 218 (7) (2015) 577–589.
[71] J.C. Chow, et al., Comparison of particle light scattering and fine particulate matter mass in central California, J. Air Waste Manage. Assoc. 56 (4) (2006) 398–410.
[72] J.G. Watson, J.C. Chow, D.H. Lowenthal, K.L. Magliano, Estimating aerosol light scattering at the Fresno supersite, Atmos. Environ. 42 (6) (2008) 1186–1196.
[73] D.K. Mikel, R. Merrill, J. Colby, EPA Handbook: Optical Remote Sensing for Measurement and Monitoring of Emissions Flux, US Environmental Protection Agency, North Carolina, 2011.
[74] G.M. Hidy, et al., Remote sensing of particulate pollution from space: have we reached the promised land? J. Air Waste Manage. Assoc. 59 (10) (2009) 1130–1139.
[75] R.M. Hoff, S.A. Christopher, Remote sensing of particulate pollution from space: have we reached the promised land? J. Air Waste Manage. Assoc. 59 (6) (2009) 645–675.
[76] J. Slanina, P. De Wild, G. Wyers, The application of denuder systems to the analysis of atmospheric components, Adv. Environ. Sci. Technol. 24 (1992) 129–154.
[77] Y. Ye, I. Galbally, I. Weeks, Emission of 1, 3-butadiene from petrol-driven motor vehicles, Atmos. Environ. 31 (8) (1997) 1157–1165.
[78] P.R. Westlin, W.W. Balcke, Compliance assurance monitoring rule- status of final rulemaking, in: Air & Waste Management Association 90. Annual Meeting.[np]. 1997, 1997.
[79] Part I e New Source Review Q&A, Available: http://www.epa.state.il.us/air/new-source-review/new-sourcereview, 2013.
[80] D. Vallero, Air pollution, in: Kirk-Othmer Encyclopedia of Chemical Technology, John Wiley & Sons, Inc., Hoboken, New Jersey, 2015.
[81] D.A. Vallero, Excerpted and Revised From: Fundamentals of Air Pollution, fifth ed., Elsevier Academic Press, Waltham, MA, 2014. p. 999 pages cm.
[82] C. Secrest, D. Hoyt, New technology applications, in: Presented at the National Multimedia Inspection/Enforcement Workshop, Dallas, Texas, May 15 and 16, 2012, 2012.
[83] E. Murray, J. Van der Laan, Remote measurement of ethylene using a CO_2 differential-absorption lidar, Appl. Opt. 17 (5) (1978) 814–817.
[84] E. Browell, Lidar remote sensing of tropospheric pollutants and trace gases-Programs of NASA Langley Research Center, in: Sensing of Environmental Pollutants, 1978, pp. 395–402.
[85] W.B. Grant, Effect of differential spectral reflectance on DIAL measurements using topographic targets, Appl. Opt. 21 (13) (1982) 2390–2394.
[86] G. Russwurm and J. Childers, n.d. FT-IR Open-Path Monitoring Guidance Document, ManTech Environmental Technology, Inc, Research Triangle Park, NC, EPA/600/R-96/0401996.

[87] S. EPA, Air Quality Criteria for Ozone and Related Photochemical Oxidant: Volume I of III, EPA/600/P-93/004AF, US Environmental Protection Agency Office of Research and Development, Washington, DC, 1996.

[88] S. Sarkar, S.M. Parihar, A. Dutta, Fuzzy risk assessment modelling of East Kolkata wetland area: a remote sensing and GIS based approach, Environ. Model Softw. 75 (2016) 105–118.

[89] E. Lutsch, et al., Identifying long-range transport of wildfire emissions to the Arctic using a network of ground-based FTIR spectrometers, satellite observations, and transport models, in: 51st CMOS Congress-Canadian Meteorological and Oceanographic Society, 2017.

[90] U.S. Environmental Protection Agency, Trace atmospheric gas analyzer (TAGA), 20 November 2013. Available: http://www.epa.gov/region06/6lab/taga.htm, 2013.

[91] J. H. Gilliam and E. S. Hall, n.d. Reference and Equivalent Methods Used to Measure National Ambient Air Quality Standards (NAAQS) Criteria Air Pollutants-Volume I.

[92] S.-O. Monitoring, SO2 NAAQS Designations Source-Oriented Monitoring Technical Assistance Document, 2013.

[93] W. Dailey, G. Fertig, A novel NDIR analyzer for NO, SO_2 and CO analysis, Anal. Instrum. 15 (1977).

[94] R. Stevens, J. Hodgeson, Applications of chemiluminescent reactions to measurement of air pollutants-instrumentation, Anal. Chem. 45 (4) (1973) A443–A449.

[95] R.J. Paur, F.F. McElroy, Technical Assistance Document for the Calibration of Ambient Ozone Monitors, Environmental Monitoring and Support Laboratory, Office of Research and Development, US Environmental Protection Agency, 1979.

[96] K. Ashley, NIOSH manual of analytical methods 5th edition and harmonization of occupational exposure monitoring, Gefahrstoffe, Reinhaltung der Luft= Air quality control/Herausgeber, BIA und KRdL im VDI und DIN 2015 (1–2) (2015) 7.

[97] P.M. Eller, M.E. Cassinelli, NIOSH Manual of Analytical Methods, Diane Publishing, 1994.

[98] P.A. Solomon, G. Norris, M. Landis, M. Tolocka, Chemical analysis methods for atmospheric aerosol components, in: Aerosol Measurement: Principles, Techniques, and Applications, 2001, pp. 261–294.

[99] B. Effectivity, n.d. Department of Environment and Natural Resources.

[100] Primary National Ambient Air Quality Standard for Sulfur Dioxide; Final Rule, 2010.

[101] C.D. Hollowell, Current Instrumentation for Continuous Monitoring for SO2, 1972.

[102] H. Okabe, P.L. Splitstone, J.J. Ball, Ambient and source SO2 detector based on a fluorescence method, J. Air Pollut. Control Assoc. 23 (6) (1973) 514–516.

[103] K.L. Rubow, Filtration: fundamentals and applications, in: Aerosol and Particle Measurement, University of Minnesota, Minneapolis, Minnesota, 2004.

[104] D.A. Vallero, J.R. Kominsky, M.E. Beard, O.S. Crankshaw, Efficiency of sampling and analysis of asbestos fibers on filter media: implications for exposure assessment, J. Occup. Environ. Hyg. 6 (1) (2008) 62–72.

[105] R. Heindryckx, R. Dams, Chemical analysis of atmospheric aerosols by instrumental neutron activation, Prog. Nucl. Energy 3 (3) (1979) 219–252.

[106] R. Dams, A critical review of nuclear activation techniques for the determination of trace elements in atmospheric aerosols, particulates and sludge samples (Technical Report), Pure and Applied Chemistry 64 (1992) 991.

[107] B.Y. Liu, D. Pui, A. Kapadia, Electrical aerosol analyzer: history, principle, and data reduction, Minnesota University, Department of Mechanical Engineering, Minneapolis, USA, 1976.

[108] K. Whitby, B. Cantrell, Atmospheric aerosols – characteristics and measurement, in: International Conference on Environmental Sensing and Assessment, Las Vegas, Nev, 1976, p. 1.

[109] T.G. Dzubay, X-Ray Fluorescence Analysis of Environmental Samples, Ann Arbor Science Publishers Ann Arbor, MI, 1977.

[110] E. Sawicki, J.D. Mulik, E. Wittgenstein, Ion Chromatographic Analysis of Environmental Pollutants, Ann Arbor Science Publishers, 1978.

[111] R.D. Cadle, The Measurement of Airborne Particles, Wiley-Interscience, New York, 1975. 353 p.

[112] B.Y. Liu, Fine Particles: Aerosol Generation, Measurement, Sampling, and Analysis, Elsevier, 2012.

[113] R. Hashmonay, Particulate Matter Measurements Using Open-Path Fourier Transform Infrared Spectroscopy, 20020001768, vol. 2001, 2001.

[114] C.M. Simonescu, Application of FTIR spectroscopy in environmental studies, in: Advanced Aspects of Spectroscopy, InTech, 2012.

[115] B.B. Hicks, M.L. Wesely, J.L. Durham, Critique of methods to measure dry deposition workshop summary, Argonne National Lab., IL, USA; Environmental Protection Agency, Research Triangle Park, NC, USA, 1980.

[116] M.R. Ekhtera, G.A. Mansoori, M.C. Mensinger, A. Rehmat, B. Deville, Supercritical fluid extraction for remediation of contaminated soil, in: M.A. S. Abraham (Ed.), Supercritical Fluids: Extraction and Pollution Prevention, vol. 670, American Chemical Society, Washington, DC, 1997, pp. 280–298.

FURTHER READING

[117] T. Barzyk, D.A. Vallero (Ed.), Nitric Oxide Model, Research Triangle Park, North Carolina, U.S. Environmental Protection Agency, 2013.

[118] A. Venkatram, M. Snyder, V. Isakov, S. Kimbrough, Impact of wind direction on near-road pollutant concentrations, Atmos. Environ. 80 (2013) 248–258.

[119] P. Visibility, E. An, Report to Congress, EPA-450/5-79-008, US Environmental Protection Agency, 1979.

[120] D.A. Vallero, Fundamentals of Air Pollution, fourth ed., Elsevier, Amsterdam; Boston, 2008. pp. xxiii, 942 p.

[121] K. James, R. Gregory, Eye and Brain: The Psychology of Seeing, 1976.

[122] W.C. Malm, Introduction to Visibility, Cooperative Institute for Research in the Atmosphere, NPS Visibility Program, Colorado State University, 1999.

[123] Interagency Monitoring of Protected Visual Environments, 22 November 2013. Available: http://vista.cira.colostate.edu/improve/Education/Glossary/Haze.htm, 2013.

[124] A. Waggoner, N. Ahliquist, R. Charlson, Measurement of the aerosol total scatter–backscatter ratio, Appl. Opt. 11 (12) (1972) 2886–2889.

[125] X. Pang, M.D. Shaw, A.C. Lewis, L.J. Carpenter, T. Batchellier, Electrochemical ozone sensors: a miniaturised alternative for ozone measurements in laboratory experiments and air-quality monitoring, Sensors Actuators B Chem. 240 (2017) 829–837.

[126] D.A. Vallero, T.M. Letcher, Unraveling Environmental Disasters, Newnes, 2012.

[127] J. Schmaltz, MODIS: Moderate Resolution Imaging Spectroradiometer, National Aeronautics and Space Administration, 2018.

[128] C. Seftor, Tracking Dust Across the Atlantic, E. O. National Aeronautics and Space Administration, 2013.

[129] U.S. Environmental Protection Agency, EnviroAtlas Interactive Map: User's Guide, EnviroAtlas, 2016. July 2016. Available: http://www.epa.gov/enviroatlas/enviroatlas-data-layer-matrix.

[130] A.C. Stern, Fundamentals of Air Pollution, Elsevier, 2014.

[131] E. P. A. U.S., Managing the Quality of Environmental Data in EPA Region 9, 13 August 2013. Available: http://www.epa.gov/region9/qa/pdfs/aircrf.pdf, 2013.

[132] D.C. Mussatti, R. Srivastava, P.M. Hemmer, R. Strait, EPA air pollution control cost manual, in: Air Quality Strategies and Standards Division of the Office of Air Quality Planning and Standards, vol. 27711, US Environmental Protection Agency, Research Triangle Park, NC, 2002.

[133] J. Moreno, S. Moreno-Grau, and A. Garcıa-Sánchez, n.d. Lidar probes air pollution.

[134] R.J. Bryan, Ambient air quality surveillance, Air Pollution 3 (1976) 343–392.

[135] P.M. Giever, Particulate matter sampling and sizing, in: Air Pollution: Measuring, Monitoring, and Surveillance of Air Pollution, vol. 3, 1976, p. 1.

[136] G.J. Stensland, et al., NADP Quality Assurance Report: Central Analytical Laboratory, 1980.

[137] EPA-625/SR-96/010b, Project summary: Compendium of Methods for the Determination of Toxic Organic Compounds in Ambient Air-Second Edition, (1999).

[138] D.A. Vallero, J.J. Peirce, Engineering the Risks of Hazardous Wastes, Butterworth-Heinemann, Burlington, MA, 2003, p. xxi. 306 p.

Chapter 12

Air pollution risk calculations

12.1 INTRODUCTION

Arguably, this entire book is about risk. Risk is the proverbial bottom line of air quality management. Scientists, engineers, and managers use risk metrics to determine the success or failure of an operation or project. This chapter introduces key risk concepts, along with examples of hazards, exposures, and risk characterizations. Air pollution calculations for the most part support some aspect of assessing and addressing risks.

12.1.1 Risk, reliability and resilience

The systems deployed by air quality managers must sufficiently reduce risk and be reliable and resilient. Reliability, that is, the probability that the system will perform its intended function for a specified time and under specified conditions [1, 2], is covered explicitly in Chapter 16. However, keeping with much of the engineering economics literature, the calculations in Chapter 16 focus on the item, which is the fundamental unit of reliability engineering. Consideration of risk from a systems engineering must characterize and predict the cumulative success of all the items and the externalities that will affect the system. This calls for the acknowledgment of the interrelationship between risk and reliability. For example, the greater the reliability of an air pollution control system, the better it is at reducing the emissions of harmful substances, which translates into lower risks to humans and the environment. Reliability, like risk, is an expression that incorporates probability, but instead of the likelihood of an adverse and undesirable outcome, it expresses the likelihood of a desired outcome. Since both are probabilities, their values range between zero and unity, often expressed as a percentage ranging from 0% to 100%. Complete reliability, that is, 100%, and, conversely, 0% risk are the ideals.

Reliability is the extent to which something can be trusted. This trust is conditional, including the length of time that a system, process, or item is trusted. They are reliable so long as they perform the designed functions under the specified conditions during a certain period. For example, every sampling and analytic device discussed in Chapter 11 require a level of reliability to be useful to air quality management. Often, items within a system are known to fail at some rate, for example, x number of failures per hour of operation. If a pump is expected to fail, on average, once every 1000 h of operation and a study is using 10 pumps in the field for 11,000 h, they may not be sufficiently reliable. At a minimum, it would be wise to have at least one replacement extra pumps.

Reliability is a metric of accountability for risk management, that is, reliability means that the systems selected and managed will not fail during their proposed operating periods. Or stated more positively, reliability is the mathematical expression of success. Reliability is the probability that something that is in operation at time 0 (t_0) will still be operating until the designed life (time $t=(t_t)$). As such, it is also a measure of the engineer's social accountability. Neighborhoods, municipalities, and other clients want to be assured that that air they breathe is healthful, which includes assurances that air pollution controls are working as designed and will not fail, that is, will continue to remove pollutants as designed.

The probability of a failure per unit time is the hazard rate or failure density [$f(t)$]. This is a function of the likelihood that an adverse outcome will occur. It is not a function of the severity of the outcome [3]. The $f(t)$ is not affected by whether the outcome is very severe, for example, lung cancer and the loss of an entire species in an ecosystem, or relatively benign, for example, slight odors or minor leaf damage. The likelihood that something will fail at a given time interval is found by integrating the hazard rate over a defined time interval:

$$P\{t_1 \le T_f \le t_1\} = \int_{t_1}^{t_2} f(t)dt \qquad (12.1)$$

where T_f = time of failure.

Air Pollution Calculations. https://doi.org/10.1016/B978-0-12-814934-8.00012-0

Thus, the reliability function $R(t)$ of a system at time t is the cumulative probability that the system has not failed in the time interval from t_0 to t_t:

$$R(t) = P\{t_1 \leq T_f \leq t_1\} = 1 - \int_0^t f(x)dx \tag{12.2}$$

It is not a question of "if" but "when" an engineered system will fail. An engineer or manager may be able to extend the optimal time of an air pollution control or prevention technology, that is, increasing t_t to make the system more resilient under conditions that would otherwise lead to failure. For example, proper engineering design of a fabric filter system should not only work well during "normal" conditions but also resist failure during hostile conditions, for example, a time period of highly corrosive particulate matter (PM) loading. A design flaw occurs if the wrong material is used, for example, filters not matched to worst-case corrosivity PM emissions. However, even when the proper materials are used, the failure is not eliminated entirely. Selecting the right materials simply protracts the time before the failure occurs, that is, increases T_f [4].

In addition to reliability, risk is also related to resilience. Reliable systems are also generally resilient systems. Like many other environmental systems, engineers can learn from nature. For example, ecosystems must be resilient. Ecosystem resilience and engineering resilience are mirror images of one another. The less resilient a system, the more that the engineer and manager must do to make it reliable. A very resilient natural system can withstand a wide array of insult. A system that lacks resilience is a sensitive and vulnerable system, needing controls to remain operational when conditions change.

Resilient systems have wide tolerances, whereas sensitive system is easily harmed outside of very narrow tolerance ranges. Thus, an ecologically resilient system is in less need of engineering controls against insult than a nonresilient system. For example, atmospheric deposition of a pollutant onto a resilient system may cause little impact and/or require a relatively short time to recover. Conversely, the same deposition onto a nonresilient system could be devastating.

Unfortunately, human activities can transform resilient systems into nonresilient ones. For example, a wild, biodiverse area may be quite resilient in the face of drought and insect infestation. Human activities, for example, building roads and infrastructure around and near this area, can reduce biodiversity and change biological processes, making it less able to withstand these insults. Thus, a natural system may have been able to withstand lower pH rainfall caused by sulfur dioxide (SO_2) emissions, but other human activities may have made the system more vulnerable to the acid rain.

Engineered systems, including air pollution equipment, must also be resilient, that is, able to return to the original form or state after being bent, compressed, or stretched or the ability to recover or return to the desired state readily following the application from some form of stress [5–7].

12.1.2 Risk and precaution

In evidence-based risk assessments, the onus is on the air quality regulator to show that an action or agent is unsafe. This is the predominant perspective of most US health and environmental regulations, that is, it is up to the agency to stop an action, such as a new chemical being used in a product, only if it has sufficient information to support this decision. However, the agency usually may require the applicant to provide such information. Chemical risk assessment is a scientific approach to answering three basic questions [8]:

1. How much of chemical is in each environmental medium or compartment, for example, in the soil, water, air, carpet, and walls?
2. What is the exposure to the chemical, that is, the amount of contact with a receptor in each medium and compartment?
3. How toxic is the chemical?

Many nations take a different view of environmental problems, especially those that have the potential of large-scale and irreversible damage. Under this approach, any application is denied unless the applicant can provide sufficient information showing that the action or product is safe [9–13]. Unlike traditional risk assessments, which assume governance holds the burden of proof that something is harmless, the precautionary onus assumes that the burden of proof is entirely on those who anticipate the new action, and if "reasonable suspicion" arises of a severe and potentially irreversible outcome, the action as proposed should be denied. This requires an objective, well-structured, comprehensive analysis of alternatives to provide a needed service, including a "no action" if any of the alternatives are worse than doing nothing new.

Increasingly, evidence-based risk assessments are being augmented or even supplanted by precaution, especially if the decision involves a reasonable likelihood that an adverse effect is severe and irreversible [14]. The precautionary approach also calls for systems thinking and sustainable solutions. As discussed in Chapter 2, new screening models and tools are needed, including multicriteria decision analysis (MCDA), which allows for the consideration of numerous variables, from various information sources. New ways to communicate risk and reliability include expert elicitation [15], that is, gathering insights from a swath of the scientific community on newly emerging or otherwise poorly understood challenges to air quality.

Among the important gaps in data and tools needed for risk-based and precaution-based decision-making are more reliable means of estimating and predicting exposures to stressors. For example, human health characterization factors (CFs) in LCAs have benefited from improved hazard, especially toxicity, information [16]. Given that health risk is a function of hazard and exposure, both traditional risk assessment and LCA's human health *CF* must be based on reliable exposure predictions [17]. For instance, in early life stages a chemical compound that, when used, could result in the formation and release of pollutants at some later stage. Thus, eliminating the chemical compound before it can be used in the household or changing chemical synthesis or product manufacturing approaches in early stages will have eliminated the exposure *and* the risk [18].

12.1.3 Emerging focus on near-field exposures

Chemical risk assessment was codified for federal agencies in 1983 by the National Research Council (NRC) of the National Academy of Sciences [19]. Air was among the media addressed. Air pollution engineers and managers may be tempted to perceive their roles as limited to laws and rules specifically targeted to address air quality. In the United States, the main statutory authority rests with the Clean Air Act and its amendments, which predominantly considers pollutants from a "far-field" perspective. That is, a substance is released, and exposure occurs after some time and within some space beyond the source. Thereafter, risk is calculated based on the toxicity of substances and the likely exposure to those substances. However, these laws and rules are not the only ones that drive risk.

Recently, the Toxic Substances Control Act and other laws have authorized greater attention to "near-field" exposure scenarios, that is, exposure occurs when a person uses a product or engages in some other activity within a home or other microenvironment.

Several NRC reports subsequent to the 1983 red book have since provided guidance on ways to assess risks of chemicals, including reports that highlighted the disparity between the rate of deployment of new anthropogenic chemicals and assessment of their potential risks to public health [20, 21]. Central to these evolving recommendations has been to replace the current practice of extensive animal-based characterization of chemical hazard, dose-response relationships, and extrapolation to human health with high-throughput in vitro tests, in silico models, and evaluations of efficacy at the human population level. In addition, human and ecosystem risk assessments require reliable approaches for exposure to these chemicals. Noting that risk is a function of both hazard and exposure, the NRC added recommendations for advancing risk-based science that underpins environmental and human health decision-making [22]. Among these recommendations is the introduction of credible ways to screen and prioritize chemical substances before these chemicals become ingredients in products and components in articles that reach the marketplace. Such an exposure-based prioritization approach will depend on the high throughput (HTP) and other tools not only that are rapidly deployed but also that are scientifically sound.

As mentioned, the Clean Air Act and its amendments have driven the selection of air pollutants of concern in the United States, which fall into two categories, that is, "criteria" and "hazardous" air pollutants. The criteria pollutants are lead (Pb), tropospheric ozone (O_3), carbon monoxide (CO), nitrogen oxides (NO_X), sulfur dioxide (SO_2), particulate matter with aerodynamic diameters of 10 μm or less (PM_{10}), and particulate matter with diameters of 2.5 μm or less ($PM_{2.5}$). The 187 hazardous pollutants include organic and inorganic compounds, including compounds of mercury (Hg), hydrochloric acid (HCl) and other acid gases, and heavy metals such as nickel and cadmium, and hazardous organic compounds such as benzene, formaldehyde, and acetaldehyde are included among these HAPs (see Table 17.1 in Chapter 17 for the complete listing).

Although this may appear to be a large number of compounds regulated under air pollution laws, it is dwarfed by other regulations, especially those regulated under the Toxic Substances Control Act (TSCA) and its amendments. Unlike the Clean Air Act, which regulates emissions, TSCA considers the potential risks from potential exposure to ingredients in yet-to-be-released products and estimated risks for products already in use. If the risks are unacceptable, new products may not be released as formulated, or the uses will be strictly limited to applications that meet minimum risk standards. For products already in the marketplace, the risks are periodically reviewed.

Another product-related development in recent years is the growth in the importance of screening and prioritizing chemicals for possible harm and exposure prior to their appearance in the marketplace. For example, research suggests a link between exposure to certain chemicals and damage to the endocrine system humans and wildlife. In the United States, the

Endocrine Disruptor Screening Program focuses on methods and procedures to detect and to characterize the endocrine activity of pesticides and other chemicals [23].

TSCA gives the EPA the authority to track thousands of industrial chemicals currently produced or imported into the United States. This is accomplished through screen of the chemicals and requiring that reporting and testing be done for any substance that presents a hazard to human health or the environment. If chemical poses a potential or actual risk that is unreasonable, the EPA may ban the manufacture and import of that chemical.

Governments in North America, Europe, and Asia track thousands of new chemicals being developed by industries each year, if those chemicals have either unknown or dangerous characteristics. This information is used to determine the type of control that would be needed to protect human health and the environment from these chemicals. Manufacturers and importers of chemical substances first submit information about chemical substances already on the market during an initial inventory. Since the initial inventory was published, commercial manufacturers or importers of substances not on the inventory have been subsequently required to submit notices to the EPA, which has developed guidance about how to identify chemical substances to assign a unique and unambiguous description of each substance for the inventory. The categories include the following:

- Polymeric substances
- Certain chemical substances containing varying carbon chain
- Products containing two or more substances, formulated and statutory mixtures
- Chemical substances of unknown or variable composition, complex reaction products, and biological materials (UVCB substance)

Historically, air pollution has mainly been concerned with so-called "far-field" exposure scenarios, that is, those where a pollutant is released and finds its way to the receptor. Indoor air pollution is a "near-field" exposure scenario in which pollutants are either generated indoors or penetrated from far-field pollutants. More recently, exposure from product use is increasingly the focus of risk assessors and researchers. As evidence, the European Commission and individual nations are strengthening their methods for screening and prioritizing chemicals based on precautions before and after these substances enter the marketplace. The US Congress recently amended TSCA to prioritize and evaluate the risks of existing chemical substances. The law contains deadlines and minimum requirements for the number of chemicals that must undergo risk evaluation and lays out a process and the criteria by which prioritization and risk evaluation must be conducted. A chemical designated as low priority indicates a risk evaluation is not warranted at that time. Final designation of a chemical or chemical category as a high priority immediately initiates the risk evaluation process [24]. TSCA requires that high-priority chemicals undergo risk evaluation to determine whether a chemical presents an unreasonable risk of injury to health or the environment, without consideration of costs or other nonrisk factors, including an unreasonable risk to a potentially exposed or susceptible subpopulation.

Meanwhile, systems sciences have been rapidly advancing across the disciplines, for example, systems biology and systems chemistry. The system approach has been employed not only in research and development but also in practice. For example, engineering and biomedicine rely less on reductionist thinking and increasingly call for translational science from one discipline to another. This requires that biological, chemical, and physical principles be meshed with the social sciences to explain why and how systems, including the human body and ecosystems, respond to stress. Engineers and physicians, of course, have long recognized that the real world cannot be reduced to the sum of its parts and that failure analysis and disease diagnosis and treatment almost always has included "black boxes." This is the quarry of system thinking, that is, explaining how these black boxes work and applying these lessons to air quality.

Of the thousands of chemicals in the marketplace, many find their way into the waste streams discussed in this book. Some chemical compounds are inherently toxic, so extremely so. Some chemicals are nearly ubiquitous. They can be found in myriad manufacturing processes, consumer products, and the environment. Some of these are particularly persistent and bioaccumulate in the environment and the food chain [25, 26].When a chemical compound is both harmful *and* likely to reach people and ecosystems, they present special challenges.

Prioritizing these chemicals based on the harm they may cause is now an international concern [27–31]. However, such prioritization is complex in that many chemicals may be toxic parent compounds that degrade into other compounds. Some of these may be even more toxic, persistent, and/or biodegradable than the parent compounds [32]. Even parent compounds that are not so toxic may degrade into toxic compounds. It must also consider the likelihood of contact with receptors. For example, a chemical used in an industrial process may be relatively safe within an industrial life stage if workers are wearing proper personal protection equipment, but in downstream, life cycle stages may become problematic. During their residence time in solid waste, wastewater or other environmental media may allow them to combine and react with other substances within these substrates to become more hazardous. In nature and engineered bioreactors, for example, landfills, microbial populations help to degrade these substances [33–36] but can also generate increasingly toxic degradation products, that is, bioactivation [37, 38].

Most chemical screening, until relatively recently, has been based on the inherent properties of chemical. The screens were built from historical data from animal and epidemiological studies, often based on pure doses. Recently, screening has been based on both hazard, especially toxicity, and exposure information. For example, exposure prioritization can complement and/or be integrated into decision tools, such as EPA's Chemistry Dashboard [39], which includes individual chemical structures for over 700,000 compounds and combines bioassay screening data, exposure modes, and product categories. Screening tools can be beneficial in identifying analytics associated with data-poor and emerging substance, for example, nanomaterials, by showing rankings of chemicals based on hazard and exposure potentials [15]. Such screening tools can also support the evaluation of a hypothetical portfolio of products (e.g., cleaning products and cosmetics) for various life stages of a product. A portfolio of products and an accompanying set of their chemical ingredients can allow decision-makers to rank products according to potential risk, including the likelihood of the formation and transformation of pollutants [40].

12.2 ACCOUNTABILITY BASED ON RISK REDUCTION AND SUSTAINABILITY

Risk is generally understood to be the likelihood that an unwelcome event will occur. Risk is also investigated scientifically within well-established frameworks. Risk analysis addresses the factors that lead to a risk. Risk management finds and implements ways to reduce this risk, for example, removing pollutants before they are emitted, changing processes to reduce the likelihood pollutants will form, or reducing exposure to air pollutants by recommending certain activities such as staying inside during episodes of ozone pollution. Risk management is often differentiated from risk assessment, which is composed of the scientific consideration hazards and exposures that make up a risk. As such, risk is a common metric for air quality management. Both traditional risk management and precaution-driven management rely on scientifically credible risk assessments to underpin air quality decisions [19].

One means of determining whether a pollutant is handled properly is if it has exceeded some level of acceptable risk. This measure of risk, then, is an expression of operational success or failure. Too much risk means the air quality management has failed society. Societal expectations of acceptable risk are mandated by the standards and specifications of certifying authorities. These include health codes and regulations, zoning and building codes and regulations, design principles, canons of professional engineering and medical practice, national standard-setting bodies, and standards promulgated by international agencies (e.g., ISO or the International Organization for Standardization). In the United States, for example, standards can come from a federal agency, such as air quality standards and guidelines of the US Environmental Protection Agency or specifications for equipment, such as those of the National Institute of Standards and Testing (NIST). Standards are also articulated by private groups and associations, such as those from ASTM (formerly the American Society for Testing and Materials).

A useful measure of risk-based accountability is whether the risk presented by an alternative is "as low as reasonably practicable" (ALARP), a concept coined by the UK Health and Safety Commission [41]. The range of possibilities fostered by this standard falls within three domains (see Fig. 12.1). In uppermost domain, the risk is clearly unacceptable. The bottom indicates generally acceptable risk. However, the size of these domains varies considerably on perspective. Reaching unanimity on the "acceptable" level of risk is nearly impossible, but consensus can be reached for many, if not most, environmental and public health management actions.

The ALARP depends on a defensible margin of safety that is both protective and reasonable. Hence, reaching ALARP necessitates qualitative and/or quantitative measures of the amount of risk reduced and costs incurred with the design decisions. The ALARP principle assumes that it is possible to compare marginal improvements in safety, for example, marginal risk decreases, with the marginal costs of the increases in reliability [41]. Quantitating the ALARP regions is challenging. For example, defining benchmark doses and thresholds for "tolerability" is fraught with subjectivity. Indeed, the quantification may have to fall into ranges of protection, for example, distance to a container or filling point and the types of structures and activities allowed in the event of fugitive emissions, explosions, or other chemical releases. The more hazardous and the greater the likelihood of exposure, the further residences, offices, and other human structures must be sited [42].

To ascertain possible risks, the first step is to identify general potential threat and the scenario(s) of events that could realize the threat and lead to an undesirable outcome. To assess the importance of a given scenario, the severity of the effect and the likelihood, that is, risk, that it will occur in that scenario are calculated. The ALARP literature sometimes uses the term "hazard" synonymously with threat; however, hazard as understood in air pollution and most environmental literature

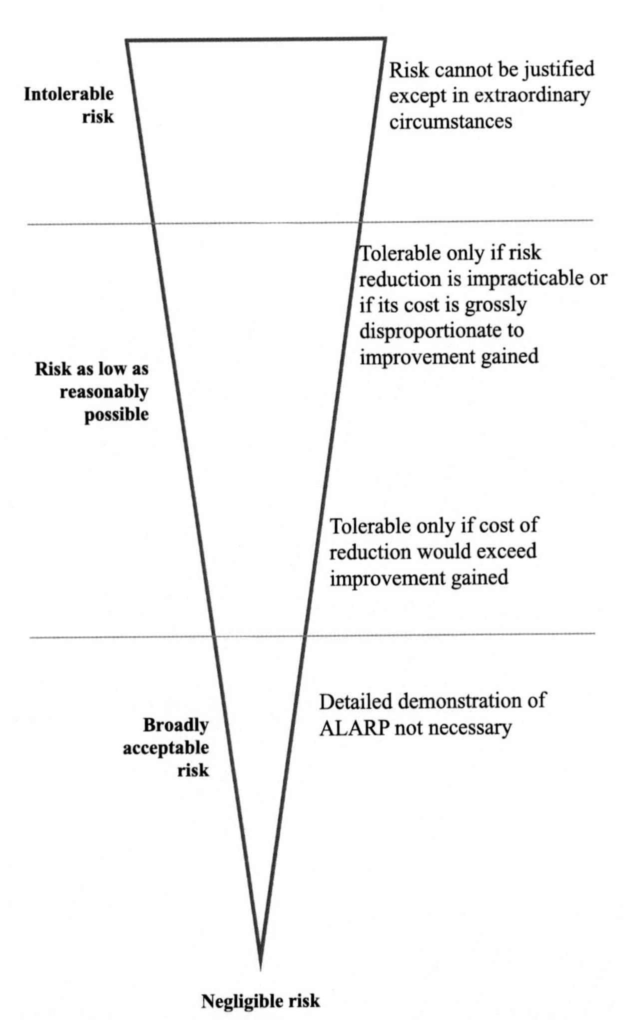

FIG. 12.1 Regions of risk tolerance. *(Modified from Health and Safety Executive, Guidance on ALARP Decisions in COMAH. (2017), 14 May 2017. Available: http://www.hse.gov.uk/foi/internalops/hid_circs/permissioning/spc_perm_37/.)*

is the inherent property of a substance, for example, its potency in producing a toxic effect such as cancer, birth defects, and the loss of ecosystem biodiversity. Sometimes, hazard is extended to abiotic harm, for example, the corrosion of metals.

ALARP is a semiquantitative risk assessment approach. Fig. 12.2 indicates the risk of fatality with distance from a gas pipeline in South America. You are responsible for approving the zoning plan for your town. Using ALARP, where would you recommend excluding residences?

The highest individual risk is immediately above or at the pipeline but is still below 10^{-5} per year, that is, 1 expected fatality per 100,000 people per year, which is the limit to risks considered acceptable in this region, that is, state of São Paulo [43]. Risks to pipelines are considered unacceptable above 10^{-4} per year, that is, 1 expected fatality per 10,000 people per year. Risks between 10^{-4} and 10^{-5} are considered negotiable, that is, the risks should be examined, and active measures taken to reduce them as much as possible, that is, ALARP.

An ALARP argument could be made to keep all development at least 110 m away from the pipeline as one of the active measures to reduce risks. However, this would assume that the data are correct and that the standard is sufficiently protective.

Fig. 12.3 compares societal risks calculated during the risk assessment phase of the implemented pipeline and the alternative route, which avoids crossing highly populated areas. According to the risk acceptance criteria adopted by the state of São Paulo, the estimated societal risks fell within the ALARP region, that is, between the "acceptable" and "unacceptable" zones and where existing risks should be actively managed and reduced where reasonably practicable. From the societal risk assessment of the alternative route, merely two changes to the pipeline routing in critical urban areas lowered risks across the whole system. This an example of combining quantitative and qualitative information in a risk assessment to inform decision-makers, in this case land use planners, of the need to lower risks.

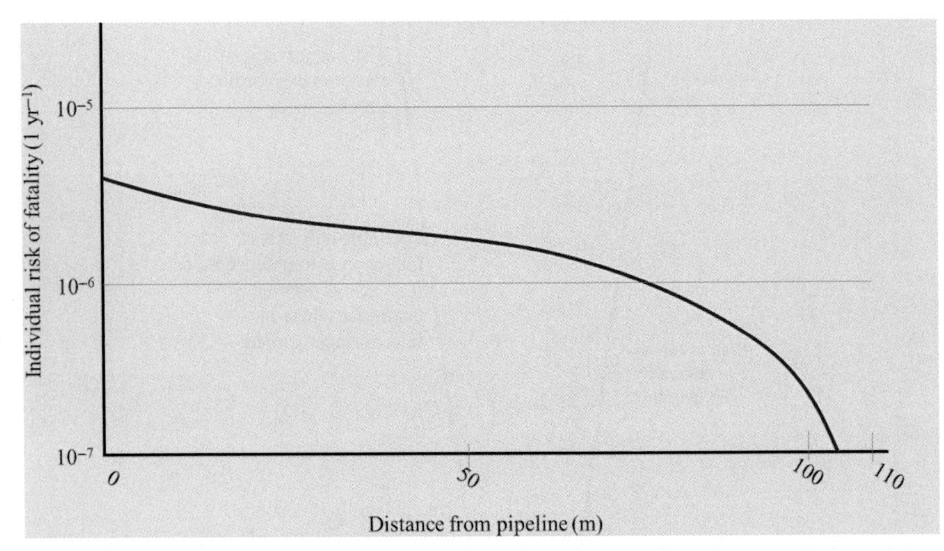

FIG. 12.2 Individual risk associated with a South American natural gas pipeline. *(Data from D. Kirchhoff, B. Doberstein, Pipeline risk assessment and risk acceptance criteria in the State of Sao Paulo, Brazil, Impact Assessment Project Appraisal, 24 (3) (2006) 221–234.)*

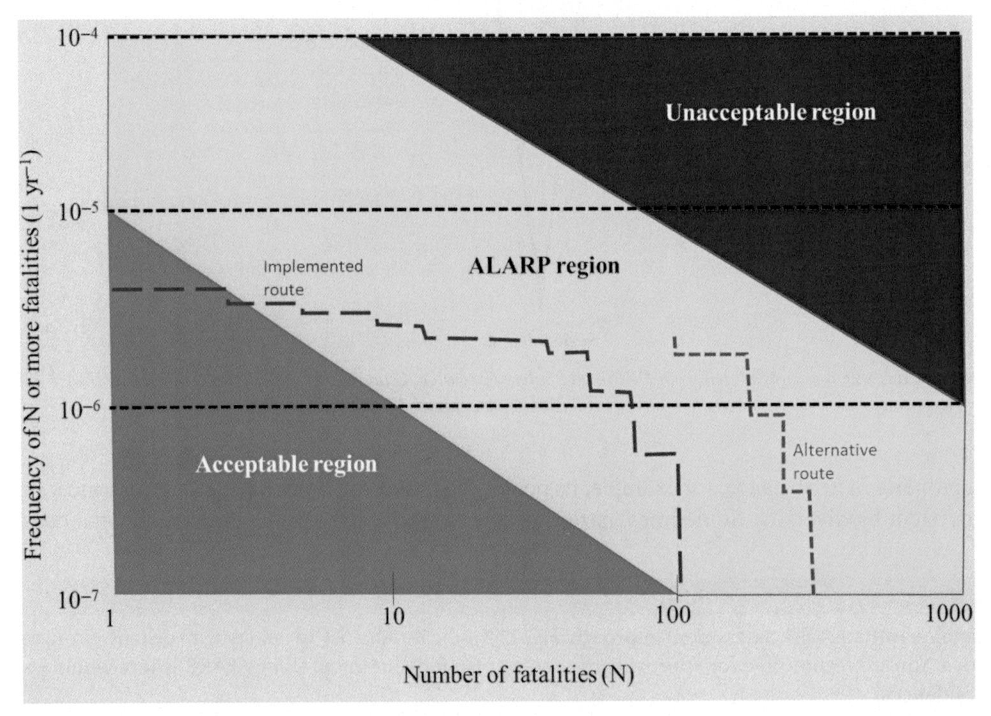

FIG. 12.3 Comparison of frequency and number of fatalities to identify the ALARP region, that is, the region in need of actions to reduce risk. In this case, the alternative route would reduce the risks substantially compared with the original pipeline locations. Note: This is not an exact depiction of the actual calculation of this real-world study, but a generalization to demonstrate an ALARP-based calculation. *(Based on the information from D. Kirchhoff, B. Doberstein, Pipeline risk assessment and risk acceptance criteria in the State of Sao Paulo, Brazil, Impact Assessment Project Appraisal, 24 (3) (2006) 221–234.)*

12.2.1 Hazard

Hazard is an inherent trait. In air pollution risk assessments, hazard is the physical, chemical, or biological agent of harm. Often, the risk assessment begins by identifying the hazard, followed by an assessment of its potency, that is, the dose-response. Thus, the hazard exists and can manifest itself in different domains of space and time. The hazard exists as soon as it is formed, such as a component of a manufacturing process. For example, if 1,1,1-trichloroethane (TCE) is used as solvent in a chemical processing plant, it is inherently hazardous to the workers because it is carcinogenic. It is also hazardous when it exits the facility via vents and stacks or through the walls.

The hazard of a substance does not change, but the risk changes depending on the exposure. Thus, the decision to use a hazardous substance is seldom made solely on its toxicity or other hazards but almost always also considers the likelihood of exposure to the hazard. Conversely, even a relatively nontoxic substance can present unacceptable risks if the potential exposure is sufficiently high.

Synthesizing substances to be less inherently hazardous, that is, green chemistry [40, 44–46], falls outside the province of most air pollution engineers and managers. Collectively, however, the environmental and public health management profession and its associations may call upon manufacturers and users of hazardous chemicals to use safer substances that provide the same function.

12.2.2 Exposure

The second component of the risk assessment is the potential of exposure to the hazard. Risk assessment combines the hazard with the receptor's contact with that hazard, that is, exposure. The types of receptors range in scale and complexity; for example, the exposed receptor may be

- an individual organism, for example, a human or other species;
- a subpopulation, for example, asthmatic children or endangered plant species in a habitat;
- an entire population, for example, all persons in a city, nation, or the world;
- a system, for example, a forest ecosystem.

In the previous TCE example, people can contact the substance in occupational and environmental settings. Thus, the exposure to TCE varies by activities (high for workers who use it, less for workers who may not work with TCE but are nearby and breathe the vapors, and even less for other workers). Usually, worker exposure is based on a 5-workday exposure (e.g., 8 or 10h), whereas environmental exposures, especially for chronic diseases like cancer, are based on lifetime, 24h per day exposures. Thus, environmental regulations are often more stringent than occupational regulations when aimed at reducing exposure to a substance.

The explanation of exposure to chemicals and other agents requires a complete understanding of the chemical space, that is, physicochemical descriptors that explain a mechanism or action, along with usage space, that is, how something is used. The integration of chemical and usage space can be thought of as exposure space.

In recent years, risk assessors have broadened the concept of exposure beyond a simple integration of chemical concentration with respect to time to one that includes biology, geography, and activity, that is, the exposome. Analogous to the genome, that is, a person's complete set of genetic material, the exposome is the person's complete biological makeup, activities, and environmental setting. A person, for example, can be exposed to physical, chemical, and biological agents differentially, depending on ethnicity; genetic predisposition and other biological makeup; lifestyle (e.g., diet, habits, and exercise); and the time spent indoors, outdoors, at work, and the activities in these locations (environment). Collectively, these constitute that person's exposome [47].

12.2.3 Air pollution risk characterization

To ascertain possible risks from substances, the first step is to identify general hazard (a potential threat) and then to develop a scenario of events that could take place to unleash the potential threat and lead to an effect. To assess the importance of a given scenario, the severity of the effect and the likelihood that it will occur in that scenario is calculated. This combination of the hazard particular to that scenario constitutes the risk.

The relationship between the severity and probability of a risk follows a general equation [48]:

$$R = f(S, P) \tag{12.3}$$

where risk (R) is a function (f) of the severity (S) and the probability (P) of harm. The risk equation can be simplified to be a product of severity and probability:

$$R = S \times P \tag{12.4}$$

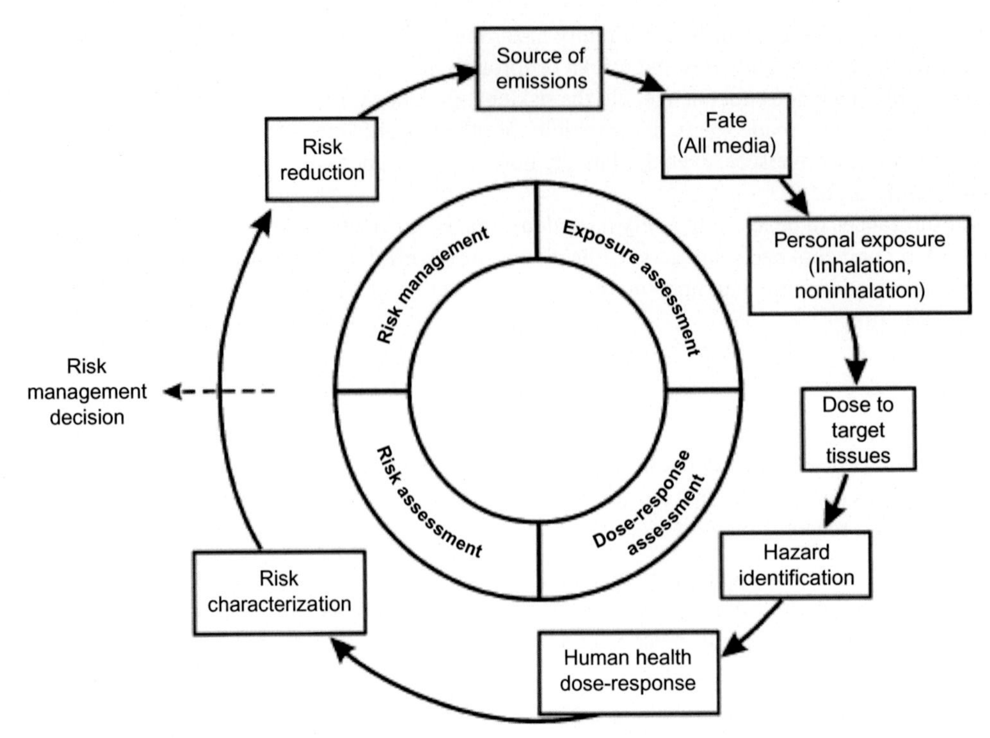

FIG. 12.4 Risk assessment and management paradigm as employed by environmental agencies in the United States. The *inner circle* includes the steps recommended by the National Research Council. *(Source: National Academy of Science, National Research Council, Risk Assessment in the Federal Government: Managing the Process. National Academy Press, Washington, DC, 1983.)*

The traditional chemical risk assessment paradigm (see Fig. 12.4) is a step-wise process. It begins with the identification of a hazard, which is composed of a summary of an agent's physicochemical properties and routes and patterns of exposure and a review of toxic effects. The tools for hazard identification account for the chemical structures that are associated with toxicity, metabolic and pharmacokinetic properties, short-term animal and cell tests, long-term animal (in vivo) testing, and human studies (e.g., epidemiology, such as longitudinal and case-control studies). These comprise the core components of hazard identification; however, additional hazard identification methods have been emerging that increasingly provide improved reliability of characterization and prediction.

Almost all environmentally important substances are mixtures. The chemical definition distinguishes mixtures from compounds. Within an environmental and public context, mixtures are combinations of constituents having unique hazards. Some of the constituents are relatively nontoxic but are the substrates and matrices within which the hazardous constituents reside. These substrates and matrices may also make the substances more dangerous. For example, relatively nonreactive substances may become bioavailable in fatty or low pH substrates. People and ecosystems are exposed to an array of compounds simultaneously [49]. Until relatively recently, toxicologists studied mixtures in a step-wise manner, adding substances one at a time to ascertain the response of an organism with each iteration. Recent toxicological studies have begun to look at multicomponent mixtures. From an exposure perspective, a mixture is a coexposure. Key coexposure considerations include how the individual constituents' physical and chemical properties affect those of other constituents and vice versa and whether receptors experience additive, synergistic, or antagonistic effects when exposed to different substances simultaneously.

Thus, characterizing the inherent properties of an air pollutant is but the first step in chemical risk assessment. A growing number of tools have emerged to assist in this characterization. Risk assessors now can apply biomarkers of genetic damage (i.e., toxicogenomics) for more immediate assessments and improved structure-activity relationships (SAR), which have incrementally been quantified in terms of stereochemistry and other chemical descriptions, that is, using quantitative structure-activity relationships (QSAR) and computational chemistry. For the most part, however, health-effects research has focused on early indicators of outcome, making it possible to shorten the time between exposure and observation of an adverse effect [50].

How does exposure to benzo(a)pyrene differ demographically?

This type of information is now available from public websites, especially the US EPA's Chemistry Dashboard. Fig. 12.5 provides the current rates of exposure to benzo(a)pyrene from a query from this dashboard [51], which derived the exposure data from heuristics [52]. Younger and thinner people are more highly exposed, as are reproductive age females. The differences are higher than they may at first appear because the exposures in milligrams per kg bodyweight per day ($mg\,kg^{-1}\,d^{-1}$) are shown as logarithms. Since benzo(a)pyrene is a polycyclic aromatic hydrocarbon (PAH) and a product of incomplete combustion (PIC), air pollution is likely to be the predominant source of these exposures.

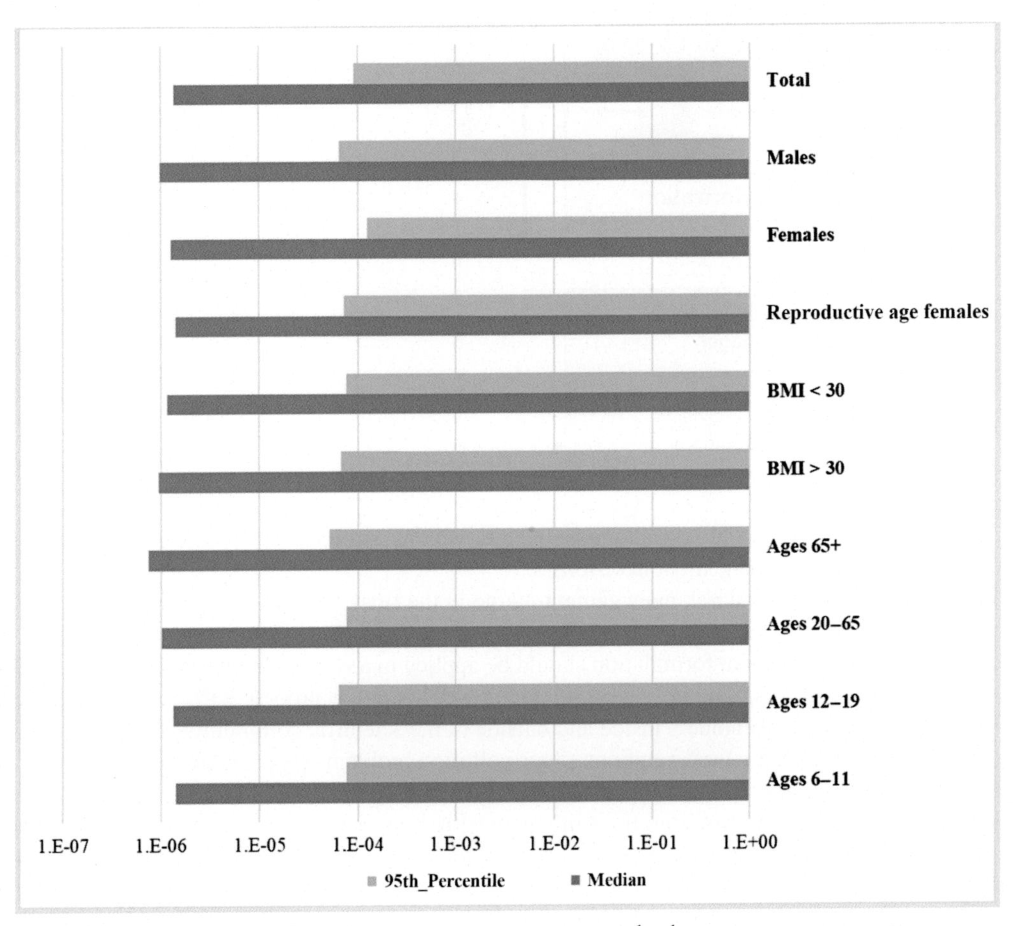

FIG. 12.5 Median and 95th percentile exposure to benzene in the United States ($mg\,kg^{-1}\,d^{-1}$). Note: Longer bar means lower exposure. *BMI*, body mass index.

Risk can be extrapolated from available knowledge to similar chemical or biological agents with similar characteristics or to yet untested but similar environmental conditions (e.g., a field study's results in one type of field extrapolated to a different agricultural or environmental remediation setting). In chemical hazard identification, this is accomplished by structural activity relationships.

In the United States, ecological risk assessment paradigms have differed from human health risk assessment paradigms. The ecological risk assessment framework (see Fig. 12.6) is based mainly on characterizing exposure and ecological effects. Both exposure and effects are considered during problem formulation [53].

Interestingly, the ecological risk framework is driving current thinking in human risk assessment. The process shown in the inner circle of Fig. 12.4 does not target the technical analysis of risk so much as it provides coherence and connections between risk assessment and risk management. In the early 1980s, risk assessment and management were conflated. A share of the criticism of federal response to environmental disasters, such as those in Love Canal, New York, and Times Beach, Missouri, belonged to the mixing of scientifically sound studies (risk assessment) with the decisions on whether to pursue certain actions (risk management) [54]. When carried out simultaneously, decision-making may be influenced by the need

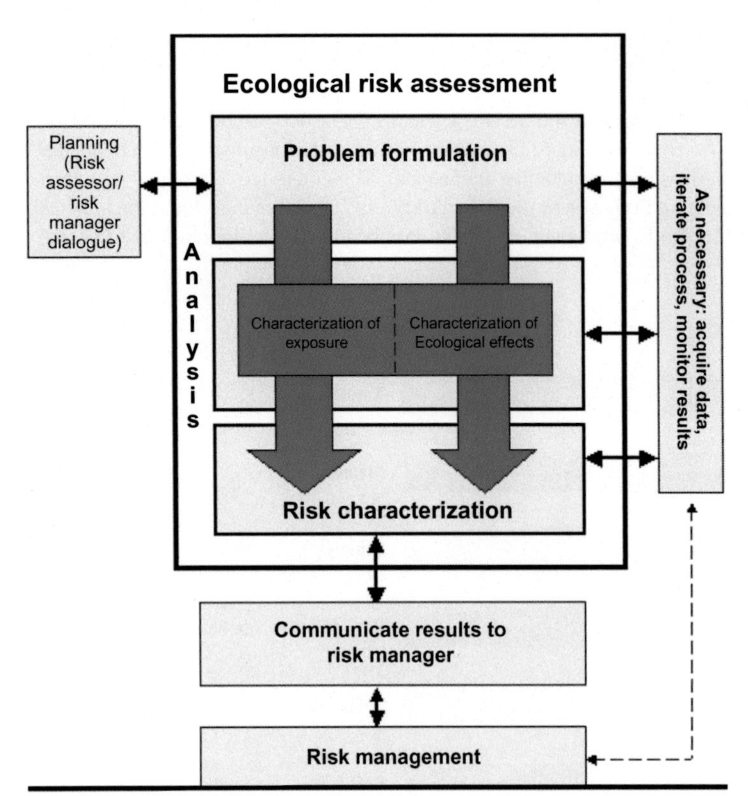

FIG. 12.6 Framework for integrated human health and ecological risk assessment. *(Reproduced with permission from U.S. Environmental Protection Agency, Guidelines for Human Exposure Assessment. (2017), 11 May 2018. Available: https://www.epa.gov/risk/guidelines-human-exposure-assessment. Adapted from US Environmental Protection Agency, Framework for Ecological Risk Assessment. USEPA Risk Assessment Forum, Washington, DC, n.d., EPA 630/R-92/0011992; World Health Organization, Report on Integrated Risk Assessment. World Health Organization, Geneva, Switzerland, 2001, WHO/IPCS/IRA/01/12; G.W. Suter II, Ecological Risk Assessment. CRC Press, 2016.)*

for immediacy, convenience, or other political and financial motivations, as opposed to a rational and scientifically credible assessment that would underpin management decisions.

At some point, risk assessment and risk management merge in the final step, risk characterization [55], which applies "both quantitative and qualitative elements of risk analysis and, of the scientific uncertainties in it, should be fully captured by the risk manager" [20]. The problem formulation should be applied in any risk characterization effort, both human and ecological, since it has the advantage of providing an analytic-deliberative process early on, since it combines sound science with input from various stakeholders inside and outside of the scientific community. This can be helpful, not only in siting facilities but also in nonstructural solutions to air pollution problems (e.g., waste minimization, changes in use scenarios, product substitution, and life cycle perspectives).

As mentioned, the term "hazard" has been used in chemical risk assessments to connote either intrinsic effects of a stressor or a margin of safety by comparing a health effect with an estimate of exposure concentration. Thus, the term becomes ambiguous when applied to most nonchemical toxic agents, such as those encountered in biological systems.

Specific investigations are needed in the laboratory and field when adverse outcomes may be substantial and small changes may lead to very different functions and behaviors from unknown and insufficiently known physical, chemical, or biological agents. For example, a chemical compound may have only been used in highly controlled experiments with little or no information about how it would behave inside of an organism. Often, the proponents of a product will have done substantial research on the benefits and operational aspects of the chemical constituents, but the regulatory agencies and the public may call for more and better information about unintended and yet-to-be-understood consequences and side effects [56].

Even when much is known, there often remain large knowledge gaps when trying to estimate environmental impacts. Physical, chemical, and biological factors can influence risk; for example, downstream effects can be even more difficult to predict than direct effects, since they occur not only within variable space but also in variable time regimes.

12.3 DEFINING RISK

Engineers design systems to reduce risk and look for ways to enhance the reliability of these systems. Every environmental manager deals directly or indirectly with risk and reliability. Air pollution control systems must keep operating rain or shine, no matter the conditions. Thus, systems must be sufficiently resilient to continue to meet the risk and reliability standards.

The discussion thus far may imply that the meaning of risk enjoys consensus. Disconcertingly for most engineers and technologists who loathe ambiguity, the term "risk" has numerous meanings that depend on the context. Risk is defined by the International Organization for Standardization (ISO), as the effect of uncertainty on objectives and that such effect can be either positive or negative [57]. The Council of Supporting Agencies has adopted a similar core definition but holds that risks are negative effects and "opportunities" are positive effects. On its face, that may not appear to be particularly important; however, the ISO 31000 definition is commonly held within the technical risk management and infrastructure project management communities, including air quality management.

The ability to communicate effectively can be hampered from the outset primarily because technical professionals often cannot communicate what risk means in terms that laypeople understand and because they may not share a common definition among themselves. Although risk has some very precise definitions within a specific scientific community, for example, biomedical or environmental engineering, the various scientific disciplines have divergent concepts of risk (see Table 12.1).

TABLE 12.1 Comparison of definitions of risk in technical publications versus social vernacular

1. Possibility of loss, injury, disadvantage, or destruction; to expose to hazard or danger; to incur risk of danger
2. An expression of possible loss over a specific period of time or number of operational cycles
3. Consequence per unit time = frequency (events per unit time) × magnitude (consequences per event)
4. Measure of the probability and severity of adverse effects
5. Conditional probability of an adverse effect (given that the necessary causative events have occurred)
6. Potential for unwanted negative consequences of an event or activity
7. Probability that a substance will produce harm under specified conditions
8. Probability of loss or injury to people and property
9. Potential for realization of unwanted, negative consequences to human life, health, or the environment
10. Product of the probability of an adverse event times the consequences of that event were it to occur
11. Function of two major factors: (a) probability that an event, or series of events of various magnitudes, will occur and (b) the consequences of the event(s)
12. Probability distribution over all possible consequences of a specific cause that can have an adverse effect on human health, property, or the environment
13. Measure of the occurrence and severity of an adverse effect to health, property, or the environment
1. Probability of an adverse event amplified or attenuated by degrees of trust, acceptance of liability, and/or share of benefit
2. Opportunity tinged with danger
3. A code word that alerts society that a change in the expected order of things is being precipitated
4. Something to worry about/have hope about
5. An arena for contending discourses over institutional relationships, sociocultural issues, and political and economic power distributions
6. A threat to sustainability/current lifestyles
7. Uncertainty
8. Part of a structure of meaning based in the security of those institutional settings in which people find themselves
9. The general means through which society envisages its future
10. Someone's judgment on expected consequences and their likelihood
11. What people define it to be—something different to different people
12. Financial loss associated with a product, system, or plant
13. The converse of safety

Source: S. Macgill, Y. Siu, A new paradigm for risk analysis, Futures, 37 (10) (2005) 1105–1131.

Managers with engineering backgrounds usually express risk as an equation, which is the multiplicative product of the consequences of failure and the likelihood of failure. Within the public health and security community, risk is often understood to be synonymous with vulnerability, which is expressed as the multiplicative product of the consequences of failure, the hazard rate, and the likelihood of experiencing the hazard. In this sense, risk is indirectly proportional to reliability. Increasing a system's reliability decreases risk. Environmental and public health risk, as mentioned, is the product of hazard and exposure [48]. There is a common thread running through all these conceptions. We want to be aware of what has happened, is happening, and may happen under various scenarios. Thus, technical professionals may argue that any of these methods define "risk." Rather, these and many other methods do not *define*, but simply *express* risk.

Engineers and environmental managers dislike uncertainty, so they extend the domain of the known domain to the domain of the lesser known. Historical data show that something failed under specified conditions and did not fail under different, specified conditions. If this is all the information with which to design, the gap between the two delineates the region of uncertainty. This is where factors of safety must be applied. Therefore, risk is the effect of uncertainty on objectives, which translates into something unwanted. Indeed, environmental and public health agencies define risk as a probability of an *adverse outcome* (e.g., mortality, morbidity, the loss of habitat, and diminished ecological diversity).

12.3.1 Modeling to support accountability

Risk management depends on models to estimate exposures. Such models range from "screening level" to "high tiered." Screening models included generally overpredict exposures because they are based on conservative default values and assumptions. They provide a first approximation that screens out exposures not likely to be of concern [30, 58–62]. Conversely, higher-tiered models typically include algorithms that provide specific site characteristics and time-activity patterns and are based on relatively realistic values and assumptions. Such models require data of higher resolution and quality than the screening models and, in return, provide more refined exposure estimates [58].

Risk involves a stressor, a receptor, and an outcome. Environmental stressors can be modeled in a unidirectional and one-dimension fashion. A conceptual framework can link exposure to environmental outcomes across levels of biological organization (see Fig. 12.7). Thus, environmental exposure and risk assessment considers coupled networks that span multiple levels of biological organization that can describe the interrelationships within the biological system. Mechanisms can be derived by characterizing and perturbing these networks (e.g., behavioral and environmental factors) [63]. This can

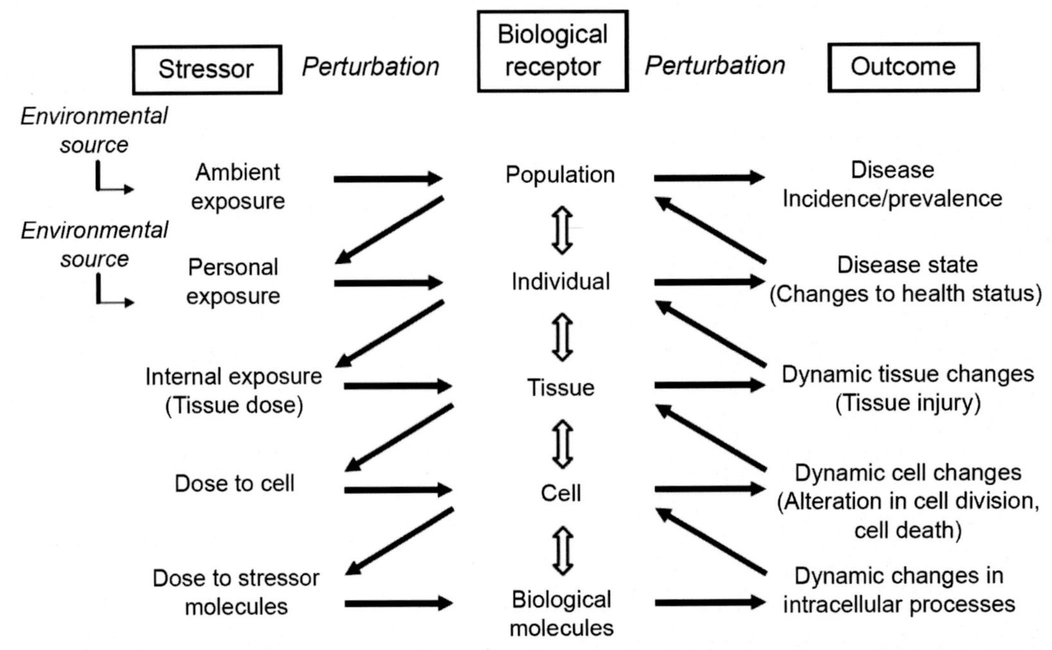

FIG. 12.7 Systems cascade of exposure-response processes. In this instance, scale and levels of biological organization are used to integrate exposure information with biological outcomes. The stressor (chemical or biological agent) moves both within and among levels of biological organization, reaching various receptors, thereby influencing and inducing outcomes. The outcome can be explained by physical, chemical, and biological processes (e.g., toxicogenomic mode of action information). *(Source: E.A.C. Hubal et al., Exposure science and the US EPA National Center for computational Toxicology, J. Expos. Sci. Environ. Epidemiol. 20 (3) (2010) 231–236.)*

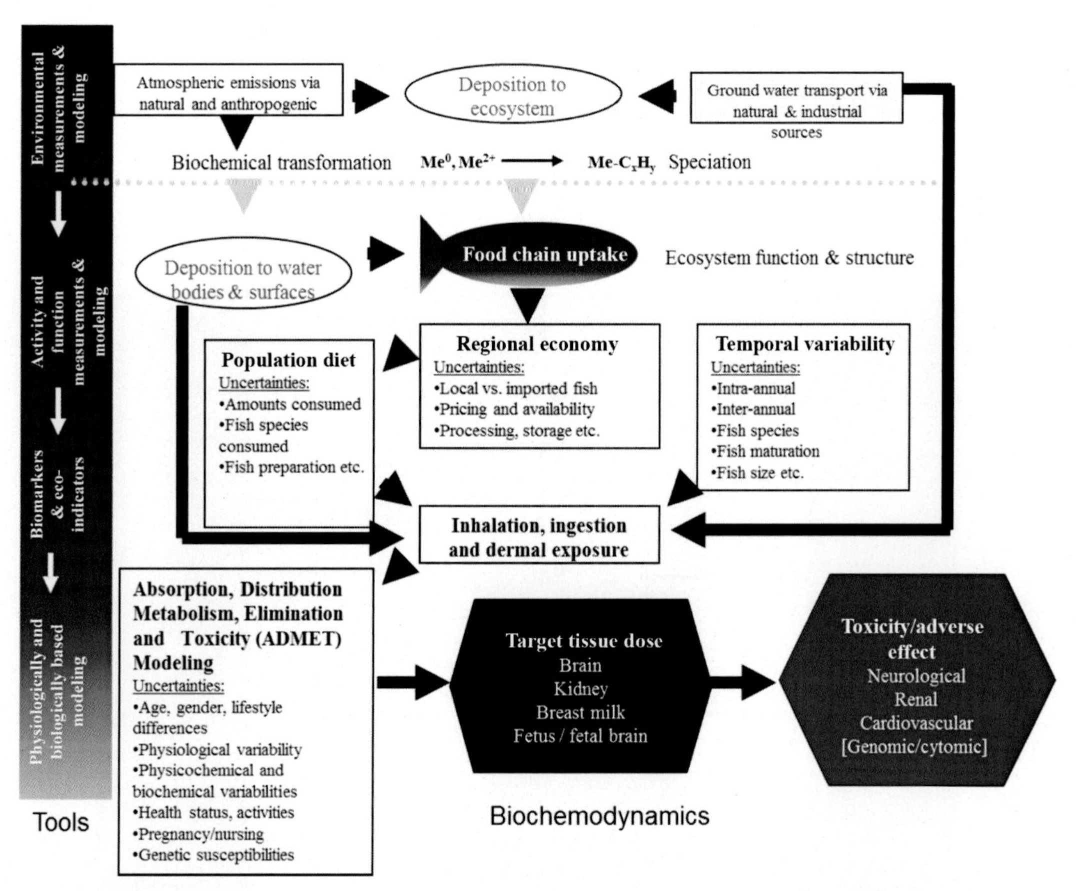

FIG. 12.8 Biochemodynamic pathways for a single, hypothetical metal (Me) in two oxidation states Me^0 and Me^{2+}, which is alkylated to $Me\text{-}C_xH_y$ within organisms. The fate is mammalian tissue. This depicts a combined human-ecosystem food web model. Various modeling tools are available to characterize the movement, transformation, uptake, and fate of the compound. Similar biochemodynamic paradigms can be constructed for multiple chemicals (e.g., mixtures) and microorganisms. *(Reproduced with permission from D. Vallero, Environmental Biotechnology: A Biosystems Approach, Elsevier Science, 2015; D. Vallero, Discussion of Exposure Within the Context of Food Chains, in: D. Mangis (Ed)., Research Triangle Park, NC, 2005.)*

apply to a food chain or food web model (see Fig. 12.8) or a model within an organism, for example, a human kinetic model (see Fig. 12.9), or numerous other modeling platforms.

To assess the risks associated with an air quality management decision, three questions must be asked [48]:

1. What are the specific environmental concerns or harm that will or can occur?
2. What is the probability that the concerns will be realized or that harm will occur?
3. What are the adverse outcomes (e.g., to health and the environment) when the harm occurs, including how widespread in time and space?

The three questions can begin with an analysis of existing data, that is, by observing patterns and analogies from events that are similar to the potential threat being considered. From there, scenarios can be developed to follow various paths to beneficial, adverse, and indifferent outcomes. This decision tree provides an estimate of the importance of each scenario and selecting the one with the most acceptable risk. This could involve developing a benefit-to-risk ratio or relationship or benefit-to-cost ratio or relationship. The challenge is how to quantify many of the benefits and risks, since risk is a function of likelihood and severity of an adverse outcome from that hazard (Eq. 12.4). Of all the environmental hazards; the most attention has been devoted to toxicity. Other important environmental hazards are shown in Table 12.2. Hazards can be expressed according to the physical and chemical characteristics, as in Table 12.2, and in the ways they may affect living things. For example, Table 12.3 summarizes some of the expressions of biologically based criteria of hazards. Other hazards, such as flammability, are also important to environmental engineering. However, the chief hazard in most environmental situations has been toxicity.

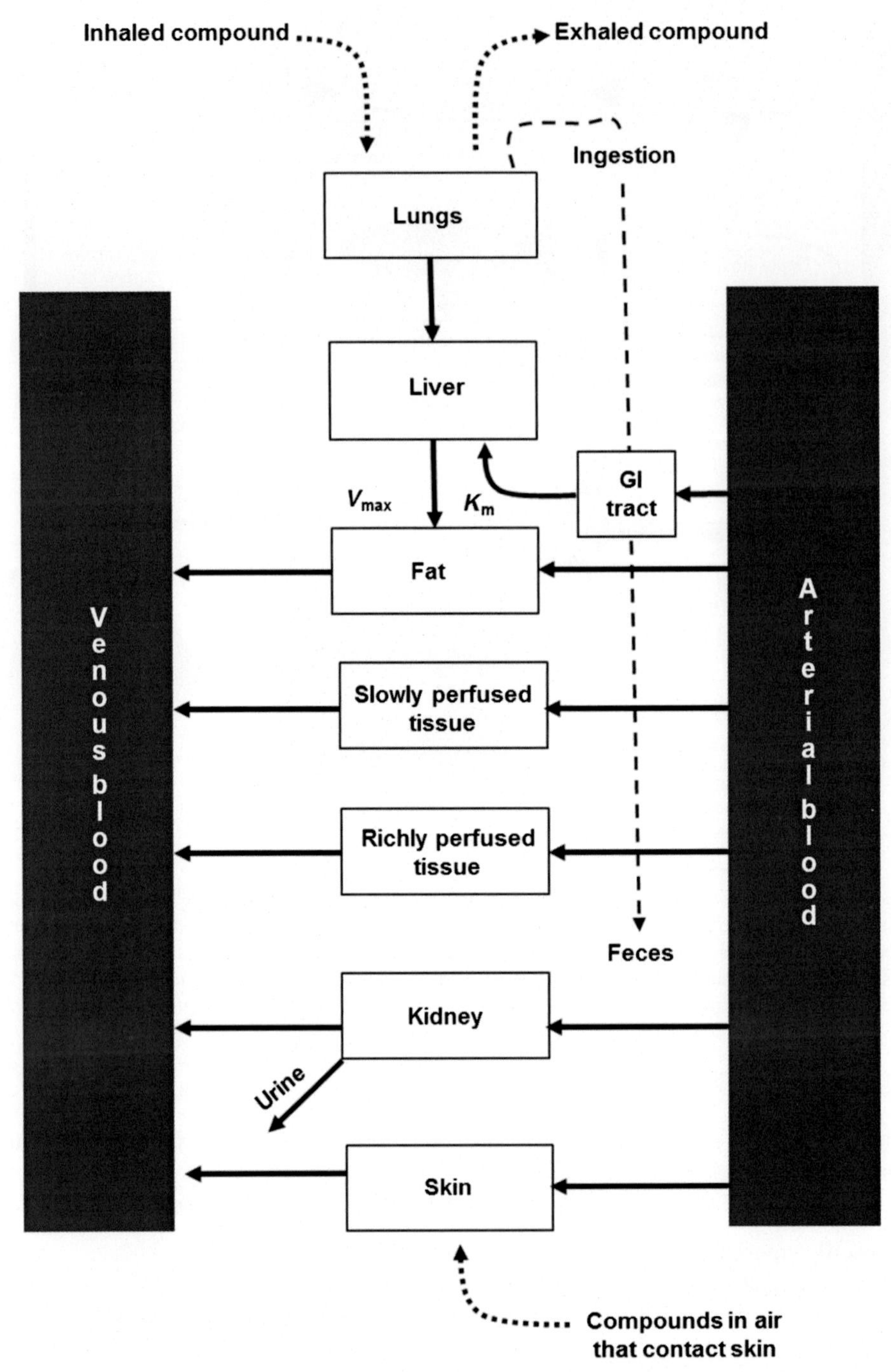

FIG. 12.9 Toxicokinetic model used to estimate dose as part of an environmental exposure. This diagram represents the static lung, with each of the compartments (brain, carcass, fat, kidney, liver, lung tissue, rapidly and slowly perfused tissues, spleen, and the static lung) having two forms of elimination, an equilibrium binding process, and numerous metabolites. Notes: V_{max} is the maximum velocity at which the enzyme catalyzes a reaction, and K_m is the substrate concentration that is required for the reaction to occur at one-half of V_{max}. *(Reproduced with permission from D. Vallero, Environmental Biotechnology: A Biosystems Approach, Elsevier Science, 2015; modified from D.A. Vallero, V.G. Zartarian, S.I. Isukapalli, T.R. Mccurdy, T. Mckone, P. Georgopoulos, C.C. Dary, Modeling and predicting pesticide exposures, in: R. Krieger (Ed), Hayes Handbook of Pesticide Toxicology, Elsevier Science, New York, NY, 2010, pp. 995–1020; X. Zhang et al., A physiologically based pharmacokinetic/pharmacodynamic model for carbofuran in Sprague-Dawley rats using the exposure-related dose estimating model, Toxicol. Sci. 100 (2) (2007) 345–359.)*

TABLE 12.2 Hazards defined by the Resource Conservation and Recovery Act

Hazard type	Criteria	Physical/chemical classes in definition
Corrosivity	A substance with an ability to destroy tissue by chemical reactions	Acids, bases, and salts of strong acids and strong bases. The waste dissolves metals, other materials, or burns the skin. Examples include rust removers, waste acid, alkaline cleaning fluids, and waste battery fluids. Corrosive wastes have a pH of <2.0 or >12.5. The US EPA waste code for corrosive wastes is "D002"
Ignitability	A substance that readily oxidizes by burning	Any substance that spontaneously combusts at 54.3°C in air or at any temperature in water, or any strong oxidizer. Examples are paint and coating wastes, some degreasers, and other solvents. The US EPA waste code for ignitable wastes is "D001"
Reactivity	A substance that can react, detonate, or decompose explosively at environmental temperatures and pressures	A reaction usually requires a strong initiator (e.g., an explosive like TNT and trinitrotoluene), confined heat (e.g., salt peter in gunpowder), or explosive reactions with water (e.g., Na). A reactive waste is unstable and can rapidly or violently react with water or other substances. Examples include wastes from cyanide-based plating operations, bleaches, waste oxidizers, and waste explosives. The US EPA waste code for reactive wastes is "D003"
Toxicity	A substance that causes harm to organisms. Acutely toxic substances elicit harm soon after exposure (e.g., highly toxic pesticides causing neurological damage within hours after exposure). Chronically toxic substances elicit harm after a long period of time of exposure (e.g., carcinogens, immunosuppressants, endocrine disruptors, and chronic neurotoxins)	Toxic chemicals include pesticides, heavy metals, and mobile or volatile compounds that migrate readily, as determined by the toxicity characteristic leaching procedure (TCLP), or a "TC waste." TC wastes are designated with a waste codes "D004" through "D043"

From D. Vallero, Environmental Biotechnology: A Biosystems Approach, Elsevier Science, 2015.

TABLE 12.3 Biologically based classification criteria for chemical substances

Criterion	Description
Bioconcentration	The process by which living organisms concentrate a chemical contaminant to levels exceeding the surrounding environmental media (e.g., water, air, soil, or sediment)
Lethal dose (LD)	A dose of a contaminant calculated to expect a certain percentage of a population of an organism (e.g., minnow) exposed through a route other than respiration (dose units are mg (contaminant) kg^{-1} body weight). The most common metric from a bioassay is the lethal dose 50 (LD_{50}), wherein 50% of a population exposed to a contaminant is killed
Lethal concentration (LC)	A calculated concentration of a contaminant in the air that, when respired for four hours (i.e., exposure duration = 4 h) by a population of an organism (e.g., rat), will kill a certain percentage of that population. The most common metric from a bioassay is the lethal concentration 50 (LC_{50}), wherein 50% of a population exposed to a contaminant is killed. (Air concentration units are mg (contaminant) L^{-1} air)

Source: United Nations Environment Programme, Rio declaration on environment and development, United Nations Environment Programme, Nairobi, Kenya, 1992.

12.4 DOSE-RESPONSE CURVES

The first means of determining exposure is to identify *dose*, the amount (e.g., mass) of a contaminant that comes into contact with an organism. The applied dose is the amount administered to an organism. The internal dose is the amount of the contaminant that enters the organism. The absorbed dose is the amount of the contaminant that is absorbed by an organism over a certain time interval. The biologically effective dose is the amount of the contaminants or its metabolites that reaches a particular "target" organ, such as the amount of a hepatotoxin (a chemical that harms the liver) that finds its way to liver cells or a neurotoxin (a chemical that harms the nervous system) that reaches the nerve or other nervous system cells. Theoretically, the higher the concentration of a hazardous substance or microbe that contacts an organism, the greater the response will be, that is, the expected adverse outcome. The pharmacological and toxicological gradient is the so-called "dose-response" curve (Fig. 12.10). Generally, increasing the amount of the dose means a greater incidence of the adverse outcome.

Dose-response assessment generally follows a sequence of five steps [64]:

1. Fitting the experimental dose-response data from animal and human studies with a mathematical model that fits the data reasonably well
2. Expressing the upper confidence limit (e.g., 95%) line equation for the selected mathematical model
3. Extrapolating the confidence limit line to a response point just below the lowest measured response in the experimental point (known as the "point of departure"), that is, the beginning of the extrapolation to lower doses from actual measurements
4. Assuming the response is a linear function of dose from the point of departure to zero response at zero dose
5. Calculating the dose on the line that is estimated to produce the response

The risk assessor can use published physical and chemical hazard characteristics for all the chemicals used in the life cycle of a process. This may not be completely applicable to all air pollutants. For example, if an airborne microbe is harmful to particular type of cell (e.g., a nerve), it may follow the steps just as a neurotoxic chemical. However, if the microbial modifications change microbial populations in an organism or in an ecosystem, the dose-response may become much more complex than a single, abiotic chemical hazard.

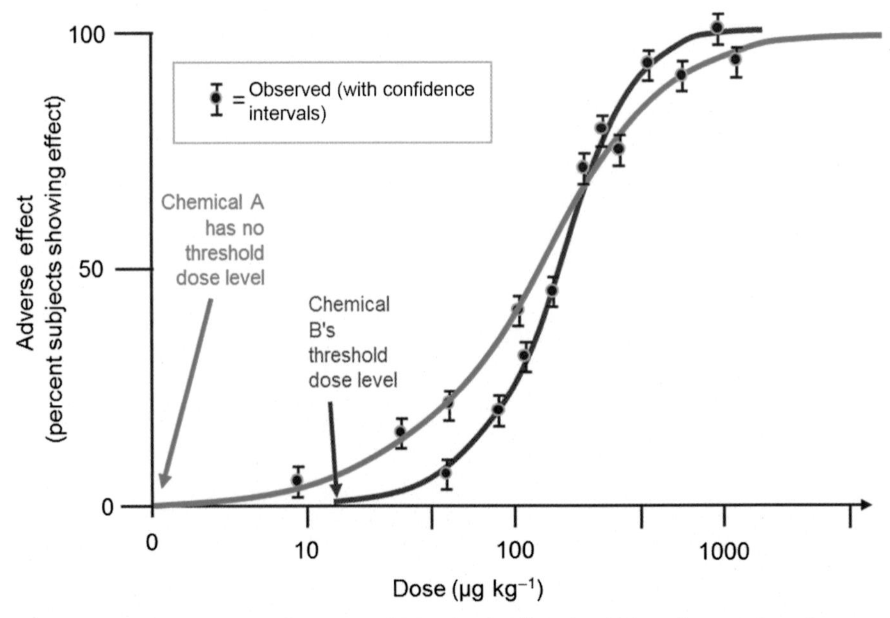

FIG. 12.10 Prototypical dose-response curves. Curve A represents the "no-threshold" curve, which predicts a response (e.g., cancer) even if exposed to a single molecule (one-hit model). As shown, the low-dose end of the curve, that is, below which experimental data are available, is linear. Thus, Curve A represents a linearized multistage model. Curve B represents toxicity above a certain threshold, that is, the no observable adverse effect level (NOAEL) or, if inhaled, the no observable adverse effect concentration (NOAEC). These are the levels or concentrations below which no response is expected. Another threshold is the no observable effect concentration (NOEC), which is the highest concentration where no effect on survival is observed ($NOEC_{survival}$) or where no effect on growth or reproduction is observed ($NOEC_{growth}$). Note that both curves are sigmoidal in shape because of the saturation effect at high dose (i.e., less response with increasing dose). *(Modified from D. Vallero, Environmental Contaminants: Assessment and Control. Academic Press, 2010.)*

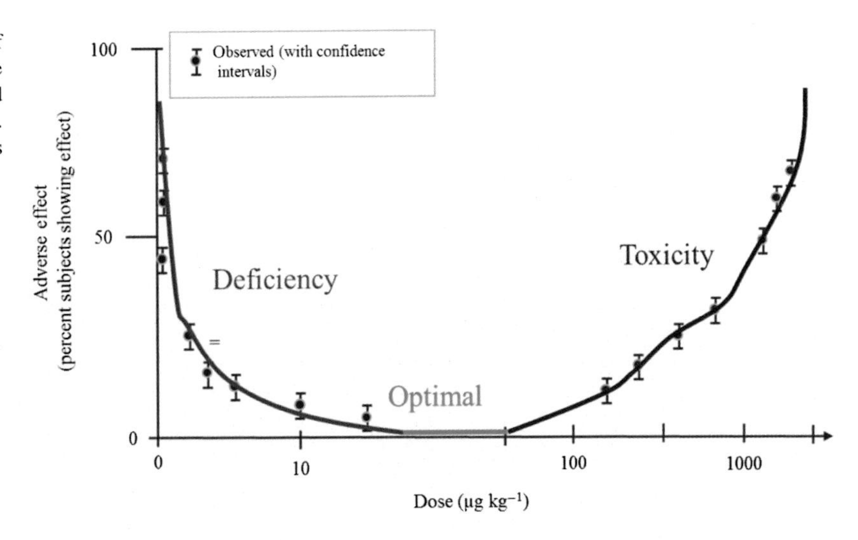

FIG. 12.11 U-shaped curve with dose regions of deficiency and toxicity. Between them is the optimal range, that is, sufficiently high to avoid deficiency and sufficiently low to avoid toxicity. This applies to nutrients and vitamins, as well as some metals, for example, trivalent chromium.

The curves in Fig. 12.10 represent those generally found for toxic chemicals [16]. Once a substance is suspected of being toxic, the extent and quantification of that hazard is assessed. Another curve could be shown for essential compounds like vitamins and certain metallic compounds. In such a curve (Fig. 12.11), the left-hand side (low dose or low exposure) of the curve would represent deficiency and the right-hand side (high dose or exposure) would represent toxicity, with an optimal, healthy range between these two adverse responses.

This step is frequently referred to as a dose-response evaluation because this is when researchers study the relationship between the mass or concentration (i.e., dose) and the damage caused (i.e., response). Many dose-response studies are ascertained from animal studies (in vivo toxicological studies), but they may also be inferred from studies of human populations (epidemiology). To some degree, "petri dish" (i.e., in vitro) studies, such as mutagenicity studies like the Ames test [17] of bacteria, complement dose-response assessments but are mainly used for screening and qualitative or, at best, semi-quantitative analysis of responses to substances. The actual name of the test is the "Ames *Salmonella*/microsome mutagenicity assay," which shows the short-term reverse mutation in histidine-dependent *Salmonella* strains of bacteria. Its main use is to screen for a broad range of chemicals that induce genetic aberrations leading to genetic mutations. The process works by using a culture that only allows those bacteria whose genes revert to histidine interdependence to form colonies. As a mutagenic chemical is added to the culture, a biological gradient can usually be determined. That is, the greater the amount of the chemical that is added, the greater the number of microbes and the larger the size of colonies on the plate. The test is widely used to screen for mutagenicity of new or modified chemicals and mixtures. It is also a "red flag" for carcinogenicity, since cancer is a genetic disease and a manifestation of mutations.

The toxicity criteria include both acute and chronic effects and include both human and ecosystem effects. These criteria can be quantitative. For example, a manufacturer of a new chemical may have to show that there are no toxic effects in fish exposed to concentrations below $10\,\text{mg}\,\text{L}^{-1}$. If fish show effects at $9\,\text{mg}\,\text{L}^{-1}$, the new chemical would be considered to be toxic.

A contaminant is acutely toxic if it can cause damage with only a few doses. Chronic toxicity occurs when a person or ecosystem is exposed to a contaminant over a protracted period of time, with repeated exposures. The essential indication of toxicity is the dose-response curve. The curves in Fig. 12.10 are sigmoidal because toxicity is often concentration-dependent. As the doses increase, the response cannot mathematically stay linear (e.g., the toxic effect cannot double with each doubling of the dose). So, the toxic effect continues to increase but at a decreasing rate (i.e., decreasing slope). Curve A is the classic cancer dose-response, that is, any amount of exposure to a cancer-causing agent may result in an expression of cancer at the cellular level (i.e., no safe level of exposure). Thus, the curve intercepts the x-axis at 0.

Curve B is the classic noncancer dose-response curve. The steepness of the three curves represents the potency or severity of the toxicity. For example, Curve B is steeper than Curve A, so the adverse outcome (disease) caused by chemical in Curve B is more potent than that of the chemical in Curve A. Obviously, potency is only one factor in the risk. For example, a chemical may be very potent in its ability to elicit a rather innocuous effect, like a headache, and another chemical may have a rather gentle slope (lower potency) for a dreaded disease like cancer.

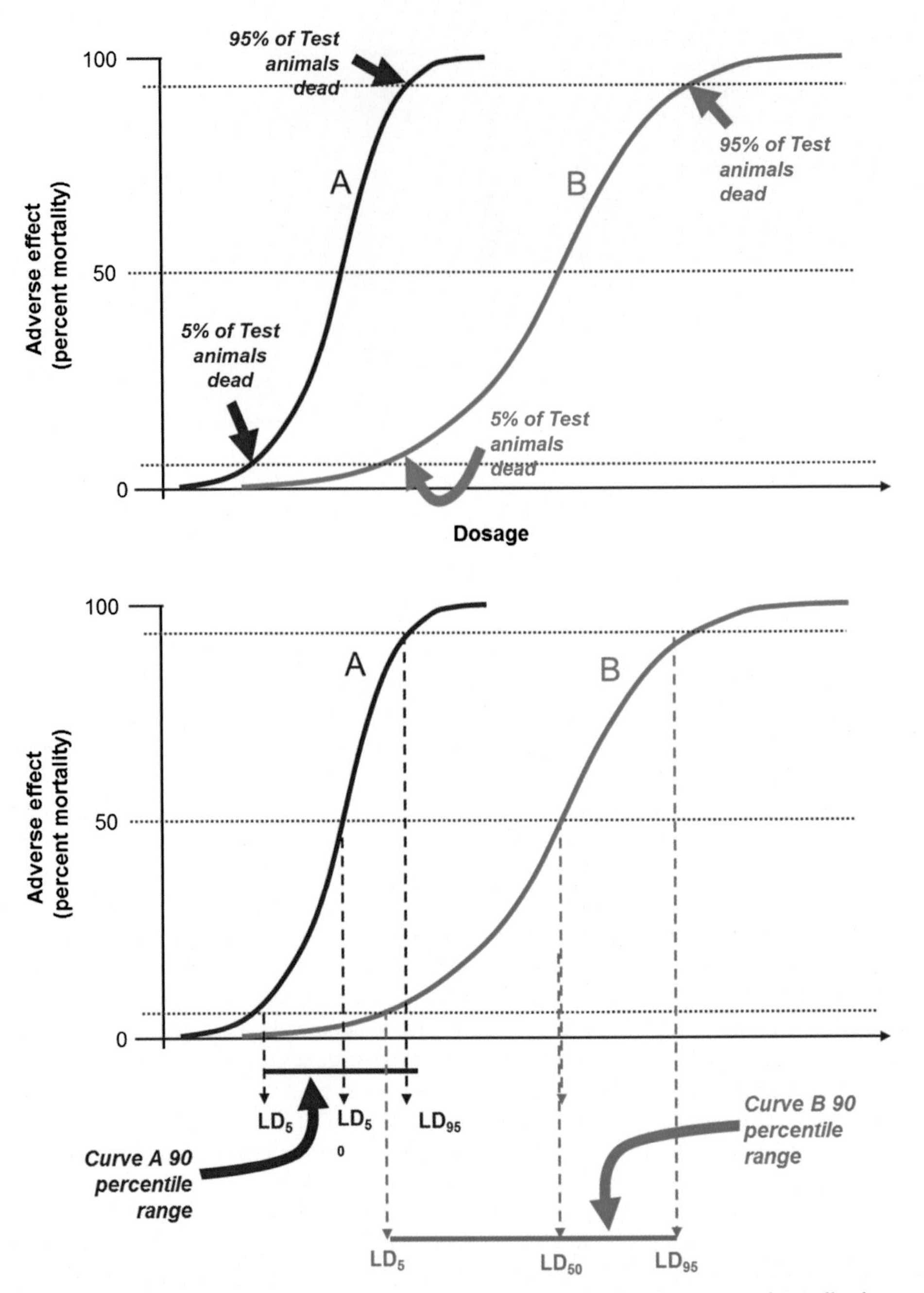

FIG. 12.12 The greater the potency or severity of response (i.e., steepness of the slope) of dose-response curve, the smaller the range of toxic response (90 percentile range shown in bottom graph). Also, note that both curves have thresholds and that Curve B is less acutely toxic based upon all three reported lethal doses (LD_5, LD_{50}, and LD_{95}). In fact, the LD_5 for Curve A is nearly the same as the LD_{50} for Curve B, meaning that about the same dose, contaminant A kills nearly half the test animals but contaminant B has only killed 5%. Thus, contaminant A is much more acutely toxic. *(Source: D. Vallero, Translating Diverse Environmental Data into Reliable Information: How to Coordinate Evidence from Different Sources. Academic Press, 2017.)*

With increasing potency, the range of response decreases. As shown in Fig. 12.12, a severe response represented by a steep curve will be manifested in greater mortality or morbidity over a smaller range of dose. For example, an acutely toxic contaminant's dose that kills 50% of test animals (i.e., the LD_{50}) is closer to the dose that kills only 5% (LD_5) and the dose that kills 95% (LD_{95}) of the animals. The dose difference of a less acutely toxic contaminant will cover a broader range, with the differences between the LD_{50} and LD_5 and LD_{95} being more extended than that of the more acutely toxic substance.

12.5 UNCERTAINTY AND FACTORS OF SAFETY

Air quality management is an endeavor steeped in uncertainty. The amount and type of substances that need to be handled change in space and time. The adopted practices vary. What we know, or believe we know, is never 100% certain. One definition of risks embraced by many engineers is that it is the effect of uncertainty on our ability to meet our objectives. In this instance, the "effect" is the a deviation from the "expected" outcome [6]. Of course, this begs the question, who is expecting the outcome? Policy makers and the public may care most about traffic flow or availability of affordable products and simply "assume" that the air quality manager is ensuring that the environment and public health are protected.

Air quality decisions fall into two broad categories [65, 66]. The first is a decision under risk and the second a decision under uncertainty. This is but another example of a twist on the term "risk," since a decision under risk is one where the probability of all possible outcomes is known. When these outcomes cannot be known or are only incompletely known, the decision is said to be under risk. Optimizing for best approaches to handling a substance depends on the amount and quality of available data to construct an event tree, that is, probability that each possible outcome can be drawn (see Fig. 12.13). Often, optimization is less about selecting the "good" versus "bad" options but about the "good enough, with fewer side effects" versus "not good enough, with too many side effects." Typically, even the best alternatives can come with additional costs and risks. For example, increasing the desired beneficial outcomes can simultaneously increase detrimental outcomes (see Fig. 12.14). Obviously, identifying every contributing event is an impossible task for all but the simplest decisions; so, such event trees often include very broad assimilations of possible outcomes. Even if there is a wealth of information about possible outcomes, these outcomes occurred in the past. None of the scenarios and contingencies that led to an outcome will ever again occur in the same way.

Risk management uncertainty results from temporal, spatial, and operational variability and the lack of knowledge [67]. Perhaps, more than most scientific venues, the factors that lead to environmental risk are highly variable, given the large number of habitats, sources of pollution, diseases in populations, geographic diversity, and many other environmental circumstances. For example, within a square meter of soil, the microbial populations, soil texture, organic matter, and chemical makeup can be highly variable. Extending this to the effect of dredging near a 10 ha wetland or trying to determine the impact of a leaking underground gasoline tank propagates these uncertainties. This is compounded by the measurement imprecision, voids, and gaps in observations, practical obstacles (e.g., no access to private property and physical barriers),

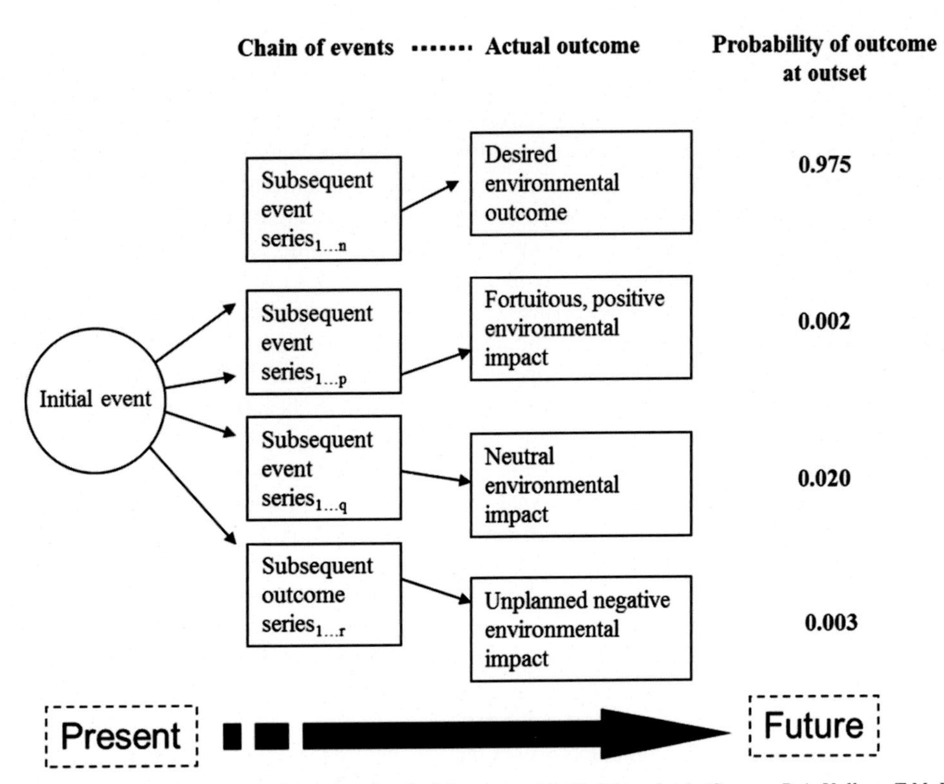

FIG. 12.13 Hypothetical example of an event tree representing decision under risk (fictitious data). *(Source: D.A. Vallero, T.M. Letcher, Unraveling Environmental Disasters. Newnes, 2012.)*

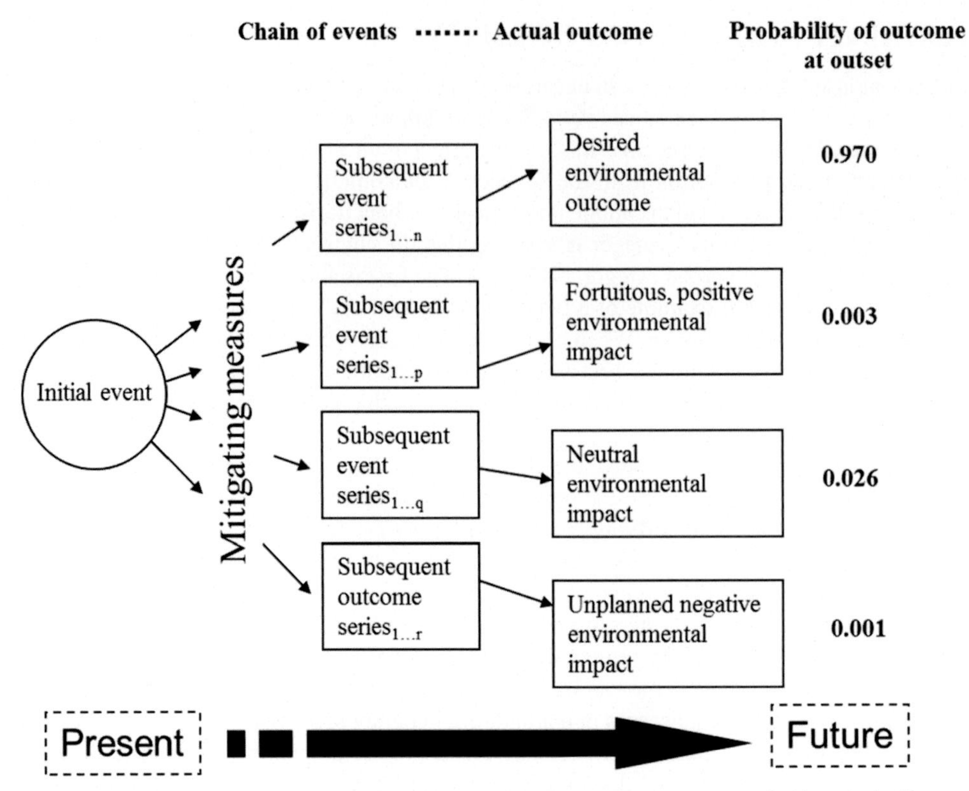

FIG. 12.14 Event tree where probabilities in Fig. 12.13 increase for both desirable and adverse outcomes (fictitious data). *(Source: D.A. Vallero, T.M. Letcher, Unraveling Environmental Disasters. Newnes, 2012.)*

the lack of consistency in measured and modeled results, and simply not being able to understand what the information means, that is, structural uncertainty [68].

Environmental and health risk distinguish carcinogenic from noncancer outcomes. The term "noncancer" is commonly used to separate cancer outcomes (e.g., bladder cancer, leukemia, or adenocarcinoma of the lung) from other maladies, such as neurotoxicity, immune system disorders, and endocrine disruption. The policies of many regulatory agencies and international organizations treat cancer differently than noncancer effects, particularly in how the dose-response curves are drawn. As discussed in the dose-response curves, there is no safe dose for carcinogens. Cancer dose-response is almost always a nonthreshold curve, that is, no safe dose is expected (Curve A in Fig. 12.10). Theoretically at least, noncancer outcomes can have a dose below which the adverse outcomes do not present themselves (Curve B in Fig. 12.10). So, for all other diseases, safe doses of compounds can be established. These are known as reference doses (RfD), usually based on the oral exposure route. If the substance is an air pollutant, the safe dose is known as the reference concentration (RfC), which is calculated in the same manner as the RfD, using units that apply to air (e.g., $\mu g\ m^{-3}$). These references are calculated from thresholds below which no adverse effect is observed in animal and human studies. If the models and data were perfect, the safe level would be the threshold, known as the no observed adverse effect level (NOAEL) or concentration (NOAEC).

The term "noncancer" has a completely different meaning than the term "anticancer" or "anticarcinogens." Anticancer procedures that include radiation and drugs are those that are used to attack tumor cells. Anticarcinogens are chemical substances that work against the processes that lead to cancer, such as antioxidants and essential substances that help the body's immune, hormonal, and other systems to prevent carcinogenesis.

Any hazard identification or dose-response research is never perfect, which means that the data derived from these investigations are often beset with various forms of uncertainty. Chief reasons for this uncertainty include variability among the animals and people being tested and differences in response to the compound by different species (e.g., one species may have decreased adrenal gland activity, while another may show thyroid effects). Whereas this is usually associated with chemical risk, these uncertainties can also be part of microbial data sets. For example, certain immunocompromised subpopulations may respond adversely to microbial exposures that are below thresholds for the general population.

Sometimes, studies only indicate the lowest concentration of a contaminant that causes the effect, that is, the lowest observed adverse effect level (LOAEL) or concentration (LOAEC), but the actual threshold is unknown. If the LOAEC is used, one is less certain how close this is to a safe level where no effect is expected. Often, there is temporal incongruence, such as most of the studies taking place in a shorter timeframe than in the real world. Thus, in lieu of long-term human studies, hazards and risks may have to be extrapolated from acute or subchronic studies of the same or similar agents. Likewise, routes and pathways of exposure used to administer the agent to subjects may differ from the likely real-world exposures. For example, if the dose of substance in a research study is administered orally but the pollutant is more likely to be inhaled by humans, this route-to-route extrapolation adds uncertainty. This is particularly problematic for microbial exposures; for example, inhalational anthrax (*Bacillus anthracis*) is more virulent in human populations than is ingestional anthrax.

Finally, the hazard and exposure data themselves may be weak because the studies from which they have been gathered lack sufficient quality or the precision, accuracy, completeness, and representativeness, or they may not be directly relevant to the risk assessment at hand.

The factors underlying the uncertainties are quantified as specific uncertainty factors (UFs). The uncertainties in the RfD are largely due to the differences between results found in animal testing and expected outcomes in human population. As in other bioengineering operations, a factor of safety must be added to calculations to account for UFs. So, for environmental risk analyses and assessments, the safe level is expressed in the RfD or in air the RfC. This is the dose or concentration below which regulatory agencies do not expect a specific unacceptable outcome. Thus, all the uncertainty factors adjust the actual measured levels of no effect (i.e., the threshold values, e.g., NOAELs and LOAELs) in the direction of a zero concentration. This is calculated as

$$\text{RfD} = \frac{\text{NOAEL}}{\text{UF}_{\text{inter}} \times \text{UF}_{\text{intra}} \times \text{UF}_{\text{iother}}}. \tag{12.5}$$

The calculation of the RfC is identical:

$$\text{RfC} = \frac{\text{NOAEC}}{\text{U}_{\text{inter}} \times \text{U}_{\text{intra}} \times \text{U}_{\text{other}}} \tag{12.6}$$

The first of the three types of uncertainty are those resulting from the difference in the species tested and that of *Homo sapiens* (UF_{inter}). Humans may differ from other species in their sensitivity to substances, which means their response to be exposed to the compound will differ from that of the tested species. This difference is reflected in UF_{inter}. The second uncertainty factor is due to the fact that certain human subpopulations are more sensitive to the effects of a compound than the general human population. These are known as intraspecies uncertainty factors (UF_{intra}). The third type of uncertainties (UF_{other}) results when the available data and science are lacking, such as when a LOAEL is used rather than a NOAEL. That is, data show a dose at which an effect is observed, but the "no effect" threshold must be extrapolated. Since the UFs are in the denominator, the greater the uncertainties, the closer the safe level (i.e., the RfD) is to zero, that is, the threshold is divided by these factors. The UFs are usually multiples of 10, although the UF_{other} can range from 2 to 10.

A particularly sensitive subpopulation is children, since they are growing and tissue development is much more prolific than in older years. To address these sensitivities, the Food Quality Protection Act (FQPA) now includes what is known as the "10 ×" rule. This rule requires that the RfD for products regulated under FQPS, for example, pesticides, must include an additional factor of 10 of protection of infants, children, and females between the ages of 13 and 50 years old. This factor is included in the RfD denominator along with the other three UF values. The RfD that includes the UFs and the 10 × protection is known as the population adjusted dose (PAD). A risk estimate that is $<100\%$ of the acute or chronic PAD does not exceed the agency's risk concern.

What is the reference dose for chlorpyrifos?

The US EPA's decision-making regarding the reregistration of the organophosphate pesticide, chlorpyrifos, demonstrates the RfD process. The acute dietary scenario had a NOAEL of $0.5\,\text{mg}\,\text{kg}^{-1}\,\text{d}^{-1}$, and the three UF values equaled 100.

Thus, the acute RfD $= 5 \times 10^{-3}$ mg kg^{-1} d^{-1}.

However, the more protective acute PAD $= 5 \times 10^{-4}$ mg kg^{-1} d^{-1}.

The chronic dietary scenario is even more protective, since the exposure is long term. The chronic NOAEL was found to be $0.03\,\text{mg}\,\text{kg}^{-1}\,\text{d}^{-1}$.

Thus, the chronic RfD for chlorpyrifos $= 3 \times 10^{-4}$ mg kg^{-1} d^{-1}.

And the more protective acute PAD $= 5 \times 10^{-5}$ mg kg^{-1} d^{-1}.

Therefore, had the NOAEL threshold been used alone without the safety adjustment of the RfD, the allowable exposure would have been three orders of magnitude higher [18].

Uncertainty can also come from error. As discussed in Chapter 11, two errors can occur when information is interpreted improperly or in the absence of sound science. The first is the "false negative" reporting that there is no problem when one in fact exists. This not only can occur due to improper sampling and analysis [69] but also can be due to many other steps in a risk assessment. The need to address this problem is often at the core of the positions taken by environmental and public health agencies and advocacy groups. They may ask questions like the following:

- What if a monitor's level of detection is above that needed to show that an air pollutant is, in fact, being emitted?
- What if the leak detector registers zero, but in fact, toxic substances are being released from the tank?
- Is there a way to determine quantitatively the amount of substance in space with known levels of confidence?
- What if this substance really does cause cancer but the tests are unreliable?
- What if people are being exposed to a contaminant from a product, but via a pathway other than the ones being studied?
- What if there is a relationship that is different from the laboratory when this substance is released into the "real world," such as the difference between how a chemical behaves in the human body by itself as opposed to when other chemicals are present (i.e., the problem of "complex mixtures")?

The other concern is, conversely, the "false positive" [70]. This can be a major challenge for public health agencies with the mandate to protect people from exposures to environmental contaminants. For example, what if previous evidence shows that an agency had listed a compound as a potential endocrine disruptor, only to find that a wealth of new information is now showing that it has no such effect? This can happen if the conclusions were based upon faulty models or models that only work well for lower organisms, but subsequently, developed models have taken into consideration the physical, chemical, and biological complexities of higher-level organisms, including humans. False positives may force public health officials to devote inordinate amounts of time and resources to deal with so-called nonproblems. Of course, simply labeling a matter as nonproblematic is not the same as it being safe. Indeed, this can be a ploy, so air quality managers must be certain. False positives also erroneously frighten people about potentially useful products, that is, known as an opportunity risk. False positives, especially when they occur frequently, create credibility gaps between engineers and scientists and the decision-makers. In turn, the public, who have entrusted the scientists, engineers, and managers with providing clean air, begins to lose confidence.

Environmental risk assessment calls for high quality, scientifically based information. Put in engineering language, the risk assessment process is a "critical path" in which any unacceptable error or uncertainty along the way will decrease the quality of the risk assessment and, quite likely, will lead to a bad environmental decision.

12.6 EXPOSURE ESTIMATION

An exposure is any contact with an agent [71]. For chemical and biological agents, this contact can come about from several exposure pathways, that is, routes taken by a substance, beginning with its source to its endpoint (i.e., a target organ, like the liver, or a location short of that, such as in fat tissues). The principal exposure route for air pollutants is the respiratory systems; however, exposure can occur via other routes and pathways, for example, ingesting food upon which air pollutants have been deposited.

Exposure results from sequential and parallel processes in the environment, from release to environmental partitioning to movement through pathways to uptake and fate in the organism (see Fig. 12.15). The substances often change to other chemical species because of the body's metabolic and detoxification processes. Certainly, genetic modifications can affect such processes. New substances, known as degradation products or metabolites, are produced as cells and use the parent compounds as food and energy sources. These metabolic processes, such as hydrolysis and oxidation, are the mechanisms by which chemicals are broken down.

Physical agents, such as electromagnetic radiation, ultraviolet (UV) light, and noise, do not follow this pathway exactly. The contact with these sources of energy can elicit a physiological response that may generate endogenous chemical changes that behave somewhat like the metabolites. For example, UV light may infiltrate and damage skin cells. The UV light helps to promote skin tumor promotion by activating the transcription factor complex activator protein-1 (AP-1) and enhancing the expression of the gene that produces the enzyme cyclooxygenase-2 (*COX2*). Noise, that is, acoustic energy, can also elicit physiological responses that affect an organism's chemical messaging systems, that is, endocrine, immune, and neural. It is possible that genetically modified organisms will respond differently to these physical agents than their unmodified counterparts.

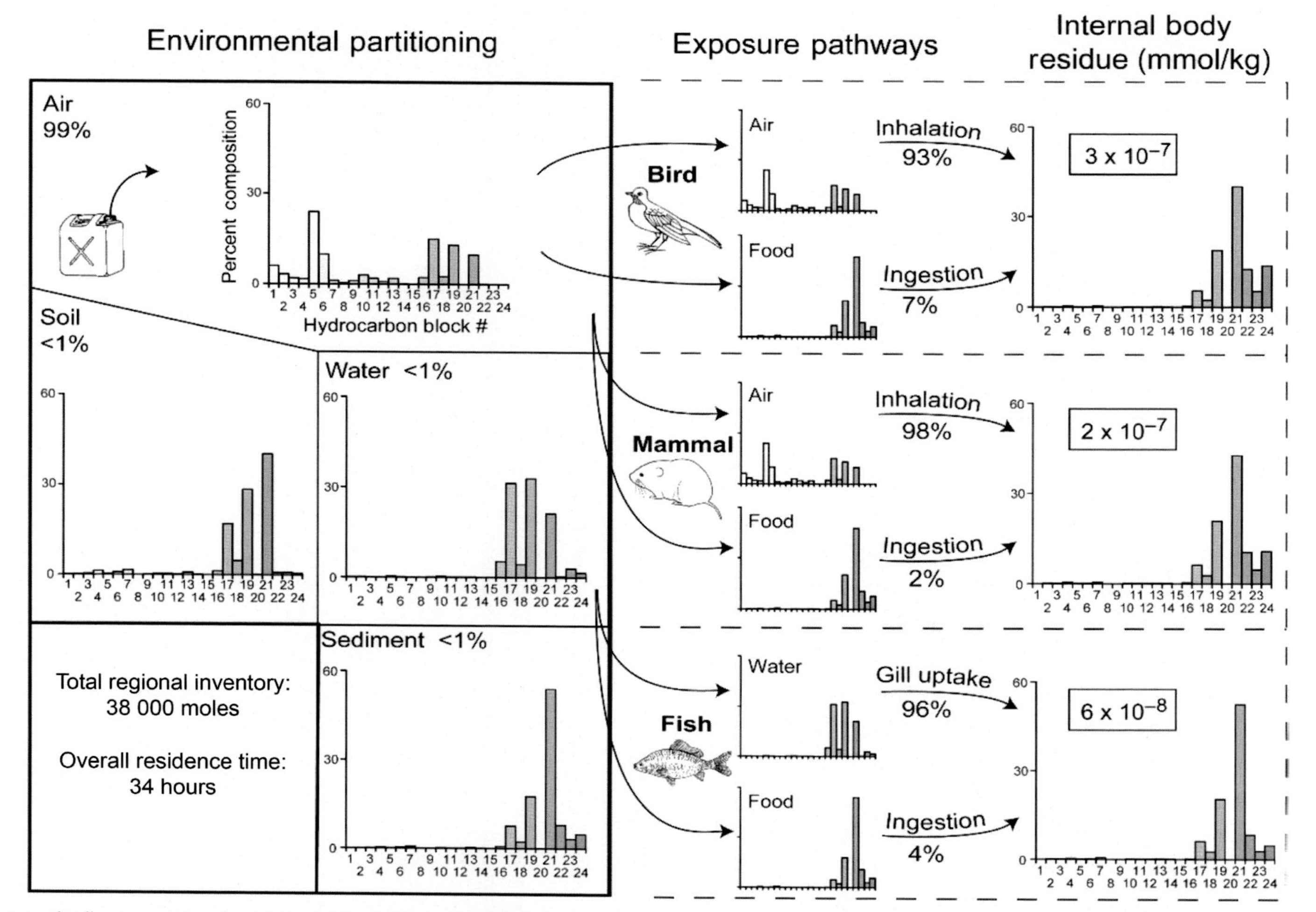

FIG. 12.15 Processes leading to organismal uptake and fate of chemical and biological agents after release into the environment. In this instance, the predominant sources are air emissions, and predominant pathway of exposure is inhalation. However, due to deposition to surface waters and the agent's affinity for sediment, the ingestion pathways are also important. Dermal pathways, in this case, do not constitute a large fraction of potential exposure. *(Source: T. McKone, W. Riley, R. Maddalena, R. Rosenbaum, D. Vallero, Common issues in human and ecosystem exposure assessment: the significance of partitioning, kinetics, and uptake at biological exchange surfaces, Epidemiology 17 (6) (2006) S134; R.K. Rosenbaum et al., USEtox human exposure and toxicity factors for comparative assessment of toxic emissions in life cycle analysis: sensitivity to key chemical properties, Int. J. Life Cycle Assess. 16 (8) (2011) 710–727.)*

The exposure pathway also includes the ways that humans and other organisms can contact the hazard. The pathway has five parts:

1. The source of contamination (e.g., fugitive dust or leachate from a landfill)
2. An environmental medium and transport mechanism (e.g., soil with water moving through it)
3. A point of exposure (such as a well that is used for drinking water)
4. A route of exposure (e.g., inhalation, dietary ingestion, nondietary ingestion, dermal contact, and nasal)
5. A receptor population (those who are exposed or who are where there is a potential for exposure)

If all five parts are present, the exposure pathway is known as a completed exposure pathway. In addition, the exposure may be short term, intermediate, or long term. Short-term contact is known as an acute exposure, that is, occurring as a single event or for only a short period of time (up to 14 days). An intermediate exposure is one that lasts from 14 days to <1 year. Long-term or chronic exposures are >1 year in duration.

Determining the exposure for a neighborhood can be complicated. For example, even if we do a good job identifying all the contaminants of concern and its possible source (no small task), we may have little idea of the extent to which the receptor population has come into contact with these contaminants (steps 2 through 4). Thus, assessing exposure involves not only the physical sciences but also the social sciences, for example, psychology and behavioral sciences. People's activities greatly affect the amount and type of exposures. That is why exposure scientists use several techniques to establish activity patterns, such as asking potentially exposed individuals to keep diaries, videotaping, and using telemetry to monitor vital information, for example, heart and ventilation rates.

General ambient measurements, such as air pollution monitoring equipment located throughout cities, are often not good indicators of actual population exposures. For example, metals and their compounds comprise the greatest mass of toxic substances *released* into the US environment. This is largely due to the large volume and surface areas involved in metal extraction and refining operations. However, this does not necessarily mean that more people will be exposed at higher concentrations or more frequently to these compounds than to others. A substance that is released or even that if it resides in the ambient environment is not tantamount to its coming in contact with a receptor. Conversely, even a small amount of a substance under the right circumstances can lead to very high levels of exposure (e.g., handling raw materials and residues at a waste site).

A recent study by the Lawrence Berkley Laboratory demonstrates the importance of not simply assuming that the concentrations of a released pollutant or even background concentrations are a good indicator of actual exposure [20]. The researchers were interested in how sorption may affect microenvironments, so they set up a chamber constructed of typical building materials and furnished with actual furniture like that found in most residential settings. Several air pollutants were released into the room and monitored (see Fig. 12.16). With the chamber initially sealed, the observed decay of xylene, a volatile organic compound, in vapor-phase concentrations results from adsorption onto surfaces (walls, furniture, etc.).

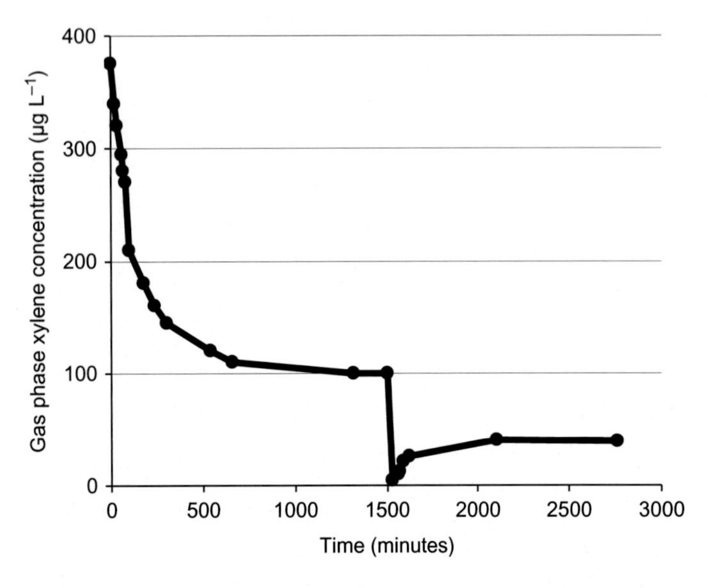

FIG. 12.16 Concentrations of xylene measured in its vapor phase in a chamber sealed during adsorption and desorption periods. *(Data from B. Singer, A tool to predict exposure to hazardous air pollutants, Environ. Energy Technol. Div. News 4 (4) (2003) 5.)*

The adsorption continues for hours, with xylene concentrations reaching a quasi-steady state. At this point, the chamber was flushed with clean air to free the vapor-phase xylene. The xylene concentrations shortly after the flush began to rise again until reaching a new steady state. This rise must be the result of desorption of the previously sorbed xylene, since the initial source is gone. Sorption is one of the biochemodynamic processes that must be considered to account for differences in the temporal pattern of microenvironmental (e.g., occupational) and ambient concentrations.

The simplest quantitative expression of exposure is.

$$E = D/t. \tag{12.7}$$

where E is the human exposure during the time period t [units of concentration ($mg\,kg^{-1}d^{-1}$)], D is the mass of pollutant per body mass ($mg\,kg^{-1}$), and t is time (day). D, the dose of a pollutant, is usually measured near the interface of the person and the environment, during a specified time period. This measurement is sometimes referred to as the potential dose (i.e., the chemical has not yet crossed the boundary into the body but is present where it may enter the person, such as on the skin, at the mouth, or at the nose).

A one-minute measurement with a sensor monitor shows 1.75 ppm carbon monoxide. If this represents the average concentration, what is the CO exposure for a man weighing 90 kg who breathes 8 L air per minute?

Based on the density of CO, 1 ppm = 0.873, so this is $2\,mg\,m^{-3} = 0.002\,mg\,L^{-1}$.

So, the man is breathing $0.0016\,mg\,min^{-1}$, and $(1440)(0.0016) = 2.3\,mg\ d^{-1}$.

Therefore, $E = \frac{2.3\,mg\,d^{-1}}{90kg} = 0.0256\,mg\ CO\ kg^{-1}\ d^{-1}$.

Expressed quantitatively, exposure is a function of the concentration of the agent and time. It is an expression of the magnitude and duration of the contact. That is, exposure to a contaminant is the concentration of that contact in a medium integrated over the time of contact:

$$E = \int_{t=t_1}^{t=t_2} C(t)dt \tag{12.8}$$

where E is the exposure during the time period from t_1 to t_2 and $C(t)$ is the concentration at the interface between the organism and the environment, at time t.

The concentration at the interface is the potential dose (i.e., the agent has not yet crossed the boundary into the body but is present where it may enter the receptor). Since the amount of a chemical agent that penetrates from the ambient atmosphere into a control volume affects the concentration term of the exposure equation, a complete mass balance of the contaminant must be understood and accounted for; otherwise, exposure estimates will be incorrect. Recall that the mass balance consists of all inputs and outputs and chemical changes to the contaminant:

$$\text{Accumulation or loss of contaminant A} = \text{Mass of A transported in} - \text{Mass of A transported out} \pm \text{Reactions} \tag{12.9}$$

The reactions may be either those that generate substance A (i.e., sources) or those that destroy substance A (i.e., sinks). Thus, the amount of mass transported in is the inflow to the system that includes pollutant discharges, transfer from other control volumes and other media (e.g., if the control volume is soil, the water and air may contribute mass of chemical A), and formation of chemical A by abiotic chemistry and biological transformation. Conversely, the outflow is the mass transported out of the control volume, which includes uptake, by biota, transfer to other compartments (e.g., volatilization to the atmosphere), and abiotic and biological degradation of chemical A. This means the rate of change of mass in a control volume is equal to the rate of chemical A transported in less the rate of chemical A transported out, plus the rate of production from sources and minus the rate of elimination by sinks. Stated as a differential equation, the rate of change contaminant A is

$$\frac{d[A]}{dt} = -v \cdot \frac{d[A]}{dx} + \frac{d}{dx}\left(\Gamma \cdot \frac{d[A]}{dx}\right) + r \tag{12.10}$$

where v is the fluid velocity, Γ is a rate constant specific to the environmental medium, $\frac{d[A]}{dx}$ is the concentration gradient of chemical A, and r refers to the internal sinks and sources within the control volume.

TABLE 12.4 Effect of extrinsic conditions on chemicals, as represented by preservation and holding time requirements for anion sampling and analysis

Analyte	Preservation	Holding time
Part A: Common anions		
Bromide	None required	28 days
Chloride	None required	28 days
Fluoride	None required	28 days
Nitrate-N	Cool to 4°C	48 h
Nitrite-N	Cool to 4°C	48 h
Ortho-phosphate-P	Cool to 4°C	48 h
Sulfate	Cool to 4°C	28 days
Part B: Inorganic disinfection by-products		
Bromate	50 mg L^{-1} EDA	28 days
Bromide	None required	28 days
Chlorate	50 mg L^{-1} EDA	28 days
Chlorite	50 mg L^{-1} EDA, cool to 4°C	14 days

Source: U.S. Environmental Protection Agency, Method 300.1: Determination of Inorganic Anions in Drinking Water by Ion Chromatography, 1997.

Reactive compounds can be particularly difficult to measure. For example, many volatile organic compounds in the air can be measured by first collecting in stainless steel canisters and analyzed by chromatography in the lab. However, some of these compounds, like the carbonyls (notably aldehydes like formaldehyde and acetaldehyde), are prone to react inside the canister, meaning that by the time the sample is analyzed, a portion of the carbonyls is degraded (underreported). Therefore, other methods, such as trapping the compounds with dinitrophenylhydrazine (DNPH), treated silica gel tubes that are frozen until being extracted for chromatographic analysis. The purpose of the measurement is to see what is in the air, water, soil, sediment, or biota at the time of sampling, so any reactions before the analysis give measurement error.

If the released chemical is reactive, some or all of it may have changed into another form (i.e., *speciated*) by the time it is measured. Even relatively nonreactive compounds may speciate between when the sample is collected (e.g., in a water sample, an air canister, a soil core, or a bag) and when the sample is analyzed. In fact, each contaminant has unique characteristics that vary according to the type of media in which it exists and extrinsic conditions like temperature and pressure (see Table 12.4).

The general exposure Eq. (12.10) is rewritten to address each route of exposure, accounting for chemical concentration and the activities that affect the time of contact. The exposure calculated from these equations is the chemical intake (I) in units of concentration (mass per volume or mass per mass) per time, such as mg kg^{-1} d^{-1}:

$$I = \frac{C \cdot CR \cdot EF \cdot ED \cdot AF}{BW \cdot AT} \quad (12.11)$$

where C is the chemical concentration of contaminant (mass per volume), CR is the contact rate (mass per time), EF is the exposure frequency (number of events and dimensionless), and ED is the exposure duration (time).

The factors in Eq. (12.11) are further specified for each route of exposure, such as the lifetime average daily dose (LADD) as shown in Table 12.5. Exposure can be calculated for various scenarios that will change the exposure duration and frequency. These include average daily dose (ADD), LADD, and acute dose rate (ADR). The averaging time (AT) differs for these exposure scenarios. The ADD averages exposure over the period of time during which the exposure occurred, making it amenable to risk calculations for many noncancer effects. The LADD is used for cancer risk assessments, with LADD usually described in terms of lifetime probabilities, although this level of exposure does not occur over the entire lifetime. Thus, for LADD, the AT is replaced with lifetime. The ADR assumes a much shorter AT, that is, 1 day [72].

Acute and subchronic exposures require different equations, since the exposure duration (ED) is much shorter. For example, instead of LADD, acute exposures to noncarcinogens may use maximum daily dose (MDD) to calculate exposure. However, even these exposures follow the general model given in Eq. (12.11).

TABLE 12.5 Equations for calculating lifetime average daily dose (LADD) for various routes of exposure

Route of exposure	Equation LADD (in mg kg^{-1} d^{-1}) =	Definitions
Inhaling aerosols (particulate matter)	$\frac{(C)\cdot(PC)\cdot(IR)\cdot(RF)\cdot(EL)\cdot(AF)\ (ED)\cdot(10^{-6})}{(BW)\cdot(TL)}$	C = concentration of the contaminant on the aerosol/ particle (mg kg^{-1}) PC = particle concentration in air (gm m^{-3}) IR = inhalation rate (m^{-3} h^{-1}) RF = respirable fraction of total particulates (dimensionless, usually determined by aerodynamic diameters, e.g., 2.5 μm) EL = exposure length (h d^{-1}) ED = duration of exposure (d) AF = absorption factor (dimensionless) BW = body weight (kg) TL = typical lifetime (d) 10^{-6} is a conversion factor (kg to mg)
Inhaling vapor-phase contaminants	$\frac{(C)\cdot(IR)\cdot(EL)\cdot(AF)\cdot(ED)}{(BW)\cdot(TL)}$	C = concentration of the contaminant in the gas phase (mg m^{-3}) Other variables the same as above
Drinking water	$\frac{(C)\cdot(CR)\cdot(ED)\cdot(AF)}{(BW)\cdot(TL)}$	C = concentration of the contaminant in the drinking water (mg L^{-1}) CR = rate of water consumption (L d^{-1}) ED = duration of exposure (d) AF = portion (fraction) of the ingested contaminant that is physiologically absorbed (dimensionless) Other variables are the same as above
Contact with soilborne contaminants	$\frac{(C)\cdot(SA)\cdot(BF)\cdot(FC)\cdot(SDF)\cdot(ED)\cdot(10^{-6})}{(BW)\cdot(TL)}$	C = concentration of the contaminant in the soil (mg kg^{-1}) SA = skin surface area exposed (cm^{-2}) BF = bioavailability (percent of contaminant absorbed per day) FC = fraction of total soil from contaminated source (dimensionless) SDF = soil deposition, the mass of soil deposited per unit area of skin surface (mg cm^{-1} d^{-1}) Other variables are the same as above

Source: M.A. Hollinger, M.J. Derelanko, Handbook of Toxicology. CRC Press, 2002.

In the process of synthesizing pesticides over an 18-year period, a polymer manufacturer has contaminated the soil on its property with vinyl chloride. The plant closed 2 years ago, but vinyl chloride vapors continue to reach the neighborhood surrounding the plant at an average concentration of 1 mg m^{-3}. Assume that people are breathing at a ventilation rate of 0.5 m^3 h^{-1} (about the average of adult males and females over 18 years of age [21]). The legal settlement allows neighboring residents to evacuate and sell their homes to the company. However, they may also stay. The neighbors have asked for advice on whether to stay or leave, since they have already been exposed for 20 years

Vinyl chloride is highly volatile, so its phase distribution will be mainly in the gas phase rather than the aerosol phase. Although some of the vinyl chloride may be sorbed to particles, we will use only vapor-phase LADD equation, since the particle phase is likely to be relatively small. Also, we will assume that outdoor concentrations are the exposure concentrations. This is unlikely; however, since people spend very little time outdoors compared with indoors, so, this may provide an additional factor of safety. To determine how much vinyl chloride penetrates living quarters, indoor air studies would have to be conducted. For a scientist to compare exposures, indoor air measurements should be taken.

Find the appropriate equation in Table 12.5 and insert values for each variable. Vinyl chloride is well absorbed, and since it is a gas at room temperature, inhalation is the major route of exposure. Much of the mass is retained in the lungs [73], so for a worst case, we can assume that AF = 1. However, more specific absorption data can be found on vinyl chloride, for example, in the US EPA Chemistry Dashboard [74].

To simplify, we will also assume that the person that stays in the neighborhood is exposed at the average concentration 24 h per day (EL = 24) and that a person lives the remainder of his/her entire typical lifetime exposed at the measured concentration.

TABLE 12.6 Commonly used human exposure factors

Exposure factor	Adult male	Adult female	Child (3–12 years of age) [23]
Body weight (kg)	70	60	15–40
Total fluids ingested ($L\ d^{-1}$)	2	1.4	1.0
Surface area of the skin, without clothing (m^2)	1.8	1.6	0.9
Surface area of the skin, wearing clothes (m^2)	0.1–0.3	0.1–0.3	0.05–0.15
Respiration/ventilation rate ($L\ min^{-1}$)—resting	7.5	6.0	5.0
Respiration/ventilation rate ($L\ min^{-1}$)—light activity	20	19	13
Volume of air breathed ($m^3\ d^{-1}$)	23	21	15
Typical lifetime (years)	70	70	NA
National upper-bound time (90th percentile) at one residence (years)	30	30	NA
National median time (50th percentile) at one residence (years)	9	9	NA

Sources: J. Moya et al., Exposure Factors Handbook, US Environmental Protection Agency, Washington, 2011; E.J. Calabrese, ATSDR Public Health Assessment Guidance Manual. CRC Press, 1992.

Although the ambient concentrations of vinyl chloride may have been higher when the plant was operating, the only measurements we have are those taken recently. Thus, this is an area of uncertainty that must be discussed with the clients. The common default value for a lifetime is 70 years [75], so we will assume the longest exposure would be 70 years (25,550 days). Table 12.6 provides commonly used default values in exposure assessments. Of course, longer lifetimes, for example, 75 years, are used in some models and calculations. If the person is now 20 years of age and has already been exposed for that time and lives the remaining 50 years exposed at 1 $mg\,m^{-3}$:

$$\begin{aligned} LADD &= \frac{(C)\cdot(IR)\cdot(EL)\cdot(AF)\cdot(ED)}{(BW)\cdot(TL)} \\ &= \frac{(1)\cdot(0.5)\cdot(24)\cdot(1)\cdot(25550)}{(70)\cdot(25550)} \\ &= 0.2\ mg\ kg^{-1}\ day^{-1} \end{aligned}$$

If the 20-year-old person leaves today, the exposure duration would be for the 20 years that the person lived in the neighborhood. Thus, only the ED term would change, that is, from 25,550 days to 7300 days (i.e., 20 years).

Thus, the LADD falls to 2/7 of its value:

$LADD = 0.05\,mg\,kg^{-1}\,day^{-1}$.

Based on the hazard and exposure calculations, risks can be characterized quantitatively. There are two general ways that such risk characterizations are used in air pollution decisions, that is, direct risk assessments and risk-based treatment targets.

12.7 DIRECT CANCER RISK CALCULATIONS [3]

Although in its simplest form, risk is the product of the hazard and exposure, but assumptions can greatly affect risk estimates. For example, cancer risk can be defined as the theoretical probability of contracting cancer when continually exposed for a lifetime (e.g., 70 years) to a given concentration of a carcinogen. The probability is usually calculated as an upper confidence limit. The maximum estimated risk may be presented as the number of chances in a million of contracting cancer.

Two measures of risk are commonly reported. One is the individual risk, that is, the probability of a person developing an adverse effect (e.g., cancer) due to the exposure. This is often reported as a "residual" or increased probability above background. For example, if we want to characterize the contribution of all the power plants in the United States to increased cancer incidence, the risk above background would be reported. The second way that risk is reported is population risk, that is, the annual excess number of cancers in an exposed population. The maximum individual risk might be calculated from exposure estimates based upon a "maximum exposed individual" or MEI. The hypothetical MEI lives an entire lifetime outdoors at the point where pollutant concentrations are highest. Assumptions about exposure will greatly affect the risk estimates. For example, the cancer risk from power plants in the United States has been estimated to be 100- to 1000-fold lower for an average exposed individual than that calculated for the MEI.

For cancer risk assessments, the hazard is generally assumed to be the cancer slope factor (CSF) and the long-term exposure (E):

$$\text{Cancer risk} = \text{CSF} \times E \tag{12.12}$$

Cancer is a chronic illness, that is, often occurs after many years of exposure or at least after a latency period following episodic exposures. Therefore, the appropriate value of E should also be chronic (e.g., LADD). Thus, the estimated risk is the product of LADD and potency (SF).

Using the lifetime average daily dose value from the vinyl chloride exposure calculation in the previous section, estimate the direct risk to the people living near the abandoned polymer plant. What information needs to be communicated?

Insert the calculated LADD values and the vinyl chloride inhalation slope factor of 3.00×10^{-1} from Appendix 2. The two LADD values under consideration, the cancer risk to the neighborhood exposed for an entire lifetime (exposure duration = 70 years), gives us $0.2\,\text{mg}\,\text{kg}^{-1}\,\text{d}^{-1} \times 0.3\,(\text{mg}\,\text{kg}^{-1}\,\text{d}^{-1})^{-1} = 0.06$. This is an incredibly high risk! The threshold for concern is often 1 in a million (0.000001), while this is a probability of 6%.

Even at the shorter duration period (20 years of exposure instead of 70 years), the risk is calculated as $0.05 \times 0.3 = 0.017$ or nearly a 2% risk. The combination of a very steep slope factor and very high lifetime exposures leads to a very high risk. Vinyl chloride is a liver carcinogen, so unless corrective actions significantly lower the ambient concentrations of vinyl chloride, the prudent course of action is that the neighbors accept the buyout and leave the area.

Incidentally, vinyl chloride has relatively high aqueous solubility and can be absorbed to soil particles, so ingestion of drinking water (e.g., people on private wells drawing water from groundwater that has been contaminated) and dermal exposures (e.g., children playing in the soil) are also conceivable. The total risk from a single contaminant like vinyl chloride is equal to the sum of risks from all pathways (e.g., vinyl chloride in the air, water, and soil):

$$\text{Total Risk} = \sum \text{risks from all exposure pathways} \tag{12.13}$$

Cancer risk varies by the potency of the carcinogen and the exposure to that carcinogen. As for any air pollution risk, cancer risk is a function of hazard and exposure. As mentioned, since cancer is a chronic disease, the exposure must be long term or at least must address the latency between exposure and cancer outcome. Thus, cancer risk assessments generally express hazard potency for the dose-response curves from animal research and human epidemiology. The quasilinear segment of these curves provides what is known as the *CSF*, and the long-term exposure is the lifetime average daily dose (LADD).

Usually, cancer occurs after prolonged exposure to a carcinogen. As such, the exposure setting must be characterized. This includes both the physical environment and the activities of the potentially exposed populations. From this assessment, sources of the air pollutant can be identified along with the possible routes of exposure. To quantify exposure, the concentration of the carcinogen and the intake variables must be assigned. Daily intake (DI) must be estimated or modeled:

$$\text{DI}\ (\text{mg/kg} - \text{day}) = \frac{\text{Concentration}\ (\text{mg volume}^{-1}) \times \text{Intake}\ (\text{volume day}^{-1})}{\text{body mass (kg)}} \tag{12.14}$$

Then, chronic daily intake (CDI) can be estimated by averaging DI over long-term exposure (e.g., 70 years for LADD).

Exposure factors vary. For example, an average lifetime may be assumed to be 70 years, but the exposure duration may be for 25 years (e.g., the source may have only been operating for 25 years). Body weight may be assumed to be 70 kg; frequency of exposure may be <365 days a year (e.g., 250 days per year). The intake rate must account for time period and the amount of the air pollutant. For example, inhalation rates will vary by activity and person, so an average exposure inhalation rate may be used, for example, $2.5\,\text{m}^3\,\text{hr.}^{-1} \times 8\,\text{h day}^{-1}$. However, as indicated in Table 12.7, inhalation rates can be highly variable.

An exposure equation must account for both chemical concentration of the airborne carcinogen and a person's activities that affect the time of contact. The chemical intake (I) is expressed in units of concentration (mass per volume or mass per mass) per time, such as $\text{mg}\,\text{kg}^{-1}$-day. Since the intake of air pollutants is almost entirely via inhalation, the contaminant concentration in the air can be used exclusively:

$$I = C_A \times \frac{IR}{BW} \times \frac{ET \times EF \times ED}{AT} \tag{12.15}$$

TABLE 12.7 Tracheobronchial region default values for humans of different ages and activity patterns

	Minute volume[a] [V_e] ($L\,min^{-1}$)	Surface area[a] [SA] (cm^2)	(Ve/SATB) (L/min-cm^2)	Tracheobronchial Regional gas dose ratio	NOAEL (human equivalent concentration) (mg m^3)
Outdoor worker M	17.5	2660	0.0066	0.94	0.15
Sedentary worker M	15.4	2660	0.0058	1.1	0.18
Sedentary worker F	12.6	2640	0.0048	1.3	0.21
15 Year M	14.0	2520	0.0056	1.1	0.18
15 Year F	10.9	2250	0.0048	1.3	0.21
10 Year	10.6	1830	0.0058	1.1	0.18
5 Year	6.1	1340	0.0046	1.4	0.22
1 Year	3.6	857	0.0042	1.5	0.24
3 Months	2.0	712	0.0028	2.2	0.35
Human equivalent concentration—default	13.8	3200	0.0043	1.4	0.22

[a]*These values are from International Commission on Radiological Protection (ICRP) (1994). ICRP Publication 66: Human Respiratory Tract Model for Radiological Protection.* Annals of the ICRP, *Volume 24:1–3, Tables B-16A and B-16B. Note:* M, *male;* F, *female.*

Data from U.S. Environmental Protection Agency, Office of Emergency and Remedial Response, Risk Assessment Guidance for Superfund: Human Health Evaluation Manual, Development of Risk-Based Preliminary Remediation Goals, 1991; M. Bailey et al., Risk Assessment Guidance for Superfund: Volume I: Human Health Evaluation Manual (Part F, Supplemental Guidance for Inhalation Risk Assessment). Environmental Protection Agency, 2009.

where

C_A = contaminant concentration in air ($mg\,m^{-3}$)
IR = inhalation rate ($m^3\,hr^{-1}$)
BW = body weight (kg)
ET = exposure time ($hrs\,day^{-1}$)
EF = exposure frequency ($days\,yr^{-1}$)
ED = exposure duration (yrs)
AT = averaging time (days over which exposure is averaged)

Exposure equations vary in their acceptability for use in risk assessments. When estimating risks presented by inhalation, the concentration of the contaminant in the air should be used as an exposure metric, that is, mass per volume (e.g., $mg\,m^{-3}$) instead of inhalation intake of the air pollutant based on *BW* and *IR*. This means that Eq. (12.15) is not consistent with EPA's Methods for Derivation of Inhalation Reference Concentrations and Application of Inhalation Dosimetry [76], which considers the amount of the contaminant that reaches the target site in a person's body not simply a function of *BW* and *IR*. Rather, the interaction of the inhaled contaminant with the tissue in the respiratory track is determined by other factors, for example, biological, species-specific relationships of the exposure concentration to dose and the physicochemical properties of the chemical contaminant [77]. Thus, although the exposure equations are credible from a scientific perspective, regulators vary in what they consider to be acceptable equations. For example, statutes may specify equations, as may other legally binding requirements, for example, court orders.

An exposure concentration (EC) is a time-weighted average concentration derived from measured or modeled concentrations of a contaminant in the air but adjusted according to the characteristics of the exposure scenario. For an acute exposure, the EC and CA values will be almost the same. However, estimating cancer risks from chronic exposure involves both the *CA* measured at a point of exposure point and scenario-specific parameters, such as the exposure duration and frequency:

$$EC = \frac{C_A \times ET \times EF \times ED}{AT} \quad (12.16)$$

The value for AT for chronic exposures is ED in years × 365 days yr^{-1} × 24 h day^{-1}. Since this gives AT in hours, it is multiplied by 24 for an AT in days. Thus, EC usually is C_A that has been time-weighted over the duration of exposure and incorporates information on human activity patterns. The US EPA cancer risk guidelines apply a linear extrapolation from exposures observed in animal studies or human occupational studies [64].

For air toxics, the unit risk estimate (URE) is often used instead of the Eq. (12.12). The URE is an upper-bound estimate of a person's chance of contracting cancer over a lifetime of exposure when exposed to 1 microgram of the carcinogen per cubic meter of (1 $\mu g\, m^{-3}$) air. Risks from exposures to concentrations other than one microgram per cubic meter can be calculated by multiplying the actual concentration to which a person is exposed by the URE.

If a person breathes air in which the average concentration of a carcinogenic air toxic is $10^{-5}\,\mu g\, m^{-3}$, what is the URE?

Since cancer is a chronic disease, the lifetime exposure is assumed, that is, 70 years. The person who breathes air containing an average of 1 microgram per cubic meter for 70 years would have a URE of 1 chance in 10,000, that is, an addition 10^{-5} risk of contracting cancer as a result.

It is common to combine the *EC* with the inhalation-related URE, that is, the inhalation unit risk (IUR), which like the *CSF* is a predictive cancer risk potency estimate. The IUR is the upper-bound excess lifetime cancer risk estimated to result from continuous exposure to an agent at a concentration of 1 $\mu g\, m^{-3}$ in air.

The IUR in a city is found to be 2×10^{-6} per $\mu g\ m^{-3}$; what are the predicted excess cancer cases?

Two excess cancer cases (upper-bound estimate) are expected to develop per 1,000,000 people if exposed daily for a lifetime to 1 μg of the chemical in 1 cubic meter of air.

In the United States, the default approach for predicting cancer risk is a linear extrapolation from exposures observed in animal or human occupational studies. A nonlinear approach should be selected when there are sufficient supporting data to ascertain the carcinogen's mode of action (MOA) to conclude that it is not linear at low doses and the carcinogen does not demonstrate mutagenic or other activity consistent with linearity at low doses [77]. This approach involves drawing a straight line from the point of departure (POD) to the origin on the dose-response curve. The POD is the dose-response point that marks the beginning of a low-dose extrapolation, which can be the lower boundary of the dose for an estimated incidence or a change in response level from a dose-response model (i.e., the benchmark dose, BMD) or a NOAEC or LOAEC for an observed incidence or change in level of response. The default linear extrapolation approach is generally considered to be conservatively protective of public health, including sensitive subpopulations [64]. The slope of this line is the IUR. Thus, the IUR can be presented as a linear extrapolation from a POD of 10% response (LEC_{10}):

$$IUR = \frac{0.1}{LEC_{10(HEC)}}$$

where the $LEC_{10(HEC)}$ is the lowest effective dose using a 10% response concentration, dosimetrically adjusted to a human equivalent concentration.

The pollutant concentration in air (*CA*) must include all physical phases, that is, both the pollutant in vapor phase and those in aerosol phase. The aerosol phase can simply be the same compound in liquid and solid phase and any of the pollutant to this sorbed to particulate. For example, at equilibrium benzene that will exist in both gaseous and liquid phase, it can be sorbed on the surface of a particle (adsorbed), and it can be sorbed within the particle matrix (absorbed). The *CA* value must account for all these phases. These values must be normalized, for example, all expressed in units of mass per volume (e.g., mg m^{-3}) or mass per mass (e.g., mg kg^{-1}).

Many carcinogens and other toxic air pollutants are organic compounds that are highly diverse in their physical characteristics and chemical composition. Volatile organic compounds (VOCs), for example, exist in the ambient air almost entirely in the gas phase since their vapor pressures in the environment are usually $>10^{-2}$ kilopascals. For VOCs, then, CA may be almost equal to vapor-phase concentrations. Conversely, relatively nonvolatile organic compounds (NVOCs) with vapor pressures $<10^{-5}$ kilopascals are predominantly found in and on particles, unless significant energy is added to increase their volatility. Thus, the CA for NVOCs can be estimated to equal the total amount of the air pollutant in particulate matter collected. However, semivolatile organic compounds (SVOCs), with vapor pressures between 10^{-2} and 10^{-5} kilopascals, can exist as in both the gas and particle phases in the ambient air.

Requirements and measures of success are seldom if ever as straightforward as the earlier vinyl chloride example. In fact, the bioengineer would be ethically remiss if the only advice given is to the local community, that is, whether to accept the buyout. Of course, one of the engineering canons is to serve a "faithful agent" to the clientele. However, the first engineering canon is to hold paramount the health and safety of the public. Thus, the engineer must balance any proprietary information that the client wants to be protected with the need to protect public health. In this case, the engineer must inform the client and prime contractors, for example, that the regulatory agencies need to know that even if the current neighbors are moving, the threat to others, including future populations, are threatened. In other words, a systematic approach is needed; since the current population may be moved from harm's way, remediation is still likely needed to reduce the vinyl chloride concentrations to acceptable levels. This is because biochemodynamic processes are complex. The contaminant may remain in untreated or poorly treated areas, which may later be released, for example, future excavation, long-term transport from groundwater to surface waters and the atmosphere or from groundwater to drinking water sources. Thus, postclosure monitoring should be designed and operated based on worst-case scenarios.

The risk of adverse outcome other than cancer (so-called noncancer risk) is generally called the "hazard quotient" (HQ). It is calculated by dividing the maximum daily dose (MDD) by the acceptable daily intake (ADI):

$$\text{Noncancer risk} = \text{HQ} = \frac{\text{MDD}}{\text{ADI}} = \frac{\text{Exposure}}{\text{RfD}} \tag{12.17}$$

Note that this is an index, not a probability, so it is an indication of relative risk. If the noncancer risk is >1, the potential risk may be significant, and if the noncancer risk is <1, the noncancer risk may be insignificant. Thus, the reference dose, RfD, and reference concentration, RfC, are types of ADI.

The chromic acid (Cr^{6+}) mist dermal chronic RfD of 6.00×10^{-3} mg kg^{-1} day^{-1}. If the actual dermal exposure of people living near a metal processing plant is calculated (e.g., by intake or LADD) to be 4.00×10^{-3} mg kg^{-1} day^{-1}, calculate the hazard quotient for the noncancer risk of the chromic acid mist to the neighborhood near the plant and interpret the meanings

From Eq. (12.17), $\frac{\text{Exposure}}{\text{RfD}} = \frac{4.00 \times 10^{-3}}{6.00 \times 10^{-3}} = 0.67$. Since this is <1, one would not expect people chronically exposed at this level to show adverse effects from skin contact. However, at this same chronic exposure, that is, 4.00×10^{-3} mg kg^{-1} d^{-1}, to hexavalent chromic acid mists via oral route, the RfD is 3.00×10^{-3} mg kg^{-1} d^{-1}, meaning the HQ = 4/3 or 1.3. The value is >1, so we cannot rule out adverse noncancer effects.

If a population is exposed to more than one contaminant, the hazard index (HI) can be used to express the level of cumulative noncancer risk from pollutants 1 through n:

$$\text{HI} = \sum_{1}^{n} \text{HQ} \tag{12.18}$$

The HI is useful in comparing risks at various locations, for example, benzene risks in St. Louis, Cleveland, and Los Angeles. It can also give the cumulative (i.e., additive risk) in a single population exposed to more than one contaminant. For example, if the HQ for benzene is 0.2 (not significant), toluene is 0.5 (not significant), tetrachloromethane is 0.4 (not significant), and the cumulative risk of the three contaminants is 1.1 (potentially significant).

Realistic estimates of the hazard and the exposures are desirable in such calculations. However, precaution is the watchword for risk. Estimations of both hazard (toxicity) and exposure are often worst-case scenarios, because the risk calculations can have large uncertainties. Models usually assume effects to occur even at very low doses. Human data are usually gathered from epidemiological studies that, no matter how well they are designed, are fraught with error and variability (science must be balanced with the rights and respect of subjects, populations change, activities may be missed, and confounding variables are ever present). Uncertainties exist in every phase of risk assessment, from the quality of data, to limitations and assumptions in models, to natural variability in environments and populations.

Ecosystem hazard and exposure calculations differ somewhat from human health risk calculations. The risk quotient (RQ) is a deterministic approach, that is, dividing a point estimate of exposure by a point estimate of effects [78]. The RQ is often applied for comparing risks with ecosystems:

$$\text{RQ} = \frac{\text{Exposure}}{\text{Ecotoxicity}} \tag{12.19}$$

Ecotoxicity is an effect level or endpoint that is derived from toxicity tests (e.g., bioassays) that yield values such as LC_{50} or NOEC. Commonly in ecotoxicity studies, the effect level is the median effective concentration, that is, EC_{50}, which is the concentration of test substance that reduces biological activity, especially algal growth or growth rate or *Daphnia* immobilization, by 50% [79]. Since these organisms are waterborne, the EC50 units are mg L^{-1}. Thus, ecosystem exposure can be expressed as the estimated environmental concentration (EEC):

$$RQ = \frac{EEC}{EC_{50}} \tag{12.20}$$

The RQ is a relative value, that is, its purpose is to associate the exposure and toxicity ratio to a level of concern (LoC). After the risk quotient is calculated, it is compared with the LoC. An LoC is a policy tool that is used to interpret the RQ to analyze potential risk to nontarget organisms and inform decisions on the need for regulatory action [78]. A few ecological LoCs are listed in Table 12.8.

TABLE 12.8 Comparison of risk quotients (RQ) and levels of concern (LoC)

Risk presumptions for birds		
Risk presumption	**RQ**	**LoC**
Acute risk	EEC/LC_{50} or LD_{50}/ft^2 or LD_{50}/day	0.5
Acute restricted use	EEC/LC_{50} or LD_{50}/ft^2 or LD_{50}/day or $LD_{50} < 50\,mg\,kg^{-1}$	0.2
Acute endangered species	EEC/LC_{50} or LD_{50}/ft^2 or LD_{50}/day	0.1
Chronic risk	EEC/NOEC	1.0
Risk presumptions for wild mammals		
Risk presumption	**RQ**	**LoC**
Acute high risk	EEC/LC_{50} or LD_{50}/ft^2 or LD_{50}/day	0.5
Acute restricted use	EEC/LC_{50} or LD_{50}/ft^2 or LD_{50}/day or $LD_{50} < 50\,mg\,kg^{-1}$	0.2
Acute endangered species	EEC/LC_{50} or LD_{50}/ft^2 or LD_{50}/day	0.1
Chronic risk	EEC/NOEC	1.0
Risk presumptions for aquatic animals		
Risk presumption	**RQ**	**LoC**
Acute high risk	EEC/LC_{50} or EC_{50}	0.5
Acute restricted use	EEC/LC_{50} or EC_{50}	0.1
Acute endangered species	EEC/LC_{50} or EC_{50}	0.05
Chronic risk	EEC/NOAEC	1.0
Acute endangered species	EEC/EC_{05} or NOEC	1.0
Risk presumptions for aquatic plants		
Risk presumption	**RQ**	**LoC**
Acute high risk	EEC/EC_{50}	1.0
Acute endangered species	EEC/EC_{05} or NOEC	1.0
Risk presumptions for terrestrial and semiaquatic plants		
Risk presumption	**RQ**	**LoC**
Acute high risk	EEC/EC_{25}	1.0
Acute endangered species	EEC/EC_{05} or NOEC	1.0

Source of information: U.S. Environmental Protection Agency. Technical Overview of Ecological Risk Assessment: Risk Characterization. (2017), 5 September 2018. Available: https://www.epa.gov/pesticide-science-and-assessing-pesticide-risks/technical-overview-ecological-risk-assessment-risk.

12.8 RISK-BASED TREATMENT TARGETS

For most of the second half of the 20th century, environmental protection was based on two types of controls, that is, technology-based and quality-based. Technology-based controls are set according to what is "achievable" from the current state of the science and engineering. These are feasibility-based standards. The Clean Air Act has called for best achievable control technologies (BACT) and more recently for maximum achievable control technologies (MACT). Both standards reflect the reality that even though from an air quality standpoint it would be best to have extremely low levels of pollutants, technologies are not available or are not sufficiently reliable to reach these levels. Requiring unproved or unreliable technologies can even exacerbate the pollution, such as in the early days of wet scrubbers on coal-fired power plants. Theoretically, the removal of sulfur dioxide could be accomplished by venting the power plant flue through a slurry of carbonate, but the technology at the time is unproved and unreliable, allowing all-too-frequent releases of untreated emissions while the slurry systems were being repaired. Selecting a new technology over older proved techniques is unwise if the tradeoff of the benefit of improved treatment over older methods is outweighed by the numerous failures (i.e., no treatment).

Technology-based standards are a part of most environmental programs. Air emission controls, wastewater treatment, groundwater remediation, soil cleaning, sediment reclamation, drinking water supply, and hazardous waste site cleanup all are in part determined by availability and feasibility of control technologies.

Quality-based controls are those that are required to ensure that an environmental resource is in good enough condition to support a particular use. For example, a stream may need to be improved so that people can swim in it and so that it can be a source of water supply. Certain streams may need higher levels of protection than others, such as the so-called wild and scenic rivers. The parameters will vary but usually include minimum levels of dissolved oxygen and maximum levels of contaminants. The same goes for air quality, where ambient air quality must be achieved, so that concentrations of contaminants listed as National Ambient Air Quality Standards and certain toxic pollutants are below levels established to protect health and welfare.

Risk-based approaches to environmental protection, especially contaminant target concentrations, are designed to require engineering controls and preventive measures to ensure that risks are not exceeded. The risk-based approach embodies elements of both technology-based and quality-based standards. The technology assessment helps determine how realistic it will be to meet certain contaminant concentrations, while the quality of the environment sets the goals and means to achieve cleanup. Engineers are often asked, "How clean is clean?" When do we know that we have done a sufficient job of cleaning up a spill or hazardous waste site? It is often not possible to have nondetectable concentrations of a pollutant. Commonly, the threshold for cancer risk to a population is one in a million excess cancers. However, one may find that the contaminant is so difficult to remove that we almost give up on dealing with the contamination and put in measures to prevent exposures, that is, fencing the area in and prohibiting no access. This is often done as a first step in remediation but is an unsatisfying and controversial (and usually politically and legally unacceptable). Thus, even if costs are high and technology unreliable, the engineer must find suitable and creative ways to clean up the mess and meet risk-based standards.

Risk-based target concentrations can be calculated by solving for the target contaminant concentration in the exposure and risk equations. Since risk is the hazard (e.g., slope factor) times the exposure (e.g., LADD), a cancer-risk-based cleanup standard that can be found by enumerating the exposure equations within the risk equation (in this instance, the drinking water equation from Table 12.5) gives

$$\text{Risk} = \frac{\text{C} \cdot \text{CR} \cdot \text{EF} \cdot \text{ED} \cdot \text{AF} \cdot \text{SF}}{\text{BW} \cdot \text{AT}} \tag{12.21}$$

and solving for C,

$$\text{C} = \frac{\text{Risk} \cdot \text{BW} \cdot \text{AT}}{\text{CR} \cdot \text{EF} \cdot \text{ED} \cdot \text{AF} \cdot \text{SF}} \tag{12.22}$$

This is the target concentration for each contaminant needed to protect the population from the specified risk, for example, 10^{-6}. In other words, this is the concentration that must not be exceeded to protect a population having an average body weight and over a specified averaging time from an exposure of certain duration and frequency that leads to a risk of one in a million. While one-in-a-million added risk is commonly used benchmark, cleanup may not always be required to achieve this level. For example, if a site is considered to be a "removal" action, that is, the principal objective is to get rid of a sufficient amount of contaminated soil to reduced possible exposures, that risk reduction target may be as high as one additional cancer per 10,000 (i.e., 10^{-4}).

The decision regarding the actual cleanup level, including whether to approach a contaminated site or facility as a removal or a remedial action, is not risk assessment but falls within the province of risk management. The engineer or manager will have input to the decision but will not be the only party in that decision. The risk assessment data and information will comprise much of the scientific underpinning of the decision, but legal, economic, and other societal drivers will also be considered to arrive at cleanup levels. It is not unusual, for example, for a legal document, such as a consent decree, to prescribe cleanup levels more protective than typical removal or remedial concentration levels.

Solvents-R-We's chemical plant has been emitting tetrachloromethane (CCl_4) for >20 years. Over the years, air pollution studies by the county and state have found varying concentrations of CCl_4 downwind from the plant. A well that serves as the principal water supply for the town of Apple Chill has found been found to contain 80 mg L^{-1} CCl_4. If the average adult in the town drinks 2 L d^{-1} of water from the well and lives in the town for an entire lifetime, what is the lifetime cancer risk to the population if no treatment is added? What concentration is needed to ensure that the population cancer risk is below 10^{-6}?

The lifetime cancer risk added to Apple Chill's population can be estimated using the LADD and slope factor for CCl_4. In addition to the assumptions given, we will use default values from Table 12.5. We will also assume that people live in the town for their entire lifetimes and that their exposure duration is equal to their typical lifetime.

Thus, ED and TL terms are canceled, leaving the abbreviated LADD $= \frac{(C)\cdot(CR)\cdot(AF)}{(BW)}$.

Since we have not specified male or female adults, we will use the average body weight, assuming that there are about the same number of males as females. We look up the absorption factor for CCl_4 and find that it is 0.85, so the adult lifetime exposure is

$$\text{LADD} = \frac{(80)\cdot(2)\cdot(0.85)}{(65)} = 4.2\,\text{mg kg}^{-1}\,\text{d}^{-1}$$

Using the midpoint value between the default values ($\frac{15+40}{2} = 27.5$kg) for body weight and default CR values (1 L d^{-1}), the children lifetime exposure is the following:

LADD $= \frac{(80)\cdot(1)\cdot(0.85)}{(27.5)} = 2.5\,\text{mg kg}^{-1}\,\text{d}^{-1}$ for the first 13 years and the adult exposure of 4.2 mg kg^{-1} d^{-1} thereafter.

The oral SF for CCl_4 is 1.30×10^{-1} kg d^{-1}, so the added adult lifetime risk from drinking the water is.

$$4.2\times(1.30\times10^{-1}) = 5.5\times10^{-1}$$

And the added risk to children is.

$$2.5\times(1.30\times10^{-1}) = 3.3\times10^{-1}$$

Some subpopulations are more vulnerable to exposures than others. For example, for children, environmental and public health agencies recommend an additional factor of safety beyond what would be used to calculate risks for adults. This is known as the "10×" rule, that is, children need to be protected 10 times more than adults because they have longer life expectancies (so latency periods for cancer need to be accounted for) and their tissue is developing prolifically and changing. So, in this case, with the added risk, our reported "risk" would be 3.3. While this is statistically impossible (i.e., one cannot have a probability greater than one because it would mean that the outcome is >100% likely, which of course is impossible!), it is actually an adjustment to the cleanup concentration. Since the combination of a very high slope of the dose-response curve and a very high LADD increases the risk, children need a measure of protection beyond the general population. This is accomplished by removing either the contaminants from the water or the provision of a new water supply. In any event, the city public works and/or health department should mandate another source of drinking water (e.g., bottled water) immediately.

The cleanup of the water supply to achieve risks below 1 in a million can be calculated from the same information and reordering the risk equation to solve for C:

$$\text{Risk} = \text{LADD}\times\text{SF}$$

$$\text{Risk} = \frac{(C)\cdot(CR)\cdot(AF)\cdot(SF)}{(BW)}$$

$$C = \frac{(BW)\cdot\text{Risk}}{(CR)\cdot(AF)\cdot(SF)}$$

Based on adult LADD, the well water must be treated so that the tetrachloromethane concentrations are below:

$$C = \frac{(65)\cdot10^{-6}}{(2)\cdot(0.85)\cdot(0.13)} = 2.9\times10^{-4}\,\text{mg L}^{-1} = 290\,\text{ng L}^{-1}$$

Based on children's LADD and the additional "10×," the well water must be treated so that the tetrachloromethane concentrations are below:

$$C = \frac{(27.5)\cdot10^{-7}}{(1)\cdot(0.85)\cdot(0.13)} = 2.5\times10^{-5}\,\text{mg L}^{-1} = 25\,\text{ng L}^{-1}$$

The town must remove the contaminant so that the concentration of CCl_4 in the finished water will be at a level six orders of magnitude less than the untreated well water, that is, lowered from $80\,mg\,L^{-1}$ to $25\,ng\,L^{-1}$. Cleanup standards are part of the arsenal needed to manage risks. However, other considerations needed to be given to a contaminated site, such as how to monitor the progress in lowering levels and how to ensure that the community stays engaged and is participating in the cleanup actions, where appropriate. Even when the engineering solutions are working well, the risk manager must allot sufficient time and effort to these other activities, otherwise skepticism and distrust can arise.

It is important to note that these example calculations are instructive, but not sufficient. There are numerous often better ways to calculate risk-based cleanup concentrations. As mentioned, assume lifetime exposure durations are conservative estimation of long-term exposures, especially for chronic diseases, for example, cancer and cardiovascular ailments. Regarding toxicity, the literature is constantly advancing for both cancer and noncancer endpoints, with slope factors and RfDs updated. Every cleanup is unique, so there will also likely be additional requirements beyond exposure and hazard factors in the risk equation.

By the way, since at least part of the contamination appears to be linked to air emissions by Solvents-R-We, source apportionment modeling is in order, that is, to identify the relative contributions of this and other sources to the town's CCl_4 contamination. For the air emissions, receptor models are in order. The atmospheric contribution can then be added to other pollutant transport models, for example, groundwater, to determine the total CCl_4 transport.

This is an instance of an air pollution problem that results in an oral exposure route. One caveat for these calculations is that inhalation and other routes may also add to the CCl_4 dose.

Some general risk assessment principles have been almost universally adopted by regulatory agencies, especially those concerned with cancer risks from environmental exposures (see Table 12.9).

As mentioned, many risks are not directly tied to a single agent but occur in the milieu of mixtures and coexposures. In addition, the risks to public health, such as cancer and noncancer endpoints, are only one class of outcomes of concern. Others include ecosystem changes from gene flow and other ecological endpoints and opportunity risks.

TABLE 12.9 General principles applied to health and environmental risk assessments conducted by regulatory agencies in the United States

Principle	Explanation
Human data are preferable to animal data	For purposes of hazard identification and dose-response evaluation, epidemiological and other human data better predict health effects than animal models
Animal data can be used in lieu of sufficient, meaningful human data	While epidemiological data are preferred, agencies can extrapolate hazards and generate dose-response curves from animal models
Animal studies can be used as basis for risk assessment	Risk assessments can be based upon data from the most highly sensitive animal studies
Route of exposure in animal study should be analogous to human routes	Animal studies are best if from the same route of exposure as those in humans, for example, inhalation, dermal, or ingestion routes. For example, if an air pollutant is being studied in rats, inhalation is a better indicator of effect than if the rats are dosed on the skin or if the exposure is dietary
Threshold is assumed for noncarcinogens	For noncancer effects, for example, neurotoxicity, endocrine dysfunction, and immunosuppression, there is assumed to be a safe level under which no effect would occur (e.g., "no observed adverse effect level," not only NOAEL, which is preferred, but also "lowest observed adverse effect level," LOAEL)
Threshold is calculated as a reference dose or reference concentration (air)	Reference dose (RfD) or concentration (RfC) is the quotient of the threshold (NOAEL) divided by factors of safety (uncertainty factors and modifying factors; each usually multiples of 10): $\mathrm{RfD} = \frac{\mathrm{NOAEL}}{\mathrm{UF} \times \mathrm{MF}}$
Sources of uncertainty must be identified	Uncertainty factors (UFs) address: • interindividual variability in testing, • interspecies extrapolation, • LOAEL-to-NOAEL extrapolation, • Subchronic-to-chronic extrapolation, • Route-to-route extrapolation, • Data quality (precision, accuracy, completeness, and representativeness) Modifying factors (MFs) address uncertainties that are less explicit than the UFs

Continued

TABLE 12.9 General principles applied to health and environmental risk assessments conducted by regulatory agencies in the United States—cont'd

Principle	Explanation
Factors of safety can be generalized	The uncertainty and modifying factors should follow certain protocols, for example, 10=for extrapolation from a sensitive individual to a population, 10=rat to human extrapolation, 10=subchronic-to-chronic data extrapolation, and 10=LOAEL used instead of NOAEL
No threshold is assumed for carcinogens	There is no safe level of exposure assumed for cancer-causing agents
Precautionary principle applied to cancer model	A linear, no-threshold dose-response model is used to estimate cancer effects at low doses, that is, to draw the unknown part of the dose-response curve from the region of observation (where data are available) to the region of extrapolation
Precautionary principle applied to cancer exposure assessment	The most highly exposed individual is generally used in the risk assessment (upper-bound exposure assumptions). Agencies are reconsidering this worst-case policy and considering more realistic exposure scenarios

Source: U.S. Environmental Protection Agency, General Principles for Performing Aggregate Exposure and Risk Assessment, 2001, EPA/600/R-03/036.

An air quality decision must be comprehensive. Even when one aspect seems to be favorable for a particular site, the overall risk of one process or operation may be greater than that of another (e.g., less reliance of toxic raw materials and decreased likelihood of release of chemical toxins). Thus, a complete life cycle must be constructed, and the composite and individual risks are compared among various alternatives.

12.9 CAUSALITY

Zero risk can be calculated either when the hazard (e.g., toxicity) does not exist or when the exposure to that hazard is zero. A substance found to be associated with cancers based upon animal testing or observations of human populations can be further characterized. Linking causes to outcomes begins with the weight-of-evidence and statistical association. The association of two factors, such as the level of exposure to a compound and the occurrence of a disease, does not necessarily mean that one necessarily "causes" the other. Often, after study, a third variable explains the relationship. However, it is important for science to do what it can to link causes with effects. Otherwise, corrective and preventive actions cannot be identified. So, strength of association is a beginning step toward cause and effect. A major consideration in strength of association is the application of sound technical judgment of the weight of evidence. For example, characterizing the weight of evidence for carcinogenicity in humans consists of three major steps [24]:

1. Characterization of the evidence from human studies and from animal studies individually
2. Combination of the characterizations of these two types of data to show the overall weight of evidence for human carcinogenicity
3. Evaluation of all supporting information to determine if the overall weight of evidence should be changed

Note that none of these steps is certain.

Air quality managers and engineers need tools to support the risk analysis, but none or all these tools are sufficient. They are too complicated and complex for simple risk calculations. There are numerous ways to evaluate the performance of an air quality system. Does it "work" (effectiveness)? Is it the best way to reach the end for which we strive (efficiency)? If it works and if it is the best means of providing the outcome, what is the probability of benefit from the system for a societal (medical, industrial, agricultural, or environmental) problem under prescribed conditions [27].

Engineers add a few steps. We must consider whether the technology will likely continue to "work as desired" (reliability), and further, we must consider the hazards that can come about as the new technology is used. Risk is a function of likelihood that the hazard will in fact be encounter, so we must also try to predict the adverse implications that society might face (risk). Thus, the "risk" refers to the possibility and likelihood of undesirable and possibly harmful effects. Errors in risk prediction can range from not foreseeing outcomes that are merely annoying (e.g., occasional odors) to those that are devastating (e.g., the release of carcinogens into the environment) [28].

Air quality engineers and managers must often rely in part on semiquantitative or qualitative risk assessments. Retrospective failure analyses can be more quantitative, since there are forensic techniques available to tease out and assign weights to the factors that led to the outcome. There are even methods to calculate outcomes that had other steps been taken.

Air quality management success is a function of the amount of risk that has been reduced or avoided. Decisions must be based on scientifically credible information. This information is part of the risk assessment that informs waste management decisions. Thus, the success of all waste operations is, to some extent, a reflection of the operation's hazards and potential exposures to these hazards. Thus, the engineer and air quality manager are accountable for reducing risks, which means that the systems to do so are reliable and resilient in the face of dynamic operating and environmental conditions.

REFERENCES

[1] R. Vandoorne, P.J. Gräbe, Stochastic modelling for the maintenance of life cycle cost of rails using Monte Carlo simulation, Proc. Inst. Mech. Eng. Part F: J. Rail Rapid Transit 232 (4) (2018) 1240–1251.
[2] P. O'Connor, A. Kleyner, Practical Reliability Engineering, John Wiley & Sons, 2012.
[3] D.A. Vallero, Excerpted and Revised From: Fundamentals of Air Pollution, fifth ed., Elsevier Academic Press, Waltham, MA, 2014. p. 999 pages cm.
[4] S. Macgill, Y. Siu, A new paradigm for risk analysis, Futures 37 (10) (2005) 1105–1131.
[5] E. Hollnagel, D.D. Woods, N. Leveson, Resilience Engineering: Concepts and Precepts, Ashgate Publishing, Ltd, 2007.
[6] J.D. Solomon, Communicating Reliability, Risk and Resiliency to Decision Makers, October 12, 2017, 2017.
[7] J.D. Solomon, D.A. Vallero, From Our Partners – Communicating Risk and Resiliency: Special Considerations for Rare Events, 1 May 2017. Available: https://cip.gmu.edu/2016/06/01/partners-communicating-risk-resiliency-special-considerations-rare-events/, 2016.
[8] U.S. Environmental Protection Agency, An Examination of EPA Risk Assessment Principles and Practices, EPA/100/B-04/001, 2004, Available at: http://www.epa.gov/OSA/pdfs/ratf-final.pdf.
[9] E. Persson, What are the core ideas behind the precautionary principle? Sci. Total Environ. 557 (2016) 134–141.
[10] C. Singh, 19_The Precautionary Principle and Environment Protection, 2016.
[11] C.G. Turvey, E.M. Mojduszka, C.E. Pray, The precautionary principle, the law of unintended consequences, and biotechnology, in: Presented at the Agricultural Biotechnology: Ten Years After, Ravello, Italy, July 6–10, 2005, 2005.
[12] P. Harremoës, et al., Late Lessons from Early Warnings: The Precautionary Principle 1896–2000, Office for Official Publications of the European Communities, 2001.
[13] Science & Environmental Health Network, Wingspread Conference on the Precautionary Principle, 1998.
[14] United Nations Environment Programme, Rio declaration on environment and development, United Nations Environment Programme, Nairobi, Kenya, 1992.
[15] M.D. Wood, K. Plourde, S. Larkin, P.P. Egeghy, A.J. Williams, V. Zemba, I. Linkov, D.A. Vallero, Advances on a decision analytic approach to exposure-based chemical prioritization. Risk Anal. (2018), https://doi.org/10.1111/risa.13001.
[16] P.A. Schulte, et al., Occupational safety and health, green chemistry, and sustainability: a review of areas of convergence, Environ. Health 12 (1) (2013) 1.
[17] D.A. Vallero, Air pollution monitoring changes to accompany the transition from a control to a systems focus, Sustainability 8 (12) (2016) 1216.
[18] A.M. Gauthier, et al., Chemical assessment state of the science: evaluation of 32 decision-support tools used to screen and prioritize chemicals, Integr. Environ. Assess. Manag. 11 (2) (2015) 242–255.
[19] National Academy of Science, National Research Council, Risk Assessment in the Federal Government: Managing the Process, National Academy Press, Washington, DC, 1983.
[20] National Research Council, Science and Decisions: Advancing Risk Assessment, The National Academies Press, Washington, DC, 2009, p. 424.
[21] National Research Council, Toxicity Testing in the 21st Century: A Vision and a Strategy, National Academies Press, 2007.
[22] National Research Council, Exposure Science in the 21st Century: A Vision and a Strategy, The National Academies Press, Washington, DC, 2012, p. 196.
[23] U.S. Environmental Protection Agency, Endocrine Disruptor Screening Program (EDSP) Overview, 8 March 2018. Available: https://www.epa.gov/endocrine-disruption/endocrine-disruptor-screening-program-edsp-overview, 2017.
[24] U.S. Environmental Protection Agency, Final Rule, "Procedures for Chemical Risk Evaluation Under the Amended Toxic Substances Control Act" 2018.
[25] Environmental Health Analysis Center, PBT Profiler Methodology (2.000 ed.), Available: http://www.pbtprofiler.net/Methodology.asp, 2012.
[26] Environmental Health Analysis Center, PBT Profiler, 2.000 ed, U.S. Environmental Protection Agency, 2012.
[27] D.J. Dix, K.A. Houck, M.T. Martin, A.M. Richard, R.W. Setzer, R.J. Kavlock, The ToxCast program for prioritizing toxicity testing of environmental chemicals, Toxicol. Sci. 95 (1) (2007) 5–12.
[28] P.P. Egeghy, D.A. Vallero, E.A.C. Hubal, Exposure-based prioritization of chemicals for risk assessment, Environ. Sci. Pol. 14 (8) (2011) 950–964.
[29] S. Gangwal, et al., Incorporating exposure information into the toxicological prioritization index decision support framework, Sci. Total Environ. 435 (2012) 316–325.
[30] R.S. Judson, et al., In vitro screening of environmental chemicals for targeted testing prioritization: the ToxCast project, Environ. Health Perspect. 118 (4) (2010) 485.

[31] J.F. Wambaugh, et al., High-throughput models for exposure-based chemical prioritization in the ExpoCast project, Environ. Sci. Technol. 47 (15) (2013) 8479–8488.
[32] D. Mackay, A. Fraser, Bioaccumulation of persistent organic chemicals: mechanisms and models, Environ. Pollut. 110 (3) (2000) 375–391.
[33] V. Boonyaroj, C. Chiemchaisri, W. Chiemchaisri, K. Yamamoto, Enhanced biodegradation of phenolic compounds in landfill leachate by enriched nitrifying membrane bioreactor sludge, J. Hazard. Mater. 323, pp (2017) 311–318.
[34] Y. Long, Y.-Y. Long, H.-C. Liu, D.-S. Shen, Degradation of refuse in hybrid bioreactor landfill, Biomed. Environ. Sci. 22 (4) (2009) 303–310.
[35] R. Muñoz, S. Villaverde, B. Guieysse, S. Revah, Two-phase partitioning bioreactors for treatment of volatile organic compounds, Biotechnol. Adv. 25 (4) (2007) 410–422.
[36] U.S. Environmental Protection Agency, Landfill Bioreactor Performance: Second Interim Report, Outer Loop Recycling & Disposal Facility, Louisville, Kentucky, EPA/600/R-07/060, 2007.
[37] J.G. Sims, J.A. Steevens, The role of metabolism in the toxicity of 2, 4, 6-trinitrotoluene and its degradation products to the aquatic amphipod Hyalella azteca, Ecotoxicol. Environ. Saf. 70 (1) (2008) 38–46.
[38] D.P. Williams, B.K. Park, Idiosyncratic toxicity: the role of toxicophores and bioactivation, Drug Discov. Today 8 (22) (2003) 1044–1050.
[39] A.L. Karmaus, D.L. Filer, M.T. Martin, K.A. Houck, Evaluation of food-relevant chemicals in the ToxCast high-throughput screening program, Food Chem. Toxicol. 92 (2016) 188–196.
[40] P.T. Anastas, R.L. Lankey, Life cycle assessment and green chemistry: the yin and yang of industrial ecology, Green Chem. 2 (6) (2000) 289–295.
[41] Health and Safety Executive, Guidance on ALARP Decisions in COMAH, 14 May 2017. Available: http://www.hse.gov.uk/foi/internalops/hid_circs/permissioning/spc_perm_37/, 2017.
[42] B. Ale, Tolerable or acceptable: a comparison of risk regulation in the United Kingdom and in the Netherlands, Risk Anal. 25 (2) (2005) 231–241.
[43] Companhia Ambiental do Estado de São Paulo, CETESB. 2001, in: Relatório de qualidade das águas interiores do estado de São Paulo, 2000.
[44] D.T. Allen, D.R. Shonnard, Green Engineering: Environmentally Conscious Design of Chemical Processes, Pearson Education, 2001.
[45] S. Billatos, Green Technology and Design for the Environment, CRC Press, 1997.
[46] D.A. Vallero, C. Brasier, Sustainable Design: The Science of Sustainability and Green Engineering, John Wiley & Sons, 2008.
[47] C.P. Wild, The exposome: from concept to utility, Int. J. Epidemiol. 41 (1) (2012) 24–32.
[48] D. Vallero, Environmental Biotechnology: A Biosystems Approach, Elsevier Science, 2015.
[49] A. Kortenkamp, T. Backhaus, M. Faust, State of the art report on mixture toxicity, Contract 70307 (2007485103) (2009) 94–103.
[50] National Academy of Science, National Research Council, Biosolids Applied to Land: Advancing Standards and Practices, National Academies Press, 2002.
[51] U.S. Environmental Protection Agency, Chemistry Dashboard: Benzo(a)pyrene, 5 September 2018. Available: https://comptox.epa.gov/dashboard/dsstoxdb/results?search=DTXSID2020139#exposure-predictions, 2018.
[52] J.F. Wambaugh, et al., High throughput heuristics for prioritizing human exposure to environmental chemicals, Environ. Sci. Technol. 48 (21) (2014) 12760–12767.
[53] US Environmental Protection Agency, n.d. Framework for Ecological Risk Assessment. Washington, DC: USEPA Risk Assessment Forum, EPA 630/R-92/0011992.
[54] D. Vallero, Paradigms Lost: Learning from Environmental Mistakes, Mishaps and Misdeeds, Butterworth-Heinemann, 2005.
[55] E.K. Silbergeld, Risk assessment and risk management: an uneasy divorce, in: D.G. Mayo, R.D. Hollander (Eds.), Acceptable Evidence: Science and Values in Risk Management, 1994, pp. 99–114.
[56] P. Doblhoff-Dier, et al., Safe biotechnology 10: DNA content of biotechnological process waste. The safety in biotechnology working party on the European Federation of Biotechnology, Trends Biotechnol. (2000).
[57] International Organization for Standardization, ISO 31000: 2009 Risk Management—Principles and Guidelines, ISO, Geneva, Switzerland, 2009.
[58] U.S. Environmental Protection Agency, Guidelines for Human Exposure Assessment: Peer Review Draft, 2017, Available at: https://www.epa.gov/sites/production/files/2016-02/documents/guidelines_for_human_exposure_assessment_peer_review_draftv2.pdf.
[59] A. Guy, C. Gauthier, G. Griffin, Adopting alternative methods for regulatory testing in Canada, in: Proceedings of the 6th World Congress on Alternatives & Animal Use in the Life Sciences. AATEX, 14 2008, pp. 322–327.
[60] X. Zhang, J.A. Arnot, F. Wania, Model for screening-level assessment of near-field human exposure to neutral organic chemicals released indoors, Environ. Sci. Technol. 48 (20) (2014) 12312–12319.
[61] Chemical Computing Group, Molecular Operating Environment: Chemoinformatics and Structure Based Tools for High Throughput Screening, Chemical Computing Group, Montreal, Canada, 2013.
[62] D.C. Hilton, R.S. Jones, A. Sjödin, A method for rapid, non-targeted screening for environmental contaminants in household dust, J. Chromatogr. A 1217 (44) (2010) 6851–6856.
[63] E.A.C. Hubal, et al., Exposure science and the US EPA National Center for computational Toxicology, J. Expos. Sci. Environ. Epidemiol. 20 (3) (2010) 231–236.
[64] Proposed Guidelines for Carcinogen Risk Assessment, Available: http://cfpub.epa.gov/ncea/cfm/recordisplay.cfm?deid=2833, 2015.
[65] M. Peterson, An Introduction to Decision Theory, Cambridge University Press, 2009.
[66] M.D. Resnik, Choices: An Introduction to Decision Theory, University of Minnesota Press, 1987.
[67] M.B. Van Asselt, Perspectives on uncertainty and risk, in: Perspectives on Uncertainty and Risk, Springer, 2000, pp. 407–417.
[68] D. Vallero, Translating Diverse Environmental Data into Reliable Information: How to Coordinate Evidence from Different Sources, Academic Press, 2017.

[69] M.D. Adams, P.S. Kanaroglou, A criticality index for air pollution monitors, Atmos. Pollut. Res. 7 (3) (2016) 482–487.
[70] P. Boffetta, J.K. McLaughlin, C. La Vecchia, R.E. Tarone, L. Lipworth, W.J. Blot, False-positive results in cancer epidemiology: a plea for epistemological modesty, J. Natl. Cancer Inst. 100 (14) (2008) 988–995.
[71] D. Vallero, Environmental Contaminants: Assessment and Control, Academic Press, 2010.
[72] U.S. Environmental Protection Agency, Example Exposure Scenarios, EPA/600/R-03/036, 2004.
[73] J. Krajewski, M. Dobecki, J. Gromiec, Retention of vinyl chloride in the human lung, Occup. Environ. Med. 37 (4) (1980) 373–374.
[74] U.S. Environmental Protection Agency, Chemistry Dashboard: Vinyl Chloride ADME, 5 September 2018. Available: https://comptox.epa.gov/dashboard/dsstoxdb/results?search=DTXSID8021434#adme-ivive-subtab, 2018.
[75] U.S. Environmental Protection Agency, E-FAST Glossary, 5 September 2018. Available: https://www.epa.gov/tsca-screening-tools/e-fast-glossary#ladd, 2016.
[76] U.S. Environmental Protection Agency, Methods for derivation of inhalation reference concentrations and application of inhalation dosimetry, US EPA, Washington, DC, 1994.
[77] M. Bailey, et al., Risk Assessment Guidance for Superfund: Volume I: Human Health Evaluation Manual (Part F, Supplemental Guidance for Inhalation Risk Assessment), Environmental Protection Agency, 2009.
[78] U.S. Environmental Protection Agency, Technical Overview of Ecological Risk Assessment: Risk Characterization, 5 September 2018. Available: https://www.epa.gov/pesticide-science-and-assessing-pesticide-risks/technical-overview-ecological-risk-assessment-risk, 2017.
[79] ChemSafetyPRO, How to Calculate Hazard Quotient (HQ) and Risk Quotient (RQ), 5 September 2018. Available: https://www.chemsafetypro.com/Topics/CRA/How_to_Calculate_Hazard_Quotients_(HQ)_and_Risk_Quotients_(RQ).html, 2018.

FURTHER READING

[80] D. Kirchhoff, B. Doberstein, Pipeline risk assessment and risk acceptance criteria in the State of Sao Paulo, Brazil, Impact Assessment Project Appraisal 24 (3) (2006) 221–234.
[81] U.S. Environmental Protection Agency, Guidelines for Human Exposure Assessment, 11 May 2018. Available: https://www.epa.gov/risk/guidelines-human-exposure-assessment, 2017.
[82] World Health Organization, Report on Integrated Risk Assessment, World Health Organization, Geneva, Switzerland, 2001, WHO/IPCS/IRA/01/12.
[83] G.W. Suter II, Ecological Risk Assessment, CRC Press, 2016.
[84] D. Mangis, D. Vallero (Ed.), Discussion of Exposure Within the Context of Food Chains, Research Triangle Park, NC, 2005.
[85] D.A. Vallero, V.G. Zartarian, S.I. Isukapalli, T.R. Mccurdy, T. Mckone, P. Georgopoulos, C.C. Dary, Modeling and predicting pesticide exposures, in: R. Krieger (Ed.), Hayes Handbook of Pesticide Toxicology, Elsevier Science, New York, NY, 2010, pp. 995–1020.
[86] X. Zhang, et al., A physiologically based pharmacokinetic/pharmacodynamic model for carbofuran in Sprague-Dawley rats using the exposure-related dose estimating model, Toxicol. Sci. 100 (2) (2007) 345–359.
[87] D.A. Vallero, T.M. Letcher, Unraveling Environmental Disasters, Newnes, 2012.
[88] T. McKone, W. Riley, R. Maddalena, R. Rosenbaum, D. Vallero, Common issues in human and ecosystem exposure assessment: The significance of partitioning, kinetics, and uptake at biological exchange surfaces, Epidemiology 17 (6) (2006) S134.
[89] R.K. Rosenbaum, et al., USEtox human exposure and toxicity factors for comparative assessment of toxic emissions in life cycle analysis: sensitivity to key chemical properties, Int. J. Life Cycle Assess. 16 (8) (2011) 710–727.
[90] B. Singer, A tool to predict exposure to hazardous air pollutants, Environ. Energy Technol. Div. News 4 (4) (2003) 5.
[91] U.S. Environmental Protection Agency, Method 300.1: Determination of Inorganic Anions in Drinking Water by Ion Chromatography, 1997.
[92] M.A. Hollinger, M.J. Derelanko, Handbook of Toxicology, CRC Press, 2002.
[93] J. Moya, et al., Exposure Factors Handbook, US Environmental Protection Agency, Washington, 2011.
[94] E.J. Calabrese, ATSDR Public Health Assessment Guidance Manual, CRC Press, 1992.
[95] U.S. Environmental Protection Agency, Office of Emergency and Remedial Response, Risk Assessment Guidance for Superfund: Human Health Evaluation Manual, Development of Risk-Based Preliminary Remediation Goals,1991.
[96] U.S. Environmental Protection Agency, General Principles for Performing Aggregate Exposure and Risk Assessment, 2001, EPA/600/R-03/036.

Chapter 13

Air pollution control technologies

13.1 INTRODUCTION

The best means of addressing air pollutants is to prevent their existence through waste minimization and prevention. These techniques have improved, but unfortunately, the most common way to control air pollution remains pollution removal and treatment. However, the life cycle approach discussed in Chapter 1 is becoming more common. For example, steps can be taken early in the life cycle to reduce the amounts and concentrations of contaminants that reach the stack, for example, precleaning technologies like mechanical collectors. These do not completely eliminate the pollutant, but decrease the size and volume, inlet loading, of substances that need to be controlled at the stack, especially particulate matter (PM) inlet loading [1].

For pollution removal to be accomplished, the polluted carrier gas must pass through a control device or system, which collects or destroys the pollutant and releases the cleaned carrier gas to the atmosphere. The control device or system selected must be specific for the pollutant of concern. If the pollutant is an aerosol, the device used will, in most cases, be different from the one used for a gaseous pollutant. If the aerosol is a dry solid, a different device must be used than for liquid droplets.

Not only the pollutant itself but also the carrier gas, the emitting process, and the operational variables of the process affect the selection of the control system. Table 13.1 illustrates the large number of variables that must be considered in controlling pollution from a source. As such, each emission requiring controls is to at least some extent unique. Therefore, the calculations presented in this chapter are merely examples, which need to be customized according to these variables and other specific conditions and factors of the pollutant source.

After the control system is installed, its operation and maintenance become a major concern. Important reasons for an operation and maintenance (O&M) program are (1) the necessity of continuously meeting emission regulations, (2) prolonging control equipment life, (3) maintaining productivity of the process served by the control device, (4) reducing operation costs, (5) promoting better public relations and avoiding community alienation, and (6) promoting better relations with regulatory officials [2].

The O&M program has the following minimum requirements: (1) an equipment and record system with equipment information, warranties, instruction manuals, etc.; (2) lubrication and cleaning schedules; (3) planning and scheduling of preventive maintenance; (4) a storeroom and inventory system for spare parts and supplies; (5) listing of maintenance personnel; (6) costs and budgets for O&M; and (7) storage of special tools and equipment.

13.2 PARTICULATE MATTER CONTROLS

As discussed in Chapter 10, air pollutants are generally classified as either aerosol phase or gas phase. The air pollution control technologies differ substantially between these two classes. Removing and controlling particulate matter (PM) emissions take advantage of physical principles. Unlike gases, which flow with the air, particles deposit differentially within the airway by mechanical processes. Pollution control equipment to remove PM makes use of several physical mechanisms:

- Gravitational settling
- Inertial impaction
- Centrifugal inertial force
- Brownian motion

Air Pollution Calculations. https://doi.org/10.1016/B978-0-12-814934-8.00013-2

TABLE 13.1 Key characteristics of air pollution control devices and systems

Factor	Characteristic of concern
General	Collection efficiency
	Legal limitations such as best available technology
	Initial cost
	Lifetime and salvage value
	Operation and maintenance costs
	Power requirement
	Space requirements and weight
	Materials of construction
	Reliability
	Reputation of manufacturer and guarantees
	Ultimate disposal/use of pollutants
Carrier gas	Temperature
	Pressure
	Humidity
	Density
	Viscosity
	Dew point of all condensables
	Corrosiveness
	Inflammability
	Toxicity
Process	Gas flow rate and velocity
	Pollutant concentration
	Variability of gas and pollutant flow rates, temperature, etc.
	Allowable pressure drop
Pollutant (if gas phase)	Corrosiveness
	Inflammability
	Toxicity
	Reactivity
Pollutant (if aerosol phase)	Size range and distribution
	Particle shape
	Agglomeration tendencies
	Corrosiveness
	Abrasiveness
	Hygroscopic tendencies
	Stickiness
	Inflammability
	Toxicity
	Electric resistivity
	Reactivity

Information sources D.A. Vallero, Fundamentals of Air Pollution, 5th ed., Elsevier Academic Press, Waltham, MA, 2014, p. 999 pages cm; L. Theodore, A.J. Buonicore, Air Pollution Control Equipment: Selection, Design, Operation and Maintenance, 1982; L. Theodore, A.J. Buonicore, Air Pollution Control Equipment, 1988; A.C. Stern, H.C. Wohlers, R.W. Boubel, W.P. Lowery, Fundamentals of Air Pollution, Academic Press, New York and London, 1973; S. Oglesby, G.B. Nichols, Electrostatic precipitators. in: A.C. Stern (Ed.), Air Pollution, 1977.

- Electrostatic attraction
- Thermophoresis
- Diffusiophoresis

The importance of each varies according to the class of equipment. Each mechanism applies a combination of forces to a particle that makes it move.

13.3 FORCES

The design of the control equipment takes advantage of this motion, that is, moving the particle toward a collecting surface. Obviously, engineering must apply physical principles within a domain of uncertainty, so the more that is known about the mass and energy relationships of the particle, the airstream, and the surfaces of the control equipment, the more that can be included in a design. These basic relationships are summarized as

$$\sum F = m_p \cdot a_p = m_p \frac{dv_p}{dt} \tag{13.1}$$

where $\sum F$ = the sum of all forces acting on a particle (g cm sec^{-2}), m_p = particle mass (g), a_p = particle acceleration (cm sec^{-2}), v_p = particle velocity, and t = time (sec). In English units, these relationships are expressed as

$$\sum F = \frac{m_p \cdot a_p}{g_c} \tag{13.2}$$

where $\sum F$ = the sum of all forces acting on a particle (foot-pounds, lb_f), m_p = particle mass (international avoirdupois pound, lb_m), and a_p = particle acceleration (ft sec^{-2}). The foot-pound is the product of lb_m and g_0 (i.e., the standard gravitational field). That is, 1 lb_f is equal to the force exerted by 1b_m in a gravitational field. The gravitational constant (g_c) is expressed as

$$g_c = 32.2 \frac{\text{lb}_\text{m} \cdot \text{ft}}{\text{lb}_\text{f} \cdot \text{sec}^2} \tag{13.3}$$

All these mechanisms are involved in particle movement. In practice, much of the motion that accounts for efficient particle removal involves three basic processes:

1. Initial capture of particles on surfaces
2. Gravity settling of solids into the hopper
3. Removal of solids from the hopper

For example, high-efficiency particulate control systems, for example, fabric filters and electrostatic precipitators, apply these processes, although the initial step of capturing varies considerably (e.g., one uses filtration, whereas the other uses electrostatics). Also, the means of using gravity will vary among devices. Thus, this discussion differentiates the particle collection mechanisms that control the effectiveness of initial capture of the incoming particles and gravity settling of the collected solids.

The distribution of PM sized is the first information that is needed when designing, selecting, and installing equipment. The type of source, operation conditions, and other factors lead to significant differences in the particle size ranges, even from within the same industry or other source categories. In addition, the type of control equipment may use the same physical mechanisms as other equipment but in varying proportions.

Indeed, technologies even within the same class, for example, filtration, are not limited to filter mechanisms. For example, a pulse-jet fabric filter uses inertial impaction, Brownian motion, and electrostatic attraction to capture particles in the size range of 100 μm to <0.01 μm onto the accumulated layers of dust on the exterior surfaces of the bags. Thus, the bags must be cleaned regularly. Large aggregates of dust cake become dislodged from the filter surface and drop into the hopper. These visible agglomerations of solids are several orders of magnitude larger than the individual particle, that is, between 10,000 and 50,000 μm (1.0–5.0 cm). Thus, the aggregate mass causes them to readily fall into the hopper. Unfortunately, poor O&M, for example, improper filter cleaning, will allow for dislodging much smaller agglomerations, thus greatly decreasing the settling rate. This means that even with a very efficient capture of PM, the control equipment can become much less efficient if not followed by well-maintained gravity settling systems.

The principle of pressure differential is crucial to particulate removal. The laws of potentiality state that flow moves from high to low pressure. So, if pressure can decrease significantly below that of the atmosphere, air will move to that pressure trough. If there is a big pressure difference between air outside and inside, the flow will be quite rapid. So, the "vacuum" (it is really a pressure differential) is created inside the vacuum cleaner using an electric pump. When the air rushes to the low-pressure region, it carries particles with it. Increasing velocity is proportional to increasing mass and numbers of particles that can be carried. This is the same principle as the "competence" of a stream, which is high (i.e., can carry heavier loads) in a flowing river, but the competence declines rapidly at the delta where stream velocity approaches zero. This causes the river to drop its particles in descending mass, that is, sedimentation of heavier particles first, but colloidal matter remaining suspended for much longer times.

Two of the most common types of particle collection systems in industry are cyclones and fabric filter systems. Both the cyclone and the fabric filter are all designed to remove particles. In the United States, air quality standards were first directed at total suspended particulates (TSP) as measured by a high-volume sampler, that is, a device that collected a large range of sizes of particles (aerodynamic diameters up to 50 μm), with added emphasis on smaller particles with time. Presently, ultrafine particles (aerodynamic diameter ≤ 100 nm) are a major new emphasis. Some of the older PM removal systems will fall short in removing them but will likely still need to employ the same principles mentioned above but in different ways. For example, coarse PM will still need to be removed, but after this process, other mechanisms will be needed for the ultrafines.

Particle size is also important because of the relationship between the diameter of a sphere and the sphere's surface area. For example, in a multistage impactor, each stage removes particles of progressively smaller diameter. The PM collected on each stage is an indication that inertial forces can be used to collect PM. As shown in Table 13.2, with each order of magnitude decrease in diameter, the surface area increases by two orders of magnitude. Thus, a 1 μm particle has six orders of magnitude greater surface area than a 1 mm particle. This difference in size-to-surface-area ratio affects sorption, which means that smaller particles will be more likely to sorb materials, including toxic compounds.

13.3.1 Gravity

After initial capture, gravity is the next important mechanism. Gravity settling was among the first particulate control devices, working on the principal that with expanding volume, gas velocity decreases along with the gas stream's ability to hold the particle. Thus, the forces exerted on the particle must be calculated to estimate the extent to which a particle or clumps formed from particle aggregates can be collected by gravitational settling. These forces are the gravitational force (F_G), buoyant force (F_B), and drag force (F_D). The gravitational force causes the PM mass to fall. This is expressed as

$$F_G = m_p \cdot g = \rho_p \cdot V_p \cdot g \quad (13.4)$$

where F_G = gravitational force (g cm sec^{-2}), m_p = particle mass (g), g = particle acceleration due to gravity (980 cm sec^{-2}), ρ_p = particle density (g cm^{-3}), and V_p = particle volume (cm^3).

TABLE 13.2 Relationship between spherical particle diameter, volume, and surface area

Diameter (μm)	Volume (cm^3)	Area (cm^2)
0.1	5.23×10^{-16}	3.14×10^{-10}
1.0	5.23×10^{-13}	3.14×10^{-8}
10.0	5.23×10^{-10}	3.14×10^{-6}
100.0	5.23×10^{-7}	3.14×10^{-4}
1000.0	5.23×10^{-4}	3.14×10^{-2}

Particle settling is directly proportional to mass. However, since PM is usually expressed as size (diameter), mass is assumed to be proportional to diameter. Theoretical "cut size" is the size above which all particles will be collected. Usually, the cut size or cut diameter describes the particle diameter that is removed with 50% efficiency. The 2.5 μm cut size captures half of the $PM_{2.5}$ and allows half to pass through the device. If the particle is assumed to be spherical, the volume term in Eq. (13.6) can be simplified as

$$V_p = \frac{\pi d_p^3}{6} \quad (13.5)$$

where d_p is the physical diameter of the particle (cm). Thus, substituting this term in Eq. (13.6) gives

$$F_G = \frac{\pi d_p^3 \rho_p g^p}{6} \quad (13.6)$$

13.3.2 Buoyancy

A force in the opposite direction of gravity is buoyancy. Buoyancy force (B_F) resists gravity, so the greater the buoyancy, the more likely a particle will remain suspended in the air. This can be expressed as

$$F_B = m_g \cdot g = \rho_g \cdot V_p \cdot g \quad (13.7)$$

where F_B = buoyancy force (g cm sec^{-2}), m_g = mass of displaced gas (g), g=particle acceleration due to gravity (980 cm sec^{-2}), ρ_g = carrier gas density (g cm^{-3}), and V_p = particle volume (cm^3). Again, assuming a spherical particle, this equation can also be simplified:

$$F_B = \frac{\pi d_p^3 \rho_g g^g}{6} \quad (13.8)$$

Note that F_B depends on the density of the gas, which is usually air. Even in very polluted carrying gases, the gas is predominantly air. Thus, ρ_g is on the order of 10^{-2} lb_m ft.$^{-3}$, whereas the F_G depends on ρ_p, which is on the order of 10^2 lb_m ft.$^{-3}$. This density difference means that in most practical situations, the buoyant force is orders of magnitude smaller than the gravitational force and is usually neglected.

13.3.3 Drag

As the particle begins to move downward due to F_G, it encounters a resistive force that increases with increasing downward velocity. This is the third important force acting on a particle, that is, drag force:

$$F_D = \frac{A_g \cdot \rho_g \cdot V_p^2 \cdot C_D}{2} = \frac{\pi d_p^2 \rho_g v^g p^2 \cdot C_D}{8} \quad (13.9)$$

where F_D = drag force (g cm sec^{-2}); A_p=cross-sectional area of the particle (cm), which equals $\frac{\pi d_p^2}{4}$; ρ_g=gas density (g cm^{-3}); and C_D=drag coefficient (dimensionless).

The drag force results from the gas in front of the particle that is being displaced as the particle moves. This imparts momentum on the gas. Thus, F_D equals the momentum per unit time imparted by the gas on the particle. Some of the particle velocity (v_p) is imparted to the gas as gas velocity (v_g). The amount of energy that v_p gives to v_g is related to a factor of friction, which is the drag coefficient (C_D) (see Fig. 13.1).

The drag coefficient is related to particle velocity and the flow pattern of the gas around the particle. This flow is determined by the Reynolds number (N_R or Re).[1] As discussed in Chapter 10, at some point in the velocity increase, a fluid's flow ceases to be laminar and becomes turbulent. The Re is expressed as the ratio of inertial to viscous forces in a fluid:

$$\text{Re} = \frac{\text{Inertial Forces}}{\text{Viscous Forces}} \quad (13.10)$$

1. For general fluid dynamic aspects of the Reynolds number is N_R. The symbol *Re* is commonly used in air pollution science and engineering, so it is used here. The underlying theory and practice is the same; only the symbols differ.

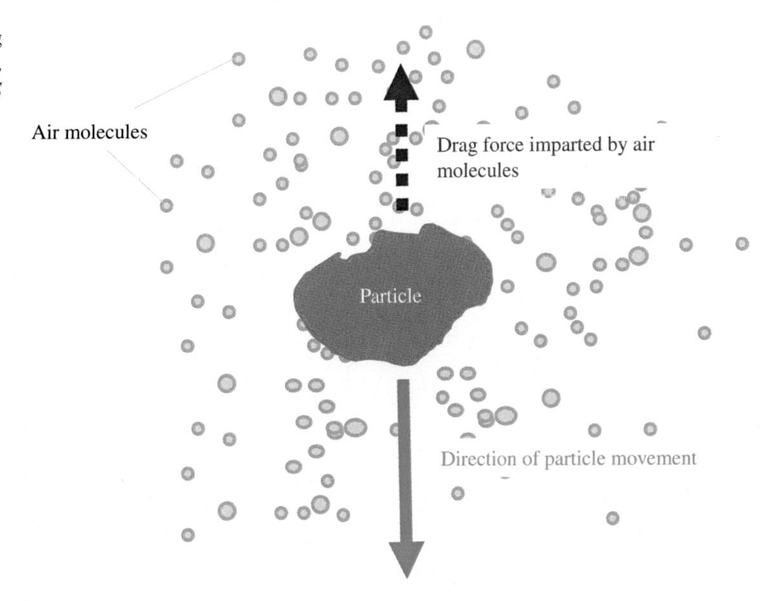

FIG. 13.1 Drag force of air imparted onto a particle moving because of gravitational force. *(Modified from J.R. Richards, Control of gaseous emissions, 3rd ed., Student Manual: US Environmental Protection Agency APTI Course 415, 2000.)*

The inertial forces are proportional to the velocity and density of the fluid and to the diameter of the conduit in which the fluid is moving. An increase in any of these factors will lead to a proportional increase in the momentum of the flowing fluid. The coefficient of viscosity or absolute viscosity (μ) represents the total viscous force of the fluid, so N_R can be calculated as

$$\mathrm{Re} = \frac{d_e v \rho}{\mu} \tag{13.11}$$

where D_e is the conduit's equivalent diameter, which evaluates the fluid flow as a physical length. Recall that $\mu\, \rho^{-1}$ is the kinematic viscosity υ, so the Reynolds number can be stated as the relationship between the size of the conduit, the average fluid velocity v, and υ:

$$\mathrm{Re} = \frac{d_e v}{\upsilon} \tag{13.12}$$

At very low velocities, the mass moves in discrete layers parallel to one another. The only movement across the fluid layers is molecular motion, which creates viscosity. Such a flow is laminar. The dimensionless particle Reynolds ($\mathrm{Re_p}$ or N_{Rp}) number can also be expressed in terms like those of the forces but adding gas viscosity:

$$\mathrm{Re}_p = \frac{d_p v_p \rho_g}{\mu_g} \tag{13.13}$$

where μ_g is the gas viscosity ($\mathrm{g\,cm^{-1}\,s^{-1}}$).

As mentioned, three particle flow regions exist: laminar, transition, and turbulent (see Fig. 13.2). For low values of the particle Reynolds number ($\mathrm{Re_p} < 1$), the flow is considered laminar.[2] Laminar flow is defined as flow in which the fluid moves in layers smoothly over an adjacent particle surface. For much greater values of the particle Reynolds number (e.g., $\mathrm{Re_p} > 1000$), the flow is turbulent.[3] Turbulent flow is characterized by erratic motion of fluid, with a violent interchange of momentum throughout the fluid near the particle surface. For particle Reynolds numbers between 1 and 1000, the flow is in the transition region, that is, the flow can be either laminar or turbulent, depending on local conditions. In most air pollution control applications, particles $<100\,\mu\mathrm{m}$ are in the laminar flow region. Transition and turbulent flow conditions are relevant primarily to the gravity settling of large agglomerates in fabric filters and electrostatic precipitators.

The experimental data that are the basis for Fig. 13.2 can be used to relate the values of C_D and $\mathrm{Re_p}$ as expressed in equations for each particle Reynolds region [3]. The drag coefficient in the laminar flow region ($\mathrm{Re}_p < 1$) is

2. The laminar region is also known as the Stokes region.
3. The turbulent region is also known as the Newton region.

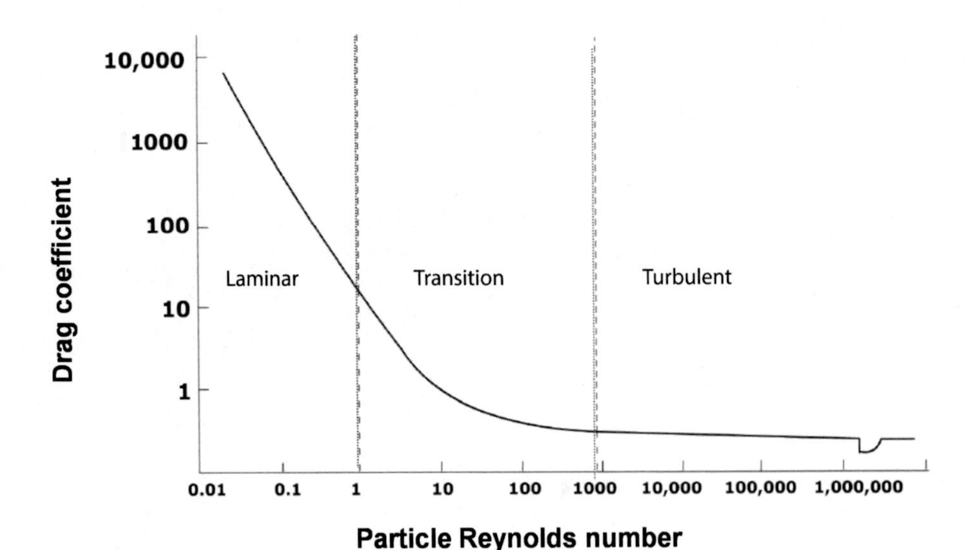

FIG. 13.2 Relationship between drag coefficient and particle Reynolds number. *(Source: US Environmental Protection Agency, APTI 413: Control of particulate matter emissions. Chapter 4: Particle collection mechanisms, 2012.)*

$$C_D = \frac{24}{Re_p} \tag{13.14}$$

The drag coefficient in the transition flow region ($1 < Re_p < 1000$) is

$$C_D = \frac{18.5}{Re_p^{0.6}} \tag{13.15}$$

The drag coefficient in the turbulent flow region ($Re_p > 1000$) is

$$C_D = 0.44 \tag{13.16}$$

These equations apply to particles with diameters >3 μm, that is, those where the gas is continuous around the particle. For smaller particles, the gas appears as distinct molecules. Such small-diameter particles can slip between the gas molecules, which means they fall faster than relationships developed for continuous media predict. English mathematician Ebenezer Cunningham deduced in 1910 that the drag coefficient should be reduced for small particles [4]. This means that the drag coefficient equation for the laminar region must be modified to include a term called the Cunningham slip correction factor (C_c):

$$C_D = \frac{24}{Re_p \cdot C_C} \tag{13.17}$$

An approximation of the slip correction (see Fig. 13.3) can be based on the temperature and particle diameter:

$$C_C = 1 + \frac{6.21 \times 10^{-4} T}{d_p} \tag{13.18}$$

where T = absolute temperature (K) and d_p = particle diameter (μm).

Applying the slip corrections and substituting terms allow drag forces to be calculated for each flow region. The drag force for the laminar region is

$$F_D = \frac{3\pi \mu_g v_p d_p}{C_C} \tag{13.19}$$

The drag force for the transition flow region is

$$F_D = 2.30 \left(d_p \cdot v_p \right)^{1.4} \cdot \mu_g^{0.6} \cdot \rho_g^{0.4} \tag{13.20}$$

The drag force for the turbulent flow region is

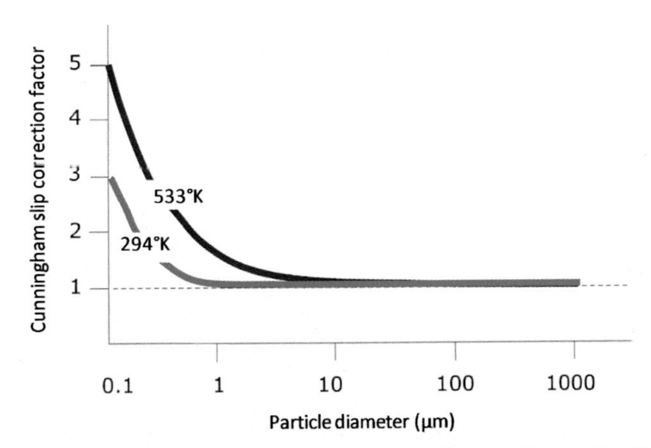

FIG. 13.3 Cunningham slip correction factor as a function of temperature and particle diameter. *(Source: D.A. Vallero, Fundamentals of Air Pollution, 5th ed., Elsevier Academic Press, Waltham, MA, 2014, p. 999 pages cm; data from U.S. Environmental Protection Agency, Chapter 4: particle collection mechanisms, 3rd ed., in: J.R. Richards (Ed.), APTI Course 413: Control of particulate matter emissions, ICES Ltd, Research Triangle Park, NC, 2000. [Online]. Available: http://www.4cleanair.org/APTI/413Combined.pdf.)*

$$F_D = 0.05\pi\left(d_p \cdot v_p\right)^2 \cdot \rho_g \tag{13.21}$$

where μ_g = gas viscosity ($g\,cm^{-1}\,s^{-1}$), d_p = particle diameter (cm), ρ_g = gas density ($g\,cm^{-3}$), vp = particle velocity relative to the gas ($cm\,sec^{-1}$), and C_C = Cunningham slip factor (dimensionless).

With buoyancy not a factor, the two forces dominating particle deposition are gravity and drag. As the particle accelerates, its velocity increases. As shown in Fig. 13.1, gravity and drag work against one another, either keeping the particle suspended or moving it toward settling. Thus, another way of considering Eq. (13.3) is to show when this balance is achieved:

$$\sum F = F_G - F_D = m_p \cdot a_p = m_p \frac{dv_p}{dt} = 0 \tag{13.22}$$

The drag force on the particle also increases with increasing velocity. At some point, velocity will become high enough that the drag force will be as large as the gravitational force. At this point, the gravitational and drag force balance, that is, the net force is zero, and the particle will no longer accelerate, reaching a constant velocity. This constant velocity is known as the terminal settling velocity (v_t).

Terminal velocities differ for the three regions and can be derived by substituting the terms:

$$\text{Laminar flow}: v_t = \frac{g \cdot C_C \cdot \rho_p \cdot d_p^2}{18\mu_g} \tag{13.23}$$

$$\text{Transition flow}: v_t = \frac{0.153 g^{0.71} \cdot \rho_p^{0.71} \cdot d_p^{1.14}}{\mu_g^{0.43} \cdot \rho_g^{0.29}} \tag{13.24}$$

$$\text{Turbulent flow}: v_t = 1.74\left(\frac{g \cdot \rho_p \cdot d_p}{\cdot \rho_g}\right)^{0.5} \tag{13.25}$$

where v_t = terminal particle settling velocity ($cm\,sec^{-1}$), g = particle acceleration due to gravity ($980\,cm\,sec^{-1}$), C_C = Cunningham slip correction factor (dimensionless), μ_g = gas viscosity ($g\,cm^{-1}\,s^{-1}$), d_p = particle diameter (cm), ρ_p = particle density ($g\,cm^{-3}$), ρ_g = gas density ($g\,cm^{-3}$), and v_p = particle velocity relative to the gas ($cm\,sec^{-1}$).

The slip correction and Reynolds numbers can now be used to determine the flow region in which the particles are settling in the pollution control equipment. This is found by using a characteristic value, that is, the K value, for each type of flow:

$$K = d_p\left(\frac{g \cdot \rho_p \cdot \rho_g}{\mu_g^2}\right)^{0.33} \tag{13.26}$$

The units for the variables are the same as the previous equations, with the gravitational acceleration $(g) = 980\,cm\,sec^{-2}$.

Each flow region type has a K-value range:
Laminar region: $K < 2.62$
Transitional region: $2.62 < K < 69.12$
Turbulent region: $K > 69.12$

At 20°C, what is the K value of a 10 μm particle in air (density = 1.2×10^{-3} g cm^{-3} and viscosity = 1.8×10) with a particle density of 1.0 g cm^{-3}? What is the flow region for this system?

$$K = 1.0 \times 10^{-3}\,\text{cm} \left(\frac{(980\ \text{cm sec}^{-2}) \times (1.0\ \text{g cm}^{-3}) \times (1.2\ \text{g} \times 10^{-3}\ \text{cm}^{-3})}{(1.80 \times 10^{-4}\ \text{g cm}^{-1}\ \text{sec}^{-1})^2} \right)^{0.33} \approx \underline{\mathbf{0.3}}$$

Thus, this is in the laminar flow region.

Note that for more viscous gases, the K value would be much smaller, since the term for gas viscosity in the denominator is squared (squaring numbers <1 decreases the product). Thus, more viscous gases have larger laminar flow regions (i.e., flow is less likely to be turbulent) than less viscous gases.

Assuming $C_C = 1.0$, what is the terminal velocity for the particle in the previous calculation?

$$V_t = \frac{g \cdot C_C \cdot \rho_p \cdot d_p^2}{18\mu_g} = \left(\frac{(980\ \text{cm sec}^{-2}) \times (1.0\ \text{g cm}^{-3}) \times (10 \times 10^{-4}\ \text{cm})^2}{18 \times (1.80 \times 10^{-4}\ \text{g cm}^{-1}\ \text{sec}^{-1})} \right) \approx 0.3\ \text{cm sec}^{-1}.$$

This particle, that is, diameter = 10 μm, is considered by regulators to be a coarse particle in the ambient air but is relatively small with regard to gravity-based settling equipment.

What is the K value, flow region, and terminal velocity for a larger particle with an aerodynamic diameter = 100 μm, and which is denser particle, that is, $\rho_p = 2$ of 1.0 g cm^{-3}?

$$K = 1.0 \times 10^{-2}\,\text{cm} \left(\frac{(980\ \text{cm sec}^{-2}) \times (2.0\ \text{g cm}^{-3}) \times (1.2\ \text{g} \times 10^{-3}\ \text{cm}^{-3})}{(1.80 \times 10^{-4}\ \text{g cm}^{-1}\ \text{sec}^{-1})^2} \right)^{0.33} \approx \underline{\mathbf{38}}$$

This is well within the transitional flow regime.
The terminal velocity would be calculated using Eq. (13.26):

$$V_t = \frac{0.153 g^{0.71} \cdot \rho_p^{0.71} \cdot d_p^{1.14}}{\mu_g^{0.43} \cdot \rho_g^{0.29}} = \frac{[0.153 \times (980\ \text{cm sec}^{-2}]^{0.71} \times (2.0\ \text{g cm}^{-3})^{0.71} \times (1.0 \times 10^{-2}\text{cm})^{1.14}}{(1.80 \times 10^{-4}\ \text{g cm}^{-1}\ \text{sec}^{-1})^{0.43} \times (1.2 \times 10^{-3}\ \text{g cm}^{-3})^{0.29}} \approx \underline{\mathbf{86\ cm\ sec^{-1}}}$$

Obviously, with this greater terminal velocity, the denser and larger particle would be much more amenable to gravitational collection equipment. These equations have been used to calculate the terminal settling velocities of a wide range of particles with physical diameters from 10^{-2} to 10^5 μm. Table 13.3 indicates that terminal settling velocities are negligible for the very fine PM, that is, diameters <2.5 μm, and even for much of the coarse PM fraction, that is, <10 μm. This indicates that PM control devices that employ only gravitational settling are primarily used for very coarse PM and are useful for initial separation and as precleaners designed to decrease the large-particle fraction before entering a next stage of treatment or in front of air movers and other ancillary equipment. They are particularly useful for gravity settling to remove large agglomerated masses or clumps of dust that have been collected on fabric filter, precipitator plates, or other collection surfaces. These large clumps of material have high terminal settling velocities (e.g., the bottom rows in Table 13.3).

Another important force that acts in similar way as gravitation is centrifugal force. This is discussed in detail later in this chapter.

TABLE 13.3 Terminal settling velocities for spherical, unit-density particles in air at 25°C

Particle size (μm)	Terminal settling velocity at 25°C (cm sec^{-1})	Flow condition
0.1	0.000087	Laminar
1.0	0.0035	Laminar
2.5	0.2	Laminar
10.0	0.304	Laminar
50.0	7.5	Laminar
80.0	19.3	Laminar
100.0	31.2	Transitional
200.0	68.8	Transitional
1000.0	430.7	Transitional
10,000.0	1583	Turbulent
100,000.0	5004	Turbulent

Note: $PM_{2.5}$ based on an interpolation.
Source: U.S. Environmental Protection Agency, Chapter 4: particle collection mechanisms, 3rd ed., in J.R. Richards (Ed.), APTI Course 413: Control of particulate matter emissions, ICES Ltd, Research Triangle Park, NC, 2000. [Online]. Available: http://www.4cleanair.org/APTI/413Combined.pdf.

13.4 PARTICLE MORPHOLOGY [5]

The assumption in these force equations that a particle is spherical, however, is not actually the case in most air pollutant scenarios. Indeed, another particle characteristic that influences removal is the relationship between aerodynamic diameter and shape or morphology. For example, the settling velocities in Table 13.3 are assumed to be spheres, but these values would differ for other shapes. The unit-density sphere mentioned in the table's caption is a correction for both density (i.e., $\rho_p = 1\ g\ cm^{-3}$ = the density of water at STP) and the condition of the air (i.e., calm) but should also be corrected for the different morphology of a particle compared with that of a sphere.

Few particles are spheres, however, so particles that appear to be different in physical size and shape can have the same aerodynamic diameter, as shown in Table 13.4. Conversely, some particles that appear similar can have somewhat different aerodynamic diameters. Particles can be spheres, hollow spheres, irregular shapes, flakes, fibers, aggregates, and condensation flocs [6].

13.5 MECHANISMS OF PARTICULATE REMOVAL

The principal PM removal technologies are inertial and gravitational separation, fabric filtration, and electromagnetics. Filtration is an important technology in every aspect of environmental engineering (i.e., air pollution, wastewater treatment, drinking water, and even hazardous waste and sediment cleanup). Basically, filtration consists of four mechanical processes: (1) diffusion, (2) interception, (3) inertial impaction, and (4) electrostatics (see Fig. 10.17 in Chapter 10).

A fifth mechanism can be added when discussing PM removal, that is, gravitational settling.

Recall from Chapter 5 that diffusion is important only for very small particles (<0.1 μm diameter) because the Brownian motion allows them to move in a "random walk" away from the airstream. Interception works mainly for particles with diameters between 0.1 and 1 μm. The particle does not leave the airstream but comes into contact with the filter medium (e.g., a strand of fiberglass or fabric fiber). Inertial impaction, as explained in the cyclone discussion, collects particles sufficiently large to leave the airstream by inertia (diameters >1 μm).

Electrostatics consist of electric interactions between the atoms in the filter and those in the particle at the point of contact (van der Waals force) and electrostatic attraction (charge differences between particle and filter medium). These are the processes at work in large-scale electrostatic precipitators (ESPs) that are employed in coal-fired power plant stacks around the world for particle removal. Other important factors affecting filtration efficiencies include the thickness and pore diameter or the filter; the uniformity of particle diameters and pore sizes; the solid volume fraction; the rate of particle loading onto the filter (e.g., affecting particle "bounce"); the particle phase (liquid or solid); capillarity and surface tension

TABLE 13.4 Morphology's effect on particle diameter (d) and density (ρ_p) for three particles with the same aerodynamic diameter (d_p)

	Shape	Density (ρ_p) and diameter (d)	Aerodynamic diameter (d_p)
	Solid sphere	$\rho_p = 2.0\ \text{g cm}^{-3}$ $d = 1.4\ \mu m$	
	Hollow sphere	$\rho_p = 0.5\ \text{g cm}^{-3}$ $d = 2.8\ \mu m$	$d_p = 2.0\ \mu m$
	Irregular	$\rho_p = 2.3\ \text{g cm}^{-3}$ $d = 1.3\ \mu m$	

Reproduced with permission from US Environmental Protection Agency, APTI 413: Control of particulate matter emissions. Chapter 3: Particle sizing, 2012.

(if either the particle or the filter media are coated with a liquid); and characteristics of air or other carrier gases, such as velocity, temperature, pressure, and viscosity.

Environmental engineers have been using filtration to treat air and water for several decades. Air pollution controls employing fabric filters (i.e., baghouses) remove particles from the airstream by passing the air through a porous fabric. The fabric filter is efficient at removing fine particles and can exceed efficiencies of 99%. Based solely on an extrapolation of air pollution control equipment, filtration should be better a cyclone. However, this does not take into operational efficiencies and effectiveness, which are very important to the consumer and the engineer. Changing the bag and insuring that it does not exceed its capacity must be monitored closely by the user. Also, the efficiency of the equipment is only as good as the materials being used. For example, a filter's efficiency depends on interception, inertial impaction, electromagnetics, and to a lesser extent diffusion. However, the cyclone only requires optimization for inertia.

Selecting the correct control device is a matter of optimizing efficiencies. In some environments, for example, research "clean rooms" in laboratories and assembly operations (e.g., semiconductors and high-efficiency particulate air (HEPA) filters) are fitted to equipment to enhance removal efficiency. Efficiency is often expressed as a percentage. So, a 99.99% HEPA filter is efficient enough to remove 99.99% particles from the airstream. This means that if 10,000 particles enter the filter, on average, only one particle would pass all the way through the filter. This is the same concept that we use for incinerator efficiency, but it is known as destruction and removal efficiency (DRE), since it depends on at least two forms of energy: (1) thermal destruction of pollutants into simpler, less toxic compounds and (2) motion for the physical removal of the breakdown products. For example, in the United States, federal standards require that hazardous compounds can only be incinerated if the process is 99.99% efficient, and for the more toxic compounds (i.e., "extremely hazardous wastes"), the so-called "rule of six nines" applies (i.e., DRE >99.9999%). The HEPA and DRE calculations are simply restatements of Eq. (13.1):

$$\text{DRE or HEPA efficiency} = \frac{M_{in} - M_{out}}{M_{in}} \times 100 \qquad (13.27)$$

Thus, if $10\ \text{mg min}^{-1}$ of a hazardous compound is fed into the incinerator, only $0.001\ \text{mg min}^{-1} = 1\ \mu\text{g min}^{-1}$ is allowed to exit the stack for a hazardous waste. If the waste is an extremely hazardous waste, only $0.00001\ \text{mg min}^{-1} = 0.01\ \mu\text{g min}^{-1} = 10\ \text{ng min}^{-1}$ is allowed to exit the stack. This is the same concept that is used throughout environmental engineering to calculate treatment and removal efficiencies. For example, assume that raw wastewater enters a treatment facility with $300\ \text{mg L}^{-1}$ biochemical oxygen demand (BOD_5), $200\ \text{mg L}^{-1}$ suspended solids (SS), and $10\ \text{mg L}^{-1}$ phosphorous (P). If the plant must meet effluent standards of $<10\ \text{mg L}^{-1}$ BOD_5, $<10\ \text{mg L}^{-1}$ SS, and $<1\ \text{mg L}^{-1}$ P, the removal rates of these contaminants must be

97% for BOD_5, 95% for SS, and 90% for P, respectively. The pure efficiency values may be misleading because the ease of removal can vary significantly with each contaminant. In this case, gravitational settling in the primary stages of the treatment plant can remove most of the SS, and the secondary treatment stage removes most of the BOD, but more complicated, tertiary treatment is needed for removing most of the nutrient P.

Although the fine and ultrafine PM has received the largest focus recently, there remain important reasons to remove the mass in the coarse fraction of particles, that is, ranging between 2.5 and 10 μm aerodynamic diameters. These may consist of potentially toxic components, for example, resuspended road dust; brake lining residues; industrial by-products; tire residues; heavy metals; and aerosols generated by organisms and their materials, for example, pollen, cysts, and spores (known as "bioaerosols"). A large fraction of these coarse particles may deposit to the upper airways, causing health scientists to link them to asthma. And since asthma appears to be increasing in children, their role in triggering asthma and other respiratory diseases needs to address a compliment of ultrafine, fine, and coarse particles.

Thus, equipment selection has been complicated by technical specifications. Efficiency is an important part of effectiveness, although the two terms are not synonymous. As we discussed, efficiency is simply a metric of what you get out of a system compared with what you put in. However, you can have a bunch of very efficient systems that may *in toto* be ineffective. They are all working well as designed, but they may not be working on the right things, or their overall configuration is not optimal to solve the problem at hand. So, the correct control device is not only the one that gives optimal efficiency but also the one that effectively addresses the specific pollution problem at hand.

13.6 AEROSOL REMOVAL TECHNOLOGIES

The efficiency of removal and treatment technologies depends on the type of particle, especially the size and physicochemical properties. It also depends on the characteristics of gas stream carrying the particles. Notably, the moisture content and stickiness of the airstream and particles are very important factors on equipment selection. Generally, dry aerosols differ so much from the carrier gas stream that their removal should present no major difficulties. The aerosol is different physically, chemically, and electrically from the gas in which it is suspended. It has vastly different inertial properties than the carrying gas stream and can be subjected to an electric charge. It may be soluble in a specific liquid. With such a variety of removal mechanisms that can be applied, it is not surprising that PM, such as mineral dust, can be removed by a filter, wet scrubber, or ESP with equally satisfactory results.

13.6.1 Precleaners

Mechanical collectors sited within a manufacturing process to reduce the inlet loading of PM are known as precleaners. Usually, these remove large-diameter, that is, > 10 μm, and abrasive aerosols.

There are five major types of mechanical collectors [7]. Settling chambers and elutriators capture aerosols by gravity settling. Momentum or impingement separators, also known as baffle or knockout chambers, remove PM by gravity settling and inertia. Mechanically aided collected and inertial centrifugal collectors remove PM by inertia [1, 7].

The collection efficiency of mechanical collection generally improves with increasing particle size and density and with gas stream velocity and number of turns, baffles, or other sharp direction changes to gas flow. Fractional collection efficiencies are 5% or less for a particle diameter of 5 μm, 10%–20% for a particle diameter of 10 μm, and up to 99% for particle diameters of 90 μm or larger [7]. Precleaning is mainly aimed at treating low gas flows and low volumes compared with the treatment technologies in the next sections.

In gravitational settling chamber, large-diameter particles are removed from the gas stream by slowing the gas stream's flow. The design of settling chambers may be either a single expansion chamber or a multiple-tray settling chamber (see Fig. 13.4). The multiple-tray configuration collects smaller-diameter aerosols than the single-chamber configuration because the latter is a series of horizontal plates, so a particle falls a shorter distance to reach collecting surfaces.

An elutriator consists of a configuration of vertical towers or tubes in series. Particles with terminal settling velocities that exceed the upward gas velocity, that is, about 13 cm sec^{-1} or higher, become separated and collected at the chamber's bottom [8]. As is true for gravitational settling chambers, the finer particles, that is, those with lower terminal settling velocities, remain in the gas flow and exit the elutriator. Like gravitational settling, elutriator collection efficiency increases with particle diameter and density and the number of tubes or towers. Efficiency also increases with decreasing gas velocity.

Elutriators are used for precleaning gas streams in granulated plastics, secondary metals, and agricultural and petrochemical processes. They are usually used for gas flows and volumes less than other mechanical collectors [8].

In momentum separators, particles become separated from the gas stream by inducing a sharp directional change, for example, 90–180 degree turn, that diverts the gas stream into a hopper, so that the momentum, that is, inertia, drives the particles from the gas streamlines to the hopper (see Fig. 13.5). The number of directional changes can be increased by adding baffles (Fig. 13.5B), which slightly increases collection efficiencies.

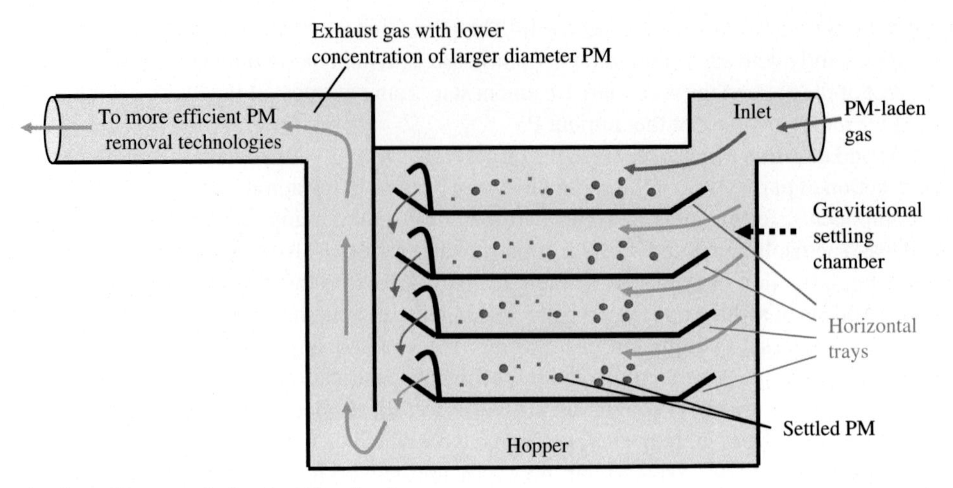

FIG. 13.4 Schematic of a multitray gravitational settling chamber.

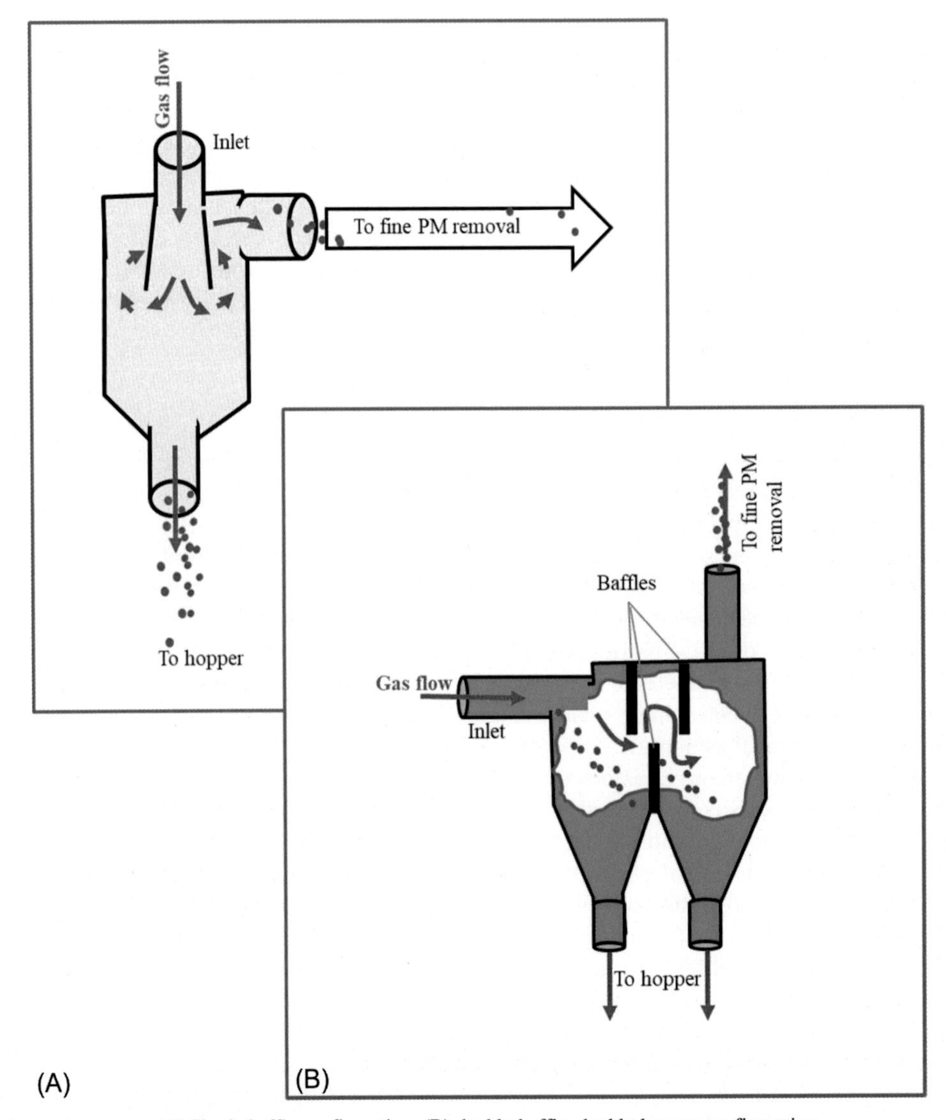

FIG. 13.5 Momentum separators. (A) Single baffle configuration; (B) double-baffle, double-hopper configuration.

In a louver configuration, the momentum collector employs a series of blades, that is, flat plates, at an angle to the gas stream. Like the baffles, the blades lengthen and slow the path of the gas stream around the blades, so that the momentum of the particles is parallel to the louver surface and across the gas stream. This results in the larger-diameter and denser particles separating into a confined volume of the gas stream.

Estimate the collection efficiency of a 50 μm-diameter particle that has a density of 4.6 g cm^{-3} in a gravitational settling chamber that is 5 m wide (*W*) × 2 m high (*H*) × 10 m long (*L*) when the gas velocity (v_g) is 0.3 m sec^{-1}. Assume a particle and gas stream conditions of 20°C and 1 atm.

A few equations need to be introduced to solve this problem [9].

Collection efficiency (η_i) is

$$\eta_i = 1 - e^{-X} \tag{13.28}$$

Where $X_i = \frac{t_r}{t_{si}}$

And t_r is the chamber residence time and t_{si} is the particle settling time. Since the residence time of a particle in the chamber is the length times gas velocity

$$t_r = \frac{L}{v_g} \tag{13.29}$$

Since $v_g = \frac{Q}{W \times H}$, volumetric gas flow rate is

$$Q = \frac{v_g}{W \times H} \tag{13.30}$$

Therefore

$$t_r = \frac{LWH}{Q} \tag{13.31}$$

In the worst case, that is, longest path to the bottom of chamber, the particle settling time (t_s) is

$$t_r = \frac{H}{v_t} \tag{13.32}$$

Where v_t is the particle terminal settling velocity

The collection efficiency can now be stated as

$$\eta_i = 1 - e^{-\frac{v_i LW}{Q}} \tag{13.33}$$

Assuming the Cunningham slip factor is 1, the terminal settling velocity of a particle in a laminar region, $\mathrm{Re_p} < 1$, is given by

$$\mathrm{v_t} = \frac{g\rho_\mathrm{p} \mathrm{d^2 p_i}}{18\mu_\mathrm{g}} \tag{13.34}$$

where

$\mathrm{v_t}$ = particle terminal settling velocity (ft/s)

G = acceleration of particle due to gravity (32.17 ft./s)

ρ_p = particle density ($\mathrm{lb_m/ft^3}$)

μ_g = gas viscosity ($\mathrm{lb_m}$/(ft · sec))

$\mathrm{d_p}$ = physical particle diameter (ft)

Particles that are smaller than 100 μm generally are in this region, but the linear relationship for drag coefficient can be extended into the transition region, up to a particle Reynolds number of about 5–10, without introducing significant error. So, this means that Eq. (13.34) is applied to particles considerably larger than 100 μm. The error introduced by this assumption should be determined, and if it is substantial, the terminal settling velocity for the transition region should be used. Substituting Eq. (13.34) into Eq. (13.33) yields the collection efficiency for particles in the laminar region

$$\eta_i = 1 - e^{-\left(\frac{g\rho LWN_c}{18\mu_g Q}\right) d_{pi}^2} \tag{13.35}$$

Thus, the volumetric gas flow rate is

$$Q = v_g WH = \left(0.3\frac{m}{\mathrm{sec}}\right)(5m)(2m) = 3.0\frac{m^3}{\mathrm{sec}} = 3.0x10^6\frac{cm^3}{\mathrm{sec}}$$

The collection efficiency is

$$\eta_i = 1 - e^{-\left(\frac{g\rho_L W N_c}{18\mu_g Q}\right) d_{pi}^2} = 1 - e^{-\left[\frac{\left(980^{ft}/_{sec^2}\right)\left(120^{lbm}/_{ft^3}\right)(1000ft)(500ft)(1)}{18\left(1.80x10^{-4g}/_{cm \cdot sec}\right)\left(3.0x10^6 \frac{cm^3}{sec}\right)}\right]\left(50x10^{-4}ft\right)^2}$$

$$= 0.997 = 99.7\% \text{collection efficiency.}$$

This appears to be a very good efficiency rating. However, this is the removal of a large-diameter particle, that is, 50 μm. Air permits and regulations are generally targeted to much smaller particle, that is, PM_{10} and $PM_{2.5}$. Therefore, this may be acceptable for precleaning, but is definitely not acceptable for emissions leaving the stack.

After pretreatment, unacceptable concentrations of PM must be removed by more efficient and sophisticated equipment, especially to collect finer PM remaining in the airstream. These are described in the following sections [7].

13.6.2 Inertial collectors

Inertial collectors, which include cyclones, baffles, louvers, and rotating impellers, operate on the principle that the aerosol material in the carrying gas stream has a greater inertia than the gas. Since the drag forces on the particle are a function of the diameter squared and the inertial forces are a function of the diameter cubed, it follows that as the particle diameter increases, the inertial (removal) force becomes relatively greater. Inertial collectors, therefore, are most efficient for larger particles. The inertia is also a function of the mass of the particle, so that heavier particles are more efficiently removed by inertial collectors. These facts explain why an inertial collector will be highly efficient for the removal of 10 μm rock dust and very inefficient for 5 μm wood particles. It would be very efficient, though, for 75 μm wood particles.

The most common inertial collector is the cyclone, which is used in two basic forms: the tangential inlet and the axial inlet (see Fig. 13.6). In actual industrial practice, the tangential inlet type is usually a large (1–5 m in diameter) single cyclone, while the axial inlet cyclone is relatively small (about 20 cm in diameter and arranged in parallel units for the desired capacity).

For any cyclone, regardless of type, the radius of motion (i.e., curvature), the particle mass, and the particle velocity determine the centrifugal force exerted on the particle. This centrifugal force may be expressed as

$$F = MA \tag{13.36}$$

where F = force (centrifugal), M = mass of the particle, and A = acceleration (centrifugal) and

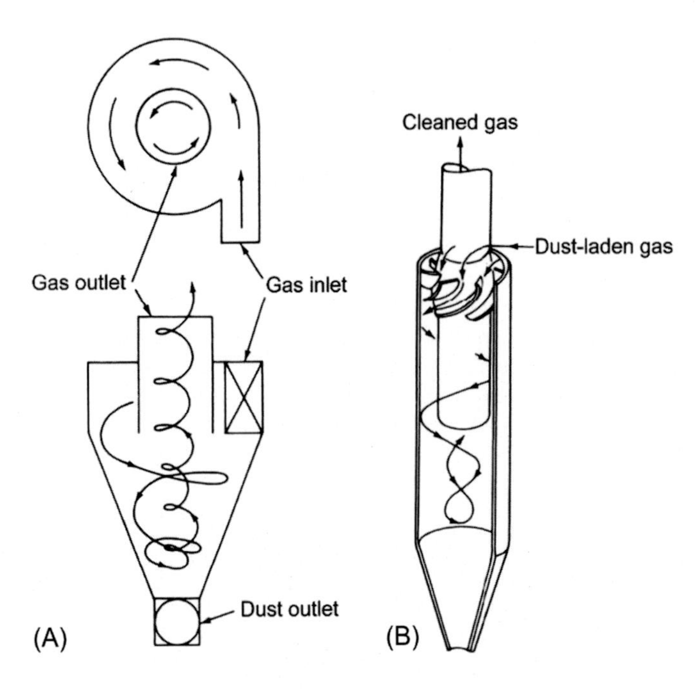

FIG. 13.6 Two common cyclone configurations: (A) tangential inlet and (B) axial inlet. *(Source: D.A. Vallero, Fundamentals of Air Pollution, 5th ed., Elsevier Academic Press, Waltham, MA, 2014, p. 999 pages cm.)*

$$A = \frac{V^2}{R} \tag{13.37}$$

where V = velocity of particle and R = radius of curvature. Therefore,

$$F - \frac{MV^2}{R} \tag{13.38}$$

Other types of inertial collectors that might be used for particulate separation from a carrying gas stream depend on the same theoretical principles developed for cyclones. Table 13.5 summarizes the effect of the common variables on inertial collector performance.

Although decreasing the radius of curvature and increasing the gas velocity both result in increased efficiency, the same changes cause increased pressure drop through the collector. Design of inertial collectors for maximum efficiency at minimum cost and minimum pressure drop is a problem that lends itself to computer optimization. Unfortunately, many inertial collectors, including most of the large single cyclones, have been designed to fit a standard-sized sheet of metal rather than a specific application and gas velocity. As tighter emission standards are adopted, the major use of inertial collectors will be to preclean in front of more sophisticated fine PM and gas control devices.

In a cyclone (Fig. 13.7), air is rapidly circulated causing suspended particles to change directions. Due to their inertia, the particles continue in their original direction and leave the airstream (see Fig. 13.8). This works well for larger particles because of their relatively large masses (and greater inertia), but very fine particles are more likely to remain in the airstream and stay suspended. The dusty air is introduced in the cyclone from the top through the inlet pipe tangential to the cylindrical portion of the cyclone. The air whirls downward to form a peripheral vortex, which creates centrifugal forces. As a result, individual particles are hurled toward the cyclone wall and, after impact, fall downward where they are collected in a hopper. When the air reaches the end of the conical segment, it will change direction and move upward toward the outlet.

TABLE 13.5 Effect of independent variables on inertial collection efficiency

Independent variable of concern	Increase or decrease to improve efficiency
Radius of curvature	Decrease
Mass of particle	Increase
Particle diameter	Increase
Particle surface/volume ratio	Decrease
Gas velocity	Increase
Gas viscosity	Decrease

Source: D.A. Vallero, Fundamentals of Air Pollution, 5th ed., Elsevier Academic Press, Waltham, MA, 2014, p. 999 pages cm.

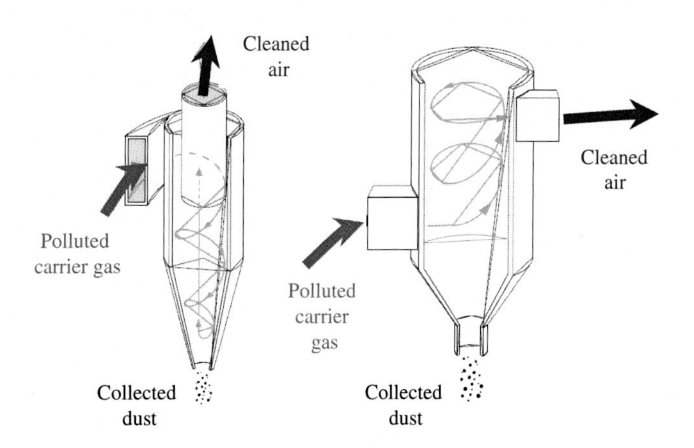

FIG. 13.7 Schematic of simple cyclone separators: (A) top inlet type and (B) bottom inlet type. *(Source: D.A. Vallero, Fundamentals of Air Pollution, 5th ed., Elsevier Academic Press, Waltham, MA, 2014, p. 999 pages cm.; reproduced with permission from U.S. Environmental Protection Agency, Air Pollution Control Orientation Course. (2013), 30 November 2013. Available: http://www.epa.gov/air/oaqps/eog/course422/ce6.html.)*

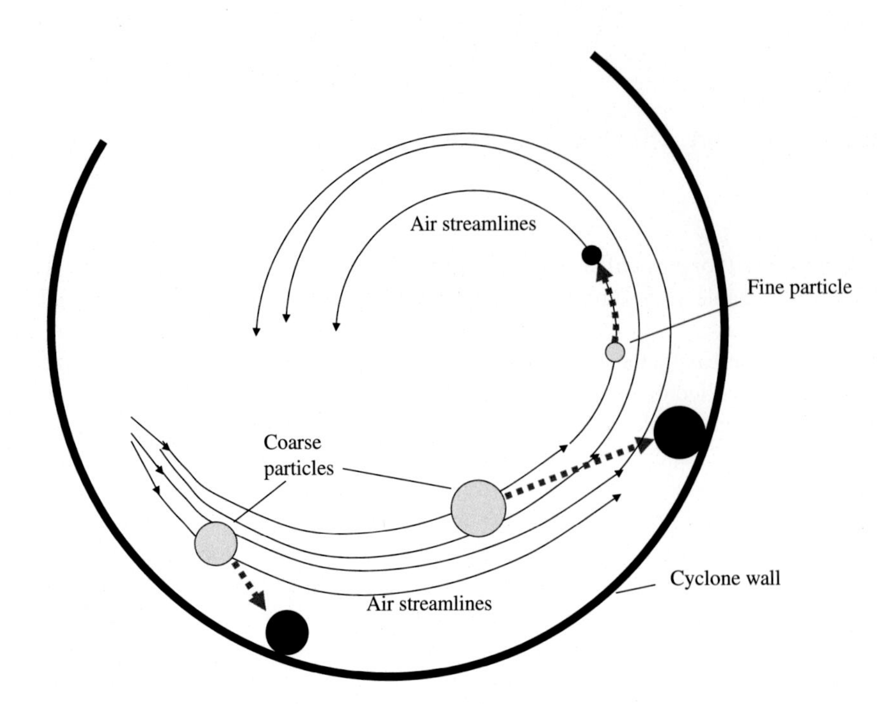

FIG. 13.8 Inertial forces in a cyclone separator. The inertia of heavier particles (mainly coarse fraction, with some very dense fine particles) is high enough that they leave the airstream of the vortex created by the cyclone inside configuration. The particle next collides with the cyclone wall. The fine and ultrafine particles, having smaller mass, continue moving with the airstream, thus remaining suspended and not collected until the stream velocity decreases to a point where it can no longer carry the particle. If this velocity is never reached, the particle will exit the cyclone. *(Source: D.A. Vallero, Fundamentals of Air Pollution, 5th ed., Elsevier Academic Press, Waltham, MA, 2014, p. 999 pages cm.)*

This forms an inner vortex. The upward airflow against gravitation allows for additional separation of particles. The cyclone vacuum cleaner applies the same inertial principles, with the collected dust hitting the sides of the removable cyclone separator and falling to its bottom.

13.6.2.1 *Centrifugal force in a cyclone*

The geometry of pollution control equipment is important. For example, circular designs are usually preferable to avoid pockets of zero to very low airflows, which can attract PM and increase the frequency of cleaning. In addition, inertial forces are often very effective when PM-laden air flows in a circular route within a cylinder (see Fig. 13.9).

The approach used to estimate terminal settling velocity due to gravity can be used for the motion of a particle due to centrifugal force (F_C). The drag and centrifugal forces balance at some point:

$$\sum F = F_C - F_D = 0 \tag{13.39}$$

$$F_C = \frac{m_p \nu_\tau^2}{R} \tag{13.40}$$

where F_C=centrifugal force (g cm sec^{-2}), m_p=particle mass (g), ν_τ=tangential velocity of the gas (cm sec^{-1}), and R=radial position of the particle (cm). The combined term $\frac{\nu_\tau^2}{R}$ is similar to the g (gravitational constant) term used to calculate gravitational settling.

The particle mass can also be expressed in terms of particle density and volume:

$$F_C = \frac{\pi d_p^3 \rho_p \nu_\tau^2}{6R} \tag{13.41}$$

where d_p=physical particle density (cm) and ρ_p=particle density (g cm^{-3}). And for fine PM with laminar Reynolds numbers, the drag force is

$$F_D = \frac{3\pi \mu_g \nu_p d_p}{C_C} \tag{13.42}$$

Eqs. (13.29), (13.30) can be substituted into Eq. (13.27) to find particle's velocity in the cylinder:

$$v_p = \frac{C_C d_p^2 \rho_p \nu_\tau}{18 \mu_g R} \tag{13.43}$$

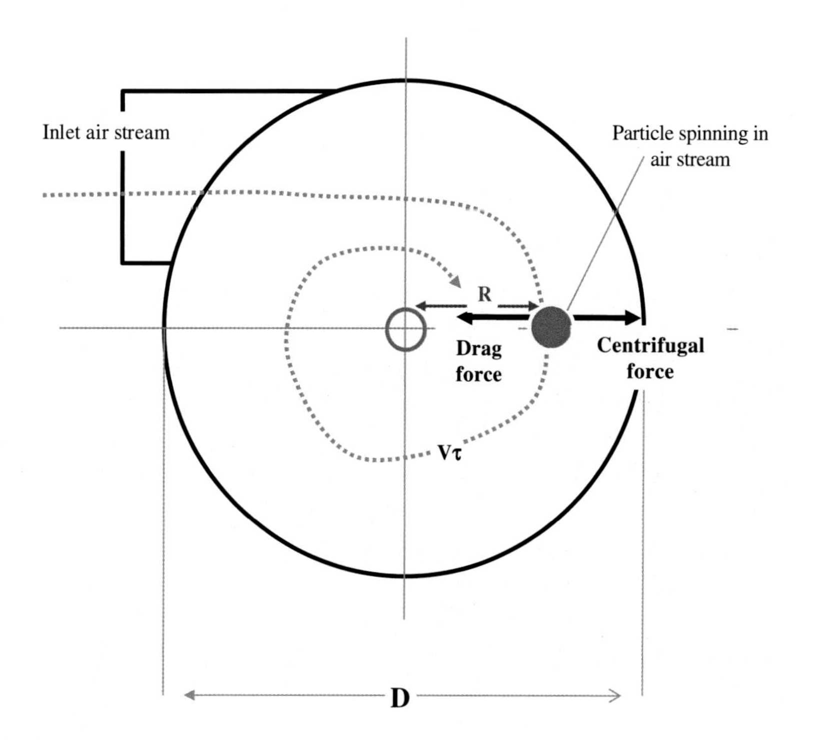

FIG. 13.9 Centrifugal force demonstrated by the top-view cutaway schematic of a cyclone. v_τ = tangential velocity of the air, D = cyclone inside diameter, R = radial position of the particle, and C_C = Cunningham slip factor. *(Source: D.A. Vallero, Fundamentals of Air Pollution, 5th ed., Elsevier Academic Press, Waltham, MA, 2014, p. 999 pages cm; reproduced with permission from U.S. Environmental Protection Agency, Chapter 4: particle collection mechanisms, 3rd ed., in: J.R. Richards (Ed.), APTI Course 413: Control of particulate matter emissions, ICES Ltd, Research Triangle Park, NC, 2000. [Online]. Available: http://www.4cleanair.org/APTI/413Combined.pdf.)*

Thus, the particle velocity across the air streamlines and its velocity moving toward the walls of the cyclone are proportional to the square of the particle's diameter. As was the case for settling velocities, the larger the particle, the more effective the cyclone will be at PM removal. Also, no matter the size, the radial velocity of the particle is proportional to the square of the airstream tangential velocity and inversely proportional to the radius of the cyclone. Thus, when selecting a cyclone system, these factors need to be known, since they determine the amount of spin within the cyclone. Higher airstream velocities will increase radial velocities, which increase the amount of PM collected. It is important not to "oversize" a cyclone, since a small radius causes a sharper turn of the airstream, which improves cyclone PM collection efficiencies.

13.6.3 Inertial impaction

A moving particle in the airstream can come into contact with surfaces of stationary and slower-moving objects. The airstream flows around the object, but due to the particle's inertia, the particle is displaced across the air streamlines. As such, it moves into the object and, if there is sufficient inertia, contacts the object and is captured (see Fig. 10.18 in Chapter 10). The object can be a plate or other target surfaces.

Note the similarities of the inertial force competing with the drag force to the gravitational and centrifugal forces discussed earlier. Indeed, impaction efficiencies may be evaluated with the same approaches used for gravitational and centrifugal forces. For particles with $Re < 1$, impaction effectiveness can relate to a dimensionless inertial impaction parameter (Ψ), calculated as

$$\Psi = \frac{C_C d_p^2 3 y_p \rho_p}{18 \mu_g D_C} \tag{13.44}$$

where C_C = Cunningham slip correction factor (dimensionless), d_p = physical particle diameter (cm), y_p = difference in velocity and collection object (cm sec^{-1}), D_C = diameter of the collection object (cm), ρ_p = density of particle, and μ_g = gas velocity (g cm^{-1} s^{-1}). If the Stokes number (Stk) is used as the inertial impaction parameter rather than Ψ, the parameter will be twice the value calculated in Eq. (13.44) since, rather than a coefficient of 18 in the denominator, the Stk denominator has a coefficient of 9.

With increasing Ψ values, particles are more likely to move radially toward a collection surface. Conversely, smaller Ψ values mean that the particle is likely to remain in the airstream and miss the target collection surface. Therefore, larger-diameter, faster-moving, and denser particles are more likely to be captured by inertial impaction. Also, smaller collection targets (e.g., fine wires or filter fibers) increase the likelihood of inertial capture.

13.6.4 Brownian motion

As mentioned, molecular diffusion is only an important collection mechanism for very small particles (e.g., $< 0.2\,\mu m$ diameters). These ultrafine PM or nanoparticles can be deflected slightly by coming into contact with gas molecules. This deflection is the result of kinetic energy from fast-moving gas molecule imposed upon the small particle. The dimensionless diffusional collection parameter (Ψ_D) is analogous to the inertial impaction parameter discussed in the previous section:

$$\Psi_D = \frac{X_p}{D_C \nu_p} \tag{13.45}$$

where X_p = particle diffusivity (cm^2 sec^{-1}), D_C = collection target's diameter (cm), and ν_p = relative velocity between the particle and the target object. Particle diffusivity is expressed as

$$X_p = \frac{C_C kT}{3\pi\mu_g d_p} \tag{13.46}$$

where k = Boltzmann's constant (g cm^2 sec^{-1} K^{-1}), C_C = Cunningham slip correction factor (dimensionless), d_p = particle's physical diameter, T = absolute temperature (K), and μ_g = gas viscosity (g cm^{-1} s^{-1}).

As Ψ_D increases, so does the tendency of a particle to be collected by applying Brownian motion. Thus, very small particles in low-viscosity gases are the best candidates for collection by molecular diffusive processes. However, the additional collection provided by diffusion becomes negligible above $0.3\,\mu m$ diameter.

13.6.5 Electrostatic mechanisms

Field charging and diffusion (or ion) charging are employed in PM collection equipment. Field charging works best for particles larger than $2\,\mu m$, with collection rapidly diminishing with decreasing diameter. The opposite is true for diffusion charging. Diffusion charging is most effective for particles with diameters $<0.4\,\mu m$ and improves with diminishing diameters.

Diffusion charging occurs from collisions between particles and unipolar ions in the airstream. The collisions result from random walks (Brownian) of both ions and very small particles. The diffusion charging continues until the surface gains particles that have sufficient charge to repel the ions. Diffusion charging needs no electric field, but if one exists, it will influence particles that have acquired a charge. The charged particles will move according to the electric field lines in the direction of the lower field strength. Thus, the collection surface would ideally have a large charge differential with the particles.

Field charging results from the creation of a strong electric field having a high unipolar ion concentration. The ions need a large enough mass to cause local disruptions of the electric field lines. The ions travel toward the particle surface along the field lines intersecting the particle. The electric charge transfer will continue until the particle has an electric field strength sufficient to repel the electric field (i.e., it has reached a saturation charge).

The number of charges imparted to a particle from field charging is a function of the square of the particle diameter. This charge level is reached rapidly, and once charged, the particles are strongly influenced by the electric field that makes for a force toward the collection surface. The particle charge is often expressed as the product of the number of charges (n) times the smallest unit of charge (i.e., the charge of an electron, which is 4.8×10^{-10} statcoulombs). The force on a particle with n units of charge in an electric field (E) is expressed as

$$F_E = n \cdot e \cdot E \tag{13.47}$$

where F_E = electrostatic charge (dyne), n = the number of charges, e = charge of electron (4.8×10^{-10} statcoulombs), and E = electric field strength (statvolt cm^{-1}).

Electric fields create forces that can be orders of magnitude greater than gravity. The particle velocities within electric fields can be estimated in the same way as gravity settling and centrifugal force:

$$F_E - F_D = 0 \tag{13.48}$$

$$n \cdot e \cdot E - \frac{3\pi\mu_g d_p \nu_p kT}{C_C} = 0 \tag{13.49}$$

The migration or drift velocity toward the collecting service (ω) applies to particles in the laminar flow region. For $Re_p > 1$, the mathematics become much more complicated. This migration is expressed as

$$\omega = \frac{n \cdot e \cdot E \cdot C_C}{3\pi\mu_g d_p \nu_p} \tag{13.50}$$

What is the migration velocity of a 2 μm unit-density particle carrying 800 units of charge in an electric field of 2 kV cm^{-1}? The gas temperature is 20°C.

- 300 V = 1 statvolt
- 1 statvolt = 1 statcoulomb cm^{-1}
- 1 dyne = 1 statcoulomb^2 cm^{-2} = 1 g cm sec^{-2}
- $C_C = 1.09$ (as calculated in Example 4–2)

The electric field in centimeter-gram-second units is

$$E = 2\frac{kV}{cm} = 2{,}000\frac{V}{cm}\left(\frac{statvolt}{300\ volts}\right) = 6.67\frac{statvolt}{cm}$$

$$= 6.67\frac{statcoulom\ bs}{cm^2}$$

$$\omega = \frac{neEC_c}{3\pi\mu_g d_p} = \frac{(800)(4.8x10^{-10} statcoulom\ bs)\left(6.67\frac{statcoulom\ bs}{cm^2}\right)(1.09)}{3\pi\left(1.8x10^{-4}\ g/_{cm \cdot sec}\right)(2x10^{04} cm)}$$

$$= 8.23^{cm/_{sec}}$$

The migration velocity of the 2 μm particle is much greater than the settling velocity of the same sized particle. Again, this illustrates that electric fields can provide much larger force than gravity.

13.6.6 Thermophoresis and diffusiophoresis

The remaining forces on particles that can be deployed for pollutant collection are thermophoresis and diffusiophoresis, both relatively weak compared with those already discussed [10].

Temperature differences on either side of a particle can lead to particle movement, that is, thermophoresis. The particle is deflected toward the colder side, since energy is transported from higher to lower potentials.

Similarly, a particle can move due to differences in concentrations of molecules in one part of a particle than another. This difference leads to an increase in molecular collisions, with the particle moving toward the region of lower concentration. Diffusiophoresis is the spontaneous motion of small particles suspended and induced by a concentration gradient of a different substance. The motion of one chemical species responds to a concentration gradient in another chemical species. The major means of achieving diffusiophoresis is to use water vapor that is migrating with the bulk airstream toward colder liquid, for example, scrubber liquid, onto which the water condenses. An airstream with 15% H_2O vapor by volume and liquid with relatively constant low temperature can enhance collection efficiency substantially. For example, a study found that after a 30 in WC pressure drop, venturi downstream from an industrial incinerator had removed 90% of the PM and cooled the gases to 77°C water and saturated the airstream with H_2O vapor, further cooling the airstream to 27°C in a 3 m countercurrent packed-bed cooling tower cooled and removed another 60% of fine PM [11].

Thus, if condensation or evaporation of water is taking place, phoresis can make for particle movement. High moisture conditions lead to temperature and concentration gradients. However, typical temperature and concentration conditions in a gas stream lead to very little phoresis, so extra measures, such as the packed-bed cooling tower mentioned above, need to be provided.

13.6.7 Net effects of forces

For most dry pollutant collection scenarios, particle size is the key factor. Particles with diameters >100 μm are generally effectively collected with inertial impaction and electrostatics and to a lesser extent from gravitational settling. Particles with diameters ranging from 10 to 100 μm respond well to inertial and electric forces, that is, proportional to the square of the particle diameter. However, the efficiencies diminish dramatically with smaller diameters to a point where molecular diffusion becomes a force of choice for very small particles (< 0.3 μm diameters) [10]. Of course, a quick perusal of the equations in this chapter indicated that no single parameter dictates the choice of equipment. For example, gas velocity and electric field strengths can be modulated to improve PM collection substantially.

13.6.8 Filtration

A filter removes PM from the carrying gas stream because the particulate impinges on and then adheres to the filter material. As time passes, the deposit of PM becomes greater, and the deposit itself then acts as a filtering medium. When the deposit becomes so heavy that the pressure necessary to force the gas through the filter becomes excessive or the flow reduction severely impairs the process, the filter must be either replaced or cleaned.

The filter medium can be fibrous, such as cloth; granular, such as sand; a rigid solid, such as a screen; or a mat, such as a felt pad. It can be in the shape of a tube, sheet, bed, fluidized bed, or any other desired form. The material can be natural or man-made fibers, granules, cloth, felt, paper, metal, ceramic, glass, or plastic. It is not surprising that filters are manufactured in an infinite variety of types, sizes, shapes, and materials.

The theory of filtration of aerosols from a gas stream is much more involved than the sieving action that removes particles in a liquid medium. The mechanisms mentioned in Chapter 10 and shown in Fig. 10.17 account for the filtration of most particles, although diffusion is important only for very small particles. Note that the particles are often depicted as spheres (see Fig. 13.7). In practice, the particles and filter elements are seldom spheres or cylinders. Thus, decisions on the optimal equipment should simply be based on not only size and mass but also morphology of the particles.

Direct interception occurs when the fluid streamline carrying the particle passes within one-half of a particle diameter of the filter element. Regardless of the particle's size, mass, or inertia, it will be collected if the streamline passes sufficiently close. Inertial impaction occurs when the particle would miss the filter element if it followed the streamline, but its inertia resists the change in direction taken by the gas molecules, and it continues in a sufficiently direct course to be collected by the filter element. Electrostatic attraction occurs because the particle, the filter, or both possess sufficient electric charge to overcome the inertial forces; the particle is then collected instead of passing the filter element. Note that size separation ("sieving") plays little or no role in filtration.

13.6.8.1 *Filter efficiency*

Particles can be measured as either mass or count. Particle count is the number of particles in a given band of mass, such as particles with aerodynamic diameters >10 μm (coarse fraction), those with diameters <10 μm but >2.5 μm (PM_{10} fraction), and those with diameters <2.5 μm ($PM_{2.5}$ fraction, also known as the fine fraction). However, the bands can be further subdivided. For example, there has been concern recently about the so-called nanoparticles. These have diameters <100 nm. Filtration is the most common method used to measure particles in the air. So, a sample taken from a filter that could show bands within the fine fraction may resemble that is shown in Table 13.6.

TABLE 13.6 Mass of particles collected on a filter (fictitious data)

Size range (mm)	Count (number of particles)	Mass (mg)	Flow rate (l min^{L3})	Integration time (min)	Mass concentration (mg m^{L3})	Description
>10	2	100	16	60	96	Reentrained dust
>2.5 < 10	20	10	16	60	9.6	Tailpipe emission
>0.01 < 2.5	200	1	16	60	0.96	Suspended colloids
>0.01	20,000	0.1	16	60	0.096	Nanoparticles; mainly carbon (fullerenes = C-60)

Since filtration is important in both measuring and controlling particle matter, expressions of filter efficiency are crucial to air pollution technologies. Eq. (13.3) provides the overall efficiency of any air pollution removal equipment. The efficiency (E) equation can be restated specifically for particles:

$$E = \frac{N_{in} + N_{out}}{N_{in}} \tag{13.51}$$

$$E = \frac{C_{in} + C_{out}}{C_{in}} \tag{13.52}$$

where N is the number of particles (count) and C is the mass concentration (subscripts, in = entering and out = exiting).

Pollution control equipment often characterizes efficiency in terms of the fraction entering versus that exiting the filter known as particle penetration (P):

$$P = \frac{N_{out}}{N_{in}} = 1 - E \tag{13.53}$$

$$P = \frac{C_{out}}{C_{in}} = 1 - E \tag{13.54}$$

Since P is the inverse of E, an inefficient filter is one that allows a large number or mass of particles to penetrate the filter. Thus, the air pollution engineer needs to specify the tolerances for filtration in any design of measurement or control technologies. Inherent to penetration calculations is the velocity of air entering the system. The front (entry side) of the filter is known as the face, so face velocity (U_0) is the air's velocity just before the air enters the filter:

$$U_0 = \frac{Q}{A} \tag{13.55}$$

where Q is the volumetric flow and A is the area of the cross section through which the air is passing. However, since the flow is restricted to the void spaces of the filter, the actual velocity in the filter itself is higher than the face velocity (same air mass through less volume). This is true for flow through any porous medium, such as polyurethane traps (see Fig. 10.2 in Chapter 10) and columns of sorbent granules, like XAD resins (see Fig. 10.3 in Chapter 10). Thus, the velocity within the filter (U_{filter}) is

$$U_{filter} = \frac{Q}{A(1-\alpha)} \tag{13.56}$$

where α is the packing density (solidity),[4] which is inverse to the filter (or trap) porosity:

$$\alpha = \frac{V_{ff}}{V_{total}} = 1 - \varphi \tag{13.57}$$

where V_{ff} = filter fiber volume, V_{total} = total filter volume, and φ = porosity.

The filter configuration is similar to that of a sieve; however, filters are quite different from sieves, given that size separation is not even listed as one of the mechanisms for PM removal. In fact, fibrous filters are more akin to numerous microscopic layers of filters, each with a specific probability of catching a particle, depending on the particle's shape and size. Therefore, the efficiency is enhanced with filter thickness. Thus, size capture is not one of the most important mechanisms for collection, compared with inertial impaction, interception, and electrostatics.

Other less important mechanisms that result in aerosol removal by filters are (1) gravitational settling due to the difference in mass of the aerosol and the carrying gas, (2) thermal precipitation due to the temperature gradient between a hot gas stream and the cooler filter medium that causes the particles to be bombarded more vigorously by the gas molecules on the side away from the filter element, and (3) Brownian deposition as the particles are bombarded with gas molecules that may cause enough movement to permit the aerosol to come into contact with the filter element. Brownian motion may also cause some of the particles to miss the filter element because they are moved away from it as they pass by.

Regardless of the mechanism that causes the aerosol to come in contact with the filter element, it will be removed from the airstream only if it adheres to the surface. Aerosols arriving later at the filter element may then, in turn, adhere to the

4. Packing density is analogous to bulk density in soil science, that is, inversely proportional to porosity. Bulk density is calculated as the dry weight of soil divided by its volume. The volume includes that of the soil particles and the pores among particles, often expressed in g cm^{-3}.

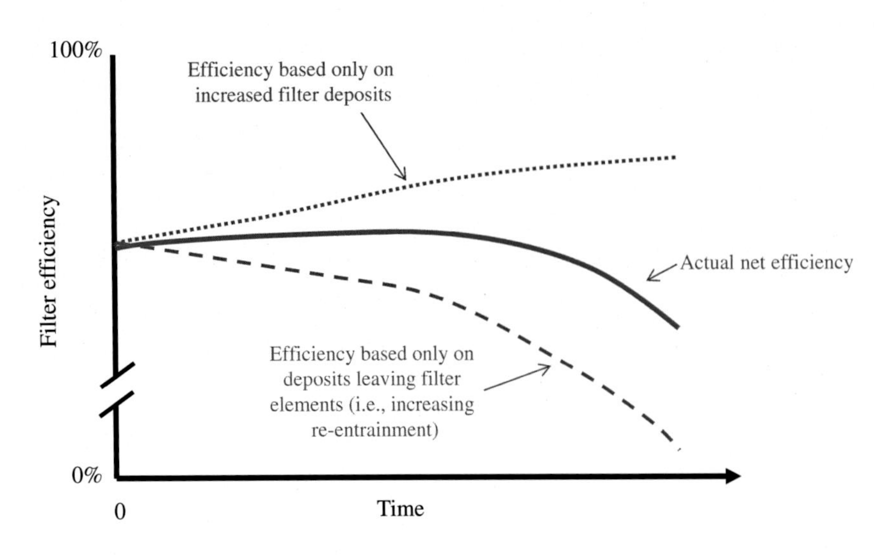

FIG. 13.10 Change of filtration efficiency with time.

collected aerosol instead of the filter element. The result is that actual aerosol removal seldom agrees with theoretical calculations. One should also consider that certain particles do not adhere to the filter element even though they touch it. As time passes, the heavier deposits on the filter surface will be dislodged more easily than the light deposits, resulting in increased reentrainment. Because of plugging of the filter with time, the apparent size of the filter element increases, causing more interception and impaction.

The contravening effects of all these variables on the particle buildup and reentrainment are shown in Fig. 13.10. In this hypothetical example, the net aerosol collection efficiency actually increases slightly compared with that of the clean filter (time = 0), due to the combination of attractions to filter element and newly deposit mass onto the filter elements. After time, however, the reentrainment of deposits decreases the filter efficiency, leading to the time when filters become much less efficient and need cleaning and replacement.

13.6.8.2 Fabric filter system configurations

The types of industrial filtration systems vary considerably. Fabric filters, commonly known as baghouses, can be highly efficient in removing PM. Filtration technologies have continuously improved from the early versions in the mid-twentieth century, with some systems efficiently removing ultrafine (~100 nm diameter) particles. The mechanisms include those discussed under inertia separation systems, for example, impaction, electrostatics, and molecular (Brownian) diffusion.

Like other air pollution control equipment, filtration is only as efficient as the O&M allows, with filters susceptible to chemical degradation, physical damage (e.g., punctures, flexing, and abrasion), and thermal breakdown. Chemical, physical, and thermal susceptibility vary by filter materials (see Table 13.7). In addition, dust cake fissures can allow excessive penetration. Also, problems can result during cleaning and bag replacement, such as excessive pressure drop.

A common configuration of a fabric filter system, that is, the shaker filter, is shown in Fig. 13.11. The filter bags are fabricated from woven material, with the material and weave selected to fit the specific application. Cotton and synthetic fabrics are used for relatively low temperatures, and glass cloth fabrics can be used for elevated temperatures, up to 290°C.

The basic operating principle of all fabric filters is that they capture particles directly in the filter material or in the dust layers that form on these materials. Thus, the first mechanism at work is inertial impaction, which is quite effective for capturing PM with diameters $\geq 1\,\mu m$, given the tortuosity of the pathways taken by particles and the many opportunities to contact objects, that is, myriad situations like those depicted in Fig. 10.18C in Chapter 10.

Static charge also builds in the fabric and dust cake, so electrostatic attraction also adds PM capture efficiency. However, since the charges can be either negative or positive, particles will oscillate between attracted and repelled depending on the net charge differential between the filter and particles. Opposite or neutral polarity leads to particle attraction. Size separation or sieving can also add to PM capture after a dust layer forms [12].

Brownian diffusion can add to the attraction of very small particles, especially given the protracted time and length required to traverse the filter and dust cake. In addition, the pore spaces are quite small, so the distance to surfaces is small enough that the diffusion path is very short.

The filtration collection curve is somewhat U-shaped, that is, filters collect large and very small particles. Filtration is very efficient for PM_{10} and larger particles and reasonably efficient for ultrafines (<100 nm), but not nearly as efficient for

TABLE 13.7 Thermal and acid resistance of common fabric filter materials

Filter material	Common or trade name	Maximum temperature (°F)		Acid resistance	Resistance to abrasion and flex
		Continuous	Surges		
Natural fiber, cellulose	Cotton	180	225	Poor	Good
Polyolefin	Polyolefin	190	200	Good to excellent	Excellent
Polypropylene	Polypropylene	200	225	Excellent	Excellent
Polyamide	Nylon	200	225	Excellent	Excellent
Acrylic	Orlon	240	260	Good	Good
Polyester	Dacron	275	325	Good	Excellent
Aromatic polyamide	Nomex	400	425	Fair	Excellent
Polyphenylene sulfide	Ryton	400	425	Good	Excellent
Polyimide	P-84	400	425	Good	Excellent
Fiberglass	Fiberglass	500	550	Fair	Fair
Fluorocarbon	Teflon	400	500	Excellent	Fair
Stainless steel	Stainless steel	750	900	Good	Excellent
Ceramic	Nextel	1300	1400	Good	Fair

Source: US Environmental Protection Agency, APTI 413: Control of particulate matter emissions. Chapter 7: Fabric filters, 2012.

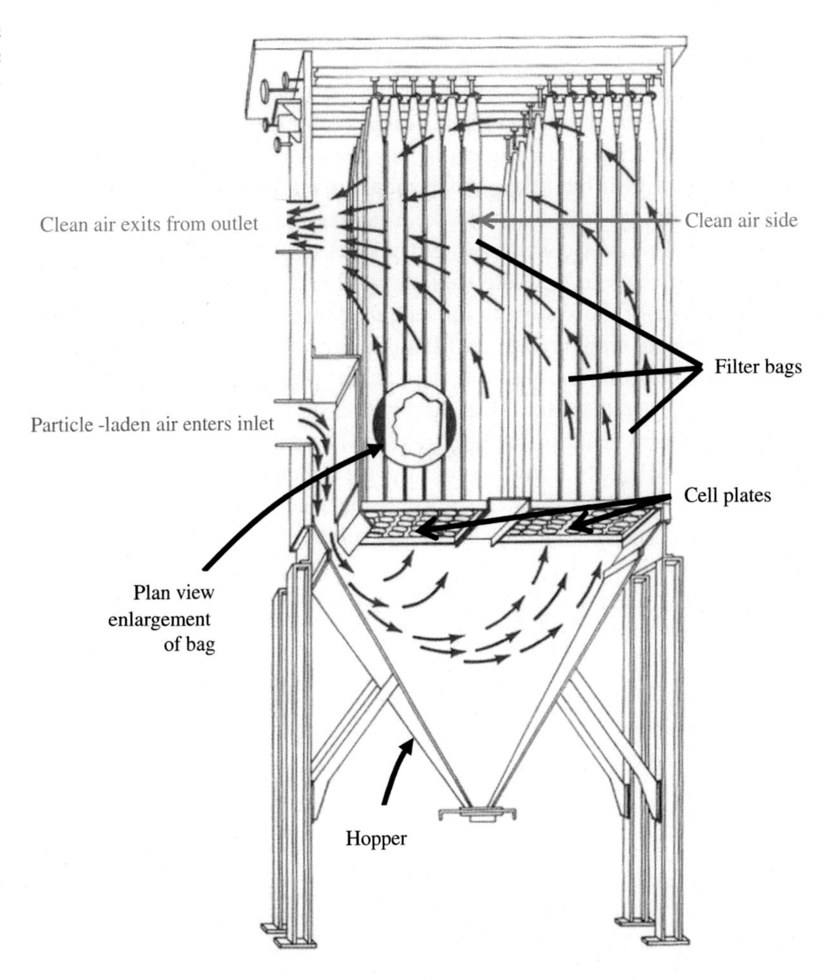

FIG. 13.11 Common baghouse configuration (shaker filter). *(Source: US Environmental Protection Agency, APTI 413: Control of particulate matter emissions. Chapter 7: Fabric filters, 2012.)*

those particles in between. And these efficiencies only occur after a sufficient dust layer forms on the filter material. Thus, efficiency is decreased substantially after cleaning or installation of new fabric. Thus, the cleaning schedule must be optimized between preventing excessive blockage and destroying efficient dust cake layers.

Other types of industrial filtration systems include (1) fixed beds or layers of granular material such as coke or sand, some of the original designs for cleaning large quantities of gases from smelters and acid plants involved passing the gases through such beds; (2) plain, treated, or charged mats or pads (common throwaway air filters used for hot air furnaces and for air conditioners are of this type); (3) paper filters of multiple plies and folds to increase filter efficiency and area (the throwaway dry air filters used on automotive engines are of this type); (4) rigid porous beds that can be made of metal, plastic, or porous ceramic (these materials are most efficient for the removal of large particles such as the 30 μm particles from a wood sanding operation); and (5) fluidized beds in which the granular material of the bed is made to act as a fluid by the gas passing through it. Most fluidized beds are used for heat or mass transfer. Their use for filtration has not been extensive.

13.6.8.3 Fabric filter operation, maintenance and performance evaluation

The filter ratio for baghouses, also called the gas-to-cloth ratio or air-to-cloth (*A/C*) ratio [12], varies from 0.6 to 1.5 m^3 of gas per minute per square meter of fabric. The air-to-cloth (*A/C*) ratio [12], which is an expression of the airstream flow rate through the fabric cloth, is

$$A/C = \frac{Q_g}{A_{fabric}} \tag{13.58}$$

where *A/C* is expressed as length time^{-1}(L T^{-1}), Q_g is the actual gas flow rate (L^2 T^{-1}), and A_{fabric} is the fabric surface area (L^2).

With increasing *A/C* ratios, the velocities of the gas moving through filter media and dust layers also increase. At high *A/C*, ultrafine particles may pass unscathed and not be captured or reentrained from loosely retained after capture, known as filter bleed-through. In addition to particle size, bleed-through is also increased with the amount of flex and movement of the bags and decreased with the thickness of dust layers on the fabric. Other inherent material properties may also affect bleed-through, for example, pore collapse.

One means of knowing whether the dust layer is optimal is by measuring pressure drop. The pressure drop across the fabric is a function of the filter ratio; it ranges from about 80 mm of water for the lower filter ratios to about 200 mm of water for the higher ratios. Before selecting any bag filter system, a thorough engineering study should be made, followed by a consultation with different bag and baghouse manufacturers. Lower than optimal static pressure drop may mean that there is insufficient thickness of the dust cake, whereas higher than optimal pressure drop may indicate that the gas flow is overly constricted. This flow constriction can be caused by mechanical and electric failures in air moving equipment. Thus, monitoring equipment can be located to indicate flange-to-flange static pressure drop (at the outlet and inlet ducts) and/or at the unfiltered side of the system to indicate media static pressure drop.

The pressure drop through the filter is a function of two separate effects. The clean filter has some initial pressure drop. This is a function of filter material; depth of the filter; the superficial gas velocity, which is the gas velocity perpendicular to the filter face; and the viscosity of the gas. Pressure drop increases from the initial clean filter's resistance as particles begin to adhere to form a cake on the filter surface. This cake increases in thickness as approximately a linear function of time, and the pressure difference necessary to cause the same gas flow also becomes a linear function with time. Usually, the pressure available at the filter is limited so that as the cake builds up, the flow decreases. Timing of filter cleaning can be based, therefore, on (1) increased pressure drop across the filter, (2) decreased volume of gas flow, or (3) time elapsed since the last cleaning.

Put simply, media static pressure drop can be expressed as the sum of the pressure drop across the filter media plus the pressure drop across the dust cake:

$$\Delta P_{total} = \Delta P_{media} + \Delta P_{dust-cake} \tag{13.59}$$

Pressure drop is often expressed as inches of water column (in. WC).[5] The drop in static pressure across the filter media (ΔP_{media}) is approximated as

$$\Delta P_{media} = k_1 \cdot v_f \tag{13.60}$$

5. 1 WC = 0.25 kPa.

where k_1 = the filter media resistance constant ($\frac{\text{in.WC}}{(\text{ft min}^{-1})(\text{lb}_m \text{ft}^{-2})}$) and v_f = velocity through the filter (ft min^{-1}). The drop in static pressure through the dust cake ($\Delta P_{\text{dust-cake}}$) is approximated as

$$\Delta P_{\text{dust-cake}} = k_2 \cdot c_i \cdot v_f^2 \cdot t \tag{13.61}$$

where k_2 = the filter media resistance constant ($\frac{\text{in.WC}}{(\text{ft min}^{-1})(\text{lb}_m \text{ft}^{-2})}$), c_i = the PM concentration (lb$_m$ ft.$^{-3}$), and t = filtration time (min).

With substitutions, the total pressure drop is

$$\Delta P_{\text{total}} = (k_1 \cdot v_f) + \left(k_2 \cdot c_i \cdot v_f^2 \cdot t\right) \tag{13.62}$$

The filter drag (S_f) can be found by dividing both sides of Eq. (13.62) by the velocity (v_f):

$$S_f = \frac{\Delta P_{total}}{v_f} = k_1 + (k_2 \cdot c_i \cdot v_f \cdot t) \tag{13.63}$$

where S_f = filter drag force (in. WC ft.$^{-1}$ min^{-1}).

These and other general predictions of pressure drop are gross approximations given the variability in dust cake thickness and other real-world conditions.

The bags must be periodically cleaned to remove the accumulated PM (note in Fig. 13.11 the general downward trend in efficiency with time). Bag cleaning methods vary widely with the manufacturer and with baghouse style and use. Methods for cleaning include (1) mechanical shaking by agitation of the top hanger, (2) reverse flow of gas through a few of the bags at a time, (3) continuous cleaning with a reverse jet of air passing through a series of orifices on a ring as it moves up and down the clean side of the bag, and (4) collapse and pulsation cleaning methods.

Common fabric filter cleaning mechanisms include shaking, reverse air, and pulse jets. Shaker systems are cleaned after stopping the gas flow, allowing the bags to relax (usually <1 min) and shaking the bags for 10–100 cycles with a rocker-arm assembly in 1–5 cycles per second. The cleaning cycles are usually controlled by a timing device that deactivates the section being cleaned. The dusts removed during cleaning are collected in a hopper at the bottom of the baghouse and then removed, through an air lock or star valve, to a bin for ultimate disposal.

The reverse air filtration system construction and operation resemble that of shaker collectors, including a tube sheet that separates the bags in the upper portion of the collector from the hoppers. The open bottoms of the bags are attached to the tube sheet, and the closed tops are attached to an upper support. Gases enter through the hopper and pass up through the filter bag, depositing the dust cake inside the bag. Cleaner air exits through an outlet duct. Reverse air collectors usually employ woven fabrics, as well as membrane bags and felted bags. Their typical *A/C* ratio is 0.8–1.8 cm sec^{-1} [13].

There are two basic types of pulse-jet collectors: top access and side access, with the top-access pulse jet more common (see Fig. 13.12). In pulse-jet collectors, the tube sheet is located near the top of the unit, and the bags are suspended from it. This free-hanging design is needed to facilitate bag replacement, to direct gas stream movement upward, and to avoid abrasive surfaces near the bottom of the bags. The pulse-jet cleaning occurs from compressed air that passes through a drier, if gas has too much moisture, which can condense upon cooling and increase blinding. The compressed air is filtered to remove oil droplets that condense from lubricating oils in the compressor, heated, and delivered by compressed tubes to the bags [13].

Cartridge filter systems are similar to pulse-jet fabric filter systems, with filter elements supported on a tube sheet that is usually mounted near the top of the filter housing. Filtering is performed by the filter media and the dust cake supported on the exterior of the filter media. The filter media is usually cellulose, polypropylene, or other flex-resistant material covered with felt. Cartridges are usually shorter than pulse-jet bags. They not only can be cylindrical but also may have other shapes to increase surface area. Their short filter element makes it less susceptible to being scratched by suspended matter in the airstream and allows cleaning with compressed air in a process similar to pulse-jet cleaning.

The actual performance of filtration systems is a function of numerous factors, so the evaluation of performance is multifactorial as well. These factors include the type of fabric, *A/C* ratio, approach velocity, filter bag spacing and size, bag access, design of cleaning system and hopper, bypass dampening, and the type and locations of monitoring instrumentation.

The *A/C* ratio is the most useful means of sizing fabric filters. Gross *A/C* is the actual gas flow rate at maximum operating conditions divided by the total area of fabric in the system:

$$(A/C)_{gross} = \frac{Q_{\max}}{A_{total}} \tag{13.64}$$

where $(A/C)_{gross}$ = gross air-to-cloth ratio [(ft^3 min^{-1})ft.$^{-2}$], Q_{max} = maximum actual flow rate (f^3 min^{-1}), and A_{total} = total fabric area (ft^2).

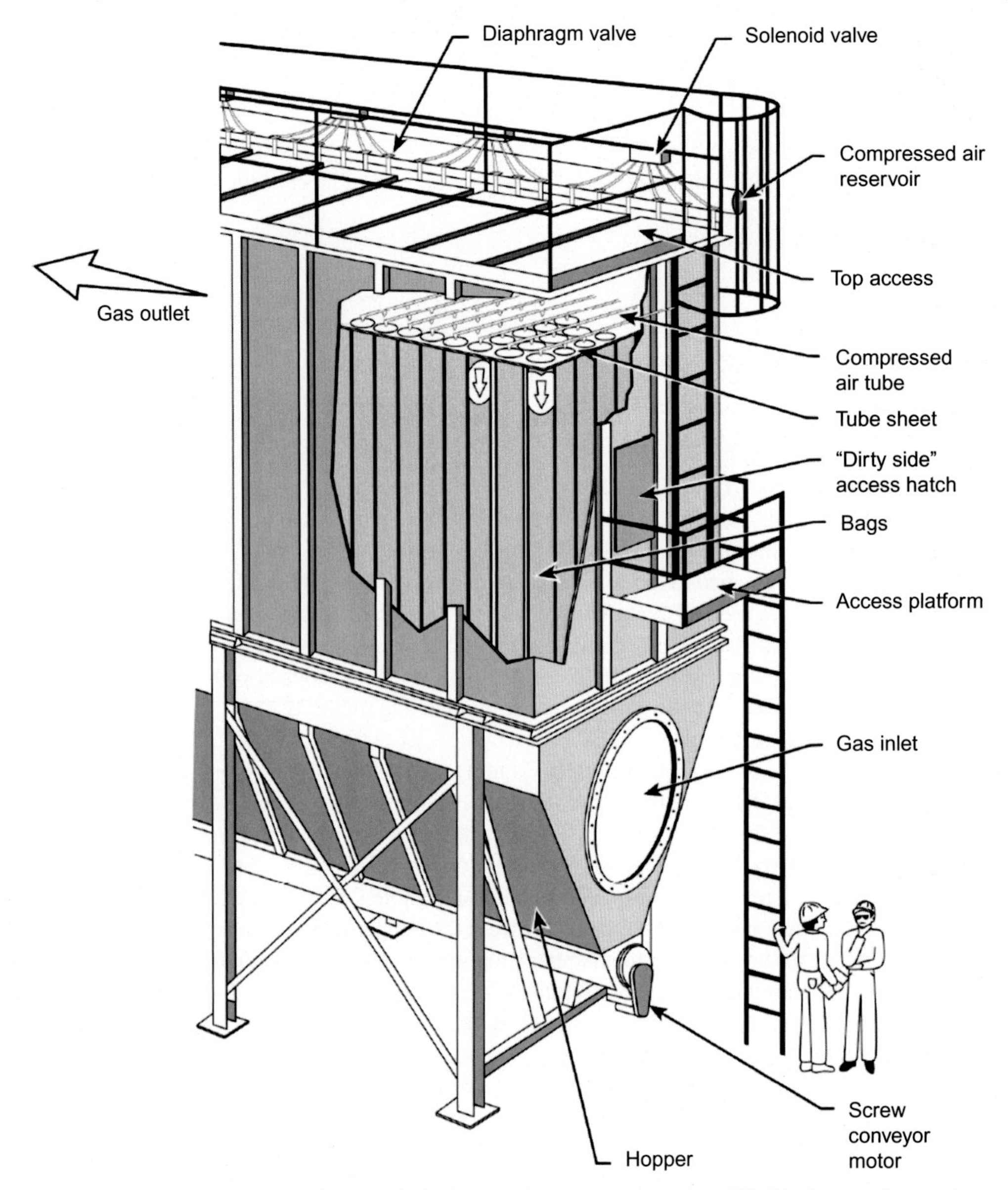

FIG. 13.12 Pulse-jet fabric filter with top access. *(Source: US Environmental Protection Agency, APTI 413: Control of particulate matter emissions. Chapter 7: Fabric filters, 2012.)*

The net *A*/*C* ratio is commonly used for systems that have multiple compartments, with one or more of them separated from the gas flow during cleaning or maintenance or as a regularly scheduled part of a modular system:

$$(A/C)_{gnet} = \frac{Q_{\max}}{A_{net}} \tag{13.65}$$

where $(A/C)_{nets}$ = net air-to-cloth ratio [(ft^3 min^{-1})ft.$^{-2}$], Q_{max} = maximum actual flow rate (f^3 min^{-1}), and A_{net} = isolated compartment's fabric area (ft^2).

Bag area of a cylindrically shaped bag or shaker is

$$A = \pi DL \tag{13.66}$$

where A/C = bag surface area (ft^2), D = bag diameter (ft), and L = bag length (ft). This assumes that filtration only occurs on the side of the bag, not on the top or bottom. If the bag is a pleated cylindrical cartridge, the area is

$$A = 2ndh \tag{13.67}$$

where A = cartridge surface area (ft^2), n = number of pleats, d = depth of pleat (ft), and h = pleat height (ft).

A reverse air baghouse with 20 compartments has 360 bags in each compartment. Each bag has a diameter of 11 in. and is 30 ft long. At an actual gas flow rate of 1.2 million cubic feet per minute and assuming that two compartments are presently out of service, calculate the gross and net air-to-cloth ratios

Solution

$$\text{Bag area} = \pi DL$$

$$\text{Area/bag} = \pi\,(11\,\text{in.})\,(\text{ft}/12\,\text{in.})\,30\,\text{ft.} = 86.35\,\text{ft}^2/\text{bag}$$

The gross air-to-cloth ratio is calculated assuming that all the bags are in service.

$$\text{Total number of bags} = (360\,\text{bags/compartment})\,(20\,\text{compartments}) = 7200\,\text{bags}$$

$$\text{Total fabric area} = (7200\,\text{bags})\left(86.35\,\text{ft}^2/\text{bag}\right) = 621{,}720\,\text{ft}^2$$

$$(A/C)_{gross} = \frac{1.2x10^{6\,ft^3}/\text{min}}{621{,}720\,ft^2} = \frac{1.93^{ft^3}/\text{min}}{ft^2}$$

The net air-to-cloth ratio is calculated by subtracting the compartments that are not in filtering service.

$$\text{Total number of bags} = (360\,\text{bags/compartment})\,(18\,\text{compartments}) = 6480\,\text{bags}$$

$$\text{Total fabric area} = (6480\,\text{bags})\left(86.35\,\text{ft}^2/\text{bag}\right) = 559{,}548\,\text{ft}^2$$

$$(A/C)_{gross} = \frac{1.2x10^{6\,ft^3}/\text{min}}{559{,}458\,ft^2} = \frac{2.14^{ft^3}/\text{min}}{ft^2}$$

A cartridge baghouse has 4 compartments, 16 cartridges per compartment, a cartridge length of 2 ft, and a cartridge diameter of 8 in. Use a pleat depth of 1.5 in. and a total of 36 pleats in the cartridge. Use an actual gas flow rate of 4000 ft min^{-1}. Assume one compartment is out of service. Calculate the gross and net air-to-cloth ratios

Solution

$$\text{Cartridge area} = 2\text{ndh}$$

$$\text{Area/cartridge} = 2(36\,\text{pleats})(1.5\,\text{in.}/(12\,\text{in.peu ft}))(2\,\text{ft}) = 18\,\text{ft}^2$$

The gross air-to-cloth ratio is calculated assuming that all the bags are in service.

$$\text{Total number of cartridges} = (16\ \text{cartridges/compartment})(4\ \text{compartments}) = 64\ \text{cartridges}$$

$$\text{Total fabric area} = (64\,\text{cartridges})\,(18\,\text{ft2/cartridge}) = 1152\,\text{ft2}$$

$$(A/C)_{gross} = \frac{4000^{ft^3}/\text{min}}{1152\,ft^2} = \frac{3.47^{ft^3}/\text{min}}{ft^2}$$

The net air-to-cloth ratio is calculated by subtracting the compartments that are not in filtering service.

$$\text{Total number of cartridges} = (16\ \text{cartridges/compartment})(3\ \text{compartments}) = 48\ \text{cartridges}$$

$$\text{Total fabric area} = (48\,\text{cartridges})\left(18\,\text{ft}^2/\text{cartridge}\right) = 864\,\text{ft}^2$$

$$(A/C)_{gross} = \frac{4000^{ft^3}/\text{min}}{864\,ft^2} = \frac{4.62^{ft^3}/\text{min}}{ft^2}$$

TABLE 13.8 Air-to-cloth ratios for selected industrial processes

Process	Shaker	Reverse air	Pulse jet
Basic oxygen furnaces	2.5–3.0	1.5–2.0	6–8
Brick manufacturers	2.5–3.2	1.5–2.0	9–10
Coal-fired boilers	1.5–2.5	1.0–2.0	3–5
Electric arc furnaces	2.5–3.0	1.5–2.0	6–8
Ferroalloy plants	2.0	2.0	9
Gray iron foundries	2.5–3.0	1.5–2.0	7–8
Lime kilns	2.5–3.0	1.5–2.0	8–9
Municipal incinerators	1.5–2.5	1.0–2.0	2.5–4.0
Phosphate fertilizer	3.0–3.5	1.8–2.0	8–9
Portland cement kilns	2.0–3.0	1.2–1.5	7–10

Source: US Environmental Protection Agency, APTI 413: Control of particulate matter emissions. Chapter 7: Fabric filters, 2012.

The air-to-cloth ratio used for a system depends on the particle size distribution, fabric characteristics, particulate matter loadings, and gas stream conditions. Low values for a design air-to-cloth ratio are generally used when the particle size distribution includes a significant fraction of very fine PM or if particulate loading is high. The air-to-cloth ratios for cartridge filters are usually maintained at values less than approximately 4 (ft^3 min^{-1}) ft^2. General *A/C* ratio values for shaker, reverse air, and pulse-jet fabric filters in various settings are provided in Table 13.8. However, this is merely a general guideline, since sites and situations vary. In addition, air-to-cloth ratio requirements have been decreasing in recent decades, commensurate with more stringent PM emission limits.

Other aspects of filter system performance should be part of the evaluation. This includes inspection of the extent of gravity settling of dust cake agglomerates, sheet, and particles that can be released during cleaning. Collection efficiency and static pressure drops can result when fine PM matter, with its inherently low settling velocities, returns to the filter surface. Thus, a sufficient amount of time is needed to allow for particles that have been resuspended to resettle. Also, gas entering the hopper instead of the collector side during online cleaning (e.g., pulse jets) can block particle and aggregates from settling.

The gas approach velocity of a pulse-jet system is directly proportional to the *A/C* ratio. Thus, large dust agglomerates or sheets rather than small aggregates or individual particles should be released during cleaning.

Calculate the total bag area and gas approach velocity for two identical pulse-jet systems with the following design characteristics:

Design characteristic	First unit	Second unit
Compartment area, ft^2	130	130
Number of bags	300	300
Bag diameter, in.	6	6
Bag height, ft	10	10
Air-to-cloth ratio, (ft3/min)/ft2	5	8

The bag area is calculated using the product of the bag circumference times bag length:

$$\text{Bag area} = \pi DL = \pi(6\,\text{in.})(1\,\text{ft.}/12\,\text{in.})(10\,\text{ft}) = 15.7\,\text{ft}^2/\text{bag}$$

$$\text{Total bag area} = (300\,\text{bags})\left(15.7\,\text{ft}^2/\text{bag}\right) = 4710\,\text{ft}^2$$

$$\text{Total gas flow rate of first unit} = \frac{5\left(ft^3/\text{min}\right)}{ft^2}\left(4710\,ft^2\right) = 23,550\,ft^3/\text{min}$$

$$\text{Total gas flow rate of second unit} = \frac{8\left(ft^3/\text{min}\right)}{ft^2}\left(4710\,ft^2\right) = 37,680\,ft^3/\text{min}$$

The area for gas flow at the bottom of the pulse-jet bags is identical in both units.

Area for flow = total area – bag projected area
= total area – (number of bags)(circular area of bag at bottom)
$= 130\,\text{ft}^2 - (300)(\pi D^2/4)$
$= 130\,\text{ft}^2 - 58.9\,\text{ft}^2$
$= \mathbf{71.1\,ft^2}$

$$\text{Gas approach velocity for first unit} = \frac{23,550\,ft^3/\text{min}}{71.1\,ft^2} = 331\,ft/\text{min}$$

$$\text{Gas approach velocity for second unit} = \frac{37,680\,\text{ft}^3/\text{min}}{71.1\,\text{ft}^2} = 530\,\text{ft}/\text{min}$$

The approach velocity in pulse-jet systems is directly proportional to bag length. An increase in the second unit's length from 10 to 16 ft., for example, would increase the approach velocity substantially to about 850 ft. min^{-1}, so long as the *A/C* ratio is constant. This combined with the longer path that fine PM would have to travel in the more elongated bag would increase problems caused by additional gravity settling.

The type of instrumentation should be properly matched to the equipment. As a minimum, for fabric filters, this likely includes instrumentation for static pressure drop measurement, inlet and outlet gas temperature measurements, and opacity monitors.

It should be noted that often several different air pollution control devices are used in sequences and series. For example, when addressing a very toxic compound, particulates and gases must be removed and/or treated. In these configurations, both inertial and filtration equipment may be deployed, along with scrubbers and sorbing media for gases (see Fig. 13.13).

13.6.9 Electrostatic precipitation

High-voltage ESPs have been widely used throughout the world for particulate removal since Fredrick Cottrell invented the first one early in the twentieth century. Most of the original units were used for the recovery of process materials, but today, gas cleaning for air pollution control is often the main reason for their installation. The ESP has distinct advantages over other aerosol collection devices:

(1) It can easily handle high-temperature gases, which makes it a likely choice for boilers, steel furnaces, etc.
(2) It has an extremely small pressure drop, so fan costs are minimized.
(3) It has an extremely high collection efficiency if operated properly on selected aerosols (many cases are on record, however, in which relatively low efficiencies were obtained because of unique or unknown dust properties).
(4) It can handle a wide range of particulate sizes and dust concentrations (most precipitators work best on particles smaller than 10 μm, so that an inertial precleaner is often used to remove the large particles).
(5) If it is properly designed and constructed, its operating and maintenance costs are lower than those of any other type of particulate collection system.

Three of the disadvantages of ESPs are the following:

(1) The initial cost is the highest of any particulate collection system.
(2) A large amount of space is required for the installation.
(3) ESPs are not suitable for combustible particles such as grain or wood dust.

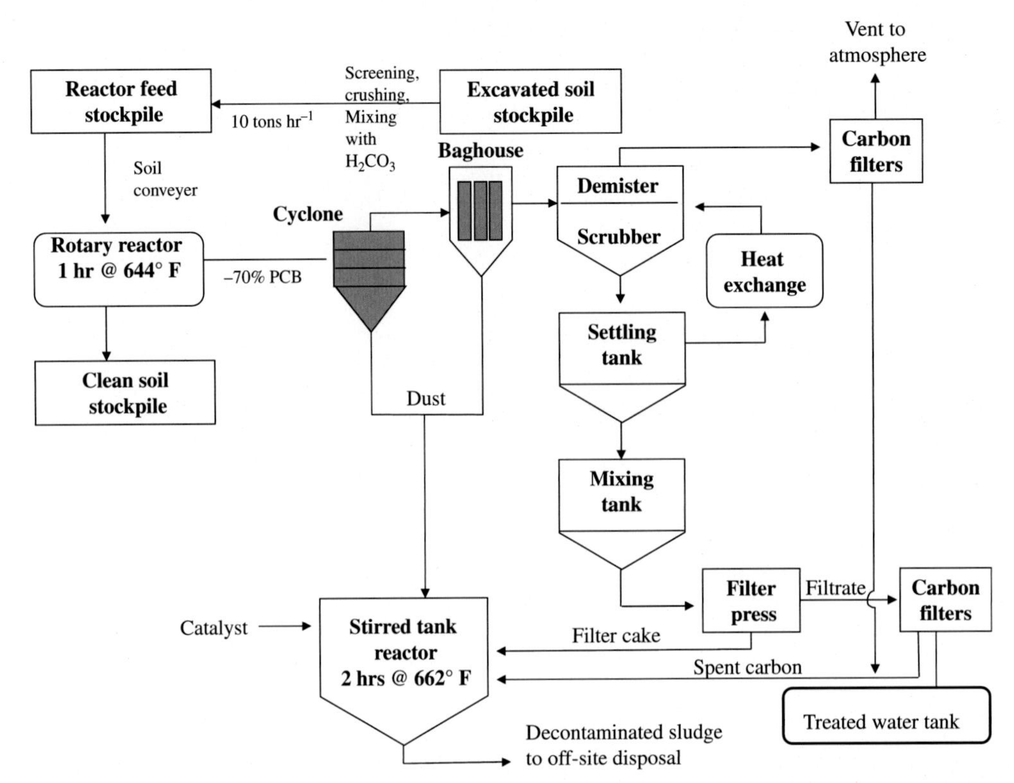

FIG. 13.13 Schematic of a treatment system for treating soil that had been contaminated with polychlorinated biphenyls (PCBs). PCBs are semivolatile compounds that can exist both as aerosols and vapors. After incineration with a rotary kiln, the dust removal relies on both a cyclone and a baghouse to capture various sizes of PM. The vapors are captured after dust removal, with a final stage employing sorption onto granulated activated carbon. *(Source: Federal Remediation Technologies Roundtable, Screening Matrix and Reference Guide, 4th ed., Washington, DC, 2002.)*

The particles move advectively with the gas stream that is moving through the ESP unit (see Fig. 13.14). The particles become charged and then are attracted by the charge differential to either side. The particles migrate toward the collection surface, where they are captured. The accumulated solids or liquids are deposited into the hoppers (or sumps for liquids) that are below the electric zone.

The ESP works by charging dust with ions and then collecting the ionized particles on a surface. The collection surfaces may consist of either tubular or flat plates. For cleaning and disposal, the particles are then removed from the collection surface, usually by rapping the surface.

The alternating power supplied to the primary control cabinet is at a constant 480 V and 60 cycles per second. A transformer-rectifier (T-R) set provides this power. The primary alternating power is converted to a secondary pulse-type direct power in the T-R set. The relatively low primary voltage is stepped up to a secondary voltage of >50,000 V. The voltage applied to the discharge electrodes is called the secondary voltage because the electric line is on the transformer's secondary side, that is, the high-voltage-generating side [14]. With increasing supply of primary voltage, the secondary voltage applied to the discharge electrodes increases. A high-voltage (30 kV or more) DC field is established between the central wire electrode and the grounded collecting surface.

The voltage is high enough that a visible corona can be seen at the surface of the wire. The result is a cascade of negative ions in the gap between the central wire and the grounded outer surface. Any aerosol entering this gap is both bombarded and charged by these ions. The aerosols then migrate to the collecting surface because of the combined effect of this bombardment and the charge attraction. When the particle reaches the collecting surface, it loses its charge and adheres because of the attractive forces existing. It should remain there until the power is shut off, and it is physically dislodged by rapping, washing, or sonic means.

The electric discharges from the precipitator discharge electrodes are termed corona discharges and are needed to charge the particles. Within the negative corona discharge, electrons are accelerated by a very strong electric field and strike and ionize gas molecules. Each collision of a fast-moving electron with a gas molecule produces an additional electron and a positively charged gas ion. The corona discharges can be described as an "electron avalanche" since large numbers of electrons are generated during multiple electron-gas molecule collisions [14].

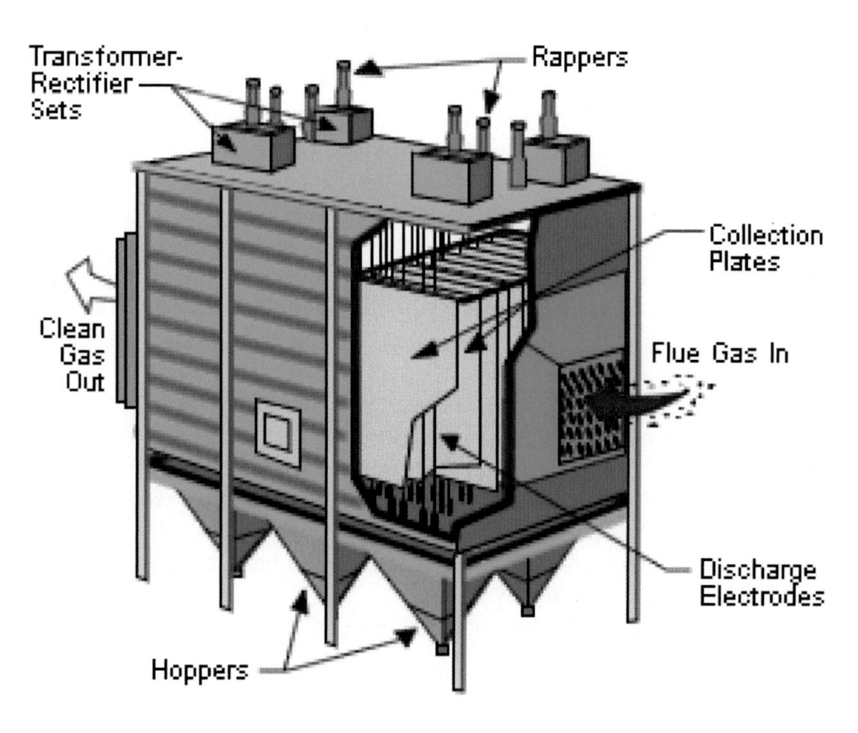

FIG. 13.14 Commercial plate-type electrostatic precipitator. *(Source: U.S. Environmental Protection Agency, Basic Concepts in Environmental Sciences—Module 6: Air Pollutants and Control Techniques. (2007), 30 June 2007. Available: http://www.epa.gov/eogapti1/module6/matter/control/control.htm.)*

The positive gas ions generated in the ionization process turn back toward the discharge electrode. Some positive gas ions deposit on particles within the corona, adding a positive charge. The positively charged particles collect on negative discharge electrodes and must be cleaned sufficiently often to continue inducing charge.

Electrons released in the corona discharges are captured by gas molecules, further from the discharge electrode, where the electric field strength is lower. The negatively charged gas ions move toward the grounded collection plates. Some are captured by particles, inducing a negative charge. The particles quickly reach a maximum charge, that is, saturation charge, at which the electrostatic field created by the captured ions is sufficiently strong to deflect additional gas ions approaching the particle [14]. The magnitude of the saturation charge is dependent on the particle size:

$$q = (3\pi\varepsilon_0 E)d_p^2 \tag{13.68}$$

where q = charge on particle, ε_0 = permittivity of free space (8.85×10^{-12} farads m^{-1}), E_0 = strength of electric field (volts m^{-1}), and d_p = physical diameter of particle.

Since gas ions have small surfaces on which to deposit, small particles have low saturation charges. Thus, saturation charge increases with surface area and with the square of the particle diameter. Large particles accumulate higher electric charges on their surface and, therefore, are more strongly affected by the applied electric field.

Upon attaching ions, particles are influenced by the strong, nonuniform electric field between the discharge electrode and the grounded collection plate. As a result, the charged particles migrate toward the grounded plates at a velocity given by

$$\omega = \frac{qE_pC_C}{3\pi d_p\mu_g} \tag{13.69}$$

where ω = migration velocity, q = charge on particle, E_0 = electric field near collector plate, C_C = Cunningham slip correction factor, μ_g = gas viscosity, and d_p = physical diameter of particle.

Simultaneously to the migration from the charge differential, drag forces, which depend on the particle mass or the cube of the particle diameter, are inducing particle to move unabated through the precipitator. Thus, very small particles deposit near the inlet, while increasingly larger particles deposit deeper into the precipitator. Often, particles larger than about 30 μm diameter can be removed with precleaning. Otherwise, the ESP would have to be much longer. Small-sized particles charge more slowly, but once they acquire a charge, they move rapidly to the collection plate.

Particles with $>1.0\,\mu m$ diameters accumulate charged gas ions by locally disrupting the electric field. As a result, the gas ions are momentarily directed to the particle surface rather than the collection plate, known as contact charging. Particles with $<0.1\,\mu m$ diameters have insufficient mass to disrupt the electric field. Rather, very small particles accumulate charges as in their random walks through the gas ions, known appropriately as diffusional or ion charging. The combined effect of contact and diffusion charging creates a particle size-collection efficiency relationship that is efficient for larger PM but diminishing for very small particles, dropping to a low of <70% for $\sim 0.5\,\mu m$ diameter particles and increasing in efficiency, nearing 100% for PM_{10}. The high collection efficiencies for $PM > 1.0\,\mu m$ diameter due are the result of improved effectiveness of contact charging for large particles. Increased diffusion charging is the reason for good collection efficiencies for particles smaller than $0.1\,\mu m$ diameter. The suboptimal range between 0.1 and $1.0\,\mu m$ diameter results from size-dependent limitations of both of these charging mechanisms. Electrostatic systems are notoriously weak in collecting particles in this size range.

In a tube-type ESP, the tubes are 8–25 cm in diameter and 1–4 m long. They are arranged vertically in banks with the central wires, about 2 mm in diameter, suspended in the center with tension weights at the bottom. Many innovations, including square, triangular, and barbed wires, are used by different manufacturers.

A plate-type ESP is similar in principle to the tubular type except that the air flows across the wires horizontally, at right angles to them. The particles are collected on vertical plates, which usually have fins or baffles to strengthen them and prevent dust reentrainment. Fig. 13.14 illustrates a large plate-type precipitator. These precipitators are usually used to control and collect dry dusts.

Problems with ESPs develop because the final unit does not operate at ideal conditions. Indeed, ESP operation is vulnerable to particle characteristics. Gas channeling through the unit can result in high dust loadings in one area and light loads in another. The end result is less than optimum efficiency because of much reentrainment. The chemical composition of the dust cake affects conductance and resistance. Actually, electrons can flow through the dust cake in two ways. Electrons can pass directly through each dust particle until reaching the metal surface (bulk conductance), or they can be conducted surface to surface (surface conductance); for example, vapor-phase conductive compounds sorb to the particles. Obviously, bulk conductance requires that the particle material has sufficient conductivity. The resistivity of the dust greatly affects its reentrainment in the unit. If a high-resistance dust collects on the plate surface, the effective voltage across the gap is decreased.

It should also be noted that resistance is temperature-dependent. For example, resistivity of inorganic oxides is orders of magnitudes lower at gas temperatures <150°C than at 180°C. Resistivity is also orders of magnitudes lower at gas temperatures >200°C than at 180°C. Thus, the range between 150 and 200°C is to be avoided, with temperatures between 250 and 350 nearly optimal, depending on the specific dust material [14].

Some power plants burning high-ash, low-sulfur coal have reported very low efficiency from the precipitator because the ash needed more SO_2 to decrease its resistivity. The suggestion that precipitator efficiency could be greatly improved by *adding* SO_2 or SO_3 to the stack gases has met, not surprisingly, with skepticism given these compounds' reputations as potent air pollutants. However, combustion of coal and other fossil fuels often generates sufficient H_2SO_4 and other particle surface conditioning agents that lower dust layer resistivities to allow for conductance through dust layers.

Electrostatic precipitators work well in many settings; however, some need to be avoided or at least customized to address certain unwanted properties of substances in gas streams. These include fly ash with extremely low resistivities, sticky PM, and flammable and explosive substances. In addition, ESPs produce ozone (O_3), so operators must ensure that waste streams do not produce unacceptable amounts of this pollutant, that is, trade-off between PM and O_3.

13.6.10 Wet collection

Scrubbers, or wet collectors, have been used as gas-cleaning devices for many years. They predominantly collect PM by applying two basic physical mechanisms already discussed, that is, inertial impaction and Brownian motion. The wet scrubbing process uses two processes to remove an aerosol from the gas stream. The first is to wet the particle by the scrubbing liquid. As shown in Fig. 13.15, this process is essentially the same whether the system uses a spray to atomize the scrubbing liquid or a diffuser to break the gas into small bubbles. In either case, it is assumed that the particle is trapped when it travels from the supporting gaseous medium across the interface to the liquid scrubbing medium. Some relative motion is necessary for the particle and liquid-gas interface to come in contact. In the spray chamber, this motion is provided by spraying the droplets through the gas so that they impinge on and make contact with the particles. In the bubbler, inertial forces and severe turbulence achieve this contact. In either case, the smaller the droplet or bubble, the greater the collection efficiency. In the scrubber, the smaller the droplet, the larger will be the surface area for a given weight of liquid, with a greater likelihood of wetting the particles. In a bubbler, smaller bubbles mean not only that more interface area is available but also that the particles have a shorter distance to travel before reaching an interface where they can be wetted.

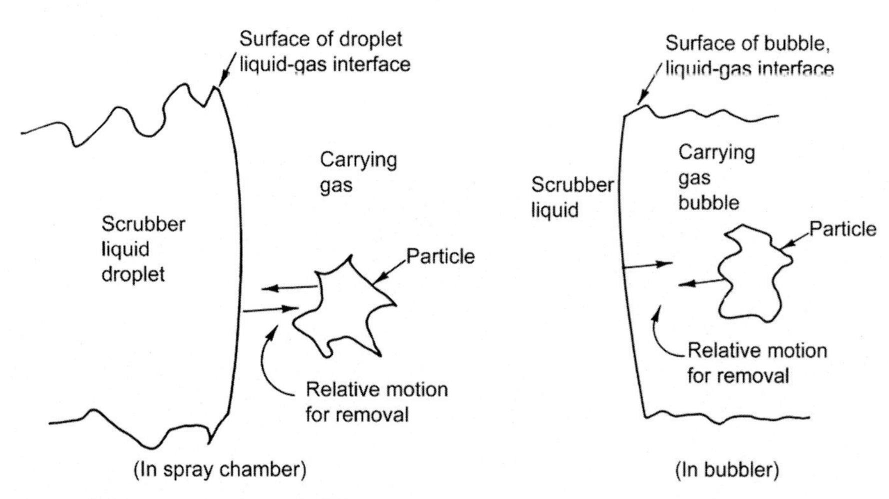

FIG. 13.15 Wetting of aerosols in a spray chamber or bubbler.

The second process employed by wet collectors is to remove the wetted particles on a collecting surface, followed by their eventual removal from the device. The collecting surface can be in the form of a bed or simply a wetted surface. One common combination follows the wetting section with an inertial collector that then separates the wetted particles from the carrying gas stream. Thus, collection has two phases: First, the particle is captured within a droplet; then, the droplets aggregate into a slurry that must be removed and treated.

Increasing either the gas velocity or the liquid droplet velocity in a scrubber will increase the efficiency because of the greater number of collisions per unit time. The ultimate scrubber in this respect is the venturi scrubber, which operates at extremely high gas and liquid velocities with a very high pressure drop across the venturi throat. Fig. 13.16 illustrates a configuration of a venturi scrubber unit.

Large amounts of water are involved in scrubbings, so the scrubbing liquid is recirculated to minimize the amount of liquid needing to be treated and discharged. The scrubbing liquid is collected in the sump of the scrubber and mist eliminator delivered by gravity to a recirculation tank having a liquid residence time of several minutes. This provides sufficient time to introduce alkali additives and to adjust the pH back to the proper range, if necessary. A centrifugal pump recirculates the liquid back to the scrubber vessel from the recirculation tank. Some scrubber systems use spray nozzles to atomize the scrubbing liquid. To project droplets across an entire circular area, a full-cone nozzle is used most frequently. This improves the effective gas-liquid contact. Full-cone nozzles usually provide mean diameters of 100–1000 μm that are lognormally distributed [15].

The liquid-to-gas ratio (L/G) is the rate of liquid flow to a scrubber, expressed as volume per volume (e.g., gallons of liquid per 1000 actual cubic feet of gas flow). To collect PM, wet scrubber systems generally operate with L/G ratios between 4 and 20 gal/1000 actual cubic feet (acf). Higher ratios do not appear to improve performance. Plus higher L/G changes the droplet distribution within the scrubber, which adversely affects collection efficiencies. Low L/G can dramatically decrease efficiency by decreasing the number of available target objects for particle impaction. At low L/G ratio conditions, a portion of the particle-laden gas stream may pass through the collection zone without encountering a liquid target. The L/G ratio can be defined based on either inlet or outlet gas flow rates or based on either actual or standard gas flow rates. A practical definition of L/G is

$$\mathrm{L:G} = \frac{Q_{i-l}}{Q_{o-g}} \tag{13.70}$$

where $Q_{i\text{-}l}$ = the inlet liquid flow (gallons per minute, gpm) and $Q_{o\text{-}g}$ = the outlet gas flow (actual $\mathrm{ft^3\,min^{-1}}$, acfm).

$Q_{o\text{-}g}$ is used because it is already measured as part of an emission test program and obtaining an accurate flow measurement is easy at the scrubber outlet, because this is where PM being loaded into the gas stream is decreased and where sampling ports are available. Indeed, most scrubber inlet ducts are not well suited for testing gas flow rates.

L/G ratio needs to be above the minimum level necessary to ensure proper gas-liquid distribution. In most gas-atomized scrubbers, this minimum value is approximately 4 gal/1000 acf. In other types of scrubbers, the value can be as low as 2 gal/1000 acf.

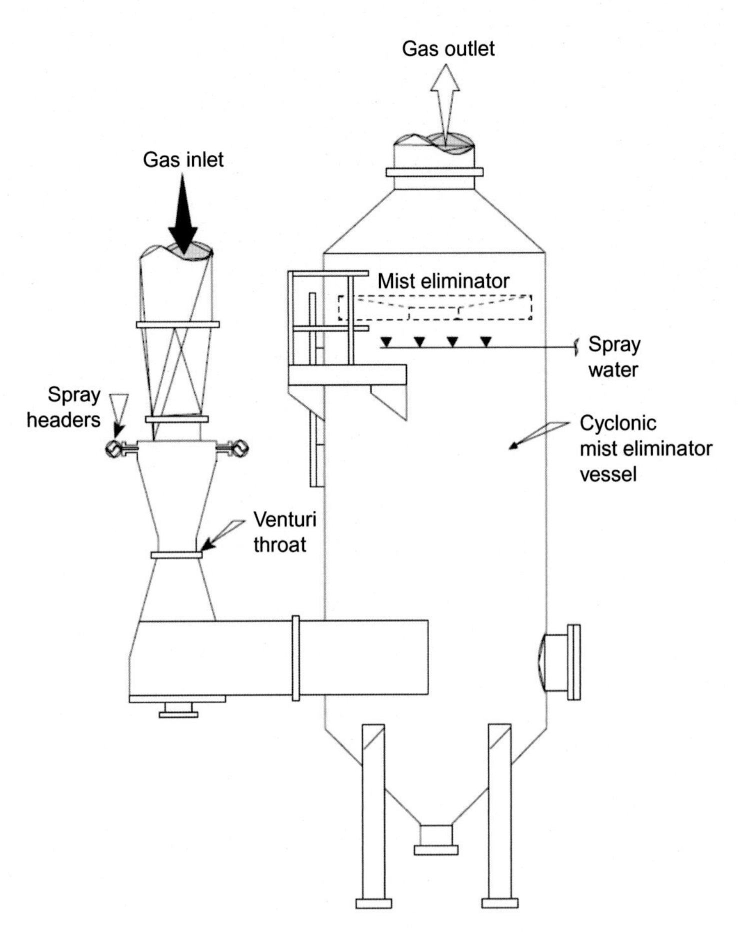

FIG. 13.16 Venturi scrubber and mist eliminator vessel. *(Source: J.R. Richards, Control of gaseous emissions, 3rd ed., Student Manual: US Environmental Protection Agency APTI Course 415, 2000.)*

What is the design liquid-to-gas ratio for a scrubber system that has an outlet gas flow rate of 10,000 acfm, a pump discharge rate of 110 gpm, and a liquid purge rate of 20 gpm? The purge stream is withdrawn from the pump discharge side

$$\text{Solution}: \text{L}:\text{G} = \frac{Q_{i-l}}{Q_{o-g}}$$

$$Q_{i-l} = 110 - 10\,\text{gpm} = 90\,\text{gpm}$$

$$\text{Therefore}, \text{L}:\text{G} = \frac{90\,\text{gpm}}{10^4\,\text{acfm}} = 9 \times 10^{-3}\frac{\text{gal.}}{\text{acf}} = \frac{9\,\text{gal.}}{1000\,\text{acf}}$$

The chemical composition of the particles and gases in the airstream often requires pretreatment. For example, if acid gases, for example, H_2SO_4, HCl, and HF, are present, an alkali addition system is used on wet scrubber systems that collect acidic PM or to treat gas streams that have acidic gases or vapors that could absorb in the liquid stream. In addition, CO_2 formed in most combustion processes can also be absorbed, with the $H_2CO_3{}^*$ lowering the pH. The most common alkalies used in scrubbers are liming agents (e.g., $CaCO_3$ or $Ca(OH)_2$), soda ash [Na_2CO_3], and sodium hydroxide [NaOH].

The amount of alkali needed depends on the quantities of acidic gases captured and the molar ratios necessary for the following reactions:

$$SO_3 + Ca(OH)_2 \rightarrow CaSO_4 + H_2O \tag{13.71}$$

$$2HCl + Ca(OH)_2 \rightarrow CaCl_2 + 2H_2O \tag{13.72}$$

$$2HF + Ca(OH)_2 \rightarrow CaF_2 + 2H_2O \tag{13.73}$$

Assuming an HCl removal efficiency of 95%, what is the amount of calcium hydroxide (lime) needed to neutralize the HCl absorbed from a gas stream with 50 ppmV HCl and a flow rate of 10,000 scfm? This can be calculated using Reaction (13.72)

Solution

$$50\,\text{ppmv} = \frac{50\text{ft}^3\text{HCl}}{10^6\text{ft}^3\text{total}} = 0.00005\frac{50\text{ft}^3\text{HCl}}{\text{ft total}} = 0.00005\frac{\text{lb}-\text{mole HCl}}{\text{lb}-\text{mole total}}$$

$$\text{HCl absorbed} = 10{,}000\text{scfm}\left(\frac{\text{lb}-\text{mole}}{385.4\text{scf}}\right)\left(0.00005\frac{\text{lb}-\text{mole HCl}}{\text{lb}-\text{mole total}}\right)(0.95)$$

$$= 0.00123\frac{\text{lb}-\text{mole}}{\text{min}}$$

$$\text{Ca(OH)}_2\text{required} = \left(\frac{1\text{ lb}-\text{mole Ca(OH)}_2}{2\text{ lb}-\text{mole HCl}}\right)\left(0.00123\frac{\text{lb}-\text{mole HCl}}{\text{min}}\right)$$

$$= 0.00062\frac{\text{lb}-\text{mole}}{\text{min}}\left(74\frac{\text{lb Ca(OH)}_2}{\text{lb}-\text{mole}}\right)\left(60\frac{\text{min}}{\text{hr}}\right)$$

$$= 2.75\frac{\text{lb}}{\text{hr}}$$

13.6.11 Mist elimination

Scrubber vessels generate relatively large water droplets that become entrained in the gas stream. Most contain captured particles that must be removed by mist eliminators from the gas stream before being emitted to the atmosphere. In addition to minimizing the carryover of PM-containing droplets to the atmosphere, mist eliminators also protect downstream equipment, for example, fans and other air movers, from PM-laden droplets and minimize water losses. Solids accumulate because of the impaction of solid-laden water droplets and due to the chemical precipitation of dissolved solids from the scrubbing liquid. Mist eliminators usually have at least one spray nozzle to remove these accumulated solids. The four most common types of mist eliminators are chevrons, mesh or woven pads, tube banks, and cyclones.[6] Static pressure drops across their surfaces range from 0.5 in. WC to >4 in. WC [15].

13.6.11.1 Scrubber performance

Scrubber manufacturers have extensive databases regarding the performance of their commercial brands of scrubbers for removing pollutants from different types of industrial sources. While useful, these data must be augmented by additional site-specific and specific PM characterization information before installing a system, beginning with the site-specific inlet particle size distribution and uncertainties around this distribution (e.g., does the distribution vary according to certain operations and maintenance schedules?). Scrubber performance depends on many factors, including the following:

1. Particle size distribution
2. Mean and maximum gas flow rates
3. Mean and maximum inlet gas temperatures
4. Chemical composition of PM and gas stream, including concentrations of corrosive materials present in the inlet gas stream
5. Concentrations of potentially explosive materials present in the inlet gas stream
6. Makeup water availability
7. Purge liquid treatment and disposal requirements
8. Process type, raw materials, and fuels
9. Source operating schedule
10. Area available for scrubber and wastewater treatment
11. Alkali supply requirements

6. For a detailed discussion of these four systems, see [15]. U.S. Environmental Protection Agency, "Chapter 8: Wet scrubbers," in *Control of Particulate Matter Emissions: Student Manual: Apti Course 413*. North Carolina State University, 2012.

12. Emission test data

This site-specific information together with the historical performance data informs the decision of whether a scrubber is applicable to the pollutant stream and, if so, which type. The data also provide a basis for the design of the scrubber system components, including estimates of static pressure drops across the system.

Scrubbers have distinct advantages and disadvantages. For example, venturi scrubbers have high collection efficiencies, can handle mists, and are capable of handling flammable and explosive dusts (a distinct advantage over ESPs). Their design is simple and relatively easy to install and maintain. They provide cooling of hot gases and can neutralize corrosive PM and gases [16]. However, venturi scrubbers, like all wet scrubbers, produce large amounts of wastewater and sludge that must be treated. The final exhaust must be reheated. In addition, these systems have high pressure drops and require protection from weather, for example, freezing [17].

13.6.11.2 Dry scrubbers

Gravel bed filters that recirculate the gravel filter medium using an external cleaning or washing system are known as dry scrubbers. Some units also use an electrostatic field across the gravel bed to enhance the removal of the particulate material. The dry scrubber may have to be followed by a baghouse for the effluent to achieve acceptable emission standards. The advantage of dry scrubbers is their ability to remove large quantities of particulate pollutants, such as fly ash, from hot gas streams.

13.6.11.3 Bio-scrubbers

Bioscrubbers are an example of an integrated environmental control system. They have a suspended microbial population in a flowing water phase. Absorption takes place in typical wet scrubbing equipment, for example, spray towers, plate towers, or packed beds. The effluent from the wet scrubber is then transferred to a separate aerated vessel, where degradation of the contaminants is performed by suspended microbes. Nutrients, acids, or bases are added to the recirculated water to maintain conditions conducive to microbial populations that lead to optimal pollutant removal. This can be an enhancement to a scrubber and particularly useful in the enhancement of not only removing but also biodegrading hazardous air pollutants in both gas and particulate phases.

13.7 REMOVAL OF LIQUID DROPLETS AND MISTS

The term mist generally refers to liquid droplets from submicron size to about 10 μm. If the diameter exceeds 10 μm, the aerosol is usually referred to as a *spray* or simply as *droplets*. Mists tend to be spherical because of their surface tension and are usually formed by nucleation and the condensation of vapors [18]. Larger droplets are formed bursting bubbles, by entrainment from surfaces, by spray nozzles, or by splash-type liquid distributors. The large droplets tend to be elongated relative to their direction of motion because of the action of drag forces on the droplets [17].

Mist eliminators are widely used in air pollution control systems to prevent free moisture from entering the atmosphere. Usually, such mist eliminators are found downstream from wet scrubbers. The recovered mist is returned to the liquid system, resulting in lowered liquid makeup requirements.

As is the case for dry particles, mist aerosols and droplets differ substantially physically for the carrying gas stream, including diameter, density, and phase (e.g., gas vs liquid). Indeed, some particles in mist collection systems are the size of ordinary raindrops. Thus, most of the removal mechanisms are like those employed for the removal of dry particulates.

Mist collection is further simplified because the particles are spherical and tend to resist reentrainment. In addition, mists tend to agglomerate after coming in contact with the surface of the collecting device. Control devices developed particularly for condensing mist are discussed separately.

13.7.1 Filtering liquids

Filters for mists and droplets have larger pores and open areas than those used for dry particles. If a filter is made of many fine, closely spaced fibers, it will become wet due to the collected liquid. Such wetting leads to matting of the fibers and retention of greater amounts of liquid, which leads eventual to a blocked filter. Water droplets in the dust cake can severely increase the resistance to gas flow. At the very least, the water fills voids in the dust cake that would normally be the route of the carrying gas. At some point, the water-filled pore space in the dust cake becomes sufficient to pack the PM or to form a layer of mud, rendering the filter impervious to airflow. The problem is known as fabric blinding. Indeed, other substances besides water cause fabric blinding. Moisture can come from numerous sources, including entrained condensed droplets and droplets carried in the compressed air, for example, in pulse-jet fabric filters, or from excessive gas cooling in baghouses; for example, those serving combustion sources that produce high vapor concentrations of water and other compounds can cause water condensation in the dust cakes.

Water is the most common but not the exclusive liquid that can cause blinding. In addition to liquids forming from previously volatilized vapors mentioned above can be liquids used in the pollution control equipment itself. For example, oil droplets can deposit in the upper, clean side surfaces of the fabric filters and prevent airflow. This means that proper O&M requires filtering of the entire inlet gas stream in the unaffected lower portions of the part of the filter (e.g., jet bag). Indeed, blinding agents do not even have to be "wet." High-velocity carrying gas laden with ultrafine particles can be driving these very small particles into the filter fabric before a protective residual dust layer has formed. Newly installed bags in baghouse compartment with several seasoned bags are particularly vulnerable to such blinding. This is because the seasoned bags already have a residual dust cake layer; thus, the gas velocities through the new bag are excessively high. The carrying gas will follow the path of least resistance, that is, through the clean pores of the newly installed bags. Ultrafine particle blinding is possible after installing new bags at sources that generate high concentrations of PM $\leq$ µm aerodynamic diameter. To ameliorate this problem, the new bags can be conditioned prior to service by passing through air containing resuspended coarse particulate [19].

This is a major difference between dry and wet filtration. Thus, rather than closely spaced fiber weaves, wet filtration system is usually composed of either knitted wire or wire mesh packed into a pad. This looser filtration medium yields a lower pressure drop than the media of the filters used for dry PM. The reported pressure drop across wire mesh mist eliminators is 1–2 cm of water at face velocities of $5\,m\,s^{-1}$. The essential collection mechanisms employed for filtration of droplets and mists are inertial impaction and, to a lesser extent, direct interception.

13.7.2 Electrostatic precipitators for mists and droplets

ESPs for liquid droplets and mists are essentially of the wetted wall type. Fig. 13.17 shows a wet-wall precipitator with tubular collection electrodes [20]. In wet ESPs, the upper ends of the tubes form weirs, and water flows over the tube ends to irrigate the collection surface. The collecting electrode becomes coated with a water film. The PM collected in the wet-type electrostatic precipitator is discharged along with the collecting electrode washing liquid. Thus, wastewater treatment is necessary.

Fig. 13.18 shows an alternative type of wet precipitator with plate-type collection electrodes. In this design, sprays located in the ducts formed by adjacent collecting electrodes serve to irrigate the plates [20]. These are often supplemented by overhead sprays to ensure that the entire plate surface is irrigated. The design of such precipitators is similar to that of conventional systems except for the means of keeping insulators dry, measures to minimize corrosion, and provisions for removing the slurry.

13.7.3 Inertial collection of mists and droplets

Inertial collectors for mists and droplets are widely used. They include cyclone collectors, baffle systems, and skimmers in ductwork. Inertial devices can be used as primary collection systems, precleaners for other devices, or mist eliminators. The systems are relatively inexpensive and reliable and have low pressure drops.

Cyclone mist eliminators and collectors have virtually the same efficiency for both liquid aerosols and solid particles. To avoid reentrainment of the collected liquid from the walls of the cyclone, an upper limit is set to the tangential velocity that can be used. The maximum tangential velocity should be limited to the inlet velocity. Even at this speed, the liquid film may creep to the edge of the exit pipe, from which the liquid is then reentrained.

Baffle separators of the venetian blind, V, W, and wave types are widely used for spray removal. They have small space requirements and low pressure drops. They operate by diverting the gas stream and ejecting the droplets onto the collector baffles. Efficiencies of single stages may be only 40%–60%, but by adding multiple stages, efficiencies approaching 100% may be obtained.

13.7.4 Scrubbers for mists and droplets

A widely used type of scrubber for mists and droplets is the venturi scrubber. It has been used for the collection of sulfuric acid and phosphoric acid mists with very high efficiency. The scrubbing contact is made at the venturi throat, where very small droplets of the scrubbing liquid (usually water) are injected. At the throat, gas velocities as high as $130\,m\,s^{-1}$ are used to increase collision efficiencies. Water, injected for acid mist control, ranges from 0.8 to $2.0\,L\,m^{-3}$ of gas. Collection

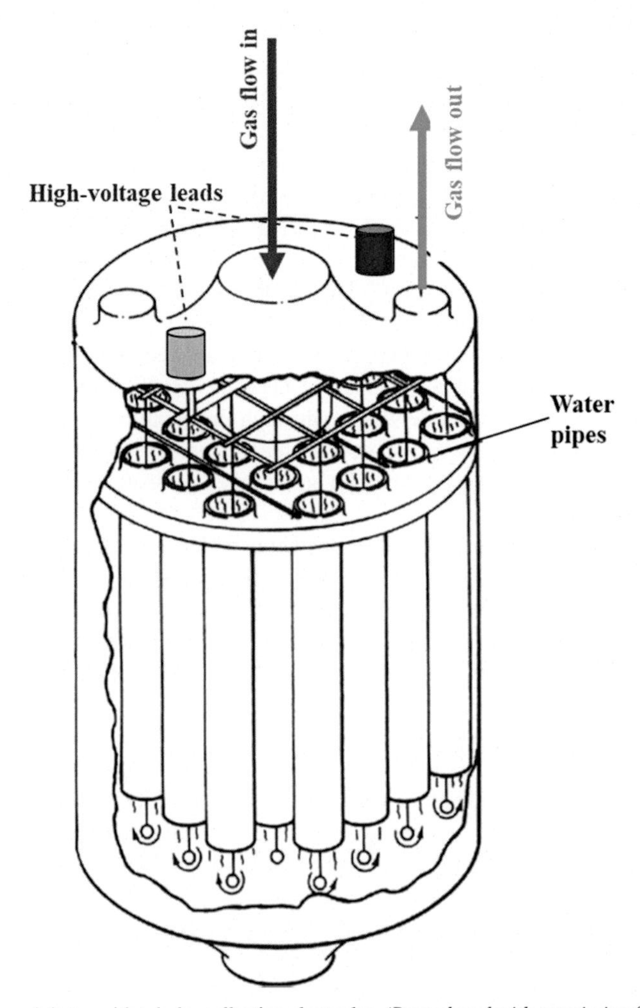

FIG. 13.17 Wet-wall electrostatic precipitator with tubular collection electrodes. *(Reproduced with permission from L. Theodore, A.J. Buonicore, Air Pollution Control Equipment: Selection, Design, Operation and Maintenance, 1982; S. Oglesby, G.B. Nichols, Electrostatic precipitators, in: A.C. Stern (Ed.), Air Pollution, 1977.)*

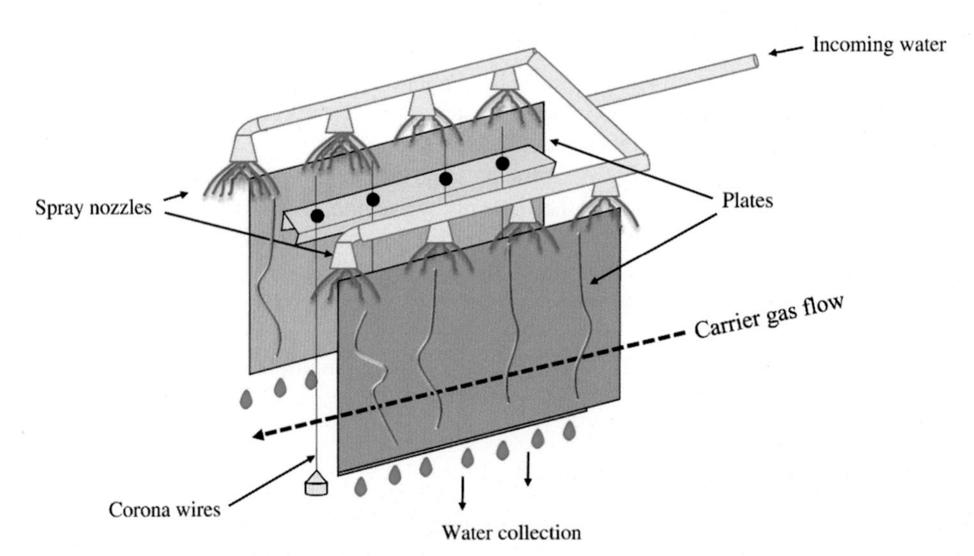

FIG. 13.18 Wet electrostatic precipitator with plate collection electrodes. *(Reproduced with permission from L. Theodore, A.J. Buonicore, Air Pollution Control Equipment: Selection, Design, Operation and Maintenance, 1982; S. Oglesby, G.B. Nichols, Electrostatic precipitators, in: A.C. Stern (Ed.), Air Pollution, 1977.)*

efficiencies approaching 100% are possible, but high efficiencies require a gas pressure drop of 60–90 cm of water across the scrubber. Normal operation, with a submicron mist, is reported to be in the 90%–95% efficiency range [20].

One problem in using scrubbers to control mists and droplets is that the scrubber also acts as a condenser for volatile gases. For example, a hot plume containing volatile hydrocarbon gases, such as the exhaust from a gas turbine, may be cooled several hundred degrees by passing through a scrubber. This cooling can cause extensive condensation of the hydrocarbons, resulting in a plume with a high opacity. Jet engine exhaust has been cooled in a test cell by the use of water sprays that can result in droplet loadings 10%–100% greater than those measured at the engine exhaust plane because of the condensation of hydrocarbons that were normally exhausted as gases [21].

13.7.5 Other systems

Many unique systems have been proposed and some used to control the release of mists and droplets. Ceramic candles are thimble-shaped, porous, acid-resistant ceramic tubes. Although efficiencies exceeding 98% have been reported, the candles have high maintenance requirements because they are very fragile.

Electric cyclones utilize an electrode in the center of the cyclone to establish an electric field within the cyclone body. This device is more efficient than the standard cyclone. It is probably more applicable to mists and droplets than to dry particulates, due to possible fire or explosion hazards with combustible dusts.

Sonic agglomerators have been used experimentally for sulfuric acid mists and as mist eliminators. Commercial development is not projected at this time because the energy requirements are considerably greater than those for venturi scrubbers of similar capacity.

13.7.6 Comparison of particulate removal systems

When selecting a system to remove particulate from a gas stream, many choices concerning equipment can be made. The selection could be made based on cost, gas pressure drop, efficiency, temperature, resistance, etc. Table 13.9 summarizes these factors for comparative purposes. The tabular values must not be considered absolute because great variations occur between types and manufacturers. No table is a substitute for a qualified consulting engineer or a reputable manufacturer's catalog.

13.8 GAS-PHASE POLLUTANT CONTROLS

Gaseous pollutants may be easier or more difficult to remove from the carrying gas stream than aerosols, depending on the individual situation. Certainly, the fact that the airstream and the pollutants are of the same physical phase removes some of the complexities of PM removal. Of course, any separation system relying on differences in inertial properties will not work for gases.

Gases may be reactive to other chemicals, which is an important chemical property that can be used to collect vapor-phase pollutants.

Four general methods of separating gaseous pollutants are currently in use. These are (1) absorption in a liquid, (2) adsorption on a solid surface, (3) condensation to a liquid, and (4) conversion into a less polluting or nonpolluting gas.

13.8.1 Absorption

Absorption of pollutant gases is accomplished by using a selective liquid in a wet scrubber, packed tower, or bubble tower. Pollutant gases commonly controlled by absorption include sulfur dioxide, hydrogen sulfide, hydrogen chloride, chlorine, ammonia, oxides of nitrogen, and low-boiling hydrocarbons.

The scrubbing liquid must be chosen with specific reference to the gas being removed. The gas solubility in the liquid solvent should be high so that reasonable quantities of solvent are required. The solvent should have a low vapor pressure to reduce losses; be noncorrosive, inexpensive, nontoxic, nonflammable, and chemically stable; and have a low freezing point. It is no wonder that water is the most popular solvent used in absorption devices. The water may be treated with an acid or a base to enhance the removal of a specific gas. If carbon dioxide is present in the gaseous effluent and water is used as the scrubbing liquid, a solution of carbonic acid will gradually replace the water in the system.

In many cases, water is a poor scrubbing solvent. Sulfur dioxide, for example, is only slightly soluble in water, so a scrubber of very large liquid capacity would be required. SO_2 is readily soluble in an alkaline solution, so scrubbing solutions containing ammonia or amines are used in commercial applications.

TABLE 13.9 Comparison of particulate removal systems

Type of collector	Particle size range (μm)	Removal efficiency	Space required	Maximum temperature (°C)	Pressure drop (cm H_2O)	Annual cost (US$ per year m^{-3})[a]
Baghouse (cotton bags)	0.1–0.1	Fair	Large	80	10	28.00
	1.0–10.0	Good	Large	80	10	28.00
	10.0–50.0	Excellent	Large	80	10	28.00
Baghouse (Dacron, nylon, Orlon)	0.1–1.0	Fair	Large	120	12	34.00
	1.0–10.0	Good	Large	120	12	34.00
	10.0–50.0	Excellent	Large	120	12	34.00
Baghouse (glass fiber)	0.1–1.0	Fair	Large	290	10	42.00
	1.0–10.0	Good	Large	290	10	42.00
	10.0–50.0	Good	Large	290	10	42.00
Baghouse (Teflon)	0.1–1.0	Fair	Large	260	20	46.00
	1.0–10.0	Good	Large	260	20	46.00
	10.0–50.0	Excellent	Large	260	20	46.00
Electrostatic precipitator	0.1–1.0	Excellent	Large	400	1	42.00
	1.0–10.0	Excellent	Large	400	1	42.00
	10.0–50.0	Good	Large	400	1	42.00
Standard cyclone	0.1–1.0	Poor	Large	400	5	14.00
	1.0–10.0	Poor	Large	400	5	14.00
	10.0–50.0	Good	Large	400	5	14.00
High-efficiency cyclone	0.1–1.0	Poor	Moderate	400	12	22.00
	1.0–10.0	Fair	Moderate	400	12	22.00
	10.0–50.0	Good	Moderate	400	12	22.00
Spray tower	0.1–1.0	Fair	Large	540	5	50.00
	1.0–10.0	Good	Large	540	5	50.00
	10.0–50.0	Good	Large	540	5	50.00
Impingement scrubber	0.1–1.0	Fair	Moderate	540	10	46.00
	1.0–10.0	Good	Moderate	540	10	46.00
	10.0–50.0	Good	Moderate	540	10	46.00
Venturi scrubber	0.1–1.0	Good	Small	540	88	112.00
	1.0–10.0	Excellent	Small	540	88	112.00
	10.0–50.0	Excellent	Small	540	88	112.00
Dry scrubber	0.1–1.0	Fair	Large	500	10	42.00
	1.0–10.0	Good	Large	500	10	42.00
	10.0–50.0	Good	Large	500	10	42.00

[a] *Includes costs for water and power, operation and maintenance, capital equipment, and insurance (in 1994 US$).*

Chlorine, hydrogen chloride, and hydrogen fluoride are examples of gases that are readily soluble in water, so water scrubbing is very effective for their control. For years, hydrogen sulfide has been removed from refinery gases by scrubbing with diethanolamine. More recently, the light hydrocarbon vapors at petroleum refineries and loading facilities have been absorbed, under pressure, in liquid gasoline and returned to storage. All the gases mentioned have economic importance when recovered and can be valuable raw materials or products when removed from the scrubbing solvent.

13.8.2 Adsorption devices

Adsorption of pollutant gases occurs when certain gases are selectively retained on the surface or in the pores or interstices of prepared solids. The process may be strictly a surface phenomenon with only molecular forces involved, or it may be combined with a chemical reaction occurring at the surface once the gas and adsorber are in intimate contact. The latter type of adsorption is known as chemisorption.

The solid materials used as adsorbents are usually very porous, with extremely large surface-to-volume ratios. Activated carbon, alumina, and silica gel are widely used as adsorbents depending on the gases to be removed. Activated carbon, for example, is excellent for removing light hydrocarbon molecules, which may be odorous. Silica gel, being a polar material, does an excellent job of adsorbing polar gases. Its characteristics for the removal of water vapor are well known.

Solid adsorbents must also be structurally capable of being packed into a tower, resistant to fracturing, and capable of being regenerated and reused after saturation with gas molecules. Although some small units use throwaway canisters or charges, the majority of industrial adsorbers regenerate the adsorbent to recover not only the adsorbent but also the adsorbate, which usually has some economic value.

The efficiency of most adsorbers is very near 100% at the beginning of operation and remains extremely high until a break point occurs when the adsorbent becomes saturated with adsorbate. At this break point, the slope of the percentage of mass of gaseous fluid that is not sorbed increases dramatically with time. It is at the break point that the adsorber should be renewed or regenerated. This is shown graphically in Fig. 13.19.

Industrial adsorption systems are engineered so that they operate in the region before the break point and are continually regenerated by units. Fig. 13.20 shows a schematic diagram of such a system, with steam being used to regenerate the saturated adsorbent. Fig. 13.21 illustrates the actual system shown schematically in Fig. 13.20.

13.8.3 Condensers

In many situations, the most desirable control of vapor-type discharges can be accomplished by condensation. Condensers may also be used ahead of other air pollution control equipment to remove condensable components. The reasons for using condensers include (1) the recovery of economically valuable products, (2) removal of components that might be corrosive or damaging to other portions of the system, and (3) reduction of the volume of the effluent gases.

Although condensation can be accomplished either by reducing the temperature or by increasing the pressure, in gas-removal practice, it is usually done by temperature reduction only.

Condensers may be of one or two general types depending on the specific application. Contact condensers operate with the coolant, vapors, and condensate intimately mixed. In surface condensers, the coolant does not come in contact with either the vapors or the condensate. The usual shell-and-tube condenser is of the surface type. Fig. 13.22 illustrates a contact condenser that might be used to clean or preclean a hot corrosive gas.

Table 13.10 lists several applications of condensers currently in use. For most operations listed, air and noncondensable gases should be kept to a minimum, as they tend to reduce condenser capacity.

13.9 AIR POLLUTANT TREATMENT TECHNOLOGIES

The techniques discussed thus far are mainly physical removal processes, rather than transforming the compounds in the emission to other, safer substances. Gases and aerosols are captured before they are released to the atmosphere. However, processes are available to form new compounds in addition to physical collection mechanisms.

A widely used system for the control of organic gaseous emissions is oxidation of the combustible components to water and carbon dioxide. Thermal processes to all phases of hazardous substances are both aerosol and gas phase. Mixtures of phases, however, can introduce control problems. For example, thermal processes like incineration often work much better for homogeneous gas-phase reactions as opposed to heterogeneous reactions involving solids, liquids, and gases.

Other systems such as the oxidation of H_2S to SO_2 and H_2O are also used even though the SO_2 produced is still considered a pollutant. The trade-off occurs because the SO_2 is much less toxic and undesirable than the H_2S. The odor

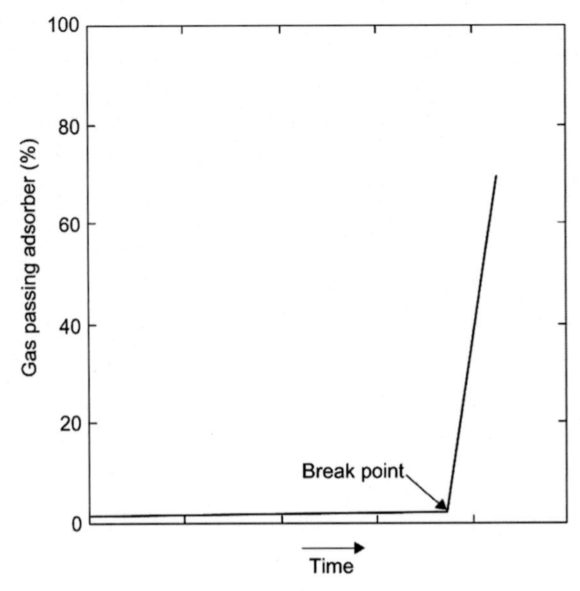

FIG. 13.19 Adsorbent break point at saturation with adsorbate. *(Source: D.A. Vallero, Fundamentals of Air Pollution, 5th ed., Elsevier Academic Press, Waltham, MA, 2014, p. 999 pages cm.)*

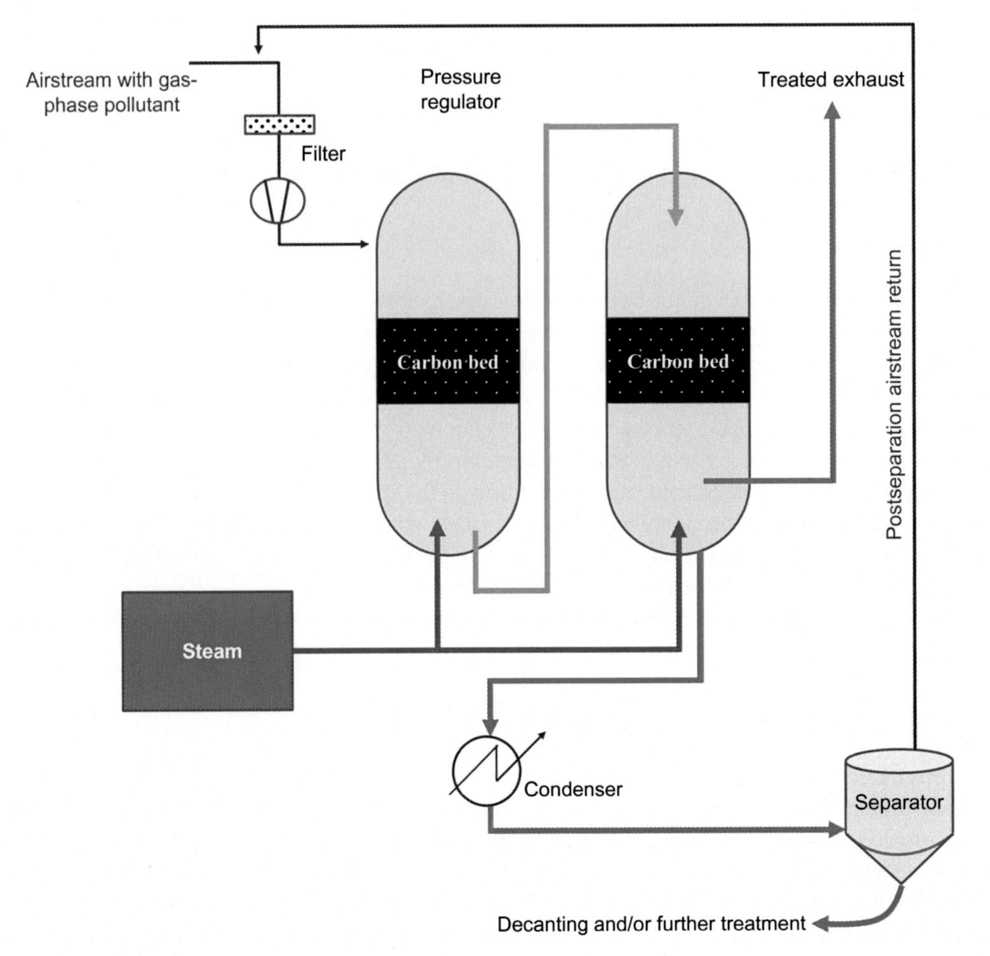

FIG. 13.20 Flow diagram of absorber.

FIG. 13.21 Pollution control facility with an absorption tower in Milford foundry, New Hampshire. Gaseous emissions are introduced at the base of the treatment column, where a gas diffuser ensures that they are evenly distributed through the system. The gas then contacts purification liquid in the packing of sorbents (e.g., zeolite) in the absorption zone. A second diffuser ensures that the purification liquid is evenly spread. *(Source: Environ-Access Inc, Environmental Fact Sheet. Treatment of air and gas: Treatment of gaseous emissions through a wet scrubber, vol. F2-03-96, M.E. Inc, Sherbrooke, Quebec, Canada, 1996.)*

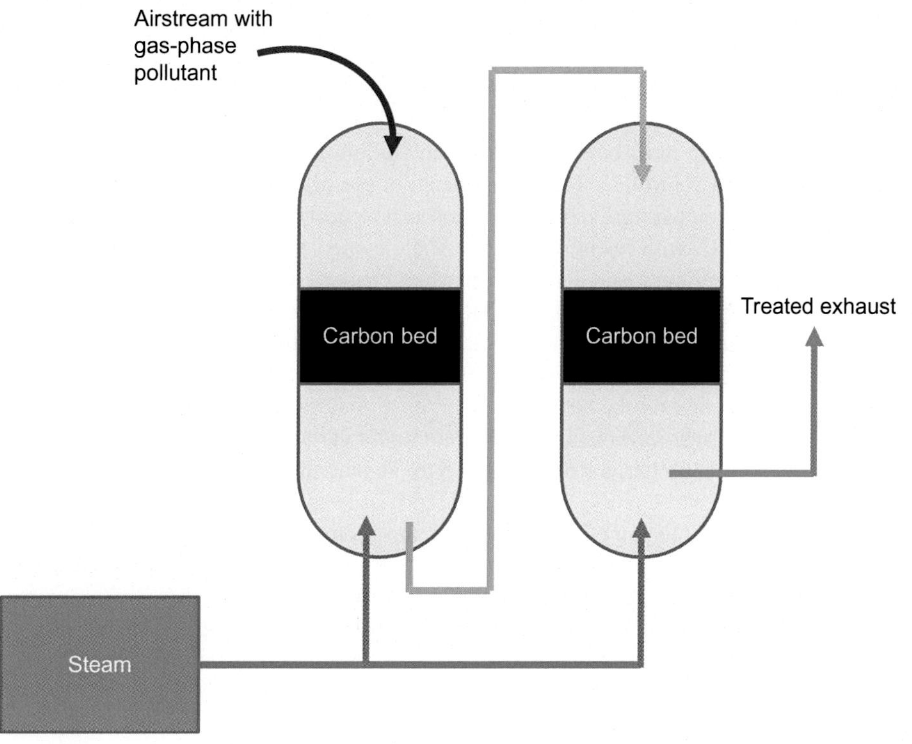

FIG. 13.22 Contact condenser.

TABLE 13.10 Selected condenser applications for gas-phase air pollutant removal

	Petrochemical	Basic chemical	Miscellaneous
Petroleum refining	Manufacturing	Manufacture	Industries
Gasoline accumulator	Polyethylene gas vents	Ammonia	Dry cleaning
Solvents	Styrene	Chlorine solutions	Degreasers
Storage vessels	Copper naphthenates		Tar dipping
Lube oil refining	Insecticides		Kraft paper
	Phthalic anhydride		
	Resin reactors		
	Solvent recover		

threshold for H_2S is about three orders of magnitude less than that for SO_2. For oxidation of H_2S to SO_2, the usual device is simply an open flare with a fuel gas pilot or auxiliary burner if the H_2S is below the stoichiometric concentration. If the SO_2 is above emission or other operation limits, it will also have to be treated, for example, by scrubbing.

13.9.1 Thermal treatment

Thermal treatment can be classified according to the extent of oxidation involved. Combustion is oxidation at high temperatures, whereas pyrolysis also occurs at high temperatures but at lower O_2 concentrations. Thermal processes are commonly used for rapid decomposition of organic compounds, for example, the organic content of municipal wastes. At near-zero air supply, the organic compounds in wastes are treated pyrolytically, that is, thermal cracking, which generates hydrogen gas (H_2), CO, and N_2, and liquid and solid substances, for example, coke, oils, and water. Thermal treatment at partial air is known as gasification, producing H_2, CO, CO_2, CH_4, water, slag, and ash. At excess air, known as incineration, the by-products are CO_2, O_2, N_2, water, slag, and ash [22].

Higher O_2 concentrations and excess air are common in treating many gas-phase pollutants. Afterburners are widely used as control devices for oxidation of undesirable combustible gases. The two general types are (1) direct-flame afterburners, in which the gases are oxidized in a combustion chamber at or above the temperature of autogenous ignition, and (2) catalytic combustion systems, in which the gases are oxidized at temperatures considerably below the autogenous ignition point.

Direct-flame afterburners are the most commonly used air pollution control device in which combustible aerosols, vapors, gases, and odors are to be controlled. The components of the afterburner include the combustion chamber, gas burners, burner controls, and exit temperature indicator. Usual exit temperatures for the destruction of most organic materials are in the range of 650–825°C, with retention times at the elevated temperature of 0.3–0.5 s.

Direct-flame afterburners are efficient and economical when properly operated. Costs to operate and maintain these systems are similar to those of the auxiliary gas fuel systems. Operating and maintenance costs are essentially those of the auxiliary gas fuel. For large industrial applications, the overall cost of the afterburner operation may be considerably reduced by using heat recovery equipment as shown in Fig. 13.23. This configuration includes a fume inlet to an insulated forced draft fan, a regenerative shell-and-tube heat exchanger, an automatic bypass around the heat exchanger for temperature control (required for excess hydrocarbons in fume steam under certain process conditions), a refractory-lined combustion chamber, refractory, with a discharge stream leaving the regenerative heat exchanger and then the ventilating air heat exchanger for further waste heat recovery. If the recovered heat is used for other purposes, for example, in chemical reactors it is known as cogeneration. In some innovative schemes, heat recovery can provide heat for reactors in neighboring industries. Boilers and kilns also provide efficient pollutant destruction of volatile organic compounds and other vapor-phase pollutants in numerous industrial settings.

Catalytic afterburners are used primarily in industry for the control of solvents and organic vapor emissions from industrial ovens. Since the 1970s, they were miniaturized and used as emission control devices for gasoline-powered automobiles.

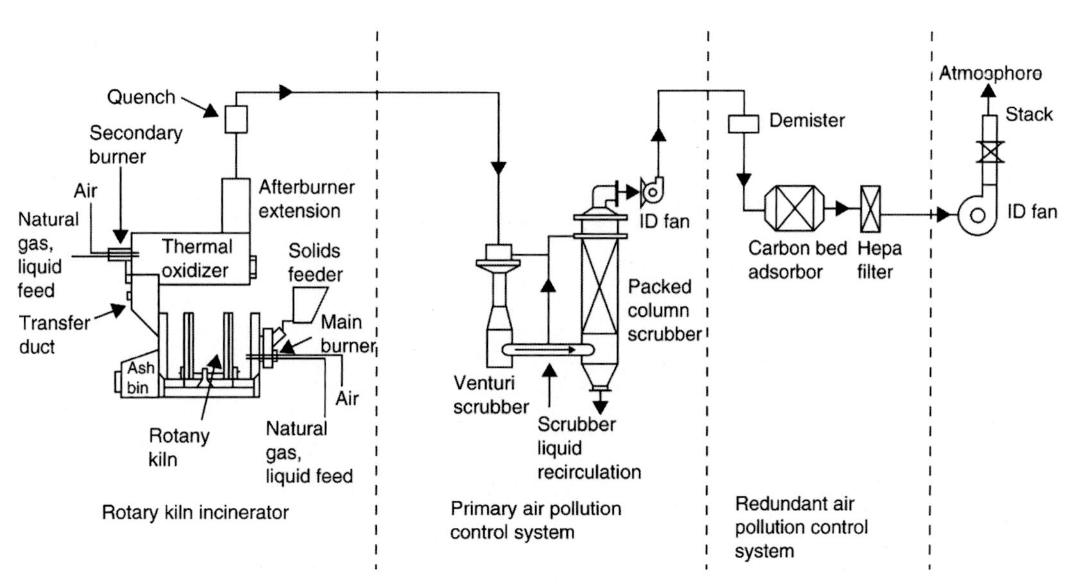

FIG. 13.23 Afterburner with heat recovery. *(Source: H.C. Engineers, Fundamentals of Air Pollution, in: D.A. Vallero (Ed.), Academic Press, 2014.)*

The main advantage of the catalytic afterburner is that the destruction of the pollutant gases can be accomplished at a temperature range of about 315–485°C, which results in considerable savings in fuel costs. However, the installed costs of the catalytic systems are higher than those of the direct-flame afterburners because of the expense of the catalyst and associated systems, so the overall annual costs tend to balance out.

In most catalytic systems, there is a gradual loss of activity due to contamination or attrition of the catalyst, so the catalyst must be replaced at regular intervals. Other variables that affect the proper design and operation of catalytic systems include gas velocities through the system, amount of active catalyst surface, residence time, and preheat temperature necessary for complete oxidation of the emitted gases.

13.9.2 Biological control systems

Waste streams with low to moderate concentrations of volatile organic compounds (VOCs) may be treated with biological systems. The VOC concentrations must be sufficiently high to serve as food sources for microbes, but not so high as to be toxic to them. These are similar to biological systems used to treat wastewater, classified as three basic types: (1) biofilters, (2) biotrickling filters, and (3) bioscrubbers.

Biofilms of microorganisms (bacteria and fungi) are grown on porous media in biofilters and biotrickling systems. The air or other gas containing the VOCs is passed through the biologically active media, where the microbes break down the compounds to simpler compounds, eventually to carbon dioxide (if aerobic), methane (if anaerobic), and water. The major difference between biofiltration and trickling systems is how the liquid interfaces with the microbes. The liquid phase is stationary in a biofilter (see Fig. 13.24), whereas liquids move through the porous media of a biotrickling system (i.e., the liquid "trickles").

A particularly valuable form of biofiltration uses compost as the porous media. Compost contains numerous species of beneficial microbes that are already acclimated to organic wastes. Industrial compost biofilters have achieved removal rates at the 99% level. Biofilters are also the most common method for removing VOCs and odorous compounds from airstreams. In addition to a wide array of volatile chain and aromatic organic compounds, biological systems have successfully removed vapor-phase inorganics, such as ammonia, hydrogen sulfide, and other sulfides including carbon disulfide and mercaptans.

The operational key is the biofilm. The gas must interface with the film. In fact, this interface may also occur without a liquid phase (see Fig. 13.25). According to Henry's law, the compounds partition from the gas phase (in the carrier gas or airstream) to the liquid phase (biofilm). Compost has been a particularly useful medium in providing this partitioning.

The bioscrubber is a two-unit setup. The first unit is an adsorption unit, as discussed earlier. This unit may be a spray tower, bubbling scrubber, or packed column. After this unit, the airstream enters a bioreactor with a design quite similar to an activated sludge system in a wastewater treatment facility. Bioscrubbers are much less common than biofiltration systems.

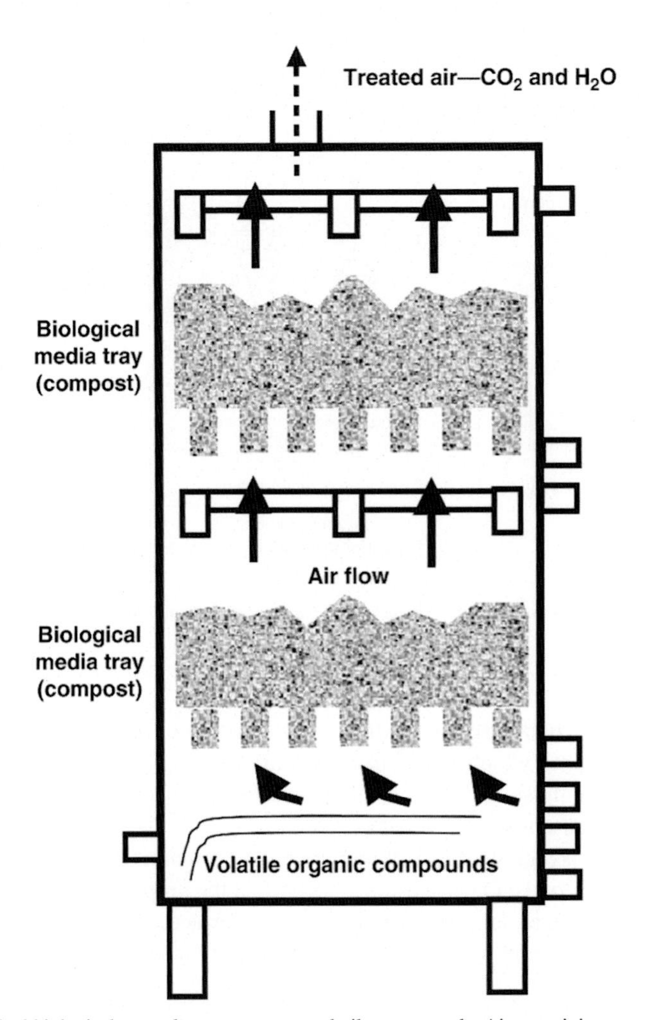

FIG. 13.24 Schematic of packed-bed biological control system to treat volatile compounds. Air containing gas-phase pollutants (C_G) traverse porous media. The soluble fraction of the volatilized compounds in the airstream partition into the biofilm (C_L) according to Henry's law:$C_L = \frac{C_G}{H}$ where H is the Henry's law constant (see Chapters 4 and 5). *(Source: D.A. Vallero, Fundamentals of Air Pollution, 5th ed., Elsevier Academic Press, Waltham, MA, 2014, p. 999 pages cm; reproduced with permission from S.J. Ergas, K.A. Kinney, Biological control systems, in: Air Pollution Engineering Manual, 2000, p. 55.)*

All three types of biological systems have relatively low operating costs since they are operated near ambient temperature and pressure conditions. Power needs are generally for air movement, and pressure drops are low ($<10\,\text{cm}\ H_2O\ \text{m}^{-1}$ or <1.2 in. WC ft.$^{-1}$ of packed bed). Other costs include amendments (e.g., nutrients) and humidification. Another advantage is the usual small amount of toxic by-products and low rates of emissions of greenhouse gases (oxides of nitrogen and carbon dioxide), compared with thermal systems. Success is highly dependent on the degradability of the compounds present in the airstream, their fugacity and solubility needed to enter the biofilm (see Fig. 13.25), and pollutant loading rates. Care must be taken in monitoring the porous media for incomplete biodegradation, the presence of substances that may be toxic to the microbes, excessive concentrations of organic acids and alcohols, and pH. The system should also be checked for shock and the presence of dust, grease, or other substances that may clog the pore spaces of the media [23].

13.9.3 Comparison of gas-phase removal systems

As with particulate removal systems, it is apparent that many choices are available for the removal of gases from effluent streams. Table 13.11 presents some of the factors that should be considered in selecting equipment.

For the control of SO_2, several systems are currently in development and use. Table 13.12 briefly explains these systems. Likewise, numerous systems are available for removing oxides of nitrogen (NO_x). In this case, the removal includes of gas-phase N species other than NO and NO_2, including the greenhouse gas, N_2O. Generally, oxides of N can be reduced by lowering peak operating temperatures and decreasing the time at which operations are at peak temperatures. Also, these oxides can be reduced by burning and reburning and by chemical reduction reactions, both catalyzed and

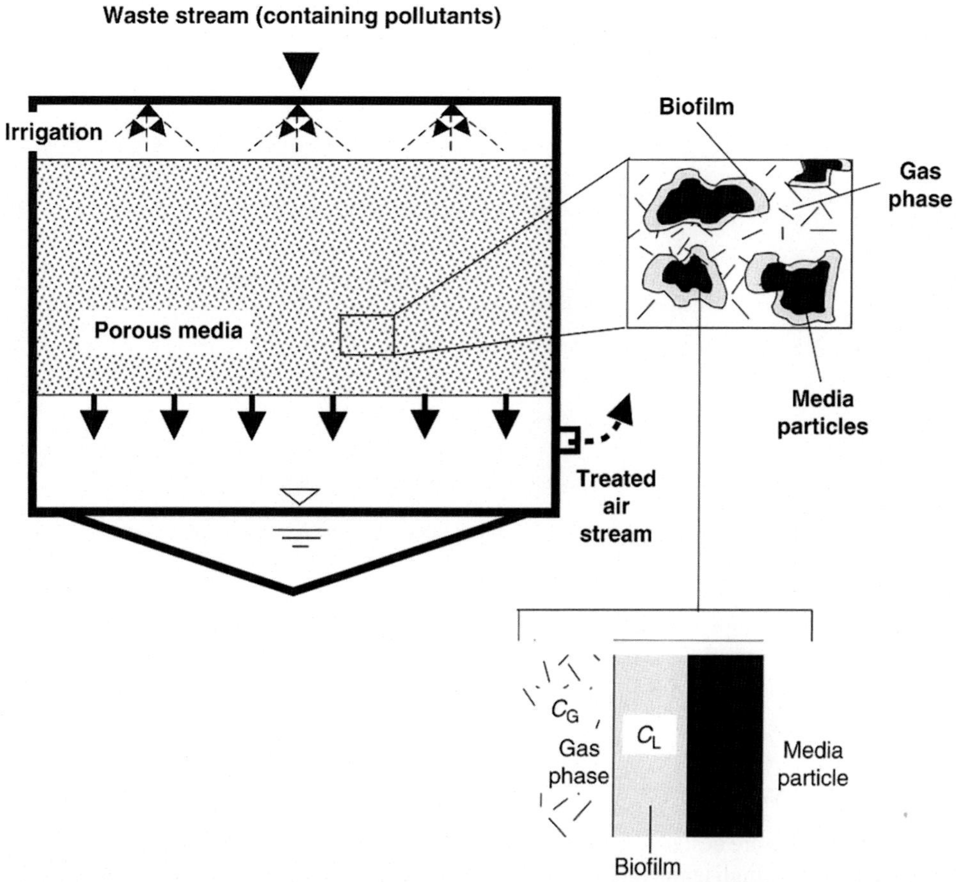

FIG. 13.25 Biofiltration without a liquid phase used to treat vapor-phase pollutants. Air carrying the volatilized contaminants upward through porous media (e.g., compost) containing microbes acclimated to break down specific organic contaminants. The wastes at the bottom of the system can be heated to increase the partitioning to the gas phase. Microbes in the biofilm surrounding each individual compost particle metabolize the contaminants into simpler compounds, eventually converting them into carbon dioxide and water vapor. *(Source: D.A. Vallero, Fundamentals of Air Pollution, 5th ed., Elsevier Academic Press, Waltham, MA, 2014, p. 999 pages cm.)*

TABLE 13.11 Comparison of gaseous pollutant removal systems

Type of equipment	Pressure drop (cm H_2O)	Installed cost (1990 US$ per m^3)	Annual operating cost (1990 US$ per m^3)
Scrubber	10	9.80	14.00
Absorber	10	10.40	28.00
Condenser	2.5	28.00	7.00
Direct-flame afterburner	1.2	8.20	8.40 + Gas
Catalytic afterburner	2.5	11.60	28.00 + Gas
Biological control systems	Low (e.g., <1 in compost)	Variable (low to moderate)	Variable (low to moderate)

Source: D.A. Vallero, Fundamentals of Air Pollution, 5th ed., Elsevier Academic Press, Waltham, MA, 2014, p. 999 pages cm.

TABLE 13.12 Potential sulfur dioxide control approaches

Method	Remarks
Limestone or lime injection (dry)	Calcined limestone or lime reacts with sulfur oxides. They are then removed with a dry particulate control system
Limestone or lime injection (wet)	Calcined limestone or lime reacts with sulfur oxides, which are then removed by wet scrubbers
Sodium carbonate	Sodium carbonate reacts with sulfur oxides in a dry scrubber to form sodium sulfite and CO_2. Sodium sulfite is then removed with a baghouse
Citrate process	Citrate is added to scrubbing water to enhance SO_2 solution into water. Sulfur is then removed from the citrate solution
Copper oxide adsorption	Oxides of sulfur react with copper oxide to form copper sulfate. Removal with a dry particulate control system follows
Caustic scrubbing	Caustic neutralizes sulfur oxides. This method is used on small processes

Source: D.A. Vallero, Fundamentals of Air Pollution, 5th ed., Elsevier Academic Press, Waltham, MA, 2014, p. 999 pages cm.

noncatalyzed. Other methods include substituting O_2 for air in combustion processes, absorption, oxidation followed by sorption, and various combinations of these approaches. Sequestration, that is, storing gases in natural reservoirs like salt domes, is used to lower CO_2 emissions [24].

13.9.4 Mercury removal

Most gas-phase control systems are designed for organic compounds. However, metals and their compounds also can exist in the vapor phase, especially at high temperatures. Often, these quickly condense to become aerosols, but mercury (Hg) is the exception, with substantial amounts in the atmosphere in various forms, including vapor, aerosol, and ions.

Gas-phase Hg is a particularly important and difficult substance to remove from gas streams, including emissions from coal-fired power plants. When Hg-laden coal is burned, the Hg is vaporized in the furnace and exits as Hg^0. Up to 98% of the Hg is then oxidized to Hg^{2+}, which has high aqueous solubility.

Factors that enhance mercury control are a low temperature in the system ($<150°C$), the use of an effective mercury sorbent, and a method to collect the sorbent. Generally, high concentrations of carbon in the fly ash enhance mercury sorption onto PM, with the ash then removed by the PM control device, for example, an electrostatic precipitator, wet flue gas desulfurization system (wFGD), and combinations of these and other technologies. Additionally, the presence of hydrogen chloride (HCl) in the flue gas stream can result in the formation of mercuric chloride ($HgCl_2$), which is readily adsorbed onto carbon-containing PM. Conversely, sulfur dioxide (SO_2) in flue gas can act as a reducing agent to convert oxidized mercury to elemental mercury, which is more difficult to collect [25].

The most common techniques for removing Hg are carbon filter beds, wet scrubbing, selenium filters, and activated carbon injection. Gas-phase oxidized Hg can be collected by absorption to aqueous solution in the wFGD and removed with wFGD waste. Once captured from the gas phase, Hg^{2+} can persist in wFGD liquids in a water-soluble form, partition to the solid phase, or be reduced to Hg^0.

13.9.4.1 Carbon filter beds

Three carbon filter designs have been used to remove Hg: cross flow, countercurrent, and counter-cross flow. The cross flow design has been applied to utility boilers and waste combustors, the countercurrent design to waste combustors, and the counter-cross flow design to utility boilers. In the cross flow filters, the flue gas flows horizontally through the filter bed, while the adsorbent migrates through the filter from top to bottom. The bed is about 1 m thick and is divided into three layers. The first layer removes PM, heavy metals (including mercury), organic compounds, and SO_2. Removal of HCl occurs primarily in the second layer. The third layer acts as a safety layer. The three layers are separated by perforated metal sheets. On the outlet side, there is a slotted sheet designed to prevent particles from being carried out of the filter with the flue gas. Additionally, an impact separator is located downstream of the slotted sheet, functioning as a safety barrier against particle emissions.

Pressure drop is the primary rate-limiting factor for carbon removal from the bed. The designed pressure drop across the bed is 1.5–1.9 cm (6–7.5 in.) of water. The pressure drop across the whole system including fans and ducting is about 30 cm (12 in.) of water. Because of PM collection in and compaction of each layer, approximately 10–13 cm (4–5 in.) of material is periodically sloughed from each layer. Because of greater vibration of particles and, thus, greater settling of the particles within the first layer, carbon is sloughed from this layer on the shortest time interval, typically once every 6–8 h. The second layer is sloughed once per day, and the third layer is sloughed once every 2 weeks. Based on these removal rates and bed thicknesses, the mass of carbon in the filter is fully replaced at an average rate of once per year [25].

Spent carbon can be disposed of by combustion if the unit is equipped with a wet scrubbing system. The combustion process degrades organic compounds captured in the carbon, and the wet scrubber collects the heavy metals and acid gases.

13.9.4.2 Wet scrubbing

Wet scrubbing controls acid gases; metals; PM; and semivolatile organic compounds (SVOCs), for example, chlorinated dioxins and furans. Single-stage scrubbers can be used to remove acid gases. Two-stage scrubbers can be used for acid gases and metals. Three-stage systems having a two-stage scrubber followed by a high-efficiency wet particulate control system are designed for improved control of fine particulates, metals, and SVOCs. Single-stage scrubbers can also be installed following other pollution controls for increased pollutant control (i.e., polishing scrubbers).

Hydrophilic Hg compounds, for example, HgC1 and HgO, can be effectively captured (>90%) in the wet scrubber. If there is significant Hg^0 in the flue gas, however, collection efficiencies will be limited. The captured mercury is precipitated out during wastewater treatment with additives (i.e., sodium sulfide and dithiocarbamate). Three-stage systems have shown Hg removal efficiencies of over 90%. The amount of hydrophilic Hg species in the gas stream is a major factor determining effectiveness of this control approach [25].

13.9.4.3 Selenium filters

Selenium (Se) filters operate based on the affinity between Hg and Se. The Hg-laden flue gas passes through the filter, which consists of ceramic grains impregnated with metallic selenium. The gas pathway through the filter is tortuous, which increases the contact between the mercury and the selenium, forming mercury selenite (HgSe). Selenium filters are effective on flue gas streams with inlet mercury concentrations of up to 9 mg scm^{-1} (3900 g per million scf). At very low Hg concentrations, removal efficiency decreases due to the dramatic decrease in Hg-Se molecular collisions. At higher Hg concentrations, the lifetime of the filter is short. Because the removal of mercury in the filter is based on the formation of HgSe, the selenium in the filter is eventually exhausted. The selenium filter is designed to convert approximately 50 kg of mercury to HgSe per cubic meter of filter material (3 lbs. $ft.^{-3}$). The combination of pressure drop, Hg content in the flue gas, and the mechanical construction of the filter drive filter life. On average, the filter lifetime is 5 years, after which the filter element is replaced. Once the lifetime of the filter mass has expired, the HgSe must be properly disposed as a hazardous waste. Se filter Hg removal effectiveness has been shown for metal smelters and crematories, but not yet for a utility boiler.

13.9.4.4 Activated carbon injection

Powdered activated carbon can be injected into the gas upstream of air pollution control devices. Since activated carbon has many internal pores and large surface area, it can adsorb a broad range of trace contaminants, including Hg. After injection into the flue gas and adsorption of Hg and other contaminants, the activated carbon is captured in the PM control device. The factors affecting the performance of activated carbon injection are the temperature of the flue gas, the amount of activated carbon injected, the concentration of each Hg species in the flue gas, the extent of contact between the carbon and Hg, and the type of carbon used. Flue gas temperature is crucial because Hg volatilizes at temperatures above 150–200°C. The flue gas temperature needs to be within or preferably below this range for the mercury to adsorb onto the carbon. The combustion device and the corresponding composition of the flue gas will affect this temperature range. In a municipal waste combustor (MWC), where there is a substantial amount of HCl resulting in the formation of $HgCl_2$, temperatures within and below the noted range have proved to be effective when injecting carbon. Test data from an MWC retrofitted with activated carbon injection indicate mercury removals >95%. Pilot testing on a coal-fired utility boiler indicated that a temperature under 90–120°C was necessary for effective mercury removal. With activated carbon injection, efficient distribution of the carbon in the flue gas is also important. The amount of carbon needed to achieve a specific level of mercury removal will vary depending on the fuel being burned, the amount of carbon inherent to the system, and the type of PM control device. At a given carbon feed rate, a fabric filter provides more Hg control than an ESP because of the additional Hg adsorption that occurs on the bags of the fabric filter (due to the increased gas contact time). Mercury is predominately removed upstream of an ESP-equipped facility; however, the nominal residence time of 1 s or less limits the capture. In addition, Hg is not effectively collected across the ESP, further requiring substantially higher carbon feed rates than the fabric filter-equipped facilities [25].

13.10 REMOVAL OF ODORS

An odor can be described as a physiological response to the activation of the sense of smell [20]. It can be caused by a chemical compound (e.g., H_2S) or a mixture of compounds (e.g., coffee roasting). Generally, if an odor is objectionable, any perceived quantity greater than the odor threshold will be cause for complaint. The control of odors, therefore, becomes a matter of reducing them to less than their odor thresholds, preventing them from entering the atmosphere, or converting them to a substance that is not odorous or has a much higher odor threshold. Odor masking is not recommended for a practical, long-term odor control system.

13.10.1 Odor reduction by dilution

If the odor is not a toxic substance and has no harmful effects at concentrations below its threshold, dilution may be the least expensive control technique. Dilution can be accomplished either by using tall stacks or by adding dilution air to the effluent. Tall stacks may be costlier if only capital costs are considered, but they do not require the expenditure for energy that is necessary for dilution systems. In addition, if the emission contains other pollutants, taller stacks will increase the distance traveled by the pollutant and, thus, will contribute to the long-range transport and the potential cumulative effects of these pollutants.

The odor threshold for most atmospheric pollutants may be found in the literature [20]. By properly applying the diffusion equations, one can calculate the height of a stack necessary to reduce the odor to less than its threshold at the ground or at a nearby structure. A safety factor of two orders of magnitude is suggested if the odorant is particularly objectionable.

Odor control by the addition of dilution air involves a problem associated with the breakdown of the dilution system. If a dilution fan, motor, or control system fails, the odorous material will be released to the atmosphere. If the odor is objectionable, complaints will be noted immediately. Good operation and maintenance of the dilution system become an absolute requirement, and redundant systems should be considered.

13.10.2 Odor removal

It is sometimes possible to close an odorous system to prevent the release of the odor to the atmosphere. For example, a multiple-effect evaporator can be substituted for an open contact condenser on a process emitting odorous, noncondensable gases.

Another possible solution to an odor problem is to substitute a less noxious or more acceptable odor within a process. An example of this type of control is the substitution of a different resin in place of a formaldehyde-based resin in a molding or forming process.

Many gas streams can be deodorized by using solid adsorption systems to remove the odor before the stream is released to the atmosphere. Such procedures are often both effective and economical.

13.10.3 Odor conversion

Many odorous compounds may be converted to compounds with higher odor thresholds or to nonodorous substances. An example of conversion to another compound is the oxidation of H_2S, odor threshold 0.5 ppb, to SO_2, odor threshold 0.5 ppm. The conversion results in another compound with an odor threshold three orders of magnitude greater than that of the original compound.

An example of conversion to a nonodorous substance would be the passage of a gas stream containing butyraldehyde, $CH_3CH_2CH_2CHO$, with an odor threshold of 40 ppb, through a direct-fired afterburner that converts it to CO_2 and H_2O, both nonodorous compounds. It should be noted that using a direct-fired afterburner, particularly one without heat recovery, to destroy 40 ppb is not an economical use of energy, and some other odor control system may be more desirable.

This chapter has introduced several examples and calculations of the theoretical basis for controlling and removing air pollutants. These are guidelines, but air pollution sources vary. The engineer and manager must consider the specific conditions in selecting the best technologies to meet required emission limits.

REFERENCES

[1] U.S. Environmental Protection Agency, Momentum separators, in: Air Pollution Control Technology Fact Sheet, 2018.

[2] L. Theodore, A.J. Buonicore, Air Pollution Control Equipment: Selection, Design, Operation and Maintenance, 1982.

[3] U.S. Environmental Protection Agency, Chapter 4: particle collection mechanisms, in: J.R. Richards (Ed.), APTI Course 413: Control of particulate matter emissions, third ed., ICES Ltd, Research Triangle Park, NC, 2000 [Online]. Available: http://www.4cleanair.org/APTI/413Combined.pdf.

[4] E. Cunningham, On the velocity of steady fall of spherical particles through fluid medium, Proc. R. Soc. Lond. A 83 (563) (1910) 357–365.

[5] D.A. Vallero, Excerpted and Revised From: Fundamentals of Air Pollution, fifth ed., Elsevier Academic Press, Waltham, MA, 2014. p. 999 pages cm.

[6] U.S. Environmental Protection Agency, Chapter 3: Particle Sizing, in: Control of Particulate Matter Emissions: Student Manual: Apti Course 413, North Carolina State University, 2012.

[7] EPA-450/3-81-005a, Control Techniques for Particulate Emissions from Stationary Sources, vol. 1, 1982.

[8] U.S. Environmental Protection Agency, Elutriators, in: Air Pollution Control Technology Fact Sheet, 2018.

[9] U.S. Environmental Protection Agency, Chapter 5: Settling chambers, in: J.R. Richards (Ed.), APTI Course 413: Control of Particulate Matter Emissions, third ed., ICES Ltd, Research Triangle Park, NC, 2000 [Online]. Available: https://www.apti-learn.net/lms/register/display_document.aspx?dID=262.

[10] U.S. Environmental Protection Agency, Chapter 4: Particle collection mechanisms, in: Control of Particulate Matter Emissions: Student Manual: Apti Course 413, North Carolina State University, 2012.

[11] C. Menoher, An incinerator scrubber that works: a case study, in: Third Symposium on the Transfer and Utilization of Particulate Control Technology: Volume III. Particulate Control Devices, Research Triangle Park, North Carolina, 1983.

[12] U.S. Environmental Protection Agency, Chapter 7: Fabric filters, in: Control of Particulate Matter Emissions: Student Manual: Apti Course 413, North Carolina State University, 2012.

[13] E. P. A. U.S, Fabric filters, in: C. o. G. E. S. M. A. C. 413, North Carolina State University, 2012.

[14] U.S. Environmental Protection Agency, Chapter 9: Electrostatic precipitators, in: Control of Particulate Matter Emissions: Student Manual: Apti Course 413, North Carolina State University, 2012.

[15] U.S. Environmental Protection Agency, Chapter 8: Wet scrubbers, in: Control of Particulate Matter Emissions: Student Manual: Apti Course 413, North Carolina State University, 2012.

[16] C.D. Cooper, F.C. Alley, Air Pollution Control: A Design Approach, Waveland Press, 2010.

[17] H. Perry, Perry's Chemical Engineering Handbook, eighth ed., McGraw-Hill, New York, 2007.

[18] C.G. Bell, W. Strauss, Effectiveness of vertical mist eliminators in a cross flow scrubber, J. Air Pollut. Control Assoc. 23 (11) (1973) 967–969.

[19] J. Richards, Control of Gaseous Emissions: Student Manual: Apti Course 415, North Carolina State University, 2000.

[20] A.C. Stern, Air Pollution: Vol. 2. The Effects of Air Pollution, Academic Press, 1977.

[21] A. Teller, Gas turbine emission control: a systems approach, J. Air Pollut. Control Assoc. 27 (2) (1977) 148–149.

[22] M.J. Quina, J.C. Bordado, R.M. Quinta-Ferreira, Air pollution control in municipal solid waste incinerators, in: The Impact of Air Pollution on Health, Economy, Environment and Agricultural Sources, InTech, 2011.

[23] S.J. Ergas, K.A. Kinney, Biological control systems, in: Air Pollution engineering manual, 2000, p. 55.

[24] R.J. Gehl, C.W. Rice, Emerging technologies for in situ measurement of soil carbon, in English, Climatic Change 80 (1–2) (2007) 43–54.

[25] L. De Rose, Introduction to Air Toxics: Student Manual, 2009.

FURTHER READING

[26] D.A. Vallero, Fundamentals of Air Pollution, fifth ed., Elsevier Academic Press, Waltham, MA, 2014. p. 999 pages cm.

[27] L. Theodore, A.J. Buonicore, Air Pollution Control Equipment, 1988.

[28] A.C. Stern, H.C. Wohlers, R.W. Boubel, W.P. Lowery, Fundamentals of Air Pollution, Academic Press, New York and London, 1973.

[29] S. Oglesby, G.B. Nichols, Electrostatic precipitators, in: A.C. Stern (Ed.), Air Pollution, 1977.

[30] U.S. Environmental Protection Agency, Air Pollution Control Orientation Course, 30 November 2013. Available: http://www.epa.gov/air/oaqps/eog/course422/ce6.html, 2013.

[31] U.S. Environmental Protection Agency, Basic Concepts in Environmental Sciences—Module 6: Air Pollutants and Control Techniques, 30 June 2007. Available: http://www.epa.gov/eogapti1/module6/matter/control/control.htm, 2007.

[32] Environ-Access Inc, Environmental Fact Sheet. Treatment of air and gas: Treatment of gaseous emissions through a wet scrubber, vol. F2-03-96, M. E. Inc, Sherbrooke, Quebec, Canada, 1996.

[33] Los Angeles Air Pollution District, in: R.W. Boubel, D.L. Fox, B. Turner, A.C. Stern (Eds.), Fundamentals of Air Pollution, Academic Press, 1994.

[34] H.C. Engineers, in: D.A. Vallero (Ed.), Fundamentals of Air Pollution, Academic Press, 2014.

Chapter 14

Air pollution dispersion models

14.1 INTRODUCTION

Models provide a means for representing a complicated system in an understandable way. They take many forms. At the beginning of any air pollution design or decision, "conceptual models" help to explain how the systems involved are expected to work, including all the factors and parameters of how a pollutant moves in the air after an emission. Conceptual models are a first and important step in identifying the major influences on where a chemical species is likely to be found in the environment and, as such, need to be developed to help target sources of data needed to assess an environmental problem. Thus, the subject matter in every chapter in this book is meaningful in conceptual modeling.

Sometimes, conceptual models are not formally developed, but understood by all users. This may present problems, since the users may incorrectly believe all agree on what needs to be known to make a decision. Ideally, the essential and desirable inputs, algorithms, and outputs should be documented. From these, the users and decision-makers may find that the intuitive choice of model may be inadequate.

In general, developing a model requires answering the following key questions:

1. What is the domain of interest? The models needed in this text all fall within the atmospheric domain. However, this can include highly diverse physical, chemical, biological, and psychosocial domains. Defining and characterizing the domain delineates the specific needs, that is, the modeling objectives, and constrains the information and algorithms needed in the model.
2. What are the processes and mechanisms involved? The boundary conditions must represent the influencing environment surrounding the study domain. Scientists often develop "physical" or "dynamic" models to estimate the location where a chemical would be expected to move under controlled conditions, only on a much smaller scale. For example, the US Environmental Protection Agency (EPA) employs chambers or full-size test houses to model the fate of pollutants indoors. Like all models, the dynamic model's accuracy is dictated by the degree to which the actual conditions can be simulated and the quality of the information that is used [1].
3. What are the metrics to generate modeling results needed? For air pollution, the major distinction is whether the results must comply with regulations, to screen for future modeling or measurement; to advance the state of the science; or to be entered into other models, tools, or dashboards. Regulatory models may be defined; for example, a state may be required to use a specific model, for example, AERMOD, to characterize and predict concentrations of air pollutants as part of the State Implementation Plan (SIP) or to meet air toxics rules [2–4].

Numerical models apply mathematical expressions to approximate a system. There are three thermodynamic systems:

1. Isolated systems, in which no matter or energy crosses the boundaries of the system (i.e., no work can be done on the system)
2. Closed systems, in which energy can exchange with surroundings, but no matter crosses the boundary
3. Open systems, in which both matter and energy freely exchange across system boundaries

Isolated systems are usually only encountered in highly controlled reactors, so are important in modeling air pollution precursors, for example, pesticide formulation and manufacturing, but are not directly pertinent to air pollution exposure and risk modeling. In fact, most microenvironmental systems are open, but with simplifying assumptions, some subsystems can be treated as closed.

14.1.1 Box models

Box models provide a simple way to model air pollution, in which an airshed is treated as a rectangular box in which the mass of pollutant is conserved (see Fig. 14.1). The box is oriented so that the wind velocity (u) is incident to and normal to

Air Pollution Calculations. https://doi.org/10.1016/B978-0-12-814934-8.00014-4

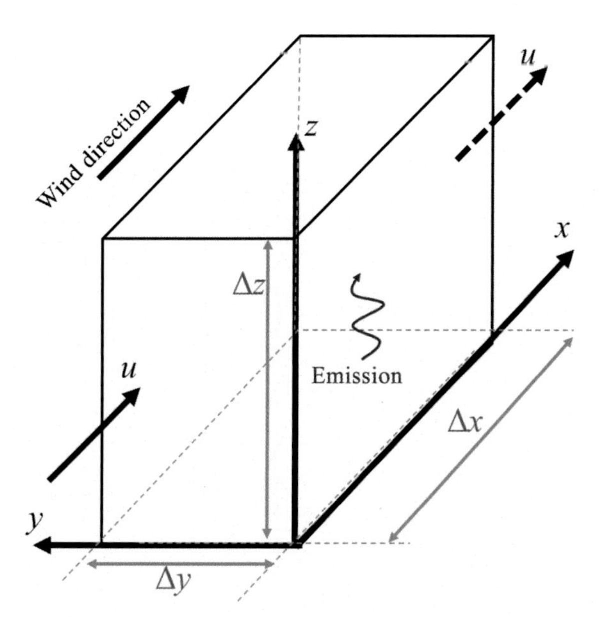

FIG. 14.1 Simple box model with dimensions $\Delta x \times \Delta y \times \Delta z$. *Note*: u is wind speed. *(Based on information from M. Markiewicz, Modelling of the air pollution dispersion, in: MANHAZ. Models and Techniques for Health and Environmental Hazards Assessment and Management. IEA, Otwock-Swierk, Poland, 2006, pp. 303–347; M. Borysiewicz, M. Borysiewicz, Models and techniques for health and environmental hazard assessment and management, Swierk Poland Inst. Atomic Energy. 20 (2006) 2010.)*

one box face. For further simplicity, box models assume that the pollutant is completely mixed within the box volume (V), that is, the concentration (C) is homogeneous. This allows for a mass balance [5]:

$$\frac{dyCV}{dt} = QA + (\bar{u} \cdot C_{in} \cdot \mathrm{W} \cdot H_{mix}) - (\bar{u} \cdot C_{in} \cdot \mathrm{W} \cdot H_{mix}) \tag{14.1}$$

where C_{in} is the concentration of the pollutant entering the box, Q is the pollutant emission rate per unit area of sources within the box, $\bar{u}$ is the average wind speed normal to the box, W is the width of the box, A is the horizontal area (length $\times$W), and H_{mix} is the mixing depth.

The integration of this equation estimates the pollutant concentration at steady state, so long as the dynamics of mixing depth are nearly stationary, and Q is constant. The box model does not resolve the spatial or temporal distribution of the air pollutant within the box, that is, the results are always average concentrations of large airsheds, since they assume homogeneity and instantaneous mixing [5]. However, if H_{mix} is allowed to be a function of downwind distance, x these models may be applied to smaller airsheds.

Usually, the simplifications that must be assumed for a single box are unacceptable for many applications. Therefore, the atmospheric domain of interest can be better characterized with multiple boxes, that is, an assemblage of boxes wherein the mass of the pollutant is exchanged (see Fig. 14.2). The mass balance of each box is calculated as discussed above, but the equations are linked using mass-in and mass-out terms [6].

Given the numerous simplifications, the applications of box models are limited. However, box models can be useful early in the decision-making processes and to screen based on average pollutant concentrations. These can be followed by runs of more sophisticated models.

14.1.2 Model frames

Pollutant dispersion models can be classified as statistical and/or deterministic. Statistical models include the pollutant dispersion models, such as the Lagrangian models, which follow the movement of a control volume starting from the source to the receptor locations. These often assume idealized Gaussian distributions of pollutants from a point of release, that is, the pollutant concentrations are normally distributed in both the vertical and horizontal directions from the source. The Lagrangian approach is common for atmospheric releases. However, recent models based on the Lagrangian approach have incorporated descriptions of complex turbulence.

Stochastic models are statistical models that assume that the events affecting the behavior of a chemical in the environment are random, so such models are based on probabilities. These are being commonly adopted in the modeling of

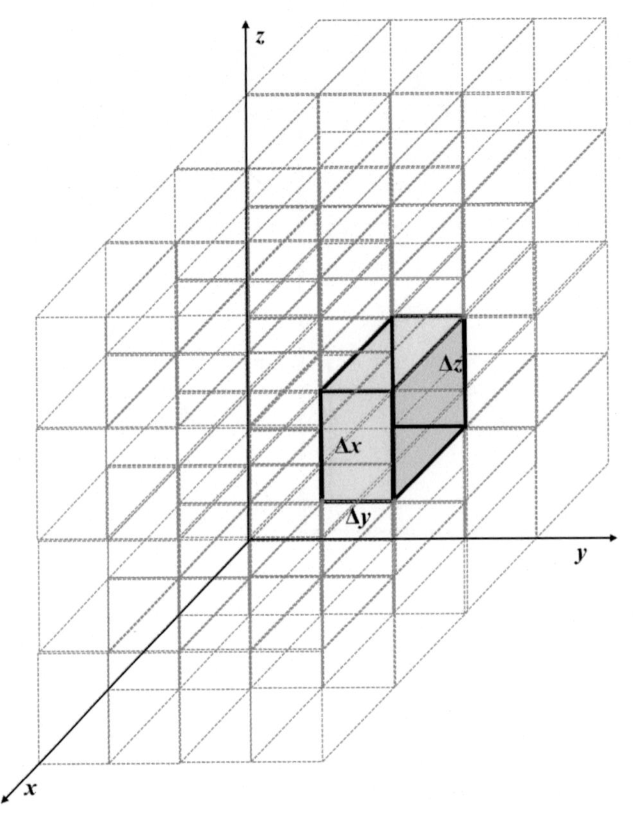

FIG. 14.2 Calculation grid for interconnected boxes in multiple box model. *(Based on information from M. Markiewicz, Modelling of the air pollution dispersion, in: MANHAZ. Models and Techniques for Health and Environmental Hazards Assessment and Management. IEA, Otwock-Swierk, Poland, 2006, pp. 303–347.)*

human exposures. For example, a stochastic model may develop a "virtual person" and randomize that person across an entire population or subpopulation. The randomization routine assumes that the population will behave like the single virtual person by generating distributions from that person's information. Obviously, nothing can be known about any other single person, but based on the randomization, the population can be described [7–10]. Stochastic models can apply to many air pollution scenarios, such as traffic routing and origin-destination studies [9] and even the behavior of an absorbed chemical within an organism, for example, stochastic pharmacokinetic models [11].

Deterministic models are used when the physical, chemical, and other processes are sufficiently understood to be incorporated to reflect the movement and fate of chemicals. One type of deterministic model used in air pollution estimation is the process-oriented model, which is based on the description of the physicochemical processes involved in air pollution, from the emission of the pollutant and its precursors, to the advection and dispersion within the atmosphere, to transformation, to deposition [12]. This type model is very difficult to develop because each process must be represented by a set of algorithms in the model. Also, the relationship between and among the systems, such as the kinetics and mass balances, must also be represented. Thus, the modeler must parameterize every important event following a pesticide's release to the environment.

Often, hybrid models using both statistical and deterministic approaches are used, for example, when one part of a system tends to be more random, while another has a very strong basis in physical principles. Numerous models are available to address the movement of chemicals through a single environmental media, but increasingly, environmental scientists and engineers have begun to develop multimedia models, such as compartmental models that help to predict the behavior and changes to chemicals as they move within and among reservoirs (e.g., carpet), in the air as dust and vapors, and in exchanges with surfaces [1, 13].

14.2 METEOROLOGICAL DATA AND GRAPHICS

Air pollutants reach receptors by being transported and perhaps transformed in the atmosphere (Fig. 14.3). As such, the outputs from air pollutant models depend on meteorology. The necessary variables and parameters vary by scale; for

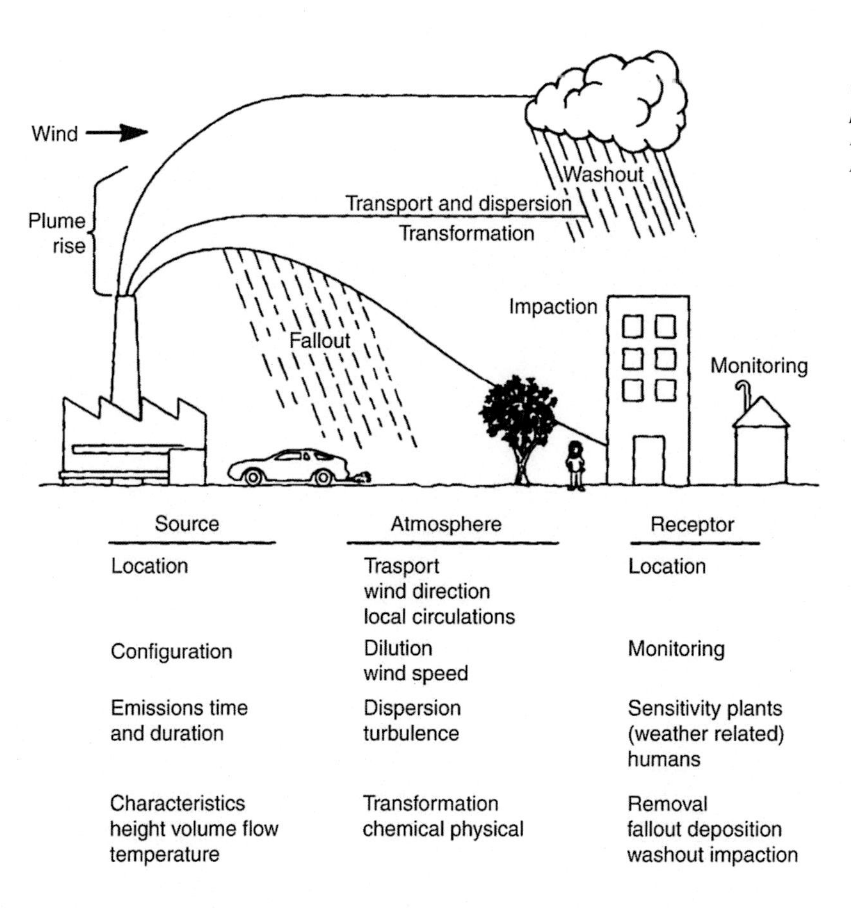

FIG. 14.3 The atmosphere's role in air pollution. *(Source: D A Vallero et al., Modeling and predicting pesticide exposures, in: R. Krieger (Ed.), Hayes Handbook of Pesticide Toxicology, Elsevier Science, New York, NY, 2010, pp. 995–1020.)*

example, large-scale models may use climatological data, whereas local and indoor air models employ site-specific and temporally explicit data and algorithms, for example, to calculate penetration of outdoor pollutants into a microenvironment.

The location of receptors relative to sources and atmospheric influences affect pollutant concentrations, and the sensitivity of receptors to these concentrations determines the effects. The location, height, and duration of release, as well as the amount of pollutant released, are also of importance. Some of the influences of the atmosphere on the behavior of pollutants, primarily the large-scale effects, are discussed here, as well as several effects of pollutants on the atmosphere.

If air movement past a continuous pollutant source is slow, pollutant concentrations in the plume moving downwind will be much higher than they would be if the air were moving rapidly past the source. If polluted air continues to have pollution added to it, the concentration will increase. Generally, a source emits into different volumes of air over time. However, there can be a buildup of concentration over time even with significant air motion if there are many sources.

Low- and high-pressure systems differ in ventilation, that is, total air volume moving past a location. Air generally moves toward the center of a low-pressure system, that is, cyclone, (see Fig. 14.4) in the lower atmosphere, due in part to the frictional turning of the wind toward low pressure. This convergence causes vertical motion near the center of the low. Usually, winds near the center of the low are light, and winds away from the center are moderate, resulting in increased ventilation rates. Note the increased wind in the area to the west of the low in Fig. 14.4. Low-pressure systems generally cover relatively small areas, although the low-pressure system shown in Fig. 14.4 covers an extensive area. They seldom remain the same at a given area for a significant period of time. Lows are frequently accompanied by cloudy skies, which may cause precipitation. The cloudy skies minimize the variation in atmospheric stability from day to night. Primarily because of moderate horizontal wind speeds and upward vertical motion, low-pressure systems are accompanied by an increase in ventilation.

High-pressure systems, that is, anticyclones, are characteristically opposite of lows. The winds flow outward from the high-pressure center, so subsiding air compensates for the horizontal transport of mass. This sinking air causes a subsidence inversion. Skies are usually clear due in part to the subsiding vertical motion. This increases incoming solar radiation during the day and decreases outgoing radiation at night, inducing atmospheric instability during the day and stability at night, with

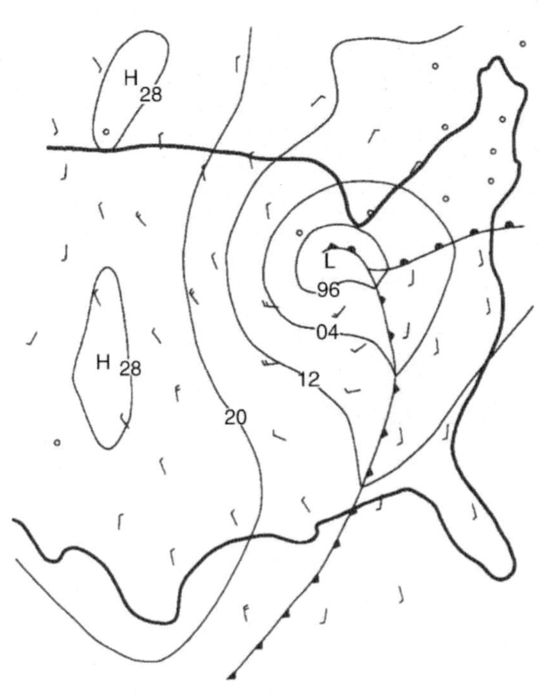

FIG. 14.4 Surface chart. *Note*: Contours are isobars of atmospheric pressure; numbers are 1000 millibars (mb) plus the number; for example, 12 is 1012 mb or 0.9988 atm. *Line with triangles* is a cold front; *line with semicircles* is a warm front; *line with both triangles and semicircles* is occluded front, that is, cold front that has caught up with a warm front. Wind direction is with the *arrow*; wind speed is 10 knots for 1 barb and 5 knots for one-half barb. *Small station circles* indicate calm. *H*, center of high pressure; *L*, center of low pressure. 1 knot = 1 nautical mile per hour = 1.15 mph = 1.9 km h^{-1}. *(Source: D.A. Vallero, Fundamentals of Air Pollution, 5th ed., Elsevier Academic Press, Waltham, MA, 2014, p. 999 pages cm.)*

frequent radiation inversions. Highs generally occupy large areas, although they are usually slow-moving. Winds over large areas are generally light; note the winds to the south of the high in the lower left corner of Fig. 14.4.

Stagnation occurs when ventilation is low. The lack of motion often occurs in the week pressure gradient near the center of the high. If the high has a warm core, air movement is stifled near the center. This produces very light winds, few clouds, and the surface-based inversions at night. During the day, the mixing height (see Chapter 1) does not reach high altitudes because of the subsidence inversion above the high-pressure system, even though the clear skies induce instability below the mixing height.

Air pollution concentrations and gradients are directly affected by meteorology. Pollutants move along with airflow, that is, advection. Concentrations in pollutant plumes can increase or decrease as a result of meteorologic conditions. For example, particulate matter (PM) concentrations may build up during stagnation and inversions. This decreases incoming solar radiation and visibility, which in turn increase the concentrations of CO_2 and other gases. This increases surface temperatures and the greenhouse effect locally, which can induce episodes of high concentrations of oxides of sulfur and nitrogen, CO, and other gas-phase pollutants as well as PM.

Another example is wind speed, which plays a major role in dilution after a pollutant is released from a stack. Dilution occurs in the direction of plume transport but with different vertical profiles (see Fig. 14.5).

What is the effect of wind speed for an elevated source with an emission of 6 mass units per second if the wind speed falls from 6 m s^{-1} to 2 m s^{-1}?

According to Fig. 14.5, there is 1 unit between the vertical parallel planes 1 m apart for a wind speed of 6 m s^{-1}. When the wind is slowed to 2 m s^{-1}, there are 3 units between those same vertical parallel planes 1 m apart. Note that this dilution by the wind takes place at the point of emission. Because of this, wind speeds used in estimating plume dispersion are generally estimated at stack top.

Fate and transport models employ algorithms based on the information discussed in Chapters 3–7. For example, the extent and degree of turbulence in the atmosphere is part of these models. Wind speed and atmospheric stability are affected by temperature and pressure. The most important mixing process that causes pollutant dispersion is eddy diffusion. The eddies in the atmosphere break up atmospheric parcels, so that the air with higher pollutant concentrations mixes with less

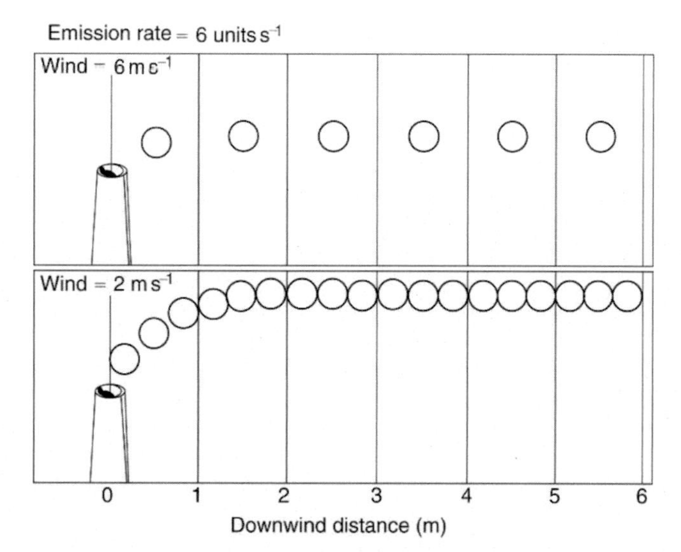

FIG. 14.5 Dilution by wind speed. *(Source: D.A. Vallero, Fundamentals of Air Pollution, 5th ed., Elsevier Academic Press, Waltham, MA, 2014, p. 999 pages cm.)*

polluted air. Eddy diffusion is most efficient if the eddy's scale approaches that of the pollutant puff or plume. The vertical temperature profile is compared with the shape and expansion of a plume in Fig. 14.6. Some of these features are due to the adiabatic lapse rate, that is, temperature change not caused any gain or loss of heat entering or exiting the system. However, the shape and extent of the plume are also the result of nonadiabatic processes of heat exchange.

As discussed in Chapter 6, dispersion is the mixing of the pollutant within the atmosphere, which can be predicted within two basic frames: Lagrangian and Eulerian. A Lagrangian model applies statistical theory of turbulence, assuming that turbulent dispersion is a random process described by a distribution function. The Lagrangian model follows the individual random movements of molecules released into the plume, using statistical properties of random motions that are characterized mathematically.

Conversely, the Eulerian model bases the mass balance around a differential volume. The velocity field is very important in environmental modeling, especially in modeling plumes in the atmosphere and in groundwater, since the velocity field is a way to characterize the motion of fluid particles and provides the means for computing these motions.

Both Lagrangian and Eulerian model frameworks begin with a release of a substance and tracking the substance in time and space. The key difference is that the Eulerian approach is based on a three-dimensional Cartesian grid as the frame of reference, whereas the Lagrangian approach employs a probability theory as a moving frame of reference. Put simply, Eulerian models have an observer standing in one location who is watching the air pass by, whereas the Lagrangian models have the observer moving along with the parcel of air as it takes a random walk through the atmosphere. Within the Lagrangian frame, the path each parcel of air takes during a specified time is an ensemble mean field related to the displacement probabilities of the chemical species of interest (see Eq. (6.15) in Chapter 6).

Both types of models are used by air pollution regulatory agencies. The advantages of Lagrangian models are that they are usually easier to code, run, and evaluate; they can handle and readily modify explicit transport mechanisms; and they are appropriate for source apportionment since the random walk from the source to the receptor is identified. However, they are not well suited for interactions among parcels, that is, puffs, and when puffs become numerous, the computation becomes increasing expensive. Examples of Lagrangian models include CALPUFF and the Reactive Plume Model (RPM).

The advantages of the Eulerian models include that they provide information about weather and transport in three-dimensional space and time; they are more real-world than Lagrangian models; concentrations can be defined over large spatial and temporal domains; and plume-in-grid approaches are available for major sources. The disadvantages are computational expense and difficulty handling source apportionment. Examples include the Urban Airshed Model (UAM), CALGRID, Regional Oxidant Model (ROM), and the Community Multiscale Air Quality (CMAQ) model.

14.3 GAUSSIAN MODELS

Gaussian plume models are based on random statistics, that is, the change in plume size is determined by standard deviations in cross-sectional, that is, x and y, and altitudinal, that is, z, dimensions (see Fig. 14.7). They have aspects of both

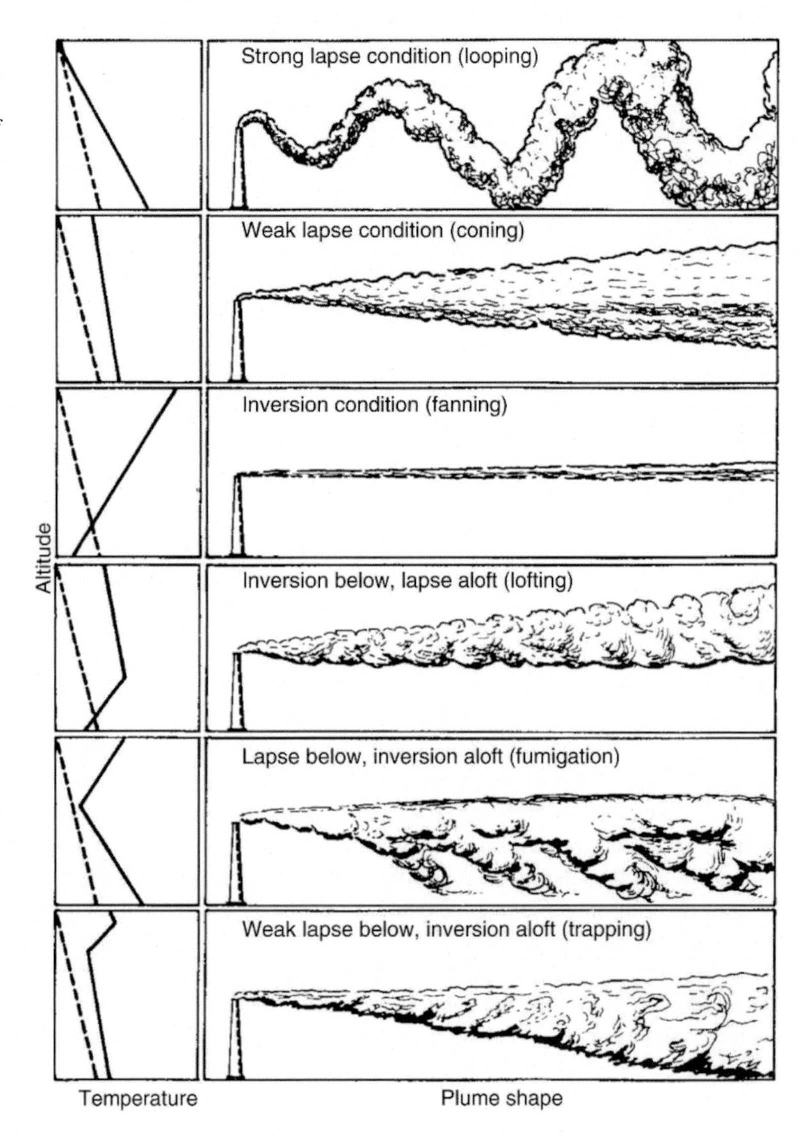

FIG. 14.6 Effect of vertical temperature profile on vertical expansion and shape change of continuous plumes after emission. *Note*: *Dashed lines* correspond to dry adiabatic lapse rate. *(Source: D.A. Vallero, Fundamentals of Air Pollution, 5th ed., Elsevier Academic Press, Waltham, MA, 2014, p. 999 pages cm.)*

Eulerian and Lagrangian models; for example, the observer's frame of reference is watching the pollutant move within a Cartesian (x,y,z) grid, but the pollutant's concentrations are randomly distributed.

Pollutant concentrations from a continuously emitting source are proportional to the emission rate. These concentrations are diluted by the wind at the point of emission at a rate inversely proportional to the wind speed. The time-averaged (about 1 h in Fig. 14.7) pollutant concentrations crosswind and vertically near the source are well described by Gaussian, that is, normal, distributions. The standard deviations of plume concentration in these two directions are empirically related to the levels of turbulence in the atmosphere and increase with distance from the source. In its simplest form, the Gaussian model ignores chemical transformations and other removal processes during transport from the source [14].

Gaussian equations are based on the coordinates shown in Fig. 14.7. The origin is assumed to be at the ground, x is the downwind dimension, y is the crosswind, and z is the vertical dimension.

The models produced values for concentration (C) based on *the* emission rate (Q) expressed as mass per time. Other variables include the following:

- Wind speed (u, length per time)
- The cross-sectional dispersion coefficient (σ_y length), that is, standard deviation of horizontal distribution of plume concentration, which is evaluated at the downwind distance x and for the appropriate stability
- Dispersion coefficient in the vertical direction (σ_z length), that is, standard deviation of vertical distribution of plume concentration, evaluated at the downwind distance x and for the appropriate stability

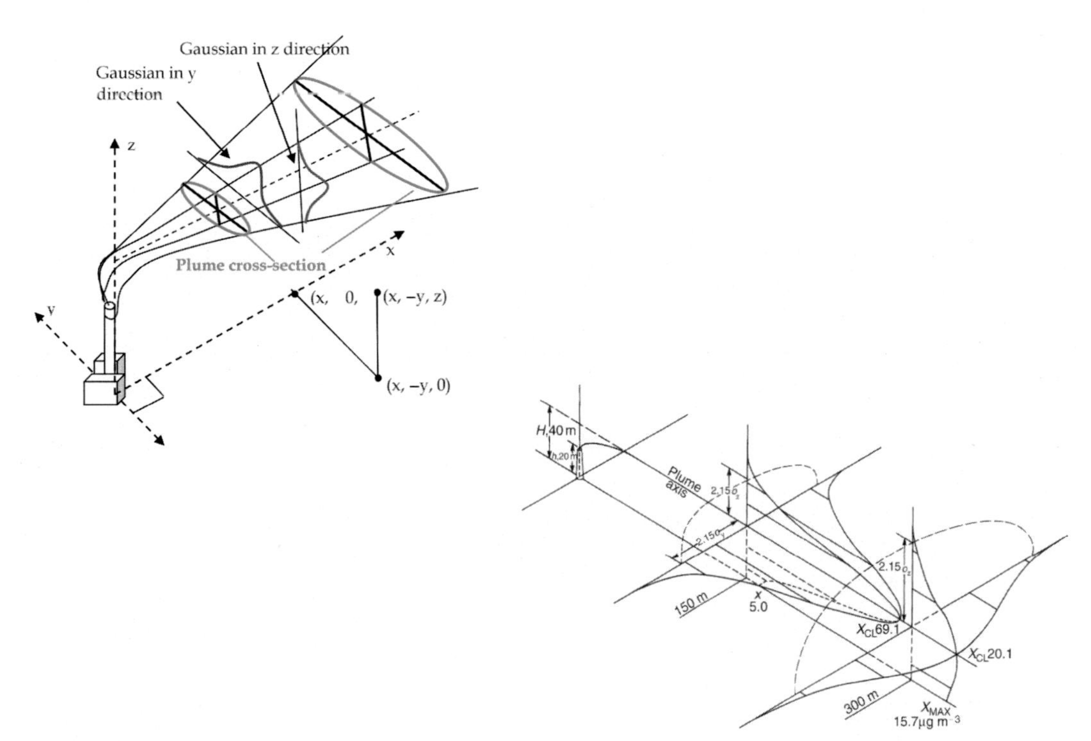

FIG. 14.7 *Top*: Gaussian plume aloft, that is, not yet reaching the ground in the two cross sections. *Bottom*: Two cross-sectional areas of a Gaussian plume showing the predicted total mass of a pollutant released at an effective height of 40m. The *cross section* under the *dashed-line curves*, with the maximum concentrations along the centerline. *(Source: D.A. Vallero, Fundamentals of Air Pollution, 5th ed., Elsevier Academic Press, Waltham, MA, 2014, p. 999 pages cm.)*

- Mixing height (*L*, length)
- Physical stack height (*h*, length)
- Effective height of emission (*H*, length), which is the sum of h + plume rise (Δh)
- Downwind distance (*x*, length)
- Crosswind distance (*y*, length)
- Receptor height above ground (*z*, length)

The concentration at location *x*, *y*, *z* from a point source located at (0, 0, *H*) is given by one of the three following equations.

The equation for stable conditions or unlimited vertical mixing (a very high mixing height) is

$$C = Q(1/u)\left\{g_1/\left[(2\pi)^{0.5}\sigma_y\right]\right\}\left\{g_2/\left[(2\pi)^{0.5}\sigma_z\right]\right\} \tag{14.2}$$

where

$$g_1 = \exp\left(-0.5y^2/\sigma_y^2\right)$$
$$g_2 = \exp\left[-0.5(H-z)^2/\sigma_z^2\right] + \exp\left[-0.5(H+z)^2/\sigma_z^2\right]$$

Note that if $y = 0$, or $z = 0$, or both z and H are 0, this equation is simplified. For locations in the vertical plane containing the plume centerline, $y = 0$ and $g_1 = 1$.

For unstable or neutral conditions where $\sigma_z > 1.6L$, the steady-state equation is

$$C = Q(1/u)\left\{g_1/\left[(2\pi)^{0.5}\sigma_y\right]\right\}(1/L) \tag{14.3}$$

For these large σ_z values, eddy reflection has occurred repeatedly at the ground and mixing height. Eddy reflection refers to the circular eddies of air moving away from the earth's surface, since they cannot penetrate that surface. Thus, the vertical extension of the plume becomes mixed uniformly though the mixing height, that is, $1/L$.

For unstable or neutral conditions where $\sigma_z < 1.6L$, the steady-state equation is

$$C = Q(1/u)\left\{g_1/\left[(2\pi)^{0.5}\sigma_y\right]\right\}\left\{g_3/\left[(2\pi)^{0.5}\sigma_z\right]\right\} \tag{14.4}$$

where

$$g_3 = \sum_{N=-\infty}^{\infty}\left\{\exp\left[-0.5(H - z + 2NL)^2/\sigma_z^2\right] + \exp\left[-0.5(H + z + 2NL)^2/\sigma_z^2\right]\right\}$$

Recently, these equations have been converted to calculators. Fig. 14.8 shows screenshots of the Atmospheric Dispersion Calculator [15], which summarizes the Gaussian equations based on information from Vesilind et al. [16]:

$$C(x, y, z) = \frac{Q}{2\pi u \sigma_y \sigma_z} e^{-\frac{y^2}{2\sigma_y^2}}\left(e^{-\frac{(z+H)^2}{2\sigma_z^2}} + e^{-\frac{(z-H)^2}{2\sigma_z^2}}\right) \tag{14.5}$$

The values are input into the model calculator.

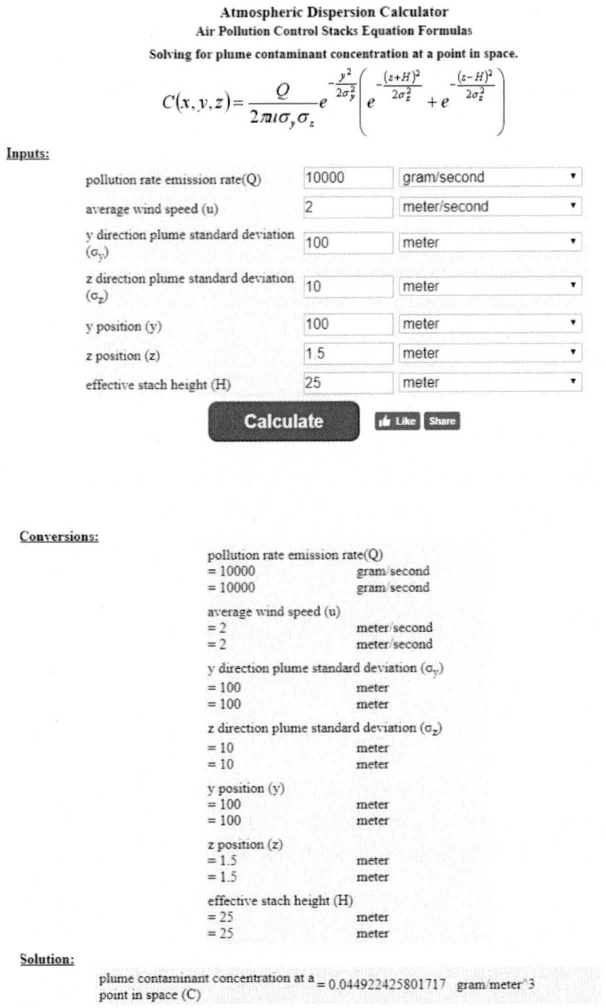

FIG. 14.8 Screenshots of Gaussian dispersion model.

Atmospheric Dispersion Calculator
Air Pollution Control Stacks Equation Formulas
Solving for plume contaminant concentration at a point in space.

$$C(x,y,z) = \frac{Q}{2\pi u \sigma_y \sigma_z} e^{-\frac{y^2}{2\sigma_y^2}}\left(e^{-\frac{(z+H)^2}{2\sigma_z^2}} + e^{-\frac{(z-H)^2}{2\sigma_z^2}}\right)$$

Inputs:

pollution rate emission rate(Q)	10000	gram/second
average wind speed (u)	2	meter/second
y direction plume standard deviation (σ_y)	100	meter
z direction plume standard deviation (σ_z)	10	meter
y position (y)	100	meter
z position (z)	1.5	meter
effective stach height (H)	25	meter

Calculate Like Share

Conversions:

pollution rate emission rate(Q)
= 10000 gram/second
= 10000 gram/second

average wind speed (u)
= 2 meter/second
= 2 meter/second

y direction plume standard deviation (σ_y)
= 100 meter
= 100 meter

z direction plume standard deviation (σ_z)
= 10 meter
= 10 meter

y position (y)
= 100 meter
= 100 meter

z position (z)
= 1.5 meter
= 1.5 meter

effective stach height (H)
= 25 meter
= 25 meter

Solution:

plume contaminant concentration at a point in space (C) = 0.044922425801717 gram/meter^3

What is the pollutant concentration of a pollutant at 1.5 m above the ground, at a location 100 m downwind from a stack (effective stack height = 25 m) with an emission rate of 1 kg s^{-1} and average wind speed of 2 m s^{-1}, if the standard deviation in cross-sectional standard deviation 100 m and in the vertical direction is 10 m?
Plugging these values into the calculator gives a concentration of 0.045 g m^{-3} = 45 mg m^{-3} for this pollutant 100 m downwind.

What is the pollutant concentration of a pollutant at 1.5 m above the ground, at a location 500 m downwind from the same stack and the same conditions?
Plugging these values into the calculator gives a concentration of 2.8 × 10^{-7} g m^{-3} = 0.28 μg m^{-3} for this pollutant 100 m downwind. Thus, at 400 m further downwind, the concentration decreased by over six orders of magnitude.

How would the concentration at 500 m downwind change had the stack been at 50 m effective height? Is it good practice to increase stack height?
Plugging this *H* into the calculator, the concentration is found to be 2.8 × 10^{-11} g m^{-3}. Doubling the stack height increased the area and volume downwind over which the pollutant is distributed, resulting in a decrease in concentration of four orders of magnitude. Of course, raising the stack height has local value; it also increases the amount of the pollutant that reaches winds aloft, which can cause regional and continental scale problems, for example, acid rain.

14.4 STABILITY

Fluctuation measurements can be used for dispersion estimates [17]. The first element is a scheme that includes the important effects of thermal stratification to yield broad categories of stability. The necessary parameters for the scheme consist of wind speed, insolation, and cloudiness, which are basically obtainable from routine observations (Table 14.1).

Pasquill's dispersion parameters were restated in terms of σ_y and σ_z by Gifford [18] to allow their use in the Gaussian plume equations. The parameters σ_y and σ_z are found by estimation from the graphs (see Fig. 14.9) as a function of the distance between source and receptor, from the appropriate curve, one for each stability class. Alternatively, σ_y and σ_z can be calculated using the equations given in Tables 14.2 and 14.3. These parameter values are most applicable for releases near the ground, for example, within about 50 m.

TABLE 14.1 Pasquill stability categories

Surface wind speed (m s^{-1})	Isolation			Night	
	Strong	Moderate	Slight	Thinly overcast or ≥4/8 low cloud	≤3/8 Cloud
<2	A	A–B	B	–	–
2–3	A–B	B	C	E	F
3–5	B	B–C	C	D	E
5–6	C	C–D	D	D	D
>6	C	D	D	D	D

Notes: 1. Strong insolation corresponds to sunny midday in midsummer in England and slight insolation to similar conditions in midwinter; 2. night refers to the period from 1 h before sunset to 1 h after sunrise; and 3. the neutral category D should also be used, regardless of wind speed, for overcast conditions during day or night and for any sky conditions during the hour preceding or following night as defined above. For hyphenated classes, use the average value; for example, A–B, take the average of values for A and B.
Source: F. Pasquill, Atmospheric Dispersion Parameters in Gaussian Plume Modeling. 2. Possible Requirements for Change in the Turner Workbook Values, 1976.

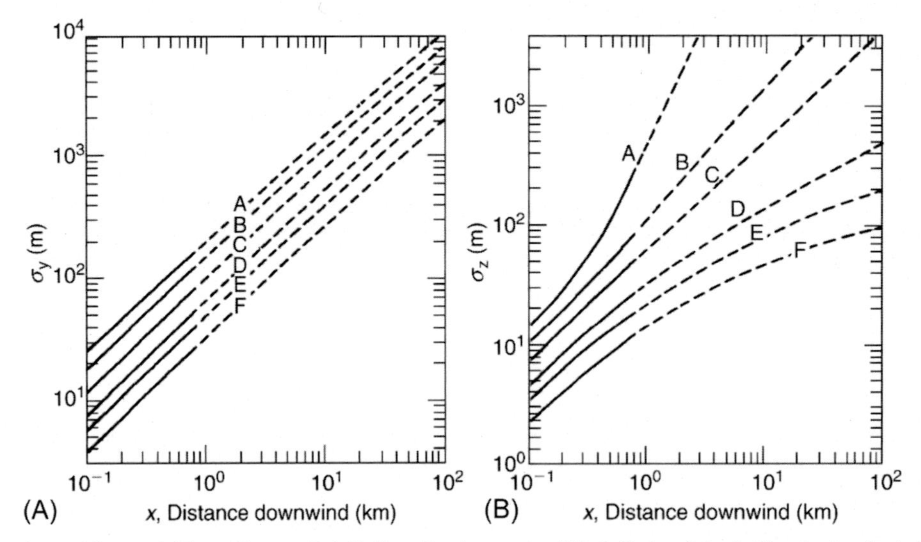

FIG. 14.9 Stability classes (A) σ_y and (B) σ_z. *(Source: D.A. Vallero, Fundamentals of Air Pollution, 5th ed., Elsevier Academic Press, Waltham, MA, 2014, p. 999 pages cm; from F.A. Gifford Jr., Use of routine meteorological observations for estimating atmospheric dispersion, Nucl. Saf. 2 (1961) 47–51.)*

TABLE 14.2 Horizontal dispersion parameters

Stability	Parameter		
A	$T=24.167$	−2.5334	ln *x*
B	$T=18.333$	−1.8096	ln *x*
C	$T=12.5$	−1.0857	ln *x*
D	$T=8.3333$	−0.72382	ln *x*
E	$T=6.25$	−0.54287	ln *x*
F	$T=4.1667$	−0.36191	ln *x*

Note: σ_y (m) = 465.116 × tan *T*; *x* is downwind distance in kilometers; *T* is one-half Pasquill's θ (degrees).
Source: D.A. Vallero, Fundamentals of Air Pollution, 5th ed., Elsevier Academic Press, Waltham, MA, 2014, p. 999 pages cm; from F.A. Gifford Jr., Use of routine meteorological observations for estimating atmospheric dispersion, Nucl. Saf. vol. 2 (1961) 47–51.

Other estimations of σ_y and σ_z for two different situations for each Pasquill stability class, as a function of distance between source and receptor. Tables 14.4 and 14.5 give these for urban and rural conditions, respectively.

Unless the source is at ground level, the maximum concentration (C_{max}) of a pollutant along the *x*-axis occurs at some point downwind, that is, after emission the chemical species require distance and time to reach ground level (see Fig. 14.10).

A factory is releasing an organic chemical species at a rate of 0.37 g s^{-1} from a stack with an effective height of 40 m. The wind speed is averaging 2 m s^{-1}. Where will the maximum concentration of this chemical be found downwind if the Pasquill stability class is B?

Assuming that the ratio of $\frac{\sigma_z}{\sigma_y}$ remains constant with distance, the maximum concentration of an emitted substance is found at the distance where

$$\sigma_{z^*} = \frac{H}{\sqrt{2}} \tag{14.6}$$

Thus, $\sigma_{z^*} = 28.3$ m. At stability class B, this occurs at $x = 0.28$ km, so at this downwind position, using the values in Table 14.2, $\sigma_y = 49.0$ m.

TABLE 14.3 Pasquill-Gifford vertical dispersion parameter

Stability	Distance (km)	*a*	*b*	σ_z[a]
A	>3.11			5000 m
	0.5–3.11	453.85	2.1166	
	0.4–0.5	346.75	1.7283	104.7
	0.3–0.4	258.89	1.4094	71.2
	0.25–0.3	217.41	1.2644	47.4
	0.2–0.25	179.52	1.1262	37.7
	0.15–0.2	170.22	1.0932	29.3
	0.1–0.15	158.08	1.0542	21.4
	<0.1	122.8	0.9447	14.0
B	≥35			5000 m
	0.4–35	109.30	1.0971	
	0.2–0.4	98.483	0.98332	40.0
	<0.2	90.673	0.93198	20.2
C	All *x*	61.141	0.91465	
D	>30	44.053	0.51179	
	10–30	36.650	0.56589	251.2
	3–10	33.504	0.60486	134.9
	1–3	32.093	0.64403	65.1
	0.3–1	32.093	0.81066	32.1
	<0.3	34.459	0.86974	12.1
E	>40	47.618	0.29592	
	20–40	35.420	0.37615	141.9
	10–20	26.970	0.46713	109.3
	4–10	24.703	0.50527	79.1
	2–4	22.534	0.57154	49.8
	1–2	21.628	0.63077	33.5
	0.3–1	21.628	0.75660	21.6
	0.1–0.3	23.331	0.81956	8.7
	<0.1	24.260	0.83660	3.5
F	>60	34.219	0.21716	
	30–60	27.074	0.27436	83.3
	15–30	22.651	0.32681	68.8
	7–15	17.836	0.4150	54.9
	3–7	16.187	0.4649	40.0
	2–3	14.823	0.54503	27.0
	1–2	13.953	0.63227	21.6
	0.7–1.0	13.953	0.68465	14.0
	0.2–0.7	14.457	0.78407	10.9
	<0.2	15.209	0.81558	4.1

Note: σ_z (m) $= ax^b$; *x* is downwind distance in kilometers.
[a]*σ_z at boundary of distance range for all values except 5000 m.*
Source: D.A. Vallero, Fundamentals of Air Pollution, 5th ed., Elsevier Academic Press, Waltham, MA, 2014, p. 999 pages cm; from F.A. Gifford Jr., Use of routine meteorological observations for estimating atmospheric dispersion, Nucl. Saf. 2 (1961) 47–51.

TABLE 14.4 Urban dispersion parameters by Briggs (for distances between 100 and 10,000 m)

Pasquill type	σ_y (m)	σ_z (m)
A–B	$0.32x(1+0.0004x)^{-0.5}$	$0.24x(1+0.001x)^{-0.5}$
C	$0.22x(1+0.0004x)^{-0.5}$	$0.20x$
D	$0.16x(1+0.0004x)^{-0.5}$	$0.14x(1+0.0003x)^{-0.5}$
E–F	$0.11x(1+0.0004x)^{-0.5}$	$0.08x(1+0.0015x)^{-0.5}$

Source: D.A. Vallero, Fundamentals of Air Pollution, 5th ed., Elsevier Academic Press, Waltham, MA, 2014, p. 999 pages cm; from F.A. Gifford Jr., Use of routine meteorological observations for estimating atmospheric dispersion, Nucl. Saf. 2 (1961) 47–51.

TABLE 14.5 Rural dispersion parameters by Briggs (for distances between 100 and 10,000 m)

Pasquill type	σ_y (m)	σ_z (m)
A	$0.22x(1+0.0001x)^{-0.5}$	$0.20x$
B	$0.16x(1+0.0001x)^{-0.5}$	$0.12x$
C	$0.11x(1+0.0001x)^{-0.5}$	$0.08x(1+0.0002x)^{-0.5}$
D	$0.08x(1+0.0001x)^{-0.5}$	$0.06x(1+0.0015x)^{-0.5}$
E	$0.06x(1+0.0001x)^{-0.5}$	$0.03x(1+0.0003x)^{-1}$
F	$0.04x(1+0.0001x)^{-0.5}$	$0.016x(1+0.0003x)^{-1}$

Source: D.A. Vallero, Fundamentals of Air Pollution, 5th ed., Elsevier Academic Press, Waltham, MA, 2014, p. 999 pages cm; from F.A. Gifford Jr., Use of routine meteorological observations for estimating atmospheric dispersion, Nucl. Saf. 2 (1961) 47–51.

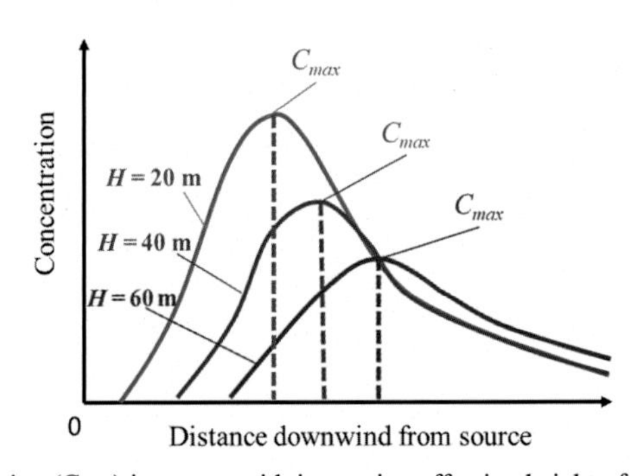

FIG. 14.10 Distance of maximum concentration (C_{mas}) increases with increasing effective height of stack for hypothetical gas-phase pollutant.

C_{max} can be estimated from

$$C_{max} = \frac{2Q}{\pi \cdot u \cdot e \cdot H^2} \cdot \frac{\sigma_z}{\sigma_y} \tag{14.7}$$

Euler's number (e) is the base of natural logarithms and equals approximately 2.718, so

$$C_{max} = \frac{2 \times 0.37}{\pi \cdot 2 \cdot e \cdot (40)^2} \cdot \frac{28.3}{49.0} = 1.56 \times 10^{-5}$$

To check whether the distance to C_{max} is reasonable, we can use the following equation at three reasonable distances for the standard deviations

$$C = \left[Q/\left(\pi u \sigma_y \sigma_z\right)\right] \exp\left[-0.5\left(H/\sigma_z\right)^2\right] \tag{14.8}$$

This yields the following values

x (km)	σ_z (m)	σ_y(m)	C (gm^{-3})
0.26	26.2	45.9	1.53×10^{-5}
0.28	28.2	49.0	1.56×10^{-5}
0.30	30.1	52.2	1.56×10^{-5}

Notes: σ_z values obtained from Table 14.3; σ_y values obtained from Table 14.4.

These verify that this simple case does indeed show C_{max} to occur at 0.28 km downwind, ± 20 m.

Puff models assume that the source is episode, that is, a puff of the pollutant is emitted at a specific time. Plume models assume that the source is continuous. For plumes, the previous equations can be stated as a mean concentration ($\overline{C}$) based on a mean emission rate ($\overline{Q}$)

$$\overline{C}(x, y, z) = \frac{\overline{Q}}{\pi \sigma_y \sigma_z u} \exp\left[-\frac{1}{2}\left(\frac{y^2}{\sigma_y^2} + \frac{z^2}{\sigma_z^2}\right)\right] \tag{14.9}$$

A tire fire is emitting NO_x at a rate of 8 gs^{-1} in a rural area. Calculate the mean NO_x concentration expected downwind along the centerline 3 km from the fire on an overcast night and wind speed of 7 ms^{-1} and the stability is class D. The background concentration of NO_x upwind from the fire is 1 μgm^{-3}

Using Table 14.5 for stability class D

$$\sigma_y = 0.08(3000)[1 + (0.0001 \times 3000)]^{-0.5} = 236.48$$

$$\sigma_z = 0.06(3000)[1 + (0.0015 \times 3000)]^{-0.5} = 990.00$$

Since the calculation is based on the centerline, that is, y = 0, assuming that the fire is at ground level, that is, z = 0, the calculation is

$$\overline{C}(x, 0, 0) = \frac{\overline{Q}}{\pi \sigma_y \sigma_z u} = \frac{8}{\pi \cdot 236.48 \cdot 990 \cdot 7} = 1.5 \times 10^{-6} \text{g m}^{-3} = 1.5 \mu\text{g m}^{-3} \text{NO}_x$$

Thus, the average concentration at this location would have to also add the background, so the modeled concentration would be 1 + 1.5 μgm^{-3} = *2.5μgm^{-3} NO_x*.

There are many other ways to estimate and predict air pollution concentrations and movement and dilution within plumes besides the Gaussian methods, arguably the simplest and most commonly used air pollution model type. The outputs from Gaussian models depend on the accuracy in calculating the dispersion coefficients, σ_y and σ_z.

14.5 EULERIAN MODELS

Eulerian models can be of numerous types, including box, analytic, numerical, and eddy simulation types [6]. As mentioned, the Eulerian model bases the mass balance around a differential volume. The velocity field is used to calculate the motion of the air. From the Eulerian perspective, the velocity field is a function of space and time:

$$\vec{\nu} = \vec{\nu}\ (\text{x, y, z, t}) \tag{14.10}$$

where $\vec{\nu}$ is the velocity field vector.

Thus, if the fluid properties and the flow characteristics at each position do not vary with time, that is, t is removed from the equation, the air flow is simply a function of space, that is, a steady or time-independent vector field:

$$\vec{\nu} = \vec{\nu}\ (x, y, z) \text{ and } \frac{\partial \vec{\nu}}{\partial x} = 0 \tag{14.11}$$

On the other hand, a time-dependent flow is considered to be an unsteady flow. Any flow with unchanging magnitude and direction of the velocity vector $\vec{v}$ is considered to be a uniform flow.

Laminar flow is in layers, while turbulent flow has random movements of fluid particles in all directions. In incompressible flow, the variations in density are assumed to be constant, while the compressible flow has density variations, which must be included in flow calculations. Viscous flows must account for viscosity, while inviscid flows assume viscosity is zero.

The velocity field is very important in modeling plumes in the atmosphere, since the velocity field is a way to characterize the motion of fluid particles and provides the means for computing these motions.

The ambient atmosphere is seldom laminar, so the Eulerian conditions are assumed to be turbulent, which means that dependent variables consist of both average and fluctuating components [5, 19].

14.6 MODEL LEVELS

Environmental models are of three general types:

Level 1 Model: This model is based on an equilibrium distribution of fixed quantities of contaminants in a closed environment (i.e., conservation of contaminant mass). No chemical or biological degradation, no advection, and no transport among compartments (such as sediment loading or atmospheric deposition to surface waters).

A Level 1 calculation describes how a given quantity of a contaminant will partition among the water, air, soil, sediment, suspended particles, and fauna, but does not taken into account chemical reactions. Early Level 1 models considered an area of 1 km^2 with 70% of the area covered in surface water. Larger areas are now being modeled (e.g., about the size of the state of Ohio).

Level 2 Model: This model relaxes the conservation restrictions of Level 1 by introducing direct inputs (e.g., emissions) and advective sources from air and water. It assumes that a contaminant is being continuously loaded at a constant rate into the control volume, allowing the contaminant loading to reach steady state and equilibrium between contaminant input and output rates. Degradation and bulk movement of contaminants (advection) is treated as a loss term. Exchanges between and among media are not quantified.

Since the Level 2 approach is a simulation of a contaminant being continuously discharged into numerous compartments and that achieves a steady-state equilibrium, the challenge is to deduce the losses of the contaminant due to chemical reactions and advective (nondiffusive) mechanisms.

Reaction rates are unique to each compound and are published according to reactivity class (e.g., fast, moderate, or slow reactions), which allows modelers to select a class of reactivity for the respective contaminant to insert into transport models. The reactions are often assumed to be first order, so the model will employ a first-order rate constant for each compartment in the environmental system (e.g., x mol hr^{-1}in water, y mol hr^{-1} in air, and z mol hr^{-1} in soil). Much uncertainty is associated with the reactivity class and rate constants, so it is best to use rates published in the literature based upon experimental and empirical studies, wherever possible.

Advection flow rates in Level 2 models are usually reflected by residence times in the compartments. These residence times are commonly set at one hour in each medium, so the advection rate (G_i) is volume of the compartment divided by the residence time (t):

$$G_i = Vt^{-1} \tag{14.12}$$

Level 3 Model: Same as Level 2, but does not assume equilibrium between compartments, so each compartment has its own fugacity. Mass balance applies to the whole system and each compartment within the system. It includes mass transfer coefficients, rates of deposition and resuspension of contaminant, rates of diffusion, soil runoff, and area covered. All these factors are aggregated into an intermedia transport term (D) for each compartment.

The assumption of equilibrium in Level 1 and 2 models is a simplification and often a gross oversimplification of what occurs in environmental systems. When the simplification is not acceptable, kinetics must be included in the model. Numerous diffusive and nondiffusive transport mechanisms are included in Level 3 modeling. For example, values for the various compartments' unique intermedia transport velocity parameters (in length per time dimensions) are applied to all contaminants being modeled (these are used to calculate the *D values*).

It is important to note that models are only as good as the information and assumptions that go into them. For example, neighborhood-scale effects can modify estimates from transport models or from measurement interpolations (barriers, channeling, local flows, or trapping). This applies to all transport models, whether highly computational or simplified. Site-specific differences can greatly affect predicted outcomes, such as whether terrain is complex or simple, whether there are frequent inversions, whether sources emit the contaminant continuously or haphazardly, etc. This is true at all scales. For example, microenvironmental models must account for variability in room configurations, movements, and activities of occupants and seasonal changes.

14.7 MODEL UNCERTAINTY

The processes that lead to the ultimate fate of air pollutants are complicated [20]. Their relationships with one another are complex. These complexities must be captured in a model for at least two reasons. First, they must document what is going on during a given period of time. Almost always, actual measurement data are not available to characterize the movement and change of materials in the environment. Second, models provide a means of predicting outcomes based on currently available information. Fig. 14.11 illustrates the steps that should be taken to develop an environmental model.

The model generators are to the left, and the users (stakeholders) are to the right. This illustrates the connection between the scientific bases for biochemodynamic factors discussed in this chapter with their applications to decision-making. Each arrow indicates the connection between processes; each factor and process introduces information to the model but simultaneously adds uncertainty.

Uncertainty lies in estimating the values to be assigned to each compartment of a model, including theoretical uncertainty (e.g., whether Henry's law published values are relevant to a particular scenario) and measurement uncertainty, for example, limitations in the ability to measure in the real world, such as the need for destructive methods or analytic problems, such as storing samples before they can be analyzed. However, models add another dimension to uncertainty (see Fig. 14.12). Uncertainty increases with each addition of information. Combining models adds even more uncertainty.

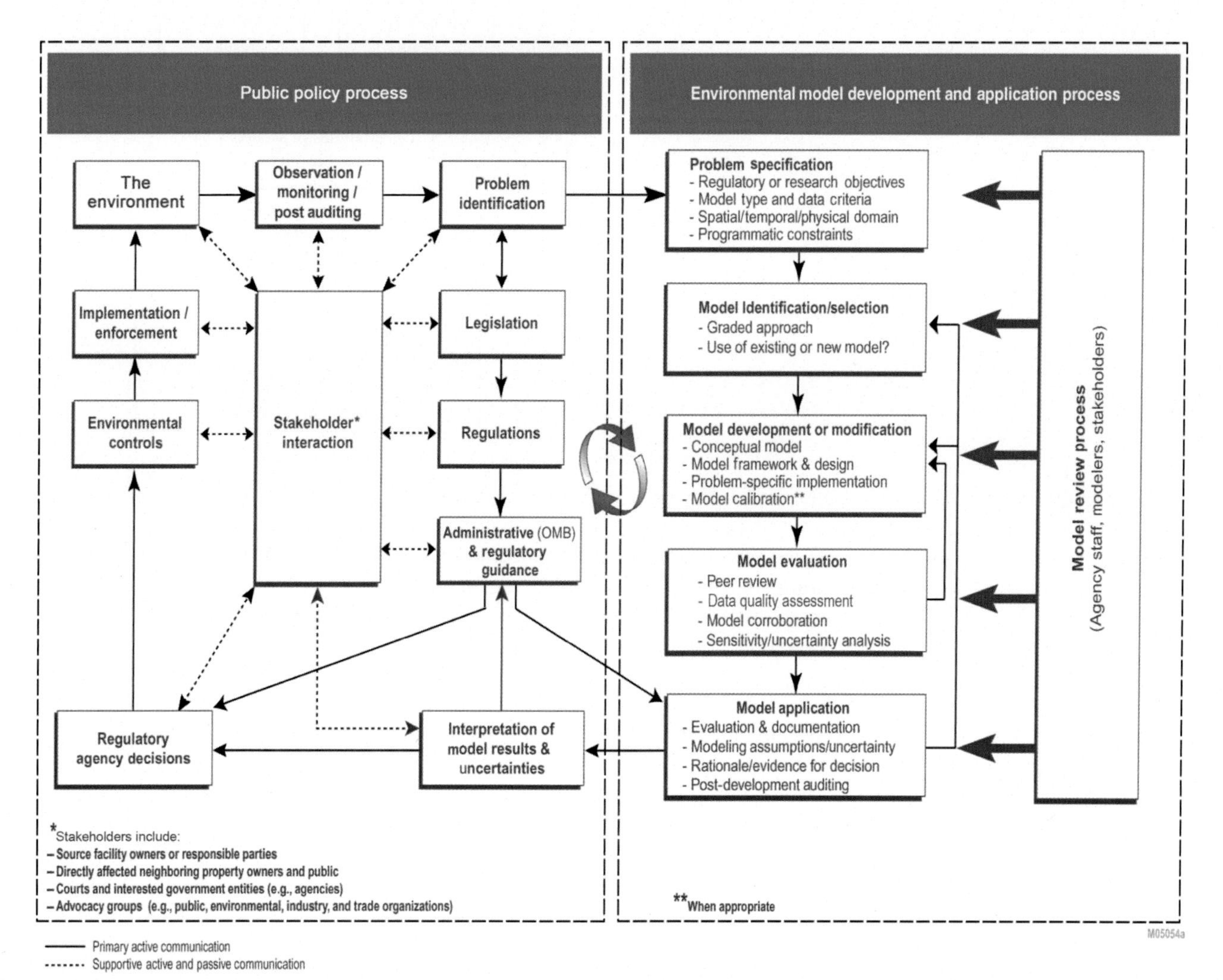

FIG. 14.11 Steps needed to develop and implement an environmental decision model from inception to completion. These include problem specification, model identification and selection (a site-specific model may be generated de novo or based on an existing model framework), model development (including problem- and site-specific model conceptualization, model formulation and design, and model calibration), model evaluation (e.g., based on peer review, data quality assessment, model code verification, model confirmation/corroboration, sensitivity analysis, and uncertainty analysis), model use (diagnostic analysis, solution, and application support for decision-making), and review after use. *(Source: S.L. Johnson, Review of Agency Draft Guidance on the Development, Evaluation, and Application of Regulatory Environmental Models and Models Knowledge Base by the Regulatory Environmental Modeling Guidance Review Panel of the EPA Science Advisory Board, EPA-SAB-06-009, Washington, DC, 20460, 2006.)*

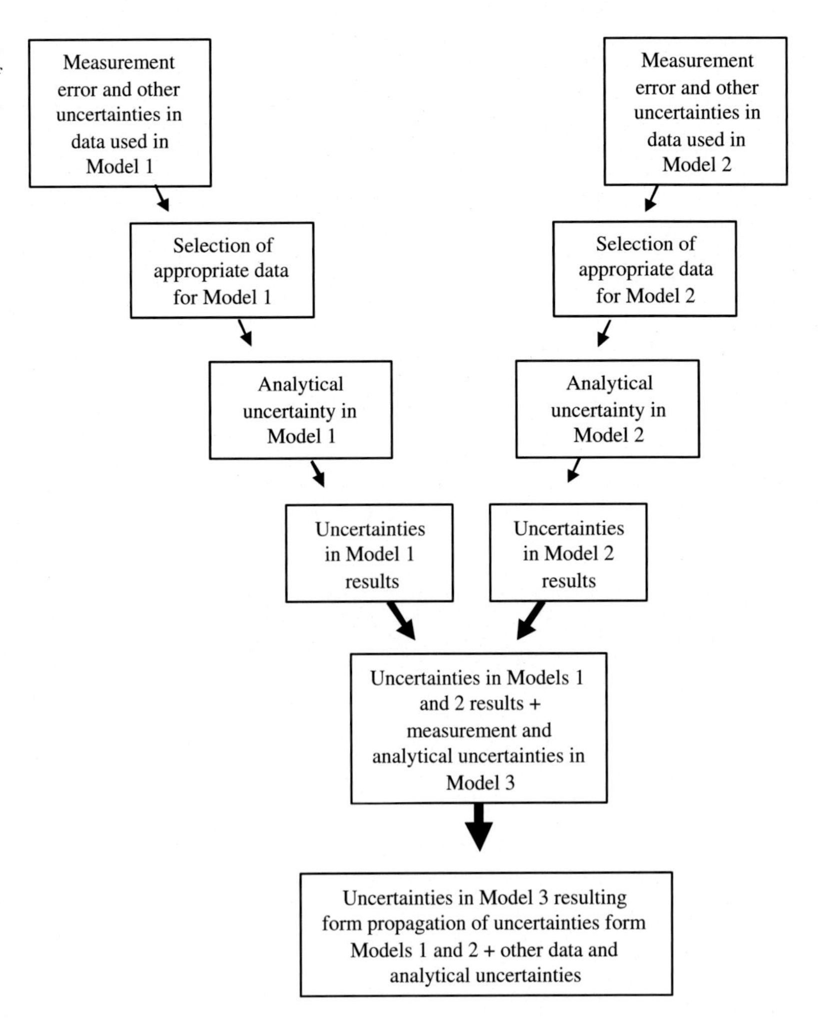

FIG. 14.12 Propagation of uncertainty in environmental models. *(Source: D.A. Vallero, Fundamentals of Air Pollution, 5th ed., Elsevier Academic Press, Waltham, MA, 2014, p. 999 pages cm.)*

Uncertainty can be classified as either aleatory or epistemic. Aleatory uncertainty is random uncertainty or stochastic uncertainty that is impossible to predict. All environmental systems have inherent, random uncertainty that cannot be reduced by additional observations. On the other hand, epistemic uncertainty arises from insufficient knowledge about the system. Thus, as more reliable information becomes available regarding the processes described in this chapter, one would expect epistemic uncertainty to fall. Thus, from a scalar perspective, any model outcome (*Y*) is affected by a function of the probability of these two types of uncertainty [21]:

$$Y = h(U, V) \tag{14.13}$$

where U = all epistemic uncertainties (uncertain parameters), V = aleatory uncertainties (stochastic variables), and h is the computational model considered as a deterministic function of both uncertainties.

An example of aleatory uncertainty would be a forecast of failure to contain a toxic substance within the physical confinement of a laboratory or a cleanup site, for example, where the occurrence of failure occurs randomly over time and the actual time of failure cannot be predicted, no matter the size of the data set. Conversely, epistemic uncertainty includes uncertainties inherent to a variable or parameter and uncertainties in the model's algorithms. For example, a model may miss a possible route by which an air pollutant can be released because wrong information about the molecule's affinity to a certain type of aerosol or soil particle. With increasing information about these relationships, the epistemic uncertainty should decrease, and the predictive capability of the model would commensurately increase.

A means of predicting this type of uncertainty is by hierarchical holographic modeling (HHM), which addresses large-system complexities by as follows:

Identifying the components and processes of all subsystems and suggesting ways in which they might interact with each other based on established/supportive information. The technique decomposes the system by looking at it from many different perspectives including the functions, activities, geopolitical boundaries, or structures of the system. HHM can be used in one of two ways—as a hazard identification tool or as a comprehensive analytic modeling tool. The analyst constructs an HHM by first identifying the most appropriate perspectives for the problem in hand. These are used to define the subsystems that in turn are further decomposed into components, processes, functions, or activities, which may or may not overlap with other subsystems. The analyst can investigate the quantitative properties of the system if the functions, activities, components, or processes of the system can be described by a series of overlapping models, subject to overall system constraints [22]. The analyst(s) can also identify hazards by comparing potential interactions between the subsystems in a qualitative fashion. This is best achieved by a team whose members are expert in one or more of the chosen perspectives. Once the hazard and exposure calculations are done, risks can be characterized risk quantitatively.

14.8 LINKS TO EXPOSURE AND DOSE MODELS

The outputs from dispersion models and air pollutant measurements are ultimately needed to characterize, quantify, and predict risks. Therefore, dispersion models and exposure models need to be linked (see Chapter 12).

Exposure modeling begins with the initial collection and organization of information to focus on the cumulative factors that lead to the exposure (see Fig. 14.13). The steps in Fig. 14.13 are often not sequential and require iterations to parameterize and input data regarding population vulnerabilities, public health information, toxicological and epidemiological data, completed exposure pathways, differential exposures, and contact with the air pollutant [23]. This effort can provide population profiles, listing of relevant chemicals, chemical groups for use in risk analysis and characterization, and a conceptual model. The final step of interface exposure assessment results with the dose-response assessment.

Rather than employing single values to generate a discrete estimate of exposure as is done in deterministic approaches, a probabilistic approach can be used to better depict the uncertainty and variability in influential input variables. Fundamental to a probabilistic approach is using statistical distributions for input variables, parameterizing these distributions, and characterizing the conditions or probabilities associated with the use of particular distributions. The complexity needed in these models depends on the variables that use statistical distributions, whether multiple variables correlate with one another, for example, body mass, fat-free body mass, overall fitness, and the number of individuals to which the model will be applied. The probability model will provide a statistical distribution of the estimated exposure or doses for the receptors.

Models used to estimate and predict exposures to air pollutants have advanced in application and reliability in recent years. Although many of the physical and chemical factors, parameters, and algorithms are similar to those discussed in this chapter, the differences are substantial in terms of scale and weighting, as well as in the behavioral and biological features. Some of the kinetics within an organism, that is, toxicokinetics, has been discussed in this chapter, but this is after intake and exposure, that is, physiologically based pharmacokinetic models. Exposure models that lead to this contact and intake depend on the types, intensities, and extent of activities, as well as biological factors like breathing rates and health status. As such, exposure and dose models are discussed in greater detail in Chapter 12.

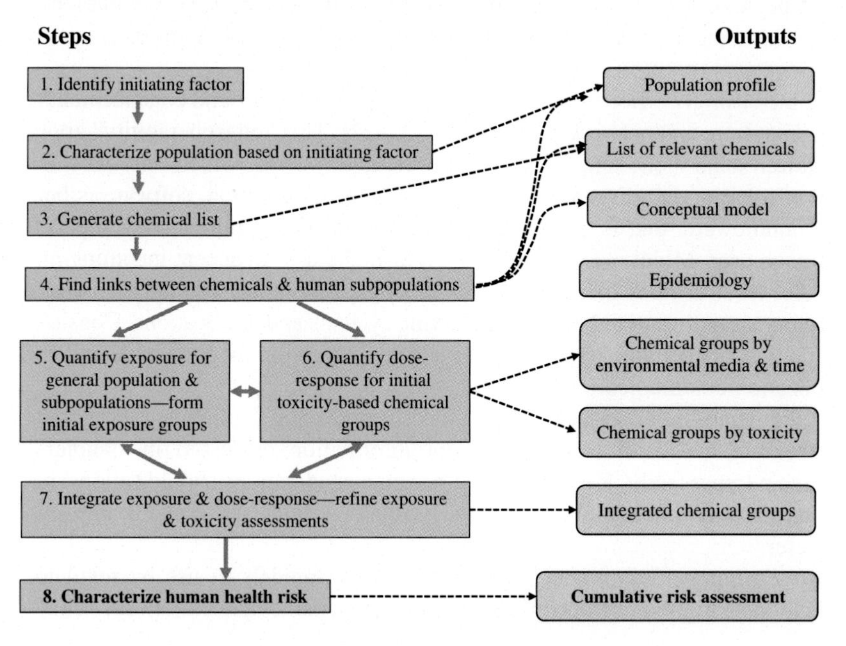

FIG. 14.13 Key steps in a cumulative risk assessment. The interdependence of exposure and toxicity assessments is indicated by *blue arrows*. *(Information from US Environmental Protection Agency, Concepts, Methods and Data Sources for Cumulative Health Risk Assessment of Multiple Chemicals, Exposures and Effects: A Resource Document. EPA/600/R-06/013F. US Environmental Protection Agency, Washington, DC, 2007. http://cfpub.epa.gov/ncea/cfm/recordisplay.cfm?deid=190187 (Accessed 1 October 2013).)*

There are many ways to classify air pollution models. In addition to the semiempirical box, Lagrangian, Gaussian, and Eulerian classes, modelers employ specialized models, such as to predict plume rise by calculating the vertical displacement and postemission behavior of pollutants [12]. Receptor models assist in linking measured concentrations of a pollutant or a tracer to a source or sources, for example, back-trajectory models. In this sense, they are the mirror image of dispersion models.

14.9 RECENT ADVANCES

Like other environmental models, air pollution models are becoming more computational and modular. Recent and rapid advances in high-performance computing hardware and software are leading to increasing applications of numerical simulation models that characterize an atmospheric plume as a system. These methods simulate spatially and temporally resolved details of the pathways of air pollutants from source emissions to pollutant concentrations within the local "virtual" microenvironment. This approach to air pollution modeling applies numerical representation of fluid flow in a fluid continuum. Such an approach is known as computational fluid dynamics (CFD), which determines a numerical solution to the governing equations of fluid flow while advancing the solution through space or time to obtain a numerical description of the complete flow field of interest. CFD models are based on the first principles of physics; they start with single-phase flow based on the Navier-Stokes equations. The Navier-Stokes equations are deterministic. They consider simultaneously the conservation of energy, momentum, and mass [24].

Practical CFD solutions require a subgrid-scale model for turbulence. As computing capacities advance, the scale where turbulence is modeled can be reduced, and the application of higher-order numerical methods can support increasingly representative and accurate turbulence models. The transport and dispersion of air pollutants is part of the fluid flow solution. CFD methods have been developed and routinely applied to aerospace and automotive industrial applications and are now being extended to environmental applications [24].

High-fidelity fine-scale CFD simulation of pollutant concentrations within microenvironments (e.g., near roadways or around buildings) is feasible now that high-performance computing has become more accessible. Fine-scale CFD simulations have the added advantage of being able to account rigorously for topographical details such as terrain variations and building structures in urban areas as well as their local aerodynamics and turbulence. Thermal heat fluxes may be added to terrain and building surfaces to simulate the thermal atmospheric boundary layer their influences on pollution transport and dispersion. The physics of particle flow and chemistry can be included in CFD simulations. The results of CFD simulations can both be directly used to better understand specific case studies and be used to support the development of better-simplified algorithms for adoption into other modeling systems. For example, CFD simulations with fine-scale physics and chemistry can enhance and complement photochemical modeling with CMAQ. Also, detailed CFD simulation for a complex site study can be used to develop reliable parameterizations to support simplified and rapid application air pollution model [24].

Data-intensive applications are becoming increasingly efficient by taking advantage of advances and enhancements of computational tools, storage capacity, grid computing, and network technologies [25–27]. This "interoperability" applies modular databases and models, leading to fewer stand-alone and expensive proprietary systems, which improves cost-effectiveness, efficiency, and the likelihood of a larger and diverse user community [28]. In addition, software is being developed to support data users, especially middleware, that is, software that merges existing programs developed for various reasons into a package designed for each user. Middleware architecture allows the user to accept incoming messages from applications and use them; provides a set of rules for encoding into a single format, for example, extensible markup language—XML; and provides directory structures to aid in interconnecting systems and applications. Consistent and standard middleware architecture will likely support future application developers by giving them a single and consistent interface for both inter- and intraapplication communications [29].

The desire for modularity is evident in Lagrangian, Eulerian, and hybrid models' increasing deployment of chemical transformation modules that incorporate the rate laws and other theoretical information discussed in Chapter 3. The examples in this chapter are for illustrative purposes only. Regulatory and design considerations should follow specified guidance and standards, so the appropriate air pollution modeling manuals, handbooks, and texts should be consulted for the specific conditions and needs of the user.

The fifth edition of *Fundamentals of Air Pollution* [30] provides information about models in use by regulatory agencies, including CMAQ, AERMOD, and models to predict tropospheric ozone and air toxic concentrations.

REFERENCES

[1] D.A. Vallero, et al., Modeling and predicting pesticide exposures, in: R. Krieger (Ed.), Hayes Handbook of Pesticide Toxicology, Elsevier Science, New York, NY, 2010, pp. 995–1020.
[2] L.G. Luther, The National Environmental Policy Act: Background and Implementation, Congressional Research Service, Library of Congress, 2005.
[3] New York Department of Environmental Conservation, State Implementation Plans, Albany, New York, 2009.
[4] U.S. Environmental Protection Agency, Final Rule: Implementation of the 2015 National Ambient Air Quality Standards for Ozone: Nonattainment Area Classifications Approach, 2018.
[5] R.S. Collett, K. Oduyemi, Air quality modelling: a technical review of mathematical approaches, Meteorol. Appl. 4 (3) (1997) 235–246.
[6] M. Markiewicz, Modelling of the air pollution dispersion, in: MANHAZ. Models and Techniques for Health and Environmental Hazards Assessment and Management. IEA, Otwock-Swierk, Poland, 2006, pp. 303–347.
[7] M. Albareda-Sambola, E. Fernández, G. Laporte, Heuristic and lower bound for a stochastic location-routing problem, Eur. J. Oper. Res. 179 (3) (2007) 940–955.
[8] S.A. Csiszar, A.S. Ernstoff, P. Fantke, O. Jolliet, Stochastic modeling of near-field exposure to parabens in personal care products, J. Expos. Sci. Environ. Epidemiol. (2016).
[9] U. Ritzinger, J. Puchinger, R.F. Hartl, A survey on dynamic and stochastic vehicle routing problems, Int. J. Prod. Res. 54 (1) (2016) 215–231.
[10] R. Vandoorne, P.J. Gräbe, Stochastic modelling for the maintenance of life cycle cost of rails using Monte Carlo simulation, Proc. Inst. Mech. Eng. Part F: J. Rail Rapid Transit 232 (4) (2018) 1240–1251.
[11] P.C. Bressloff, J.M. Newby, Stochastic models of intracellular transport, Rev. Modern Phys. 85 (1) (2013) 135.
[12] N. Moussiopoulos, Ambient Air Quality, Pollutant Dispersion and Transport Models, European Environment Agency, 1997.
[13] D. Vallero, Environmental Contaminants: Assessment and Control, Academic Press, 2010.
[14] D.A. Vallero, Fundamentals of Air Pollution, fifth ed., Elsevier Academic Press, Waltham, MA, 2014. p. 999 pages cm.
[15] J. Raymond, Atmospheric Dispersion Calculator: Air Pollution Control Stacks Equation Formulas, 28 September 2018. Available: https://www.ajdesigner.com/phpdispersion/point_space_equation.php, 2018.
[16] J.J. Peirce, R. Weiner, R. Matthews, P.A. Vesilind, Environmental Engineering, Butterworth-Heinemann, 2003.
[17] F. Pasquill, P. Michael, Atmospheric diffusion, Phys. Today 30 (1977) 55.
[18] F.A. Gifford Jr., Use of routine meteorological observations for estimating atmospheric dispersion, Nucl. Saf. 2 (1961) 47–51.
[19] R.A. Brown, Fluid Mechanics of the Atmosphere, Academic Press, 1991.
[20] P.R. Williams, B.J. Hubbell, E. Weber, C. Fehrenbacher, D. Hrdy, V. Zartarian, An overview of exposure assessment models used by the US Environmental Protection Agency, Modell. Pollut. Complex Environ. Syst. 2 (2010) 61–131.
[21] E. Hofer, M. Kloos, B. Krzykacz-Hausmann, J. Peschke, M. Woltereck, An approximate epistemic uncertainty analysis approach in the presence of epistemic and aleatory uncertainties, Reliab. Eng. Syst. Saf. 77 (3) (2002) 229–238.
[22] G. Linder, E. Little, L. Johnson, C. Vishy, B. Peacock, H. Goeddecke, Risk and Consequence Analysis Focused on Biota Transfers Potentially Associated with Surface Water Diversions between the Missouri River and Red River Basins, 2005.
[23] U.S. Environmental Protection Agency, Concepts, Methods and Data Sources for Cumulative Health Risk Assessment of Multiple Chemicals, Exposures and Effects: A Resource Document, National Center for Environmental Assessment, Office of Research and Development Cincinnati, OH, USA, 2007.
[24] D.A. Vallero, Excerpted and revised from: Fundamentals of air pollution, fifth ed., Elsevier Academic Press, Waltham, MA, 2014. p. 999 pages cm.
[25] B. Allcock, et al., Data management and transfer in high-performance computational grid environments, Parallel Comput. 28 (5) (2002) 749–771.
[26] G. Singh, et al., A metadata catalog service for data intensive applications, in: Supercomputing, 2003 ACM/IEEE Conference, IEEE, 2003, p. 33.
[27] D. Vallero, Translating Diverse Environmental Data into Reliable Information: How to Coordinate Evidence from Different Sources, Academic Press, 2017.
[28] A.L. Washington, The Interoperability of US Federal Government Information: Interoperability, in: Big Data: Concepts, Methodologies, Tools, and Applications: Concepts, Methodologies, Tools, and Applications, 2016, p. 210.
[29] 2002-P-00017, EPA Management of Information Technology Resources Under The Clinger-Cohen Act, Available:https://www.epa.gov/sites/production/files/2015-10/documents/clingercohen.pdf, 2002.
[30] H. C. Engineers, D.A. Vallero (Ed.), Fundamentals of Air Pollution, Academic Press, 2014.

FURTHER READING

[31] M. Borysiewicz, M. Borysiewicz, Models and techniques for health and environmental hazard assessment and management, Swierk Poland Inst. Atomic Energy 20 (2006) 2010.
[32] F. Pasquill, Atmospheric Dispersion Parameters in Gaussian Plume Modeling. 2. Possible Requirements for Change in the Turner Workbook Values, (1976).
[33] S.L. Johnson, Review of Agency Draft Guidance on the Development, Evaluation, and Application of Regulatory Environmental Models and Models Knowledge Base by the Regulatory Environmental Modeling Guidance Review Panel of the EPA Science Advisory Board, EPA-SAB-06-009, Washington, DC, 20460, (2006).

Chapter 15

Economics and project management

15.1 INTRODUCTION

The previous chapters discussed the scientific, engineering, and technological aspects of air pollution. Ultimately, these are the basis for environmental and health risk-based decisions and are essential elements in determining whether a project is acceptable. The acceptability, however, must also consider economics and feasibility. Any project must be reliable in the design and operation of the project, resilient to external stresses, and reasonable in the expectations that the project can be operationalized and managed by governance. Furthermore, project planning must consider the appropriate means of communicating these decisions to and working with various stakeholders.

Certainly, these are crucial criteria that must be underpinned with sound science and engineering. In addition to these, the project must also be cost-effective and feasible. Privately funded and managed projects share many feasibility aspects with public projects. Indeed, many of the contracted and private concerns have public interests and visibility, including state implementation plans (SIPs) to achieve air quality standards, technology selections to remove hazardous air pollutants, and emission permits. Public works and engineering departments usually do not and certainly should not approach air quality programs separate from other environmental, transportation, waste collection, hazardous materials handling, treatment, engineering, and remediation activities. Public agencies must search for economies of scale that solve and prevent multiple environmental, public health, and other societal problems.

This chapter draws upon engineering economics to introduce approaches for evaluating the acceptability of proposed, new projects and upgrades and enhancements to existing projects.

15.2 EMISSION STRATEGIES

At the outset, it is important to highlight the relationship between technical and cost efficiencies. For example, the first requirement is technical efficiency, that is, choosing the best alternative that protects public health and the environment. Thus, a cheaper alternative is only acceptable if it meets the technical specification necessary to meet a standard or other air quality target.

The Clean Air Act requires steps to be taken to limit the concentrations of criterion pollutants, that is, National Ambient Air Quality Standards (NAAQS), in ambient air. Nonattainment areas are those parts of the United States that exceed the NAAQS for any of the six criterion pollutants. Each state in the United States establishes an SIP to engage control strategies for reducing air pollution in nonattainment areas. These plans are periodically reviewed for effectiveness.

The SIP addresses both mobile and stationary sources, calling for the establishment of baselines and projection inventories of pollutants from sources. The inventories are tools in targeting control strategies for on-road and off-road mobile sources. The SIP also addresses stationary sources by requiring control strategies to address pollutants from industrial facilities and other stationary sources. As for mobile sources, the SIP also requires stationary source inventories based on emission data from major regulated facilities and calculated emissions from minor stationary sources (i.e., area source calculations) [1].

15.2.1 Strategies to prevent degradation of air quality

The Clean Air Act includes provisions to keep clean air clean under the new source review (NSR). The NSR embodies sets of rules that apply to a proposed construction, either new or substantial modifications to an existing site. The two categories of NSR rules are (1) nonattainment NSR and (2) Prevention of Significant Deterioration (PSD).

The nonattainment NSR rules are designed to assist in efforts to attain and maintain compliance with the National Ambient Air Quality Standards (NAAQS), applying to pollutants for which an area has been designated as not meeting one or more of the NAAQS, that is, a nonattainment area. These standards are summarized in Table 15.1. The nonattainment NSR rules assumes that minor sources and minor modifications do not significantly affect the attainment status,

Air Pollution Calculations. https://doi.org/10.1016/B978-0-12-814934-8.00015-6

TABLE 15.1 US National Ambient Air Quality Standards under the Clean Air Act Amendments

Pollutant		Primary/ secondary	Averaging time	Level	Form
Carbon monoxide (CO)		Primary	8 h	9 ppm	Not to be exceeded More than once per year
			1 h	35 ppm	
Lead (Pb)		Primary and secondary	Rolling 3-month average	0.15 μg/m^{3a}	Not to be exceeded
Nitrogen dioxide (NO_2)		Primary	1 h	100 ppb	98th percentile of 1 h daily maximum concentrations, averaged over 3 years
		Primary and secondary	1 year	53 ppb[b]	Annual mean
Ozone (O_3)		Primary and secondary	8 h	0.070 ppm[c]	Annual fourth-highest daily maximum 8 h concentration, averaged over 3 years
Particulate matter (PM)	$PM_{2.5}$	Primary	1 year	12.0 μg/m^3	Annual mean, averaged over 3 years
		Secondary	1 year	15.0 μg/m^3	Annual mean, averaged over 3 years
		Primary and secondary	24 h	35 μg/m^3	98th percentile, averaged over 3 years
	PM_{10}	Primary and secondary	24 h	150 μg/m^3	Not to be exceeded more than once per year on average over 3 years
Sulfur dioxide (SO_2)		Primary	1 h	75 ppb[d]	99th percentile of 1 h daily maximum concentrations, averaged over 3 years
		Secondary	3 h	0.5 ppm	Not to be exceeded more than once per year

[a]The level of the annual NO_2 standard is 0.053 ppm. It is shown here in terms of ppb for the purposes of clearer comparison with the 1 h standard level.
[b]Final rule signed on 1 October 2015 and effective on 28 December 2015. The previous (2008) O_3 standards additionally remain in effect in some areas. Revocation of the previous (2008) O_3 standards and transitioning to the current (2015) standards will be addressed in the implementation rule for the current standards.
[c]The previous SO_2 standards (0.14 ppm 24 h and 0.03 ppm annual) will additionally remain in effect in certain areas: (1) any area for which it is not yet 1 year since the effective date of designation under the current (2010) standards and (2) any area for which an implementation plan providing for attainment of the current (2010) standard has not been submitted and approved and that is designated nonattainment under the previous SO_2 standards or is not meeting the requirements of a state implementation plan (SIP) call under the previous SO_2 standards (40 CFR 50.4(3)). An SIP call is an EPA action requiring a state to resubmit all or part of its SIP to demonstrate attainment of the required NAAQS.
[d]In areas designated nonattainment for the Pb standards prior to the promulgation of the current (2008) standards and for which implementation plans to attain or maintain the current (2008) standards have not been submitted and approved, the previous standards (1.5 μg/m3 as a calendar quarter average) also remain in effect.

Source: U.S. Environmental Protection Agency. NAAQS Table. (2016). 25 August 2018. Available: https://www.epa.gov/criteria-air-pollutants/naaqs-table.

nor do they interfere with plans to achieve compliance with the NAAQS. Some of the key provisions are lowest achievable emission rate (LAER), offsets, alternative site analysis, and compliance certification [2].

In the United States, PSD rules are designed to curtail transport of pollution to areas whose air is presently meeting standards. The Clean Air Act requires definition of how much deterioration can be considered insignificant. PSD is needed to protect public health and welfare and to preserve, protect, and enhance the air quality in national parks, national wilderness areas, national monuments, national seashores, and other areas of special national or regional natural, recreational, scenic, or historic value. PSD also requires that economic growth will occur in a manner consistent with the preservation of existing clean air resources.

The PSD provisions also ask that regulatory actions be viewed in a manner that allows for certain levels of air pollutant releases from the perspective of the cumulative impact in any area. This provision is analogous to a virtual bubble over a group of stacks and vents, addressing the total releases under the bubble instead of emissions from each of the sources. The permit applicant can conduct a netting exercise, that is, the demonstration that by aggregating emission changes that have occurred at a source over a contemporaneous time period (see Fig. 15.1), these resulting emission changes are then reviewed to determine if the proposed project must undergo NSR. The key is account for other emission decreases and

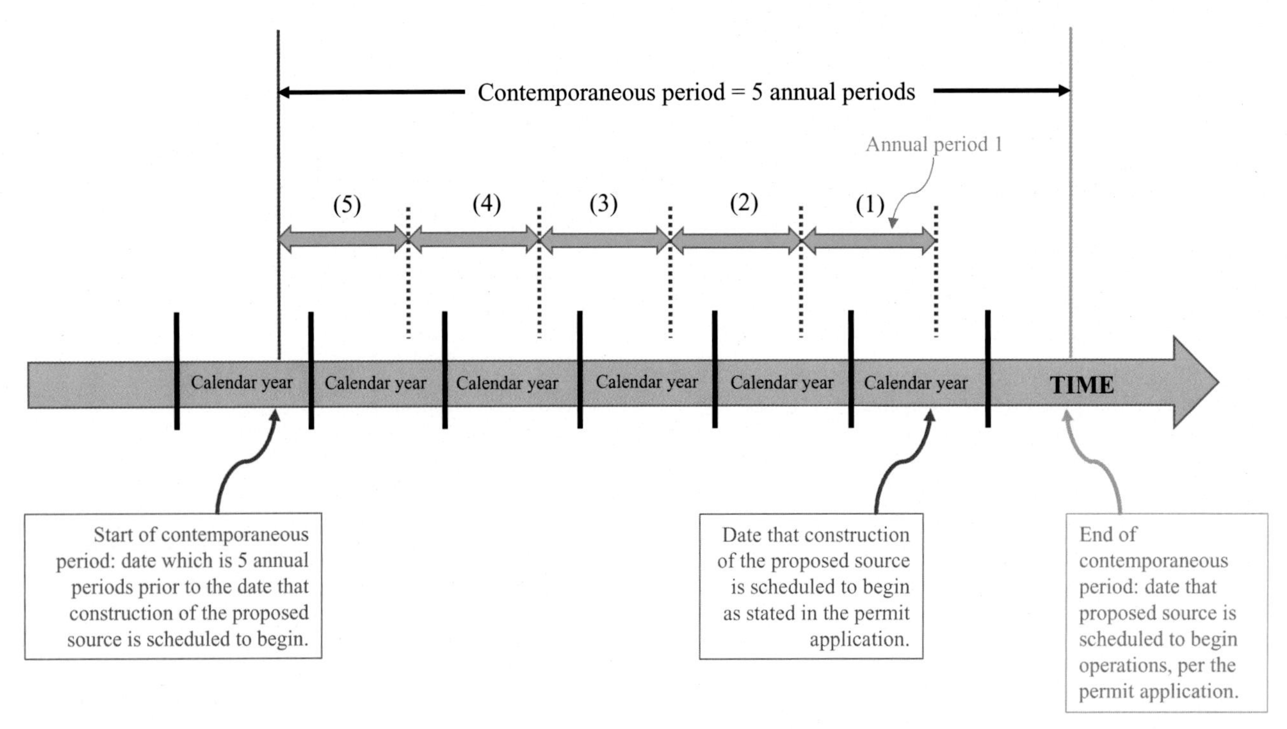

FIG. 15.1 Contemporaneous period determination for severe/marginal/moderate ozone nonattainment areas and attainment portions of the ozone transport region (volatile organic compounds or oxides of nitrogen) and for PM_{10} or $PM_{2.5}$ nonattainment areas. Notes: An annual period is 365 consecutive days; calendar year is January 1 through December 31. *(Information source: State implementation plans, 2009, Available at: http://www.dec.ny.gov/chemical/8403.)*

increases that have occurred and are expected to occur over the contemporaneous time frame. The applicant must demonstrate that the overall net increase in emissions would not be significant. The provisions governing netting differ for each attainment area; for example, netting is more stringent in serious, severe, and extreme ozone nonattainment areas [2].

On 1 October 2015, the EPA promulgated a rule that revised the primary and secondary 8 h NAAQS for ozone to a level of 0.070 ppm (ppm). These revisions started a process in which states recommend area designations, that is, as nonattainment, attainment, or unclassifiable with respect to the revised standards, to the EPA, and the EPA then evaluates air quality data and other factors prior to making final area designations. In accordance with CAA section 181(a)(1), an area designated as nonattainment for a revised ozone NAAQS must be classified, at the time of designation, as marginal, moderate, serious, severe, or extreme, depending on the severity of the ozone air quality problem in that nonattainment area. On 17 November 2016, the EPA proposed a set of nonattainment area classification thresholds and associated attainment dates and other NAAQS implementation-related provisions including submittal deadlines and specific CAA requirements for the content of nonattainment area and ozone transport region SIPs, for the 2015 ozone NAAQS. This finalized the set of nonattainment area classification thresholds and associated attainment dates, which will apply when the EPA promulgates final nonattainment area designations for the 2015 ozone ambient standard [3]. The EPA administrator signed the notice on 1 March 2018.

The PSD rules must be followed when a company adds a new source or modifies an existing source in an area designated as attainment of the NAAQS under the Clean Air Act Amendments. The PSD rules need to be addressed for the pollutants for which the area is classified as attainment with the NAAQS. The rules are a safeguard for ensuring that compliance with the NAAQS continues in time. As with nonattainment NSR, the strategy with PSD rules assumes that minor new sources and minor modifications do not significantly affect the air quality. The distinctive requirements of PSD are best available control technology (BACT); air quality analysis, modeling (allowable increments); and analysis of impacts of the project on visibility, vegetations, and soils.

15.2.2 Identifying emission sources

Preserving air quality depends on knowing which sources are contributing to the deterioration. When the results of air pollution measurements are interpreted, one of the first questions asked by scientists, engineers, and policy makers is where did it come from? Sorting out the various sources of pollution is known as source apportionment. Various models and decision tools are used to try to locate the sources of pollutants. A widely used approach is the "source-receptor model" or as it is more commonly known as the receptor model.

Using receptor models can be distinguished from the dispersion models in that dispersion models usually start from the source and estimate the direction where the plume and its contaminants are expected to move. Conversely, receptor models are based upon measurements taken in the ambient environment, and from these observations, make use of algorithms and functions to determine pollution sources.

To estimate the potential for emission of a new source or source complex, it is necessary to consider two other source categories, "fugitive" and "secondary" sources. Fugitive emissions are those from other point sources, such as unpaved plant roads, outdoor storage piles swept by the wind, and surface mining. Secondary sources are those small sources with emissions of a different character from those of a major source, necessary for the operation of the major source, or source complex.

Another consideration is the impact of indirect sources. A facility, building, structure, installation, or combination of these that generates or attracts mobile source activity is known as an indirect source of air pollution. The very existence of such a source results in emissions of any pollutant (or the precursor of an air pollutant, e.g., hydrocarbons and NO_x for ground-level O_3) for which there is a state ambient air quality standard. Examples include parking areas of shopping malls and sports arenas and other facilities that attract large numbers of motor vehicles, frequently arriving and leaving over relatively short periods of time. Of somewhat similar character are traffic interchanges, at an intersection of major highways, each highway being a line source in its own right.

15.2.3 Fuel-related strategies

Mobile sources are problematic for both conventional and toxic air pollutants. Vehicular tailpipe emissions of hydrocarbons, carbon monoxide, and oxides of nitrogen were to be reduced with the 1994 models. Standards now must be maintained over a longer vehicle life. Evaporative emission controls were mentioned as a means for reducing hydrocarbons.

Beginning in 1992, "oxyfuel" gasolines blended with alcohol began to be sold during winter months in cities with severe carbon monoxide problems. In 1995, reformulated gasolines with aromatic compounds were introduced in the nine cities with the worst ozone problems, but other cities could participate. Later, a pilot program introduced 150,000 low-emitting vehicles to California that meet tighter emission limits through a combination of vehicle technology and substitutes for gasoline or blends of substitutes with gasoline. Other states are also participating in this initiative.

15.2.4 Air toxics strategies

Air toxic emissions from vehicles and other mobile sources are a growing concern. Near-roadway environments can be substantial exposure pathways to benzene and other organic air pollutants. And in nations that have not banned leaded gasoline, near-roadway exposures can be a major contributor to Pb body burden [4]. Air toxic exposure from mobile sources, therefore, is particularly worrisome for children, given the neurological, developmental, and cancer risks.

The United States presently lists 187 hazardous air pollutants. Most of these are carcinogenic, mutagenic, and/or toxic to neurological, endocrine, reproductive, and developmental systems. Actually, this list includes many >189 chemical compounds. For example, the coke-oven emission category includes thousands of chemical species, and the metals are listed as metals and their compounds. The EPA published a list of source categories issued maximum achievable control technology (MACT) standards for each category over a specified timetable. The next step beyond MACT standards was to begin to address chronic health risks that would still be expected if the sources meet these standards, that is, residual risk reduction. The first step was to assess the health risks from air toxics emitted by stationary sources that emit air toxics after the technology-based, that is, MACT, standards are in place. The residual risk provision sets additional standards if MACT does not protect public health with an "ample margin of safety" and additional standards if they are needed to prevent adverse environmental effects.

The Clean Air Act requires an "ample margin of safety" for cancer-causing emissions. That is, if a source can demonstrate that it will not contribute to $>10^{26}$ cancer risk, then it meets the ample margin of safety requirements for air toxics. The ample margin needed to protect populations from noncancer toxins, such as neurotoxins, has varied but involves the application of some type of hazard quotient (HQ). The HQ is the ratio of the potential exposure to the substance and the level at which no adverse effects are expected. An $HQ < 1$ means that the exposure levels to a chemical should not lead to adverse health effects. Conversely, an $HQ > 1$ means that adverse health effects are possible. Due to uncertainties and the feedback that is coming from the business and scientific communities, the ample margin of safety threshold is presently ranging from HQ=0.2 to HQ=1.0. So, if a source can demonstrate that it will not contribute to greater than the threshold (whether it is 0.2, 1.0, or some other levels established by the federal government) for noncancer risk, it meets the ample margin of safety requirements for air toxics.

An air toxic source is designated as "major" if it emits 10 or more tons per year of any of the listed toxic air pollutants or 25 or more tons per year of a mixture of air toxics. In addition to releases from stacks and vents, these emission totals include equipment leaks and releases during transfer. Area sources are those facilities that emit <10 tons per year of a single air toxic or <25 tons per year of a combination of air toxics [5].

The MACT are customized for specific source categories. For example, semiconductor manufacturing, semiconductor packaging, printed wiring board manufacturing, and display manufacturing must be aware of specific air toxics that apply (see Table 15.2). In this case, the list includes both organic contaminants and metals. In addition, the manufacturer must be aware of upcoming chemicals used in its processes. The company would be encouraged to find less hazardous substances as substitutes, if possible.

TABLE 15.2 Chemicals used in semiconductor manufacturing that are scheduled for maximum achievable control technology (MACT) standards: A. Manufacturing, B. Packaging, and C. Printed wire board manufacturing

A. Semiconductor manufacturing hazardous air pollutants	
Antimony compounds	Hydrochloric acid
Arsenic compounds	Hydrofluoric acid
Arsine	Methanol
Carbon tetrachloride	Methyl isobutyl ketone
Catechol	Nickel compounds
Chlorine	Phosphine
Chromium compounds	Phosphorus
Ethyl acrylate	1,1,1-Trichloroethane
Ethylbenzene	Trichloroethylene
Ethylene glycol	Xylene
B. Semiconductor packaging hazardous air pollutants	
Chlorine	Methanol
Chromium	Methylene chloride
Ethylbenzene	Nickel compounds
Ethylene glycol	Toluene
Hydrochloric acid	1,1,1-Trichloroethane
Hydrofluoric acid	Xylene
Lead compounds	
C. Printed wire board manufacturing hazardous air pollutants	
Chlorine	Lead compounds
Dimethylformamide	Methylene chloride
Formaldehyde	Nickel compounds
Hydrochloric acid	Perchloroethylene
Hydrofluoric acid	1,1,1-Trichloroethane
Ethylene glycol	

(**Source:** US Environmental Protection Agency. Design for the Environment. Section A. Clean Air Act Amendments. (2013). http://www.epa.gov/dfe/pubs/pwb/tech_rep/fedregs/regsecta.htm (Accessed 26 September 2013).)

15.2.5 Economics and efficiency

When owners wish to build a new source, which will add a certain amount of a specific pollutant to an area that is in nonattainment with respect to that pollutant, they must, under US federal regulations, document a reduction of at least that amount of the pollutant from another source in the area. They can affect this reduction or "offset," as it is called, in another plant they own in the area or can shut down that plant. However, if they do not own another such plant or do not wish to shut down or effect such reduction in a plant they own, they can seek the required reduction or offset from another owner. Thus, such offsets are marketable credits that can be bought, sold, traded, or stockpiled ("banked") if the state or local regulatory agency legitimizes, records, and certificates these transactions. The new source will still have to meet NSPS, BACT or MACT, and/or LAER standards, whichever are applicable.

Emission trading programs generally can be accomplished by netting, offsets, bubbles, and banking. Netting allows large new sources and major modifications of existing sources to be exempted from certain review procedures if existing emissions elsewhere in the same facility are reduced by a sufficient amount. An offset allows a major new source to locate in an area that does not attain a given air quality standard, that is, a nonattainment area, if emissions from an existing source are decreased by at least as much as the new source would contribute, after installation of stringent controls. A bubble allows a regulated entity to combine the limits for several different sources into one combined limit and to determine compliance based on that aggregate limit instead of each source individually. The name alludes to an imaginary "bubble" placed over the several sources. Banking allows a regulated entity to accumulate emission credits for future use or sale by taking actions to lower emissions below the relevant standard. These approaches all have a common objective of flexibility to comply with traditional source-specific command and control standards but with the additional benefit of providing even less emissions to improve local air quality.[1]

When a new stationary source has been added to a group of existing stationary sources under the same ownership in the same industrial complex, the usual practice has been to require the new source independently to meet the offset, NSPS, BACT, and/or LAER, disregarding the other sources in the complex. Under the more recent "bubble concept" (Fig. 15.2) adopted by some states with the approval of the US EPA, the addition of the new source is allowed, whether

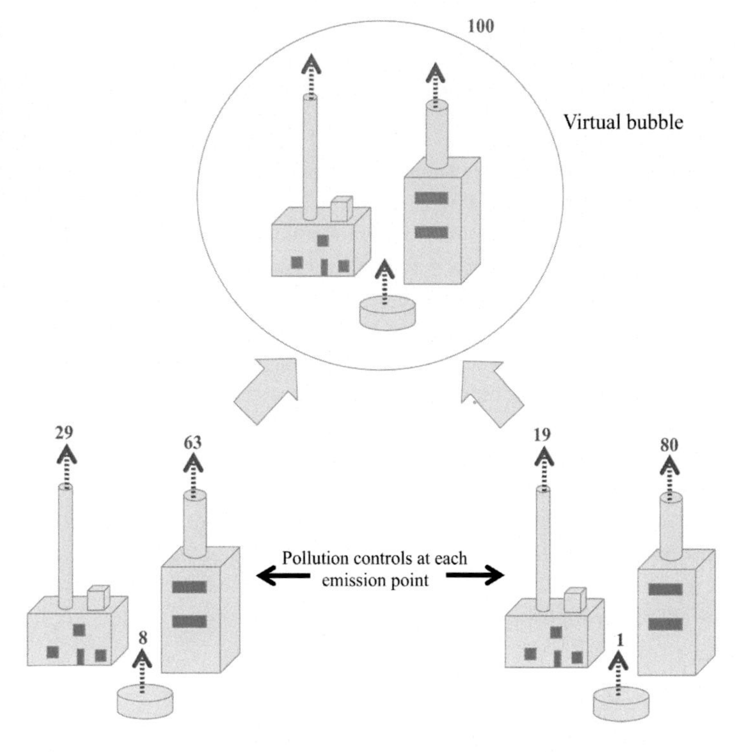

FIG. 15.2 Bubble concept. This pollution control concept places an imaginary bubble over an entire industrial plant, evaluating emissions from the facility as a whole instead of requiring control point by point on emission sources. Numbers represent emissions from individual sources, some of which can be fugitive sources, and from the entire industrial plant. Information from and adaptation of drawings by the Chemical Manufacturing Association.

1. A.D. Ellerman, P.L. Joskow, D. Harrison, Jr., Emissions trading in the U.S.: Experience, lessons, and considerations for greenhouse gases. National Economic Research Association, Inc. report. Arlington, Virginia, 2003.

or not it meets NSPS, BACT, or LAER, provided the total emission of the relevant regulated pollutants from the total complex is decreased. This can be accomplished by obtaining the required offset from one or more of the other sources within the complex, by shutdown or by improvement in control efficiency. The bubble concept has been subject to litigation and may require a ruling or challenge to make it acceptable.

15.2.6 Proportional emission predictions

The proportional modeling or rollback approach assumes that emissions and atmospheric concentrations are proportional and linearly related, that is, that a given percentage reduction in emission will result in a similar percentage reduction in atmospheric concentrations. This is most likely a valid assumption for a nonreactive gas such as carbon monoxide, whose principal source is the automobile; however, it is not recommended for many pollutants and/or those with more complicated and complex physics and chemistry, for example, ground-level O_3.[2] The simple rollback model is.

$$R = \frac{g(P) - D}{g(P) - B} \times 100 \tag{15.1}$$

where R is the required percentage reduction in emission, P is the present air quality, D is the air quality standard, B is the background concentration, and g is the growth factor in emissions (e.g., g is projected to a year in the future when emissions are expected to apply to all vehicles on the road).

The basic steps for the rollback model are as follows [6]:

1. Determine representative monitoring station(s) and the design value.
2. Determine background as the lowest pollutant concentration value recorded at an upwind monitoring location on the same day or during the same time period.
3. Prepare a microscale inventory of the sources that emit the pollutant for the time period the monitor measured.
4. Calculate the percentage for each source based upon the entire inventory.
5. Calculate the relative contribution from each source to the concentration measured for the time period.
6. Estimate the anticipated increase or decrease in emissions from each source.
7. Apply the same percentage of increase or decrease in emissions from each source to the relative contribution calculated for the same source.
8. Calculate the anticipated ambient concentration after source emissions change.

As mentioned, rollback is a substantial oversimplification of the processes involved in air pollution, so basing emission reduction on rollback is fraught with error. Indeed, proportional modeling is mentioned in the US air quality modeling only for PM and lead (Pb). Proportional models are not recommended for PM screening analysis, unless such techniques are used in conjunction with receptor modeling. Refined models are recommended for PM_{10}. However, where possible, particle size, gas-to-particle formation, and their effect on ambient concentrations may be considered. For point sources of small particles and for source-specific analyses of complicated sources, the appropriate recommended steady-state plume dispersion model should be used. Receptor models have proved useful for helping validate emission inventories and for corroborating source-specific impacts estimated by dispersion models. The chemical mass balance (CMB) model is useful for apportioning impacts from local sources. Other receptor models, for example, the positive matrix factorization (PMF) and unmix models, which do not have some of CMB's constraints, have also been applied. In regulatory applications, dispersion models have been used in conjunction with receptor models to attribute source (or source category) contributions [7].

For lead (Pb), an entire major urban area or model areas without significant sources of Pb emissions, as a minimum, a proportional model may be used for air quality analysis. However, urban or other dispersion models are encouraged in these circumstances where the use of such models is feasible. In modeling the effect of traditional line sources (such as a specific roadway or highway) on lead air quality, dispersion models applied for other pollutants can be used [7].

15.3 ENGINEERING ECONOMICS

Once the decision is made to address an air pollution problem, engineering economics, also known as engineering economy, is used to determine whether a specific project is most worthwhile compared with other candidates. Engineering

2. Preferred methods include dispersion modeling, source/receptor models such as the chemical mass balance (CMB) models, and advanced regional models (e.g., the Urban Airshed Model; link: http://www.epa.gov/scram001/photochemicalindex.htm).

economics is also used to prioritize various projects that may serve the same purpose or function and to assign value or importance to design specifications. For example, engineering economics may help the city engineer determine whether to invest in new control technologies or keep repairing the old ones; which of the commercial equipment to purchase; which improvements to public assets and operations should be funded; and whether the added cost of materials, for example, fabric filters, are worth the price.

A principal economic concept is that time represents money, that is, the so-called time value of money. Waste management is an ongoing endeavor, so funding and other resources are viewed over several years. Expensive equipment may be purchased and immediately installed yet paid for years, even if the benefit provided by the equipment ends before payoff. Therefore, the first step in evaluating a project is the cash flow, which represents the time value of money. This requires that the cash receipts and disbursements (i.e., expenses) for each project be described over a specified time period.

The costs include initial costs (e.g., purchase price, setup, and installation); operation and maintenance (O&M); overhaul; and end-of-project life costs, for example, decommissioning, disassembly, and recycling. The costs are somewhat fluid and interrelated. For example, a fly ash or scrubber sludge pit may have major decommissioning costs if hazardous substances have accumulated, such as the cases of water contamination from power plant ash ponds in North Carolina (see Discussion Box "Environmental and Economic Perils of Decommissioning: Ash Pits"). Therefore, O&M and surveillance costs upfront can prevent major costs over time. The disbursements include salvage value, that is, what the equipment or property is worth after its operational life, and revenues (e.g., sales of compost or recycled materials).

Environmental and Economic Perils of Decommissioning: Ash Pits [8]

Coal combustion residuals (CCRs) represent the largest amount of industrial wastes generated in the United States. The 10,000 t of CCRs generated per year from 470 electric generating plants consist of bottom ash, fly ash, boiler slag, and by-products from FGD. Surface impoundments (ash ponds and pits; see Fig. 15.3) and landfills hold half of the CCRs [9, 10]. If the effluent from ash is released into a surface water body in the United States, the amount and type of effluent is limited by the National Pollutant Discharge Elimination System (NPDES) and requires an NPDES permit. Since this effluent very likely contains highly toxic metals and other contaminants, the permit process is complicated beyond a single effluent limitation threshold and requires other factors, including background concentrations and characteristics of the receiving water body. For example, the US Environmental Protection Agency (EPA) applies a formula to calculate waste load allocations and to find the in-stream pollutant concentration (C_r) in mg L^{-1} in the stream reach after complete mixing occurs [11]:

$$C_r = \frac{Q_d C_d + Q_s C_s}{Q_r} \tag{15.2}$$

where Q_d = waste discharge flow in units commonly used by US environmental engineers, that is, million gallons per day (MGD) or cubic feet per second (cfs); C_d = pollutant concentration in waste discharge in milligrams per liter (mg L^{-1}); Q_s = background stream flow in mgd or cfs above point of discharge; C_s = background in-stream pollutant concentration in mg L^{-1}; and Q_r = resultant in-stream flow after discharge in MGD or cfs.

This and similar equations are used by permitting authorities as part of the process in calculating water quality-based limits. If a value of zero is used for the ambient concentration for a pollutant (C_s) in the equation, the permitting authority can establish an entire limit of discharge. The EPA is aware that the resulting limit would not account for upstream background concentrations of the pollutant [11]. Depending on the pollutant, the background concentration is almost never zero, so the limit would not prevent

FIG. 15.3 Initial stages of excavating and consolidating ash at Savannah River coal ash disposal site. *(Photo credit: U.S. Department of Energy, Office of Environmental Management, New Technology Accelerates Schedule, Reduces Cost of SRS Coal Ash Cleanup, 2017, Available at: https://www.energy.gov/em/articles/new-technology-accelerates-schedule-reduces-cost-srs-coal-ash-cleanup.)*

an in-stream exceedance of the water quality criteria. In addition, some of the released pollutant is likely to come from nonpoint and other sources besides the direct discharge of the effluent, especially seepage from the ash storage system [11]. Thus, handling CCRs requires coordination among water quality, air quality, and waste management authorities. Indeed, failures of ash storage systems can involve direct and indirect releases into surface and groundwater.

On 2 February 2014, an estimated 39,000 tons of coal ash spilled from Duke Energy's Dan River Steam Station's ash pond (Fig. 15.4) to the Dan River in Eden, NC (Fig. 15.5). The ash remained visible in a reservoir 80 miles downstream from the spill

FIG. 15.4 Breached ash pond released coal ash and coal ash wastewater into the Dan River. *(Photo credit: U.S. Fish and Wildlife Service. Natural Resource Damage Assessment and Restoration for the Dan River Coal Ash Spill. (2018). 8 March 8 2018. Available: https://files.nc.gov/ncdeq/Coal%20Ash/documents/Coal%20Ash/20140609%20FINAL%20Dan%20River%20cooperative%20NRDAR%20factsheet.pdf.)*

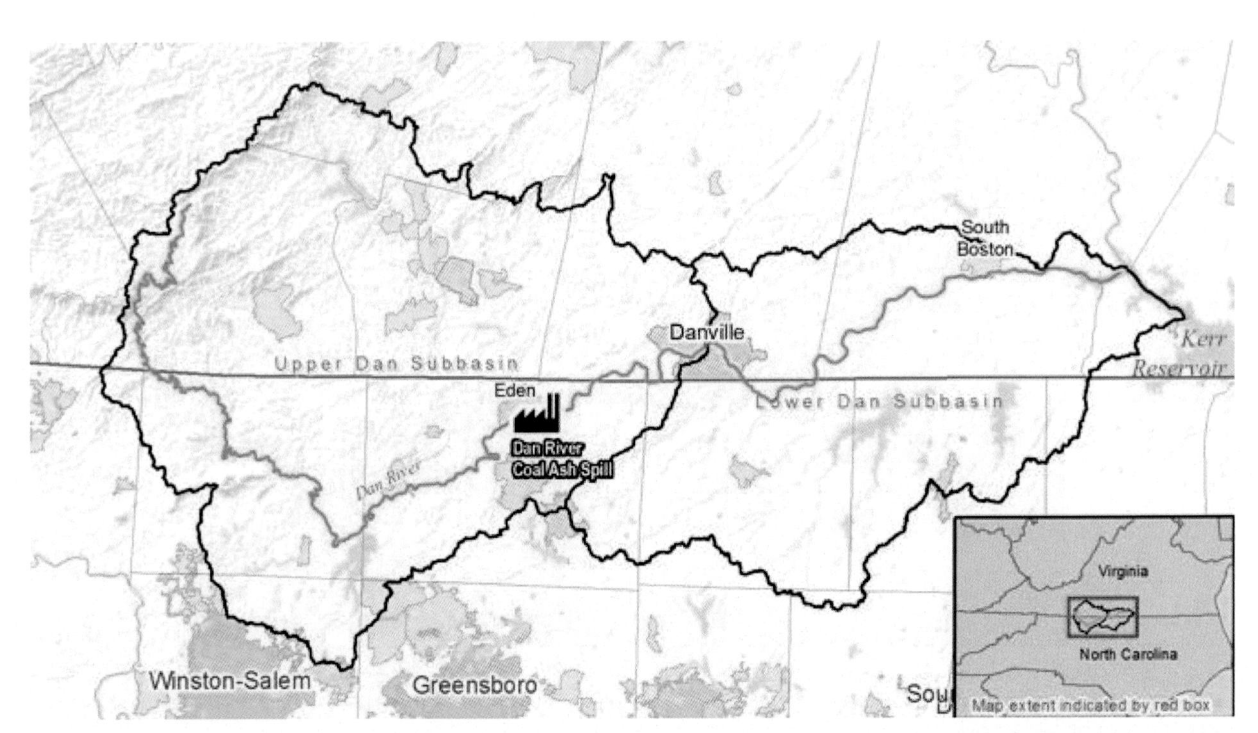

FIG. 15.5 Location of ash pond spill. *(Map credit: U.S. Fish and Wildlife Service. Natural Resource Damage Assessment and Restoration for the Dan River Coal Ash Spill. (2018). 8 March 8 2018. Available: https://files.nc.gov/ncdeq/Coal%20Ash/documents/Coal%20Ash/20140609%20FINAL%20Dan%20River%20cooperative%20NRDAR%20factsheet.pdf.)*

for days after the coal ash spill. On the day after the spill, the North Carolina Department of Environment and Natural Resources, the Virginia Department of Environmental Quality, the US Fish and Wildlife Service, and the US Environmental Protection Agency, along with the utility, Duke Energy, initiated a unified command addressing the spill, including technical assistance, oversight, data management, and environmental sampling and analysis. Other agencies eventually became involved in a massive effort to protect human health and the environment and guide the cleanup. These include the US Army Corps of Engineers, North Carolina Department of Public Health, Virginia Department of Health, and county and local agencies from communities along the Dan River and within the Kerr Reservoir watershed [10].

An extensive breach threatens both public health and ecosystems. Metals and other contaminants can reach drinking water supply intakes. The Dan River basin is a recreation area for fishing, boating, and wildlife viewing. The river is a vital habitat area for fish and wildlife, which include the endangered Roanoke logperch (*Percina rex*) and the James spinymussel (*Pleurobema collina*), along with a species currently under consideration for protection, that is, the green floater (*Lasmigona subviridis*), a freshwater mussel [12].

On 22 May 2014, EPA and Duke Energy entered into an Administrative Order on Consent (AOC). Under the requirements of the AOC, the utility removed coal ash that had accumulated at the Schoolfield Dam, Town Creek Sand Bar, and at both the Danville and South Boston water treatment facilities. The removal action was completed in July 2014, and an estimated 4000 cubic yards of coal ash was removed.

Placing the wastes in a pit or other storage systems can allow mercury (Hg), lead (Pb), arsenic (As), and other toxic metals and metalloids to migrate. For example, they can reach groundwater, a major source of drinking water for much of the world. An ash pit that is upgradient from a well can contaminate personal and municipal well water (see Fig. 15.6).Water stands in a well at a particular level, known as hydraulic head (*h*). Groundwater flows in the direction of decreasing head. The change in head measurement is the hydraulic gradient, $\frac{\partial h}{\partial x}$.

The capability of a fluid to move through underground materials is known as hydraulic conductivity (*K*). The various strata under the earth's surface may be consolidated rock or unconsolidated materials, like soil, sand, or gravel. Most commonly, however, the movement of groundwater is through the unconsolidated materials, that is, porous media. Darcy's law is an empirically derived equation for the flow of fluids through a cross section of porous media (*A*). It states that the specific discharge, *Q*, is directly proportional to the *K* and hydraulic gradient:

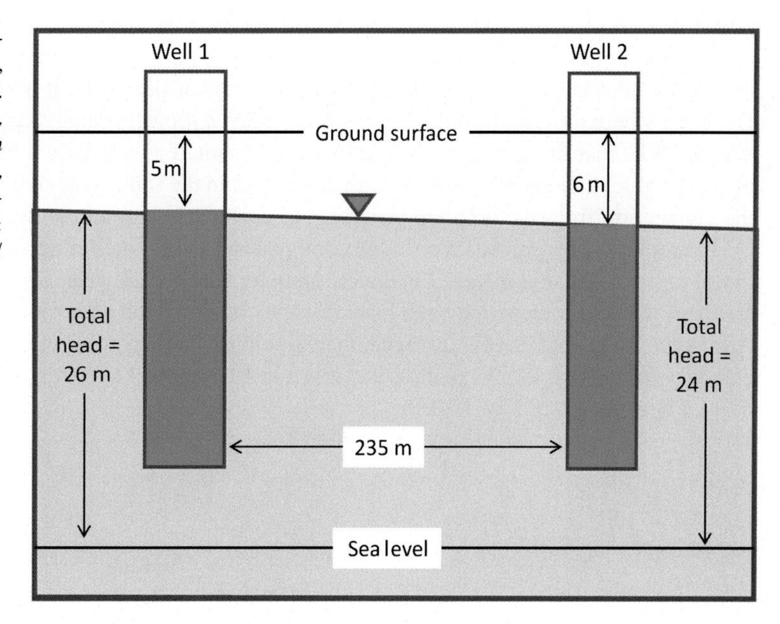

FIG. 15.6 Hydraulic gradient between two wells. The difference in head is 2 m over 235 m. Therefore, in this example, the hydraulic gradient $= \frac{\partial h}{\partial x} = \frac{26\text{m}-24\text{m}}{235\text{m}} = 0.0085\,\text{m}\,\text{m}^{-1}$. *(Graphic reproduced with permission from D.A. Vallero, Environmental impacts of energy production, distribution and transport, in: Future Energy, 2nd ed., Elsevier, 2014, pp. 551–581; data from North Carolina Department of Environment and Natural Resources. Groundwater, Aquifers & Confining Beds. (2018). 6 March 2018. Available: https://www.ncwater.org/?page.560.)*

$$\text{Darcy's law}: Q = -KA\frac{\partial h}{\partial x} \tag{15.3}$$

The law assumes that flow is laminar and that inertia can be neglected. Thus, if $K=0.5\,\text{m}\,\text{s}^{-1}$, $\text{dh/dx}=0.0085\,\text{m.m}^{-1}$, and $A=500\,\text{m}^2$, then $Q=2.125\,\text{m}^3\,\text{s}^{-1}$. Thus, if a leak has produced a plume, the aquifer is being contaminated at a rate of $2.125\,\text{m}^3\,\text{s}^{-1}$ (see Fig. 15.7).

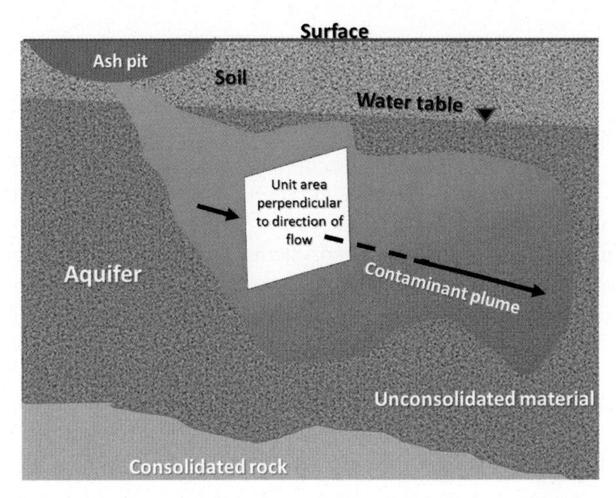

FIG. 15.7 Plume of pollutants flowing from an ash pit to a drinking water aquifer. The plane perpendicular to groundwater flow is used to calculate the movement, that is, flux, of the plume.

15.3.1 Cash flow

The cash flow diagram (CFD) is used to evaluate costs and disbursements (see Fig. 15.8). The CFD shows the amount and timing of individual cash flows and forms the basis for engineering economic analysis [13]. The CFD is developed by first drawing a segmented time-based horizontal line, divided into appropriate time unit. At each cash flow, even a vertical arrow points down to costs and up for revenues or benefits. For clarity, the CFD segments should be drawn to a relative scale.

The end of period t is the same as the beginning of period $(t+1)$. The beginning-of-period cash flows include costs for rent, lease, and insurance payments, whereas the end-of-period cash flows are O&M, salvages, revenues, and overhauls. Deciding where $t=0$ is arbitrary, for example, when a project is analyzed, when funding is approved, or when construction begins. Note that one group's cash outflow, represented as a negative value, is another group's inflow, represented as a positive value. Incidentally, two or more cash flows in the same year should be diagrammed individually to indicate clearly the connection from the problem statement to each cash flow in the diagram.

Example: A city purchases a sludge combustion facility, including an air pollution control, system for $2 million, with O&M costs of $30,000 in year 1 and increasing by $5000 each year. Some of the sludge is converted to commercial-grade fertilizer with equipment that has a local salvage company and a plastic company, providing an annual revenue of $50,000. After year 7, the converter is salvaged, that is, sold to another jurisdiction, for $110,000. To simplify, let us apply the year-end convention, that is, all expenses and revenue are assumed to occur at the end of the year. This makes the math discrete. The CFD is shown in Fig. 15.9.

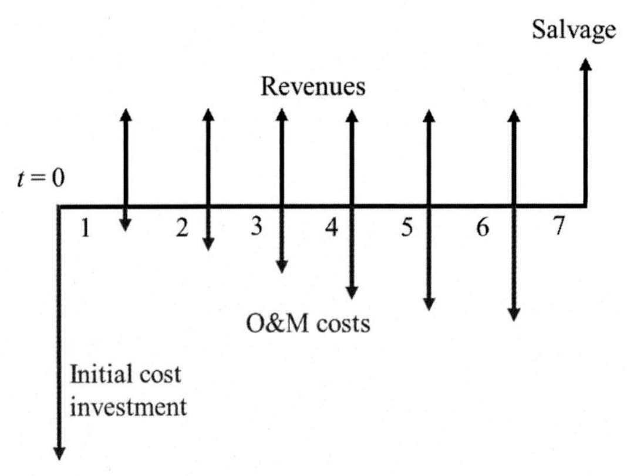

FIG. 15.8 Simple cash flow diagram. In this case, after the initial investment, revenues are constant, whereas operation and maintenance costs are increasing each year. After year 5, O&M costs exceed revenues. After year 7, the project is ended, and salvage is recovered.

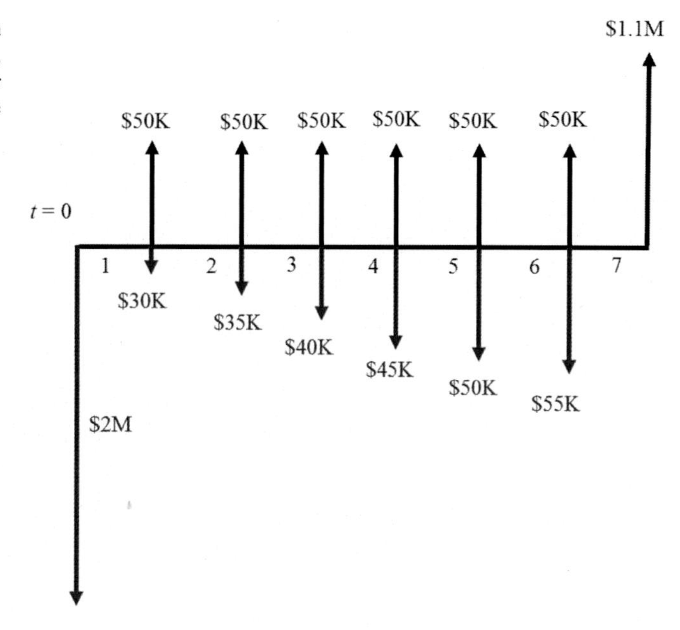

FIG. 15.9 Example cash flow diagram showing an initial $2 million expenditure. Following the initial investment, revenues are constant, whereas operation and maintenance costs are increasing each year. After year 5, O&M costs exceed revenues, so the project is ended, and salvage is recovered.

15.3.2 Discounting

Engineering economics dictates equivalence so that all values can be compared consistently. The present worth (P) is the present amount at $t = 0$. Future worth (F) is the equivalent value that will have accumulated after n periods (e.g., years) from P. The annual amount (A) is the uniform amount of money that repeats at the end of each period for n periods. If we do not assume a discrete payout, we also need to have a uniform gradient amount (G), which is the amount of money that repeats at the end of each period, beginning at the end of period 2 and ending at the end of period n.

Therefore, the CFDs can represent present value, future worth, uniform series, uniform gradient series, and exponential gradient series (see Fig. 15.10). Present value and future worth CFDs are used when determining whether a purchase is acceptable and feasible. Uniform series CFDs can be part of multiyear budgets. The two bottom CFDs in Fig. 15.10 may best apply to equipment and systems that demand greater maintenance and replacements. For example, if the reliability of a system declines with a linear slope, the uniform gradient series is used, but if the failure rates accelerate with time, the exponential (E) gradient CFD may be preferable.

Another important factor is the discount rate (i) that needs to be paid to borrow money, that is, the effective rate per time period. The annual i can be calculated as

$$i = \left(1 + \frac{r}{k}\right)^{\frac{k}{p}} - 1 \tag{15.4}$$

where $r =$ the annual interest, $k =$ the number of compounding periods per year, and $p =$ the number of periods per year corresponding to the basis for n. Thus, if payments are made monthly, $p = 12$; if daily, $p = 365$.

The discounting principle states that investing P at $t = 0$ will pay F at $t = n$. Thus, the discount formula is.

$$P = F\,(P/F, i\%, n) \tag{15.5}$$

where (P/F, $i\%$, n) is the discount factor.

Table 15.3 provides the discount factor conversion formulas for the most common conversions for $P, F, A, G,$ and E used by waste managers and engineers. Numerous online calculators are available for these conversions, as well as spreadsheet programs, especially Microsoft Excel (see far-right column of Table 15.3). Microsoft [14], for example, provides support for calculating P, F, and i at the following:

- *P=PV(i,N,A,F,Type)*: https://support.office.com/en-us/article/PV-function-23879d31-0e02-4321-be01-da16e8168cbd
- *F=FV(i,N,A,P,Type)*: https://support.office.com/en-us/article/FV-function-2eef9f44-a084-4c61-bdd8-4fe4bb1b71b3
- *i=RATE(N,A,P,F,Type,guess)*: https://support.office.com/en-us/article/RATE-function-9f665657-4a7e-4bb7-a030-83fc59e748ce

Where "type = 0" means end-of-period cash payments, "type = 1" means beginning-of-period payments, and "guess" is the value of the interest rate.

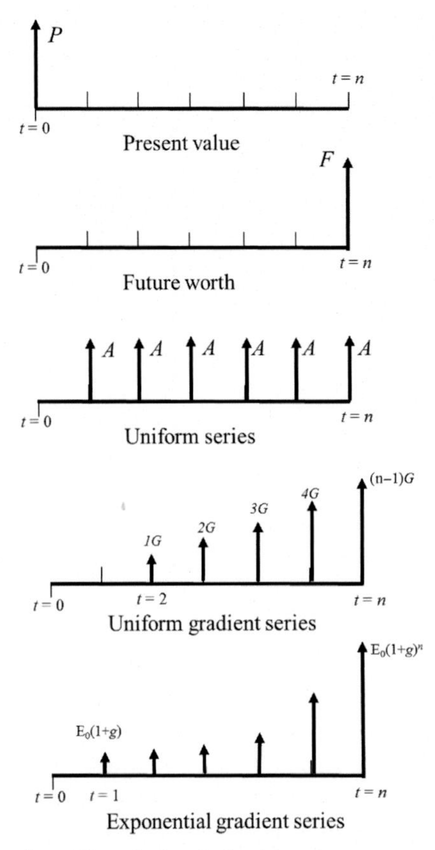

FIG. 15.10 Commonly used cash flow diagrams for public and privately funded projects.

The symbols and nomenclature are not used consistently in engineering and waste management economics. For example, public agencies may also represent present value as *PV*:

$$PV = \frac{B_{y_i}}{(1+r)^{y_i - y_0}} \tag{15.6}$$

where B = the annual benefit, r = the discount rate, y_i = the year the benefit occurs, and y_0 = the year to which future dollars are discounted.

Present value is similarly determined for annual cost (C):

$$PV = \frac{C_{y_i}}{(1+r)^{y_i - y_0}} \tag{15.7}$$

where y_i = the year the cost occurs.

Example

Assume an electric power generating station has to change to a higher-sulfur-content coal and must install a scrubber to meet sulfur dioxide standards. The scrubber is estimated to cost $5000,000. Construction and installation will occur in 2025. The year of analysis is 2020. If the discount rate is 3.6% for the year 2020, what is the present value of the construction cost?

Given: $AC = \$5000,000$; $r = 0.036$, $n = 2025 - 2020 = 5$.

Then, $PV = 5000,000/(1+0.036)^5$.

Therefore, $PV = \frac{5000000}{(1+0.036)^5} = \$4,189,587$ (see Fig. 15.11).

The $5 million is the present value. If the project can wait until 2025, the money could be put into a bank where it earns interest or put into other projects where it generates immediate benefits. The amount that needs to be deposited into the bank to have $5 million in 2025, using a 3.6% discount rate, is only $4.2 million—the present value of the cost of the project in 2020. The same logic applies to benefits.

TABLE 15.3 Conversions for waste management economics of present value (P), future worth (F), annual amount (A), gradient (G), and exponential gradient (EG)

Conversion	Symbol	Discount factor formula	Discount factor formula in Excel
Discrete compounding			
P to F	(F/P,i%,n)	$(1+i)^n$	=FV(i,n,0,-1)
F to P	(P/F,i%,n)	$(1+i)^{-n}$	=PV(i,n,0,-1)
F to A	(A/F,i%,n)	$i/((1+i)^n-1)$	=PMT(i,n,0,-1)
P to A	(A/P,i%,n)	$i*(1+i)^n/((1+i)^n-1)$	=PMT(i,n,-1)
A to F	(F/A,i%,n)	$((1+i)^n-1)/i$	=FV(i,n,-1)
A to P	(P/A,i%,n)	$((1+i)^n-1)/(i*(1+i)^n)$	=PV(i,n,-1)
G to P	(P/G,i%,n)	$((1+i)^n-1)/(i^2*(1+i)^n)-n/(i*(1+i)^n)$	{=NPV(i,(ROW(INDIRECT("1:"&n))-1))}
G to F	(F/G,i%,n)	$((1+i)^n-1)/i^2-(n/i)$	{=(P/G,i%,n) * (F/P,i%,n)}
G to A	(A/G,i%,n)	$(1/i)-n/((1+i)^n-1)$	{=(P/G,i%,n) * (A/P,i%,n)}
EG to P	(P/EG,z-1,n)	$(z^n-1)/(z^n(z-1))$, $z=(1+i)/(1+g)$	=PV(z-1,n,-1)
Discrete compounding			
F to P	(P/F,r%,n)	$e^{-r*n}=1/e^{r*n}$	=1/EXP(r*n)
P to F	(F/P,r%,n)	e^{r*n}	=EXP(r*n)
F to A	(A/F,r%,n)	$(e^r-1)/(e^{r*n}-1)$	=(EXP(r)-1)/(EXP(r*n)-1)
A to F	(F/A,r%,n)	$(e^{r*n}-1)/(e^r-1)$	=(EXP(r*n)-1)/(EXP(r)-1)
P to A	(A/P,r%,n)	$(e^r-1)/(1-e^{-r*n})$	=(EXP(r)-1)/(1–1/EXP(r*n))
A to P	(P/A,r%,n)	$(1-e^{-r*n})/(e^r-1)$	=(1–1/EXP(r*n))/(EXP(r)-1)

Note that Excel formulas for (F/G,i%,n) and (A/G,i%,n) are based on the algebraic equivalence of F/G=(P/G)*(F/P) and A/G=(P/G)*(A/P). The discount factor symbols (P/G,i%,n), (F/P,i%,n), and (A/P,i%,n) need to be replaced with the appropriate discount factor formula listed in the table.
Data from J. Wittwer, Discount Factor Table for Excel: Summary of Discount Factor Formulas for TVM Calculations in Excel®. (2018). 17 May 2018. Available: https://www.vertex42.com/ExcelArticles/discount-factors.html.

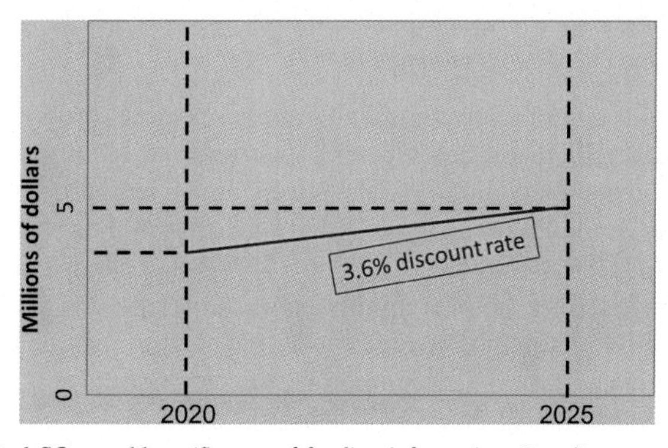

FIG. 15.11 Discount rate for hypothetical SO_2 scrubber. *(Source of funding information: Benefit cost analysis for transportation projects. (2012). Available: https://www.dot.state.mn.us/planning/program/benefitcost.html.)*

Thus, if the only thing of importance is money, you would not install the scrubber in the example. However, large public projects or, in this case, highly regulated utility fills societal needs, including increasing the capacity of existing facilities or systems and/or improving public health, environmental quality, safety, and aesthetics of neighborhoods and enhancing the performance of other existing facilities or systems. In fact, the benefit-cost and other alternative selection techniques (see the next section) must include fines, penalties, and other monetized and nonmonetized costs of not removing the SO_2.

The analysis of a large public project is conducted in the following stages:

1. Planning the analysis and defining its scope
2. Performing engineering analyses of the alternatives
3. Calculating the present value of project costs and benefits
4. Evaluating the results—benefit-cost analysis

Planning the benefit-cost analysis and performing the engineering analysis (the first two stages) require careful thought. The analysis should capture the appropriate benefits and cost differences between the base case and the identified alternatives [15]. These first two stages are the most complicated and require the most time and effort. The economic calculation stage is a relatively short and straightforward process. Evaluation and interpretation of the results require judgment and experience. The process of conducting a benefit-cost analysis can be iterative: the process may require going back to a previous stage to verify results and explore subalternatives.

15.4 COMPARING ALTERNATIVES

In most jurisdictions, the company, utility, and other regulated entities have latitude in the strategies that they employ to meet air quality standards, especially for standards that are performance-oriented. Some standards are more prescriptive and technical, but even these are often open to more than one approach on how to comply [16]. This means that expenditures are seldom a one-time expense, usually requiring the comparison of alternatives with expenditures varying in amounts and spending occurring at multiple times over a project's or equipment's useful life. Not only may the expenditures vary, but also the timing of expenditures may also be different. This section addresses the selection of appropriate implementation and funding time frames, addressing the time value of money, adjusting for prices over time, and selection of the appropriate measure of cost.

15.4.1 Social costs

It is important to keep in mind that the calculations in this chapter generally apply to monetizable costs and benefits, often the costs borne by a private or quasi-public entity. For instance, if the entity pays for the installation and operation of an air pollution control technology, all of these costs are considered to be private. Thus, selection of alternatives usually rests on minimizing these private costs. Conversely, social costs are more than the province of macroeconomics:

> *From a regulatory standpoint, social cost represents the total burden a regulation will impose on the economy. It can be defined as the sum of all opportunity costs incurred as a result of the regulation. These opportunity costs consist of the value lost to society of all the goods and services that will not be produced and consumed if firms comply with the regulation and reallocate resources away from production activities and toward pollution abatement. To be complete, an estimate of social cost should include both the opportunity costs of current consumption that will be foregone as a result of the regulation and also the losses that may result if the regulation reduces capital investment and thus future consumption [17].*

Comparing private costs of various air pollution control alternatives is more straightforward than social cost assessment because private cost comparison generally quantifies expenses incurred by a facility to purchase, finance, and operate the equipment. For example, if a governmental jurisdiction wished to encourage pollution control for a power generation industrial sector, it could provide monetary support, for example, grants, for 25% of the costs of a treatment system, for example, scrubber. The private cost for the industry would be reduced to 75% of the cost of a scrubber. Other less direct monetary support strategies have been employed, for example, eliminating or rebating sales and other taxes, which means that the private cost to the firm is lessened by the total that would otherwise need to include the tax.

The following sections stress private costs, but it is important to keep in mind that the basis for many of the costs is to provide a societal, macroeconomic value, that is, cleaner air. Thus, the private costs lead to incremental improvements or, if the air is already clean, the maintenance of existing air quality, that is, nondeterioration [16].

15.4.2 Types of cost estimates

Uncertainty is highest during the conceptual phases of a project. Cost estimate accuracy depends on the decision being made. Simply knowing that emissions have to be reduced may begin with order-of-magnitude cost estimates for air pollution control projects. For example, chemical engineers and air pollution control engineers [18] consider the order of magnitude to have an error between +50 and – 30%.

15.4.3 Proposal review

Rather than simply deciding whether to accept a proposal, it is better to decide on a functional objective and then identify ways to provide that function in the most efficient and effective way. Both benefits and costs vary, so this analysis must find alternative means of maximizing net benefits and, similarly, minimizing costs to get a desired amount of benefits. This is known as the benefit-cost analysis (BCA), which assumes the following:

1. Equivalence, that is, benefits or costs are in the same units in all projects under review. Most often, this is monetized, for example, expressed in dollars, pounds, yen, or euros.
2. The public is the "user" or "client" who will benefit from the selected project.
3. The "sponsor" is the same level of government (e.g., town, city, county, state, province, nation, or even multinational organizations).

The analysis consists of the following:

1. Identifying all user benefits (positive outcomes) and "disbenefits" (negative outcomes)
2. Quantification and comparison of these benefits and disbenefits
3. Identification and quantification of the sponsor's costs
4. Finding the equivalent net benefits and net costs at the base period, with a discount rate specific to the project
5. Considering the project "acceptable" if the equivalent user's net benefits exceed the equivalent sponsor's net costs (however, you will likely want to have a rule for exceedance before the project is accepted, e.g., some percentage above)

Basically, the BCA seeks to show whether the benefits are worth the economic investment. This can be viewed at various points in the design process. For example,

- project planning: From an economic perspective, is it better to install new pollution control equipment or repair and upgrade existing equipment?
- design and environmental study: From an economic perspective, are the benefits of location "A" worth the project costs? How does location "A" compare with "B" or "C"?
- construction planning: From an economic perspective, are the benefits of closing all industrial operations during installation of new equipment worth the delays and diversion costs, that is, compared with keeping some open?

The BCA is most valuable after a decision is made, based on risk and reliability. It is more valuable as a tool for comparing and evaluating alternatives than for whether to provide a service. For example, if a risk assessment indicates that water is being contaminated, the BCA would not override the need to clean the water, but would seek the best and most feasible means of doing so. As discussed in Chapter 35, the cleanup would have to meet a risk-based or other environmental standard. Thus, the project would have to provide water that meets this standard, and the waste manager would decide which of the options for doing this are most cost-effective.

The BCA is a tool for maximizing benefits for a certain amount of costs, for example, the basis for a waste management or public work budget. So, the BCA needs to be more than a financial comparison. Indeed, various methods are used for comparing benefits and costs for a public project. After deciding which alternatives are feasible and viable, different techniques can be applied to compare them, including the following:

Net present value (NPV): NPV is the value resulting from subtracting the expected cost from the expected benefits. Projects having an NPV less than zero should be avoided.

Net present value is an expression of economic efficiency, which is useful for private cost comparisons and for estimating social costs. In the latter case, economic efficiency is the maximization of social welfare. An efficient market allows society to maximize the NPV of benefits, that is, it is the difference between social benefits and social costs over time. The efficient level of production is "Pareto optimal," that is, it is not possible to rearrange production or to reallocate goods so that someone is better off without making someone else worse off in the process. Economic efficiency requires only that net benefits be maximized, irrespective to whom those net benefits accrue. It does not guarantee an "equitable" or "fair" distribution of benefits among consumers and producers or between subgroups of consumers or producers [17]. This is an example of a societal zero-sum game.

Internal rate of return (IRR): the IRR is the discount rate at which the present value of costs equals the present value of benefits. Alternatively, IRR is the discount rate at which NPV = 0 and BCR = 1.0. If IRR exceeds the opportunity cost of the capital, the project is economically sound and worth pursuing [2].

Benefit-cost ratio (BCR): the BCR is the ratio of the present value of benefits to the present value of costs. A BCR is a gross tool that may serve as a screen and is part of a larger analysis. In its simplest form, it is the ratio of benefits to cost of a project. For example, the city's air quality department is asked to participate in an analysis of a proposed landfill. The project receives the following scores:

	Score	Weight	Total
Benefits			
Water quality	3	5	15
Gas sales	1	3	3
Removal of open dumps	5	5	25
Customer satisfaction	3	4	12
Costs			
Expenses	5	3	15
Fugitive dust	2	2	4
Odors	3	2	6

The scores are weighted according to importance, so the most important factors in this case are water quality and closing dumps. Thus, the benefits are equal: 15+3+25+12=55. The costs are equal: 15+4+6=25. Therefore, this benefit-to-cost ratio is 55/25=2.2. This may indicate that the landfill is acceptable. However, the scores and weightings are highly subjective. Indeed, the engineer from the air quality department was overruled on the scoring and weighting of fugitive dust (she thought it should be 5 and 5, respectively, increasing the cost side of the ratio by 21, i.e., B/C=1.2). Such ratios are notorious for justifying previously ordained projects.

Ideally, the BCA would project and evaluate all possibilities, but this is not possible or even desirable because of the large uncertainties. So, project alternatives need to be screen. Expert judgment can be a first screen to differentiate infeasible alternatives to realistic options and compare them with a base case. The base case is not necessarily the "no action" alternative, although, in most BCAs, one or more of the other options have substantial improvements over the base case. Linear programming and other optimization techniques can be used to quantify the economic feasibility of a project from a life cycle/system perspective. Public projects must also differentiate economic costs and benefits from other societal costs and benefits. Thus, the benefit-cost analysis (BCA) differs from an economic impact analysis (EIA). The US Department of Transportation, for example, differentiates the two processes [19]:

> *A BCA measures the value of a project's benefits and costs to society, while an economic impact analysis measures the impact of increased economic activity within a region. Common metrics for measuring economic impacts include retail spending, business activity, tax revenues, jobs, and property values. Economic impact analyses often take a strictly positive view (i.e., increased jobs and spending) and do not examine how the resources used for a project might have benefitted alternative societal uses of the resources (i.e., they do not assess the net effect on society).*
>
> *In this sense, an economic impact analysis would consider the initial investment in infrastructure as a stimulus to the local economy, as opposed to a cost to the local government. Also, economic impact analyses are smaller scale than the usual BCA, which usually consider broad, economy-wide, and "societal" benefits and costs. Effects in one region may be offset by losses in a neighboring region, reflecting a transfer of spending or jobs that may be a net neutral summation. Thus, the two local jurisdictions would have different economic impact results, but the BCA would not change, since its focus is larger.*

15.5 REPLACEMENT COST ANALYSIS

As any waste manager can attest, things wear out. Replacement and repair will result; it is merely a matter of time. Often, replacement applies to a "supply," that is, something that is expected to be used and changed frequently. Repair is often seen as returning "capital equipment" to its designed efficiency.

These are semantic distinctions. Of course, any piece of a system needs to be replaced, including capital equipment. This is not always due to wear. Sometimes, the function has changed; for example, labor-intensive sorting equipment may be fine until a town's population reaches a threshold where the equipment needs to be replaced by higher-capacity equipment. Sometimes, there is a safety issue; for example, the older equipment may have met previous worker safety regulations, but no longer does. Often, however, economics drives the decision; for example, the mean time to repair (MTTR) may indicate it is time to replace equipment. Any type of asset eventually loses value and must be replaced due to the following:

- Obsolesce: Fast for information technology and slower for mechanical equipment
- Depletion: Supplies
- Deterioration: Mechanical equipment

Replacement cost analysis (RCA) uses specific terminology. The "defender" is the existing piece. The "challenger" is the best available replacement equipment. Current market value is the selling price of the defender in the marketplace.

There are a few ways to compare useful life, but a popular one is the equivalent uniform annual cost (EUAC).

The RCA is done in the following steps:

1. Identify alternatives (reasonable and available), at least: defender, ...best challenger.
2. Determine whether good defender marginal cost data are available.
3. If data are available, see if the marginal costs are increasing.
4. If increasing, then compare defender's next-year marginal cost with challenger's EUAC. (Use analysis Technique 1, described below.)
5. If not increasing, then must find lowest EUAC for defender and compare it with the challenger's EUAC at its minimum cost life. (Use analysis Technique 2, discussed below.)
6. If defender marginal cost data are not available, must find EUAC over given equipment life. (Use analysis Technique 3, described below.)
7. Compare defender's EUAC over its remaining life with the challenger's EUAC at its minimum cost life.
8. Calculate the EUAC for each value of the useful life (e.g., $n=1$, $n=2$, and $n=3$).
9. Time period at which the EUAC is minimized is the minimum cost life (economic useful life).

Replacement Cost Analysis Example

Assuming the conditions in Table 15.4, the engineer must decide how to address failing equipment.
Goal: Narrow replacement down to one challenger.
Approach: Evaluate each candidate challengers at its minimum cost life.
This can be done using the previously discussed alternative evaluation approaches, that is, NPW, EUAC, IRR, or BCR.
Next, find minimum cost life. An asset can have several possible lives, depending on frequency of asset replacement. The life that minimizes the asset's EUAC is known as the asset's minimum cost life.
Example:
If initial cost (P) is $7500, arithmetic gradient maintenance cost (G) is $900, uniform cost (A) is $500, and arithmetic gradient operating cost (G) is $400, at an 8% interest rate.
There is no salvage value.
The "n" is changing depending on the year, on an arithmetic gradient, that is, it is changing each year (n=1, n=2, ...).
The 15-year spreadsheet is shown in Table 15.5.
The second column can be constructed in Excel as $7500(A/P,8%,n); third column, $900(A/P,8%,n); and fourth column, $500+400 (A/P,8%,n). The fifth column is the sum of the values in columns 2, 3, and 4.
Thus, the lowest total EUAC is year 4, so this is the economic useful life, and minimum cost life is when $n=4$ years (see Fig. 15.12).

TABLE 15.4 Defender and challenger example from P. R. Rosenkrantz, Cal Poly Pomona [20]

	Where		Compare	
	Defender marginal cost	Defender marginal cost	Defender	Best challenger
1	Available	Increasing	Next-year marginal cost	EUAC at minimum cost life
2	Available	Not increasing	EUAC at minimum cost life	EUAC at minimum cost life
3	Not available		EUAC over remaining useful life	EUAC at minimum cost life

TABLE 15.5 Spreadsheet for equivalent uniform annual cost (EUAC) example from P. R. Rosenkrantz, Cal Poly Pomona [20]

Year	EUAC of capital recovery costs	EUAC of maintenance and repair costs	EUAC of operating costs	EUAC total	Interest rate
Initial year	−7500	0	−500		8%
Arithmetic gradient		−900	−400		
1	$8100.00	$0.00	$500.00	$8600.00	
2	$4205.77	$432.69	$692.31	$5330.77	
3	$2910.25	$853.87	$879.50	$4643.62	
4	$2264.41	$1263.56	$1061.58	$4589.55	<—— MIN
5	$1878.42	$1661.82	$1238.59	$4778.84	
6	$1622.37	$2048.71	$1410.54	$5081.62	
7	$1440.54	$2424.30	$1577.47	$5442.31	
8	$1305.11	$2788.67	$1739.41	$5833.19	
9	$1200.60	$3141.93	$1896.41	$6238.94	
10	$1117.72	$3484.18	$2048.53	$6650.43	
11	$1050.57	$3815.55	$2195.80	$7061.93	
12	$995.21	$4136.17	$2338.30	$7469.68	
13	$948.91	$4446.19	$2476.08	$7871.18	
14	$909.73	$4745.75	$2609.22	$8264.69	
15	$876.22	$5035.01	$2737.78	$8649.02	

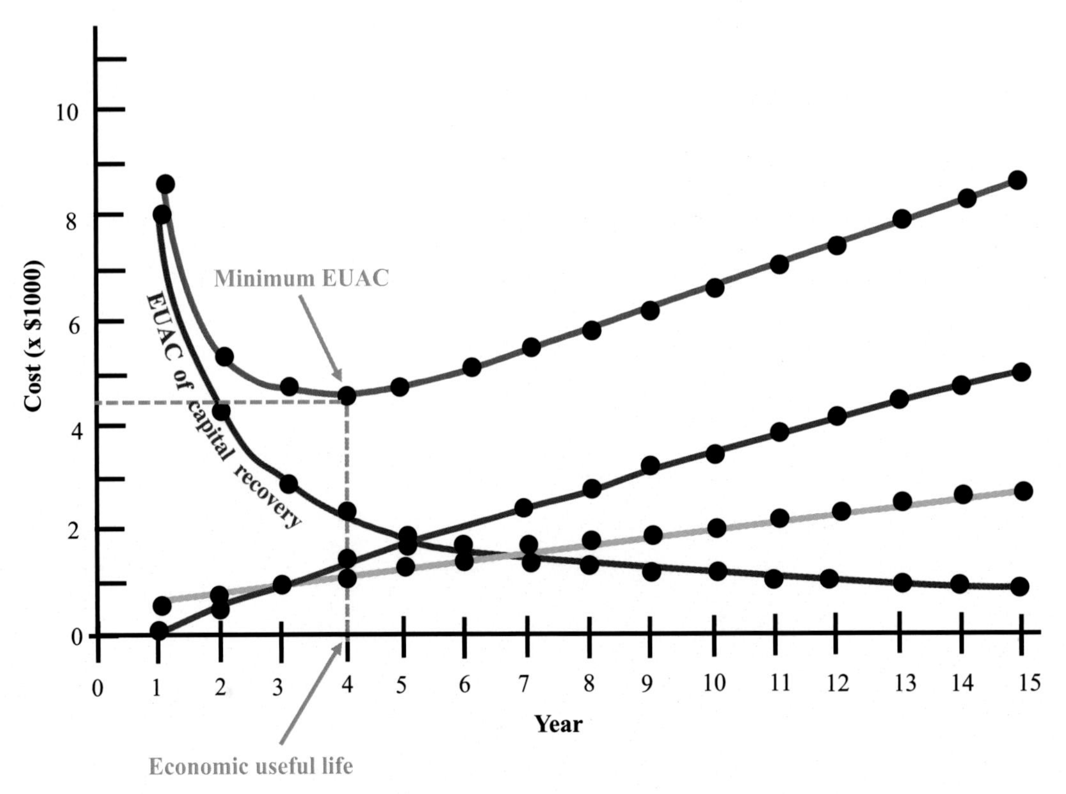

FIG. 15.12 Minimum equivalent uniform annual cost (EUAC) used to find the economic useful life in example from P. R. Rosenkrantz, California Polytechnic—Pomona [20].

Next, find the year-by-year costs for keeping an asset (i.e., marginal costs). This includes the loss in value of the asset for keeping it (usually per year), the lost interest from tying up funds in this asset. The costs are directly related to the asset, for example, expenses like insurance and O&M.
Example of marginal cost comparison:
New item:

Expense	Cost	Years	Incremental cost thereafter
Investment cost	$25,000.00	0	$ –
Annual O&M cost	$ 2000.00	1	$ 500.00
Annual insurance cost	$ 5000.00	3	$ 1500.00
Useful life (years)	7		
MARR	15%		

Year	Market value	Loss in market value	Foregone interest	O&M cost	Insurance cost	Total marginal cost
0	$ 25,000.00					
1	$ 18,000.00	$ 7000.00	$3750.00	$ 2000.00	$ 5000.00	$17,750.00
2	$ 13,000.00	$ 5000.00	$2700.00	$ 2500.00	$ 5000.00	$15,200.00
3	$ 9000.00	$ 4000.00	$1950.00	$ 3000.00	$ 5000.00	$13,950.00
4	$ 6000.00	$ 3000.00	$1350.00	$ 3500.00	$ 6500.00	$14,350.00
5	$ 4000.00	$ 2000.00	$ 900.00	$ 4000.00	$ 8000.00	$14,900.00
6	$ 3000.00	$ 1000.00	$ 600.00	$ 4500.00	$ 9500.00	$15,600.00
7	$ 2500.00	$ 500.00	$ 450.00	$ 5000.00	$11,000.00	$16,950.00

Defender:

Expense	Cost	Years	Incremental cost thereafter
Investment cost	$15,000.00	0	$ –
Annual O&M cost	$10,000.00	1	$ 1500.00
Annual insurance cost	$ –		$ –
Useful life (years)	5		
MARR	15%		

Year	Market value	Loss in market value	Foregone interest	O&M cost	Insurance cost	Total marginal cost
0	$ 15,000.00					
1	$ 14,000.00	$ 1000.00	$2250.00	$10,000.00	$ –	$13,250.00
2	$ 13,000.00	$ 1000.00	$2100.00	$11,500.00	$ –	$14,600.00
3	$ 12,000.00	$ 1000.00	$1950.00	$13,000.00	$ –	$15,950.00
4	$ 11,000.00	$ 1000.00	$1800.00	$14,500.00	$ –	$17,300.00
5	$ 10,000.00	$ 1000.00	$1650.00	$16,000.00	$ –	$18,650.00
6						
7						

In this example, the marginal costs of the defender are constantly increasing, but the marginal costs of the challenger decrease for 3 years then start to increase. By year 3, the challenger marginal costs become less than the defenders and stay the way until the end of the defender's useful life (year 5).

RCA Techniques

Technique 1: If the defender marginal costs are increasing, replace when the marginal costs of ownership is more than the EUAC of the challenger. This is appropriate approach when replacement repeatability assumptions hold the following:

- Best challenger is available for all subsequent years, and its costs will remain unchanged.
- The period of service is infinite.
- If these are too restrictive, assume a more realistic, known, and limited project life.

Technique 2: If the replacement repeatability holds, compare the EUAC of the defender asset at its minimum cost life with the EUAC of the challenger at its minimum cost life.

Technique 3: Compare the defender's EUAC over the stated life with the minimum EUAC of the challenger. There may be a problem in defining first costs of the defender and challenger. Trade-in value is not suitable; market value is appropriate.

Any cost analysis is incomplete if it neglects tax. Thus, all relevant taxes will affect the remaining life of the defender, the economic life of the challenger, and defender versus challenger comparisons. After-tax replacement depends on the depreciation Modified Accelerated Cost Recovery System (MACRS) and asset market over time [20].

15.6 LIFE CYCLE COMPARISONS

The project evaluation techniques so far have helped the waste manage select from finite alternatives. Another aspect of evaluation is improving existing systems to provide the needed function or to eliminate problems that weaken that alternative. In life cycle assessment (LCA), alternatives assessment is a formal technique to characterize the hazards of substances based on a full range of human health and environmental information [21]. Changes in the life cycle made based on these assessments ameliorate the potential for unintended consequences a priori.

The current approach for alternative assessment shown in Table 15.6 goes beyond hazard alternatives to address benefits and utility provided by a waste management option, that is, steps in LCA stages may require helium (He), which is becoming less available. A thorough alternative assessment may identify steps with a different, more sustainable and safer chemical or even use a nonchemical technique to perform the same function. Thus, the steps in Table 15.6 may be applied to other substitutions and replacements; for example, energy use, labor and skill mix, and social justice.

Alternative assessment and other life cycle approaches could supplant many of the single approach and unsustainable approaches with more systematic, sustainable solutions. Note that Table 15.6 calls for stakeholder involvement and risk communication; for example, using expert elicitation techniques with other jurisdictions, universities, and governance to identify ways to eliminate the use of toxic chemicals before they reach the waste stream [22].

Although expert judgment is almost always applied at the outset of project evaluation, for example, deciding which alternatives are not worthy of further scientific investigation, it is also useful as the evaluation continues. Project evaluation can be a protracted, lengthy, and iterative process. Indeed, there may come a point where decision-makers ask the waste manager to "settle" on a less than optimal approach. In addition, after substantial time and psychic investment in narrowing down the alternative list to a few, well-understood options, it may be difficult to be completely objective. This may be a good time to bring in someone "cold" who can view the options more systematically and advise the waste manager as to whether the original functions and objectives are being met.

Comparing air pollution control projects exists in two distinct domains: scientific credibility and economic feasibility. The chapters up to this one have addressed the former. This chapter has considered the ways to do the latter and to incorporate economic efficiencies with considerations of risk, reliability, and resilience. Economic acceptability is useless if a project fails to meet standards of performance based on public health, environmental quality, and societal acceptance.

TABLE 15.6 Key steps in alternative assessment for substituting toxic chemicals in the life cycle of a process

Process	Action	Application to finite resources
Step 1: Determine Feasibility	Consider whether alternatives 1. are commercially available and cost-effective, 2. have the potential for an improved health and environmental profile, 3. are likely to result in lasting change, 4. to what extent stakeholders are interested	Consider whether the goods and services can be provided with lesser amounts of these resources
Step 2: Collect Information	Find data about and consider 1. how well characterized the possible alternatives are; 2. how much is known about the chemical manufacturing process; 3. the range of functional uses that the chemical serves; 4. the feedstock of contaminants and residuals from the production process; 5. the work of others who are exploring alternatives for the chemical of concern, similar chemicals, and functional uses Based on the analysis of this information and preliminary stakeholder consultation, develop a proposed project scope and an approach for developing the alternative assessment	Same, except collecting market and production information
Step 3: Convene Stakeholders	Stakeholders are drawn from entire supply chain and all life cycle stages of the chemical of concern. Involvement throughout the project helps to ensure that stakeholders contribute to, understand, and support the outcome, enhancing credibility and promoting adoption of the safer alternatives. Typical stakeholders include chemical manufacturers, product manufacturers, nongovernmental organizations, government agencies, academics, retailers, consumers, and waste and recycling companies. Chemical and technology innovators are critical members of the group	Same process but include additional stakeholders, for example, from similar industries
Step 4: Identify Viable Alternatives	Collect information about viability on a range of potential alternatives. The focus is on finding alternatives that are functional with minimum disruption to the manufacturing process. To identify the most likely alternatives, it may also include viability demonstrations by chemical and product manufacturers	Same
Step 5: Conduct Hazard Assessment	Based on the best data that are available from the literature or can be modeled, assign an adjective descriptor of hazard concern level of high, moderate, or low for each alternative across a range of end points, including acute and repeated dose toxicity, carcinogenicity and mutagenicity, reproductive and developmental toxicity, neurotoxicity, sensitization and irritation, acute and chronic aquatic toxicity, and persistence and bioaccumulation. In addition, we provide a qualitative description of potential endocrine activity	Rather than an assessment solely of potential hazards, including potential substitutions and utility of these substitutions in terms of providing the same goods and services as current processes

Source: D. Vallero, Translating Diverse Environmental Data into Reliable Information: How to Coordinate Evidence from Different Sources. Academic Press, 2017; data for two left columns from U.S. Environmental Protection Agency. Design for the Environment Alternatives Assessments. (2017). 15 May 2017. Available: https://www.epa.gov/saferchoice/design-environment-alternatives-assessments and right column from D.A. Vallero, T.M. Letcher, Engineering risks and failures: lessons learned from environmental disasters, Leadersh. Manag. Eng. 12 (4) (2012) 199–209; D.A. Vallero, T.M. Letcher, Unraveling Environmental Disasters, Newnes, 2012.

REFERENCES

[1] State implementation plans, Available: http://www.dec.ny.gov/chemical/8403, 2009.

[2] Part I e New Source Review Q&A, Available: http://www.epa.state.il.us/air/new-source-review/new-sourcereview, 2013.

[3] Final Rule: Implementation of the 2015 National Ambient Air Quality Standards for Ozone: Nonattainment Area Classifications Approach, 2018.

[4] Lead Review, Available: http://www.who.int/ifcs/documents/forums/forum5/nmr_lead.pdf, 2003.

[5] Pollutants and sources, Available: http://www.epa.gov/ttn/atw/pollsour.html, 2013.

[6] N. D. o. A. Q. Clark County, State Implementation Plan for PM. Appendix K: Rollback Methodology, Available: http://www.clarkcountynv.gov/Depts/AirQuality/Documents/Planning/SIP/PM10/App_K%E2%80%93Rollback_Methodology.pdf, 2001. Accessed 25 November 2013.

[7] 70(216) 40 CFR Part 51, Revision to the Guideline on Air Quality Models: Adoption of a Preferred General Purpose (Flat and Complex Terrain) Dispersion Model and Other Revisions; Final Rule. November 9, 2005, (2005).

[8] T.M. Letcher, D. Vallero, Excerpt From: Waste: A Handbook for Management, second ed., Academic Press, 2019.

[9] J.S. Harkness, B. Sulkin, A. Vengosh, Evidence for coal ash ponds leaking in the southeastern United States, Environ. Sci. Technol. 50 (12) (2016) 6583–6592.

[10] U.S. Department of Energy, Office of Environmental Management, New Technology Accelerates Schedule, Reduces Cost of SRS Coal Ash Cleanup, (2017), Available at: https://www.energy.gov/em/articles/new-technology-accelerates-schedule-reduces-cost-srs-coal-ash-cleanup.

[11] U.S. Environmental Protection Agency, Attachment B: Water Quality-Based Effluent Limits – Coal Combustion Waste Impoundments, U.S. Environmental Protection Agency, 2018. 2017 #978, 7 March 2018.

[12] U.S. Fish and Wildlife Service, Natural Resource Damage Assessment and Restoration for the Dan River Coal Ash Spill, 8 March 8 2018. Available: https://files.nc.gov/ncdeq/Coal%20Ash/documents/Coal%20Ash/20140609%20FINAL%20Dan%20River%20cooperative%20NRDAR%20factsheet.pdf, 2018.

[13] T. Brown, Engineering Economics and Economic Design for Process Engineers, CRC Press, 2016.

[14] Microsoft Support Office, Financial functions (reference), 17 May 2018. Available: https://support.office.com/en-us/article/financial-functions-reference-5658d81e-6035-4f24-89c1-fbf124c2b1d8?ui=en-US&rs=en-US&ad=US, 2018.

[15] Benefit cost analysis for transportation projects, Available: https://www.dot.state.mn.us/planning/program/benefitcost.html, 2012.

[16] J.L. Sorrels, T.G. Walton, Cost estimation: concepts and methodology, in: EPA Air Pollution Control Cost Manual, vol. 27711, Office of Air Quality Planning and Standards, U.S. Environmental Protection Agency, Research Triangle Park, NC, 2017.

[17] E. P. A. U.S, Guidelines for preparing economic analyses, EPA 240-R-00-003, US Environmental Protection Agency, Washington, DC, 2014. Available: https://www.epa.gov/sites/production/files/2017-08/documents/ee-0568-50.pdf. Accessed 27 August 2018.

[18] K.B. Schnelle Jr., R.F. Dunn, M.E. Ternes, Air Pollution Control Technology Handbook, CRC Press, 2015.

[19] Benefit-Cost Analysis Guidance for TIGER and INFRA Applications, Available: https://www.transportation.gov/sites/dot.gov/files/docs/mission/office-policy/transportation-policy/284031/benefit-cost-analysis-guidance-2017_1.pdf, 2017.

[20] P.R. Rosenkrantz, Chapter 12: Replacement, Available: https://www.cpp.edu/~rosenkrantz/egr403site/egr403online/egr403_sv15_chapter12.ppt, 2018.

[21] C.P. Weis, The value of alternatives assessment, Environ. Health Perspect. 124 (2016) A40.

[22] M.D. Wood, K. Plourde, S. Larkin, P.P. Egeghy, A.J. Williams, V. Zemba, I. Linkov, D.A. Vallero, Advances on a decision analytic approach to exposure-based chemical prioritization. Risk Anal. (2018), https://doi.org/10.1111/risa.13001.

FURTHER READING

[23] U.S. Environmental Protection Agency, NAAQS Table, 25 August 2018. Available: https://www.epa.gov/criteria-air-pollutants/naaqs-table, 2016.

[24] D.A. Vallero, Environmental impacts of energy production, distribution and transport, in: Future Energy, second ed., Elsevier, 2014, pp. 551–581.

[25] North Carolina Department of Environment and Natural Resources, Groundwater, Aquifers & Confining Beds, 6 March 2018. Available: https://www.ncwater.org/?page=560, 2018.

[26] J. Wittwer, Discount Factor Table for Excel: Summary of Discount Factor Formulas for TVM Calculations in Excel®, 17 May 2018. Available: https://www.vertex42.com/ExcelArticles/discount-factors.html, 2018.

[27] D. Vallero, Translating Diverse Environmental Data into Reliable Information: How to Coordinate Evidence from Different Sources, Academic Press, 2017.

[28] U.S. Environmental Protection Agency, Design for the Environment Alternatives Assessments, 15 May 2017. Available: https://www.epa.gov/saferchoice/design-environment-alternatives-assessments, 2017.

[29] D.A. Vallero, T.M. Letcher, Engineering risks and failures: lessons learned from environmental disasters, Leadersh. Manag. Eng. 12 (4) (2012) 199–209.

[30] D.A. Vallero, T.M. Letcher, Unraveling Environmental Disasters, Newnes, 2012.

Chapter 16

Reliability and failure

16.1 INTRODUCTION

Addressing air pollution requires that any design, siting, construction, and operation of systems be reliable. The fundamental unit of reliability is the item, that is, a functional or structural unit of arbitrary complexity, for example, a component (such as a part or device), assembly, equipment, subsystem, or system [1]. The system in reliability differs from a thermodynamic system. A reliability system is the highest level of intergration of the item. Thus, the item is the target of success or failure.

Air pollution control and other equipment must meet four performance factors [2]:

1. Capability
2. Efficiency
3. Availability
4. Reliability

An item must be capable of meeting functional requirements. It must also be efficient, that is, effectively and feasibly meet its design objectives. The major influence on an item's capability and efficiency is its quality, that is, the extent to which an item is "fit for purpose."

Even a capable and efficient item is not reliable if it is not sufficiently available. It must be available when needed, that is, minimal downtime. For example, there are many pollution control devices that work well under laboratory conditions but when put into operation are less available than traditional technologies. This is not unique to air pollution equipment. The real world has variables that are not apparent at the outset. Only after deployment do certain problems become apparent.

Reliability is an expression of the probability that a system or item will perform as designed under specified conditions and for a specified period of time.

In addition to ensuring that simple and complex systems are reliable, they also must also be available, maintainable, and safe (RAMS). This means that reliability engineering is a part of a comprehensive protocol that ensures that these systems are cost- and time-effective, that is, integrated with quality assurance (QA), project activity tracking, and concurrent engineering.

Reliability is not an absolute quantity, but a conditional quantity. For example, consider a facility that uses two cyclones to remove particulate matter (PM), with one failing, that is, is no longer capable of functioning as designed, after 10 years and the other failing after 5 years. Is the second cyclone less reliable than the first? The answer is not necessarily. The longer-lasting one has a longer operational period but could be less reliable. Let us assume that the first one was designed to last 15 years and the second for 4 years. In this case, the first is unreliable, and the second is reliable, based on the design time period. This could occur if the first was in a facility that only operated once a month and/or collected dust that is relatively noncorrosive, whereas the second operated almost continuously for the 5-year period and/or collected corrosive dust.

Also, reliability is driven by operational efficiency. For example, if the second cyclone included a regular replacement, for example, daily, of a part that would otherwise have failed. The proper operation and maintenance (O&M) is one of the specified operational conditions. This is much like maintenance of a motor vehicle. If the crankcase oil is supposed to be changed every 5000 miles and the timing belt every 50,000 miles, these are some of the specified maintenance conditions that would lead to a stated operational time period for the engine (e.g., 10 years or 250,000 miles). If the oil is changed every 10,000 miles and the engine failed after 75,000 miles, the engine's reliability is not known, because the O&M conditions were not met.

The engineer must ensure that any engineered systems have a sufficient optimal operational time frame, where the failure rate is assumed to be at its lowest. Reliability is the probability that something that is in operation at time *0* (t_0) will still be operating until the designed life (time $t = (t_t)$). As such, it is also a measure of accountability.

Air Pollution Calculations. https://doi.org/10.1016/B978-0-12-814934-8.00016-8

Unreliable systems range from occasional and low-impact failures, such as the simple nuisance resulting from fencing that is too low to capture windblown litter, to catastrophes, for example, the loss of life from an explosion in an air pollution control device. People living near a proposed facility must be ensured the systems will work and will not fail or that in the event of a failure, the harm would be properly managed with an acceptable contingency plan [3]. The probability of a failure per unit time is known as the "hazard" rate. Engineers may recognize it as a "failure density," or $f(t)$. This is a function of the likelihood that an adverse outcome will occur, but note that it is not a function of the severity of the outcome. The $f(t)$ is not affected by whether the outcome is very severe (such as an explosion where people are hurt or killed) or relatively benign (minor leaf damage from a plume or litter on the side of a road). The likelihood of failure during a given time interval can be found by integrating the hazard rate over a defined time interval:

$$P\{t_1 \le T_f \le t_2\} = \int_{t_1}^{t_2} f(t)dt \tag{16.1}$$

where T_f = time of failure.

Thus, the reliability function $R(t)$ of a system at time t is the cumulative probability that the system has not failed in the time interval from t_0 to t_t:

$$R(t) = P\{T_f \ge t\} = 1 - \int_0^t f(x)dx \tag{16.2}$$

Reliability can be improved by extending the time (increasing t_t), thereby making the system more resistant to failure. For example, proper engineering design of a landfill barrier can decrease the flow of contaminated water between the contents of the landfill and the surrounding aquifer, for example, a velocity of a few microns per decade. However, the barrier does not completely eliminate failure, that is, $R(t) = 0$; it simply protracts the time before the failure occurs (increases T_f).

Eq. (16.2) illustrates built-in vulnerabilities, such as unscientifically sound facility siting practices or the inclusion of inappropriate design criteria, for example, cheapest land. Such mistakes or malpractice shortens the time before a failure. Thus, reliability is also a term of efficiency. Failure to recognize these inefficiencies upfront leads to premature failures (e.g., the loss of life, loss of property, law suits, and a public that has been ill-served). Risk is really the probability of failure, that is, the probability that our system, process, or equipment will fail. Thus, risk and reliability are two sides of the same coin. The common graphic representation of engineering reliability is the "bathtub" curve (see Fig. 16.1). The U-shape indicates that failure will more likely occur at the beginning, so-called infant mortality, and near the end of the life of a system, process, or equipment. Indeed, failure can occur even before acquiring equipment or starting a new route. In fact, many problems occur during the planning and idea stage. A great idea may be dismissed prematurely.

Predicting and managing desirable air quality outcomes depend on credible data and information from which to extrapolate and interpolate from the known to the less known. This is the nature of risk management models, science-based

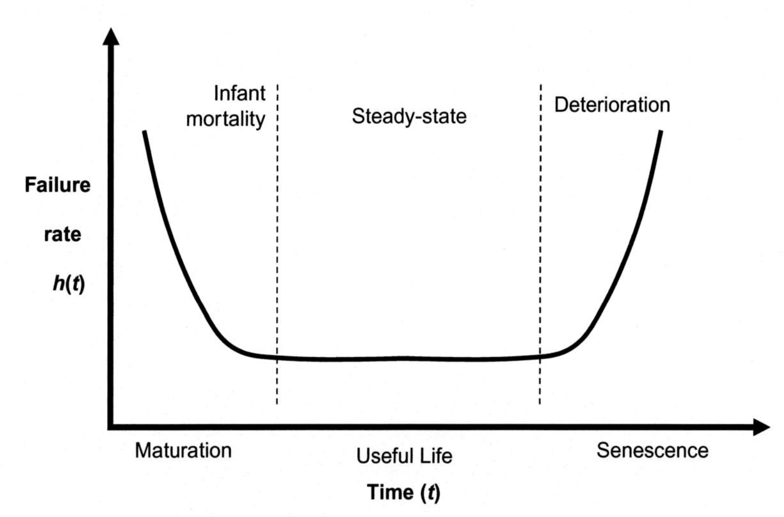

FIG. 16.1 Prototypical reliability curve. The highest rates of failure, $h(t)$, occur during the earlies and latest stages. Well-designed systems extend the useful life. The air quality manager and engineer must build in factors of safety, including closer monitoring more extensive safeguards when the failure rates are likely higher at the beginning and toward the end of the steady-state period. *(Source: D. Vallero, Paradigms Lost: Learning from Environmental Mistakes, Mishaps and Misdeeds. Butterworth-Heinemann, 2005.)*

decision-making, and forward trajectories. This is the flip side of trying to see why something happened, which is done using root cause analysis, decision trees, deconstruction/reconstruction, and failure analysis [4].

These are important tools to evaluate of reliability of air pollution control systems. For example, a common technique in determining the availability and reliability, $R(t)$, of a system is to calculate the mean time between failures (MTBF) [5]:

$$R(t) = \exp\left(-\frac{t}{\text{MTBF}}\right) = \exp(-\lambda t) \tag{16.3}$$

where t is the operation time and λ is the reciprocal of MTBF.

16.2 FAILURE CLASSIFICATION

Engineers are familiar with failure, which is effectively the opposite of reliability. Failure can be stated as a probability. Survival analysis is an approach used to look at data up to the time of interest. Since our time of interest is the time until failure, the survival function ($S(t)$) is the same as $R(t)$. A hazard is the conditional probability of failure of a currently working system. Thus, the failure can be expressed as a hazard function, that is, the instantaneous rate at which events happen if there are no previous events. The hazard function can be measured over time, for example, the lifetime of a facility's components, for example, the cells in a landfill or the kiln in an incinerator. Although Fig. 16.1 represents the classic failure curve, the change of the hazard function with time produces various patterns (see Fig. 16.2), depending on the type of facility and equipment, as well as the operating conditions.

Recently, numerous software packages have become available to the air quality manager to calculate $R(t)$; notably, reliability, availability, maintainability, and safety (RAMS) software [6]. This goes by numerous terms, but air pollution control and prevention fall under the province of "industrial ecology" [7–9]. The industrial ecology process includes identifying waste streams, the contaminants of concern, and ways to address these concerns, for example, conversion technologies. The process also evaluates, together, the economic and environmental effects of using wastes and transformations that may occur during transport, handling, and processing. Depending on the various air pollution sources, the optimal methods will vary.

The idealized patterns in Fig. 16.2 may be more applicable to engineered systems than as representative of ecosystem response to stresses. In fact, concepts like carrying capacity, resilience, and sustainability are better indicators of ecosystem reliability. For example, an ecosystem may be functioning quite well until it reaches a stress threshold, at which point it may "crash" and no longer function. This is lesson of *Tragedy of the Commons*, espoused by Garrett Hardin (1915–2003), which has become a "must-read" in every introductory ecology course. In this article, Hardin imagines an English village with a common area where everyone's cow may graze. The common is able to sustain the cows, and village life is stable, until one of the villagers figures out that if he gets two cows instead of one, the cost of the extra cow will be shared by everyone, while the profit will be his alone. So he gets two cows and prospers, but others see this and similarly want two cows. If two is possible, why not three—and so on—until the village common is no longer able to support the large number of cows, that is, the ecosystem will have exceeded its carrying capacity (see Fig. 16.3). Given the complexity of ecosystems, a crash may still occur with other changes in conditions, for example, drought, introduced and opportunistic species, dredging, landfill

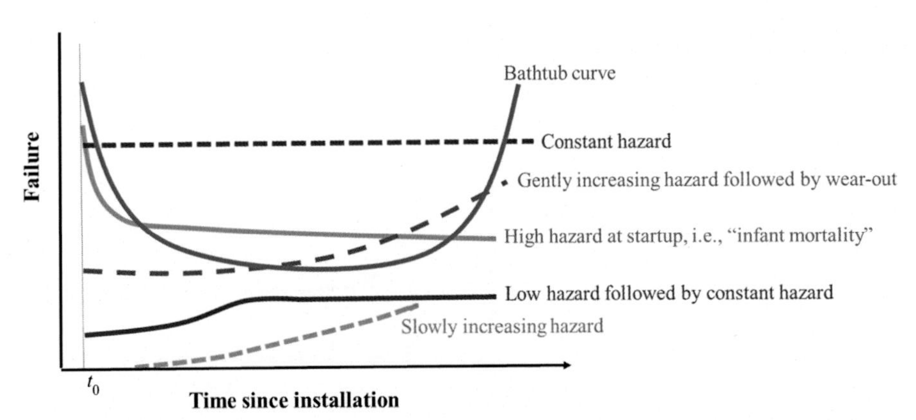

FIG. 16.2 Idealized patterns of reliability. *(Source: T. M. Letcher, D. Vallero, Excerpt From: Waste: A Handbook for Management, 2nd ed., Academic Press, 2019.)*

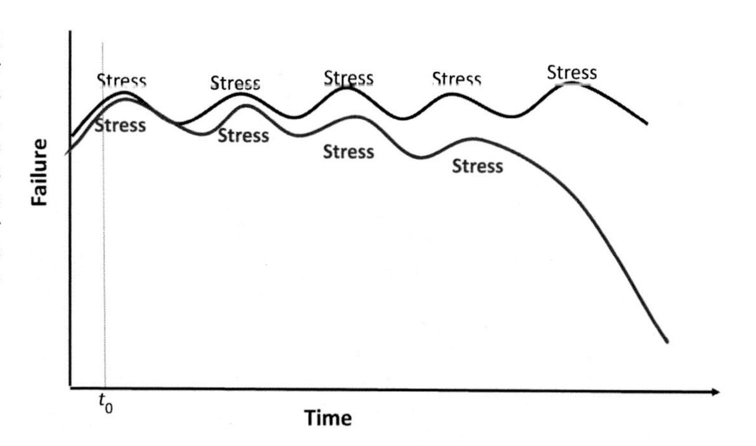

FIG. 16.3 Idealized failure of an ecosystem incremental stresses. The *black line* represents a system with relatively similar stresses over time, for example, equal number of grazing animals in Hardin's *Tragedy of the Commons*, where the system recovers. The *red line* represents a system in which the stresses increase over time, for example, an increased number of grazing animals every year. At some point, the system crashes. Note that the stresses do not have to be the same. For example, the number of grazing animals may stay the same, but drought or changes in biodiversity (e.g., from introduced plant species) may exacerbate the failure pattern, may be similar to the *red line*.

leaching, wetland destruction, and deposition of pollutants from incineration. Thus, even though the waste manager is not likely to be engaged in animal husbandry, Hardin's caution applies equally to solid waste decisions.

16.3 MARGINS OF SAFETY

The reliability of a system must be matched to an adequate margin of safety. So, if the municipal solid waste incinerator fails and needs to be replaced after 30 years, few reasonable people would likely complain. If it fails after 17 years, when stakeholders had been assured 99.9999% efficiency for 20 years, complaints may be justified, but new costs of repair may be accepted if failure rate and margin of safety had been fully disclosed and the reason for the failure adequately explained. If it fails after 10 years but its specifications require 20 years, this may very well constitute an unacceptable risk. Obviously, other factors must be considered, such as sensitive subpopulations. For example, if the failure is the excessive emissions of particulates, this may do little harm to the general population, but in certain subgroups (e.g., asthmatic children), the reliability is not acceptable. Of course, if the particulates contain carcinogens and other toxic compounds, the chronic harm would be unacceptable for everyone. The engineer and risk manager must fully disclose such limitations and recommend actions to address these failures.

Reliability engineering, a discipline within engineering, considers the expected or actual reliability of a process, system or piece of equipment to identify the actions needed to reduce failures, and once a failure occurs how to manage the expected effects from that failure. Thus, reliability is the mirror image of failure. Since risk is really the probability of failure (i.e., the probability that our system, process, or equipment will fail), risk and reliability are two sides of the same coin. The air pollution control system must have a long steady-state, useful life in the midsection of the bathtub curve compared with the regions of higher failure rates.

Another way to visualize reliability is to link potential causes to effects. Cause-and-effect diagrams (also known as *Ishikawa diagrams*) identify and characterize the totality of causes or events that contribute to a specified outcome event. The "fishbone" diagram (see Fig. 16.4) arranges the categories of all causative factors according to their importance (i.e., their share of the cause). The construction of this diagram begins with the failure event to the far right (i.e., the "head" of the fish), followed by the spine (flow of events leading to the failure). The "bones" are each of the contributing categories. This can be a very effective tool in explaining failures to clients and the public. Even better, the engineer may construct the diagrams in "real time" at a design meeting. This will help to open the design for peer review and will help get insights, good and bad, early in the design process [10].

The premise behind cause-and-effect diagrams like the fishbones and other failure logic is that all the causes must connect through a logic gate. This is not always the case, so another more qualitative tool may need to be used, such as the Bayesian belief network (BBN). Like the fishbone, the BBN starts with the failure (see Fig. 16.5). Next, the most immediate contributing causes are linked to the failure event. The next group of factors that led to the immediate causes is then identified, followed by the remaining contributing groups. This diagram helps to identify and weight the contributing factors, allowing for investigations of how one group of factors impacts the others.

As was the case for risk, the engineering and scientific communities have may define reliability in nuanced ways. Environmental engineering and other empirical sciences also use the term "reliability," to indicate quality, especially for data derived from measurements, including health data. In this use, reliability is defined as the degree to which measured results

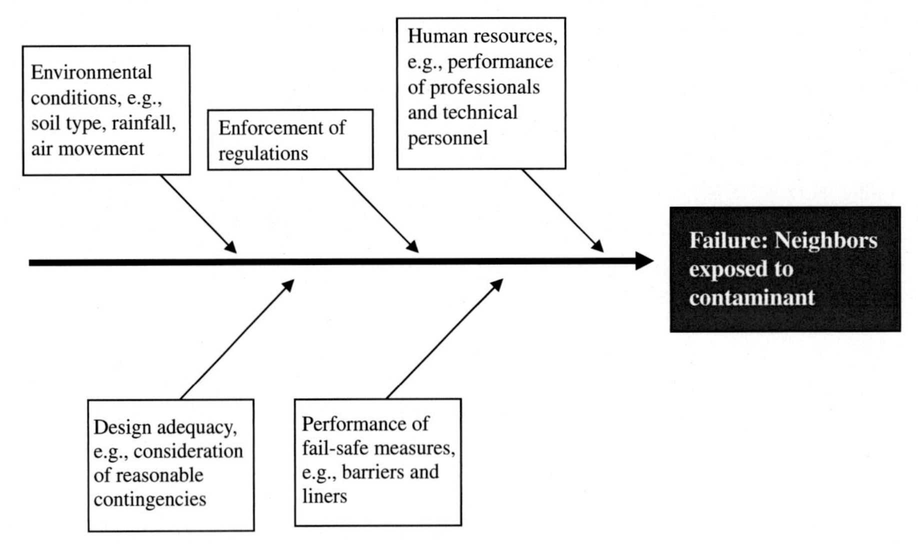

FIG. 16.4 Fishbone reliability diagram showing contributing causes to an adverse outcome (exposure to an environmental contaminant). *(Reproduced with permission from D.A. Vallero, P.A. Vesilind, Socially Responsible Engineering: Justice in Risk Management. Wiley Online Library, 2007.)*

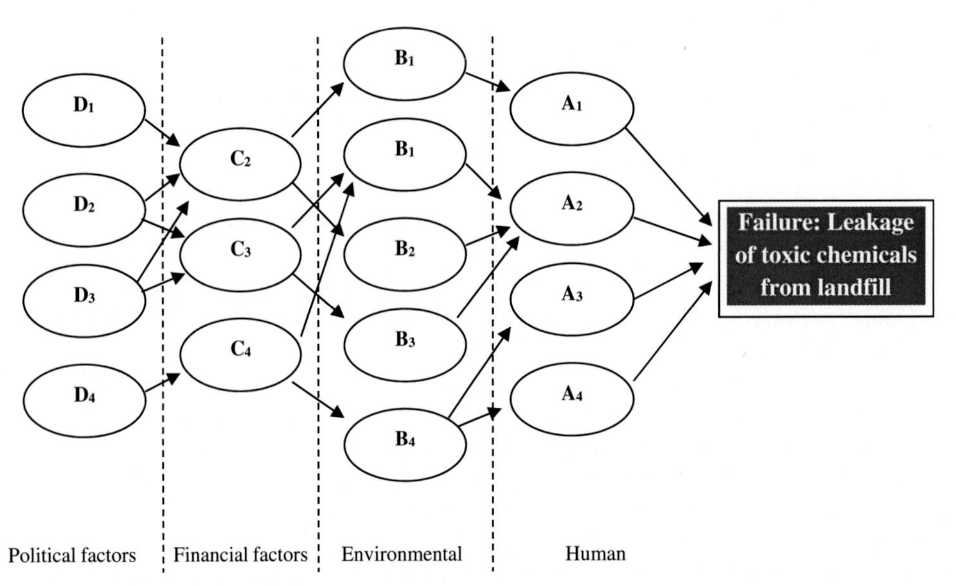

FIG. 16.5 Bayesian belief network, with three groups of contributing causes leading to a failure. *(Reproduced with permission from D.A. Vallero, P.A. Vesilind, Socially Responsible Engineering: Justice in Risk Management. Wiley Online Library, 2007.)*

are dependable and consistent with respect to the study objectives, for example, stream water quality. This specific connotation is sometimes called "test reliability" in that it indicates how consistent measured values are over time, how these values are compared with other measured values, and how they differ when other tests are applied. Test reliability, like engineering reliability, is a matter of trust. As such, it is often paired with test validity, that is, just how near to the true value (as indicated by some type of known standard) the measured value is. The less reliable and valid the results, the less confidence scientists and engineers have in interpreting and using them. This is very important in engineering communications generally and risk communications specifically.

It is important to keep in mind that, even if the waste manager is efficient at waste handling, there is always the potential for attendant, undesired outcomes. Furthermore, even if a project is acceptable, there can still be an unacceptable cumulative outcome when viewing it in combination with all the other "acceptable" projects. In other words, waste management must recognize their projects within concept of systematic success. A multistep process can be used to approach the various types of risks that may result from any engineering application [11] but particularly in view of the impact on complex ecosystems:

1. Properly accounting for uncertainty [12]
2. Basing risks on different types of information and understanding the limitations of using other peoples' data
3. Comparing risks that may result consistently among the alternative actions [13, 14]
4. Employing a tiered, intelligent, adaptive, and flexible approach to risk assessment; often beginning with a higher-tier, scoping effort and adding needed details as information is continuously reevaluated [15, 16]
5. Evaluating the value of information on which decisions will be based, including assignment of quality and relevance thresholds below which the data and information should be avoided or, at a minimum, given appropriate caveats
6. Determining the approach to validate and evaluate the models and other tools used to assess the risks [17]

Considering the cumulative risks from exposures to all types of stressors, that is, physical, chemical, and biological [18, 19], waste management decisions must consider the effect the actions will have on the integrity of ecosystems in time and space. A practical means of estimating this is to apply ecosystem indices and models [20].

Failure may occur in many forms and from many sources. A dam break or oil leak is an engineering failure, as is exposure to carcinogens in the air, water, and food. The former are examples more directly under the engineer's span of control, whereas the latter are indirect results of failures, that is, "second order" engineering failures, if you will. A system that protects one group of people at the expense of another is a type of failure.

Failure varies in kind, degree, and extent. Human-induced or human-contributed disasters can result from mistakes, mishaps, and misdeeds. The terms all include the prefix "mis-" derived from Old English, "to miss." This type of failure applies to numerous ethical failures. However, the prefix "mis-" can connote that something that is done "poorly," that is, a mistake. It may also mean that an act leads to an accident because the original expectations were overtaken by events, that is, a mishap. A mishap can occur as a result of not upholding the levels of technical competence called for by their field. Medical and engineering codes of ethics, for example, include tenets and principles related to competence, such as only working in one's area of competence or specialty. Finally, "mis-" can suggest that an act is immoral or ethically impermissible, that is, a misdeed. Interestingly, the theological derivation for the word "sin" (Greek: *hamartano*) means that when a person has missed the mark, that is, the goal of moral goodness and ethical uprightness, that person has sinned or has behaved immorally by failing to abide by an ethical principle, such as honesty and justice. Bioethical failures have come about by all three means. The lesson from Santayana is that we must learn from all of these past failures. Learning must be followed by new thinking and action, including the need to forsake what has not worked and shift toward what needs to be done.[1]

Let us consider a few types familiar to engineers, particularly with regard to their likelihood of contributing to a disaster.

16.3.1 Miscalculations

Engineers and scientists err due to their own miscalculations, such as when parentheses are not closed in computer code, leading to errors in predicting the extent of an air pollutant plume or the number of people exposed to an air pollutant. Miscalculation goes beyond math errors, such as when engineers do not correctly estimate the corrosivity that occurs from unexpected conditions. Assuming best case conditions when there is a possibility of treating corrosive pollutants could lead to disappointed clients and regulators when the actual conditions lead to premature degradation of fabric filters or electrostatic precipitators devices (e.g., not properly accounting for fatigue of materials resulting from high temperature and pH in condensed water). Such mistakes are completely avoidable if the physical sciences and mathematics are properly applied and communicated.

Best practices and good management can prevent problems by catching a miscalculation or other mistakes in the quality assurance/quality control (QA/QC) program. Even when problems do arise, reliable QA/QC programs can greatly lessen the impact of a failure, which would otherwise be disastrous.[2] An illustrative engineering case, although not an air pollution example, involved William LeMessurier, a renowned structural engineer. He was a principal designer of the Citicorp Tower in Manhattan, NY. The Citicorp Tower was constructed using LeMessurier's diagonal-bracing design that made the building unusually light for its size, completed in 1977.[3] This technique also unfortunately increased the building's tendency to sway in the wind, which was addressed by installing a tuned-mass damper (including a 400 ton concrete block

1. D. Vallero, Environmental Biotechnology: A Biosystems Approach, Elsevier Academic Press, Burlington, MA, 2009.

2. For example, Vallero made a six-order-of-magnitude mistake in his book, *Environmental Contamination.* In giving an example of units needed to calculate carbon monoxide concentrations, he had the wrong sign on an exponent, so instead of the value taken to the power of +3, it was taken to the power of −3. He thinks about this mistake frequently when doing calculations. It continues to be a lesson in humility.

3. National Academy of Engineering, Part 3: The Discovery of the Change from Welds to Bolts" Online Ethics Center for Engineering 6/23/2006. (2006). www.onlineethics.org/Topics/ProfPractice/Exemplars/BehavingWell/lemesindex/3.aspx (Accessed 02 November 2011).

floated on pressurized oil bearings) that was installed at the top to combat the expected slight swaying. During construction, without apprising LeMessurier, contractors thought that welding was too expensive and decided instead to bolt the braces. When he became aware of the change, however, LeMessurier thought that the change posed no safety hazard. He changed his mind over the next month, when new data indicatied that the switch from welds to bolts compounded another danger with potentially catastrophic consequences.

When LeMessurier recalculated the safety factor taking account of the quartering winds and the actual construction of the building, he discovered that the tower, which he had intended to withstand a thousand-year storm, was actually vulnerable to a 16-year storm. This meant that the tower could fail under meteorologic conditions common in New York on average every 16 years. Thus, the miscalculation completely eliminated the factor of safety. The disaster was averted after LeMessurier notified Citicorp executives, among others. Soon after the recalculation, he oversaw the installation of metal chevrons, welded over the bolted joints of the superstructure to restore the original factor of structural safety.

Miscalculation can be difficult to distinguish from negligence, given that in both, competence is an element of best practice. As humans, however, we all make arithmetic errors. Certainly, the authors have miscalculated far too frequently during our careers. The distinguishing features deeming the miscalculation as unacceptable are the degree of carelessness and the extent and severity of consequences.

Even a small miscalculation is unacceptable if it has the potential of either large-scale or long-lived negative consequences. At any scale, if the miscalculation leads to any loss of life or substantial destruction of property, it violates the first canon on the engineering profession to hold paramount the public's safety, health, and welfare, articulated by the National Society of Professional Engineers[4]:

Engineers, in the fulfillment of their professional duties, shall hold paramount the safety, health, and welfare of the public. To emphasize this professional responsibility, the engineering code includes this same statement as the engineer's first rule of practice.

16.3.2 Improper attention to natural circumstances

Failure can occur when factors of safety are exceeded due to extraordinary natural occurrences. Engineers can, with fair accuracy, predict the probability of failure due to natural forces like wind loads, and they design the structures for some maximum loading, but these natural forces can be exceeded. Engineers design for an acceptably low probability of failure—not for 100% safety and zero risk. However, tolerances and design specifications must be defined as explicitly as possible.

The tolerances and factors of safety have to match the consequences. A failure rate of 1% may be acceptable for a household compost pile, but it is grossly inadequate for bioreactor performance. And the failure rate of devices may spike up dramatically during an extreme natural event (e.g., power surges during storms). Equipment failure is but one of the factors that lead to uncontrolled environmental releases. Conditional probabilities of failure should be known. That way, backup systems can be established in the event of extreme natural events, like hurricanes, earthquakes, and tornados. If appropriate, contingency planning and design considerations are factored into operations, the engineer's device may still fail, but the failure would be considered reasonable under the extreme circumstances.

16.3.3 Critical path failure

No engineer can predict all of the possible failure modes of every structure or other engineered device, and unforeseen situations can occur. A classical, microbial case is the Holy Cross College football team hepatitis outbreak in 1969.[5] A confluence of events occurred that resulted in water becoming contaminated when hepatitis virus entered a drinking water system. Modeling such a series of events would probably only happen in scenarios with relatively high risks associated agents and conditions that had previously led to an adverse outcome.

In this case, a water pipe connecting the college football field with the town passed through a golf course. Children had opened a water spigot on the golf course, splashed around in the pool they created, and apparently discharged hepatitis virus into the water. A low pressure was created in the pipe when a house caught on fire and water was pumped out of the water pipes. This low pressure sucked the hepatitis-contaminated water into the water pipe. The next morning the Holy Cross football team drank water from the contaminated water line and many came down with hepatitis. The case is memorable

4. National Society of Professional Engineers, *NSPE Code of Ethics for Engineers*, Alexandria, Virginia, 2003. http://www.nspe.org/ethics/eh1-code.asp (Accessed 21 August 2005).

5. L.J. Morse, J.A. Bryan, J.P. Hurley, J.F. Murphy, T.F. O'Brien, T.F. Wacker, The Holy Cross Football Team hepatitis outbreak, J. Am. Med. Assoc. 219 (1972) 706–708.

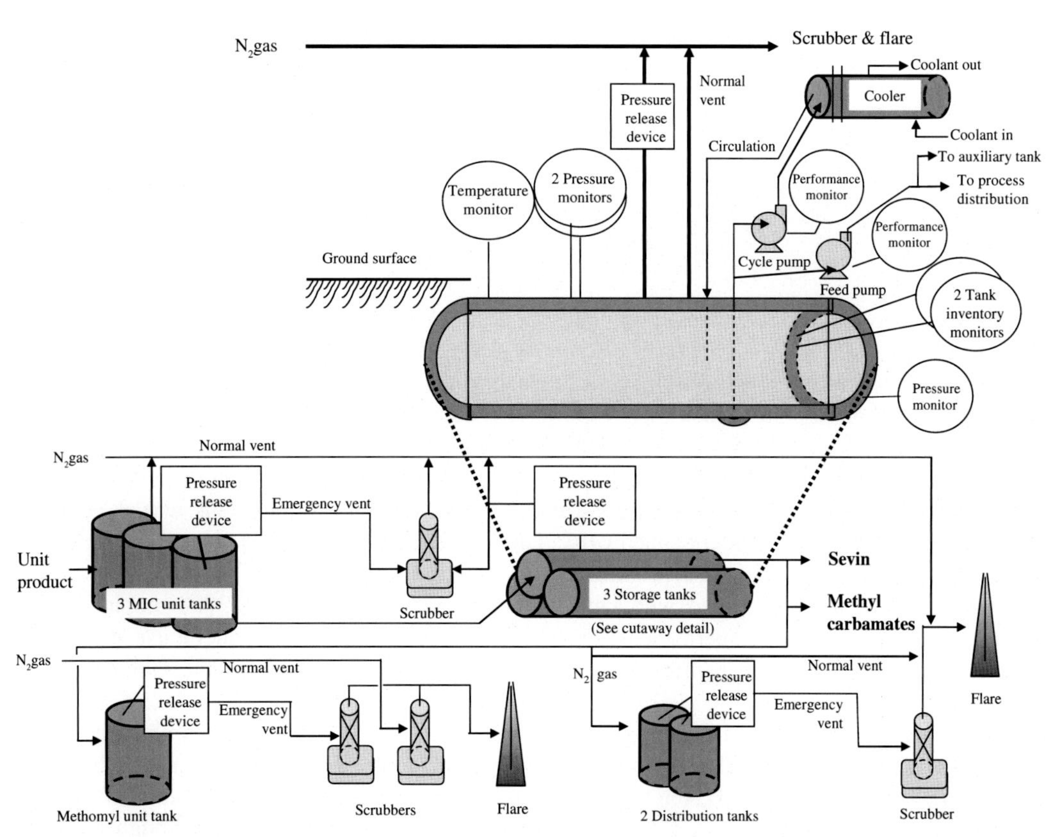

FIG. 16.6 Schematic of methyl isocyanate processes at the Bhopal, India, plant (c. 1984). *(Source: W. Worthy, Methyl isocyanate: the chemistry of a hazard, Chem. Eng. News 63 (66) (1985) 29.)*

because it was so highly unlikely—a combination of circumstances that were impossible to predict. Nevertheless, the job of engineers is to do just that, to try to predict the unpredictable and thereby to protect the health, safety, and welfare of the public.

This is an example of how engineers can fail, but may not be blamed for the failure, since such a set of factors had not previously led to an adverse action. If the public or their peers agree that the synergies, antagonisms, and conditional probabilities of the outcome could not reasonably be predicted, the engineer is likely to be forgiven. However, if a reasonable person deems that a competent engineer *should* have predicted the outcome, the engineer is to that extent accountable.

Indeed, there is always a need to consider risks by analogy, especially when related to complex, biological systems. Many complex situations are so dynamic and multifaceted that there is never an exact precedent for the events and outcomes for any real-world scenario. For example, every bioremediation project will differ from every other such project, but there are analogous situations related to previous projects that can be applied to a particular project. Are the same strains of microbes being used? Are the physical conditions, such as soil texture, and biological conditions, such as microbial ecology and plant root systems, ambient temperatures, and daily season variabilities, similar to those in previous studies? Are structurally similar compounds being degraded? Are the volumes of wastes and concentrations similar?

There are numerous examples of ignoring analogies to previous situations that led to adverse outcomes. The case of the industrial accident and air toxic cloud at Bhopal, India, illustrates this type of engineering failure. Perhaps, the biggest air pollution disaster of all time occurred in Bhopal in 1984 when a toxic cloud drifted over the city from the Union Carbide pesticide plant. This gas leak killed many people and permanently injured about tens of thousands more. Failure is often described as an outcome when not applying the science correctly (e.g., a mathematical error and an incorrect extrapolation of a physical principle). Another type of failure results from misjudgments of human systems. Bhopal had both.

The pesticide manufacturing plant in Bhopal demonstrates the chain of events that can lead to failure.[6,7] In fact, if one were to chart the Bhopal incident as a Bayesian Belief Network (Fig. 16.6), it is very nearly a worst-case scenario.

6. M.W. Martin, R. Schinzinger, Ethics in Engineering, 3rd ed., McGraw-Hill, New York, 1996.
7. C.B. Fledderman, Engineering Ethics, Prentice Hall, Upper Saddle River, NJ, 1999.

The plant, up to its closing, has produced the insecticides Sevin and Carbaryl since 1969, using the intermediate product methyl isocyanate (MIC) in its gas phase. The MIC was produced by the following reaction:

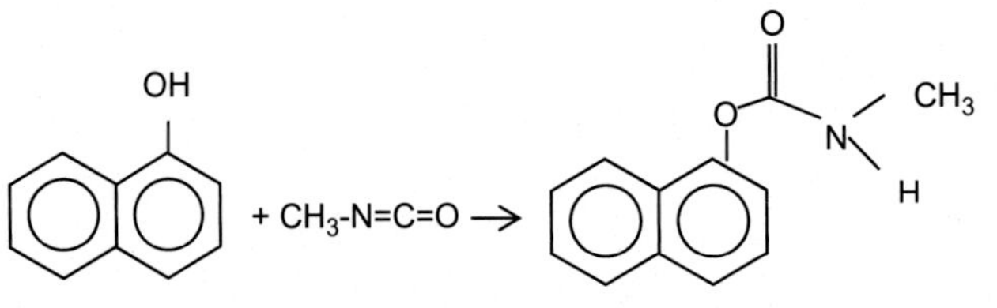

$$\text{OH} \quad + \; CH_3\text{-N=C=O} \rightarrow \quad \text{O} \quad \text{O} \quad \text{N} \quad CH_3 \quad \text{H} \tag{16.4}$$

This process was highly cost-effective, involving only a single reaction step. The schematic of MIC processing at the Bhopal plant is shown in Fig. 16.6.

MIC is highly water reactive (see Table 16.1); that is, it reacts violently with water, generating a very strong exothermic reaction that produces carbon dioxide. When MIC vaporizes, it becomes a highly toxic gas that, when concentrated, is highly caustic and burns tissues. This can lead to scalding of nasal and throat passages, blinding, and the loss of limbs, as well as death.

On 3 December 1984, the Bhopal plant operators became concerned that a storage tank was showing signs of overheating and had begun to leak. The tank contained MIC. The leak rapidly increased in size, and within one hour of the first leakage, the tank exploded and released approximately 80,000{ts}lbs. (4{ts} × $10^{4\{ts\}}$kg) of MIC into the atmosphere. The human exposure to MIC was widespread, with a half million people exposed. Nearly 3000 people died within the first few days after the exposure; 10,000 were permanently disabled. Ten years after the incident, thousands of death claims had been filed, along with 870,000 personal injury claims. However, only $90 million of the Union Carbide settlement agreement had been paid out.

The most basic physical science event tree begins with the water reactivity. That is, the combination of H_2O and MIC resulted in a highly exothermic reaction. The rapid generation of the product of this reaction, carbon dioxide (CO_2), led to an explosive increase in pressure. The next step in the event tree was the release of 40 metric tons of MIC into the atmosphere. As of 2001, many victims had received compensation, averaging about $600 each, although some claims are still outstanding.

The Indian government had required that the plant be operated exclusively by Indian workers, so Union Carbide agreed to train them, including flying Indian workers to a sister plant in West Virginia for hands-on sessions. In addition, the

TABLE 16.1 Properties of methyl isocyanate (MIC)

Common name	Isocyanic acid, methyl ester, and methyl carbylamine
Molecular mass	57.1
Properties	Melting point: −45{ts}°C; boiling point: 43–45{ts}°C Volatile liquid Pungent odor Reacts violently with water and is highly flammable MIC vapor is denser than air and will collect and stay at low areas The vapor mixes well with air and explosive mixtures are formed May polymerize due to heating or under the influence of water and catalysts Decomposes on heating and produces toxic gases like hydrogen cyanide, nitrogen oxides, and carbon monoxide
Uses	Used in the production of synthetic rubber, adhesives, pesticides, and herbicide intermediates. It is also used for the conversion of aldoximes to nitriles
Side effects	MIC is extremely toxic by inhalation, ingestion, and skin absorption. Inhalation of MIC causes cough, dizziness, shortness of breath, sore throat, and unconsciousness. It is corrosive to the skin and eyes. Short-term exposures also lead to death or adverse effects like pulmonary edema (respiratory inflammation), bronchitis, bronchial pneumonia, and reproductive effects. The Occupational Safety and Health Administration's permissible exposure limit to MIC over a normal 8-h workday or a 40-h workweek is 0.05{ts}mg{ts}m^{-3}

Sources: US Chemical Safety and Hazards Board, http://www.chemsafety.gov/lib/bhopal.0.1.htr; Chapman and Hall, Dictionary of Organic Chemistry, 5th ed., vol. 4, Mack Printing Company, USA, 1982; and T.W Graham, Organic Chemistry, 6th ed., John Wiley & Son, Inc, Canada, 1996.

company required that US engineering teams make periodic on-site inspections for safety and quality control, but these ended in 1982, when the plant decided that these costs were too high. So, instead, the US contingency was responsible only for budgetary and technical controls, but not safety. The last US inspection in 1982 warned of many hazards, including a number that have since been implicated as contributing to the leak and release.

From 1982 to 1984, safety measures declined, attributed to high employee turnover, improper and inadequate training of new employees, and low technical savvy in the local workforce. On-the-job experiences were often substituted for reading and understanding safety manuals. (Remember, this was a pesticide plant.) In fact, workers would complain of typical acute symptoms of pesticide exposure, such as shortness of breath, chest pains, headaches, and vomiting, yet they would typically refuse to wear protective clothing and equipment. The refusal in part stemmed from the lack of air conditioning in this subtropical climate, where masks and gloves can be uncomfortable.

More lenient Indian standards, rather than the US safety standards, were generally applied at the plant after 1982. This likely contributed to overloaded MIC storage tanks (e.g., company manuals cite a maximum of 60% fill).

The release lasted about two hours, after which the entire quantity of MIC had been released. The highly reactive MIC arguably could have reacted and become diluted beyond a certain safe distance. However, over the years, tens of thousands of squatters had taken up residence just outside of the plant property, hoping to find work or at least take advantage of the plant's water and electricity. The squatters were not notified of hazards and risks associated with the pesticide manufacturing operations, accepted by a local journalist who posted signs saying, "Poison Gas. Thousands of Workers and Millions of Citizens are in Danger."

This is a class instance of a "confluence of events" that led to a disaster. More than a few mistakes were made. The failure analysis found the following:

- The tank that initiated the disaster was 75% full of MIC at the outset.
- A standby overflow tank for the storage tank contained a large amount of MIC at the time of the incident.
- A required refrigeration unit for the tank was shut down 5 months prior to the incident, leading to a three- to fourfold increase in tank temperatures over expected temperatures.
- One report stated that a disgruntled employee unscrewed a pressure gauge and inserted a hose into the opening (knowing that it would do damage, but probably not nearly the scale of what occurred).
- A new employee was told by a supervisor to clean out connectors to the storage tanks, so the worker closed the valves properly, but did not insert safety disks to prevent the valves from leaking. In fact, the worker knew the valves *were* leaking, but they were the responsibility of the maintenance staff. Also, the second-shift supervisor position had been eliminated.
- When the gauges started to show unsafe pressures and even when the leaking gases started to sting mucous membranes of the workers, they found that evacuation exits were not available. There had been no emergency drills or evacuation plans.
- The primary fail-safe mechanism against leaks was a vent-gas scrubber, that is, normally, this release of MIC would have been sorbed and neutralized by sodium hydroxide (NaOH) in the exhaust lines, but on the day of the disaster, the scrubbers were not working. (The scrubbers were deemed unnecessary, since they had never been needed before.)
- A flare tower to burn off any escaping gas that would bypass the scrubber was not operating because a section of conduit connecting the tower to the MIC storage tank was under repair.
- Workers attempted to mediate the release by spraying water 100 ft. high, but the release occurred at 120 ft.

Thus, according to the audit, many checks and balances were in place, but the cultural considerations were ignored or given low priority, such as, when the plant was sited, the need to recognize the differences in land use planning and buffer zones in India compared with Western nations or the difference in training and oversight of personnel in safety programs.

In spite of a heightened awareness for years after the disaster, versions of the Bhopal incident have occurred and are likely to occur to smaller spatial extents and with, hopefully, more constrained impacts. For example, two freight trains collided in Graniteville, SC, just before 3:00 a.m. on 6 January 2005, resulting in the derailment of three tanker cars carrying chlorine (Cl_2) gas and one tanker car carrying sodium hydroxide (NaOH) liquids. The highly toxic Cl_2 gas was released to the atmosphere. The wreck and gas release resulted in hundreds of injuries and eight deaths. Some of these events are the result of slightly different conditions not recognized as vulnerable or not considered similar to Bhopal. Others may have resulted or may result due to memory extinction. Even vary large and impactful disasters fade in memory with time. This is to be expected for the lay public, but is not acceptable to engineers.

Every engineer and environmental professional needs to recognize that much of their responsibility is affected by geopolitical realities and that we work in a global economy. This means that engineers must have a respect and appreciation for how cultures differ in their expectations of environmental quality. One cannot assume that a model that works in one setting will necessarily work in another without adjusting for differing expectations. Bhopal demonstrated the consequences of ignoring these realities. Chaos theory tells us that even very small variations in conditions can lead to dramatically different outcomes, some disastrous.

Dual use and bioterrorism fears can be seen as somewhat analogous to the lack of due diligence at Bhopal. For example, extra care is needed in using similar strains and species in genetically modified microbes (e.g., substantial differences and similarities in various strains of *Bacillus* spp.). The absence of a direct analogy does not preclude that even a slight change in conditions may elicit unexpected and unwanted outcomes (e.g., weapon-grade or resistant strains of bacteria and viruses).

Characterizing as many contingencies and possible outcomes in the critical path is an essential part of many biohazards. The Bhopal incident provides this lesson. For example, engineers working with bioreactors and genetically modified materials must consider all possible avenues of release. They must ensure that fail-safe mechanisms are in place *and* are operational. Quality assurance officers note that testing for an unlikely but potentially devastating event is difficult. Everyone in the decision chain must be "on board." The fact that no incidents have yet to occur (thankfully) means that no one really knows what will happen in such an event. That is why health and safety training is a critical part of the engineering process.

16.3.4 Negligence

Engineers also have to protect the public from its members' own carelessness. The case of the woman trying to open a 2-l soda bottle by turning the aluminum cap the wrong way with a pipe wrench, having the cap fly off and into her eye, is a famous example of unpredictable ignorance. She sued for damages and won, with the jury agreeing that the design engineers should have foreseen such an occurrence. (The new plastic caps have interrupted threads that cannot be stripped by turning in the wrong direction.)

In the design of water treatment plants, engineers are taught to design the plants so that it is easy to do the right thing and very difficult to do the wrong thing. Pipes are color-coded, valves that should not be opened or closed are locked, and walking distances to areas of high operator maintenance are minimized and protected. This is called making the treatment plant "operator proof." This is not a rap exclusively on operators. In fact, such standard operating procedures (SOPs) are crucial in any operation that involves repeated actions and a flow of activities. Hospitals, laboratories, factories, schools, and other institutions rely on SOPs. Examples include mismatched transplants due to mislabeled blood types and injuries due to improper warning labels on power equipment. When they are not followed, people's risks are increased. Biosystem engineers recognize that if something can be done incorrectly, sooner or later, it will and that it is their job to minimize such possibilities. That is, both risk and reliability are functions of time.

Risk is a function of time because it is a part of the exposure equation, that is, the more time one spends in contact with a substance, the greater is the exposure. In contrast, reliability is the extent to which something can be trusted. A system, process, or item is reliable so long as it performs the designed function under the specified conditions during a certain time period. In most engineering applications, reliability means that what we design will not fail prematurely. Or stated more positively, reliability is the mathematical expression of success; that is, reliability is the probability that a system that is in operation at time *0* (t_0) will still be operating until the designed life (time $t=(t_t)$). As such, it is also a measure of the engineering accountability. People in neighborhoods near the facility want to know if it will work and will not fail. This is especially true for those facilities that may affect the environment, such as landfills and power plants. Likewise, when environmental cleanup is being proposed, people want to know how certain the engineers are that the cleanup will be successful.

Time shows up again in the so-called hazard rate, that is, the probability of a failure per unit time. Hazard rate may be a familiar term in environmental risk assessments, but many engineers may recognize it as a *failure density*, or $f(t)$. This is a function of the likelihood that an adverse outcome will occur, but note that it is not a function of the severity of the outcome. The $f(t)$ is not affected by whether the outcome is very severe (such as pancreatic cancer and the loss of an entire species) or relatively benign (muscle soreness or minor leaf damage). The likelihood that something will fail at a given time interval can be found by integrating the hazard rate over a defined time interval:

$$P\{t_1 \leq T_f \leq t_2\} = \int_{t_1}^{t_2} f(t)dt \tag{16.5}$$

where T_f = time of failure.

Thus, the reliability function $R(t)$ of a system at time t is the cumulative probability that the system has not failed in the time interval from t_0 to t_t:

$$R(t) = P\{T_f \geq t\} = 1 - \int_0^t f(x)dx \quad (16.6)$$

Engineers must be humble, since everything we design *will* fail. We can improve reliability by extending the time (increasing t_t), thereby making the system more resistant to failure. For example, proper engineering design of a landfill barrier can decrease the flow of contaminated water between the contents of the landfill and the surrounding aquifer, for example, a velocity of a few microns per decade. However, the barrier does not completely eliminate failure, that is, $R(t)=0$; it simply protracts the time before the failure occurs (increases Tf).

Hydraulics and hydrology provide very interesting case studies in the failure domains and ranges, particularly how absolute and universal measures of success and failure are almost impossible. For example, a levee or dam breach, such as the recent catastrophic failures in New Orleans during and in the wake of Hurricane Katrina, experienced failure when flow rates reached cubic meters per second. Conversely, a hazardous waste landfill failure may be reached when flow across a barrier exceeds a few cubic centimeters per decade.

Thus, a disaster resulting from this type of failure is determined by temporal dimensions. If the outcome (e.g., polluting a drinking water supply) occurs in a day, it may well be deemed a disaster, but if the same level of pollution is reached in a decade, it may be deemed an environmental problem, but not a disaster. Of course, if this is one's only water supply, as soon as the problem is uncovered, it becomes a disaster to that person. In fact, it could be deemed worse than a sudden-onset disaster, since one realizes he or she has been exposed by a long time. This was the case for some of the infamous toxic disasters of the 1970s, notably the Love Canal incident.

Love Canal is an example of a cascade of failure. The eventual exposures of people to harmful remnant waste constituent event trees were a complicated series of events brought on by military, commercial, and civilian governmental decisions. The failures involved many public and private parties who shared the blame for the contamination of groundwater and exposure of humans to toxic substances. Some, possibly most, of these parties may have been ignorant of the possible chain of events that led to the chemical exposures and health effects in the neighborhoods surrounding the waste site. The decisions by governments, corporations, school boards, and individuals in totality led to a public health disaster. Some of these decisions were outright travesties and breaches of public trust. Others may have been innocently made in ignorance (or even benevolence, such as the attempt to build a school on donated land, which tragically led to the exposure of children to dangerous chemicals). But the bottom line is that people were exposed to these substances. Cancer, reproductive toxicity, neurological disorders, and other health effects resulted from exposures, no matter the intent of the decision-maker. As engineers, neither the public nor the attorneys and company shareholders accept ignorance as an excuse for designs and operations that lead to hazardous waste-related exposures and risks.

One particularly interesting event tree is that of the public school district's decisions on accepting the donation of land and building the school on the property (see Fig. 16.7). As regulators and the scientific community learned more, a series of laws were passed and new court decisions and legal precedents established in the realm of toxic substances. Additional hazardous wastes sites began to be identified, which continue to be listed on the EPA website's National Priorities List. They all occurred due to failures at various points along a critical path.

These failures resulted from unsound combinations of scientifically sound studies (risk assessment) and decisions on whether to pursue certain actions (risk management). Many of these disasters have been attributed in part to political and financial motivation, which were perceived to outweigh good science. This was a principal motivation for the National Academy of Science's recommendation that federal agencies separate the science (risk assessment) from the policy decisions (risk management). In fact, the final step of the risk assessment process was referred to as "characterization" to mean that "both quantitative and qualitative elements of risk analysis, and of the scientific uncertainties in it, should be fully captured by the risk manager."[8]

Whereas disclosure and labeling are absolutely necessary parts of reliability in engineering, they are wholly insufficient to prevent accidents. The Tylenol tampering incident occurred in spite of a product that, for its time, was well-labeled. A person tampered with the product, adding cyanide, which led to the deaths of seven people in Chicago in 1982. The company, Johnson and Johnson took some aggressive and expensive steps to counter what would have been the critical path. They eventually recalled 31 million bottles and pioneered innovations to tamper-resistant packaging. What happened as a result was that after some large initial financial losses, the company completely recovered and avoided the business disaster many expected.[9]

8. National Research Council, Science and Decisions: Advancing Risk Assessment, National Academies Press, Washington, DC, 2009.

9. T. Jaques, Learning from past crises – do iconic cases help or hinder? Public Relat. J. 3 (1) (2009). http://www.prsa.org/SearchResults/download/6D-030103/0/Learning_from_Past_Crises_Do_Iconic_Cases_Help_or (Accessed 2 November 2011).

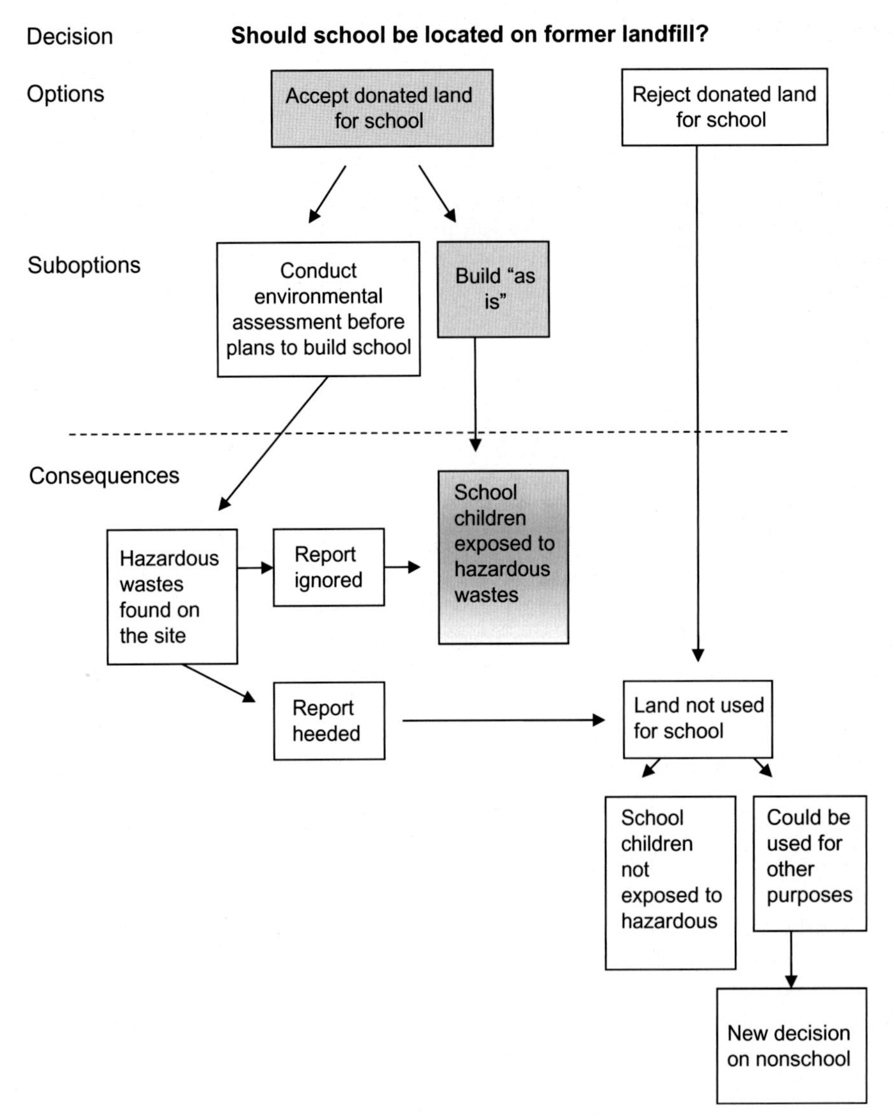

FIG. 16.7 Event tree of school-site decisions at Love Canal, NY. The *gray boxes* approximate the actual option, suboption, and consequence.

Another example of a critical path failure occurred in the early 1970s, when jet-powered airliners were replacing propeller aircraft. The fueling system at airports was not altered, and the same trucks fueled both types of craft. The nozzle fittings for both types of fuels were therefore the same. A tragic accident occurred near Atlanta, where jet fuel was mistakenly loaded into a Martin 404 propeller craft. The engines failed on takeoff, resulting in fatalities. A similar accident occurred in 1974 in Botswana with a DC-4 and again near Fairbanks, Alaska, with a DC-6.[10] The fuel delivery systems had to be modified so that it was impossible to put jet fuel into a propeller-driven airplane and vice versa.

An example of how this can be done is the modification of the nozzles used in gasoline stations. The orifice in the unleaded gas tank is now too small to take the nozzles used for either leaded fuel or for diesel fuel. Drivers of diesel-engine cars can still mistakenly pump gasoline into their cars, however. By analogy, all engineers must recognize that no amount of signs or training could prevent such tragedies.

10. Aviation Safety Network, (2002). http://aviation-safety.net/database/record.php?id=19740404-0 and http://aviation-safety.net/database/record.php?id=19760513-0 (Accessed 2 November 2011).

16.3.5 Inaccurate prediction of contingencies

Every time something fails, whether a manufactured product (a medical implant) or a constructed facility (dam breaches), it is viewed as an engineering failure. Engineers historically have been called upon to predict the problems that can occur and to design so as to minimize these adverse outcomes, protecting people from design errors, natural forces, unforeseen events, and ignorance/carelessness.

For engineers, the concept of negligence (Failure type 4) actually changed on 11 September 2001 with the attack on the United States by terrorists. The 21st century engineer is now expected to add a higher level of scrutiny and to be forward-thinking and proactive in designing ways to protect the health, safety, and welfare of the public from intentional acts. It had not occurred to most engineers that they have a responsibility to protect people from those who would want to intentionally harm other people or to destroy public facilities intentionally. This is a totally new failure mode in engineering. Such failures can be considered to be "intentional accidents," or failures resulting from intentional actions.[11]

Engineers now find themselves in the position of having to address these "intentional accidents." Military engineers of course have had to design against such destructive actions since the days of moats and castles, but those were all structures built explicitly to withstand attack. Civilian engineers have never had to think in these terms, but are now asked to design structures for this contingency. Engineering and engineers have a new challenge—to prevent such "accidents" on civilian targets by terrorists bent on harm to the public.

16.4 SUCCESS

Meeting air quality standards is bounded by numerous policies, laws, and other societal norms. Risk is an expression of operational success or failure. Management is the means of navigating among the policies, laws, and competing interests to achieve goals. Science and engineering are essential to good air quality management, but not sufficient. Proper air pollution control strategies must draw on other knowledge. Decisions must apply reasoning of all types, deductive; inductive; and, yes, intuitive. Certainly, air pollution cannot be completely managed by intuition-based decisions, but intuition is not always wrong [21]. It is important to keep in mind that many of the stakeholders in a jurisdiction apply intuitive reasoning often and, for some, almost exclusively. Intuition, in the author's opinion, is often fact-based, deductive or inductive reasoning that has long been forgotten. For example, most professions require a substantial amount of time spent working under a senior professional before one is granted a license. The senior professional may have long forgotten the specifics of why and how something works but has logged the process in memory. Indeed, the newly minted professional may know exactly why and how but has only applied this limited knowledge within the highly controlled academic or laboratory setting, whereas the seasoned professional has been applying these principles in the real world.

Having all the data needed for a *completely* informed air quality management decision is impossible. Scientific objectivity and humility dictate that decision-makers are upfront about uncertainties. Sometimes, air quality questions can only be answered by looking for patterns and analogies from events that are similar to the potential threat being considered. From there, scenarios can be developed to follow various paths to good, bad, and indifferent outcomes. This is known as a decision tree. This is not the same, necessarily, as the one with the most benefits compared with risks, that is, a benefit-to-risk ratio or relationship or benefit-to-cost ratio or relationship. However, this is indeed one of the more widely used approaches. The challenge is how to quantify many of the benefits and risks.

Humility is always the watchword for air quality management. Usually, there is more than one cause of a problem and more than one solution to the problem. Caution is in order when attributing a cause. The association of two factors, such as the level of exposure to a compound and the occurrence of a disease, does not necessarily mean that one necessarily "causes" the other. Often, after study, a third variable explains the relationship. However, the air quality manager must be diligent in linking causes with effects. Otherwise, corrective and preventive actions cannot be identified. So, strength of association is a beginning step toward cause and effect. A major consideration in strength of association is the application of sound technical judgment of the weight of evidence.

REFERENCES

[1] A. Birolini, Reliability Engineering: Theory and Practice, Springer, 2017.
[2] M. Modarres, M.P. Kaminskiy, V. Krivtsov, Reliability Engineering and Risk Analysis: A Practical Guide, CRC Press, 2016.
[3] D.A. Vallero, T.M. Letcher, Unraveling Environmental Disasters, Newnes, 2012.

11. S. Pfatteicher, Learning from failure: terrorism and ethics in engineering education, IEEE Technol. Soc. Mag. 21 (2) (2002) 8–12, 21.

[4] D.A. Vallero, T.M. Letcher, Engineering risks and failures: lessons learned from environmental disasters, Leadersh. Manag. Eng. 12 (4) (2012) 199–209.

[5] M. Eti, S. Ogaji, S. Probert, Integrating reliability, availability, maintainability and supportability with risk analysis for improved operation of the Afam thermal power-station, Appl. Energy 84 (2) (2007) 202–221.

[6] L. Sikos, J. Klemeš, RAMS contribution to efficient waste minimisation and management, J. Clean. Prod. 17 (10) (2009) 932–939.

[7] T.E. Graedel, On the concept of industrial ecology, Annu. Rev. Energy Environ. 21 (1) (1996) 69–98.

[8] R. Smith, D. Sengupta, S. Takkellapati, C. Lee, An industrial ecology approach to municipal solid waste management: I. Methodology, Resour. Conserv. Recycl. 104 (2015) 311–316.

[9] R. Smith, D. Sengupta, S. Takkellapati, C. Lee, An industrial ecology approach to municipal solid waste management: II. Case studies for recovering energy from the organic fraction of MSW, Resour. Conserv. Recycl. 104 (2015) 317–326.

[10] D.A. Vallero, P.A. Vesilind, Excerpted and Adapted from: Socially Responsible Engineering: Justice in Risk Management, Wiley Online Library, 2007.

[11] R.A. Hill, C. Sendashonga, General principles for risk assessment of living modified organisms: lessons from chemical risk assessment, Environ. Biosaf. Res. 2 (02) (2003) 81–88.

[12] ER-TG-03-1 7/00, Preparing Information on Risks, Costs and Benefits for Applications under the Hazardous Substances and New Organisms Act, 2000, p. 1996.

[13] B. Hong, et al., Model-based tolerance intervals derived from cumulative historical composition data: application for substantial equivalence assessment of a genetically modified crop, J. Agric. Food Chem. 62 (40) (2014) 9916–9926.

[14] P. Macdonald, Developing workable regulatory frameworks for the environmental release of transgenic plants, Collect. Biosaf. Rev. 6 (2012) 126–159.

[15] ER-TG-02-1 1/100, Assessment of Effects of Hazardous Substances and, 2000.

[16] R.A. Hill, From Science to Decision-Making: The Applicability of Bayesian Methods to Risk Assessment, 1996.

[17] B. Breckling, et al., Risk indication of genetically modified organisms (GMO): modelling environmental exposure and dispersal across different scales: oilseed rape in northern Germany as an integrated case study, Ecol. Indic. 11 (4) (2011) 936–941.

[18] J.G. Lundgren, J.J. Duan, RNAi-based insecticidal crops: potential effects on nontarget species, Bioscience 63 (8) (2013) 657–665.

[19] C.V. Rider, et al., Cumulative risk: toxicity and interactions of physical and chemical stressors, Toxicol. Sci. 137 (1) (2014) 3–11.

[20] Y. Li, et al., Quantifying urban ecological governance: a suite of indices characterizes the ecological planning implications of rapid coastal urbanization, Ecol. Indic. 72 (2017) 225–233.

[21] W.D. Ruckelshaus, Science, risk, and public policy, Science 221 (4615) (1983) 1026–1028.

FURTHER READING

[22] D. Vallero, Paradigms Lost: Learning from Environmental Mistakes, Mishaps and Misdeeds, Butterworth-Heinemann, 2005.

[23] T.M. Letcher, D. Vallero, Excerpt from: Waste: A Handbook for Management, second ed., Academic Press, 2019.

[24] D.A. Vallero, P.A. Vesilind, Socially Responsible Engineering: Justice in Risk Management, Wiley Online Library, 2007.

Chapter 17

Air pollution decision-making

17.1 INTRODUCTION

Numerous professions and disciplines are involved in the measurement, modeling, control, and prevention of air pollution. Researchers and academic scientists add to the knowledge base of how pollutants form, move, and change. Engineers design, install, and improve technologies for measuring, modeling, and treating air pollutants. Scientists participate in and advise policy makers on the necessary rules and regulations needed to improve air quality. These and many other practitioners must be trusted to do their part in addressing air pollution and do so in an ethical manner.

Air quality decisions, like other environmental and public health decisions, include those about releases and those about quality [1, 2].

17.1.1 Release-related decisions

Release-related decisions may be direct or indirect. A direct release-related decision involves the type of pollutant that should be regulated and the maximum amount of the pollutant that should be emitted directly from a stack, vent, storage tank, or other conveyance. An indirect release-related decision involves pollutants that reach the air after first being released into another medium, for example, from a pipe to surface water or a nonpoint release onto land or into surface water or groundwater.

The release-related decision-making process could begin when sources are known to contribute a pollutant to the atmosphere; for example, a factory is known to emit or is expected to emit a pollutant based on its manufacturing processes; such as the use a metal as catalysts to synthesize organic chemicals. This would be a clue not only that both the organic compounds and precursors may be emitted but also that the metal and its compounds may also be emitted (see Section 3 of Chapter 2). The release-related decision may also begin after a pollutant is detected in the ambient air, sometimes without first being suspected, based on the known sources. This can be because the pollutant could be coming from sources far away from the urban area, that is, long-range transport (see Chapter 6). It may also be because substances released from various sources only become pollutants after release, that is, transformation (see Chapter 3).

Regulators may consider all types of releases. For example, in the United States, modelers categorize sources into five types: point, nonpoint, on-road, nonroad, and event. An event is an intermittent release, for example, from a chemical spill, a wildfire, or prescribed burning [3]. The major distinctions among releases that form these sources are whether the pollutant is criteria air pollutant (CAP) or hazardous air pollutant (HAP). In several nations, CAPs are those regulated under National Ambient Air Quality Standards (NAAQS), whereas HAPs are known as "air toxics."

In the United States, CAPs include lead (Pb), tropospheric ozone (O_3), carbon monoxide (CO), nitrogen oxides (NO_X), sulfur dioxide (SO_2), and particulate matter 10 μm or less (PM_{10}) and particulate matter 2.5 μm or less ($PM_{2.5}$). Emission inventories and other source databases also track precursors of CAPs, for example, volatile organic compounds (VOC) and ammonia (NH_3), which are not technically CAPs, but are important precursors, that is, VOCs are precursors of O_3 and NH_3 is a precursor for PM. In the United States, HAP pollutants include the 187 remaining HAP pollutants (hydrogen sulfide was removed) from the original 188 listed in Section 112(b) of the 1990 Clean Air Act Amendments. Mercury (Hg); hydrochloric acid (HCl) and other acid gases; heavy metals such as nickel and cadmium; and hazardous organic compounds such as benzene, formaldehyde, and acetaldehyde are included among these HAPs (see Table 17.1).

17.1.2 Quality-related decisions

Quality-related decisions address the desired state of an environmental resource or the health status of a population. Obviously, the two types of environmental decisions are directly related to one another. The reason for eliminating or decreasing the emission or effluent is to transition in the direction of the desired state, that is, to enhance an ecosystem or improve the health of a human population. As shown in Fig. 17.1, the decision-maker must consider the condition of the current state, the

Air Pollution Calculations. https://doi.org/10.1016/B978-0-12-814934-8.00017-X

TABLE 17.1 Hazardous air pollutants listed under Section 112 of the Clean Air Act Amendments of 1990
Acetaldehyde
Acetamide
Acetonitrile
Acetophenone
2-Acetylaminofluorene
Acrolein
Acrylamide
Acrylic acid
Acrylonitrile
Allyl chloride
4-Aminobiphenyl
Aniline
o-Anisidine
Asbestos
Benzene (including benzene from gasoline)
Benzidine
Benzotrichloride
Benzyl chloride
Biphenyl
Bis(2-ethylhexyl)phthalate (DEHP)
Bis(chloromethyl)ether
Bromoform
1,3-Butadiene
Calcium cyanamide
Caprolactam (see modification)
Captan
Carbaryl
Carbon disulfide
Carbon tetrachloride
Carbonyl sulfide
Catechol
Chloramben
Chlordane
Chlorine
Chloroacetic acid
2-Chloroacetophenone
Chlorobenzene
Chlorobenzilate
Chloroform

TABLE 17.1 Hazardous air pollutants listed under Section 112 of the Clean Air Act Amendments of 1990—cont'd
Chloromethyl methyl ether
Chloroprene
Cresols/cresylic acid (isomers and mixture)
o-Cresol
m-Cresol
p-Cresol
Cumene
2,4-D, salts and esters
DDE
Diazomethane
Dibenzofurans
1,2-Dibromo-3-chloropropane
Dibutyl phthalate
1,4-Dichlorobenzene(p)
3,3-Dichlorobenzidine
Dichloroethyl ether (bis(2-chloroethyl)ether)
1,3-Dichloropropene
Dichlorvos
Diethanolamine
N,N-Dimethylaniline
Diethyl sulfate
3,3-Dimethoxybenzidine
Dimethyl aminoazobenzene
3,3′-Dimethyl benzidine
Dimethyl carbamoyl chloride
Dimethylformamide
1,1-Dimethyl hydrazine
Dimethyl phthalate
Dimethyl sulfate
4,6-Dinitro-o-cresol, and salts
2,4-Dinitrophenol
2,4-Dinitrotoluene
1,4-Dioxane (1,4-diethyleneoxide)
1,2-Diphenylhydrazine
Epichlorohydrin (l-chloro-2,3-epoxypropane)
1,2-Epoxybutane
Ethyl acrylate
Ethylbenzene
Ethyl carbamate (urethane)

Continued

TABLE 17.1 Hazardous air pollutants listed under Section 112 of the Clean Air Act Amendments of 1990—cont'd
Ethyl chloride (chloroethane)
Ethylene dibromide (dibromoethane)
Ethylene dichloride (1,2-dichloroethane)
Ethylene glycol
Ethylene imine (aziridine)
Ethylene oxide
Ethylene thiourea
Ethylidene dichloride (1,1-dichloroethane)
Formaldehyde
Heptachlor
Hexachlorobenzene
Hexachlorobutadiene
Hexachlorocyclopentadiene
Hexachloroethane
Hexamethylene-1,6-diisocyanate
Hexamethylphosphoramide
Hexane
Hydrazine
Hydrochloric acid
Hydrogen fluoride (hydrofluoric acid)
Hydroquinone
Isophorone
Lindane (all isomers)
Maleic anhydride
Methanol
Methoxychlor
Methyl bromide (Bromomethane)
Methyl chloride (chloromethane)
Methyl chloroform (1,1,1-trichloroethane)
Methyl ethyl ketone (2-butanone) (see modification)
Methyl hydrazine
Methyl iodide (iodomethane)
Methyl isobutyl ketone (hexone)
Methyl isocyanate
Methyl methacrylate
Methyl tert-butyl ether
4,4-Methylene bis(2-chloroaniline)
Methylene chloride (dichloromethane)
Methylene diphenyl diisocyanate (MDI)
4,4′-Methylenedianiline

TABLE 17.1 Hazardous air pollutants listed under Section 112 of the Clean Air Act Amendments of 1990—cont'd

Naphthalene
Nitrobenzene
4-Nitrobiphenyl
4-Nitrophenol
2-Nitropropane
N-Nitroso-*N*-methylurea
N-Nitrosodimethylamine
N-Nitrosomorpholine
Parathion
Pentachloronitrobenzene (quintobenzene)
Pentachlorophenol
Phenol
p-Phenylenediamine
Phosgene
Phosphine
Phosphorus
Phthalic anhydride
Polychlorinated biphenyls (Aroclors)
1,3-Propane sultone
beta-Propiolactone
Propionaldehyde
Propoxur (Baygon)
Propylene dichloride (1,2-dichloropropane)
Propylene oxide
1,2-Propylenimine (2-methyl aziridine)
Quinoline
Quinone
Styrene
Styrene oxide
2,3,7,8-Tetrachlorodibenzo-p-dioxin
1,1,2,2-Tetrachloroethane
Tetrachloroethylene (perchloroethylene)
Titanium tetrachloride
Toluene
2,4-Toluene diamine
2,4-Toluene diisocyanate
o-Toluidine
Toxaphene (chlorinated camphene)
1,2,4-Trichlorobenzene

Continued

TABLE 17.1 Hazardous air pollutants listed under Section 112 of the Clean Air Act Amendments of 1990—cont'd

1,1,2-Trichloroethane
Trichloroethylene
2,4,5-Trichlorophenol
2,4,6-Trichlorophenol
Triethylamine
Trifluralin
2,2,4-Trimethylpentane
Vinyl acetate
Vinyl bromide
Vinyl chloride
Vinylidene chloride (1,1-dichloroethylene)
Xylenes (isomers and mixture)
o-Xylenes
m-Xylenes
p-Xylenes
Antimony compounds
Arsenic compounds (inorganic including arsine)
Beryllium compounds
Cadmium compounds
Chromium compounds
Cobalt compounds
Coke-oven emissions
Cyanide compounds[a]
Glycol ethers[b]
Lead compounds
Manganese compounds
Mercury compounds
Fine mineral fibers[c]
Nickel compounds
Polycyclic organic matter[d]
Radionuclides (including radon)[e]
Selenium compounds

[a] *X'CN where X=H' or any other group where a formal dissociation may occur. For example, KCN or Ca(CN)2.*
[b] *Includes monoether and diether of ethylene glycol, diethylene glycol, and triethylene glycol R-(OCH2CH2)n -OR' where* n = *1, 2, or 3; R=alkyl or aryl groups; R'=R, H, or groups that, when removed, yield glycol ethers with the structure R-(OCH_2CH)n-OH. Polymers are excluded from the glycol category (see modification).*
[c] *Includes mineral fiber emissions from facilities manufacturing or processing glass, rock, or slag fibers (or other mineral-derived fibers) of average diameter 1 μm or less.*
[d] *Includes organic compounds with more than one benzene ring and that have a boiling point greater than or equal to 100°C.*
[e] *A type of atom that spontaneously undergoes radioactive decay.*
Notes: *For all listings above that contain the word "compounds" and for glycol ethers, the following applies: unless otherwise specified, these listings are defined as including any unique chemical substance that contains the named chemical (i.e., antimony and arsenic) as part of that chemical's infrastructure.*

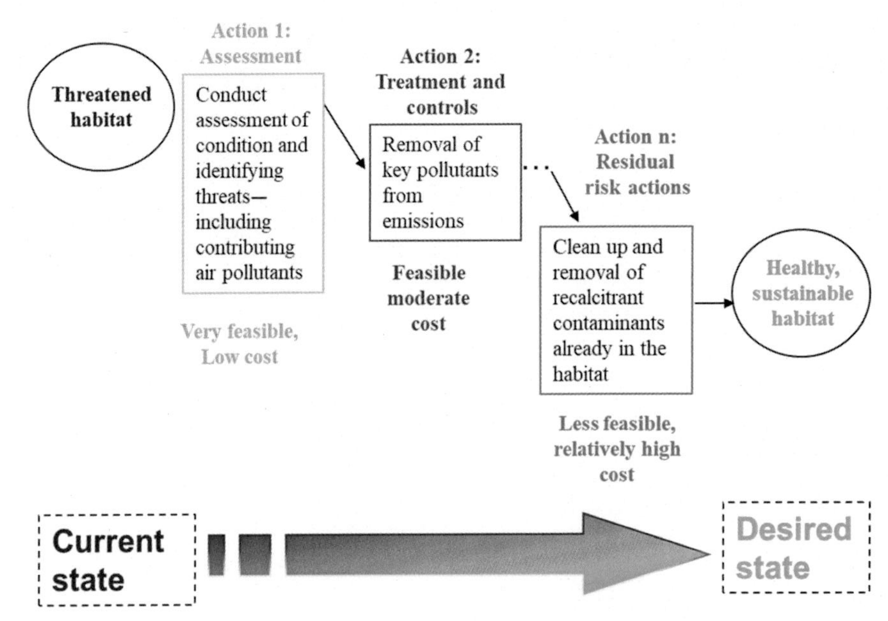

FIG. 17.1 Environmental decision-makers must decide on the actions needed to move from the current quality state to the desired state.

effectiveness of proposed actions in changing that state and moving toward the desired state, and the costs and feasibility of these actions.

17.2 DATA-BASED DECISION-MAKING

Air quality decisions are not simply "plug and play" applications. The decision-maker must select the correct information from which to employ all types of rational decision-making, that is, deductive and inductive reasoning, smattered with intuition informed by experience. Intuition and situational awareness are common traits of engineers. As evidence, plant managers and engineers had already adopted practices and technologies long before scientific consensus on the hazards associated with most air and water pollutants [4]. Further complicating decisions about air pollution policies and standards are the need to consider psychological, sociological, economic, and political factors. As discussed in Chapter 12, exposure to pollutants varies according to human activities.

Decision-making differs among nations and even at subnational levels, but to varying degrees, all pollution actions depend on reliable, timely, and spatially meaningful data [5]. Deciding the type and amount of measures needed to protect the environment from potential harm from activities that would likely provide jobs or would enhance profit-making of a key industrial sector demonstrates the diversity perspectives involved in many environmental policies. Each side provides what they deem to be evidence. Some would argue that evidentiary route is paramount and supersedes. In the case of jobs versus the environment, for example, the "jobs" advocates may predominantly go to the social sciences and economics to support their argument, whereas the environmental advocates more likely would argue that the biological sciences should hold sway. In recent decades, risk assessment came to the fore, forcing both sides to bolster the science needed to support their positions. The now includes attention be paid to all relevant scientific disciplines in what has come to be known as "science-based" environmental policy. Ostensibly, to be credible, any decision must be made within the context of the extent to which the environment or public health is protected. Lots of jobs or a better bottom line for profits does not obviate the need to weight environmental harm.

Environmental decision-making is one of the more complex types made by governments, industries, and professionals, given the need to employ almost every scientific discipline. This can be particularly challenging to professionals, especially engineers, whose paramount allegiance is the safety, health, and welfare of the public, even beyond their obligations to their clients [6]. For civil engineers, the code of ethics goes even further.

In recommending the "best" course of action, the engineer must identify the underlying science and must ensure the quality and relevance of the scientific data being used to select this action. Environmentally sound decision-making draws from multiple physical, biological, and social science disciplines, with each having unique weightings and from which

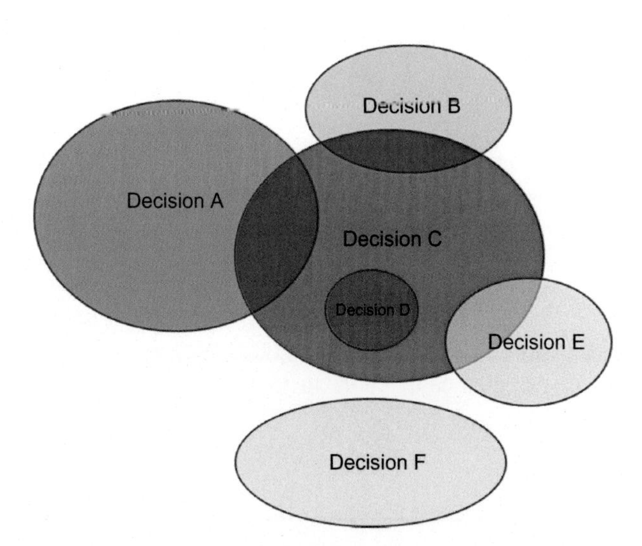

FIG. 17.2 Illustration of decision space. Decisions A, B, and E are affected by and affect Decision C, with at least one shared, available option. Decision D is subordinate and inclusive within Decision C, that is, it is an important decision, but its options fall entirely within Decision C's space (e.g., decision on road materials may fall entirely within the larger decision of whether the road is needed for access). Decision F's space is separate from the Decision A–E cluster. That is, it has unique decision space that is composed of options different from Decisions A through E. The size of the ellipse indicates the range of options available to the decision-maker. Decision D is the most constrained (less options), whereas Decision C allows for more options. *(Source: D. Vallero, Excepted and Updated from: Translating Diverse Environmental Data into Reliable Information: How to Coordinate Evidence from Different Sources. Academic Press, 2017; D. Vallero, Translating Diverse Environmental Data into Reliable Information: How to Coordinate Evidence from Different Sources. Academic Press, 2017.)*

alternatives are selected by optimizing among variables. The optimization leads to a choice among competing and available alternative courses of action.

Each decision resides with a "decision space," that is, the range of options from which the decision-maker can choose. Each option has a distribution of possible outcomes, which is a function of uncertainty about the factors needed to make the decision and the uncertainty about the various courses of action [7, 8]. Environmental courses of actions stemming from environmental decisions are nested. For example, the transition from the current state (predecision) to the desired state (following implementation of the decision) is likely to be a part of a much larger and complex decision. The stepwise progression depicted in Fig. 17.1 is merely one decision that is nested within the larger range of options available, with each evaluated, for not only how they are individually needed to provide a specific aspect of habitat condition but also how they interact with all other actions being taken to ensure a healthy and sustainable habitat. The overall decision must account for linkages among decisions and synergies and antagonisms. In other words, each individual decision space resides in a larger complex of decision space. For instance, a course of action at a scale of 1 ha may be perfectly fine, but when scaled up to 1000 ha, the interaction with other factors may eliminate it as a desired course of action. Such propagation of harm or benefit depends on an accurate accounting and weighting of factual evidence. A map of the interdependencies of decision spaces is shown in Fig. 17.2.

Few decisions boldly say that they will be made without factoring in the need to preserve or enhance environmental quality and without an eye toward how to prevent or to ameliorate environmental damage. Rather, the decisions often center around what is acceptable in terms of a value, for example, human health, ecology, and economics. As such, an air pollution decision may be a simple calculation of the amount of substance that can be emitted. The success may lie in the fact that specific actions, emission controls, are being taken and are leading to a desired outcome, for example, air quality. For example, if the emission rates of the six pollutants regulated under the Clean Air Act's National Ambient Air Quality Standards (NAAQS) drop by 63% in 35 years, a regulatory agency may deem this to be a successful transition toward the desired state (see Fig. 17.1). The success may be further articulated by tracking other environmental, economic, and social factors with these changes in emissions (see Fig. 17.3). To some extent, the correlation of cleaner air and emission controls is uncertain and indirect. They certainly are coincidental, but more statistical analysis may be needed to confirm that the relationships are causal, given the interrelationships and dependencies of engineering and social factors on air quality (see Fig. 17.4).

The first step in any environmental decision is to gather scientifically sound facts. The next step is to apply the facts to the problem at hand. The first step may not be "scientific," for example, where one looks for the facts may be a biased endeavor. Science is supposed to be objective, but scientific searches can be myopic, metaphorically looking at night for our "lost car keys" only under the area illuminated by the street light. This area of illumination is whatever paradigm

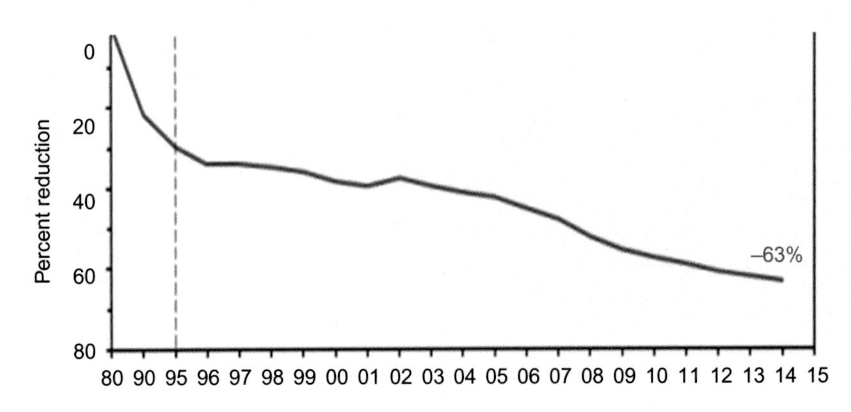

FIG. 17.3 Emission rates of the six pollutants regulated under the Clean Air Act's National Ambient Air Quality Standards (NAAQS) have continuously dropped over the past 35 years. *(Source: U.S. Environmental Protection Agency, Air Quality – National Summary. U.S. EP Agency Washington, DC, 2016; U.S. Environmental Protection Agency, NAAQS Table. in: Criteria Air Pollutants. (2016), 20 December 2016. https://www.epa.gov/criteria-air-pollutants/naaqs-table (Accessed 25 August 2018).)*

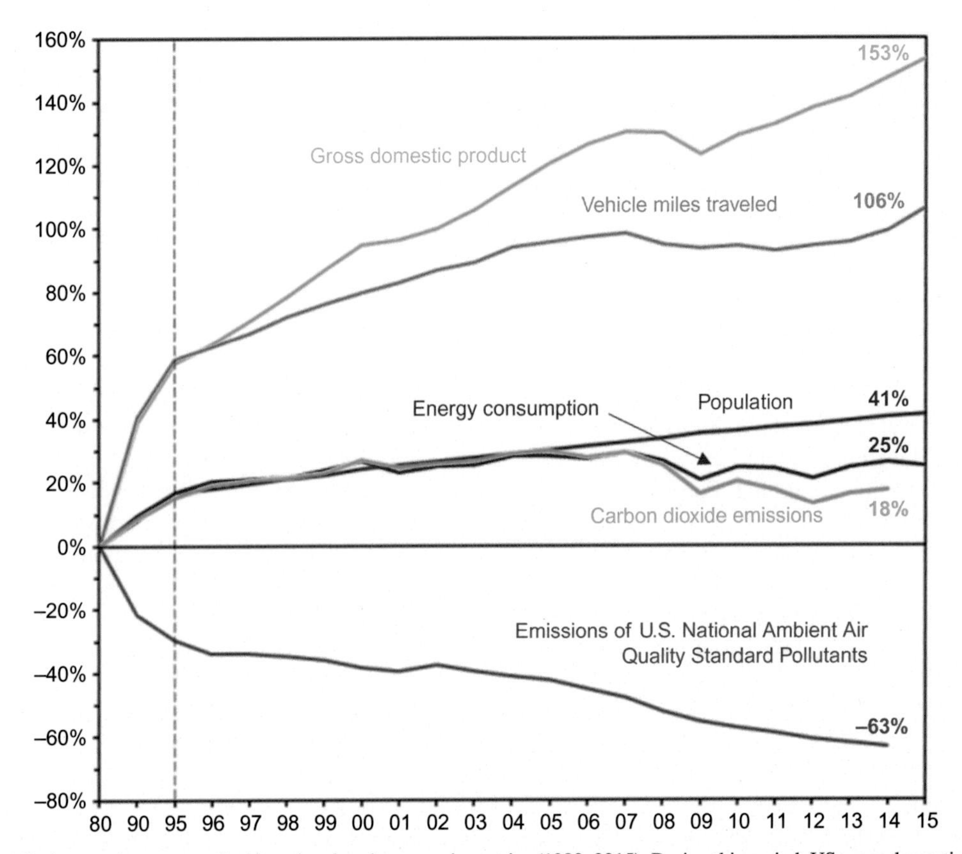

FIG. 17.4 Air pollution metrics compared with national socioeconomic metrics (1980–2015). During this period, US gross domestic product and total miles traveled by US vehicles increased substantially. The US population also increased. However, energy consumption and carbon dioxide emissions have flattened. Meanwhile, emissions of criteria air pollutants (i.e., lead, tropospheric ozone, carbon monoxide, sulfur dioxide, nitrogen dioxide, and particulate matter) continued to decrease markedly. *(Source: Source: U.S. Environmental Protection Agency, Air Quality – National Summary. U.S. EP Agency Washington, DC, 2016; U.S. Environmental Protection Agency, NAAQS Table. in: Criteria Air Pollutants. (2016), 20 December 2016. https://www.epa.gov/criteria-air-pollutants/naaqs-table (Accessed 25 August 2018).)*

that we hold dear. This is much easier than going home and returning with a flashlight, but not ideal for finding the lost keys. Thus, the "easy" method of fact finding may well be inadequate since it omits a large segment of possible solutions.

A data search limited to previous favorites may miss approaches that could work in even slightly different decision space with unique factors. A classic example of relying on a source of data exclusively when the decision demands additional perspectives is that of the "healthy worker effect" (HWE). Epidemiology is the study of the distribution and determinants of health-related phenomena in populations and the use of this information to control health problems [9, 10]. The HWE is an error in occupational epidemiology that arises from drawing conclusions about the general population from a group that has been well studied and for which data exist [11, 12]. Historically, companies and other groups have kept better tabs on the health of their workforce than most of the population. So, epidemiologists saw these data as a rich vein from which to extrapolate general disease trends. At least when they begin their jobs, workers tend to be healthier than the general population for many disease categories. Indeed, less healthy people have been screened out of many jobs, as have women and children. Conversely, workers often are exposed to contaminants and had very different lifestyles (e.g., miners breathing particulates and working in dangerous settings) than the public.

Thus, HWE is an example of systematic error or bias, which was introduced in Chapter 1. In this instance, it is "information bias," that is, the systematic difference in obtaining data from two different groups, the workers and the general population that leads to erroneous results. The bias is also the result of the difference in how the people were enrolled, that is, the so-called selection bias. The workers were enrolled because of their employment, whereas the rest of the population was enrolled by the epidemiologist trying to extrapolate from the available data. Selection bias ranged from the difficult (e.g., attempting to compare younger and older members of the population with healthy worker data) to the ridiculous (e.g., trying to tease out pregnancy or lactation information from a predominantly male database). Relying exclusively on available data that were gathered for a specific purpose is an example of selection bias. Thus, one could argue the opposite bias, that is, a type of "unhealthy worker effect" [13]. Thus, there is great uncertainty and risk associated with looking for one's keys only under the lamppost.

17.2.1 Data and information selection

Many factors go into an air quality study. It must address criteria to determine the best approach for addressing the pollution problem:

1. Overall protection of human health and environment
2. Compliance with applicable or relevant and appropriate requirements
3. Long-term effectiveness and permanence
4. Reduction of toxicity, mobility, or volume through treatment
5. Short-term effectiveness
6. Ease of implementation
7. Cost
8. State acceptance
9. Community acceptance

A sound, scientific knowledge base must support the first, third, and fourth criteria. In many countries, this includes credible and quantitative information within a risk assessment. The seventh criterion must be based on reliable economic information. Criteria 2, 3, 5, 6, 7, and 9 are the "management" components of the knowledge base. Combining these and the governmental components (criteria 8) with the scientific risk assessment is the basis for a scientific risk assessment process. This process blends quantitative physical, chemical, and biological data with semiqualitative and qualitative information, for example, surveys, illustrating the variety of data and information used in evaluating the environmental implications of pollution.

The cleanup and any science-based decision-making are not complete unless it also has a plan for communicating some rather complex information and knowledge to arrive at the best approach. Note that this is not necessarily the first choice of the scientist or engineer (who may likely heavily weight criteria 1, 3, and 4), but often the one that best arrives at criterion 9, that is, good enough to be accepted by the public *and* that protects the environment and health. The nine criteria may be viewed as sectors in a decision force field, but for the public and policy makers, other factors are even more important.

Environmental decisions are affected in varying ways by factors other than the physical and environmental sciences. The weighting of these factors determines why one alternative is chosen whereas another is rejected. This can be demonstrated with a device known as the decision force field (see Fig. 17.5).

Postrenaissance science depends on the aptly named "scientific method." The features and requirements of this centuries-old system provide a critical path to gathering and verifying new and old sources of data and applying information technologies. The writer and likely many other practitioners of science need to be reminded from time to time what this method entails.

First, the method requires objectivity. All data and information begin as a neutral commodity, which becomes increasing valuable to the user as they are transformed into information. If the data are truly a neutral commodity, they should meet the scientific requirement of objectivity. Arguably, however, this axiom only applies to data that have been completely collected and controlled by the user. It cannot be universally applied to secondary data, given that they are

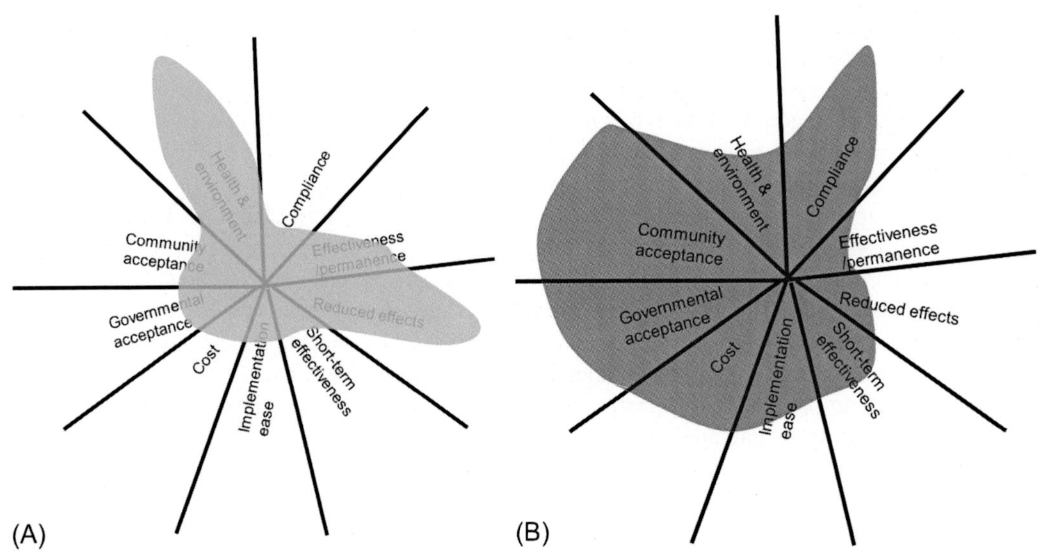

FIG. 17.5 Decision force fields. Force field A represents a decision driven mainly by objective scientific findings. Force field B is a decision influenced more by feasibility, management, and policy concerns. *(Source: D. Vallero, Excepted and Updated from: Translating Diverse Environmental Data into Reliable Information: How to Coordinate Evidence from Different Sources. Academic Press, 2017; D. Vallero, Translating Diverse Environmental Data into Reliable Information: How to Coordinate Evidence from Different Sources. Academic Press, 2017.)*

already transformed to some extent for some purpose. And this purpose is almost always something other than that of the secondary user.

For a scientific research investigation in which the results are published in peer-reviewed literature, the article must undergo review by unbiased, objective experts. This peer-reviewed process is designed to ensure that the results are of high quality and that the methodology, results, and conclusions are accurately reflected. This is a challenge when using other data collected and quality controlled by others, that is, secondary information. Each aspect of the study is reviewed somewhat independently, by more than one expert. This multipeer approach allows for more intensive scrutiny from different perspectives. For example, a study's methodology may be found to be sound by all reviews, but a person with statistical expertise may disagree with how the data are interpreted. Such criticism often does not doom publication of the study prima facie but may require a different statistical test. Other times, a study's methodology and the data being derived are not questioned by the reviewers, but they disagree with the conclusions.

Confidence in the integrity of scientific findings requires that no matter who conducts the study and no matter how many times the study is done under the conditions prescribed in the methods, the same results must be found. This is not to say that the results are necessarily identical, just consistent. This may sound oxymoronic, but it is almost always the case, given the uncertainties of any study. A theoretical physicist would have much tighter tolerances than that of a field botanist, given that the physicist can control experiment conditions to several decimal places to the right of the decimal point. However, an air pollution field study's conditions are never completely reproducible, given the variability of meteorology, emissions, and other conditions.

Selecting appropriate and sufficiently reliable data begins by identifying and accounting for the uncertainties and establishing standards for reproducibility. Quality needs to vary by type of study. A physics experiment may report a value of 100 ± 1 Å, that is, a tolerated range between 99 and 101 Å. This would be an extremely and presently impossibly tight range for an air pollution study. However, for subatomic physics studies, the range may be deemed to be too large.

Assuming the 100 ± 1 Å confidence level is appropriate, if a subsequent physics study applied the same methodology in a follow-up study that finds 102 Å, then the original study would be considered to have not been reproducible. Even the same air pollution study can include quality criteria that differ, depending on the needs of the user. For example, a first responder entering a building may be satisfied knowing that the concentrations of an explosive substance are $<1\%$, whereas the air pollution engineer may require confidence ranges that are $\pm 10\,\mu g\,m^{-1}$. Perhaps, the academic scientist conducting an indoor air study of the building may require even tighter tolerance, for example, $\pm\ 1\,\mu g\,m^{-1}$.

Quality standards and tolerances for both measured and secondary data are driven by the extent to which immediacy trumps deliberation. Indeed, some aspects of air pollution decision-making are more immediate, whereas others require greater, sometimes extensive, deliberation with a wide range of input from eclectic perspectives. For example, a toxic cloud resulting from a ruptured tanker after a derailment is an "emergency response," including evacuation, personal protective equipment, and sampling for explosiveness. These emergency steps are followed by a more deliberative cleanup.

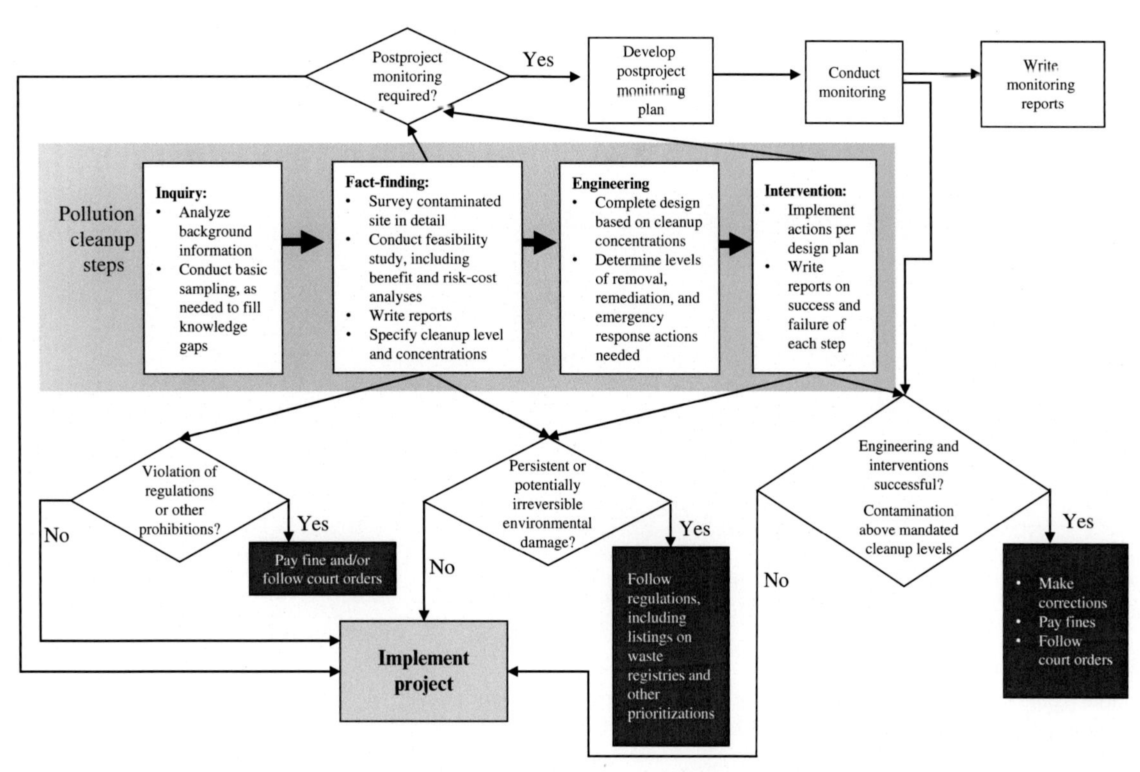

FIG. 17.6 Flow of data and information production to support cleanup actions at contaminated sites in Hungary. *(Modified from E. Almássy, National Environmental Remediation Program Guide. E. R. Program. Ministry of Environment and Water, Budapest, Hungary, 2002.)*

The data needed, therefore, evolve. Often both types of decisions depend on and add to background documents (Fig. 17.6). The main reasons for using available data are that similar sites may have been evaluated and that the existing information about these similar sites will help focus measurements at the new site, for example, sufficient number and locations of monitoring wells if groundwater depth and flow are like previous studies. Since each site is unique in terms of soil type, amounts and types of contaminants, potential exposures, etc., the available information is often complemented with site-specific sampling and modeling. This modified data set may be available to future users, although there are often restrictions imposed to protection intellectual property, confidentiality, and legal proceedings.

The data needed to support an air pollution decision evolve. These decisions often begin with background documents, but the information in these documents to be adjusted and tailored to a specific site or problem. The main reasons for using available data are that similar sites may have been evaluated and that the existing information about these similar sites will help focus measurements at the new site, for example, sufficient number and locations of meteorology and air quality are similar to those found in previous studies. Since each site is unique in terms of topography, amounts and types of contaminants, potential exposures, etc., the available information is often complemented with site-specific sampling and modeling. This modified dataset may be available to future users; although there are often restrictions imposed to protection intellectual property, confidentiality, and legal proceedings.

As discussed in Chapter 12, the traditional risk assessment process is a stepwise approach, beginning with the identification of hazard, calculation of its danger as expressed by a dose-response curve, assessing exposures to the hazard, estimating the effects, and ultimately characterizing the risks [14]. This stepwise process has weaknesses. For example, microbial and nonchemical agents do not necessarily adhere to this paradigm. It does not work well for ecological risk assessment, which requires greater integration of problem identification and analysis [15]. Even for human health risk posed by chemical agents, the linear approach is not conducive to system thinking [16, 17].

17.2.2 The nondetect problem

The selection of sampling and analysis equipment and protocols is based on these and other quality metrics. A problem that arises in using secondary data is the so-called nondetect problem.

In 1990, the fictional carcinogen benzo(a)dumbo (BAD) was measured with equipment at a level of detection (LoD) of 5 ppm with a 95% confidence but with technological advances changed to 2 ppm in 2000, 1 ppm in 2010, and 500 ppb in 2015 (Table 17.2). How does the changing LoD affect nondetects (ND)?

TABLE 17.2 Importance of unreported results in a database

	Year							
	1990		2000		2010		2015	
Site	Measured	Actual	Measured	Actual	Measured	Actual	Measured	Actual
A	ND	4.2	3.3	3.3	3.1	3.1	2.2	2.2
B	ND	4.9	ND	1.9	1.5	1.5	1.4	1.4
C	ND	4.1	ND	1.8	ND	0.8	0.6	0.6
D	ND	3.9	ND	1.5	ND	0.7	ND	0.4

Annual concentrations of a fictitious carcinogen measured in a stream at four different levels of detection (LoDs), at 95% confidence: 5.0 ppm in 1990, 2.0 ppm in 2000, 1.0 ppm in 2010, and 0.5 ppm in 2015. *ND*, nondetects below LoDs.

The table would really look like Table 17.3, since actual values are not really known. Misinterpretations of the table will occur if the data user intuitively and erroneously assumes that nondetect (ND) to be zero and/or the interpretive statistics ignore the ND values.

TABLE 17.3 Data in Table 17.2, excluding "actual" values, showing only measured values

	1990	2000	2010	2015
Site	Concentration (ppm)	Concentration (ppm)	Concentration (ppm)	Concentration (ppm)
A	ND	3.3	3.1	2.2
B	ND	ND	1.5	1.4
C	ND	ND	ND	0.6
D	ND	ND	ND	ND

17.2.3 Data manipulation

Similarly, unreported results can also lead to "data false positives," such as US Department of Housing and Urban Development (HUD)-sponsored Lead-Based Paint Hazard Control Grant Program Evaluation [18, 19]. This multisite, longitudinal study of the effects of environmental interventions on childhood blood lead levels and lead (Pb) in homes deleted over 37% of findings [20]. One recommendation from subsequent research is that data be reported, with the caveat that they do not meet all quality requirements for use in that study (e.g., CI $<$ 95%). This would allow future users to apply imputation and other techniques, which are also likely to improve with time. However, the user should document the use of these tools, so that future users are aware of the imputations and other manipulations.

The term "imputation" has a negative connotation in scientific history, that is, "making up data." There are too many examples of cooking, pruning, and trimming data to "improve" them, for example, improve correlations and other statistical metrics to meet publication standards (see Sigma Xi's excellent booklet on such practices [21]). There are three types of missing data than can be imputed [22]: missing completely at random (MCAR), missing at random (MAR), and nonrandom missing. It should be noted that imputation does address missing data, which is certainly not cheating and is obligatory if missing data were to lead to errors in generalization, that is, improperly informing deductive reasoning for proper interpretation of the data [22–24]. In this case, the only unethical imputation would be that this is not properly documented in the study reports and journal articles' text and supplemental information.

Air pollution decisions require high-quality measurement data *and* associated data, along with the metadata related to both data sets [25]. The decision to use data depends on methodological coherence, that is, the extent to which results agree when using different methodologies. Whereas reproducibility is concerned with matching exactness of methodology with concurrence of results, coherence is concerned with the agreement among different approaches with results and conclusions of different studies. Assume, for instance, that an investigator conducts a study to determine the extent to which a population has been exposed to the fictional chemical, benzo(a)dumbo (BAD). The risk associated with a chemical agent is determined by both the hazard of the agent and the amount of exposure to the agent. Actual methods for estimating exposure will be discussed later, but for now, let us simplify things. Assuming, usually incorrectly, that exposure is via only one route, inhalation. Our intrepid investigator is only able to find two data sources from which to calculate exposure estimates. One of the sources is BAD air concentrations in the sample of the male population aged 18 years and older in the city of Townsville. The other data source is the amount of BAD found in the urine of these same men. Unless one is only concerned about a single sample, a measurement, like chemical concentrations in food or urine, the information must usually be extended. The extension may be spatial, such as extending a sample to represent a larger two- or three-dimensional space, such as a sample of a small amount of soil being used to describe an entire field. The extension may be temporal, such as taking a sample of a chemical in the air for one minute to represent a whole day, week, or year. The extension may be intellectual, where a fact is used to extend knowledge.

Environmental and human health effects can occur in populations with very low doses, especially over lengthy time periods. Thus, in vivo measurement data and a growing amount of in vitro data [26–29] provide the basis for models needed to extend the range to dose-response rates in humans. In this case, the curve includes an "observed range" and an "extrapolation range." The best way to extend information from the observed data to the unknown, especially low dose, part of the response curve is to find a model that is consistent with the biological mechanisms and processes of the observed range. Biologically based models are often preferable but require large amounts of data reflecting processes like cell growth dynamics; enzyme processes; or factors that have been shown to be associated with some aspect, for example, timing, of the response, perhaps gender, age, body weight, and/or genetic variation (polymorphism). In the absence of a biologically based model, dose-response modeling usually consists of empirical curve fitting [30].

There are several ways to extrapolate from high to low doses, from animal to human responses, and from one route of exposure, but each introduces uncertainties. Risk at low exposure levels cannot be measured directly either by animal experiments or by epidemiological studies, necessitating the need for mathematical models and procedures to extrapolate from high to low doses. For both cancer slope factors and unit risks, various extrapolation models or procedures, while they may reasonably fit the observed data, may lead to large differences in the projected risk at low doses. Selecting the low-dose extrapolation method depends on the properties of the compound that affect the mechanisms of action for carcinogenesis and other relevant biological information. The choice does not solely rely on goodness of fit to the observed tumor data. With limited data and when uncertainty exists regarding the mechanisms of carcinogenic action, however, models that incorporate low-dose linearity are preferred if compatible with the information available. For example, the US EPA usually uses the linearized multistage procedure in the absence of adequate information to the contrary [31].

Building a knowledge base that accounts for adverse effects may not directly need to include dose-response information, but the information and knowledge producer should be aware of the source of the hazard data. Dose-response is commonly derived in a stepwise manner using the studies mentioned and fitting the experimental dose-response data from these studies with a mathematical model that fits the data reasonably well. The fitting must show the upper confidence limit (e.g., 95%) line equation for the selected mathematical model. The response point just below the lowest measured response in the experimental point is where the extrapolation from actual observational data begins, that is, the point of departure. The response is often assumed to be a linear function of dose from the point of departure to zero response at zero dose. Calculating the dose on the line is estimated to produce the response. The cancer dose-response curve is assumed to have no threshold (NOAEL), so the curve intercepts the y-axis at zero [32].

The first step of the linearized multistage procedure is to fit a multistage model to the data. Multistage models are exponential models approaching 100% risk at high doses, with a shape of the dose-response curve at low doses described by a polynomial function. For a first-degree polynomial, the model is equivalent to a one-hit model, which approximates a linear relationship between dose and cancer risk at low doses. The second step is to estimate an upper bound for the risk by incorporating an appropriate linear term into the statistical bound for the polynomial. When exposures are sufficiently low, any higher-order terms in the polynomial will contribute negligibly, and the graph of the upper bound will appear to be a straight line.

Other models can be used to extrapolate from the known to the unknown regions of dose-response (more commonly, the somewhat known to the lesser known regions). These include Weibull, probit, logit, one-hit, and gamma multihit models.

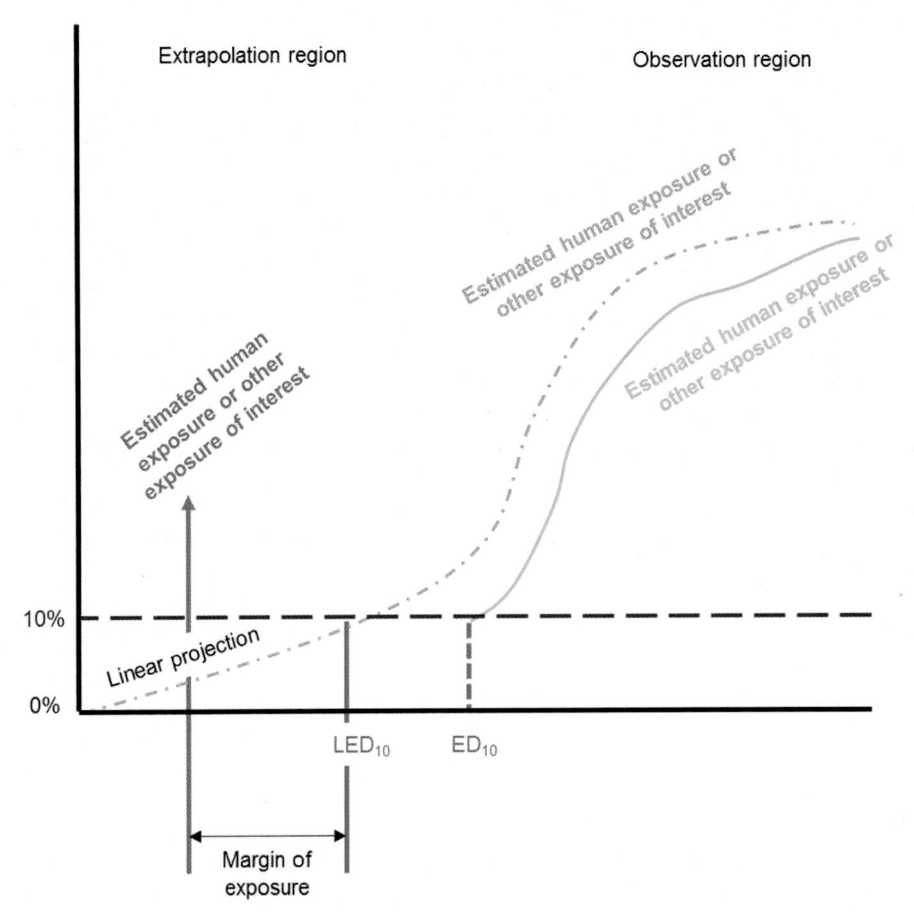

FIG. 17.7 Dose-response curve showing regions populated with observed data and data extrapolated using linearized multistage, no-threshold model. LED_{10} is the lower 95% confidence limit on dose associated with 10% risk; LED_{10} is the estimated dose needed to increase the response, for example, cancer by 10%. *(Source: Source: D. Vallero, Excepted and Updated from: Translating Diverse Environmental Data into Reliable Information: How to Coordinate Evidence from Different Sources. Academic Press, 2017; D. Vallero, Translating Diverse Environmental Data into Reliable Information: How to Coordinate Evidence from Different Sources. Academic Press, 2017.)*

Except for the one-hit model, they all tend to give characteristic, sigmoid (i.e., S-shaped) dose-response curves, with varying curvature and tail lengths. Unless corrected, their upper bounds generally follow the curvature of the models themselves, which is the case for the linearized multistage procedure. The slope factor designated in this Appendix is the slope of the straight line from the upper bound risk at zero dose to the dose producing an upper bound risk of 1%.

When building knowledge bases, scientists are confronted with the need to extend knowledge from the known domains to the unknown or lesser known domains. Certain diseases, effects, stressors, scenarios, and other phenomena are well studied or "data-rich," whereas others are understudied or "data-poor." Even if a phenomenon is studied, certain aspects are unknown, such as what happens at low concentrations. Test equipment and methods are limited; for example, limits of detection (LoD) in analytic equipment means that a substance is not detectable below a certain concentration. It is tempting to simply draw a line to the origin from the lowest measured point, but few biological phenomena behave consistently as dose increases (note the sigmoid shape and different slopes). The amount of time it takes before a response occurs indicates one aspect of the toxicity of a substance. For instance, cancer and other chronic diseases often have protracted latency periods that are impractical to duplicate with in vivo studies. For animal studies, the lethal exposure duration (LED) is used to try to bridge the known responses to the unknown responses. The LED_{10} in Fig. 17.7 is the lower 95% confidence limit on a dose associated with 10% extra risk. This is an example of such a point and in fact is often the standard point of departure. The central estimate of the LED_{10} (the estimate of a 10% increased response) also may be used to describe a relative hazard and potency ranking.

Environmental and health data sets can be based on various outcomes or effects. These include both noncancer and cancer effects. Noncancer effects include neurological, developmental, and cardiovascular diseases. Indeed, cancer effects are singled out, not only by scientists but also by policy (e.g., the decision not to allow thresholds in cancer dose-response curves). The International Agency for Research on Cancer (IARC), within the World Health Organization, promotes international collaboration in cancer research [33]. As such, they are a good source for information regarding carcinogens, including the IARC databases. The Cancer Incidence in Five Continents (CI5) series provides cancer incidence[1] data by geographic location, usually at five-year intervals, containing

1. incidence rate tables by sex, age group, and cancer site;
2. summary tables comparing registries;
3. index tables showing of the validity and completeness of the data.

A common challenge for information producers is the balance between temporal consistency and the need to update data elements and other component data sets considering societal and scientific changes. For example, if previously unrecognized cancer risk factor was discovered 25 years after the design of the data set, it would have to be added and denoted on tables; otherwise, it would appear that the risk factor contributed to zero cases 26 years ago, when it had been contributing but was previously unrecognized. This may be considered a "data false negative," that is, the phenomenon was there but the database did not previously include it. Indeed, this is not uncommon, so it behooves the user of secondary data to review the metadata and otherwise determine if the database has changed with time.

In this example, the extension is physiological, the chemical and biological fate of BAD after it is inhaled. Both data sources are merely BAD concentrations in a substrate, either air or urine. We know the BAD concentrations for a limited subpopulation. We know the BAD concentration in urine after the air is inhaled. Incidentally, the measurement of chemical concentrations in urine, feces, and other elimination product concentrations is known as biomonitoring.

When a substance measured by biomonitoring shows what is happening inside an organism, for example, state of health; disease; organ damage; contamination; and, yes, exposure, the substance is called a biomarker. In this simplistic example, our biomarker is the contaminant itself, that is, BAD. However, we could also and usually do measure BAD's metabolites and endogenous substances, for example, enzymes, that indicate exposure to BAD or damage that the BAD has done.

Among the ways to assess reproducibility, the investigator can run the forward and reverse models. For example, air pollution data may be available at the source and at the receptor. Let us say that a factory's stack has been tested and shows that 200 g of a chemical is released each hour and that 1 km downwind measurements average $2\,\mathrm{mg\,m^{-3}}$ of that same chemical. A forward model may indicate that the plume produced from the stack should produce $3\,\mathrm{mg\,m^{-3}}$. An inverse, for example, back trajectory, model estimates that the $3\,\mathrm{mg\,m^{-3}}$ downwind should come from a stack releasing $250\,\mathrm{g\,h^{-1}}$. The modeler, then, would have to determine first whether this difference is important, for example, statistically significant. If so, then the modeler would have to determine if there are other sources besides the stack for this pollutant and if the pollutant is expected to react in the plume. If so, let us further assume that the chemical indeed breaks down, so that only 80% of the emitted mass would be expected at the downwind site. The modeler also discovers that two other sources are contributing 0.5 and 1.2 g of the chemical to the plume. The subtraction for the reactions and the additions for the two new sources are then factored in, and the models are reparameterized. Now, the inverse model indicates that the original source should be releasing $220\,\mathrm{g\,h^{-1}}$, and the forward model predicts the downwind site would be expected to be $2.8\,\mathrm{mg\,m^{-3}}$. This may be acceptable or unacceptable, depending on the objectives of the study, that is, data quality and model quality objectives. The bottom line is that numerous data sets can be used to improve scientific confidence.

Data used to populate models range in extent and quality, for example, air emission data may be focused on a single source for a finite time period, whereas air quality data often include large areas and long timeframes. In addition to scientific credibility, another of the main reasons that science requires peer review is to assure that the study, as described, if replicated, would arrive at the same results.

17.3 ETHICS

All scientific disciplines have standards of behavior, but not all are codified. Professionals, for example, engineers, physicians, and attorneys, have codified norms, that is, ethical canons and codes. Air pollution draws on these, but arguably most often, it is the engineers who play prominent roles in air quality decisions. According to their codes of ethics and practice, engineers must hold paramount the public's safety, health, and welfare [34]. Engineers apply the sciences to address societal needs. Environmental engineers are particularly interested in protecting public health and ecosystem conditions [35–37].

1. Incidence is the epidemiological term for new cases for a specific time period, for example, annual incidence, whereas prevalence is the number of total cases.

Ethics may be an unexpected subject in a calculation text. However, engineers, scientists, and managers employ techniques to determine whether an action is "good" or "not so good." Or it may be better stated as "good versus good" [38]. The decision tools used to support an air pollution action have inherent weaknesses.

Certainly, decisions and actions that unnecessarily expose people and ecosystems to toxic substances, that is, bad air quality decisions, must be avoided. Mindless and misapplied science and technology is immoral. It is most likely that ethical breaches related to air pollution are predominantly unintentional. However, willful dishonest malpractice is rare yet unfortunately does occur. Often, the breach is not related to science or engineering per se, but an administrative, financial, or other ancillary aspect of air quality management, for example, inappropriate billable hours (e.g., double billing), violations of contract specifications, and inappropriate disclosures. These often fall under the fundamental canon that requires that engineers act as faithful agents and trustees [34].

Engineers and scientists may believe that if the work is sufficiently important, that is, removing carcinogens from the environment, they will be excused if they cut corners or otherwise violate an ethical principle. After all, they are result-oriented and may succumb to the fallacy that ends justify the means, even unethical means. Indeed, the author has observed the woes of very competent scientists who believe that the primacy of science will inoculate them from ethical misconduct. It certainly does not.

Even the choice of tool or model is laden with ethics. For example, an engineer may have used a model so frequently in the past that little or no thought goes into whether the model is appropriate for a given situation. The engineer has lost objectivity in the model selection process due to selection bias, that is, only choosing methods or concepts that agree with one's own perspective, avoiding those that may be better or more appropriate but that do not follow one's previously conceived paradigm. Engineers belong to a helping profession, trying to achieve outcomes that serve humankind by altering some aspect of the natural or built environment. However, the way this is done can lead to problems. Sometimes, engineers, scientists, and managers become practically myopic, solving an air pollution problem, but in the process causing other potentially worse problems. Examples are removing fly ash and sulfur dioxide from flue gases but myopically improperly storing the sludge and ash in pits that contaminate surface and groundwater; requiring additives to fuels to reduce the emissions of carbon monoxide but that themselves become air and water pollutants; and introducing cutting-edge treatment technologies that rely on and diminish the availability of finite elements that are needed elsewhere by society, for example, in medical devices. As the philosopher Georg Wilhelm Friedrich Hegel put it, "The essential tragic fact is not so much the war of good with evil as the war of good with good" [39].

In addition to "good versus good," there are also examples of "good versus good enough." Environmental scientists are challenged with the question, "How clean is clean?" Perhaps, the colloquial question, "Are we done yet?" is more appropriate.

Everything in this book aims to improve air quality, but some approaches are better than others in terms of credible science and feasibility. Indeed, the famous engineer, Norman Augustine, has stated that "…Engineers who make bad decisions don't know that they are confronting ethical issues" [40].

Air quality decisions must be underpinned by sound science and engineering. The public must be able to trust practitioners. Unfortunately, this trust is put into jeopardy when practitioners cut corners and otherwise do not adhere to the ethical standards. The most recent, notorious example of this breach of trust is "Dieselgate," in which Volkswagen has been accused in 2015 of intentionally programmed turbocharged direct injection diesel engines to activate their emission controls only during laboratory emission testing, which caused the vehicles' NO_x output to meet US standards during regulatory testing, but emit up to 40 times more NO_x in everyday driving [41]. Volkswagen is said to have deployed this programming software in about 11 million cars worldwide in model years 2009 through 2015.

17.3.1 Responsible conduct of air pollution research [35–37]

Researchers and practitioners in the biomedical and environmental sciences must balance the advancement of the respective fields with the ethical treatment of those adversely affected by exposures to myriad agents. To address air pollution, they must advance the science while simultaneously controlling and preventing the emissions of air pollutants. The amount of pollutant considered to be safe is determined from extrapolations of limited data sets using models with varying amounts of uncertainty. For example, it is likely that workers in some parts of the world are presently and routinely exposed to levels of hazardous air pollutants that would not be permitted even in short-term chamber experiments in the developed world. Is the researcher who travels to these parts of the world to conduct air pollution studies being unethical or merely realistic? [35–37].

Indeed, this ethical dilemma is more common for journalists and anthropologists than to biologists, engineers, and biomedical researchers. Journalism and anthropology expect researchers to be thrust into the lives of those they are studying. Commonly, however, they are asked to drop the notepad or camera and to help the objects of their research. Researchers in the physical and natural sciences are trained to maintain objectivity. Conversely, engineers and physicians are trained to hold paramount the public and patient, respectively. This leads to an ethical problem, then, when maintaining objectivity in some way detracts from the paramount calling of these professions.

17.3.2 Systems thinking

Environmental and biomedical ethics can converge and be scaled to address not only the technical aspects of air pollution but also the ethical aspects. For example, physicians, engineers, and research scientists each address the problems posed by lead and mercury pollution but at different scales and focuses. The physician must give primacy to the patient's well-being. The engineer must hold paramount the safety, health, and welfare of the public. The researcher must respect and ensure the consent of any subject. The professional and research ethical constructs can be interwoven within the environmental and biomedical scientific communities who must advance their individual sciences while, increasingly, take a system view to incorporate numerous perspectives. As the "bright lines" between disciplines are blurring, so must the ethical canons, that is, commensurate "system ethics" and "life cycle ethics" to address air pollution and its impacts, which range from nanoparticles to global climate change.

Ethical decision-making in medicine, public health, and environmental protection must combine technical and ethical factors. It makes use of multiplex optimization or benchmarking, where only certain outcomes are acceptable. A technically acceptable outcome may be ethically unacceptable, and an ethically acceptable outcome may be technically and fiscally infeasible. The tools needed to evaluate the health and environmental benefits and risks share aspects of most decision support tools. This means that new decision support tools must be employed to consider risks and costs over the life of the technology and beyond.

Case 1: Downstream Climate Impacts

One metric of the ethics shared by medicine and environmental professionals is whether a chemical, biological, or physical agent poses or could pose *unacceptable risk.* Risk is the likelihood of negative outcomes. Indeed, the concept of air pollution is one of risk. The same substance can be essential in one scenario but an air pollutant in another, depending on the risk. For example, there has been much recent debate in the United States as to whether carbon dioxide (CO_2) should be designated a criteria pollutant under the Clean Air Act (CAA). A criteria pollutant is one considered harmful to public health and the environment. The CAA identifies two types of national ambient air quality standards:

Primary standards provide public health protection, including protecting the health of "sensitive" populations such as asthmatics, children, and the elderly. Secondary standards provide public welfare protection, including protection against decreased visibility and damage to animals, crops, vegetation, and buildings [42, 43].

Unlike lead (Pb), carbon monoxide, sulfur dioxide, and the other criteria pollutants, CO_2 is needed for photosynthesis and is the product of respiration. Those against including CO_2 on the list of criteria pollutants argue that it is not "inherently" toxic and indeed is essential.

Those in favor of including CO_2 on the criteria pollutant list argue, however, that it is the product of complete combustion. As such, the compound is a surrogate for fossil fuel combustion, which represents the anthropogenic contribution of global greenhouse gases to the atmosphere. They argue that by extension CO_2 is a health threat, not by the typical air pollution exposure through the lungs, but because of cardiovascular and other diseases caused by higher ambient temperatures and other weather phenomena induced by higher concentrations of this greenhouse gas in the troposphere.

The risks are displaced temporally and can seem isolated from their causative agents. The prototypical cause-and-effect paradigm of air pollution is violated. That is, risk is a function of exposure and hazard. Typically, society recognizes certain agents as inherently toxic, that is, carcinogens, endocrine disruptors, and neurotoxins. However, the risk of many substances can only be defined within a given scenario. Chromium, for example, is inherently toxic only in certain oxidation states; for example, Cr^{6+} is carcinogen; Cr^{3+} is essential and might have been in your breakfast cereal this morning.

With complex, large-scale, potentially irreversible scenarios like climate change, the health effects may come because of downstream impacts; for example, warning leads to heat islands that lead to increased numbers of cardiovascular diseases and mortality. In addition, these complex problems often include psychosocial and ethical considerations. For example, any increase in the number and severity of other events brought on by climate change, such as flooding, may also increase psychological stress and concomitant cardiovascular response.

Downstream impacts, for example, warmer temperatures and altered hydrologic cycle, can also lead to changes in ecosystems that in turn affect human health. Climate change can alter seasonality of some allergenic species and disease vectors. Insect and other vector populations could lead to increases in the number and severity of vector-borne diseases. These include changes in spatial and temporal distributions of malaria, dengue, tick-borne diseases, cholera, and some other diarrheal diseases [43a]. These diseases can be expected to increase cardiovascular and other stress in human populations. Thus, opportunistic diseases brought on by environmental change that is induced by air pollutants (e.g., greenhouse gases) can result in increased cardiovascular effects in populations [44].

One of the considerations of global climate change is the extent to which the release of air pollutants may create extreme heat events. Heat waves are generally characterized by several days of warm, stable air masses, with daytime temperatures >32°C and consecutive nighttime temperatures above normal. If climate change does indeed increase the number of and severity of heat waves in the temperate zones, this will be met by greater cardiovascular effects.

Heat waves presently are the most deadly, weather-related exposure in the United States, leading to a greater number of deaths annually than those attributed to hurricanes, tornadoes, floods, and earthquakes combined. Like direct air pollutant exposures, heat-sensitive subpopulations include children and the elderly. Climate change will bring more heat waves to the United States. Increases in the number of people living in cities and population aging will further increase heat-related health risks. Studies suggest that, if current emissions hold steady, excess heat-related deaths in the United States may increase from the present average of about 700 each year as many as 5000 per year by 2050 [44a].

The indirect health effects demonstrate the complexity of causal links between emissions of air pollutants and health and ecosystem well-being. This applies not only to cardiovascular effects but also to most other adverse health and ecological outcomes [44].

Case 2: Bioaerosols and Risk Uncertainty

Whether risk is excessive is a social construct. Societal expectations of acceptable risk are mandated by codified regulations, rules, standards, and specifications, such as health codes and regulations, zoning and building codes and regulations, principles of professional engineering and medical practice, criteria in design guidebooks, and standards promulgated by international agencies (e.g., the International Standards Organization (ISO)) and national standard-setting bodies (e.g., the National Institute for Standards and Testing [45].

The same agent may be regulated differently in a biomedical setting than in an environmental setting. For example, genetically modified organisms (GMOs) are controlled to varying degrees by institutes of biomedical sciences, such as the American Medical Association and regulatory agencies, whereas food safety and environmental agencies, such as the US Food and Drug Administration, the US Department of Agriculture, the US Environmental Protection Agency, and their respective state counterpart agencies, are responsible for new biotechnologies in their respective areas. The biomedical research concern is often whether an agent can be contained within the research facility, whereas the environmental concern is whether the GMO, for example, an oil-eating bacterium, can be contained within the service area (e.g., the extent of an oil spill).

Air pollution is not the usual biotechnology topic. However, there is a growing interest in so-called bioaerosols and gene flow. Bioaerosols include any airborne organisms or their genetic material-bearing substances, such as cysts, seeds, spores, and pollen. In addition, there is concern about the atmospheric transport of these materials that can lead to horizontal diffusion of the genetic materials from GMOs, that is, gene flow.

Gene flow is an example of the biotechnological downstream risks, as evidenced by proteins from the common soil bacteria *Bacillus thuringiensis* (Bt). In Europe and North America, regulators generally require reporting of a GMO before it is introduced into the marketplace. Interestingly, in the United States, a newly modified organism is something akin to a new chemical under the Toxic Substances Control Act. In the United States, the data requirements include product characterization; mammalian toxicity; allergenicity potential; effects on nontarget organisms; environmental fate; and, for the Bt products, insect resistance management to product from losing use of both the microbial sprays and the Bt plant-incorporated protectants (PIPs). A transgene is an exogenous gene that has been introduced into the genome of another organism, and a transgenic species is one whose genome has been genetically altered. For instance, if a biotechnology is used as a PIP, the movement of transgenes from a host plant into weeds and other crops presents a concern that new types of exposures will occur. To date, *Bt* corn and potato PIPs that have been registered to date have been expressed in agronomic plant species that, for the most part, do not have a reasonable possibility of passing their traits to wild native plants. Most of the wild species in the United States cannot be pollinated by these crops (corn and potato) due to differences in chromosome number, phenology, and habitat. There is a possibility, however, of gene transfer from Bt cotton to wild or feral cotton relatives in Hawaii, Florida, Puerto Rico, and the US Virgin Islands. Where feral populations of cotton species like cultivated cotton exist, regulators have prohibited the sale or distribution of *Bt* cotton in these areas. These containment measures prevent the movement of the registered Bt endotoxin from Bt cotton to wild or feral cotton relatives.

Although, based on data submitted by the registrant and a review of the scientific literature, US regulators have concluded that there is no foreseeable risk for the specific case of unplanned pesticide production through gene capture and expression of the Colorado potato beetle control protein (Cry3A) in wild potato relatives in the United States. Tuber-bearing *Solanum* species include *Solanum tuberosum. S. tuberosum* cannot hybridize naturally with the non-tuber-bearing *Solanum* species. Three species of tuber-bearing wild species of *Solanum* occur in the United States: *S. fendleri, S. jamesii,* and *S. pinnatisectum.* But, successful gene introgression into these tuber-bearing *Solanum* species is virtually excluded due to constraints of geographic isolation and other biological barriers to natural hybridization [38]. In this case, the regulators are providing a very specific answer to a specific question, that is, will there be gene flow problems associated with this specific GMO used in this way? Unfortunately, this could open the door for other uses that may indeed cause downstream problems, for example, gene flow of similar GMOs into nonpotato crops. The altered genetic material is no longer contained.

This is a weakness of premanufacture notification, not only for GMOs but also for any new substance, either chemical or biological. That is, the notification is tightly and specifically defined, but the potential use of the substance can be large once it reaches the marketplace. Also, the decision for approval must be made in relatively short period of time, for example, 90 days. Thus, the onus falls on the regulators to decide whether it is safe, rather than the manufacturers to prove it is safe. This is the reason that many nations are moving from the evidence-based risk model to one of precaution, especially for decisions that may lead to severe and potentially irreversible impacts. Although the United States continues to hold a risk-based approach for pharmaceutical, air pollutants, and other substances, a coalition of 13 states has released their own principles asserting that chemical and product manufacturers should be required to develop and provide sufficient toxicity, exposure, and use information to regulators to conclude that chemicals and products in commerce do not endanger the public or the environment; they also call for protocols to evaluate potential alternatives to chemicals of concern. The states maintain that emerging chemicals of concern should be assessed for public and environmental safety before they go into widespread commerce and use. States acknowledge the need for a strong federal chemical regulation system while expressly preserving the authority of state and localities to implement measures to manage chemicals of concern [46].

These cases set the stage for distinguishing between research ethics and professional ethics. For example, if the researcher establishes a study design to evaluate the difference in risks between an exposed population (e.g., people too poor to leave a region experiencing increasing temperatures or to control their diet to avoid GMOs) and unexposed (e.g., people who left the region or choose to avoid GMOs), this is more or less a "natural experiment," albeit unfortunate for those being exposed. However, if the researcher is also an engineer who has found a way for people to leave or to avoid GMOs but withholds this knowledge from the experimental ("exposed") group, this is arguably unethical. If the group is assigned blindly to the control or experimental group, the scientific integrity of the study may well be satisfied, but likely not its moral integrity. Indeed, if the engineer-researcher had gained the confidence of the population, this would be a form of the therapeutic misconception experienced by physician-researchers.

The examples of climate change and biotechnologies illustrate that an air pollutant's concentration must be defined by what society considers to be an acceptable and reasonable risk (see Chapter 12). The question is where to draw the line between unacceptable and acceptable and between unreasonable and reasonable. There is little trouble building a consensus that exposing people to unacceptable risk is unethical. However, the acceptability of a pollutant concentration has both inherent and instrumental aspects. For example, dimethylmercury is inherently hazardous and is used in strictly controlled, laboratory settings. However, if none of the mercury ever leaves the lab hood, its actual risk is zero, notwithstanding its inherent properties. Thus, the use scenario drives an air pollutant's level of acceptability. As such, acceptability is value-laden. A device that destroys a tumor or that advances biomedical and environmental research may be worth the exposure to its inherently hazardous properties; so long as the risks are described accurately, the decision-making is transparent, and the ethics are considered by an objective collection of reviewers, for example, the institutional review boards who must ensure respect for persons, justice, and beneficence [46a].

As discussed in Chapter 12, a useful measure of risk acceptability is whether it is "as low as reasonably practical" (ALARP), a concept coined by the UK Health and Safety Commission and published by the UK Health and Safety Executive [46b]. The range of possibilities fostered by this standard can be envisioned to fall within three domains. Actions falling in one domain are associated with risk that is clearly unacceptable. At the other end of the risk spectrum are actions that are associated with unquestionably acceptable risk. In the middle are those actions that are more ambiguous from a risk perspective. The size of these domains varies considerably, so that reaching consensus about acceptable risk is difficult, and there is almost never unanimous agreement on the differences between acceptability and unacceptability.

Risks in the ALARP region need to be managed scientifically and ethically to produce an acceptable outcome. Thus, this process may prove useful in deciding actions related to the emission of an air pollutant. For example, it may identify which actions provide the greatest good compared with the potential harm that they may cause. For example, consider a propellant used for an asthma inhalant that can only be delivered efficiently by a gas that if present in the stratosphere would destroy ozone. It may be decided that this use should continue if it is the only reliable means of delivering the asthma medication to a particularly sensitive group. However, such single-variable assessments are uncommon and can lead to erroneous predictions of outcome. For example, biomedical research may find alternative gases that do not destroy ozone or a mechanism that does not use a propellant at all, for example, a mechanical air pump. The key is that the decision must be systematic and must consider the entire life cycle of the potential pollutant.

The ALARP depends on a defensible margin of safety that is both protective and reasonable. Hence, reaching ALARP necessitates qualitative and/or quantitative measures of the amount of risk reduced and costs incurred with the design decisions. The ALARP principle assumes that it is possible to compare marginal improvements in safety (marginal risk decreases) with the marginal costs of the increases in reliability [47, 48].

Air pollution risk assessment has numerous sources of uncertainty. For example, the dose-response relationships are difficult to extrapolate from animal studies to human population. Also, the likely exposure can be variable and highly uncertain.

Even the physical properties of an air pollutant are difficult to characterize. As evidence, engineering materials are being with dimensions <100 nm (i.e., nanoparticles) that behave very differently after being emitted. In fact, the air pollution community is concerned about the health effects of ultrafine particles (also with aerodynamic diameters <1 nm), while the biomedical, chemical, and other communities are trying to develop nanoparticles with special features. Indeed, the medical and air pollution communities are working at opposite sides of the same coins. Drug delivery researchers are looking for ways to enhance dose, while air pollution experts are seeking means to decrease biologically effective dose. Since risk is a function of hazard and the exposure to that hazard, reliable assessment of that risk depends on sound physical characterization of the hazard. However, if even the physics is not well understood due to the scale and complexity of the research, the expected hazards and potential for exposure to humans and ecosystems are even less well understood.

Indeed, the ethical uncertainty of air pollution is propagated in time and space. Nascent areas of research include ways to link protein engineering with cellular and tissue biomedical engineering applications (e.g., drug delivery and new devices); ultradense computer memory; nonlinear dynamics and the mechanisms governing emergent phenomena in complex systems; and state-of-the-art nanoscale sensors, including those based on photonics. Complicating the potential societal risks, much of this research frequently employs biological materials and self-assembly devices to design and build some strikingly different kinds of devices. Among the worst-case scenarios has to do with the replication of the "nanomachines." Advancing the state of the science to improve the quality of life (e.g., treating cancer, Parkinson's disease, and Alzheimer's disease and improving life expectancies or cleaning up contaminated hazardous wastes) can introduce difference risks [49].

Rarely is there a simple answer to the questions "How healthy is healthy enough?" And "How protected is protected enough?" Managing risks consists of balancing among alternatives. Usually, no single solution to an environmental problem is available. Whether a risk is acceptable is determined by a process of making decisions and implementing actions that flow from these decisions to reduce the adverse outcomes or at least to lower the chance that negative consequences will occur [50].

The social contract with the scientific and engineering communities expects that air pollution risks will be low, all other things being equal. Derby and Keeney [51] have stated that "acceptable risk is the risk associated with the best of the available alternatives, not with the best of the alternatives which we would hope to have available." Calculating the risks associated with these alternatives is inherently constrained by three conditions [52]:

1. The actual values of all important variables cannot be known completely and thus cannot be projected into the future with complete certainty.
2. The physical and biological sciences of the processes leading to the risk can never be fully understood, so the physical, chemical, and biological algorithms written into predictive models will propagate errors in the model.
3. Risk prediction using models depend on probabilistic and highly complex processes that make it infeasible to predict many outcomes.

17.3.3 Uncertainty

The benefit-to-cost ratio (BCR) discussed in Chapter 15 commonly supports environmental decision-making. However, the BCR is flawed or at least incomplete for many air pollution decisions. When comparing benefits with costs, values are inaccurate. Given the uncertainty, even a benefit/cost ratio that appears to be mathematically high, that is, $\gg 1$, may not provide an ample margin of safety given the risks involved. Environmental valuation is difficult since many of the values are intrinsic and not necessarily instrumental. Therefore, other values, for example, transportation efficiency, return on investment, and energy capacity, are more amenable to the BCR. Nonmonetized values like habitat preservation or protection of nongame fish spawning areas are not.

Recent advances in computational and systems sciences are affording a new look at BCRs and may be a means of preventing developing nations from repeating some of the air pollution episodes that provided painful lessons. Throughout the last quarter of the 20th century, emerging technologies were shared among Western nations and Eastern Europe and the previous Soviet Union states, with south Asia and with many other parts of the world, notwithstanding the geopolitical obstacles and cultural differences. These transfers included exponential advances in the state of the science in both conventional pollutants and air toxics.

The advances now provide data on air pollutant concentrations, air pollution exposures, and health status of populations that can be mined to provide insights that were simply not possible before. Informatics is available to interpret data and observe patterns of air pollutants to make heretofore impossible associations with effects on humans and ecosystems.

Computational methods are being applied to fluid dynamics, toxicology, and other sciences to build models that explain the movement and change of pollutants, potential exposures, and the kinetics and dynamics of these pollutants and their metabolites in humans and other organisms. Computational and informatic tools can also be applied in ways to inform and to select options for controlling the emissions, including the expected change to ambient air quality with each option. Better statistical tools are also needed to compare options, for example, Bayesian approaches that update outcomes (i.e., posterior distributions) with various control strategies.

Spatiotemporal tools are also becoming increasingly reliable. Air pollution curricula are increasing the use of geographic information systems (GIS), geostatistical methods (e.g., kriging), and land use regression techniques. Such tools can be very useful in preliminary screening of air pollution exposures and in linking potential sources to measured concentrations.

All these tools already exist, but they need to continue to be fitted to air pollution needs. This may be particularly challenging for complicated mixtures, for example, urban air toxics, and coke-oven emissions and real-world indoor pollutant mixtures. These tools are only now being applied to human activities and the use of products. Such psychosocial predictions are difficult given the many variables of humans in various microenvironments, but this is the only accurate depiction of how a person encounters an air pollutant.

The twofold challenge of addressing emerging air pollutants (e.g., bioaerosols and air toxics) and transferring lessons learned to developing economies will be called the simultaneous advancement of very low-tech and very high-tech portable, remote measurement systems. The existing system of collecting a sample and conducting laboratory analysis may not work well for remote areas. This calls for low-maintenance, adaptive technologies (e.g., passive monitors with no need for a pump) and low-maintenance, high-technology systems (e.g., solar powered and long-life battery-powered systems, with satellite links). Sophisticated analyses, for example, organic chemical analysis, will have to be conducted in many of these areas using portable and open-path technologies. These are currently in use but are limited in the number of chemicals detected and have transport and other logistic challenges, which will need to be improved for wider use.

Scientists and policy makers have recommended taking the precautionary approach to climate change. That is, even if the data are uncertain, corrective actions that do not have sufficient risks themselves should be taken. The system view reminds us that corrective actions often have their own risks; for example, fuel additives like methyl tert-butyl ether (MTBE) were effective in raising gasoline's octane rating and lowering mobile source CO emissions by oxygenation but were subsequently found to damage other parts of the environment.

Air pollution threats to human populations and ecosystems are highly complex. Thus, they must be viewed systematically. Every hazard exists within a milieu of biomedical and environmental life cycles. Identifying the factors that have contributed to the occurrence of air pollution problem is complicated. Few credible information sources are available about what was occurring during at any given time period. Almost always, measurement data are not sufficient to characterize the movement and change of materials in the environment in a way that fully explains air pollution and rely on models to extrapolate and interpolate to find meaning.

Global greenhouse gas emissions continue to rise. Numerous toxic compounds continue to be released. Risks to human health persist. Sensitive ecosystems are threatened. As mentioned, such problems are not readily resolved by the old "command and control" approaches. Innovations and market forces (e.g., incentive and emission banking and trading) must also be part of the solution.

The problems will best be addressed by new thinking that is underpinned by sound science. Philosopher Immanuel Kant is famous for the categorical imperative, which says that the right thing to do requires that a person must "...act only on that maxim whereby thou canst at the same time will that it should become a universal law" [53]. Thus, deciding on the acceptable and reasonable risk of air pollution means that it is duty of the biomedical and environmental profession to think about what would happen if everyone acted in the same way. This is the essence of sustainability. The only way to ensure that something is protected for the future is to think through all the possible outcomes and select only those that will sustain a better world.

This calls for a deep understanding of systems that operate once the design is implemented and, ideally, forms a foundation for the exploration and discovery of innovative ways to minimize risks to health and safety and increase design reliability and improve the environment. With a better understanding of sustainable processes, new strategies will emerge to supplant old ways of thinking, especially replacing those antiquated templates that depend on the subjugation of nature to achieve human ends.

It is the duty of those who have learned the hard lessons of air pollution to share with others before they repeat the mistakes of the past. The tragedies of the air pollution episodes of Donora, London, Bhopal, and elsewhere have led to advances. The advances in control and prevention that have led steady decrease in the concentrations of carbon monoxide, sulfur dioxide, nitrogen dioxide, lead, and other pollutants in North America and Europe must be shared. In addition, looming threats like climate change and stratospheric ozone destruction depend on knowledge transfer and a more sustainable system view of the atmosphere. To repeat the mistakes would be doubly tragic since we now know so much more about how not only to control their emissions but also to prevent them by removing them far upstream in the life cycle.

The biological, engineering and biomedical researcher must take care to ensure that risks are properly addressed in any study. These researchers have a rather unique challenge in balancing sound science with ethical principles, that is, they belong to professions aimed at helping individuals and the public. They must be ever mindful of not justifying noble scientific ends by using means that are available, convenient, yet unethical.

17.4 ETHICAL ANALYSIS OF DECISIONS AND ACTIONS

At the outset, it must be stated that, compared with the calculations in previous chapters, determining the extent to which an action is ethical is seldom quantitative nor even semiquantitative. At best, it is a well-documented, qualitative "calculation." This section will, however, provide some numerical approaches that may be best called "pseudoquantitative." Such calculations can give the false impression of being absolute. As such, the numerical values should never be presented without a narrative, that is, an explanation of the inherent weaknesses, limitations, and assumptions.

Arguably, the biggest air pollution engineering ethical challenges arise around decisions to deploy emerging and untested technologies. These do not enjoy a body of supporting literature, which has evolved from controlled experiments, to pilots, to limited field studies. The technology may be at the earliest stages of reliability, for example, the first phase of the bathtub curve (see Fig. 16.1 in Chapter 16). Like most newly introduced technologies, air pollution control systems may only reveal problems after applications under various conditions. Witness the advice to avoid purchasing a car the first year that major purported improvements are made. As such, the engineer may only have the documentation presented by a vendor or otherwise sources with conflicts of interest, that is, they want to see the device, method, or system installed for monetary reasons. This is certainly not inherently unethical, given that the state of the science and engineering has advanced this way. However, the decision must be as objective as possible, so the engineer must do the homework to ensure that the technology is not only technically and financially sound but also appropriate for an application.

If data are available, they may be limited in scale, such as those obtained using Likert scales, that is, a survey in which participants select a value that equates to a value or attitude, for example, 1 = strongly approve, 2 = approve, 3 = undecided, 4 = disapprove, and 5 = strongly disapprove. Such scales produce only ordinal data, that is, the size of the ranges within each category will vary, providing only the order of opinion. A score of 1 is more acceptable than a score of 2 according to the respondent. Thus, calculating measures of central tendency are to be avoided for ordinal data and restricted to interval and ratio data.

Despite a common code of practice and ethics, those who engage in the sciences have many attributes of a professional. Their credo is the scientific method. They are responsible for their own research and work. The buck stops with the individual practitioner of science. The credo of the professional is *credat emptor*, that is, let the client trust. Environmental and public health professionals are charged with responsibilities to protect the public and ecosystems. When failures occur, the professionals are accountable. When a manufacturing, transportation, or other process works well, the professional can take pride in its success. The professional is responsible for the successful project. That is why we went to school and are highly trained in our fields. We accept the fact that we are accountable for a well-running system. Conversely, when things go wrong, we are also responsible and must account for every step, from the largest and seemingly most significant to those we perceive to be the most minuscule, in the system that was in place. Professional responsibility cannot be divorced from accountability. The Greeks called this *ethike aretai* or "skill of character." It is not enough to be excellent in technical competence. Such competence must be coupled with trust gained from ethical practice.

One of the difficult tasks in writing and thinking about failures is the temptation to assign "status" to key figures involved in the episodes. Most accounts in the media and even in the scientific literature readily assign roles of villains and victims. Sometimes, such assignments are straightforward and enjoy a consensus. However, often such classifications are premature and oversimplified.

Granted, case studies have their limits, but one can strongly argue that failure has common elements, whether a reactor is nuclear, such as the meltdown in Chernobyl, or chemical, such as arguably the worst reactor disaster on record, the Bhopal toxic cloud in India. Whereas neither was a bioreactor nor biotechnological system disaster, there are profound lessons to be learned.

The Bhopal release killed thousands of people and left many more thousands injured, yet there are still unresolved disagreements about which events leading up to the disaster were most critical. In addition, the incident was fraught with conflicts of interest that must be factored into any thoughtful analysis. In fact, there is no consensus on exactly how many deaths can be attributed to the disaster, especially when trying to ascertain mortality from acute exposures versus from long-term, chronic exposures. Certainly, virtually all the deaths that occurred within hours of the methyl isocyanate (MIC) release in nearby villages can be attributed to the Bhopal plant. However, with time, the linkages between deaths and debilitations to the release become increasingly indirect and more obscure. Also, lawyers, politicians, and business people have reasons beyond good science for including and excluding deaths. Frequently, the best one can do is say that more deaths than those caused by the initial, short-term MIC exposure can be attributed to the toxic cloud. But, just how many more is a matter of debate and speculation. One important lesson to environmental biotechnology, however, is that consequences of inadequate design failure can be very protracted and persistent.

17.4.1 Life cycle methodologies

Discussions in the previous chapters have given examples of the benefits of using biology to solve societal problems, with an eye toward possible hazards from these solutions. The life cycle perspective is very valuable in identifying possible problems, even for some very beneficial environmental applications.

Environmental systems consist of intricately interconnected components that are in balance at many levels, that is, thermodynamic, fluid dynamic, trophic, and physical–chemical-biological cycling. Introducing change, sometimes seemingly small, can have profound impacts on these systems.

This life cycle view also is the first step toward preventing problems. For example, if we consider all the possible contaminants of concern, we can compare which of these must be avoided completely, which are acceptable with appropriate safeguards and controls, and which are likely to present hazards beyond our span of control. We may also ascertain certain processes that generate none of these hazards. Obviously, this is a preferable way to prevent problems. This is a case where we would be applauded for thinking first "inside the box." We can then progress toward thinking outside the box or, better yet in some cases, get rid of the box completely by focusing on function rather than processes.

As mentioned in Chapter 1, the key tool to identify and compare processes in a systematic way is the life cycle analysis (LCA). The complexity of LCA ranges from cursory attention paid, to inputs and outputs of materials and energy, to multifaceted decision fields extending deeply into time and space. The latter is preferable for decisions involving large scales, such as the cumulative buildup of greenhouse gases, or those with substantially long-term implications, such as the release of genetically altered microbes into the environment. Complex LCAs are also favored over cursory models when the effects are extensive, such as externalities and artifacts resulting in geopolitical impacts. Air pollution decisions may fall into any or all of these three categories.

The decision to increase the use of ethanol as a fuel additive and a reformulated fuel is such a decision. Thus, the recent proclamation by the US government to increase ethanol's share of refined fuel to 10% by the year 2012 provides a case study of the application of LCA, from both design and pedagogical perspectives. Ethanol had been increasingly touted as an alternative to crude oil-based fuels. This interest has been diverse, with coverage in the national media and in professional and research journals. In his 2007 State of the Union Address, US President George W. Bush set a two-part goal:

1. Setting a mandatory standard requiring 35 billion gallons of renewable and alternative fuels in the year 2017, which is approximately five times the 2012 target called for in current law. Thus, in 2017, alternative fuels will displace 15% of projected annual gasoline use.
2. Reforming the corporate average fuel economy (CAFE) standards for cars and extending the present light truck. Thus, in 2017, projected annual gasoline use would be reduced by up to 8.5 billion gallons, a further 5% reduction that, in combination with increasing the supply of renewable and alternative fuels, will bring the total reduction in projected annual gasoline use to 20%.

These and other alternative fuel standards have met with skepticism and even dissent. The viability of ethanol has been challenged from scientific and policy standpoints [54, 55]. Corn-based ethanol is indeed a biotechnology. In fact, since the presidential proclamation, dedicated corn crops and bioreactors in these states have emerged. On the other hand, geopolitical impacts, such as food versus fuel dilemmas, are being raised. Scientific challenges to any improved efficiencies and actual decreases in the demand for fossil fuels have also been voiced. Some have accused advocates of ethanol fuels of using "junk science" to support the "sustainability" of an ethanol fuel system. Notably, some critics contend that ethanol is not even renewable, since its product life cycle includes many steps that depend on fossil fuels [54, 55]. The metrics of success are often deceptively quantitative. For example, the two goals for increasing ethanol use include firm dates and percentages. However, the means of accountability can be quite subjective. For example, the 2017 target *could* be met, but if overall fossil fuel use was to increase dramatically, the percentage of total alternative use could be quite small, that is, not near the 15%. Thus, *both* absolute and fractional metrics are needed.

Another accountability challenge is accurately accounting for energy and matter losses in calculations. From a thermodynamic standpoint, the nation's increased ethanol use could actually increase demands for fossil fuels, such as the need for crude oil-based infrastructures, including farm chemicals derived from oil, farm vehicle, and equipment energy use (planting, cultivation, harvesting, and transport to markets) dependent on gasoline and diesel fuels, and even embedded energy needs in the ethanol processing facility (crude oil-derived chemicals needed for catalysis, purification, fuel mixing, and refining). A comprehensive LCA is a vital tool for ascertaining the actual efficiencies.

The questions surrounding ethanol can be addressed using a three-step methodology. First, the efficiency calculations must conform to the physical laws, especially those of thermodynamics and motion. Second, the "greenness," as a metric of sustainability and effectiveness can be characterized by life cycle analyses. Third, the policy and geopolitical options and outcomes can be evaluated by decision force field analyses. In fact, these three approaches are sequential. The first must be satisfied before moving to the second. Likewise, the third depends on the first two methods. No matter how politically attractive or favored by society, an alternative fuel must comport with the conservation of mass and energy. Further, each step in the life cycle (e.g., extraction of raw materials, value-added manufacturing, use, and disposal) must be considered in any benefit-cost or risk-benefit analysis. Finally, the societal benefits and risks must be viable for an alternative fuel to be accepted. Thus, even a very efficient and effective fuel may be rejected for societal reasons (e.g., religious, cultural, historical, or ethical).

The challenge of the scientist, engineer, and policy maker is to sift through the myriad data and information to ascertain whether ethanol truly presents a viable alternative fuel. Of the misrepresentations being made, some clearly violate the physical laws. Many ignore or do not provide correct weights to certain factors in the life cycle. There is always the risk of mischaracterizing the social good or costs, a common problem with the use of benefit–cost relationships.

Biomass-based fuel efficiencies are evaluated in terms of net energy production that is based on thermodynamics (first and second laws). The numerator of the efficiency equation includes all energy losses. However, these are dictated by the specific control volume. This volume can be of any size, from molecular to planetary. To analyze energy losses related to alternative fuels, every control volume of each step of the life cycle must be quantified.

The first two laws of thermodynamics drive this step. First, the conservation of mass and energy requires that every input and output be included. Energy or mass can be neither created nor destroyed, only altered in form. For any system, energy or mass transfer is associated with mass and energy crossing the control boundary within the control volume. If mass does not cross the boundary, but work and/or heat do, the system is a "closed" system. If mass, work, and heat do not cross the boundary, the system is an isolated system. Too often, open systems are treated as closed, or closed systems include too small control volume (Fig. 4.8 in Chapter 4). A common error is to assume that the life cycle begins at an arbitrary point conveniently selected to support a benefit-cost ratio. For example, if a life cycle for ethanol fuels begins with the corn arriving at the ethanol processing facility, none of the fossil fuel needs on the farm or in transportation will appear.

The second law is less direct and obvious than the first. In all energy exchanges, if no energy enters or leaves the system, the potential energy of the state will always be less than that of the initial state. The tendency toward disorder, that is, entropy, requires that external energy is needed to maintain any energy balance in a control volume, such as a heat engine, a waterfall, or an ethanol processing facility. Entropy is ever present. Losses must always occur in conversions from one type of energy (e.g., mechanical energy of farm equipment ultimately to chemical energy of the fuel). Thus, a series of efficiency equations must be calculated for the entire process, with losses at every step.

In the LCA process, the information is provided by the life cycle inventory (LCI). In fact, the LCA process uses the LCI data to assess the environmental implications associated with a product, process, or service, by compiling an inventory of relevant energy and material inputs and environmental releases. From this LCI, the potential environmental impacts are evaluated. The results aid in decision-making. Thus, the LCA process is a systematic, four-component process [56]:

1. *Goal definition and scoping*—Define and describe the product, process, or activity. Establish the context in which the assessment is to be made and identify the boundaries and environmental effects to be reviewed for the assessment.
2. *Inventory analysis*—Identify and quantify energy, water, and material usage and environmental releases (e.g., air emissions, solid waste disposal, and wastewater discharges).
3. *Impact assessment*—Assess the potential human and ecological effects of energy, water, and material usage and the environmental releases identified in the inventory analysis.
4. *Interpretation*—Evaluate the results of the inventory analysis and impact assessment to select the preferred product, process, or service with a clear understanding of the uncertainty and the assumptions used to generate the results.

Note that these steps track closely with the life cycle stages dictated by the laws of physics. That is, all energy and matter entering the control volume is characterized and quantified. System boundaries are drawn, and flows of energy and matter are accounted.

This stepwise process can be used to evaluate biotechnologies. First, a life cycle inventory (LCI) is constructed to define the boundaries of the possible effects of a technology (e.g., microbial populations, genetically modified organisms, and toxic chemical releases). If the technology is hypothetical, this can be done by analogy with a similar conventional process. Next, experts can participate in an expert panel to find the driving forces involved (this is known as "expert elicitation").

Then, scenarios can be constructed from these driving forces to identify which factors are most important in leading to various outcomes. This last step is known as a sensitivity analysis. The greater the weight of the factor, the greater will be the change in the outcome. For example, if one change in the production of a chemical compound lowers emissions of benzene from the current $100\,\mu g\,m^{-3}\,s^{-1}$ to $1\,\mu g\,m^{-3}\,s^{-1}$, the emission rate reduction would be 100-fold. However, if a different change of the same process lowers the benzene emission rate to $0.1\,\mu g\,m^{-3}\,s^{-1}$, that is, 1000-fold emission rate reduction, this means that the system is 10 times more sensitive the second change than to the first.

Air pollution decision-making should be part of a comprehensive, environmental system. For example, many of the control technologies discussed in Chapter 13 come at the very end of manufacturing and energy production. However, it is preferable to prevent as much of the formation of the pollutant as possible earlier in the life cycle. In fact, it may be possible to eliminate pollution even before production. For example, the extraction of crude oil produces air and water pollution, so eliminating the need for crude oil would reduce emissions of air pollutants very early in the life cycle.

The production of algal biodiesel fuel may be a way to do so. The process begins with the preparation of an inoculum by culturing an algal strain using photobioreactors, for example, indoor ponds, and ends with harvesting lipids that are converted to biofuel and transported for sale and use [57]. Through algae-based biotechnologies, inoculums can be cultured from whole organisms, which include carbohydrates, lipids, and proteins. They may also be extracted from the whole organism, for example, lipids from microalgae and carbohydrates (see Fig. 17.8).

Biofuels do have advantages but are not an air pollution prevention panacea. The LCA for the processes depicted in Fig. 17.9 must consider not only the direct inputs and pathway from harvesting the algae to combusting the fuel (the bold-lettered boxes in Fig. 13.18) but also ancillary processes that provide materials and energy to each step in the pathway (horizontal arrows and nonbolded boxes to the left of the main pathway in Fig. 13.18). Therefore, even a sustainable source of fuel, for example, algae, may rely on unsustainable steps (e.g., crude oil drilling and refining for materials needed to separate cell components) and toxic by-products (e.g., metals mined for catalysts) [57].

17.4.2 Utility and the benefit-cost analysis

Air pollution decision-making must consider the often subtle and difficult-to-measure benefits and costs, as expressed in the benefit-cost ratio (BCR). The BCR is attractive to technologists since it is, or at least appears to be, a quantitative measure of engineering success. Thus, it can also be an expression of a biotechnology's success. This utilitarian perspective is also

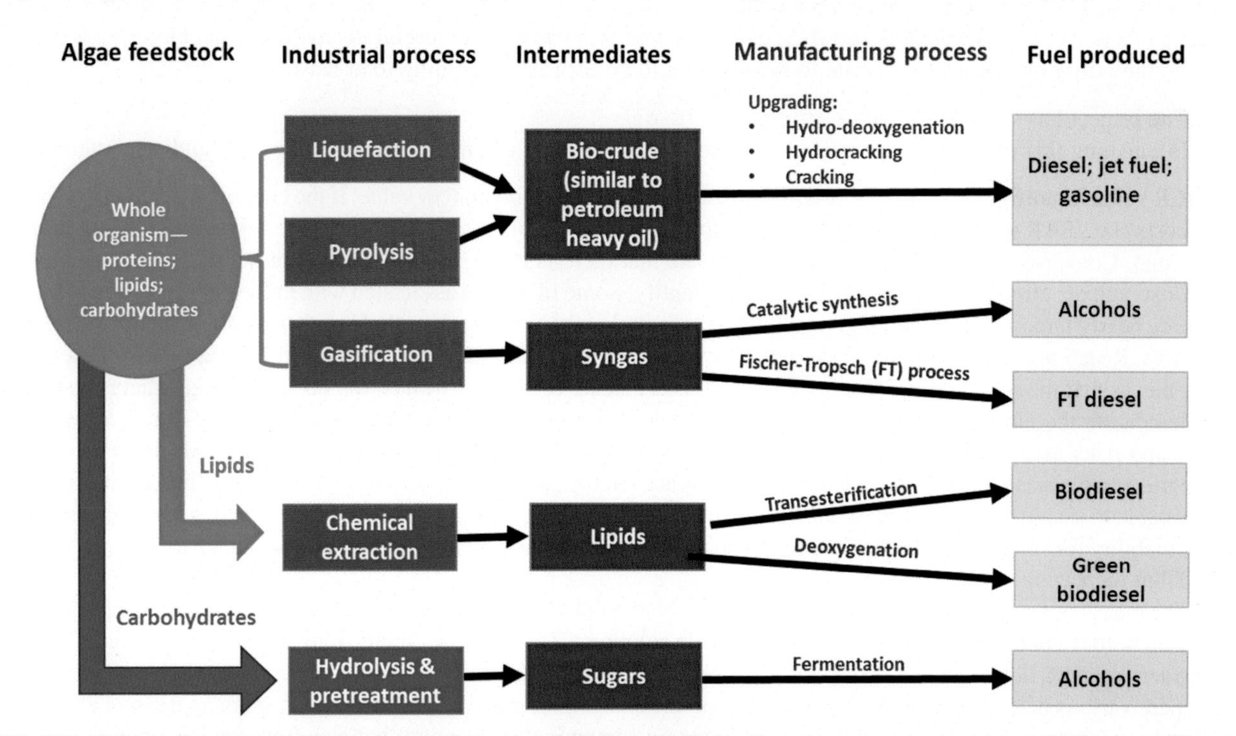

FIG. 17.8 Pathways from three algal feedstocks to produce various fuels. *Note*: The Fischer-Tropsch process converts a mixture of carbon monoxide and hydrogen to liquid-phase hydrocarbons. *(Reproduced with permission from F. Shi, et al., Recent developments in the production of liquid fuels via catalytic conversion of microalgae: experiments and simulations, RSC Adv. 2 (26) (2012) 9727–9747.)*

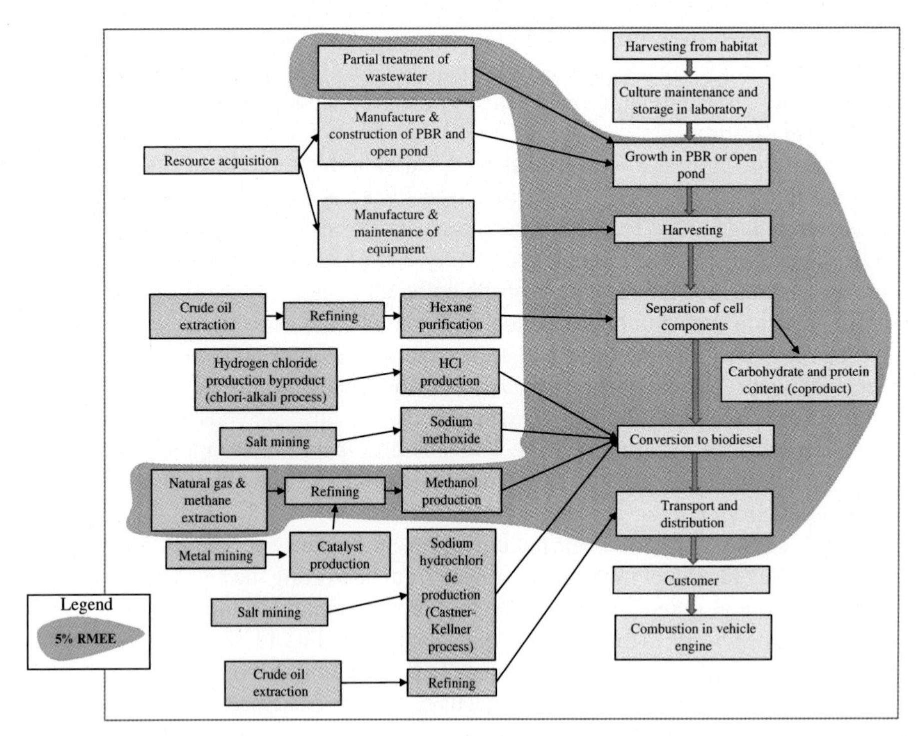

FIG. 17.9 Process flow diagram for producing fuels from algae. Primary pathway shown within dashed boundary. *(Source: F. Shi, et al., Recent developments in the production of liquid fuels via catalytic conversion of microalgae: experiments and simulations, RSC Adv. 2 (26) (2012) 9727–9747.* Note*: RMEE = relative mass-energy-economic method for system boundary selection. M. Raynolds, et al., The relative mass-energy-economic (RMEE) method for system boundary selection Part 1: A means to systematically and quantitatively select LCA boundaries, Int. J. Life Cycle Assess. 5 (1) (2000) 37–46.)*

attractive since it can compare one project to another; for example, a BCR of 2 in Project 1 means that it is a more worthwhile endeavor than Project 2, which has a BCR of 1.5. But is that true? Therefore, it is advantageous to use more than one metric of success and why a combined BCR and LCA approach can be quite useful in environmental biotechnology. Certainly, the utility of a project is crucial to the decision to go forward, according to at least two criteria:

1. The project has value based on its utility.
2. In pursuing this project, it must provide the most benefit for the greatest number (e.g., people and ecosystems).

A BCR value is more of a screening tool than an absolute measure of potential or value. If the benefits are far outweighed by the costs (i.e., $BCR \ll 1$), this is likely not a worthwhile endeavor, so long as the factors are representative and properly weighted. Conversely, a very large BCR may indicate that there is great potential for that project.

Costs and benefits vary considerably in ease to quantify. Some like those associated with the social sciences and humanities are nearly impossible to quantify and monetize accurately when compared with those that are aligned with the physical sciences. Recall also that the comparison of action versus no-action alternatives cannot always be captured within a BCR. This means that the so-called opportunity costs and risks that are associated with taking no action (e.g., what is lost by not implementing the biotechnology project) are not included in the calculations. Thus, comparisons of the *status quo* with costs and risks associated with a new technology may be biased toward the *status quo*. Thus, the decision to embrace or avoid a project is complex, involving factors that are readily quantifiable and monetizable and those that are almost impossible to assign a concrete value. This is particularly challenging for values that are long term (e.g., overall ecological sustainability for future populations) and for those decisions where two societal values are clashing, for example, banning DDT but increasing malarial risks.

The means of finding the best technological approach is a matter of optimization, which is complicated for projects that have numerous contravening risks and many possible solutions. The LCA can build from the BCR screening effort as it considers product flows, critical paths, and life cycle inventories (LCIs). If properly conducted, the LCA can be used to compare various biotechnological options, including the *status quo*, passive approaches, and more aggressive endeavors, not only in terms of the principal project objective (e.g., contamination cleanup) but also from the standpoint of various ecosystems, public health, and other societal values. In other words, the view from the past (e.g., material extraction) to the future (e.g., postproject impacts) is a critical path for each option that can be evaluated objectively.

A promising trend is the recent advances in integrating LCAs and risk assessment, including metrics for exposure and hazard [58–61]. To improve the LCA's usefulness as a predictive tool calls for enhancements of life cycle inventories (LCIs), including more harmonized formats and wider availability of LCIs. Recent progress has included the development of open-source software, making data more accessible to a larger audience and improving transferability of software. This is complemented by better exposure models and data that are also becoming more widely available. Many previous paper data sets have become digitized and searchable, for example, through the US Environmental Protection Agency's (EPA) *Exposure Factors Handbook* [62, 63].

Exposure models now make use of a wide array of data, including (a) product ingredients, (b) pharmacokinetic factors, (c) consumer product category-specific "exposure factor surrogates," and (d) time/activity estimates (i.e., human factors). These disparate data streams can then be integrated within an interface such that different exposure scenarios for target groups, for example, individuals, population, or occupational, yielding time-use profiles can be interchanged and quantitatively explored to prioritize tailored chemical exposure and ultimately dose. This allows estimates of multichemical signatures of exposure, internalized dose (uptake), remaining dose or body burden, and elimination [64]. Such models and databases should be useful for characterizing and comparing applications of synthetic biology and genetic engineering processes.

To arrive at a risk comparison, the resulting exposure data can then be mapped onto toxicity data. For example, to what extent do chemical and biological substances contribute to potential toxic outcomes at each life cycle stage? To answer this question, the LCA must consider how these substances move and change in the LCA to help determine which exposures are most likely to lead to various adverse effects, for example, cancer, neurotoxicity, development and reproductive disorders, and problems with the immune and endocrine systems in human populations, and ecosystem effects. Indeed, an LCA must ascertain the extent to which potential exposures to sentinel or endangered species differ from human exposures. Thus, the LCA may account for a direct emission from a stack; an exposure to an ingredient in a product that is intentionally applied (e.g., a pesticide); or a substance that migrates to a habitat, for example, released from a stack or outfall structure and then transported to the habitat. Thereafter, the exposure in human populations or the activities of the species in an ecosystem can be modeled (e.g., predator-prey, migration, and bioaccumulation factors), including the pharmacokinetic factors within the organism that can be modeled and estimated in the LCA [58].

One means of combining biological and chemical stressors into a cumulative risk assessment is to combine LCA with a quantitative microbial risk assessment (QMRA). Although this has principally been a tool used for waterborne pathogens, it may be useful for bioaerosols, which can carry bacteria, viruses, protozoa, or their by-products. The QMRA can be combined with risk assessments of chemical stressors. The combined tools may evaluate the potential for an array of risks, for example, food and water contamination after deposition. The bioaerosols may also contain chemicals that induce various acute and chronic effects (e.g., cancer). Pathogenic risk can be expressed as a burden of disease, for example, disability-adjusted life years (DALYs) [65]. Thus, the DALYs can be combined with risk information for the chemical stressors to depict cumulative risks, for example, to compare combined pathogenic and chemical risks from a manufacturing processes and making changes to lower those risks.

The success of an alternative technology displacing conventional manufacturing depends on the efficiency and safety with which a product can be produced and used. Complicating matters, certain materials, and costs may be embedded in the technology. As mentioned, fossil fuels are used in almost every step of ethanol production and/or operation, as is the case for all biofuels.

The LCA is useful, but it is not a panacea for biotechnological impact assessment. For example, an analysis is only as good as the availability and quality of the life cycle inventory data. Uncertainties in the inventory and in the impact assessment methodology can lead to omitted options and unforeseen impacts. The lack of agreement among elements of the impact assessment methodology, that is, internal inconsistencies, can lead to additional uncertainties and errors. Finally, many of the factors are qualitative and subjective so that differences in LCA problem formulation may in fact be due to differences in weights placed on factors. Such weighting can be biased according to the values of those conducting the LCA. For example, the previously mentioned QMRA expresses only the pathogenic burden of disease in the human populations, omitting numerous other biological adverse outcomes, such as the effects of a transgenic organism on the soil microbial community [66]. Thus, additional tools and data are needed for a comprehensive assessment of the life cycle in which air pollution is occurring.

17.5 LINKING CAUSES AND OUTCOMES

The underlying assumption in any environmental assessment is that "cause and effect" can be identified using credible science to connect exposure to a hazard and to estimate a negative outcome. As stated in most introductory statistic courses, association and causation are not synonymous. Often in linking an air pollutant to a disease or other adverse outcome involves the proverbial "third variable." Something other than the air pollution could be the reason for the relationship. Or the presence of the third variable exacerbates the outcome. In statistics classes, we are given simple examples of such occurrences:

- Studies show that people who wear shorts in Illinois eat more ice cream.
- Therefore, wearing shorts induces people to eat more ice cream.

The first statement is simply a measurement. It is stated correctly as an association. However, the second statement contains a causal link that is clearly wrong for most occurrences [12]. Something else is causing both variables, that is, the wearing of shorts and the eating of ice cream. For example, if one were to plot ambient average temperature and compare it with either the wearing of shorts or the eating of ice cream, one would see a direct relationship between the variables. That is, as temperatures increase, so does short wearing and so does the rate of ice cream eating.

The medical science community may help us deal with the causality challenge. The best that science usually can do in this regard is to provide enough weight of evidence to support or reject a suspicion that a substance causes a disease. The medical research and epidemiological communities use several criteria to determine the strength of an argument for causality, but the first well-articulated criteria were Hill's causal criteria [13].

Assigning causality when none really exists must be avoided, but if all we can say is that the variables are associated, the public is going to want to know more about what may be contributing an adverse effect (e.g., learning disabilities and blood lead levels). This was particularly problematic in early cancer research. Possible causes of cancer were being explored, and major research efforts were being directed at myriad physical, chemical, and biological agents. So, there needed to be some manner of sorting through findings to see what might be causal and what is more likely to be spurious results. Sir Austin Bradford Hill is credited with articulating key criteria that need to be satisfied to attribute cause and effect in medical research [67].

17.5.1 Strength of association [68]

Strong associations provide more certain evidence of causality than is provided by weak associations. A cause must be associated with an outcome, that is, the statistical association cannot be zero, although a cause may show up as a very low statistical association. Common metrics used to indicate association include risk ratio, odds ratio, and standardized mortality ratio.

17.5.2 Consistency

If a cause is associated with an outcome consistently under different studies using diverse methods of study of assorted populations under varying circumstances by different investigators, the link to causality is stronger. For example, if results are similar for air pollution control approach studied with scenarios, simulated with models and documented using historical information, there is greater consistency. However, if there is no agreement with the various approaches, this could mean that the results are unique to a given set of conditions; for example, if the only scenarios were towns with populations <5000, a model was parameterized only for one type of collection approach, or historical data are incomplete or erroneous.

17.5.3 Specificity

The specificity criterion holds that the cause should lead to only one outcome and that the outcome must result from only this single cause. Recall that Hill was interested in cancer and was viewing causality through a biomedical perspective, that is, this criterion appears to be based in the germ theory of microbiology, where a specific strain of bacteria and viruses elicits a specific disease. This is rarely the case for air quality management.

17.5.4 Temporality

This criterion requires that exposure to the chemical must precede the effect. For example, in a retrospective study, the researcher must be certain that the outcome was not already present before the contributing factors were present. For example, a disease may have been present before a toxic release. However, this does not eliminate the temporal link if, for example, the exacerbation of the disease occurs, as when the incidence of existing asthma events and lung disease exacerbations already present spikes. Thus, it may not mean that the agent in the release is innocuous and is not contributing to the disease. It does mean that it is not the sole cause of the disease (see Section "Specificity") or that the outcome is not new, so much as a worsening of an existing condition.

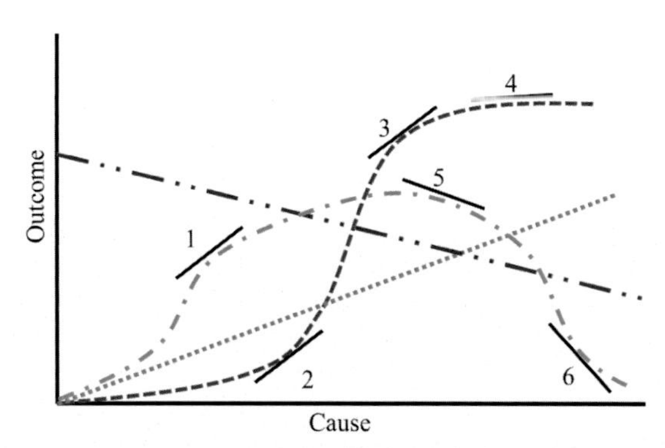

FIG. 17.10 Hypothetical cause-outcome curves. The *green* and *purple lines* have positive slopes. The *red line* has a negative slope. The *blue line* starts as a positive slope but becomes negative. The slope varies on the curves, that is, the *purple* and *blue lines*, so must be measured instantaneously at a tangent to the curve (1–6). Slopes 1, 2, 3, and 4 are positive. Slopes 1, 2, and 3 are equal. Slopes 5 and 6 are negative. The linear cause-outcomes are common expressions of scientific principles, for example, gravitation, but not in most air pollution engineering situations. *(Source: T.M. Letcher, D. Vallero, Excerpt From: Waste: A Handbook for Management, 2nd ed., Academic Press, 2019.)*

17.5.5 Gradient

If the force of a cause, for example, the intensity and extent, increases the amount of outcome, this strengthens the causal link. This is a common metric for risk assessment, known as the "dose-response" step in risk assessment (see Chapter 12). If the level, intensity, duration, or total level of chemical exposure is increased, a concomitant, progressive increase should occur in the toxic effect. However, the gradient also applies to other areas; for example, if a recycling program is intensified and the amount of recycling increases concomitantly, this is a gradient.

In air quality management, the gradient is seldom linear. Thus, depending on the shape of the curve, it must be measured at a finite time value, that is, a tangent to the cause-outcome curve (see Fig. 17.10).

17.5.6 Scientific plausibility

Hill limited this to biological plausibility, but the criterion has wider application. Generally, an association needs to follow a well-defined explanation based on a known scientific system, that is, a paradigm. A paradigm, however, is mutable. Even some of the most strongly held scientific laws have had to be revised with the evolution of knowledge [69]. Outside of research, this caution has little relevance to the practice of air quality management. Engineers and other professionals involved in air pollution control and management must apply best practices according to their profession. If results are implausible according to underlying scientific principles, they should be heeded. However, this applies to scientific and engineering principles and not one's one "favorite" practice. Simply because an approach does not follow direct from what has worked in the past does not render it implausible, but merely inconvenient. Indeed, the air quality manager must beware of Kaplan's "law of the instrument." The air quality manager's or engineer's preferred approach can be likened to a hammer, so that the entirety of air quality management is a collection of nails. There are likely to be other plausible "tools," besides one's hammer, and there are certainly myriad other problems besides the "nails." This can be particularly worrisome for modelers, who often must incorporate simplifying assumptions to make the math work. Engineering models, for example, can be accused of assuming that a chicken is spherical, but when the designed coop fails, this may be in part due to this oversimplification. After all, the real-world coop's chickens have beaks, feathers, and claws.

17.5.7 Coherence

The criterion of coherence suggests that all available evidence should form a cohesive whole. By that, the proposed causal relationship should not conflict or contradict information from experimental, laboratory, epidemiological, theory, or other knowledge sources.

17.5.8 Experimentation

Experimental evidence useful to air quality management can support a causal hypothesis. Laboratory experiments, tests, pilot studies, physical models, and natural experiments (e.g., a disaster response) are available. The experiment does not have to be specific, but can be drawn from others' experiences, for example, an informatic experiment using "big data."

17.5.9 Analogy

The term analogy implies a similarity in some respects among things that are otherwise different. It is thus considered one of the weaker forms of evidence.

For air quality management decisions, some of Hill's criteria are more important than others [68]. Reliability in a public health and environmental context relies heavily on the strength of association, for example, to establish dose-response relationships. Coherence is also very important. Animal and human data should not be extensions of one another and should not disagree. Biological gradient is crucial, since this is the basis for dose-response (the more dose, the greater the biological response).

Temporality is crucial to all scientific research, that is, the cause must precede the effect. However, this is sometimes difficult to see in some instances, such as when the exposures to suspected agents have been continuous for decades and the health data are only recently available.

Linking causes to outcome must be based on sound science and scientific judgment. Air quality managers must consider whether the selected approach will likely continue to "work" (reliability) and, further, must consider the hazards that may arise, especially in a new and untested technology that is selected. Risk is a function of likelihood that the hazard will in fact be encounter, so the air quality manager is also tasked with predicting the adverse implications that society might face. Thus, the "risk" associated with a selected approach refers to the possibility and likelihood of undesirable and possibly harmful effects. Errors in reliability and risk predictions can range from not foreseeing outcomes that are merely annoying (e.g., fuel efficiencies that are less than expected) to those that are devastating (e.g., the release of carcinogens from an incinerator) [70].

17.6 ANALYSIS OF COMPLEX AIR POLLUTION PROBLEMS

Tom and Miriam Budinger have provided an elegant approach, the "Four As," for approaching a scientific dilemma, which can be used to address possible environmental impacts that can result from a biotechnology [71]. First, one should *acquire* the facts, including the uncertainties associated with the technology. Second, the *alternative* solutions are listed and compared in parallel. Third, each solution is *assessed* with respect to principles (the authors were mainly concerned with moral theories, but other scientific theories can also be used as benchmarks). This includes a thorough risk analysis, where appropriate. Finally, a decision is made on which *action* to take. This includes a comprehensive action plan, keeping the other alternatives available if needed, with continuous adaptation and improvement and eye toward new options (since the nature of emergent technologies is that things change very rapidly).

Air Pollution Problem Analysis

Problem analysis is a stepwise process [72]:

Step 1—Scenario description

Key characters and events pertinent to the case are identified. This description includes narrative, tables, figures, maps, organization charts, critical path diagrams, and photographs that are needed to place the case in ethical context. The storyboard must be both accurate and complete. This can be challenging since almost any biotechnological endeavor includes numerous perspectives, so each perspective must be described adequately. Since this is the *descriptive stage*, no judgments about what right need be made; that is done in the next steps. Completeness means that the science and societal concepts are fully understood. For example, if the project is designed to degrade a recalcitrant chemical compound, it must be described completely in all matters that could affect the environmental decision (e.g., electron donation/acceptance of microbes in previous studies, similarities in chemical structure to compounds that have been successfully bioremediated, and problems encountered in similar projects).
A key consideration of Step 1 is assigning responsibility and accountability.

Step 2—Deductive arguments

Based upon the findings in Step 1, the validity of the decisions or lack thereof is analyzed. The syllogism includes a factual premise, a connecting fact-value premise, and an evaluative premise to reach an evaluative conclusion. Many moral (and scientific) arguments fail because of weaknesses in any of these components of the logical argument, that is, the syllogism. Depending on the case, numerous arguments must be evaluated.

Step 3—Problem-solving analysis

Once the facts are identified, articulated, and sufficiently explained, any issues must be categorized as to whether they are factual, conceptual, or related to human factors [14].

Human factors would include constraints or drivers that are not based on physical science, but more related to perceptions and expectations of the people potentially affected by the project (e.g., historical, cultural, and financial).

From descriptions in Step 1, the depth of each type of issue can be assessed. Factual issues are those that are known. This can sometimes be apparent just by reading the events, but in certain cases, the facts may not be so clear (e.g., two scientists may agree on the "fact" that carbon dioxide is a radiant gas but may disagree on whether the buildup of CO_2 in the troposphere will lead to increased global warming). Agreement on first principles of science and even the data being used may still be followed by large disagreements about the relative weightings in indices and models. This leads to a need to ascribe causality, a very difficult problem indeed.

Step 3a—Application of Hill's criteria

To begin to evaluate whether an approach is appropriate to address air pollution, oftentimes, the best that science usually can do in this regard is to provide enough weight of evidence between a cause and an effect, as articulated criteria by the previously discussed Hill's causal criteria [15] Depending on the situation, some of Hill's criteria are more relevant and important than others.

Conceptual issues involve different ways that the meaning may be understood. For example, what one considers to be "pollution" or "good lab practices" may vary (although the scientific community strives to bring consensus to such definitions).

Many engineers and scientists believe it is the job of technical societies and other collectives to try to eliminate factual and conceptual disagreements. Most agree on first principles (e.g., fundamental physical concepts like the definitions of matter and energy), but unanimity fades as the concepts drift from first principles. For example, John Ahearne, the former president of Sigma Xi, the Scientific Research Society, recently told an audience of engineers that we should not be disagreeing about the facts [16]. The progress of research and knowledge helps to resolve factual issues (eventually), and the consensus of experts aids in resolving conceptual issues. This is known as "expert elicitation," which has been employed for synthetic biology [73], invasive species intervention [74], nanotechnologies [75, 76], and other emerging technologies. Complete agreements are not generally possible even for the factual and conceptual aspects of many biotechnology decisions, so the moral or ethical issues are further complicated and often only reach consensus, not unanimity.

Step 3b—Force fields

As discussed, a simple "polar diagram" that illustrates the "forces" that pull or push the key individuals or groups toward decisions can be quite useful, at least in identifying what is important to the various stakeholders. The shape and size of the resulting diagram give an idea of what are the principal driving factors that lead to decisions. Envision a source in the outer middle of each sector pulling against the shape. A force field diagram can be drawn as a subjective assessment of each decision and for each decision-maker. For example, lawyers may proceed in one direction, while engineers another, bioreactor operators another, and the land owners another, all because of different forces.

Step 3c—Net goodness analysis

This is a subjective analysis of whether a decision will be moral or less than moral. It puts the case into perspective, by looking at each factor driving a decision from three perspectives: (1) how good or bad would the consequence be, (2) how important is decision, and (3) how likely is it that the consequence would occur. These factors are then summed to give the overall net goodness of the decision:

$$NG = \sum(\text{goodness of each consequence}) \times (\text{importance}) \times (\text{likelihood}) \tag{17.1}$$

Thus, this can be valuable in decisions that have not yet been made and in evaluating what decisions "should" have been made in a case. For example, these analyses sometimes use ordinal scales, such as 0 through 3, where 0 is nonexistence (e.g., zero likelihood or zero importance) and 1, 2, and 3 are low, medium, and high, respectively. Thus, there may be many small consequences that are near zero in importance and, since NG is a product, the overall net goodness of the decision is driven almost entirely by one or a few important and likely consequences.

There are two cautions in using this approach. First, although it appears to be quantitative, the approach is very subjective. Second, as we have seen many times in cases involving health and safety, even a very unlikely but negative consequence is unacceptable.

Step 3d—Line drawing

Graphic techniques like line drawing, flow charting, and event trees are very valuable in assessing a case. Line drawing is most useful when there is little disagreement on what the moral principles are, but when there is no consensus about how to apply them. The approach calls for a need to compare several well-understood cases for which there is general agreement about right and wrong and show the relative location of the case being analyzed. Two of the cases are extreme cases of right and wrong, respectively, that is, the positive paradigm is very close to being unambiguously moral and the negative paradigm unambiguously immoral:

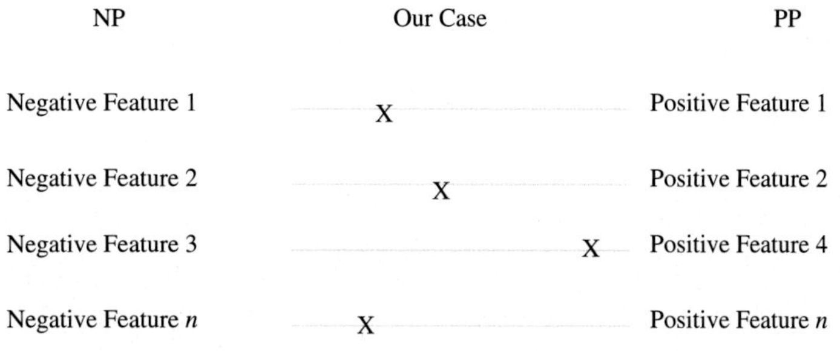

NP	Our Case	PP
Negative Feature 1	X (near NP)	Positive Feature 1
Negative Feature 2	X (middle)	Positive Feature 2
Negative Feature 3	X (near PP)	Positive Feature 4
Negative Feature *n*	X (near NP)	Positive Feature *n*

Next, our case (T) is put on a scale showing the positive paradigm (PP) and the negative paradigm (NP) and other cases that are generally agreed to be less positive than PP but more positive than NP. This shows the relative position of our case T:

This gives the sense that our case is more positive than negative but still short of being unambiguously positive. In fact, two other actual, comparable cases (2 and 3) are much more morally acceptable. This may indicate that we consider taking an approach like these if the decision has not yet been made. If the decision has been made, we will want to determine why the case being reviewed was so different from these.

Although being right of center means that our case is closer to the most moral than to the most immoral approach, other factors must be considered, such as feasibility and public acceptance. Like risk assessment, ethical analysis must account for trade-offs (e.g., security vs liberty).

Step 3e—Flow charting

Critical paths, program evaluation review technique (PERT) charts, and other flow charts are commonly used in design and engineering, especially computing and circuit design. They are also useful in ethical analysis if sequences and contingencies are involved in reaching a decision or if a series of events and ethical and factual decisions lead to the consequence of interest. Thus, each consequence and the decisions that were made along the way can be seen and analyzed individually and collectively. Fleddermann [17] shows a flow chart for the Bhopal incident. This flow chart (Fig. 17.11) deals with only one of the decisions involved in the incident, that is, where to site the plant. Other charts need to be developed for safety training, the need for fail-safe measures, and proper operation and maintenance. Thus, a "master flow chart" can be developed for all the decisions and subconsequences that ultimately led to the disaster.

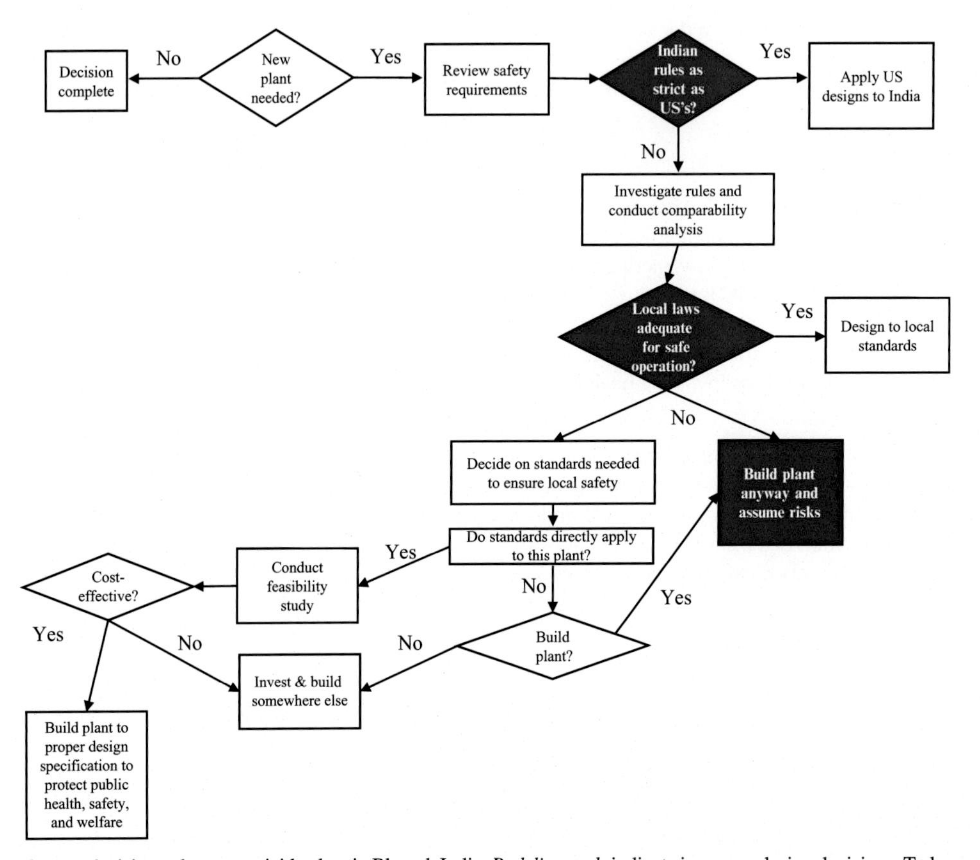

FIG. 17.11 Flow chart on decision to locate pesticide plant in Bhopal, India. *Red diamonds* indicate improper design decisions. To be more complete, the flow chart would also include operation and maintenance flows, including the inadequate training, difficult to understand safety manuals, failure to adhere to best practices in tank fill levels and flares, and improper maintenance of equipment. *(Based on information from T. Chouhan, et al., Bhopal, the Inside Story: Carbide Workers Speak out on the world's Worst Industrial Disaster, Apex Pr (1994).)*

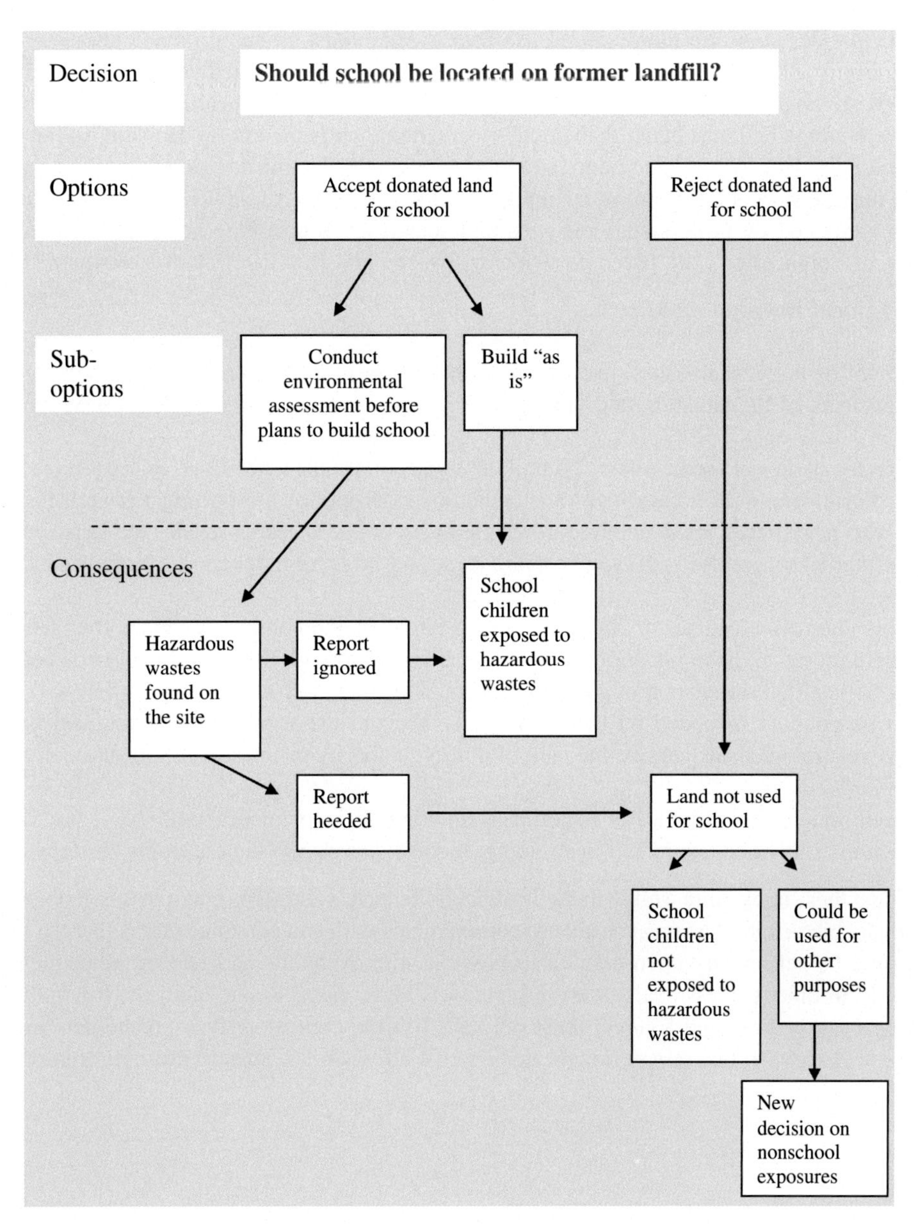

FIG. 17.12 Event tree example.

Step 3f—Event trees

Event trees or fault trees allow us to look at possible consequences from each decision. A straightforward example is provided in Fig. 17.12.

The event tree can build from all the other analytic tools, starting with the time line of key events and list of key actors. What are their interests and why were the decisions made? The event tree allows us to visualize several different paths that could have been taken and that could have led to better or worse decisions. We would do this for every option and suboption that should have been considered in our case, comparing each consequence. It may be, for example, that even in a disaster, there may have been worse consequences than what occurred. Conversely, even though something did not necessarily turn out all that badly, the event tree could point out that the outcomes were simply fortunate! In fact, the fault tree approach applies a probability to each option and suboption.

Step 4—Synthesis

Using the information from the steps above, we can begin to decide about how "right" the decision is. This is tantamount to a moral decision. That is, is the implementation of the biotechnological enterprise a moral or immoral decision compared with the other approaches that could have been taken, including the so-called "no action" alternative? If the decision has not yet been made, then the alternatives can be compared before choosing the best one.

Most environmental decisions are moral decisions. That is, they must be based on sound scientific and engineering principles, but there is usually some value being placed on one alternative compared with that of another. Thus, few environmental decisions are *amoral* (i.e., devoid of ethical content). For example, deciding whether one method of nutrient addition to improve bioremediation is better than another is predominantly an amoral decision, so long as it is completely based on undisputed facts. However, if the science is not completely driving the decision (e.g., the best nutrient addition is "too" expensive), the decision is taken on moral relevance. Deciding how to address the moral aspects of a scientific decision has been addressed by professional societies and scientific groups. One approach has been proposed by the National Academy of Engineering [18], based on work by Swazey and Bird, Weil, and Velasquez [19].

Checklist for Ethical Decision Making

- Recognize and define the ethical issues (i.e., identify what is (are) the problem(s) and who is involved or affected).
- Identify the key facts of the situation and ambiguities or uncertainties and what additional information is needed and why.
- Identify the affected parties or "stakeholders" (i.e., individuals or groups who affect, or are affected by, the problem or its resolution). For example, in a case involving intentional deception in reporting research results, those affected include those who perpetrated the deception, other members of the research group, the department and university, the funder, the journal where the results were published, and other researchers developing or conducting research on the findings.
- Formulate viable alternative courses of action that could be taken and continue to check the facts.
- Assess each alternative (i.e., its implications; whether it is in accord with the ethical standards being used and, if not, whether it can be justified on other grounds; consequences for affected parties; issues that will be left unresolved; whether it can be publicly defended on ethical grounds; the precedent that will be set; and practical constraints, e.g., uncertainty regarding consequences, the lack of ability, authority or resources, institutional, structural, or procedural barriers).
- Construct desired options and persuade or negotiate with others to implement them.
- Decide what actions should be taken, and in so doing, recheck and weigh the reasoning in steps 1–6.

It should be noted that these track quite well with the Budingers' Four A's and steps noted in this chapter. However, it is not enough to be right; the biotechnologist must be able to communicate with and convince others that the selected approach is best and needed (and be sufficiently open-minded to possible improvements and needed adjustments to the proposed approach, whether from other professionals or the lay public). Thus, based on the analysis, the findings and arguments (including all the necessary facts and figures) must be placed in the context of the stakeholders in an understandable way. The audience will vary, so one size definitely does not fit all when it comes to presenting information on possible environmental risks.

REFERENCES

[1] D. Vallero, Excepted and Updated from: Translating Diverse Environmental Data into Reliable Information: How to Coordinate Evidence from Different Sources, Academic Press, 2017.

[2] D. Vallero, Translating Diverse Environmental Data into Reliable Information: How to Coordinate Evidence from Different Sources, Academic Press, 2017.

[3] U.E.P. Agency, 2011 National Emissions Inventory, Version 1 Technical Support Document Rep, (2013).

[4] F. Bryant, Plant manager's responsibilities in air pollution control, in: Proc. Annu. Meet. Air Pollut. Control Assoc. United States, vol. 56, 1956.

[5] S.M. Capalbo, et al., Next Generation Data Systems and Knowledge Products to Support Agricultural Producers and Science-Based Policy Decision Making, Agricultural Systems, 2016.

[6] P.A. Vesilind, A.S. Gunn, Hold Paramount: The Engineer's Responsibility to Society, Cengage Learning, 2010.

[7] D. Hall, et al., Rethinking the data overload problem: Closing the gap between situation assessment and decision making, in: Proceedings of the 2007 National Symposium on Sensor and Data Fusion (NSSDF) Military Sensing Symposia (MSS), 2007.

[8] G.L. Klein, et al., Supporting a robust decision space, in: AAAI Spring Symposium: Technosocial Predictive Analytics, 2009.

[9] Centers for Disease Control and Prevention, Principles of Epidemiology in Public Health Practice: An Introduction to Applied Epidemiology and Biostatistics, U.S. Department of Health and Human Services, Atlanta, Georgia, 2012.

[10] Centers for Disease Control and Prevention, Self-study course SS1978, in: Principles of Epidemiology in Public Health Practice: An Introduction to Applied Epidemiology and Biostatistics, third ed., 2012. https://www.cdc.gov/ophss/csels/dsepd/ss1978/. Accessed 7 February 2017.

[11] J. Kirkeleit, et al., The healthy worker effect in cancer incidence studies, Am. J. Epidemiol. 177 (11) (2013) 1218–1224.

[12] R.R. Monson, Observations on the healthy worker effect, J. Occup. Environ. Med. 28 (6) (1986) 425–433.

[13] G. Wingren, Mortality in a Swedish rubber tire manufacturing plant: occupational risks or an "unhealthy worker" effect? Am. J. Ind. Med. 49 (8) (2006) 617–623.
[14] National Research Council, N. A. o. S, Risk Assessment in the Federal Government: Managing the Process, National Academy Pre, Washington, DC, 1983.
[15] G.W. Suter II, Ecological risk Assessment, CRC Press, 2016.
[16] E. Abt, et al., Science and decisions: advancing risk assessment, Risk Anal. 30 (7) (2010) 1028–1036.
[17] National Research Council, Science and Decisions: Advancing Risk Assessment, The National Academies Press, Washington, DC, 2009.
[18] W. Galke, et al., National evaluation of the US Department of housing and urban development lead-based paint hazard control grant program: study methods, Environ. Res. 98 (3) (2005) 315–328.
[19] W. Galke, et al., Evaluation of the HUD lead hazard control grant program: early overall findings, Environ. Res. 86 (2) (2001) 149–156.
[20] P.A. Succop, et al., Imputation of data values that are less than a detection limit, J. Occup. Environ. Hyg. 1 (7) (2004) 436–441.
[21] T.S.R.S. Sigma Xi, Honor in Science, Sigma Xi, Research Triangle Park, NC, 2000.
[22] T.D. Little, et al., J. Pediatr. Psychol. 39 (2) (2014) 151–162. On the Joys of Missing Data. We provide conceptual introductions to missingness mechanisms—missing completely at random, missing at random, and missing not at random—and state-of-the-art methods of handling missing data—full-information maximum likelihood and multiple imputation—followed by a discussion of planned missing designs: Multiform questionnaire protocols, 2-method measurement models, and wave-missing longitudinal designs. We reviewed 80 articles of empirical studies published in the 2012 issues of the Journal of Pediatric Psychology to present a picture of how adequately missing data are currently handled in this field. To illustrate the benefits of using multiple imputation or full-information maximum likelihood and incorporating planned missingness into study designs, we provide example analyses of empirical data gathered using a 3-form planned missing design.
[23] J.C. Little, et al., Rapid methods to estimate potential exposure to semivolatile organic compounds in the indoor environment, Environ. Sci. Technol. 46 (20) (2012) 11171–11178.
[24] T.D. Little, M. Rhemtulla, Planned missing data designs for developmental researchers, Child Dev. Perspect. 7 (4) (2013) 199–204.
[25] P.G. Georgopoulos, et al., Reconstructing population exposures to environmental chemicals from biomarkers: challenges and opportunities, J. Expos. Sci. and Environ. Epidemiol. 19 (2) (2009) 149–171.
[26] D.J. Dix, et al., The ToxCast program for prioritizing toxicity testing of environmental chemicals, Toxicol. Sci. 95 (1) (2007) 5–12.
[27] R.S. Judson, et al., In vitro screening of environmental chemicals for targeted testing prioritization: the ToxCast project, Environ. Health Perspect. 118 (4) (2010) 485.
[28] A.L. Karmaus, et al., Evaluation of food-relevant chemicals in the ToxCast high-throughput screening program, Food Chem. Toxicol. 92 (2016) 188–196.
[29] N.C. Kleinstreuer, et al., Phenotypic screening of the ToxCast chemical library to classify toxic and therapeutic mechanisms, Nat. Biotechnol. 32 (6) (2014) 583–591.
[30] EPA, U, Benchmark Dose Technical Guidance, US Environmental Protection Agency, 2012.
[31] U.S. Environmental Protection Agency, EPA's Approach for Assessing the Risks Associated with Chronic Exposure to Carcinogens, in: Risk Assessment Guidance for Superfund (RAGS) Part A, 1992. http://www.epa.gov/iris/carcino.htm. Accessed 1 May 2015.
[32] USEPA, Guidelines for Carcinogen Risk Assessment, in: Risk Assessment Forum, U.S. EP Agency, Washington, DC, 2005.
[33] International Agency for Research on Cancer, About IARC, http://www.iarc.fr/en/about/index.php, 2017. Accessed 3 May 2017.
[34] National Society of Professional Engineers, NSPE Code of Ethics for Engineers, https://www.nspe.org/resources/ethics/code-ethics, 2018. Accessed 31 August 2018.
[35] D.A. Vallero, Ethics and air pollution in the 21st century: biomedical and environmental perspectives, Ethics Biol. Eng. Med. Int. J. 4 (3) (2013).
[36] D.A. Vallero, Measurements in Environmental Engineering, in: M. Kutz (Ed.), Handbook of Measurement in Science and Engineering, 2013.
[37] D.A. Vallero, This section is excerpted and adapted from: ethics and air pollution in the 21st century: biomedical and Environmental perspectives, Ethics Biol. Eng. Med. Int. J. 4 (3) (2013).
[38] D. Vallero, Environmental Biotechnology: A Biosystems Approach, Elsevier Science, 2015.
[39] G.W.F. Hegel, et al., Hegel on tragedy, Harper & Row, New York and London, 1975.
[40] N.R. Augustine, Ethics and the second law of thermodynamics, Bridges 32 (3) (2002) 4–7.
[41] B. Chappell, It Was Installed for this Purpose, VW's US CEO Tells Congress about Defeat Device, National Public Radio, 2015.
[42] U.S. Environmental Protection Agency, Air Quality – National Summary, U.S. EP Agency, Washington, DC, 2016.
[43] U.S. Environmental Protection Agency, NAAQS Table, in: Criteria Air Pollutants, 2016. 20 December 2016, https://www.epa.gov/criteria-air-pollutants/naaqs-table. Accessed 25 August 2018.
[43a] IPCC, Climate Change 2013: The Physical Science Basis. in: T.F. Stocker, D. Qin, G.-K. Plattner, M. Tignor, S.K. Allen, J. Boschung, A. Nauels, Y. Xia, V. Bex, P.M. Midgley (Eds.), Contribution of Working Group I to the Fifth Assessment Report of the Intergovernmental Panel on Climate Change, Cambridge University Press, Cambridge, UK and New York, NY, USA, 2013, https://doi.org/10.1017/CBO9781107415324. 1535 pp.
[44] D.A. Vallero, Fundamentals of Air Pollution, Elsevier Academic Press, Waltham, MA, 2014.
[44a] S. Arbury, B. Jacklitsch, O. Farquah, M. Hodgson, G. Lamson, H. Martin, A. Profitt, Heat illness and death among workers—United States, 2012–2013, MMWR Morb. Mortal. Wkly Rep 63 (31) (2014) 661–665 Centers for Disease Control and Prevention.
[45] D. Vallero, Ethical Decisions in Emergent Science, Engineering and Technologies, InTech Open Access Publisher, 2012.
[46] P.P. Egeghy, et al., Exposure-based prioritization of chemicals for risk assessment, Environ. Sci. Pol. 14 (8) (2011) 950–964.

[46a] National Commission for the Protection of Human Subjects of Biomedical and Behavioral Research, Ethical Principles and Guidelines for the Protection of Human Subjects of Research, The Belmont Report Kennedy Inst. Ethics J. 5 (1) (1979) 83–84.
[46b] United Kingdom Health and Safety Executive, Guidance on ALARP Decisions in COMAH, 2017, May 14, 2017, Available at: http://www.hse.gov.uk/foi/internalops/hid_circs/permissioning/spc_perm_37/.
[47] R. Bell, D. Reinert, Risk and system integrity concepts for safety-related control systems, in: Proceedings of the Eighth Annual Conference on Computer Assurance, COMPASS'93, Practical Paths to Assurance, 1993, IEEE, 1993.
[48] Health and Safety Executive, Guidance on ALARP Decisions in COMAH, http://www.hse.gov.uk/foi/internalops/hid_circs/permissioning/spc_perm_37/, 2017. Accessed 14 May 2017.
[49] D.A. Vallero, P.A. Vesilind, Socially Responsible Engineering: Justice in Risk Management, Wiley Online Library, 2007.
[50] Group, R. S. S, Risk: Analysis, Perception and Management, 1992.
[51] S.L. Derby, R.L. Keeney, Risk analysis: understanding "how safe is safe enough?" Risk Anal. 1 (3) (1981) 217–224.
[52] M.G. Morgan, Risk assessment: Choosing and managing technology-induced risk: how much risk should we choose to live with? How should we assess and manage the risks we face? IEEE Spectr. 18 (12) (1981) 53–60.
[53] I. Kant, The Metaphysics of Morals (1797), 1996.
[54] D.A. Vallero, C. Brasier, Teaching green engineering: The case of ethanol lifecycle analysis, Bull. Sci. Technol. Soc. 28 (3) (2008) 236–243.
[55] D.A. Vallero, C. Brasier, Sustainable Design: The Science of Sustainability and Green Engineering, John Wiley & Sons, 2008.
[56] U.S. Environmental Protection Agency, Life Cycle Assessment: Inventory Guidelines and Principles, O. o. R. a. Development, Cincinnati, Ohio, 1993.
[57] K. Sander, G.S. Murthy, Life cycle analysis of algae biodiesel, Int. J. Life Cycle Assess. 15 (7) (2010) 704–714.
[58] J.C. Bare, D.A. Vallero, Incorporating Exposure Science into Life-Cycle Assessment, AccessScience. McGraw-Hill Education; Yearbook of Science & Technology, 2014.
[59] S.A. Csiszar, et al., A Conceptual Framework to Extend Life Cycle Assessment Using Near-Field Human Exposure Modeling and High-Throughput Tools for Chemicals, Environmental Science & Technology, 2016.
[60] O. Jolliet, et al., Health Impacts of consumer exposure during product use: near field exposure applied to risk assessment and LCA, in: International Society of Exposure Science 24th Annual Meeting, 2014.
[61] O. Jolliet, et al., Defining Product Intake Fraction to Quantify and Compare Exposure to Consumer Products, ACS Publications, 2015.
[62] J. Moya, et al., A review of physiological and behavioral changes during pregnancy and lactation: potential exposure factors and data gaps, J. Expo. Sci. Environ. Epidemiol. 24 (2014) 449–458.
[63] J. Moya, et al., Exposure Factors Handbook, 2011 edition, US Environmental Protection Agency, Washington, 2011.
[64] K.K. Isaacs, et al., SHEDS-HT: an integrated probabilistic exposure model for prioritizing exposures to chemicals with near-field and dietary sources, Environ. Sci. Technol. 48 (21) (2014) 12750–12759.
[65] R. Harder, et al., Including pathogen risk in life cycle assessment of wastewater management. 1. Estimating the burden of disease associated with pathogens, Environ. Sci. Technol. 48 (16) (2014) 9438–9445.
[66] R.A. Souza, et al., Impact of the ahas transgene and of herbicides associated with the soybean crop on soil microbial communities, Transgenic Res. 22 (5) (2013) 877–892.
[67] A.B. Hill, The environment and disease: association or causation? Proc. Roy. Soc. Med. 58 (5) (1965) 295.
[68] T.M. Letcher, D. Vallero, Excerpt from: Waste: A handbook for management, second ed., Academic Press, 2019.
[69] T.S. Kuhn, D. Hawkins, The structure of scientific revolutions, Am. J. Phys. 31 (7) (1963) 554–555.
[70] H. Petroski, To Engineer is Human, St. Martin's Press, New York, 1985.
[71] U.S. Environmental Protection Agency, Superfund cleanup process, in: Superfund, 2017. https://www.epa.gov/superfund/superfund-cleanup-process. Accessed 9 May 2017.
[72] D. Vallero, Paradigms Lost: Learning from Environmental Mistakes, Mishaps and Misdeeds, Butterworth-Heinemann, 2005.
[73] A.M. Barrett, Analyzing Current and Future Catastrophic Risks from Emerging-Threat Technologies, 2014.
[74] M.E. Wittmann, et al., Using structured expert judgment to assess invasive species prevention: Asian carp and the Mississippi Great Lakes hydrologic connection, Environ. Sci. Technol. 48 (4) (2014) 2150–2156.
[75] A. Kumar, Making a case for human health risk-based ranking nanoparticles in water for monitoring purposes, Environ. Sci. Technol. 46 (10) (2012) 5267–5268.
[76] J.E. Smith, A. Vaseashta, Advanced sciences convergence to analyze impact of nanomaterials on environment, health and safety, in: Advanced Sensors for Safety and Security, Springer, 2013, , pp. 83–91.

FURTHER READING

[77] E. Almássy, National Environmental Remediation Program Guide, E. R. Program, Ministry of Environment and Water, Budapest, Hungary, 2002.
[78] F. Shi, et al., Recent developments in the production of liquid fuels via catalytic conversion of microalgae: experiments and simulations, RSC Adv. 2 (26) (2012) 9727–9747.
[79] M. Raynolds, et al., The relative mass-energy-economic (RMEE) method for system boundary selection Part 1: A means to systematically and quantitatively select LCA boundaries, Int. J. Life Cycle Assess. 5 (1) (2000) 37–46.
[80] T. Chouhan, et al., Bhopal, the Inside Story: Carbide Workers Speak out on the world's Worst Industrial Disaster, Apex Pr, 1994.

Appendix 1

Key equations

The following are equations commonly used in air pollution calculations. See the text for the description of variables and coefficients.

ATMOSPHERIC CHEMISTRY

Conversion from ppm to µg m^{-3} (25°C and 760 mmHg):	$1\ \text{ppm (volume)} = \frac{\left(\frac{1\text{L}}{22.4}\right)\times \text{MW}\times 10^{6}\times \mu g\times \text{g}\times \text{m}^{-1}}{10^{6}\text{L}\times 298\text{K}^{\circ}\times 10^{-3}\text{m}^{3}\times \text{L}^{-1}}$
Synthesis reaction:	$A + B \rightarrow AB$
Decomposition reaction:	$AB \rightarrow A + B$
Single replacement reaction:	$A + BC \rightarrow AC + B$
Double replacement reaction:	$AB + CD \rightarrow AD + CB$
Reversal reaction (equilibrium):	$aA + bB \leftrightarrow cC + dD$
Complete combustion:	$a(CH)_x + bO_2 \xrightarrow{\Delta} cCO_2 + dH_2O$
Combustion with nitrogen:	$4CH_3NH_2\ (l) + 9O_2\ (g) \xrightarrow{\Delta} 4CO_2\ (g) + 10H_2O\ (g) + 2N_2\ (g)$
Combustion with nitrogen and sulfur:	$C_aH_bO_cN_dS_e + 4a + b - 2c) \rightarrow aCO_2 + \left(\frac{b}{2}\right) H_2O + \left(\frac{d}{2}\right) N_2 + eS$
Kinetic reaction:	$r = -\frac{\Delta(A)}{\Delta t}$
Rate law:	$r = k[A]^{m}\ [B]^{n}$
Differential rate law:	$\text{Rate} = -\frac{d[A]}{dt} = k[A]^{n}$
Integrated rate law:	$\ln[A] = -kt + \ln[A]_0$
Solubility product constant:	$K_{sp} = [C]^{c}[D]^{d}$
Generalized equilibrium constant:	$K_{eq} = \frac{k_f}{k_b} = \frac{[C]^{c}[D]^{d}}{[A]^{a}[B]^{b}}$
Indoor air concentration:	$C_{indoor} = \frac{\sum S_1 \ldots S_n}{V_{ME}\times ACH}$
Water dissociation constant:	$K_w = \frac{[H_3O^{+}][OH^{-}]}{[HOH]^{2}}$
Water dissociation constant at 25°C:	$K_w = [H_3O^{+}][OH^{-}] = 1.0 \times 10^{-14}\ \text{mol L}^{-1}$
"p" value:	$pX = -\log X$
pH:	$pH = -\log[H^{+}]$ or $pH = -\log[H_3O^{+}]$

Continued

Acid equilibrium constant for acetate:	$K_a - \frac{[H_3O^+][CH_3COO^-]}{[CH_3COOH]}$
Dissociation of carbon dioxide:	$CO_2 + H_2O \leftrightarrow H_2CO_3^*$
Henderson-Hasselbach equation:	$pK_a = pH + \log\frac{[HA]}{[A^-]}$
Arrhenius equation:	$k = A \cdot e^{\frac{E_a}{RT}}$
NO_x:	$NO_y = NO + NO_2$
NO_y:	$NO_y = NO_2 + NO + HNO_3 + PAN + 2N_2O_5 +$
	$HONO + NO_3 + NO_3^-$ compounds $+ NO_3^-$ aerosols
Time-weighted concentration:	$\overline{C} = \frac{M}{Q \times t}$
Beer-Lambert law:	$A = eb[C]$

ATMOSPHERIC PHYSICS

Stefan-Boltzmann law:	$F = \sigma T^4$
Wien's displacement law:	$v_{max} = 1.04 \times 10^{11}\ T$
Radiant flux:	$E_v\, dv = c_1 v^3\ [\exp(c_2.v \cdot T^{-1}) - 1]^{-1}\ dv$
Planck's law:	$E = hv = \frac{hc}{\lambda}$
Schrödinger equation:	$-\frac{h^2}{2m}\frac{d^2 f}{dx^2} + V(x)f = Ef$
First law of thermodynamics:	$\Delta U = q + w$
Mass balance:	$\frac{dM}{dt} = M_{in} - M_{out} \pm R$
Change of mass in control volume:	$\frac{d[A]}{dt} = -v \cdot \frac{d[A]}{dx} + \frac{d}{dx}\left(D \cdot \frac{d[A]}{dx}\right) + r$
Entropy:	$q = m.SH.\Delta T$
Boyle's law:	$V \propto P^{-1}$
Charles' law:	$V \propto T$
Gay-Lussac's law:	$P \propto T$
Ideal gas law:	$PV = nRT$
Universal gas constant:	$R = \frac{0.08212\ \text{L·atm}}{\text{mol·K}}$ or $\frac{8.314\ \text{joules}}{\text{mol·K}}$
Specific gas constant:	$R_{gas} = \frac{R}{m_{gas}}$
Flux:	$j = \frac{Q}{A};\ q = \frac{m}{A\Delta t} = \frac{m}{V} \times \frac{V}{A \cdot t} = C \cdot u$
Flux density:	$F = vC_A$
Flux density for dry deposition:	$F_d = v_d \cdot C_A$
Diffusive flux:	$j = C_A \cdot u' - C_B \cdot u' = u'\Delta C;\ j = -D\frac{\Delta C}{\Delta x}$
Fick's law:	$j = -D_O \cdot i_c;\ j = -AD\frac{\Delta C}{\Delta x};\ j = \frac{i_c}{R}$
Advective flux:	$J_{Advection} = \overline{v}\eta_e[c]$
Advective transport:	$N = QC$
Advective and diffusive flux:	$q = Cu + j$
Mean residence time:	$\tau = \frac{M}{J_{out}}$
Kinetic energy of gas molecule:	$E_g = \frac{mu^2}{2} = \frac{3RT}{2}$

Virtual temperature:	$T_v = \frac{T}{1-\left(\frac{e}{p}\right)(1-\varepsilon)}$
Newton's second law of motion:	$\mathbf{F} = \frac{d\mathbf{p}}{dt}$
Newton's third law of motion:	$\mathbf{F}_{reacting} = -\mathbf{F}_{acting}$
Velocity of flow:	$u = f(x, y, z)$
Mean mass flow:	$\frac{dm}{dt} = \dot{m} = \rho \times A \times u$
Mass continuity at steady state:	$\frac{dm}{dt} = \rho_1 \times A_1 \times u_1 = \rho_2 \times A_2 \times u_2 = \text{constant}$

PARTITIONING

Fugacity:	$Z_A = \frac{1}{RT} = \frac{C_A}{f_A}$
Fugacity capacity constant:	$C_i = Z_i \cdot f$
Total compartmental fugacity:	$f = \frac{M_{total}}{\sum_i (Z_i \cdot V_i)}$
Air-water partitioning coefficient:	$K_{AW} = \frac{C_A}{C_W} = \frac{Z_A f_A}{Z_W f_W}$
Raoult's law:	$p_i = p_i^* x_i$
Sorption:	$S = K_D \cdot C_W$
Soil or sediment partitioning:	$K_D = K_{OC} C_{org} = C_S (C_W)^{-1}$
Freundlich sorption isotherm:	$C_{sorb} = K_F C^n$
Partial pressure:	$p_a = K_H[c]$
Henry's law constant:	$K_H = \frac{P^\circ \times MW}{760 \times C_{sol}}$
Henry's law constant (unitless):	$K_{AW} = \frac{C_{air}}{C_{water}} = \frac{K_H}{R \cdot T}$
Octanol-air partitioning:	$K_{OA} = \frac{K_{ow} \cdot R \cdot T}{K_H} = \frac{K_{ow}}{K_{AW}}$; $\log K_{OA} = \log K_{ow} - \log K_{AW}$
Volumetric mass transfer rate:	$K_{1}a_{G/A}\left(\frac{S_G}{K_{G/A}}_{G/A} - S_A\right)$
Substrate partition coefficient:	$K_{G/A} = \frac{S_G}{S_A^*}$
In-stream pollutant concentration:	$C_r = \frac{Q_d C_d + Q_s C_s}{Q_r}$
Darcy's law:	$Q = -KA\frac{\partial h}{\partial x}$

METEOROLOGY

Shuttleworth's evaporation:	$E = \frac{mR_n + \gamma \cdot 6.43(1 + 0536 \times U_2) \cdot \delta e}{\lambda_v \cdot (m + \gamma)}$
Psychrometric constant:	$\gamma = \frac{0.0016286 \cdot P}{\lambda_v}$
Water use efficiency:	$WUE = \frac{[C_{CO2-fixed}]}{[C_{H2O-lost}]}$
Vapor pressure deficit:	$VPD = \mathring{p}_{saturated} - \mathring{p}_{actual}$
Actual vapor pressure:	$\mathring{p}_{actual} = \frac{RH \times \mathring{p}_{saturated}}{100}$

Continued

Scavenging ratio:	$S = C_p \cdot \rho \cdot C_{aerosol} - 1$
Hydrologic continuity equation:	$\frac{dS}{dx} = I - O$ and $\frac{dS}{dx} = P - R - E - T - I$
Humidity ratio:	$\omega = 0.62198 \cdot \frac{p_w}{p_{at} - p_w}$
Degree of saturation:	$\mu = \frac{\omega}{\omega_{sat}}$

BIOGEOCHEMISTRY

Mobilization factor:	$MF = \frac{E_h}{E_n}$
Enrichment factor (cadmium):	$EF = \frac{\left(\frac{[Cd]}{[Al]}\right)_{air}}{\left(\frac{[Cd]}{[Al]}\right)_{crust}}$

AIR POLLUTION ENGINEERING

	$N = D{\cdot}f = V[c]k = (V{\cdot}Z{\cdot}k){\cdot}f = D_R.\ f$
Fuel stoichiometric combustion reaction:	$\underbrace{C_mH_n}_{fuel} + \left(m + \frac{n}{4}\right)\underbrace{(O_2 + 3.76N_2)}_{air} \rightarrow mCO_2 + \frac{n}{2}H_2O + 3.76\left(m + \frac{n}{4}\right)N_2$
Solid/liquid fuel stoichiometric combustion reaction:	$\underbrace{C_aH_bO_cN_dS_e}_{fuel} + x\underbrace{(O_2 + 3.76N_2)}_{air} \rightarrow aCO_2 + \frac{b}{2}H_2O + \left(\frac{d}{2} + 3.76x\right)N_2 + eSO_2$
Combustion efficiency:	$CE = \left[\frac{C_{CO2}}{C_{CO2} + C_{CO}}\right] \times 100$
Destruction and removal efficiency:	$DRE = \left[\frac{W_{in} - W_{out}}{W_{in}}\right] \times 100$
Molecular momentum:	$\Delta mu = (\Delta mu_x)_f - (\Delta mu_x)_i = -(mu_x) - (mu_x) = -2\ mu_x$
Transfer rate for gas molecules:	$F_{total} = \frac{m}{l}(u_{x1} + u_{x2} + \ldots u_{xn})$
Pressure:	$p = \frac{F}{A}$
Pressure:	$P = \frac{F_{total}}{A} = \frac{m}{l^3}(u_{x1} + u_{x2} + \ldots u_{xn}) = \frac{mN}{V}\left(\frac{u_{x1} + u_{x2} + \ldots u_{xn}}{N}\right)$
Stress:	$\sigma(P) = \lim_{\delta A \to 0} \frac{\delta \mathbf{F}}{\delta \mathbf{A}}$
Density field:	$\rho = \rho(x, y, z, t)$
Velocity field:	$\vec{v} = \vec{v}\ (x, y, z, t)$
Static pressure of flow:	$p + ½\ pV^2 + pgh = \text{constant}$
Bernoulli's equation:	$\left(\frac{p}{\rho g} + \frac{V^2}{2g} + z\right)_1 = \left(\frac{p}{\rho g} + \frac{V^2}{2g} + z\right)_2$
Bernoulli's equation for pump flow:	$\left(\frac{p}{\rho g} + \frac{V^2}{2g} + z\right)_1 + h_p = \left(\frac{p}{\rho g} + \frac{V^2}{2g} + z\right)_2 + h_i$
Bernoulli's equation considering conservation of mass:	$p_1 + ½\ pV_1^2 + pgh_1 = p_2 + ½\ pV_2^2 + pgh_2$
Lagrangian velocity field:	$\vec{v} = [x(t), y(t), z(t)]$
Volumetric flow rate:	$Q = V\,t^{-1}$

Aerodynamic diameter (aerosol):	$D_{pa} = D_{ps}\sqrt{\rho_p}$
Incompressible fluid continuity:	$\nabla \cdot \boldsymbol{u} = 0$
Navier-Stokes equation:	$\rho\left[\frac{\partial \boldsymbol{u}}{\partial t} + (u \cdot \nabla)u\right] = -\nabla p + \eta \nabla^2 u + \mathrm{f}$
Continuity equation (cylinder):	$\frac{1}{r}\frac{\partial}{\partial r}(ru_r) + \frac{1}{r}\frac{\partial}{\partial \theta}u_\theta + \frac{1}{r}\frac{\partial}{\partial z}u_z = 0$
Airway volume (lung physiology):	$V_{aw} = 1.018 \times \text{Height (cm)} - 76.2$
Stokes number:	$\text{Stk} = \frac{t_0 u_0}{l_0}$; $\text{Stk} = \frac{\rho_p d_p^2 \dot{u}}{18\mu_g d}$
Impaction factor (dimensionless):	$\Psi = \frac{C_C d_p^2 3 y_p \rho_p}{18\mu_g D_C}$
Pollutant gradient:	$\frac{d[A]}{dt} = -v \cdot \frac{d[A]}{dx} + \frac{d}{dx}\left(D\frac{d[A]}{dx}\right) + r$
Maximum use concentration:	$\text{MUC} = \text{PEL} \times \text{APF}$
Sum of forces on a particle:	$\sum F = m_p \cdot a_p = m_p \frac{dv_p}{dt}$ and $\sum F = \frac{m_p \cdot a_p}{g_c}$
Gravitational constant:	$g_c = 32.2 \frac{\text{lb}_m \cdot \text{ft}}{\text{lb}_f \cdot \text{sec}^2}$
Gravitational force:	$F_G = m_p \cdot g = \rho_p \cdot V_p \cdot g$
Volume of a spherical particle:	$V_p = \frac{\pi d_p^3}{6}$
Gravitational force of a spherical particle:	$\frac{F_G = \pi d_p^3 \rho_p g^p}{6}$
Buoyancy force:	$\frac{F_B = \pi d_p^3 \rho_g g^g}{6}$
Drag force:	$\frac{F_D = \frac{A_g \cdot \rho_g \cdot V_p^2 \cdot C_D}{2} = \pi d_p^2 \rho_g v^g p^2 \cdot C_D}{8}$
Reynolds number:	$\text{Re} = \frac{\text{Inertial Forces}}{\text{Viscous Forces}}$
Reynolds number:	$\text{Re} = \frac{d_e v \rho}{\mu}$
Reynolds number:	$\text{Re} = \frac{d_e v}{\upsilon}$
Particle Reynolds number (dimensionless):	$\text{Re}_p = \frac{d_p v_p \rho_g}{\mu_g}$
Drag coefficient (laminar):	$C_D = \frac{24}{\text{Re}_p}$
Drag coefficient (transition):	$C_D = \frac{18.5}{\text{Re}_p^{0.6}}$
Drag coefficient (turbulent):	$C_D = 0.44$
Cunningham slip factor:	$C_C = 1 + \frac{6.21 \times 10^{-4} T}{d_p}$
Drag coefficient with Cunningham slip factor:	$C_D = \frac{24}{\text{Re}_p \cdot C_C}$
Drag force (laminar):	$F_D = \frac{3\pi\mu_g v_p d_p}{C_C}$
Drag force (transition):	$F_D = 2.30(d_p \cdot v_p)^{1.4} \cdot \mu_g^{0.6} \cdot \rho_g^{0.4}$
Drag force (turbulent):	$F_D = 0.05\pi(d_p \cdot v_p)^2 \cdot \rho_g$
Terminal velocity (laminar flow):	$v_t = \frac{g \cdot C_C \cdot \rho_p \cdot d_p^2}{18\mu_g}$
Terminal velocity (transition flow):	$v_t = \frac{0.153 g^{0.71} \cdot \rho_p^{0.71} \cdot d_p^{1.14}}{\mu_g^{0.43} \cdot \rho_g^{0.29}}$
Terminal velocity (turbulent flow):	$v_t = 1.74\left(\frac{g \cdot \rho_p \cdot d_p}{\cdot \rho_g}\right)^{0.5}$
Settling characteristic value:	$\text{K} = d_p\left(\frac{g \cdot \rho_p \cdot \rho_g}{\mu_g^2}\right)^{0.33}$

Continued

Collection efficiency:	$\eta_i - 1 \quad e^{-X}$ and $\eta_i = 1 - e^{-\frac{v_i I W}{Q}}$
Particle residence time:	$t_r = \frac{L}{v_g}$
Worst-case (longest path) particle residence time.	$t_r = \frac{H}{v_t}$
Volumetric gas flow:	$Q = \frac{v_g}{W \times H}$
Centrifugal force:	$F = \frac{MV^2}{R}$
Particle diffusivity:	$X_p = \frac{C_C kT}{3\pi\mu_g d_p}$
Force on a particle in electric field:	$F_E = n \cdot e \cdot E$
Face velocity at filter:	$U_0 = \frac{Q}{A}$
Air-to-cloth ratio:	$A/C = \frac{Q_g}{A_{fabric}}$
Pressure drop across filter:	$\Delta P_{total} = \Delta P_{media} + \Delta P_{dust\text{-}cake} = = (k_1 \cdot v_f) + (k_2 \cdot c_i \cdot v_f^2 \cdot t)$
Filter drag:	$S_f = \frac{\Delta P_{total}}{v_f} = k_1 + (k_2 \cdot c_i \cdot v_f \cdot t)$
Cylindrical bag area:	$A = \pi DL$
Saturation charge:	$q = (3\pi\varepsilon_0 E) d_p^2$

GEOMETRY

Area of a circle:	$A = \pi r^2 = \pi \left(\frac{D}{2}\right)^2 = \frac{\pi D^2}{4}$
Area of a cylinder:	$A = 2\pi rh + 2\pi(r)^2$
Volume of a cylinder:	$V = h\pi(r)^2$
Volume of a sphere:	$V = \frac{4}{3}\pi r^3$
Aspect ratio:	$AR = \frac{width}{height}$

STATISTICS

Root-mean-square speed:	$u_{rms} = (\overline{u}^2)^1/_2 = \left(\frac{3P}{\rho}\right)^1/_2$
Wet-dry gas concentration conversion:	$C_{dry} = \frac{C_{wet}}{1-w}$
Light interaction with atmosphere:	$-dI = b_{ext} I \cdot dx$
Extinction coefficient:	$b_{ext} = b_{rg} + b_{ag} + b_{scat} + b_{ap}$
Arithmetic mean:	$\overline{x} = \frac{\sum_{i=1}^{n} x_i}{n}$
Standard deviation:	$s = \sqrt{\left[\sum_{n=1}^{n}\left[\frac{(Y_1 - \overline{Y})^2}{n-1}\right]\right]}$
Relative percent difference:	$RPD = \frac{(C_1 - C_2)}{C_1 + C_2)}/_2 \times 100\%$
Relative standard deviation:	$RSD = \frac{s}{\overline{Y}} \times 100\%$
Percent recovery:	$\%R = \frac{S-U}{C} \times 100\%$ or $\%R = \frac{C_M}{C_{SRM}} \times 100\%$

AIR POLLUTION RISK

Risk function:	$R = f(S, P)$
Reference dose:	$\mathrm{RfD} = \frac{\mathrm{NOAEL}}{\mathrm{UF_{inter}} \times \mathrm{UF_{intra}} \times \mathrm{UF_{iother}}}$
Reference concentration:	$\mathrm{RfC} = \frac{\mathrm{NOAEC}}{\mathrm{U_{inter}} \times \mathrm{U_{intra}} \times \mathrm{U_{other}}}$
Exposure:	$E = D/t$
Exposure:	$E = \int_{t=t_1}^{t=t_2} C(t)dt$
Daily intake:	$\mathrm{DI}\ (\mathrm{mg\ kg^{-1}\ day^{-1}}) = \frac{\text{Concentration (mg volume}-1) \times \text{Intake (volume day}-1)}{\text{bodymass (kg)}}$
Chemical intake:	$I = \frac{C \cdot CR \cdot EF \cdot ED \cdot AF}{BW \cdot AT}$
Chemical intake from air pathway:	$I = C_A \times \frac{IR}{BW} \times \frac{ET \times EF \times ED}{AT}$
Exposure concentration:	$I = C_A \times \frac{IR}{BW} \times \frac{ET \times EF \times ED}{AT}$
Cancer risk:	$\mathrm{R} = \mathrm{CSF} \times E$
Noncancer risk hazard quotient:	$\mathrm{HQ} = \frac{\mathrm{MDD}}{\mathrm{ADI}} = \frac{\mathrm{Exposure}}{\mathrm{RfD}}$
Noncancer risk hazard index:	$\mathrm{HI} = \sum_1^n \mathrm{HQ}$
Eco-risk quotient:	$\mathrm{RQ} = \frac{\mathrm{Exposure}}{\mathrm{Ecotoxicity}} = \frac{\mathrm{EEC}}{\mathrm{EC_{50}}}$
Risk-based target concentration:	$\mathrm{C} = \frac{\mathrm{Risk} \cdot \mathrm{BW} \cdot \mathrm{AT}}{\mathrm{CR} \cdot \mathrm{EF} \cdot \mathrm{ED} \cdot \mathrm{AF} \cdot \mathrm{SF}}$

DISPERSION MODELING

Box model mass balance:	$\frac{dyCV}{dt} = QA + (\overline{u} \cdot C_{in} \cdot \mathrm{W} \cdot H_{mix}) - (\overline{u} \cdot C_{in} \cdot \mathrm{W} \cdot H_{mix})$
Gaussian concentration:	$C(x, y, z) = \frac{Q}{2\pi u \sigma_y \sigma_z} e^{-\frac{y^2}{2\sigma_y^2}} \left(e^{-\frac{(z+H)^2}{2\sigma_z^2}} + e^{-\frac{(z-H)^2}{2\sigma_z^2}} \right)$
Gaussian concentration (stable conditions):	$C = Q(1/u)\{g_1/[(2\pi)^{0.5}\sigma_y]\}\{g_2/[(2\pi)^{0.5}\sigma_z]\}$
Gaussian concentration (unstable or neutral conditions):	$C = Q(1/u)\{g_1/[(2\pi)^{0.5}\sigma_y]\}(1/L)$
Maximum concentration along centerline (Gaussian):	$C_{max} = \frac{2Q}{\pi \cdot u \cdot e \cdot H^2} \cdot \frac{\sigma_z}{\sigma_y}$
Mean concentration based on mean emission rate:	$\overline{C}(x, y, z) = \frac{\overline{Q}}{\pi \sigma_y \sigma_z u} \exp\left[-\frac{1}{2}\left(\frac{y^2}{\sigma_y^2} + \frac{z^2}{\sigma_z^2}\right)\right]$
Advection rate:	$G_i = V t^{-1}$
Model uncertainty:	$\mathrm{Y} = \mathrm{h(U,V)}$

RELIABILITY

Failure likelihood:	$P\{t_1 \le T_f \le t_1\} = \int_{t_1}^{t_2} f(t)dt$
Reliability function:	$R(t) = P\{t_1 \le T_f \le t_1\} = 1 - \int_0^t f(x)dx$

AIR POLLUTION ECONOMICS

Rollback model:	$R = \frac{g(P) - D}{g(P) - B} \times 100$
Annual discount rate:	$i = \left(1 + \frac{r}{k}\right)^{\frac{k}{p}} - 1$
Discount rate (spreadsheet):	$i = RATE(N,A,P,F,Type,guess)$
Discount factor:	$P/F, i\%, n$
Discount formula:	$P = F\,(P/F,\ i\%,\ n)$
Present value:	$PV = \frac{B_{y_i}}{(1+r)^{y_i - y_0}}$ and $PV = \frac{C_{y_i}}{(1+r)^{y_i - y_0}}$
Present worth (spreadsheet):	$P = PV(i,N,A,F,Type)$
Future worth (spreadsheet):	$F = FV(i,N,A,P,Type)$
Equivalent uniform annual worth:	EUAW = EUAB − EUAC
Net goodness:	NG = Σ (goodness of each consequence) × (importance) × (likelihood)

Appendix 2

Abbreviations and symbols

The following abbreviations and symbols are used in this text and/or in cited sources. The list is not completely exhaustive, since there is much variability in the use of symbols, especially since air pollution draws from numerous scientific and professional disciplines.

η_e	effective porosity (dimensionless)
Ψ	impaction factor (dimensionless)
σ *or* $\sigma°$	surface tension (forces length^{-1})
σ_y	cross-sectional dispersion coefficient
σ_z	dispersion coefficient in the vertical direction
σ **or SD**	standard deviation
σ^2	variance
ζi	mass ratio
μ	degree of saturation
μg	microgram
μm	micron or micrometer
σ_n	normal shear stress (mass length^{-1} time^{-2})
Ω	humidity ratio
Δ	energy (heat) required for reaction
$\frac{dy}{dx}$	ordinary derivative of x with respect to y
S_A^*	substrate concentration at the gas/aqueous interface
$\overline{\overline{Y}}$	source mean
τ	mean residence time
$\frac{\partial y}{\partial x}$	partial derivative of x with respect to y
[C]	molar concentration
–	mean or average (capped symbol or letter)
∇ (nabla)	gradient of a scalar field, especially in Navier-Stokes equations
∇ (nabla)	water table
≅	approximately equal to
≈	almost equal to
-	single bond
=	double bond
≡	triple bond
→	kinetic reaction
↔	equilibrium reaction
°C	degree Celsius or centigrade (temperature)
°F	degree Fahrenheit (temperature)
K	Kelvin (temperature)
°lat	latitude (degrees)
°long	longitude (degrees)
°R	degree Rankine (temperature)
μ or μ_v	dynamic viscosity (mass length^{-1} time^{-1})
μ	chemical potential (mass length^2 time^{-2})
μE	microenvironment
μ_g	gas viscosity (mass length^{-1} time^{-1})
μ_m	arithmetic mean
μM	micromole
0	initial (subscript)
A	area, surface area (length^2)
a	acceleration (length time^2)
A	activity rate (in emission calculation)
A	air (subscript)
A	aqueous phase
A	atomic weight
A	uniform cost
A/C	air-to-cloth ratio (filtration, length time^{-1})
AA	atomic absorption detection
A_{aw}	airway surface area (length^2)
ABL	atmospheric boundary layer
abs	absorbed
ABT	averaging, banking, and trading
AC	activated carbon
AC	alternating current (electricity)
ACM	asbestos-containing material
ACS	American Cancer Society
ADAF	age-dependent adjustment factor (cancer risk)
ADD	average daily dose
ADI	acceptable daily input
ADME	absorption, distribution, metabolism, and excretion (or elimination)
ADP	adenosine diphosphate
AF	absorption factor
AF	atrial fibrillation
A_{fabric}	fabric surface area (length^2)
AFO	animal feeding operation
AIRS	Aerometric Information Retrieval System (the United States)
amu	atomic mass unit
A_p	cross-sectional area of the particle (length^2) $= \frac{\pi d_p^2}{4}$

a_p particle acceleration
APCD air pollution control device
APCD air pollution control district
APCS air pollution control system
APG assigned protection factor
aq aqueous phase
AQCD Air Quality Criteria Document (the United States)
AQI air quality index
AQMA air quality maintenance area
AQS Air Quality System (part of AIRS; the United States)
ASCE American Society of Civil Engineers
ASME American Society of Mechanical Engineers
ASPECT airborne spectral photometric environmental collection technology
AT or $t_{averaging}$ averaging time for exposure (time)
atm atmosphere (unit of pressure)
ATP adenosine triphosphate
ATSDR Agency for Toxic Substances and Disease Registry (the United States)
ATUS American Time Use Survey
AWMA Air and Waste Management Association
B background pollutant concentration (rollback model)
B(a)P benzo(a)pyrene
BCA benefit-cost analysis
B/C or BCR benefit-to-cost ratio
BACT best available control technology
BAF bioaccumulation factor
b_{ag} absorption of light by NO_2 gas
b_{ap} absorption of light by particles
BCF bioconcentration factor
BCR benefit/cost ratio
b_{ext} light extinction coefficient
BF bioavailability (percent of substance absorbed per day)
BOF basic oxygen furnace
BP blood pressure
b_{rg} scattering of light by gaseous molecules (Rayleigh scattering)
b_{scat} scattering of light by particles
BTEX benzene, toluene, ethylbenzene, and xylenes (indication of gasoline contamination)
Btu or BTU British Thermal Unit (power) = 1060 Joules
BW body weight (mass)
C chemical concentration [(mass mass^{-1}), (volume volume^{-1}), or (mass volume^{-1})]
c speed of light (3×10^8 m s^{-1})
C_a airborne chemical concentration (mass volume^{-1} of air in breathing zone)
$\overline{C}$ time-weighted concentration
CAA Clean Air Act (the United States)
CAAA90 Clean Air Act Amendments of 1990 (the United States)
cal calorie
CAM compliance assurance monitoring
CARB California Air Resources Board
C_c Cunningham slip correction factor
CCR coal combustion residuals
C_D drag coefficient (dimensionless)
CDC Centers for Disease Control and Prevention (the United States)
CDD chlorinated dibenzodioxin
CDF chlorinated dibenzofuran
CEM continuous emission monitoring
CEQ Council on Environmental Quality (the United States)
CERCLA Comprehensive Environmental Response, Compensation and Liability Act (also known as Superfund; the United States)
CFC chlorofluorocarbon
CFD computational fluid dynamics
CFD cash flow diagram
CFR Code of Federal Regulations (the United States)
CHAD Consolidated Human Activity Database
CHIEF Clearinghouse for Inventories and Emission Factors (the United States)
CI confidence interval
Ci curie (radioactivity unit)
C_{ij} concentration of exposure by person i in microenvironment j
CMAQ Community Multiscale Air Quality Model (the United States)
CMB chemical mass balance (model)
CNG compressed natural gas
CNS central nervous system
COE US Army Corps of Engineers (also USACE)
COHb carboxyhemoglobinemia
COPD chronic obstructive pulmonary disease
C_{org} substrate organic matter (mass of organic carbon per mass of substrate)
cos cosine
c_p specific heat at constant pressure
cps cycles per second (1 cps = 1 Hz)
C_S equilibrium concentration of the solute in the solid phase
CSF cancer slope factor [(mass mass^{-1} time^{-1})$^{-1}$] (also SF)
c_v specific heat at constant volume
CVD cardiovascular disease
CWA Clean Water Act (the United States)
CWS Canada-wide standards
$C_xH_yO_x$ organic compound (hypothetical or representative)
D air quality standard (rollback model)
d day
d diameter (length)
D dose
D index of community diversity
D inside diameter of cyclone
D intermedia or intercompartmental transport term (fugacity models)
D mass of pollutant per body mass
D transfer coefficient or compartmental rate constant (for fugacity)
$\underline{D}$ dispersion tensor (length time^{-1})
d_0 proportionality constant (diffusion)
D_A nondiffusive transfer coefficient for fugacity
DAS data acquisition system

D_C	diameter of the collection object (length)
DC	direct current (electricity)
D_D	diffusive transfer coefficient for fugacity
DEFRA	Department for Environment, Food and Rural Affairs (the United Kingdom)
DfD	design for disassembly
DfE	design for the environment
DfR	design for recycling
DHHS	Department of Health and Human Services (the United States)
DHS	Department of Homeland Security (the United States)
d_i	percentage difference (precision)
DI	daily intake
DIAL	differential absorption lidar
dL	deciliter (volume)
DNA	deoxyribonucleic acid
DNAPL	dense (high-density) nonaqueous-phase liquid
DO	dissolved oxygen
DOAS	differential optical absorption spectroscopy
DOC	diesel oxidation catalyst
DOE	Department of Energy (the United States)
DOI	Department of the Interior (the United States)
DOI	digital object identifier
DOS	Department of State (the United States)
DOT	Department of Transportation (the United States)
DP	degradation product
d_p	physical particle diameter (length)
D_{pa} or d_{pa}	aerodynamic diameter of a particle (length)
DPF	diesel particulate filter
D_{ps} or d_{ps}	Stokes particle diameter (length)
DQO	data quality objective
DRE	destruction and removal efficiency
DRL	differential rate law
e	charge of electron (4.8×10^{-10} statcoulombs)
E	electric field strength (potential, voltage length^{-1})
E	bulk modulus (fluid stress divided by strain)
E	emission
E	energy required in a reaction
e	Euler's number = 2.71828183...
E	event (probability and statistics)
E	exposure
E	particle collection efficiency; inverse of penetration (filtration)
E_0	strength of electric field (volts length^{-1})
E10	Fuel mix of 90% gasoline and 10% ethanol
E85	Fuel mix of 15% gasoline and 85% ethanol
E_a	Arrhenius activation energy
EC	average exposure concentration
EC	elemental carbon
ECA	European Chemicals Agency
ECD	electron capture detection
ED	exposure duration (time)
EDC	endocrine-disrupting compound
EDS	energy-dispersive spectroscopy
EDSP	endocrine disruptor screening program
EEA	European Environment Agency
E_F	emission factor (in emission calculation)
EF	exposure frequency (time)
EF	enrichment factor (biogeochemistry)
EGR	exhaust gas recirculation
EIA	Energy Information Administration (the United States)
E_{ing}	ingestion exposure (mass per time)
E_{inh}	inhalation exposure (mass per time)
EIS	Emission Inventory System (the United States)
EIS	environmental impact statement
EJ	environmental justice
EL	exposure length (time per time)
ELF	extremely low frequency
EMEP	European Monitoring and Evaluation Programme
EMR	electromagnetic radiation
EPA or US EPA	Environmental Protection Agency (the United States)
E_R	percent overall emission reduction efficiency (in emission calculation)
ERDEM	Exposure-Related Dose Estimating Model
erg	energy and work unit (10^{-7} joules)
EROEI	energy return on energy investment
ERPG	Emergency Response Planning Guidelines (the United States)
ESP	electrostatic precipitator
ETS	environmental tobacco smoke
EUAB	equivalent uniform annual benefit
EUAC	equivalent uniform annual cost
EUAW	equivalent uniform annual worth
EVT	extreme value theory
exp(x)	exponential (e^x)
E_υ	energy at frequency υ
f	Coriolis parameter
F	energy flux
F	favorable
F	force (mass length time^{-2})
F	frequency (probability and statistics)
f	fugacity
f	fugacity (subscript)
F_B	buoyancy force (mass length time^{-2})
F_C	centrifugal force (mass length time^{-2})
F_D	drag force (mass length time^{-2})
F_E	electrostatic charge (force)
FFDCA	Federal Food, Drug, and Cosmetic Act (the United States)
FGD	flue gas desulfurization
FHWA	Federal Highway Administration (the United States)
FID	flame ionization detection
FIFRA	Federal Insecticide, Fungicide, and Rodenticide Act (the United States)
FOARAM	Federal Ocean Acidification Research and Monitoring Act (the United States)
f_{OC}	weight fraction of organic carbon in compartmental matrix (dimensionless)
FONSI	finding of no significant impact
FQPA	Food Quality Protection Act (the United States)
FR	Federal Register (the United States)
FRM	federal reference method (the United States)
ft	foot (length)

FTIR	Fourier-transform infrared
g	gas or vapor phase (subscript or parenthetical)
G	gas-phase emissions (mass time^{-1})
G	Gibbs free energy
g	gram
g	gravitational force
g	gravitational gas holdup
g	growth factor (rollback model)
G	nondiffusive flow rate (for fugacity)
G	gradient maintenance cost
g_0	standard gravitational field
GAC	granulated activated carbon
gal	gallon (volume)
GC	gas chromatography
g_c	gravitational constant $= 32.2\frac{lb_m \cdot ft}{lb_f \cdot sec^2} =$ 980 cm sec^{-2}
GDP	gross domestic product
GHG	global greenhouse gas
GHz	gigahertz (10^9 Hz)
G_i	advection rate (fugacity models)
GIS	geographic information system
GLP	good laboratory practice
GPS	Global Positioning System
grad	gradient; grad $f = \nabla f(x, y)$ in a plane and $\nabla f(x, y, z)$ in space
GWP	global warming potential
ΔG_f^{*0}	Gibbs free energy formation at steady state
H	hazard
h or hr	hour
h	elevation
h	physical stack height
H	effective stack height
H	enthalpy
H	headspace height
H	heat [subscript (mass length^{-1} time^{-2})]
H	Henry's Law (subscript) (also K_H)
H	horizontal (subscript)
h	Planck's constant (6.62×10^{-27} erg s)
HAP	hazardous air pollutant
HC	hydrocarbon
HEDS	Human Exposure Database System (the United States)
HEPA	high-efficiency particulate air (filter)
HHM	hierarchical holographic modeling
HI	hazard index
HMIWI	hospital/medical/infectious waste incinerators
HON	hazardous organic NESHAPs
hPa	hectopascal (unit of pressure)
HPLC	high-performance liquid chromatography
HPV	high production volume (chemical class)
HQ	hazard quotient
HR	hazard ratio
HSWA	Hazardous and Solid Waste Amendments (the United States)
HVAC	heating, ventilation, and air conditioning
HYSPLIT	Hybrid Single-Particle Lagrangian Integrated Trajectory (model)
Hz	hertz (cycles s^{-1})
$h\upsilon$	photon
i	group, species, subgroup of total (subscript)
I	immobilized
I	in situ
I	intake dose
I	intensity
I/M	inspection and maintenance program
IAEA	International Atomic Energy Agency
IARC	International Agency for Research on Cancer
ICP	inductively coupled plasma
ICRP	International Commission on Radiological Protection
ID	interior diameter
IDL	instrument detection limit
IDLH	immediately dangerous to life or health
IMPROVE	Interagency Monitoring of Protected Visual Environments (the United States)
in Hg	inches of mercury (unit of pressure)
IN	mass or energy inflow to control volume; influent (subscript)
IRL	integrated rate law
IOGAPS	integrated organic gas and particulate sampler
IPCC	Intergovernmental Panel on Climate Change
IR	intake rate [(mass time^{-1}) or (volume per time)]
IR	infrared
IR-DIAL	infrared-differential absorption lidar
IRIS	Integrated Risk Information System
IRR	internal rate of return
ISCLT	Industrial Source Complex Long Term (model)
IUR	inhalation unit risk
J	Joule (energy unit)
J	flux
k	Boltzmann's constant (1.37×10^{-16} erg K^{-1})
K	partitioning coefficient
k	reaction rate constant
$K_{1a_{G/A}}$	global volumetric mass transfer coefficient (time^{-1})
k_1	filter media resistance constant
k_2	filter dust cake resistance constant
K_a	acid equilibrium constant
K_{AW}	air-water partitioning coefficient
K_b	base equilibrium constant
K_D	soil sorption partitioning coefficient
k_e	rate at which the producers use energy (via respiration)
K_F	Freundlich isotherm constant
kg	kilogram (mass)
$K_{G/A}$	partition coefficient between the gaseous and aqueous phases
K_H	Henry's law partitioning coefficient (also H)
k_{inh}	inhalation rate (volume of air per unit time)
km	kilometer (length)
K_{oa}	octanol-air partitioning coefficient
K_{oc}	organic-carbon partitioning coefficient
K_{ow}	octanol-water partitioning coefficient
k_p	rate of chemical energy storage by primary producers
kPa	kilopascal (unit of pressure)
K_{sp}	solubility product constant
K_w	water dissociation equilibrium constant

kWh kilowatt-hour
L or l liter
l liquid phase (subscript or parenthetical)
LADD lifetime average daily dose
LAER lowest achievable emission rate
lb pound (weight)
lb_f foot-pound (force)
lb_m international avoirdupois pound (weight)
LC liquid chromatography
LC_{50} lethal concentration 50 (lethal to 50% of tested organisms)
LCA life cycle analysis or life cycle assessment
LCI life cycle inventory
LD_{50} lethal dose 50 (lethal to 50% of tested organisms)
LEL lower explosive limit
LEV low-emission vehicle
lidar remote sensing of distance using a laser (combination of light and radar)
ln(x) natural log of x
LNAPL light (low-density) nonaqueous-phase liquid
LOAEC lowest observed adverse effect concentration (also LEL or LOEL)
LOAEL lowest observed adverse effect level
LoD limit of detection (also known as DL, detection limit)
log logarithm
LoQ limit of quantitation
LOSU level of scientific understanding
LPG liquefied propane gas
LRTAP long-range transboundary air pollutants
LTP local temperature and pressure
L_v distance at which a black object is just barely visible
m mass
m meter
M mole (gram molecular weight)
M relatively nonreactive molecule in a reaction
MACRS Modified Accelerated Cost Recovery System
MACT maximum achievable control technology
max maximum (subscript)
MCE mixed cellulose ester
MCL maximum contaminant level (drinking water)
Me metal
MEI maximum exposed individual
MF mobilization factor
MFAA Millipore cellulose ester membrane
m_g mass of displaced gas
MHz megahertz (10^6 Hz)
MIC methyl isocyanate
min minimum (subscript)
min minute (time)
MIR maximum individual risk
mm Hg millimeters of mercury (unit of pressure)
mm millimeter
mM millimole
mmBTU 10^6 BTU
MMT methylcyclopentadienyl manganese tricarbonyl
mmt million metric tons
MOA mode of action
mol mole (gram molecular weight)
m_p particle mass
MQO model quality objective
MRL minimum risk level
MS mass spectroscopy
MSAT mobile source air toxic
MSDS material safety data sheet
MTBE methyl tert-butyl ether
MTTR mean time to repair
MUC maximum use concentration
mV millivolts (potential)
MW megawatt
MW molecular weight
n number of charges (electrostatics)
N concentration (number)
N Newton (unit of force)
N not significant
n number (e.g., number of moles)
N transport process rate (fugacity × compartmental rate constant)
N-100 respirator protection efficiency against particulate aerosols (99.97%), but not resistant to oil
NA nonaqueous phase
NA nonattainment
NAA neutron activation analysis
NAAQS National Ambient Air Quality Standards (the United States)
NADP National Atmospheric Deposition Program (the United States)
NAE National Academy of Engineering (the United States)
NAEI National Atmospheric Emissions Inventory (the United Kingdom)
NAMS national air monitoring station (the United States)
NAPL nonaqueous-phase liquid
NAS National Academy of Sciences (the United States)
NASA National Aeronautics and Space Administration (the United States)
NATA National Air Toxics Assessment (the United States)
ND, n.d. nondetect
NDIR nondispersive infrared
NEA Nuclear Energy Agency
NEI National Emissions Inventory (the United States)
NEPA National Environmental Policy Act (the United States)
NESHAP National Emission Standards for Hazardous Air Pollutants (the United States)
ng nanogram
NG net goodness
NHANES National Health and Nutrition Examination Survey (the United States)
N_{in} number of particles (count) entering a system (e.g., filter)

NIOSH National Institute for Occupational Safety and Health
NIST National Institute of Standards and Technology (the United States)
NLEV National Low-Emission Vehicle Program (the United States)
nM nanomole
NMHC nonmethane hydrocarbon
NMMAPS National Morbidity, Mortality, and Air Pollution Study (the United States)
NMOG nonmethane organic gases
NMVOC nonmethane volatile organic compound
NOAA National Oceanic and Atmospheric Administration (the United States)
NOAEC no observed adverse effect concentration
NOAEL no observed adverse effect level (also NEL or NOEL)
NOES National Occupational Exposure Survey (the United States)
N_{out} number of particles (count) exiting a system (e.g., filter)
NO_x oxides of nitrogen, specifically NO and NO_2
NO_y oxides of nitrogen, specifically NO_x and NO_z
NO_z HNO_3 + HONO + N_2O_5 + HO_2NO_3 + PAN + organic nitrates
NP nanoparticle
NPL National Physical Laboratory (the United Kingdom)
NPP net primary product (or productivity)
NPRI National Pollutant Release Inventory (Canada)
NPV net present value
N_R Reynolds number (also Re)
N_{Rp} dimensionless particle Reynolds number
NSF National Science Foundation (the United States)
NSR new source review
NTP National Toxicology Program (the United States)
NTP normal temperature and pressure
NTSIP National Toxic Substance Incidents Program (the United States)
NVOC nonvolatile organic compound
O octanol (subscript)
O&M operation and maintenance
O(1D) O atom in an excited singlet state
OA organic acid
OC organic carbon (subscript)
OC/EC organic carbon-to-elemental carbon ratio
OECD Organisation for Economic Co-operation and Development
OMI ozone monitoring instrument
OP organophosphorus (pesticide class)
OP-FTIR open-path Fourier-transform infrared
OSHA Occupational Safety and Health Administration (the United States)
OUT mass or energy outflow from control volume; effluent (subscript)
P* or *p pressure
p momentum
p partial
p particle (subscript)
P particle penetration; inverse of efficiency (filtration)
P particulate matter emissions (mass $time^{-1}$)
P present air quality (rollback model)
P probability
P initial cost
P^0 vapor pressure
P_1 net primary productivity (or production)
P-100 oil-proof respirator protection efficiency against particulate aerosols (99.97%)
p_a partial pressure of chemical substance gas in inhaled air
p_{a} partial pressure of gas a
Pa pascal (unit of pressure)
PA preliminary assessment
PAD population-adjusted dose
PAH polycyclic aromatic hydrocarbon (sometimes polyaromatic hydrocarbon)
PAL point, area, and line (model)
PAL-DS point, area, and line model with deposition and settling
PAMS photochemical assessment monitoring station
p_b partial pressure of chemical substance gas in blood
PBPD physiologically based pharmacodynamic
PBPK physiologically based pharmacokinetic
PBT persistent, bioaccumulative, and toxic
PC particle concentration
PC polycarbonate
PCB polychlorinated biphenyl
PCDD/F pentachlorodibenzodioxins and furans
PCE tetrachloroethane
pCi picocurie (radioactivity unit)
PCM phase contrast microscopy
PDF probability density function
PEL permissible exposure level (OSHA)
pH negative log of H^+ or H_3O^+ concentration
PIC product of incomplete combustion
PID photoionization detector
pK_a negative log of the acid dissociation constant
p_L supercooled liquid vapor pressure
PLM polarized light microscopy
PM particulate matter
PM_{10} Particulate matter, <10 μm diameter
$PM_{2.5}$ Particulate matter, <2.5 μm diameter
PNS peripheral nervous system
pOH negative log of OH^- concentration
POM polycyclic organic matter
POP persistent organic pollutant
ppb parts per billion
ppbV parts per billion by volume
PPE personal protection equipment
ppm parts per million
ppmV parts per million by volume
ppt parts per trillion
PQAO primary quality assurance organization
PQL practical quantitation limit

PRTR Pollutant Release and Transfer Register (Mexico)
PSD Prevention of Significant Deterioration (the United States)
PSI pollution standard index (replaced by AQI)
psi pounds per square inch (pressure)
PUF polyurethane foam
P_v velocity pressure
PV present value
PVC polyvinyl chloride
Q airflow
Q heat passing through boundary
q electric charge on particle
q flux
Q ionic product
QA quality assurance
QAPP quality assurance project plan
QC quality control
Q_g gas flow rate ($length^2\ time^{-1}$)
QM-A Whatman QM-A, quartz microfiber filter
QSAR quantitative structure-activity relationships
R or r correlation coefficient
r chemical production or degradation rate
R gas constant
R percentage emission reduction (rollback model)
R radial position of particle (centrifugal force calculation)
r radius
R remote
R removed
R risk
r thermodynamic loss term
r_i mole ratio
$R(t)$ reliability function
R universal gas constant (8.315 $J\ K^{-1}\ mol^{-1}$)
R-100 oil-resistant respirator protection efficiency against particulate aerosols (99.97%)
R^2 **or** r^2 coefficient of determination
RACT reasonably available control technology
RCA replacement cost analysis
RCRA Resource Conservation and Recovery Act (the United States)
Re Reynolds number (also N_R)
REACH Registration, Evaluation, Authorisation and Restriction of Chemicals (Europe)
REL recommended exposure limit
REL-TWA recommended exposure limit (time-weighted average)
Re_p dimensionless particle Reynolds number
RF radio frequency
RfC reference concentration
RfD reference dose
RH relative humidity
RIOPA Relationship Between Indoor, Outdoor, and Personal Air Study (the United States)
rms root-mean-square
RNA ribonucleic acid
RO_2 radical made of a chain of organic compounds with O_2 substituted for H
ROD record of decision
ROS reactive oxygen species
RR relative risk
RT residence time
RT retention time
RTP room temperature and pressure
RX organic compound
s or sec second (time)
S entropy
S severity of risk
s solid or sorbed phase (subscript or parenthetical)
S solid-phase concentration
S solubility
S sorption
S source or sink
S/N signal-to-noise ratio
S^2 variance
s^2 variance of the source
S_A substrate concentrations in the aqueous phase
SAR structure-activity relationships
SARA Superfund Amendments and Reauthorization Act (the United States)
SATB surface area of the tracheobronchial region
SCR selective catalytic reduction
SD or σ standard deviation
SDWA Safe Drinking Water Act (the United States)
SE standard error
SEM scanning electron microscopy
SES socioeconomic status
S_f filter drag force
SF slope factor
S_G substrate concentrations in the bulk gas phase
S_h horizontal surface
SHEDS Stochastic Human Exposure and Dose Simulation (model)
SHS secondhand smoke
sin sine
SIP state implementation plan (the United States)
SLAMS State and Local Air Monitoring Station (the United States)
SMR standard morbidity ratio
SMR standard mortality ratio
SNCR selective noncatalytic reduction
SOF soluble organic fraction
SOM soil organic matter
SOP standard operating procedure
SO_x oxides of sulfur
SPMS special purpose monitoring station
SS steady state (subscript)
Stk Stokes number
STP standard temperature and pressure
SVOC semivolatile organic compound
T temperature
t tonne
T total (subscript)
t_0 initial time
$t_{1/2}$ half-life
TAGA trace atmospheric gas analyzer
tan tangent
TC total carbon
TCDD tetrachlorodibenzo-*para*-dioxin

TCDF	tetrachlorodibenzofuran
TCE	trichloroethane
TCLP	toxicity testing leaching procedure
TCM	tetrachloromercurate
TDL	tunable diode laser
TDS	total dissolved solids
TEAM	Total Exposure Assessment Methodology Study
TEL	tetraethyl lead
TEM	transmission electron microscopy
T_f	time of failure
T_{ij}	time spent in microenvironment j by person i
TL	typical lifetime
TLV	threshold limit value
TLV-C	threshold limit value (ceiling)
TLV-STEL	threshold limit value-short-term exposure limit
TNT	trinitrotoluene
TOMS	total ozone mapping spectrometer
TPTH	triphenyltin hydroxide
tpy	tons per year
T-R	transformer-rectifier
TRI	Toxic Release Inventory (the United States)
TRS	total reduced sulfur
TS	total sulfur
TSCA	Toxic Substances Control Act (the United States)
TSP	total suspended particulate
TVA	Tennessee Valley Authority
TWA	T wave alternans
TWA	time-weighted average
u	velocity
u	wind speed
U	epistemic uncertainty
U	internal energy
U	unknown
u_{rms}	root-mean-square (rms) speed (gas molecules)
UAM	urban airshed model
UCR	unit cancer risk
UEL	upper explosive limit
$\mathbf{UF_{inter}}$	interspecies uncertainty factor
$\mathbf{UF_{intra}}$	intraspecies uncertainty factor
$\mathbf{UF_{other}}$	factor indicating RfC or RfC uncertainty other than UF_{inter} or UF_{intra}
UFP	ultrafine particles
UNEP	United Nations Environment Programme
UNFCCC	United Nations Framework Convention on Climate Change
USCG	US Coast Guard
USDA	US Department of Agriculture
USF&W	US Fish and Wildlife Services
USGS	US Geological Survey
UST	underground storage tank
UV	ultraviolet
UVCB	chemical substances of unknown or variable composition, complex reaction products, and biological materials (the United States)
UV-DOAS	differential optical absorption spectroscopy in UV range
UVF	ultraviolet fluorescence
V	fluid velocity
V	speed
V	stochastic or aleatory uncertainty
v	vapor phase (subscript or parenthetical)
V	volume
$\bar{v}$	average linear velocity (length time^{-1})
$\bar{V}$	average speed (length time^{-1})
V_{aw}	airway volume (length^3)
VC	vinyl chloride
$\mathbf{V_D}$	volume of distribution
$\mathbf{V_e}$	minute volume (volume time^{-1})
v_f	velocity through the filter (length time^{-1})
V_{ff}	filter fiber volume
v_g	gas velocity
v_g	geostrophic velocity
vis	visible light range
v_{max}	frequency of maximum radiation intensity (Wien's displacement law)
VMT	vehicle miles traveled
VOC	volatile organic compound
v_p	particle velocity
$\mathbf{V_p}$	particle volume
V_{total}	total filter volume
w	work
W	body weight
W	water (subscript)
W	watt (unit of power; J s^{-1})
WHO	World Health Organization
WOE	weight of evidence
x	at distance x from $x = 0$ (subscript)
x	distance (horizontal)
XAD	sorbing material; ion exchange resin
XRD	X-ray diffraction spectrometry
XRF	X-ray fluorescence
y	cross-sectional distance, that is, at distance y from $y = 0$ (subscript)
y	width
Y	modeling result/outcome
y	year
$\bar{y}$	sample mean
y_p	difference in velocity between the particle and collection object (length time^{-1})
z	at distance z from $z = 0$ (subscript)
Z	atomic number
z	distance (vertical)
Z	fugacity capacity
Z	zenith angle (degrees)
α	packing density (solidity)
γ	specific weight (mass length^{-3})
Δ	change
δ	solar declination (degrees)
ΔG^*	standard change in Gibbs free energy
$\Delta \mathbf{P_{dust-cake}}$	pressure drop across dust cake on filter
$\Delta \mathbf{P_{media}}$	pressure drop across filter media
$\Delta \mathbf{P_{total}}$	total pressure drop
ε_0	permittivity of free space (8.85 × 10^{-12} farads m^{-1})
η	hour angle (degrees)
θ	potential temperature

λ	average or expected counts or events per unit time (Poisson equation)
λ	degradation term in chemical decay
λ	wavelength (length, often nm)
ν_p	relative velocity between the particle and the target object
ν_τ	tangential velocity of gas
π	3.14159...
ρ	density (volume length^{-3})
ρ_0	standard density (1 g m^{-3})
ρ_g	gas density (mass length^{-3})
ρ_p	particle density (volume length^{-3})
σ	Stefan-Boltzmann constant
τ	tortuosity
τ_s	shear stress (mass length^{-1} time^{-2})
υ or υ_v	kinematic viscosity (length^2 time^{-1}) = dynamic viscosity divided by density ($\mu\ \rho^{-1}$)
υ	specific volume (reciprocal of density, length^3 mass^{-1})
φ	angle (degrees or radians)
φ	filter porosity
φ	fugacity coefficient
φ	latitude (degrees)
φ°	fraction, ratio, or coefficient
$\mathbf{X}$	leaving group from organic compound
$\mathbf{X_p}$	particle diffusivity (length^2 time^{-1})
Ψ	surface deposition
Ψ_D	diffusional collection parameter (dimensionless)
Ω	migration velocity (electrostatics)
Ω	angular speed of the earth's rotation
ω	solar azimuth (degrees)
ω	spherical particle mass

Index

Note: Page numbers followed by *f* indicate figures, *t* indicate tables, and *b* indicate boxes.

B

C

D

F

G

O

P

Q

R

S

X

Printed in the United States
By Bookmasters